Engineering Mechanics

Engineering Mechanics

D. P. Sharma
Assistant Professor
Department of Mechanical Engineering
Rajasthan Technical University, Kota

Pearson

ISBN 978-81-317-3222-9

First Impression, 2010
Eleventh Impression, 2023
Twelfth Impression, 2024

Published by Pearson India Education Services Pvt. Ltd, CIN: U72200TN2005PTC057128

Head Office: 1st Floor, Berger Tower, Plot No. C-001A/2, Sector 16B, Noida – 201 301,Uttar Pradesh, India.

Registered Office: Featherlite, 'The Address' 5th Floor, Survey No 203/10B, 200 Ft MMRD Road, Zamin Pallavaram, Chennai - 600044
Website: in.pearson.com; Email:companysecretary.india@pearson.com

Digitally printed in India by Trinity Academy for Corporate Training Ltd, New Delhi in the year of 2024.

To

Shri J. C. Sharma

To the Students

With the hope that
this book will develop their
interest in Engineering Mechanics.

Contents

Preface

The importance of mechanics for engineering education is manifold. It forms the root of machines in mechanical engineering and an initial stage work in civil engineering design work. It is a support for electrical and other branches of engineering.

I have observed that students usually learn basic theory before entering engineering colleges but need more practice at problem solving. Besides, engineering mechanics becomes interesting if the student is able to solve the problem without external guidance. This book has been organized keeping these aspects in view. It not only imparts mathematical way of problem solving but prepares the student to predict the results, forces and motions by imagining and visualizing the consequences on the bodies. This helps them to be more creative in the engineering profession.

The problems have been made as interesting as possible being practical and resembling real-life situations. Numerous diagrams and a large number of problems have been included to fulfill the need of every student.

Any error brought to our notice as well suggestions will be gratefully acknowledged.

ACKNOWLEDGEMENTS

I am thankful to colleagues, friends and students who encouraged me to write this book. I have referred to various books while writing book and acknowledge with gratitude the help taken from these books.

Finally, I am grateful to my family, especially my wife, Krishna Sharma, for her continued patience, understanding, and support throughout the preparation of this book.

D. P. Sharma

1 Statics of Particles and Rigid Bodies

INTRODUCTION

Engineering mechanics is the branch of science that considers the motion of bodies and the effects of forces on that motion. It may be divided into two parts: *statics* and *dynamics*. Dynamics is further divided into *kinematics* and *kinetics*.

Statics deals with the special case of a body at rest or a body that moves with a constant velocity. *Kinematics* is concerned with rates of change of geometrical quantities in a moving system; it does not involve the concept of force. *Kinetics* treats the causes and the nature of motion that results from specified forces.

BASIC CONCEPTS AND FUNDAMENTAL LAWS OF MECHANICS

The following concepts and fundamental laws are basic to the study of mechanics.

Time is the measure of the succession of events and is a basic quantity in dynamics. Time is not directly involved in the analysis of statics problems.

Space is the geometric region occupied by bodies whose positions are described by linear and angular measurements relative to a coordinate system.

Mass is a measure of the inertia of a body, which is its resistance to a change of velocity. The mass of a body affects the gravitational attraction force between it and other bodies.

Force is the action of one body on another. It may be exerted by actual contact or from a distance, as in the case of gravitational forces and magnetic forces. The action of a force is characterized by its magnitude, the direction of its action and its point of application. Force is represented by vector.

Particle is defined as a body whose size, in a given physical situation, is not significant in the analysis of its response to the forces that act on it. In other words, the body may be modelled as a point of concentrated mass, and the rotational motion of the body can be ignored.

Rigid body may be defined as a body that does not deform under the action of forces. Generally, the size of a rigid body influences its response to forces.

NEWTON'S LAW OF GRAVITATION

This states that two particles of mass M and m are mutually attracted with equal and opposite force $\mathbf{F}$ and $-\mathbf{F}$ of magnitude F given by the formula

$$F = \frac{GMm}{r^2},$$

where r is the distance between the two particles and G is a universal constant called the constant of gravitation.

NEWTON'S LAWS

First Law

If the resultant force acting on a particle is zero, the particle will remain at rest (if originally at rest) or will move with a constant speed in a straight line (if originally in motion).

Second Law

If the resultant force acting on a particle is not zero, the particle will have an acceleration proportional to the magnitude of the resultant force and in the direction of the resultant force.

$$\mathbf{F} = ma,$$

Third Law

The forces of action and reaction between interacting bodies are equal in magnitude, opposite in direction and collinear (they lie on the same line).

THE PRINCIPLE OF TRANSMISSIBILITY

This principle states that the conditions of equilibrium or of motion of a rigid body will remain unchanged if a force acting at a given point on the rigid body is replaced by a force of the same magnitude and direction, but acting at a different point, provided that the two forces have the same line of action.

In Figure 1.1 F is the force acting on a rigid body at point A. According to the principle of forces, this force has the same effect on the body as the force F applied at point B.

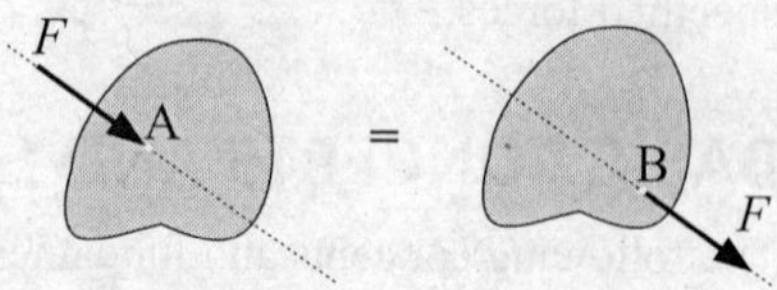

FIGURE 1.1 |

SYSTEM OF FORCES

When two or more than two forces of different magnitude and direction act upon a body, they constitute a system of forces.

CONCURRENT FORCES

Two or more forces that act at the same point are called *concurrent forces*. Concurrent forces need not have the same direction. They simply act at the same point. If they do have the same direction, they are *collinear forces*.

COPLANAR FORCES

Two or more forces whose directed arrows lie in the same plane are called *coplanar forces*. Since two concurrent forces always lie in a common plane, they are always coplanar. Three or more concurrent forces are not necessarily coplanar.

RESULTANT FORCE

It is possible to find a single force which will have the same effect as that of a number of forces acting on a body. Such a single force is called *resultant force* and the process of finding out the resultant force is called *composition of forces*. The reverse of the composition of forces is called *resolution of a force*.

PARALLELOGRAM LAW OF FORCES

If two forces acting at a point are represented, in magnitude and direction, by the sides of a parallelogram drawn from the point, their resultant force is represented, both in magnitude and in direction, by the diagonal of the parallelogram drawn through that point.

In Figure 1.2, let the two forces $\mathbf{F}_1$ and $\mathbf{F}_2$, acting at the point O be represented, in magnitude and direction, by the directed line OA and OB inclined at an angle θ with each other. Then if the parallelogram OACB be completed, the resultant force, **R**, will be represented by the diagonal OC.

From Figure 1.2, we can write

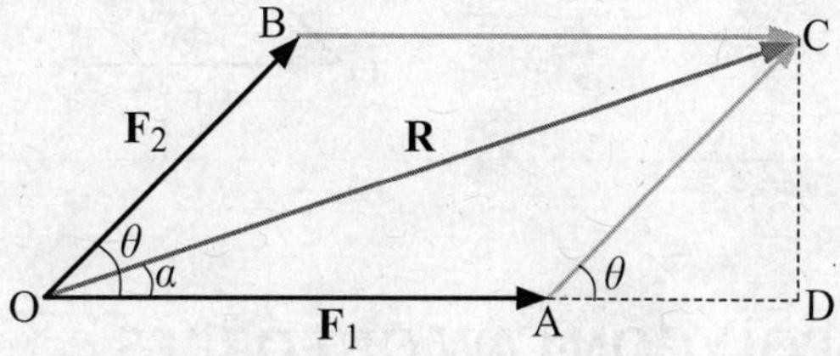

FIGURE 1.2 |

$$OA = BC = F_1; OB = AC = F_2$$

$$\angle BOA = \angle CAD = \theta; AD = F_2 \cos\theta \text{ and } CD = F_2 \sin\theta$$

In triangle OCD,

$$OC^2 = OD^2 + CD^2$$

or
$$OC^2 = (OA + AD)^2 + CD^2$$

or
$$R^2 = (F_1 + F_2 \cos\theta)^2 + (F_2 \sin\theta)^2$$

or
$$R^2 = F_1^2 + 2F_1F_2 \cos\theta + F_2^2 \sin^2\theta + F_2^2 \cos^2\theta$$

or
$$R^2 = F_1^2 + F_2^2 + 2F_1F_2 \cos\theta$$

Magnitude of the resultant **R**

$$R = \sqrt{F_1^2 + F_2^2 + 2F_1F_2 \cos\theta}$$

and

$$\tan\alpha = \frac{CD}{OD} \quad \text{or} \quad \tan\alpha = \frac{F_2 \sin\theta}{F_1 + F_2 \cos\theta}$$

Direction of the resultant

$$\alpha = \tan^{-1}\left(\frac{F_2 \sin\theta}{F_1 + F_2 \cos\theta}\right)$$

TRIANGLE LAW OF FORCES

If two forces acting simultaneously on a body are represented in magnitude and direction by two sides of a triangle in order, then the third side will represent the resultant of the two forces in the direction and magnitude taken in opposite order.

If three forces are acting on a body and they are represented by three sides of a triangle in magnitude and direction, then the body will be in equilibrium condition (Fig. 1.3). Figure 1.3

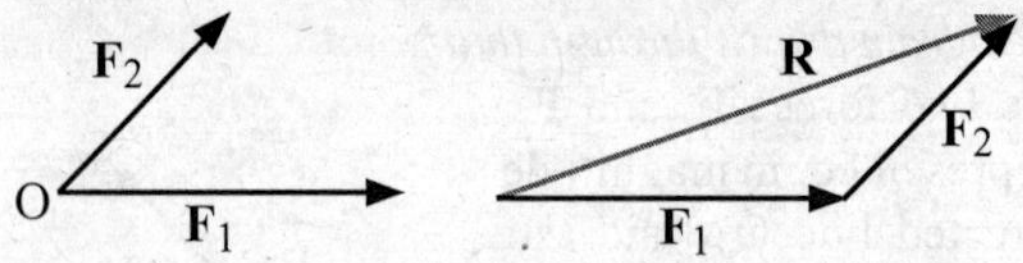

FIGURE 1.3 |

POLYGON LAW OF FORCES

When the forces acting on a body are more than two, the triangle law can be extended to polygon law.

If a number of coplanar concurrent forces acting simultaneously on a body are represented in magnitude and direction by the sides of a polygon, taken in order, then their resultant can be represented by closing side of the polygon in magnitude and direction in the opposite order.

Consider the forces $\mathbf{F}_1$, $\mathbf{F}_2$, $\mathbf{F}_3$ and $\mathbf{F}_4$ are acting at a point O as shown in Figure 1.4(a). Starting from the point O, the vector OA represents the force $\mathbf{F}_1$ in magnitude (using suitable scales) and direction. From the tip A, draw vector AB representing the force $\mathbf{F}_2$. Similarly, vector BC represents the force $\mathbf{F}_3$ and vector CD represents force $\mathbf{F}_4$. Join the starting point O to the end point D giving a vector OD in opposite order. Vector OD represents the resultant force $\mathbf{R} = \mathbf{F}_1 + \mathbf{F}_2 + \mathbf{F}_3 + \mathbf{F}_4$ in magnitude and direction as shown in Figure 1.4(b).

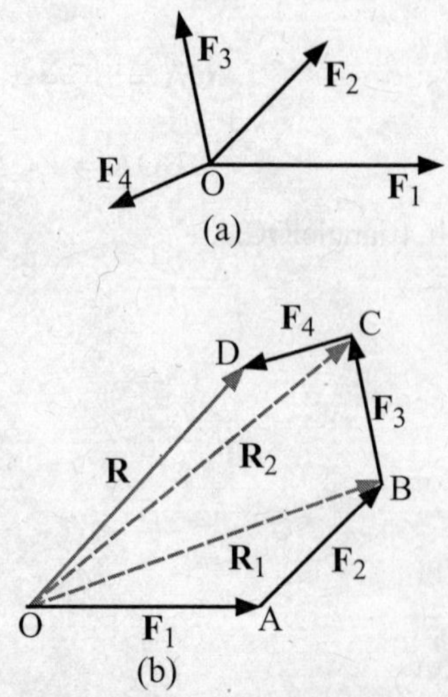

FIGURE 1.4 |

From the triangle law of forces

$$\mathbf{R}_1 = \mathbf{F}_1 + \mathbf{F}_2,$$

and

$$\mathbf{R}_2 = \mathbf{R}_1 + \mathbf{F}_3 = (\mathbf{F}_1 + \mathbf{F}_2) + \mathbf{F}_3,$$

and

$$\mathbf{R} = \mathbf{R}_2 + \mathbf{F}_4 = (\mathbf{F}_1 + \mathbf{F}_2 + \mathbf{F}_3) + \mathbf{F}_4$$

$$\mathbf{R} = \mathbf{F}_1 + \mathbf{F}_2 + \mathbf{F}_3 + \mathbf{F}_4$$

RESOLUTION OF FORCE

In resolution, a single force **F** acting on a particle may be replaced by two or more forces which together have the same effect on the particle. These forces are called *components* of the original force **F**, and the process of substituting them for **F** is called resolving the force **F** *into* components.

Consider a force F acting on a particle at O is to be resolved along OM and ON making angle α and β with the force, respectively. Components of the force **F** along OM and ON are **P** and **Q**, respectively. Draw AC parallel to OB and BC parallel to OA. Complete the parallelogram OACB as shown in Figure 1.5.

Applying the law of sine to ΔOAC,

$$\frac{OA}{\sin\beta} = \frac{AC}{\sin\alpha} = \frac{OC}{\sin[180 - (\alpha+\beta)]}$$

or

$$\frac{P}{\sin\beta} = \frac{Q}{\sin\alpha} = \frac{F}{\sin(\alpha+\beta)}$$

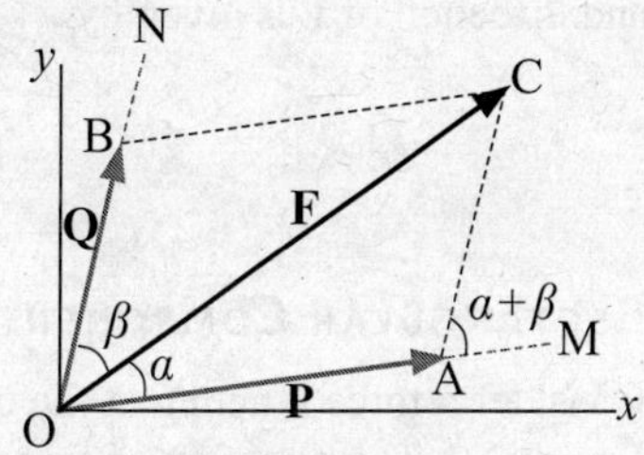

FIGURE 1.5

From which

$$\text{Component } P = F\frac{\sin\beta}{\sin(\alpha+\beta)}$$

$$\text{Component } Q = F\frac{\sin\alpha}{\sin(\alpha+\beta)}$$

RESOLUTION OF FORCE IN *XY*-PLANE

If the force **F** is resolved into two components which are perpendicular to each other, then these are called *rectangular components* F_x and F_y (Fig. 1.6).

Substituting $\alpha + \beta = 90$ and $\theta = \alpha$ in the equation

$$P = R\frac{\sin\beta}{\sin(\alpha+\beta)} \text{ and } Q = R\frac{\sin\alpha}{\sin(\alpha+\beta)}, \text{ we get}$$

$$P = F_x = F\frac{\sin(90-\theta)}{\sin 90} \text{ and}$$

$$Q = F_y = F\frac{\sin\theta}{\sin 90}$$

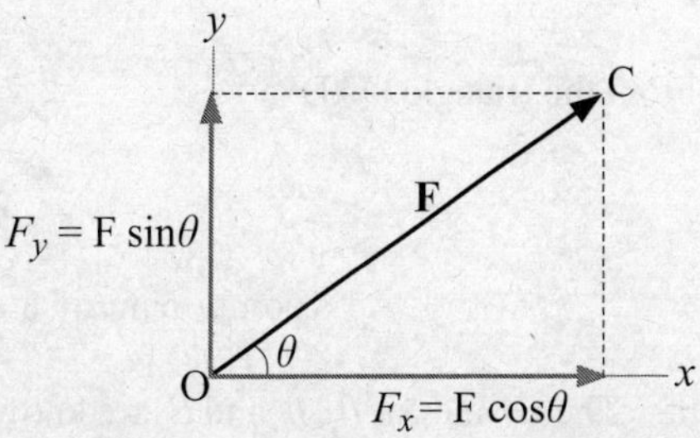

FIGURE 1.6

Component of force **F** in x-direction is $F_x = F\cos\theta$

Component of force **F** in y-direction is $F_y = F\sin\theta$

We may express **F** in the vector form

$$\mathbf{F} = F_x i + F_y j$$

$$\mathbf{F} = F\cos\theta i + F\sin\theta j$$

Magnitude of **F** is given by

$$F = \sqrt{F_x^2 + F_y^2}$$

and direction of **F** is given by

$$\theta = \tan^{-1}\frac{F_y}{F_x}$$

Rectangular Components of a Force in Space

Consider a force **F** acting at the origin O of the system of rectangular coordinates x, y, z as shown in Figure 1.7. The direction of force **F** is defined by three angles θ_x, θ_y, and θ_z. Let F_x, F_y, and F_z be the components of force F in x, y, and z directions, respectively. From the triangle OAB

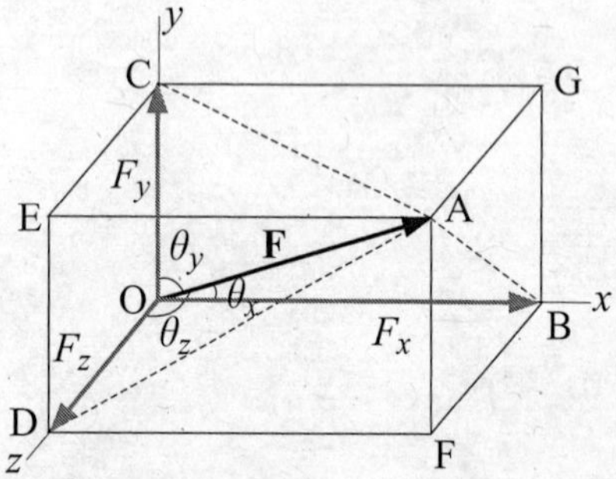

FIGURE 1.7 |

$$\cos\theta_x = \frac{OB}{OA} = \frac{F_x}{F},$$

component of **F** along x-direction $\quad F_x = F\cos\theta_x$

From the triangle OAC

$$\cos\theta_y = \frac{OC}{OA} = \frac{F_y}{F},$$

component of **F** along y-direction $\quad F_y = F\cos\theta_y$

From the triangle OAD

$$\cos\theta_z = \frac{OD}{OA} = \frac{F_z}{F},$$

component of **F** along z-direction $\quad F_z = F\cos\theta_z$

The cosines of θ_x, θ_y and θ_z are known as the direction cosine of the force **F**. The three angles θ_x, θ_y and θ_z are not independent, they are related as $\cos^2\theta_x + \cos^2\theta_y + \cos^2\theta_z = 1$. We may express **F** in the vector form

$$\mathbf{F} = F_x i + F_y j + F_z k$$

When the components F_x, F_y and F_z of the force **F** are given, the magnitude of force **F** is

$$F = \sqrt{F_x^2 + F_y^2 + F_z^2}$$

and direction of force

$$\theta_x = \cos^{-1}\left(\frac{F_x}{F}\right),\ \theta_y = \cos^{-1}\left(\frac{F_y}{F}\right),\ \text{and } \theta_z = \cos^{-1}\left(\frac{F_z}{F}\right)$$

COMPOSITION OF COPLANAR FORCES BY METHOD OF RESOLUTION

This is an analytical method of finding the resultant of multiple forces. In this method, first find the components of each force in two mutually perpendicular directions. Then the components in each direction are algebraically added to get the two components. These two component forces, which are mutually perpendicular, are combined to get the resultant force.

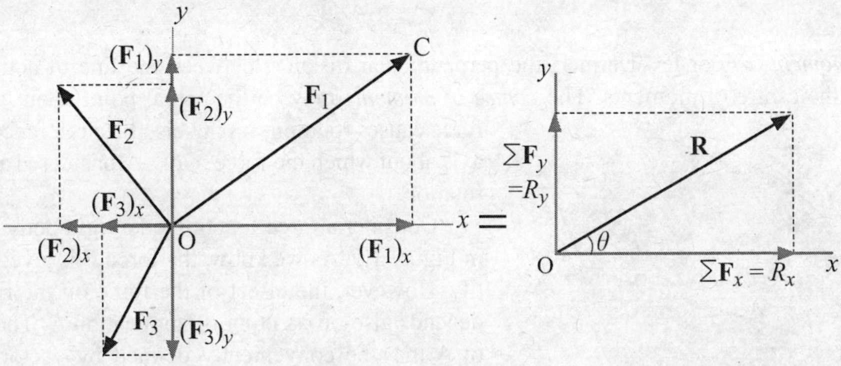

FIGURE 1.8 |

Consider, for instance, forces $\mathbf{F}_1$, $\mathbf{F}_2$, $\mathbf{F}_3$, … acting on a particle O (Fig. 1.8). Their resultant **R** is defined by the relation

$$\mathbf{R} = \mathbf{F}_1 + \mathbf{F}_2 + \mathbf{F}_3 + \cdots$$

Resolving each force into its rectangular components, we write

$$R_x i + R_y j = \{(F_1)_x i + (F_1)_y j\} + \{(F_2)_x i + (F_2)_y j\} + \{(F_3)_x i + (F_3)_y j\} + \cdots$$

From which, it follows that

$$R_x = (F_1)_x + (F_2)_x + (F_3)_x + \cdots$$

or

$$R_x = \sum F_x$$

and

$$R_y = (F_1)_y + (F_2)_y + (F_3)_y + \cdots$$

or

$$R_y = \sum F_y$$

Resultant

$$\mathrm{R} = \sqrt{R_x^{\,2} + R_y^{\,2}}$$

or

$$\mathrm{R} = \sqrt{\left(\sum F_x\right)^2 + \left(\sum F_y\right)^2}$$

and the direction of resultant

$$\alpha = \tan^{-1}\left(\frac{\sum F_y}{\sum F_x}\right)$$

MOMENT OF FORCE

The *moment* of a force is a measure of its tendency to cause a body to rotate about a specific point or axis. This is different from the tendency for a body to move, or translate, in the direction of the force. A moment is due to a force not having an equal and opposite force directly along it's line of action.

The magnitude of the moment of force acting about a point or axis is directly proportional to the distance of the force from the point or axis. It is defined as the product of the force F and the moment arm d.

$$M = Fd$$

The *moment arm* or lever arm is the perpendicular distance between the line of action of the force and the centre of moments. The *centre of moments* may be the actual point about which the force causes rotation. It may also be a reference point or axis about which the force may be considered as causing rotation.

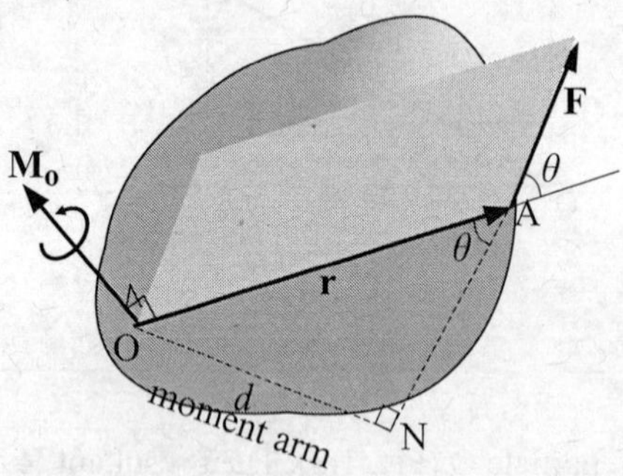

FIGURE 1.9 |

Consider a force **F** acting on a rigid body as shown in Figure 1.9. As we know, the force **F** is a vector quantity. However, the effect of the force on the rigid body depends also on its point of application A. The position of A may be conveniently defined by vector **r** which joins the fixed reference point O with A.

Magnitude of moment of force **F** about O is

$$M_o = Fd = Fr \sin \theta$$

In vector form

$$\mathbf{M}_o = \mathbf{r} \times \mathbf{F}$$

where $|\mathbf{r}| = r$, $|\mathbf{F}| = F$

According to the definition of vector product, the direction of $\mathbf{M}_o$ is perpendicular to **r** and **F**.

RECTANGULAR COMPONENTS OF THE MOMENT

Force **F** and position vector **r** in rectangular components may be written as

$$\mathbf{F} = F_x i + F_y j + F_z k$$

$$\mathbf{r} = xi + yj + zk$$

Thus,

$$\mathbf{M}_o = \mathbf{r} \times \mathbf{F} = \begin{vmatrix} i & j & k \\ x & y & z \\ F_x & F_y & F_z \end{vmatrix}$$

or
$$\mathbf{M}_o = i(yF_z - zF_y) - (xF_z - zF_x)j + (xF_y - yF_x)k$$

In case of problems involving only two dimensions, the force **F** may be assumed to lie in the xy-plane. Carrying $z = 0$ and $F_z = 0$, we obtain

$$\mathbf{M}_o = (xF_y - yF_x)k$$

VARIGNON'S THEOREM (PRINCIPLE OF MOMENTS)

Varignon's theorem states that "the moment of several concurrent coplanar forces about any point O in their plane equals the moment of their resultant about point O."

Let several forces $\mathbf{F}_1$, $\mathbf{F}_2$, $\mathbf{F}_3$, ... be applied at the same point A in plane as shown in Figure 1.10, and **r** be the position vector of A. Moment of forces about O is

$$\mathbf{M}_o = \mathbf{r} \times \mathbf{F}_1 + \mathbf{r} \times \mathbf{F}_2 + \mathbf{r} \times \mathbf{F}_3 + \cdots$$

From the property of vector, we have

$$\mathbf{M}_o = \mathbf{r} \times (\mathbf{F}_1 + \mathbf{F}_2 + \mathbf{F}_3 + \cdots)$$

or

$$\mathbf{M}_o = \mathbf{r} \times \mathbf{R},$$

where **R** is the resultant of forces $\mathbf{F}_1$, $\mathbf{F}_2$, $\mathbf{F}_3$, ..., given by $\mathbf{R} = \mathbf{F}_1 + \mathbf{F}_2 + \mathbf{F}_3 + \cdots$.

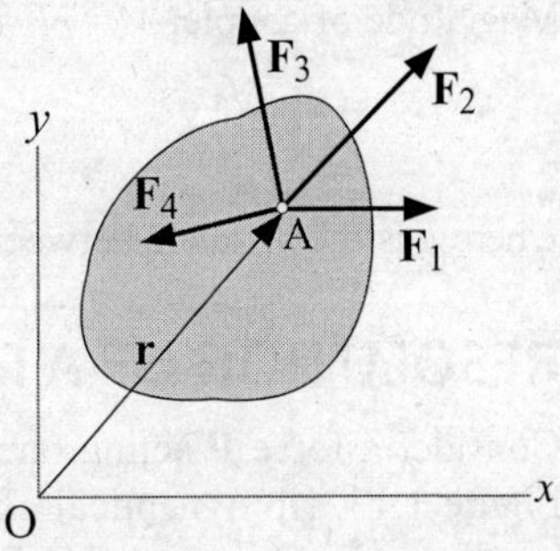

FIGURE 1.10 |

Thus,

$$\mathbf{r} \times \mathbf{F}_1 + \mathbf{r} \times \mathbf{F}_2 + \mathbf{r} \times \mathbf{F}_3 + \cdots = \mathbf{r} \times \mathbf{R}$$

This relation makes it possible to replace the direct determination of the moment of a force **F** by the determination of the moments of two or more component forces.

COUPLES

The moment produced by two equal, opposite and non-collinear forces is called couple. It does not produce any translation, but produces only rotation. The resultant force of a couple is zero.

The moment of a couple is the product of the magnitude of one of the forces and the perpendicular distance between their lines of action (Fig. 1.11).

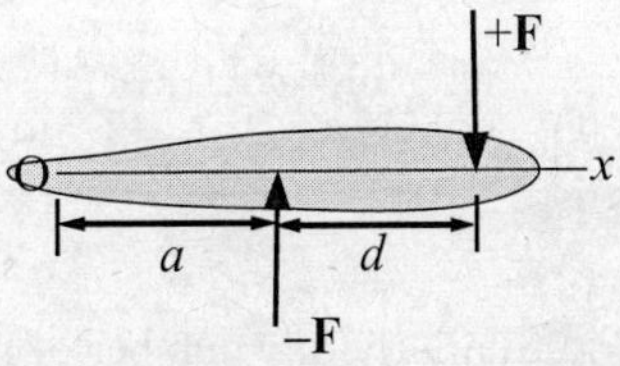

FIGURE 1.11 |

$$M = F(a + d) - Fa$$

or

$$M = Fd$$

The magnitude of the moment of a couple is the same for all points in the plane of the couple. A couple may be moved anywhere in its plane or a parallel plane without changing its external effect. The magnitude of a couple is independent of the reference point, and its tendency to create a rotation will remain constant.

Consider the couple consisting of forces +**F** and −**F** acting at points A and B as shown in Figure 1.12. Denoting by $\mathbf{r}_A$ and $\mathbf{r}_B$, respectively, the position vectors of the points A and B, we find that the sum of the moments of the two forces about O is

$$\mathbf{M} = \boldsymbol{r}_A \times \mathbf{F} + \boldsymbol{r}_B \times (-\mathbf{F}) = (\boldsymbol{r}_A - \boldsymbol{r}_B) \times \mathbf{F}$$

Let

$$\mathbf{r} = \mathbf{r}_A - \boldsymbol{r}_B,$$

$$\text{moment of couple} \quad \mathbf{M} = \mathbf{r} \times \mathbf{F}$$

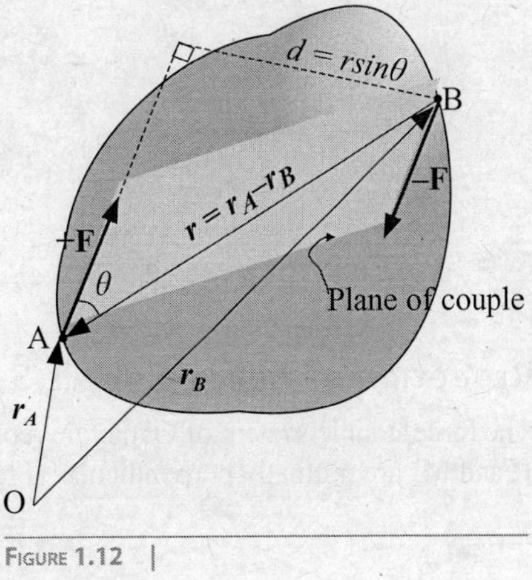

FIGURE 1.12 |

Direction of couple **M** is perpendicular to the plane containing the two forces.

Magnitude of couple, $M = rF \sin\theta = Fd$

$$M = Fd,$$

where d is the distance between the lines of action of +**F** and −**F**.

RESOLUTION OF A FORCE INTO A FORCE AND A COUPLE

Consider a force **F** acting on a rigid body at point A defined by the position vector **r** as shown in Figure 1.13. No dynamical effects are produced by adding self-equilibrating forces +**F** and −**F** at another point O as shown in Figure 1.13(b). The force **F** at A and the force −**F** at point O constitute a couple, $\mathbf{M}_o = \mathbf{r} \times \mathbf{F}$. Thus, the force **F** at point A is transferred to point O, and a couple $\mathbf{M}_o$ is introduced as shown in Figure 1.13(c). This type of combination of the force **F** and the couple **M** is called *force-couple* system.

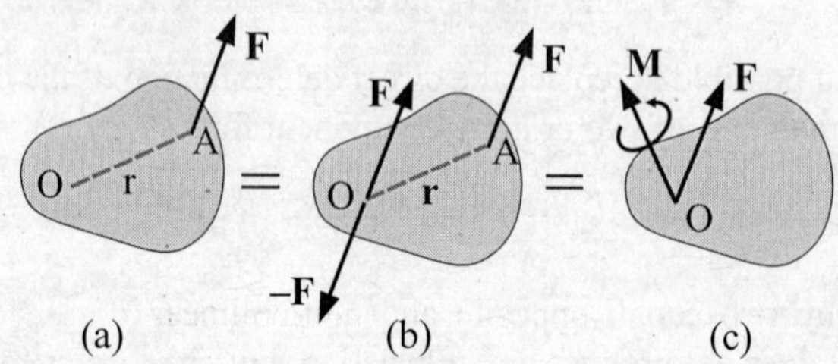

FIGURE 1.13 |

Similarly, any number of forces $\mathbf{F}_1$, $\mathbf{F}_2$, $\mathbf{F}_3$, … that act on a body may be transferred to a common point, provided that the couples $\mathbf{r}_1 \times \mathbf{F}_1$, $\mathbf{r}_2 \times \mathbf{F}_2$, $\mathbf{r}_3 \times \mathbf{F}_3$, … are introduced (Fig. 1.14). The vectors $\mathbf{r}_1$, $\mathbf{r}_2$, $\mathbf{r}_3$, … are vectors from reference point O to the points of application of the forces $\mathbf{F}_1$, $\mathbf{F}_2$, $\mathbf{F}_3$, …, respectively. The forces $\mathbf{F}_1$, $\mathbf{F}_2$, $\mathbf{F}_3$, … are now concurrent at point O, and their resultant $\mathbf{R} = \mathbf{F}_1 + \mathbf{F}_2 + \mathbf{F}_3 + \ldots$ and the resultant couple $\mathbf{M}_o = \mathbf{r}_1 \times \mathbf{F}_1 + \mathbf{r}_2 \times \mathbf{F}_2 + \mathbf{r}_3 \times \mathbf{F}_3 + \ldots$.

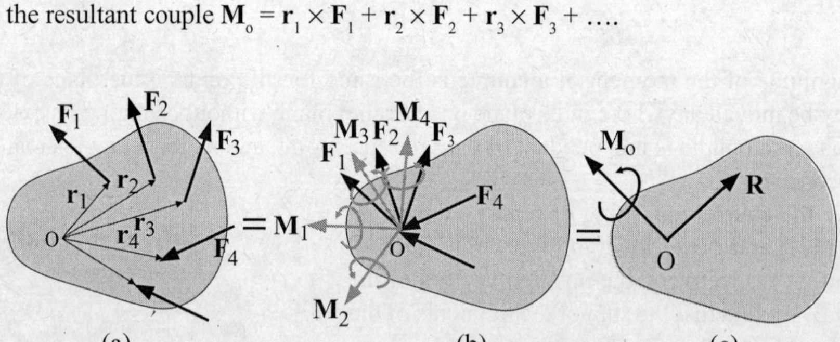

FIGURE 1.14 |

REDUCTION OF FORCE-COUPLE SYSTEM TO SINGLE FORCE

The force-couple system of O may be replaced by a single force **R** acting along a new line of action if **R** and $\mathbf{M}_o$ are mutually perpendicular (Fig. 1.15). It is applicable in coplanar force system.

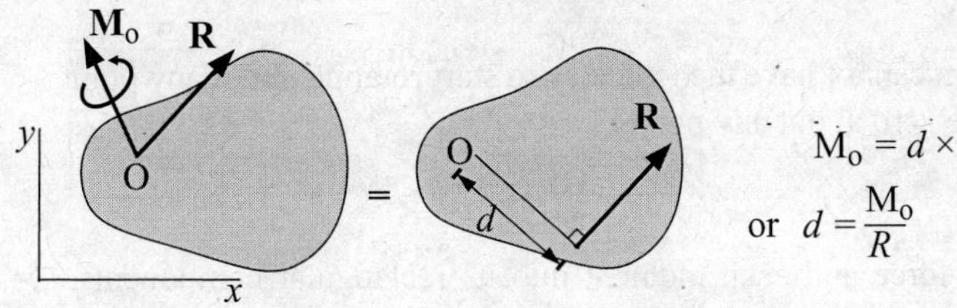

FIGURE 1.15 |

FREE BODY DIAGRAM

One of the most useful aids for solving a statics problem is the free body diagram (FBD).

A free body diagram is a sketch of the body that shows the body (by itself, free of the other parts of the system) and all the forces applied to it, that is, all forces acting on the body.

Figure 1.16 shows free body diagrams.

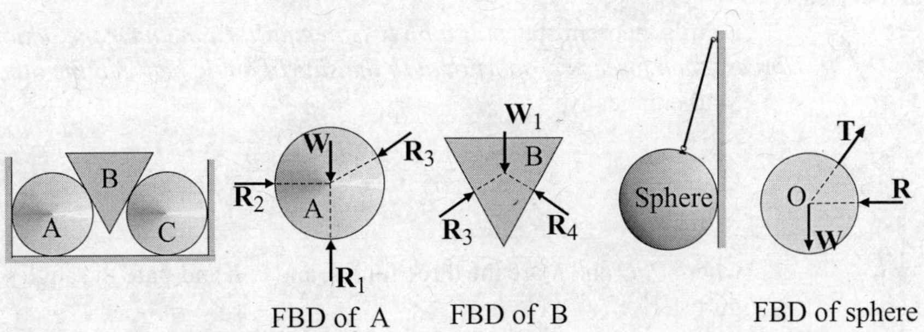

FIGURE 1.16 |

EQUILIBRIUM

The term *equilibrium* implies that either the body is at rest or it moves with a constant velocity. We shall deal with bodies at rest, or bodies in *static equilibrium*. A body is in static equilibrium when the force system acting on it tends to produce no net translation or rotation of the body.

A body is said to be in *static equilibrium* when the resultant force on it must be zero and the body must have no tendency to rotate. This second condition of equilibrium requires that the net moment about any point be zero.

CONDITIONS FOR EQUILIBRIUM

First Condition

A body is in equilibrium, that is, the body does not accelerate, if the vector sum of all the forces acting on it is zero.

$$\sum \mathbf{F} = 0$$

Second Condition

A body in equilibrium cannot have the tendency to start rotating about any point; so the sum of external moments must be zero about any point.

$$\sum \mathbf{M} = 0$$

Resolving each force and each moment into its rectangular components, we may express the necessary and sufficient conditions for the equilibrium of a body by the following six scalar equations:

$$\begin{aligned} \sum F_x &= 0 \qquad & \sum M_x &= 0 \\ \sum F_y &= 0 \qquad & \sum M_y &= 0 \\ \sum F_z &= 0 \qquad & \sum M_z &= 0 \end{aligned}$$

LAMI'S THEOREM

If a body is in equilibrium under the action of only three coplanar and concurrent forces, Lami's theorem can be used.

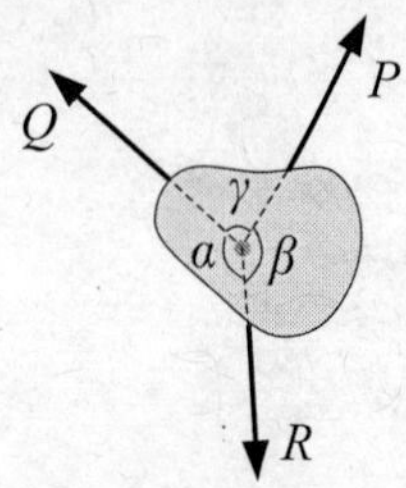

FIGURE 1.17 |

Lami's theorem states: *If a body is in equilibrium under the action of three forces, each force is proportional to the sine of angle between the other forces.*

Mathematically,

$$\frac{P}{\sin\alpha} = \frac{Q}{\sin\beta} = \frac{R}{\sin\gamma},$$

where P, Q and R are the three forces, and α, β and γ are the angles as shown in Figure 1.17.

PROOF OF LAMI'S THEOREM

Consider three coplanar forces P, Q and R acting at a point O as shown in Figure 1.18. Now complete the parallelogram OACB with OA and OB as adjacent sides as shown in the figure. Diagonal OC represents resultant of two forces P and Q.

Since these forces are in equilibrium, the resultant of the forces P and Q must be in line with OD and equal to R, but in opposite direction.

From the geometry of Figure 1.18,

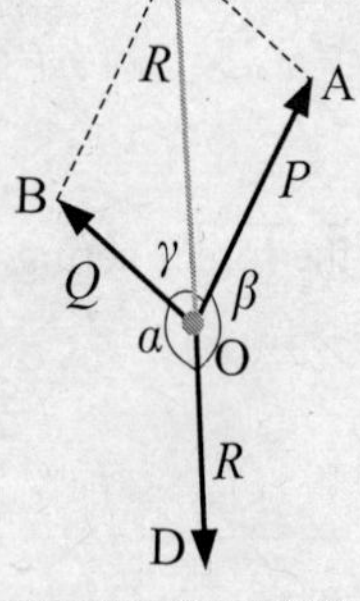

FIGURE 1.18 |

$$\text{OA} = P, \text{OB} = Q \text{ and OC} = R$$

$$\alpha + \beta + \gamma = 360° \text{ or } \angle\text{AOB} = \gamma = 360° - (\alpha + \beta)$$

$$\angle\text{AOC} = 180° - \beta, \angle\text{OAC} = 180° - \gamma$$

$$\angle\text{ACO} = 180° - \angle\text{ACO} - \angle\text{OAC} = 180 - (180 - \beta) - (180 - \gamma)$$

$$\text{or } \angle\text{ACO} = \beta + \gamma - 180 = 360 - \alpha - 180 = 180 - \alpha$$

We know that in triangle OAC

$$\frac{\text{OA}}{\sin\angle\text{ACO}} = \frac{\text{AC}}{\sin\angle\text{ACO}} = \frac{\text{OC}}{\sin\angle\text{OAC}}$$

or
$$\frac{P}{\sin(180^\circ - \alpha)} = \frac{Q}{\sin(180^\circ - \beta)} = \frac{R}{\sin(180^\circ - \gamma)}$$

$$\frac{P}{\sin \alpha} = \frac{Q}{\sin \beta} = \frac{R}{\sin \gamma} \quad \text{Lami's theorem}$$

CATEGORIES OF EQUILIBRIUM

	Force system	Free-body diagram	Equilibrium equations
1-direction	Collinear	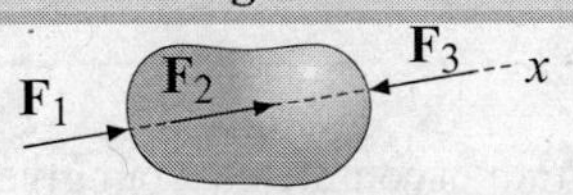	$\sum F_x = 0$
2-directions x-y plane	Concurrent at a point	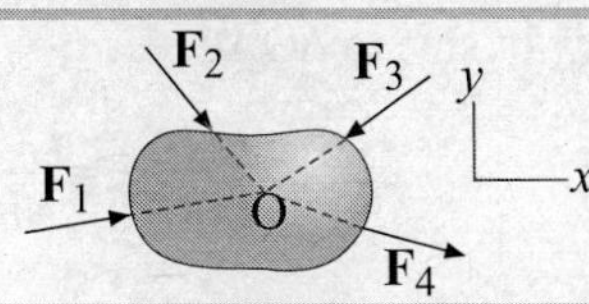	$\sum F_x = 0$ $\sum F_y = 0$
2-directions x-y plane	Parallel	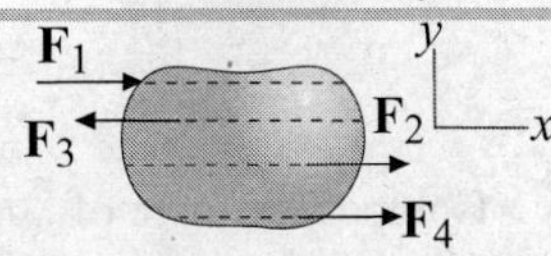	$\sum F_x = 0$ $\sum M_z = 0$
2-directions x-y plane	General	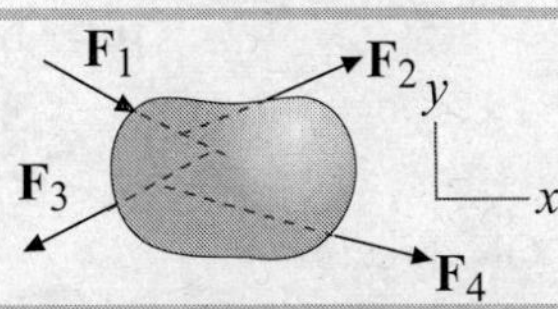	$\sum F_x = 0 \quad \sum M_z = 0$ $\sum F_y = 0$
3-directions space	Concurrent with a line	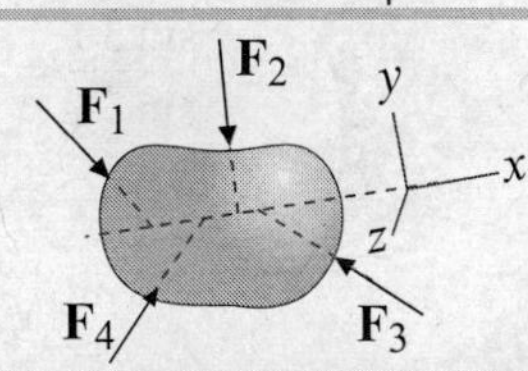	$\sum F_x = 0 \quad \sum M_y = 0$ $\sum F_y = 0 \quad \sum M_z = 0$ $\sum F_z = 0$
3-directions space	Concurrent wih a point	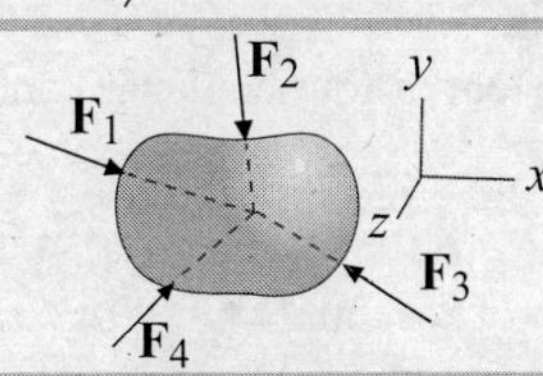	$\sum F_x = 0$ $\sum F_y = 0$ $\sum F_z = 0$
3-directions space	Parallel	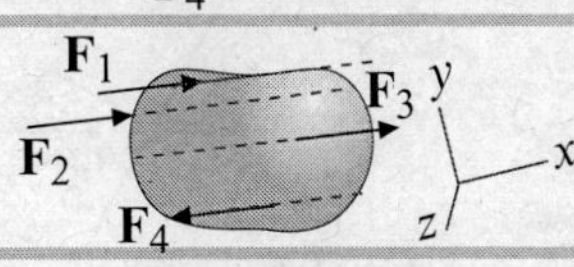	$\sum F_x = 0 \quad \sum M_y = 0$ $\sum M_z = 0$
3-directions space	General	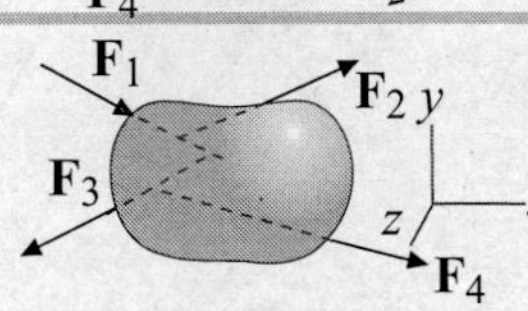	$\sum F_x = 0 \quad \sum M_x = 0$ $\sum F_y = 0 \quad \sum M_y = 0$ $\sum F_z = 0 \quad \sum M_z = 0$

SOLVED PROBLEMS

PROBLEM 1.1

Resolve the force 80 kN into two components, one along OM and the other along ON [see Fig. 1.1(a)].

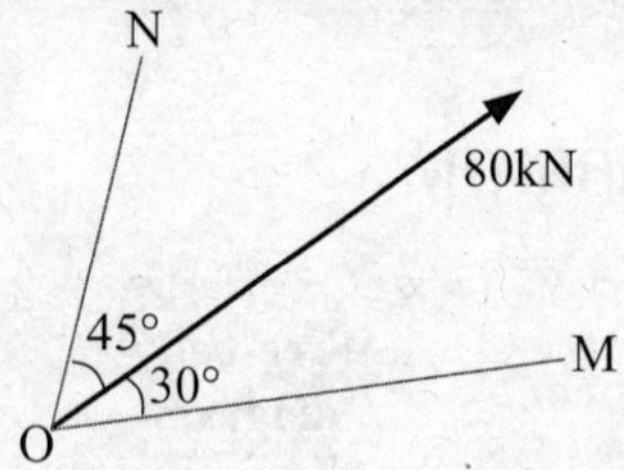

FIGURE 1.1 (a) |

Solution. The parallelogram is completed from given force 80 kN, direction line OM and ON [Fig. 1.1(b)]. Applying the law of sine to ΔOAC,

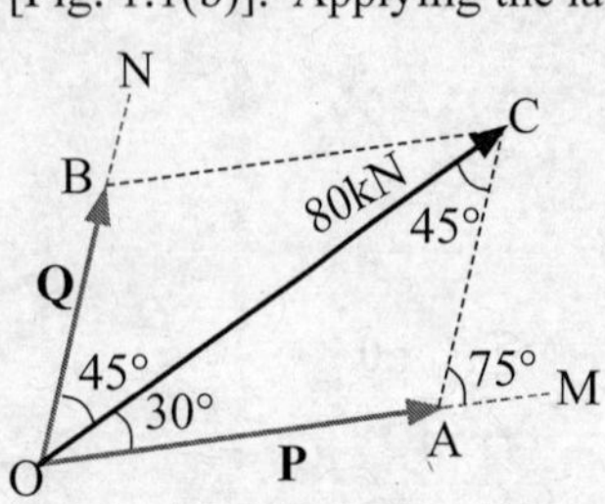

FIGURE 1.1 (b) |

$$\frac{OA}{\sin 45} = \frac{AC}{\sin 30} = \frac{80}{\sin(180 - 30 - 45)}$$

$$\frac{P}{\sin 45} = \frac{Q}{\sin 30} = \frac{80}{\sin 75}$$

Component force of 80 kN in OM direction is

$$P = 80\frac{\sin 45}{\sin 75} = 58.56$$

$$\boxed{P = 58.56\text{kN}}$$

Component force of 80 kN in ON direction is

$$Q = 80\frac{\sin 30}{\sin 75} = 41.41$$

$$\boxed{Q = 41.41\text{kN}}$$

PROBLEM 1.2

Determine the resultant of a system of four coplanar concurrent forces as shown in Figure 1.2(a) (by graphical method).

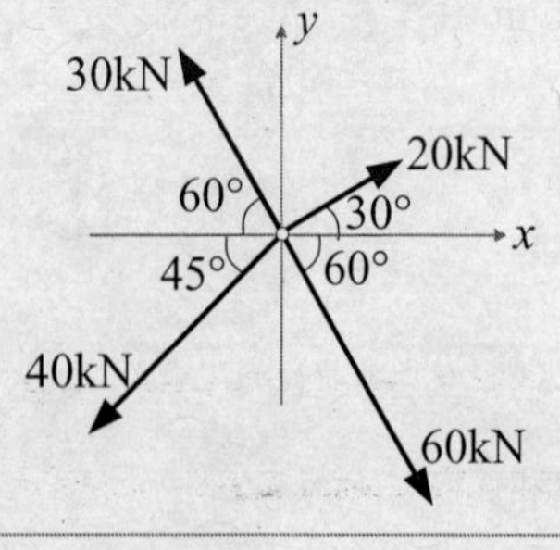

FIGURE 1.2 (a) |

Solution. Choose a suitable scale to represent the given forces. Let the scale is 10 kN = 1 cm. We construct the force polygon [Fig. 1.2(b)]. In the polygon construction, the forces may be arranged in various orders. In this problem, four different force polygons can be constructed. Note that the different polygons all provide the same result R.

Resultant $R = 44.4$ kN

$$R = 44.4 \text{ kN}$$

Direction of resultant

$$\theta = -84.78°$$

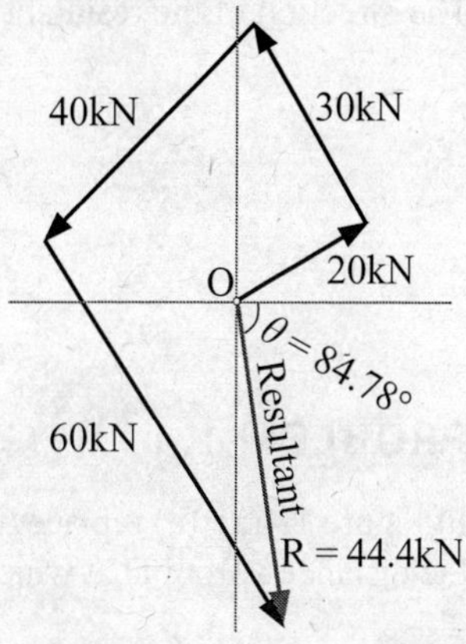

FIGURE 1.2 (b)

PROBLEM 1.3

Determine the resultant a system of four coplanar concurrent forces as shown in Figure 1.3(a) by composition of forces by method of resolution.

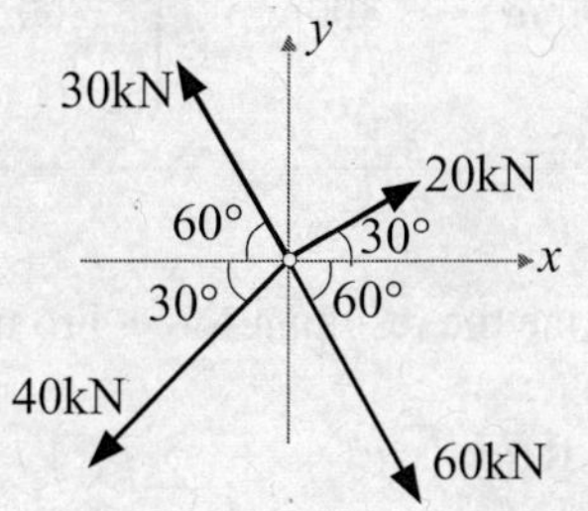

FIGURE 1.3 (a)

Solution. The x and y components of each force are determined by trigonometry as shown in the table. The x-components act to the right and y-components act upward and are represented by positive members.

Force	Magnitude (kN)	Angle q with (x-axis)	x-Components (F cos θ)	y-Components (F sin θ)
F_1	20	30	17.32	10
F_2	30	120	−15	25.98
F_3	40	225	−28.28	−28.28
F_4	60	300	30	51.96
			$\sum F_x = 4.04$kN	$\sum F_y = -44.26$kN

Thus, the resultant **R** of the four forces is

$$\mathbf{R} = 4.04i - 44.26j \text{ kN}$$

The magnitude of the resultant, $R = \sqrt{(4.04)^2 + (-44.26)^2}$

$$R = 44.4\text{kN}$$

The direction of the resultant

$$\alpha = \tan^{-1}\frac{\sum F_y}{\sum F_x} = \tan^{-1}\left(-\frac{44.26}{4.04}\right)$$

$$\alpha = -84.78^\circ$$

PROBLEM 1.4

In Figure 1.4(a), (a) replace the three forces by an equivalent force and couple at D and (b) determine the distance x from D at which the resultant of the three forces acts.

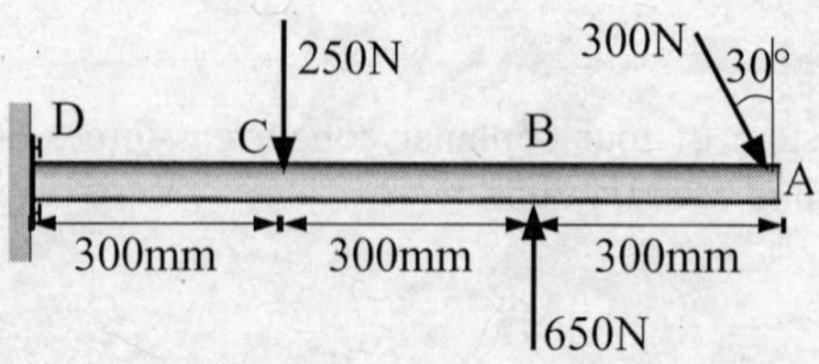

FIGURE 1.4 (a) |

Solution. (a) First we determine the resultant force. From Figure 1.4(b), we get

$$\mathbf{R} = \left(\sum F_x\right)i + \left(\sum F_y\right)j$$

or $\mathbf{R} = (300 \sin 30)i + (650 - 250 - 300 \cos 30)j$

or $\mathbf{R} = 150i + 140.2j\ N$

Magnitude of the resultant **R** is

$$R = 205.3 \text{ N}$$

and direction of the resultant

$$\theta = 43.06^\circ$$

Moment of forces about D

$$M_D = 250 \times \frac{300}{1000} - 650 \times \frac{600}{100} + 300 \cos 30 \times \frac{900}{1000}$$

$$M_D = -81.18 \text{ N m}$$

or $$M_D = 81.18 \text{ N m (CCW)}$$

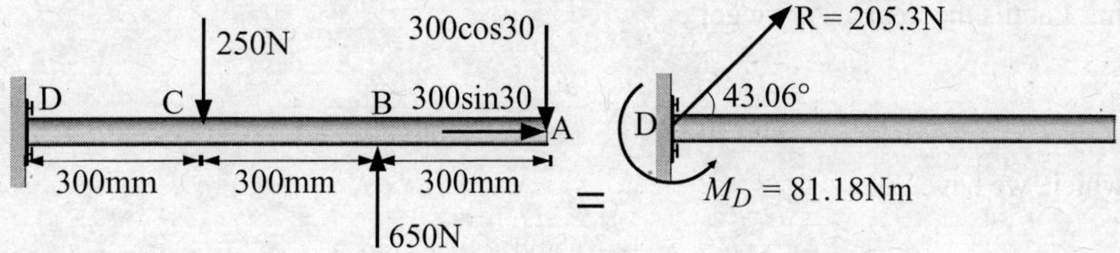

FIGURE 1.4 (b) |

(b) Let the resultant be at a distance x from D [Fig. 1.4(c)].
Moment of resultant about D is $M_D = R \sin 43.06 \times x$
Substituting the values of M_D and R, we have

$$81.18 = 205.3 \times \sin 43.06 \times x$$

$$x = \frac{81.18}{205.3 \times \sin\ 43.06} = 0.579$$

$x = 579$ mm

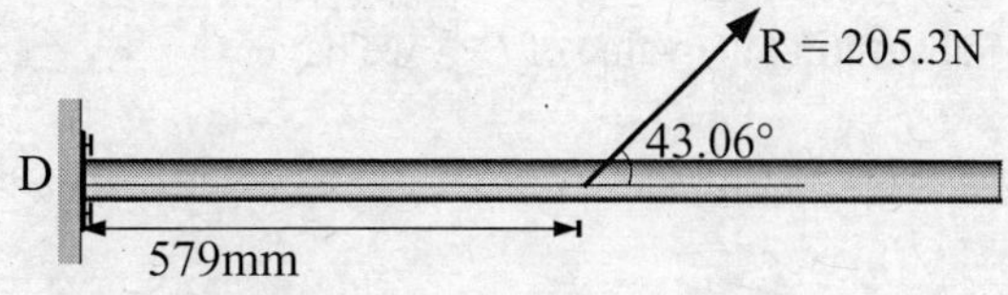

FIGURE 1.4 (c) |

PROBLEM 1.5

Figure 1.5(a) shows the slider crank mechanism. The connecting rod AB of length l supports a variable compressive force F_c. Derive an expression for the moment of F_c about the crank axis O in terms of F_c, l, r and θ.

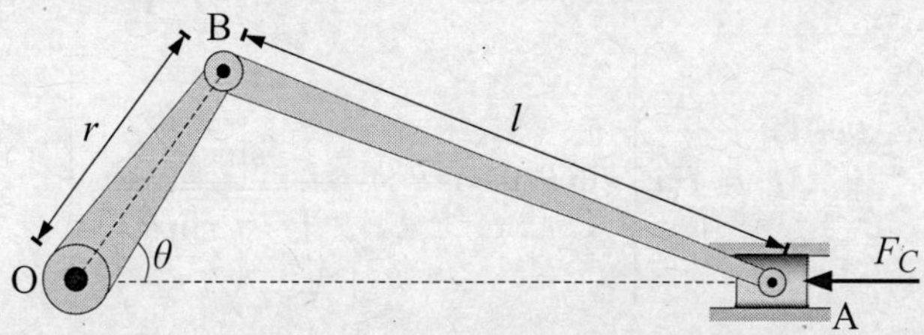

FIGURE 1.5 (a) |

Solution. The free body diagram of slider and crank is drawn as given in Figure 1.5 (b). We begin with slider A which is subjected to three forces, one is applied force F_c, the others are reactive forces F_{AB} and R.

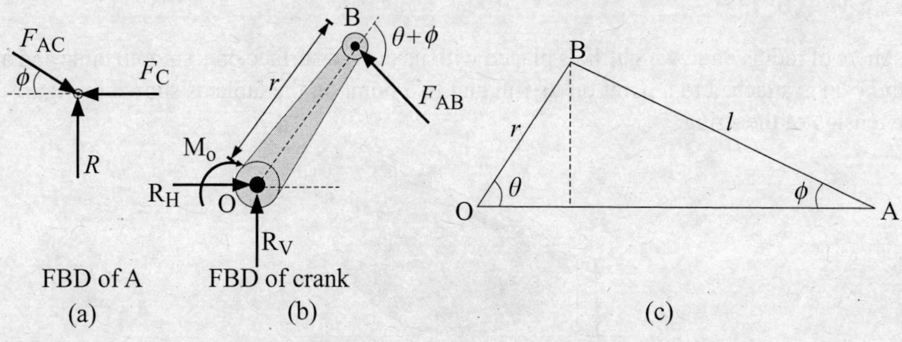

FIGURE 1.5 (b) |

Applying Lami's theorem at A, we get

$$\frac{F_c}{\sin(90+\phi)} = \frac{F_{AB}}{\sin 90} = \frac{R}{\sin(180-\phi)}$$

From which, we have

$$F_{AB} = \frac{F_c}{\cos\phi} \text{ and } R = F_c \tan\phi$$

Force F_{AB} is transferred at joint B. Figure 1.5(b) shows the free body diagram of crank OB.

$$M_o = F_{AB} \sin[180 - (\theta + \Phi)] \times r$$

Substituting the value of F_{AB}, we have

$$M_o = \frac{F_c}{\cos\phi} \sin(\theta+\phi) \times r$$

$$M_o = F_c r \left[\sin\theta + \cos\theta \times \frac{\sin\phi}{\cos\phi} \right]$$

From the geometry of figure,

$$r \sin\theta = l \sin\phi \Rightarrow \sin\phi = \frac{\sin\theta}{(l/r)} \quad \text{or} \quad \cos\phi = \sqrt{1 - \left(\frac{\sin\theta}{l/r}\right)^2}$$

or

$$M_o = F_c r \left[\sin\theta + \cos\theta \times \frac{\sin\theta}{\frac{l}{r}\sqrt{1-\left(\frac{\sin\theta}{l/r}\right)^2}} \right]$$

$$M_o = F_c r \left[\sin\theta + \frac{\sin 2\theta}{2\sqrt{(l/r)^2 - \sin^2\theta}} \right]$$

PROBLEM 1.6

A hemisphere of radius r and weight W is placed with its curved surface on a smooth table and a string of length l ($<r$) is attached to a point on its rim and to a point on the table as shown in Figure 1.6(a). Find the tension of the string.

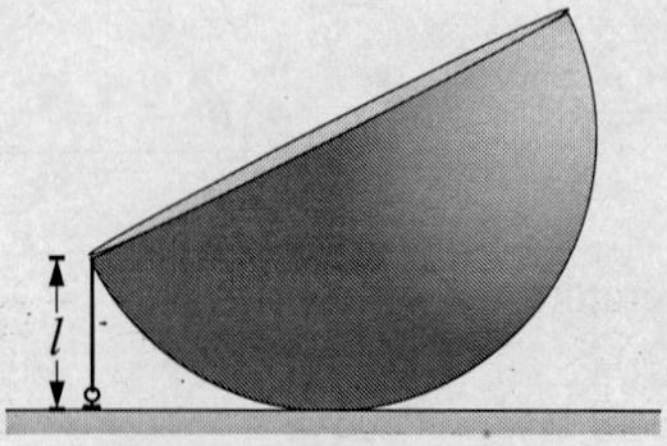

FIGURE 1.6(a) |

Solution. Free body diagram is shown in Figure 1.6(b). Let O be the centre of the base and OD the axis of the hemisphere. The centre of gravity G of the hemisphere lies on OD such that OG = 3*r*/8. The weight *W* of the hemisphere acts vertically downward at G.

Let the base of the hemisphere be inclined at an angle θ to the horizontal.

Taking the moment of the force about contact point C, we have

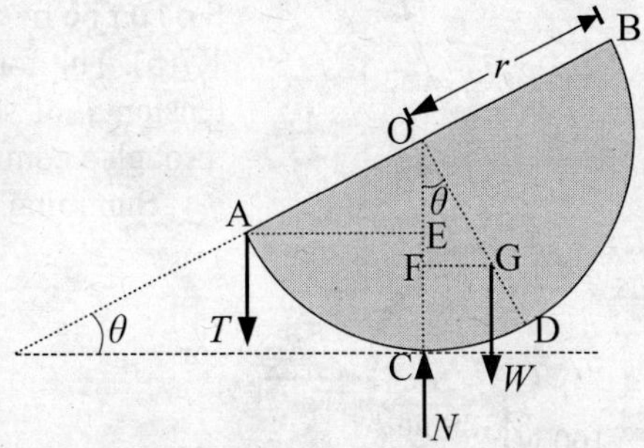

FIGURE 1.6 (b)

$$\sum M_c = 0 \Rightarrow T \times \text{AE} - W \times \text{FG} = 0$$

or $$T \times r\cos\theta - W \times \frac{3}{8} r\sin\theta = 0$$

or $$T = \frac{3W}{8}\tan\theta$$

or $$T = \frac{3W}{8} \times \frac{\sin\theta}{\cos\theta} = \frac{3W}{8} \times \frac{\sin\theta}{\sqrt{1-\sin^2\theta}}$$

or $$T = \frac{3W}{8} \times \frac{[(r-l)/r]}{\sqrt{1-[(r-l)/r]^2}}$$

$$\left(\sin\theta = \frac{\text{OE}}{\text{OA}} = \frac{r-l}{r}\right)$$

or $$T = \frac{3W}{8} \times \frac{r-l}{\sqrt{2rl-l^2}}$$

Tension of the string, $T = \dfrac{3W}{8} \times \dfrac{r-l}{\sqrt{2rl-l^2}}$

PROBLEM 1.7

A thin ring of weight 100 N and radius 300 mm is held against a smooth wall by a 200 mm long string AB [Fig. 1.7(a)]. Determine (a) the tension in the string and (b) the reaction at C.

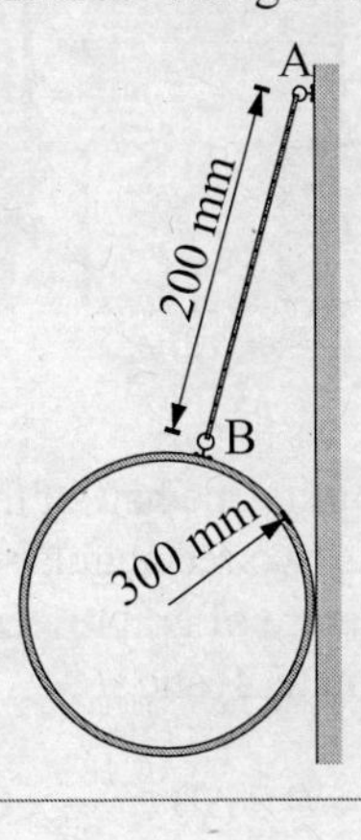

FIGURE 1.7(a)

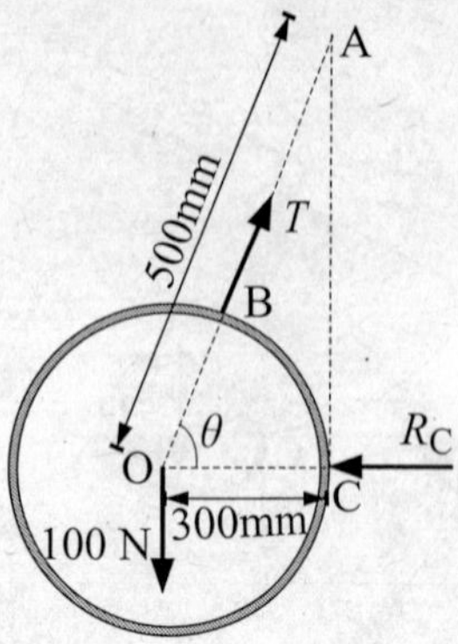

FIGURE 1.7(b) |

Solution. A free body diagram of ring is drawn as shown in Figure 1.7(b). The forces acting on the ring are its weight *mg*, reaction R_C, and tension T of string. For the equilibrium, these three forces should pass through a common point O.

Summing forces vertically, we have,

$$\sum F_y = 0 \Rightarrow T \sin\theta - 100 = 0$$

or $$T \sin\theta = 100$$

or $$T = \frac{100}{\sin\theta} = \frac{100}{400/500}$$

and tension of string, $T = 125$ N

Summing forces horizontally, we have

$$\sum F_x = 0 \Rightarrow T\cos\theta - R_C = 0$$

or $$R_C = T\cos\theta$$

or $$R_C = 125 \times \frac{300}{500}$$

$$R_C = 75\text{N}$$

PROBLEM 1.8

Two halves of a round homogeneous smooth cylinder are held together by a thread wrapped round the cylinder with two weights each equal to P attached to its ends as shown in Figure 1.8(a). The weight of complete cylinder is W. Determine the minimum value of weight P for which both halves of the cylinder will be in equilibrium.

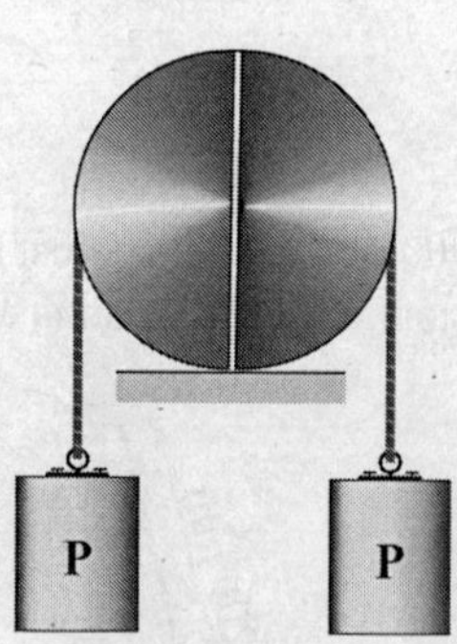

FIGURE 1.8(a) |

Solution. The free body diagram of the one half of the cylinder is shown in Figure 1.8(b) with the reaction at A represented in terms of its two rectangular components A_x and A_y. The weight of one half cylinder is $W/2$ and acts through its centre of gravity OG = $4r/3\pi$

Now taking moment about A to eliminate A_x and A_y,

$$\sum M_A = 0 \Rightarrow P \times AB - \frac{W}{2} \times OG - P \times OC = 0$$

or
$$P \times 2r - \frac{W}{2} \times \frac{4r}{3\pi} - P \times r = 0$$

or
$$Pr = \frac{2W}{3\pi} r$$

$$P = \frac{2W}{3\pi}$$

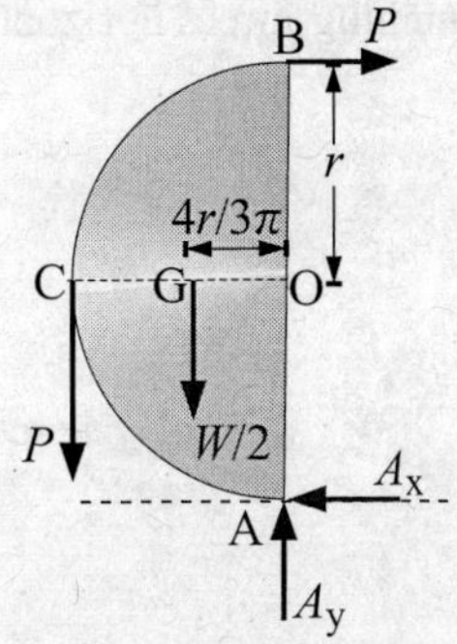

FIGURE 1.8(b)

PROBLEM 1.9

Find the length l if the mass m maintains a specified tension T in the belt for the position shown in Figure 1.9(a). Neglect the mass of the arm and central pulley compared with m. Also find the reaction at A.

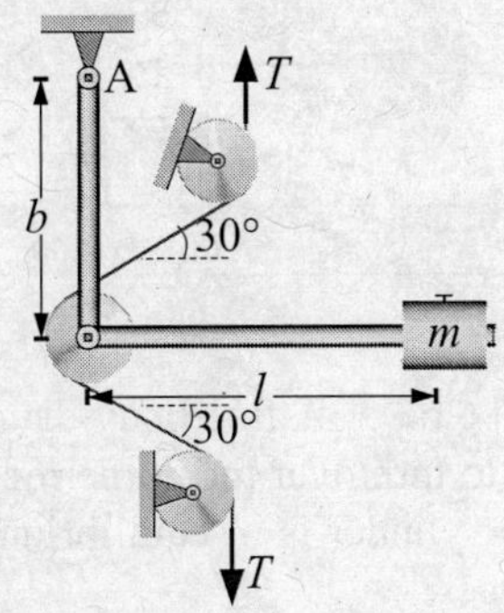

FIGURE 1.9(a)

Solution. The free body diagram of the system is shown in Figure 1.9(b) with the pin reaction at A represented in terms of its two rectangular components.

Taking moment about A

$$\sum M_A = 0 \Rightarrow$$

$$mg \times l - T\cos 30 \times b - T\cos 30 \times b = 0$$

or
$$mgl = 2Tb\cos 30 \times b$$

$$l = \frac{\sqrt{3}Tb}{mg}$$

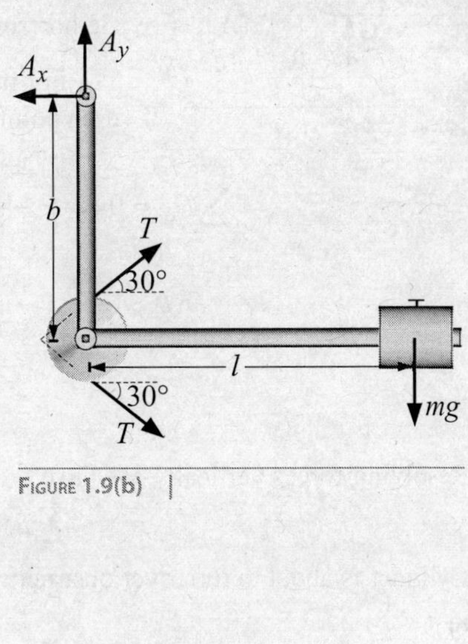

FIGURE 1.9(b)

Summing forces vertically, we have

$$\sum F_y = 0 \Rightarrow A_y - mg - T\sin 30 + T\sin 30 = 0$$

or
$$A_y = mg$$

Summing forces horizontally

$$\sum F_x = 0 \Rightarrow T \cos 30 + T \cos 30 - A_x = 0$$

or

$$A_x = 2T \cos 30$$

or

$$A_x = \sqrt{3}\,T$$

$$\text{Reaction at A} = \sqrt{A_x^2 + A_y^2}$$

$$A = \sqrt{(mg)^2 + (\sqrt{3}T)^2}$$

PROBLEM 1.10

Determine the force P required to begin rolling a uniform cylinder of mass m over the obstruction of height h as shown in Figure 1.10(a).

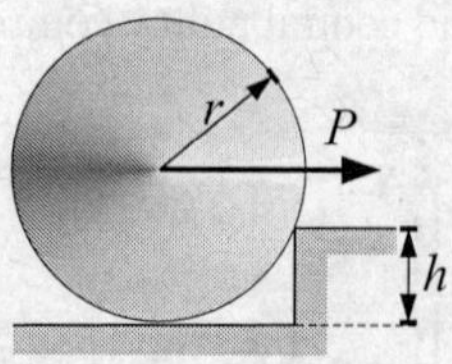

FIGURE 1.10(a)

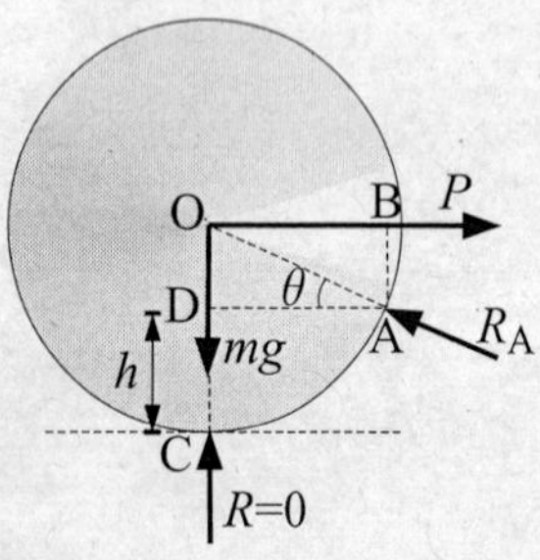

FIGURE 1.10(b)

Solution. A free body diagram of cylinder, when the cylinder is about to turn over the corner of the obstruction, is shown in Figure 1.10(b). Cylinder is in equilibrium under the action of the following three forces:

1. weight mg acting vertically downward
2. reaction R_A of the point A. The direction of R_A is unknown
3. horizontal force P.

For the equilibrium, these three forces should pass through a common point, hence reaction R_A must pass through the point O.

Resolving forces horizontally, we get

$$\sum F_x = 0 \Rightarrow P - R_A \cos\theta = 0$$

$$\left(\sin\theta = \frac{\text{OD}}{\text{OA}} = \frac{r-h}{r},\ \cos\theta = \sqrt{1-\sin^2\theta} = \frac{\sqrt{2rh-h^2}}{r}\right)$$

$$P = R_A \frac{\sqrt{2rh-h^2}}{r} \tag{i}$$

Resolving forces vertically, we have

$$\sum F_y = 0 \Rightarrow R_A \sin\theta - mg + R = 0$$

Cylinder is about to turn over obstruction; $R = 0$

or

$$mg = R_A \sin\theta$$

$$mg = R_A\left(\frac{r-h}{r}\right) \tag{ii}$$

Eliminating R_A from Equations (i) and (ii), we obtain

$$\frac{P}{mg} = \frac{\sqrt{2rh-h^2}}{r-h}$$

$$P = mg\,\frac{\sqrt{2rh-h^2}}{r-h}$$

Method of Moments

This problem can also be solved by taking moments of all the three forces about point A.

$$\sum M_A = 0 \Rightarrow P \times \text{AB} = mg \times \text{AD}$$

or $$P \times (\text{OC} - \text{CD}) = mg \times \sqrt{\text{OA}^2 - \text{OD}^2}$$

or $$P \times (r-h) = mg \times \sqrt{r^2 - (r-h)^2}$$

$$P = mg \times \frac{\sqrt{2rh-h^2}}{r-h}$$

PROBLEM 1.11

A uniform bar with end rollers has a mass of 50 kg and is supported by horizontal and vertical surfaces and by the wire CD [Fig. 1.11(a)]. Determine the tension T in the wire and reaction against the rollers at A and B.

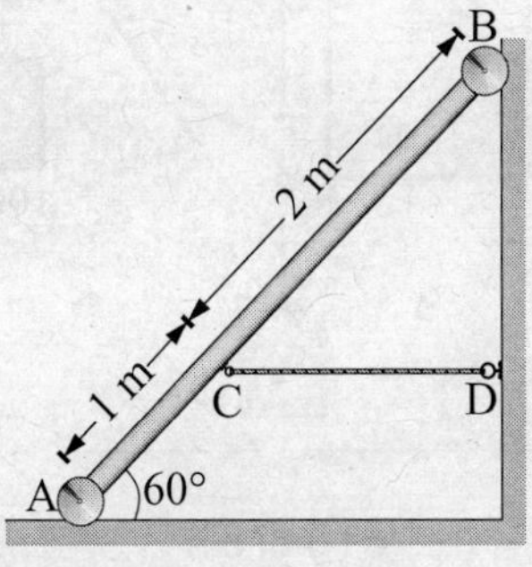

FIGURE 1.11(a) |

Solution. A free body diagram of the bar is drawn as shown in Figure 1.11(b). The reaction at each roller is perpendicular to the surface. T is the tension in wire and 490.5 N is the weight of bar which acts at the centre G. Note that there are three unknown R_A, R_B and T which may be found from the three equations of equilibrium.

Summing forces horizontally, we get

$$\sum F_x = 0 \Rightarrow T - R_B = 0$$

$$T = R_B$$

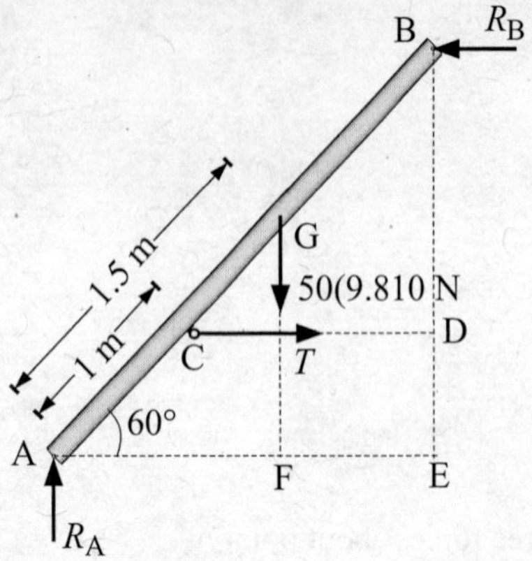

FIGURE 1.11(b) |

Summing forces vertically, we have

$$\sum F_y = 0 \Rightarrow R_A - 50(9.81) = 0$$

or

$$R_A = 50 \times 9.81 = 490.5\text{N}$$

$$R_A = 490.5\text{N}$$

Now, taking moment about A

$$\sum M_A = 0 \Rightarrow 950.5 \times \text{AF} + T \times \text{DE} - R_B \times \text{BE} = 0$$

or $950.5 \times (1.5 \cos 60) + T \times (1 \sin 60) - T \times (3 \sin 30) = 0$

or

$$367.8 + 0.866T - 1.5T = 0$$

From which, we get

$$T = \frac{367.80}{0.634} = 580.2\text{N}$$

$$T = 580.2\text{N}$$
$$R_A = T = 580.2\text{N}$$

PROBLEM 1.12

The rope pulley arrangement shown in Figure 1.12(a) is under equilibrium. Find the value of weight W. Neglect the weight of pulley and friction.

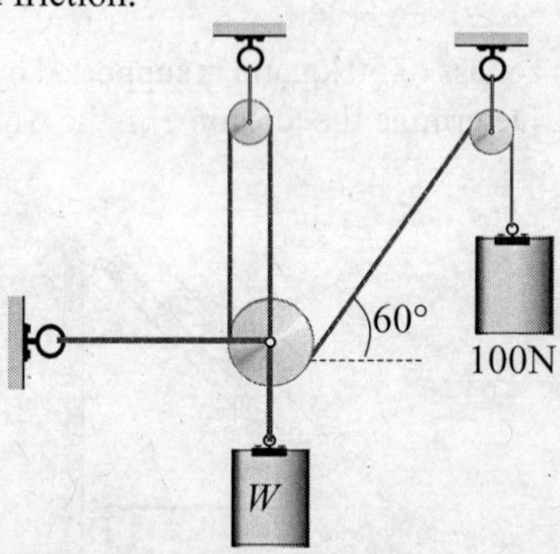

FIGURE 1.12(a) |

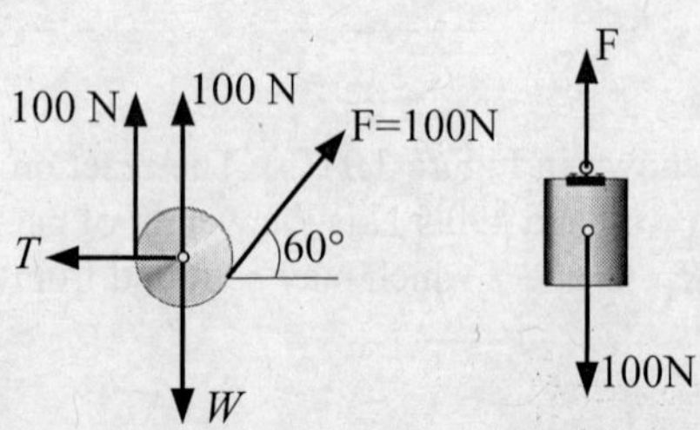

FIGURE 1.12(b) |

Solution. Free body diagrams of pulley and weight are shown in Figure 1.12(b). Five forces are acting on the pulley. Resolving the forces in horizontal and vertical direction for equilibrium, we get

$$\sum F_x = 0 \Rightarrow 100 \cos 60 - T = 0$$

or

$$T = 50\text{N}$$

$$\sum F_y = 0 \Rightarrow 100 \sin 60 + 200 - W = 0$$

$$W = 50\sqrt{3} + 200$$

or

$$W = 286.60\text{N}$$

PROBLEM 1.13

A cord supported at A and B carries a load of 20 kN at D and a load of W at C as shown in Figure 1.13(a). Find the value of W so that CD remains horizontal.

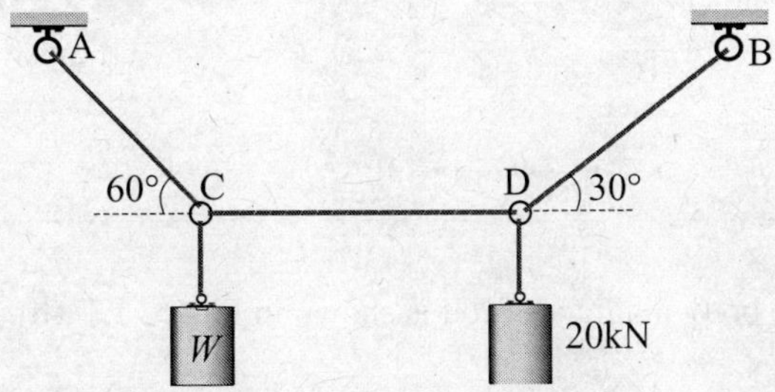

FIGURE 1.13(a)

Solution. Free body diagram of joint D is shown in Figure 1.13(b). Three forces T_1, T_2 and 20 kN are acting at point D.

Applying Lami's theorem at D, we have

$$\frac{T_1}{\sin 90} = \frac{T_2}{\sin\,(90+30)} = \frac{20}{\sin(180-30)}$$

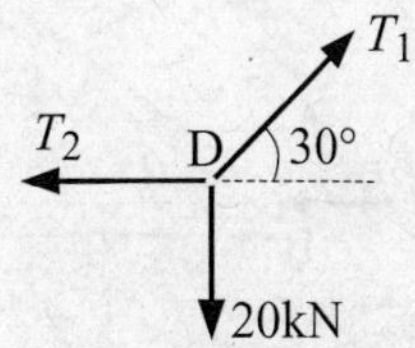

FIGURE 1.13(b)

or $$T_1 = \frac{T_2}{\cos 30} = \frac{20}{\sin 30}$$

$$T_1 = 40\text{kN}, \quad T_2 = 20\sqrt{3}\text{kN}$$

Free body diagram of joint C is shown in Figure 1.13(c).

Three forces T_2, T_3 and W are acting at point C. Applying Lami's theorem at C,

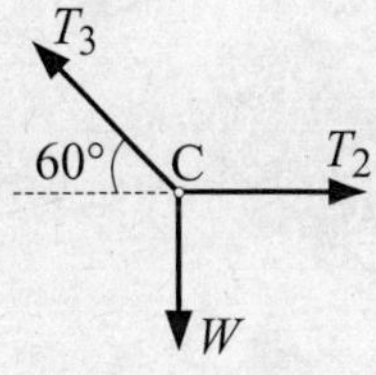

FIGURE 1.13(c)

$$\frac{T_2}{\sin\,(90+60)} = \frac{T_3}{\sin 90} = \frac{W}{\sin(180-60)}$$

or $$\frac{20\sqrt{3}}{\cos 60} = T_3 = \frac{W}{\sin 60}$$

$$T_3 = \frac{40\sqrt{3}}{\cos 60}, \quad \text{and} \quad W = 60\text{kN}$$

$$W = 60 \text{ kN}$$

PROBLEM 1.14

A rigid rod with negligible weight and small transverse carries a load W whose position is adjustable. The rod rests on a small roller at A and bears against a smooth vertical wall at B [Fig. 1.14(a)]. Determine the distance x for any given value of θ such that the rod will be in equilibrium. Neglect the friction.

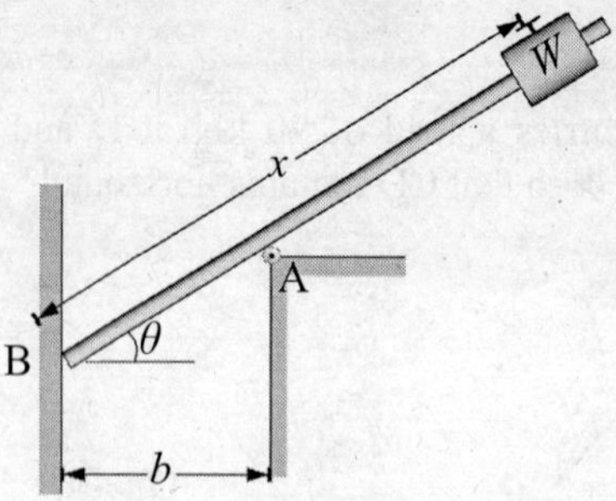

FIGURE 1.14(a) |

Solution. The free body diagram of rod is shown in Figure 1.14(b). Summing moment about B, we have

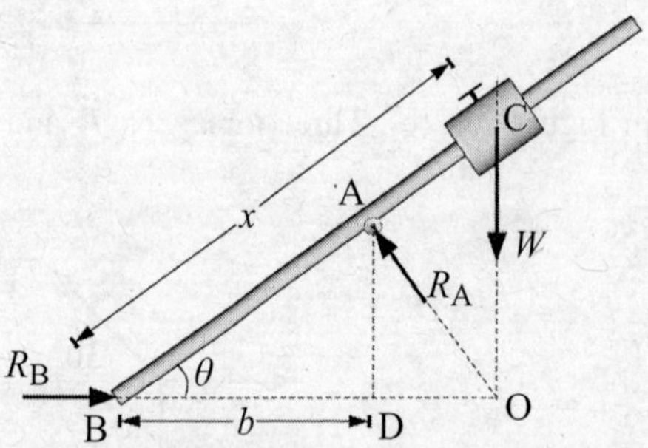

FIGURE 1.14(b) |

$$\sum M_B = 0 \Rightarrow R_A \times AB - W \times OB = 0$$

or $$R_A \times \frac{b}{\cos\theta} - W \times x\cos\theta = 0 \qquad \text{(i)}$$

Summing forces vertically, we get

$$\sum F_v = 0 \Rightarrow R_A \times \cos\theta - W = 0$$

or $$R_A = \frac{W}{\cos\theta} \qquad \text{(ii)}$$

Substituting $R_A = W/\cos\theta$ into Equation (i), we have

$$\frac{W}{\cos\theta} \times \frac{b}{\cos\theta} - W \times x\cos\theta = 0$$

or $$x = \frac{b}{\cos^3\theta}$$

$$x = b\sec^3\theta$$

PROBLEM 1.15

A uniform bar AB of weight 20 kN hinged at A is kept horizontal by supporting and setting a 50 kN weight with the help of a string tied at B and passing over a smooth peg D as shown in Figure 1.15(a). Determine the tension in the string and reaction at the supports A and D.

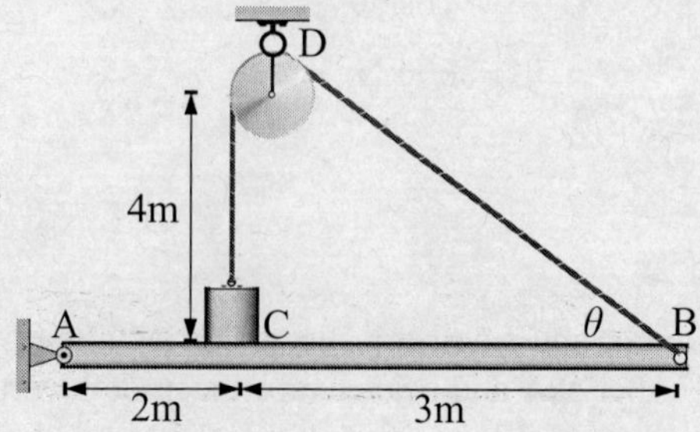

FIGURE 1.15(a) |

Solution. The free body diagram of the beam AB is shown in Figure 1.15(b) with the pin reaction at A represented in terms of its two rectangular components. The weight of the beam is 20 kN and acts through its centre G. It may be noted that the weight of 50 kN is partially resting on the bar and is partially supported by the string to the extent of tension T.

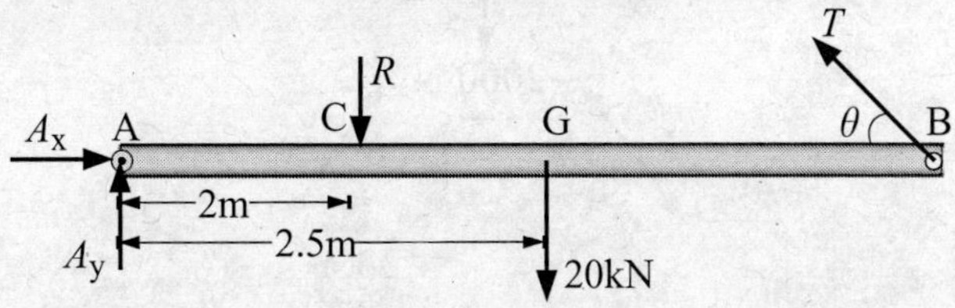

FIGURE 1.15(b) |

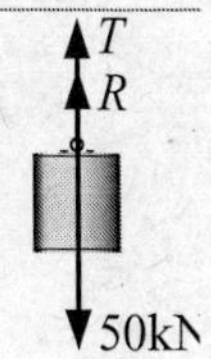

We take moments about A to eliminate A_x and A_y

$$\sum M_A = 0 \Rightarrow R \times AC + 20 \times AG - T \sin\theta \times AB = 0$$

or
$$R \times 2 + 20 \times 2.5 - T \times \frac{4}{5} \times 5 = 0$$

or
$$2R + 50 - 4T = 0 \tag{i}$$

$$\sum F_y = 0 \Rightarrow T + R - 50 = 0 \tag{ii}$$

Eliminating R from the Equations (i) and (ii),

$$2(50 - T) + 50 - 4T = 0$$

$$T = \frac{150}{6} \text{ kN}$$

PROBLEM 1.16

Two smooth cylinders, each of weight 1000 N and radius 250 mm are connected at their centres by a string of length 800 mm and rest upon a horizontal plane, supporting above a third cylinder of weight 2000 N and radius 150 mm as shown in Figure 1.16(a). Find the tension of string and reactions at all contact points of cylinder B.

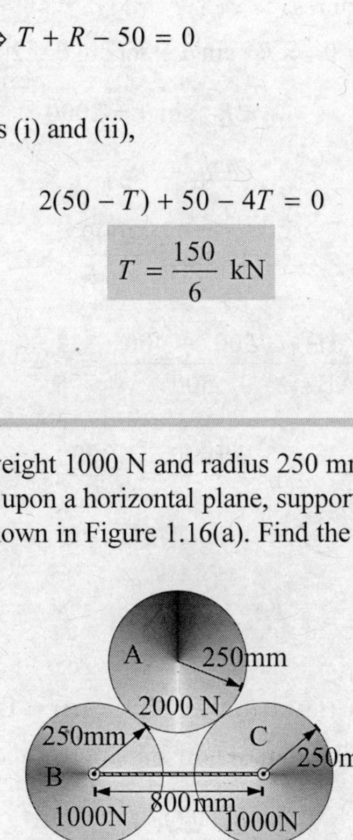

FIGURE 1.16(a) |

Solution. A free body diagram of each cylinder is drawn as shown in Figure 1.16(b) in its relative position to the others. We begin with cylinder A.

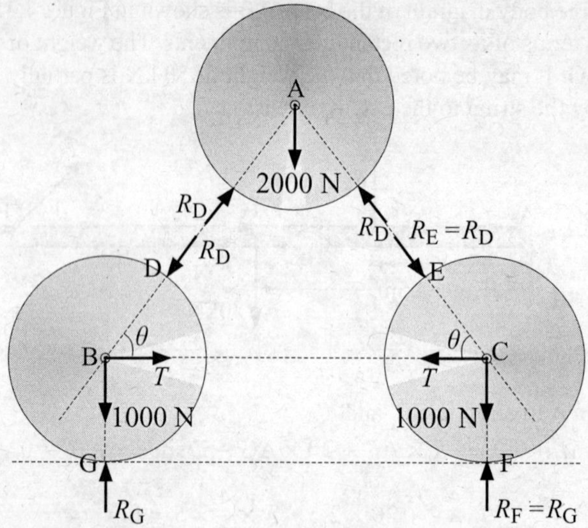

FIGURE 1.16(b) |

An equilibrium of vertical forces requires

$$\sum F_y = 0 \Rightarrow R_D \sin\theta + R_E \sin\theta - 2000 = 0$$

or

$$2R_D \sin\theta = 2000$$

$$[R_E = R_D]$$

or

$$R_D = \frac{2000}{\sin\theta}$$

$$\left[\sin\theta = \frac{AH}{AB} = \frac{\sqrt{500^2 - 400^2}}{500} = \frac{3}{5} \Rightarrow \cos\theta = \frac{4}{5}\right]$$

or

$$R_D = \frac{1000}{\frac{3}{5}} = \frac{5000}{3}\text{ N}$$

$$R_D = R_E = \frac{5000}{3}\text{ N}$$

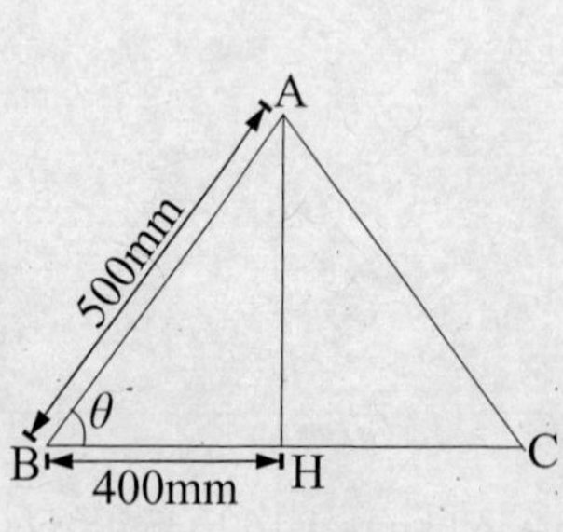

FIGURE 1.16(c) |

EQUILIBRIUM OF CYLINDER B

Equilibrium of horizontal forces requires [see Fig. 1.16(c)]

$$\sum F_x = 0 \Rightarrow T - R_D \cos\theta = 0$$

or

$$T = R_D \cos\theta = \frac{5000}{3} \times \frac{4}{5}$$

$$T = \frac{4000}{3}\text{ N}$$

Equilibrium of vertical forces requires

$$\sum F_y = 0 \Rightarrow R_G - 1000 - R_D \sin\theta = 0$$

or
$$R_G - 1000 - \frac{5000}{3} \times \frac{3}{5} = 0$$

or
$$R_G = 2000\text{N}$$

$$R_G = R_E = 2000\text{N}$$

PROBLEM 1.17

Find the value of the couple M required to roll a 50 kg roller up the incline [Fig. 1.17(a)]. Also determine the contact force at C. The surface of the incline is sufficiently rough to prevent slipping.

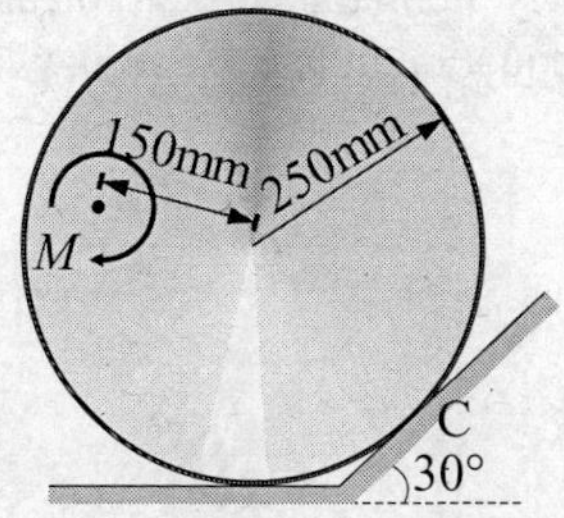

FIGURE 1.17(a)

Solution. Figure 1.17(b) shows the free body diagram of the roller, when it is just about to roll over the floor. Now taking the moment about C, we have

$$\sum M_c = 0 \Rightarrow M - 50(9.81) \times \text{CD} = 0$$

or
$$M = 50 \times 9.81 \times - \frac{250}{1000} \sin 30 = 61.3$$

$$M = 61.3\text{Nm}$$

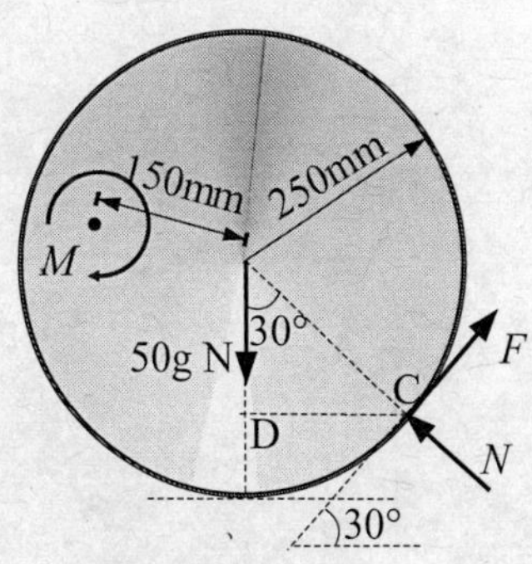

FIGURE 1.17(b)

For convenience, we choose the x-axis parallel to the incline plane and y-axis perpendicular to the incline plane.

Resolving the forces in y-direction, we get

$$\sum F_y = 0 \Rightarrow 50(9.81) \times \cos 30 - N = 0$$

or
$$N = 50 \times 9.81 \times \frac{\sqrt{3}}{2}$$

or
$$N = 424.7\text{N}$$

Resolving the forces x-direction,

$$\sum F_x = 0 \Rightarrow F - 50(9.81) \times \sin 30 = 0$$

or
$$F = 50 \times 9.81 \times \frac{1}{2}$$

or
$$F = 245.2\text{N}$$

Thus, reaction at C

$$R = \sqrt{F^2 + N^2} = \sqrt{(245.2)^2 + (424.7)^2} = 490.4\text{N}$$

$$R = 490.4\text{N}$$

PROBLEM 1.18

A collar C of weight W may move freely along the vertical rod as shown in Figure 1.18(a). The spring is unstretched when $y = 0$. Derive an equation in y, k, a and W which must be satisfied when the collar is in equilibrium.

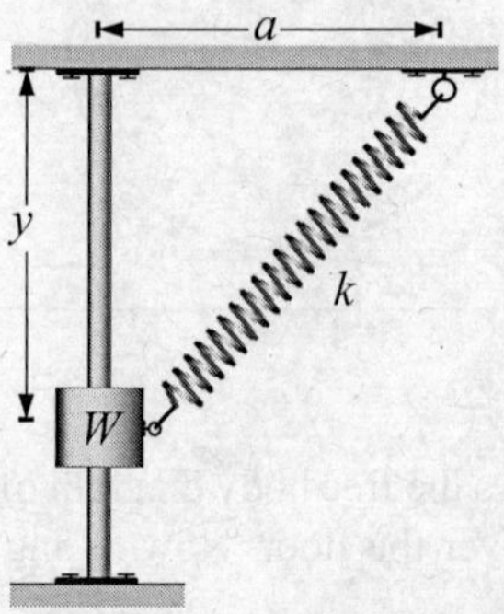

FIGURE 1.18(a) |

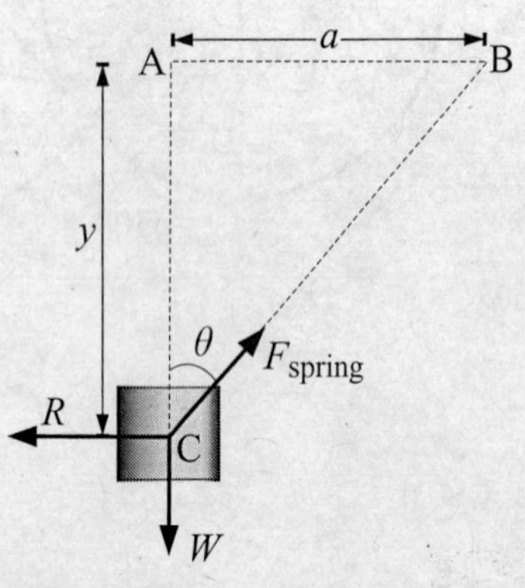

FIGURE 1.18(b) |

Solution. A free body diagram of collar is shown in Figure 1.18(b). The forces acting on the collar are its weight W, reaction R and spring force F_{spring}. Resolving the forces in vertical direction, we get

$$\sum F_v = 0 \Rightarrow F_{spring} \cos\theta - W = 0$$

Substituting $\cos\theta = \text{AC/BC} = y / \sqrt{a^2 + l^2}$ and

$F_{spring} = k\delta = k(\text{BC} - \text{AB}) = k(\sqrt{a^2 + y^2} - a)$, we have

$$k(\sqrt{a^2 + l^2} - a) \times \frac{y}{\sqrt{a^2 + l^2}} - W = 0$$

or $$ky\left[1-\frac{a}{\sqrt{a^2+l^2}}\right]=W$$

$$ky\left[1-\frac{a}{\sqrt{a^2+l^2}}\right]=W$$

PROBLEM 1.19

Two cables hold a rod AB of weight W and length l as shown in Figure 1.19(a). Knowing that cable AC is horizontal and that the rod forms an angle of 30° with the horizontal, determine (a) the angle θ that cable BD forms with the horizontal and (*b*) the tension in each cables.

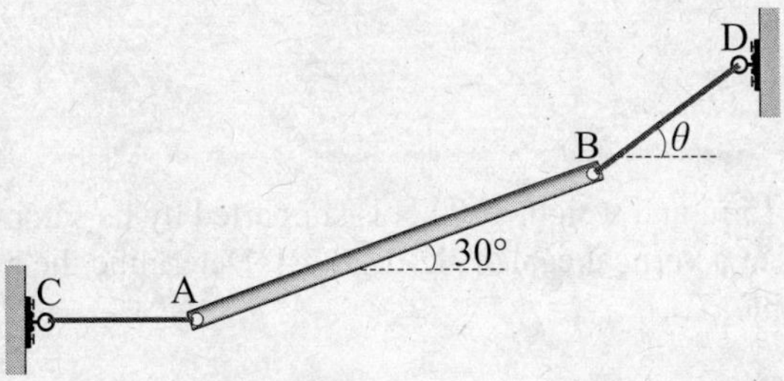

FIGURE 1.19(a)

Solution. Given: weight of rod = W, length of rod = l, angle of rod = 30°, and cable AC is horizontal.

Let F_{BD} = tension of cable BD.

Figure 1.19(b) shows the free body diagram of rod. Resolving forces in vertical direction

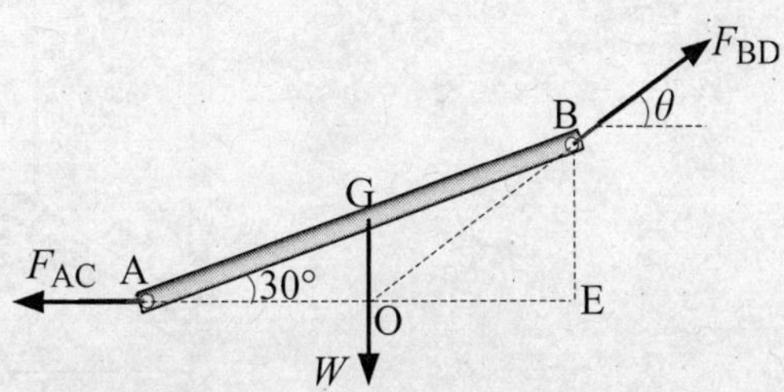

FIGURE 1.19(b)

$$\sum F_v = 0 \Rightarrow F_{BD}\sin\theta - W = 0 \qquad \text{(i)}$$

Summing moment about A.

$$\sum M_A = 0 \Rightarrow F_{BD}\cos\theta \times \text{BE} - F_{BD}\sin\theta \times \text{AE} + W \times \text{AO} = 0$$

or $$F_{BD}\cos\theta \times \text{AB}\sin 30 - F_{BD}\sin\theta \times \text{AB}\cos 30 + W \times \frac{\text{AB}\cos 30}{2} = 0$$

or $$F_{BD}(\sin\theta\cos 30 - \cos\theta\sin 30] = \frac{W}{4}$$

or $$F_{BD}\sin(\theta - 30) = \frac{W}{4} \qquad \text{(ii)}$$

Eliminating W from Equations (i) and (ii), we have

$$\frac{\sin(\theta - 30)}{\sin\theta} = \frac{1}{4}$$

On simplifying, we have

$$\tan\theta = \frac{2}{2\sqrt{3} - 1} = 0.811$$

$$\theta = \tan^{-1} 0.811 = 39°$$

or

$$\theta = 39°$$

PROBLEM 1.20

A uniform pole of length 15 m and weight 1000 N is supported by its smooth ends against a vertical wall and by the tension T in a vertical cable [Fig. 1.20(a)]. Determine the reaction at the ends of the pole and tension of the cable, T.

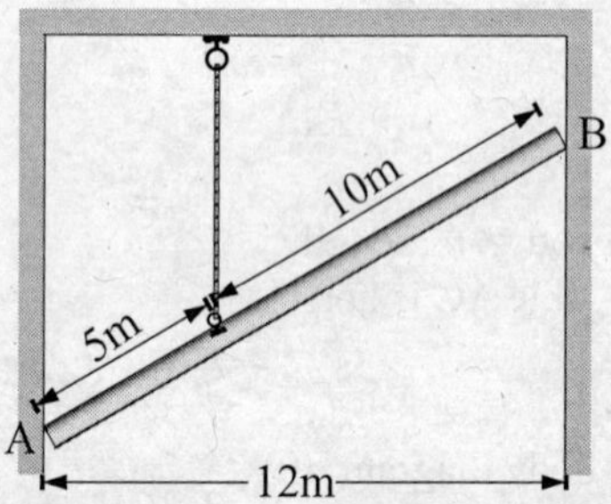

FIGURE 1.20(a) |

Solution. Figure 1.20(b) shows the free body diagram of the pole. The reactions at the ends of the rod are perpendicular to the smooth wall.

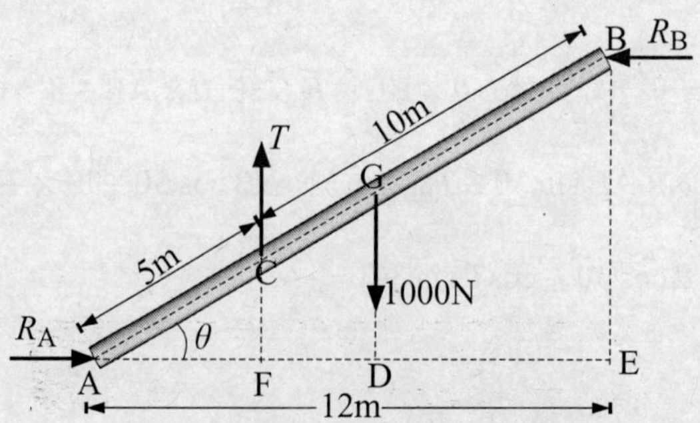

FIGURE 1.20(b) |

Let T be the tension of cable. Equilibrium equations are

$$\sum F_x = 0 \Rightarrow R_A - R_B = 0$$

or
$$R_A = R_B \tag{i}$$

$$\sum F_y = 0 \Rightarrow T - 1000 = 0$$

or
$$T = 1000\text{N} \tag{ii}$$

$$\sum M_A = 0 \Rightarrow 1000 \times \text{AD} - T \times \text{AF} - R_B \times \text{BE} = 0$$

$$1000 \times \text{AG} \cos\theta - T \times \text{AC} \cos\theta - R_B \times \text{AB} \sin\theta = 0$$

Substituting $T = 1000\text{N}$, $\text{AG} = (15/2)\text{m}$, $\text{AC} = 5\text{m}$, $\cos\theta = (12/15)$ and $\sin\theta = 9/15$, we have

$$1000 \times \frac{15}{2} \times \frac{12}{15} - 1000 \times 5 \times \frac{12}{15} - R_B \times 15 \times \frac{9}{15} = 0$$

From which, we obtain

$$R_B = \frac{2000}{9}\text{N}$$

From Equation (i),

$$R_A = R_B = \frac{2000}{9}\text{N}$$

$$T = 1000\text{N}, \quad R_A = R_B = 222.2\text{N}$$

PROBLEM 1.21

A uniform disc of mass M and radius R carries an eccentric mass m_0 at a radius r and is in an equilibrium position on the rough inclined plane as shown in Figure 1.21(a). Determine the angle θ in terms of M, R, m_0, α and r.

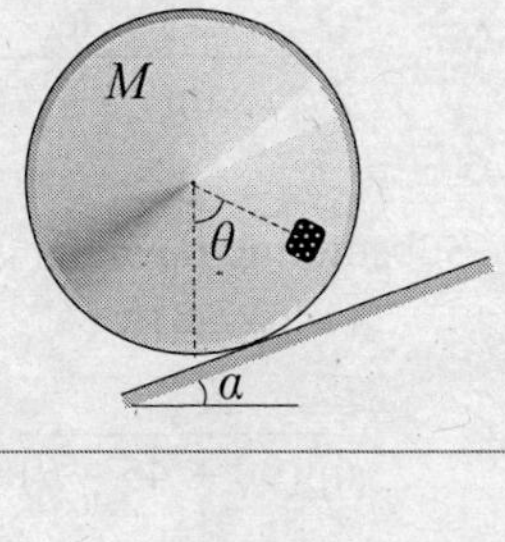

FIGURE 1.21(a)

Solution. Taking moment about C [Fig. 1.21(b)],

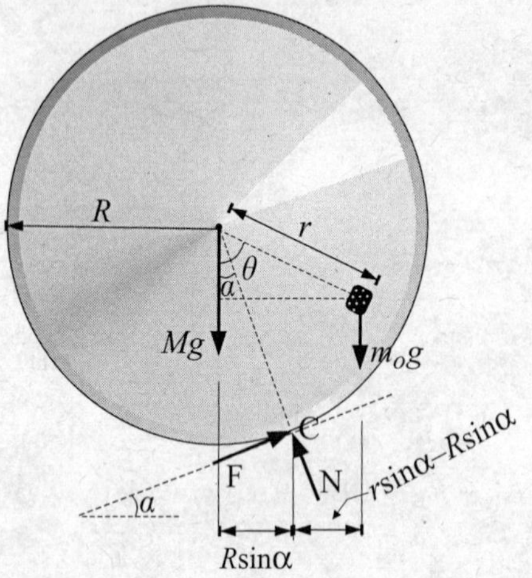

FIGURE 1.21(b) |

$$\sum M_C = 0 \Rightarrow -Mg \times R\sin\alpha + m_0 g \times (r\sin\theta - R\sin\alpha) = 0$$

or $$Mg \times R\sin\alpha = m_0 g \times r\sin\theta - m_0 g \times R\sin\alpha$$

or $$(M + m_0)gR\sin\alpha = m_0 g \times r\sin\theta$$

or $$\left(1 + \frac{M}{m_0}\right)\frac{R}{r}\sin\alpha = \sin\theta$$

$$\theta = \sin^{-1}\left[\left(1 + \frac{M}{m_0}\right)\frac{R}{r}\sin\alpha\right]$$

PROBLEM 1.22

Find the angle of tilt θ with the horizontal so that the contact force at B will be one-third of that at A for a smooth cylinder [Fig. 1.22(a)].

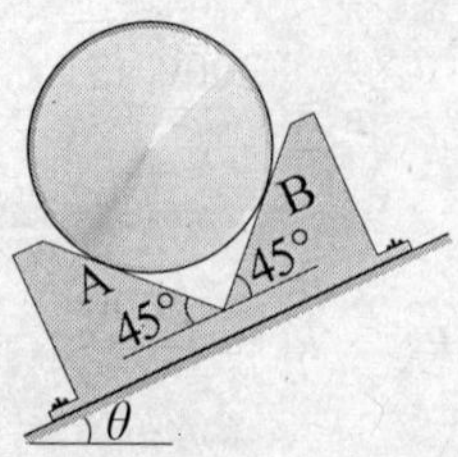

FIGURE 1.22(a) |

Solution. Ref. Figure 1.22(b), Since the cylinder is in equilibrium under the action of only three forces, mg, R_A and R_B, applying Lami's theorem, we get

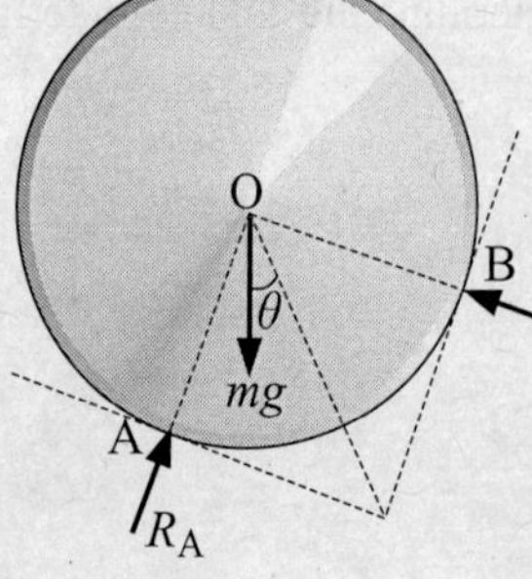

FIGURE 1.22(b) |

$$\frac{mg}{\sin 90} = \frac{R_A}{\sin(90 + 45 - \theta)} = \frac{R_B}{\sin(90 + 45 + \theta)}$$

or $$\frac{R_A}{\sin(90 + 45 - \theta)} = \frac{R_B}{\sin(90 + 45 + \theta)}$$

Substituting $R_B = R_A / 3$, we have

$$\frac{R_A}{\sin(90 + 45 - \theta)} = \frac{R_A}{3\sin(90 + 45 + \theta)}$$

or $$3\cos(45+\theta) = \cos(45-\theta)$$

or $$3\cos 45\cos\theta - 3\sin 45\sin\theta = \cos 45\cos\theta + \sin 45\sin\theta$$

or $$2\cos 45\cos\theta = 4\sin 45\sin\theta$$

or $$\tan\theta = \frac{1}{2} \Rightarrow \theta = \tan^{-1} 0.5$$

$$\theta = 26.56^\circ$$

PROBLEM 1.23

A uniform rod having a length l and weight W is placed in a smooth hemisphere bowl as shown in Figure 1.23(a). Determine the angle of inclination θ for equilibrium.

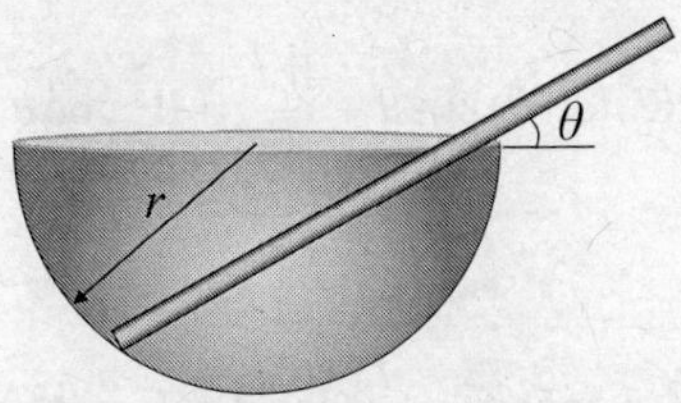

FIGURE 1.23(a) |

Solution. Free body diagram of the rod is shown in Figure 1.23(b) under the static equilibrium.

$$\sum M_A = 0 \Rightarrow -R_B \times AB + W(AG\cos\theta) = 0$$

or $$R_B = W\frac{AG\cos\theta}{AB} = W\frac{\frac{1}{2}\cos\theta}{2r\cos\theta}$$

or $$R_B = \frac{Wl}{4r} \tag{i}$$

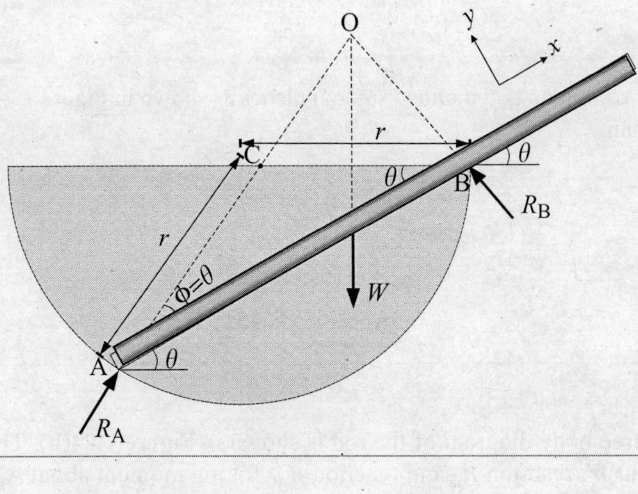

FIGURE 1.23(b) |

Resolving the forces along the rod, we get

$$\sum F_x = 0 \Rightarrow +R_A \cos\theta - W \sin\theta = 0$$

or
$$R_A = W \tan\theta \tag{ii}$$

Revolving the forces perpendicular to the rod, we get

$$\sum F_y = 0 \Rightarrow R_A \sin\theta + R_B - W \cos\theta = 0$$

or
$$R_A \sin\theta + R_B = W \cos\theta \tag{iii}$$

Substituting the value of R_A and R_B, we have

$$(W \tan\theta)\sin\theta + \frac{Wl}{4r} = W \cos\theta$$

On simplifying, we obtain

$$2\cos^2\theta - \frac{l}{4r}\cos\theta - 1 = 0$$

On solving the quadratic equation, we get

$$\cos\theta = \frac{l + \sqrt{l^2 + 128r^2}}{16r}$$

$$\theta = \cos^{-1}\left(\frac{l + \sqrt{l^2 + 128r^2}}{16r}\right)$$

PROBLEM 1.24

A uniform rod AB is resting on two inclined smooth planes as shown in Figure 1.24(a). Determine the angle θ for equilibrium.

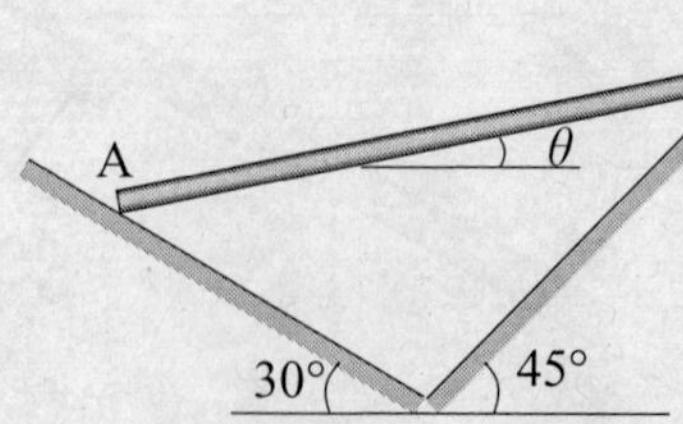

FIGURE 1.24(a) |

Solution. A free body diagram of the rod is shown in Figure 1.24(b). The forces acting on the rod are its weight W, reaction R_A and reaction R_B. Taking moment about A,

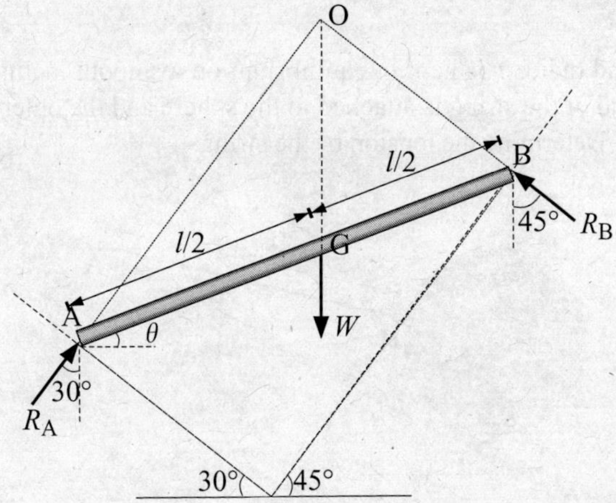

FIGURE 1.24(b) |

$$\sum M_A = 0 \Rightarrow W \times \left(\frac{l}{2}\cos\theta\right) - R_B \times \cos(45-\theta) \times l = 0$$

or $$R_B = \frac{W\cos\theta}{2\cos(45-\theta)} \tag{i}$$

Now taking moment about B

$$\sum M_B = 0 \Rightarrow -W \times \left(\frac{l}{2}\cos\theta\right) + R_A \times \cos(30+\theta) \times l = 0$$

$$R_A = \frac{W\cos\theta}{2\cos(30+\theta)} \tag{ii}$$

$$\sum F_x = 0 \Rightarrow R_A \sin 30 - R_B \sin 45 = 0$$

$$R_A \sin 30 = R_B \sin 45 \tag{iii}$$

Substituting the value of R_A and R_B from Equations (i) and (ii), we obtain

$$\left(\frac{W\cos\theta}{2\cos(30+\theta)}\right)\sin 30 = \left(\frac{W\cos\theta}{2\cos(45-\theta)}\right)\sin 45$$

or $$\sin 45\cos(30+\theta) = \sin 30\cos(45-\theta)$$

On simplifying, we get

$$\tan\theta = \frac{\sqrt{3}-1}{2} = 0.366 \Rightarrow \theta = \tan^{-1} 0.366 = 20.1°$$

$$\theta = 20.1°$$

PROBLEM 1.25

A sphere of weight W and radius r is kept in equilibrium on a smooth inclined plane by means of a string of length l. One end of the string is attached to the sphere and the other to the inclined plane as shown in Figure 1.25(a). Determine the tension of the string.

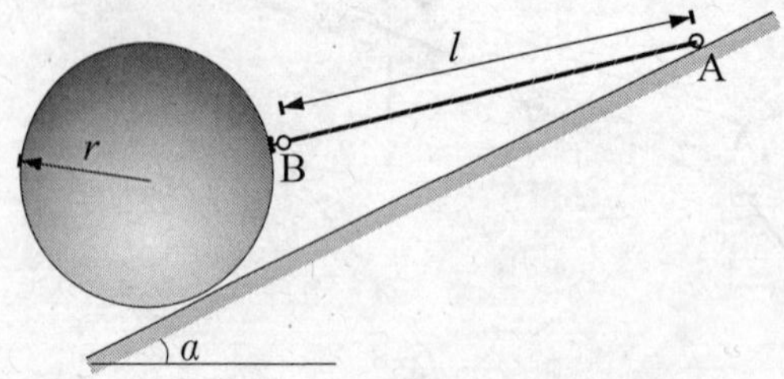

FIGURE 1.25(a) |

Solution. Figure 1.25(b) shows free body diagram of the sphere. The sphere is in equilibrium under the action of three forces: (a) its weight W, (b) tension T in the string and (c) reaction R. Since lines of action of forces R and W meet at O, so the action of the third force T will pass through O. Taking moment about C, we get

$$\sum M_c = 0 \Rightarrow T \times \text{CE} - W \times \text{CD} = 0$$

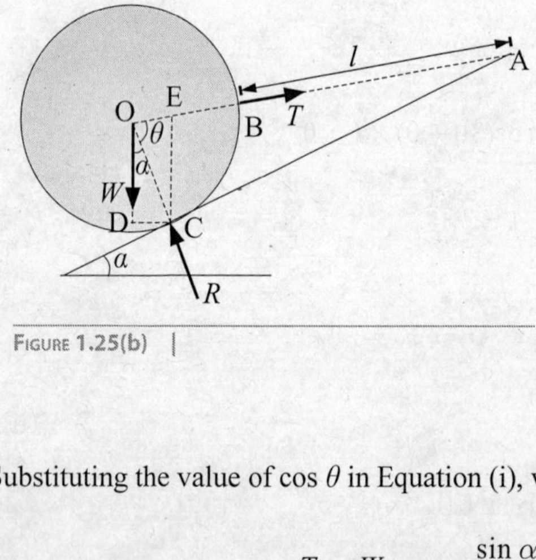

FIGURE 1.25(b) |

or $$T \times \text{OC} \sin\theta - W \times \text{OC} \sin\alpha = 0$$

or $$T = W \frac{\sin\alpha}{\sin\theta} \qquad \text{(i)}$$

Also in ΔAOC, we find that

$$\cos\theta = \frac{\text{OC}}{\text{OA}} = \frac{r}{r+l}$$

$$\Rightarrow \sin\theta = \sqrt{1-\cos^2\theta} = \frac{\sqrt{l^2+2lr}}{l+r}$$

Substituting the value of cos θ in Equation (i), we obtain

$$T = W \frac{\sin\alpha}{\sqrt{l^2+2lr}\ /\ (l+r)} = \frac{W \sin\alpha(l+r)}{\sqrt{l^2+2lr}}$$

$$T = \frac{W \sin\alpha(l+r)}{\sqrt{l^2+2lr}}$$

PROBLEM 1.26

A 100 kg weight is attached to a small pulley which may roll on the cable ABC [Fig. 1.26(a)]. The pulley and weight are held in the position shown by a second cable DE which is parallel to BC.

Determine the tension (a) in cable ABC and (b) in cable DE. Neglect the radius of the pulley and weight of cable.

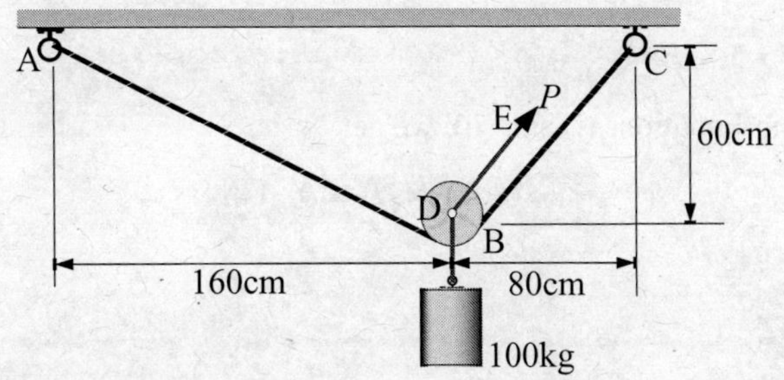

FIGURE 1.26(a)

Solution. Free body diagram is drawn as shown in Figure 1.26(b). Then we write

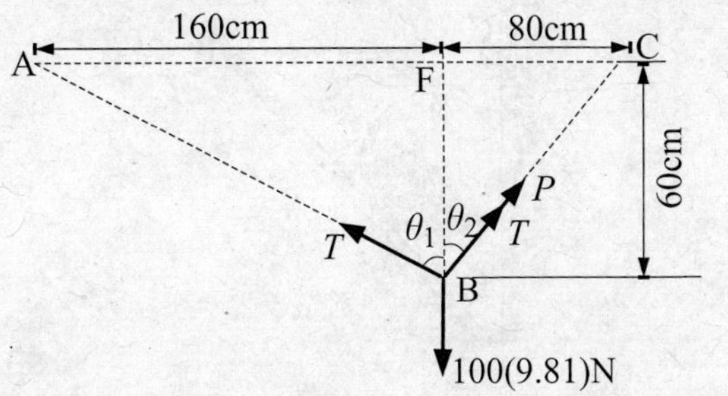

FIGURE 1.26(b)

$$\tan\theta_1 = \frac{160}{60} = 2.66 \Rightarrow \theta_1 = 69.44°$$

$$\tan\theta_2 = \frac{80}{60} = 1.33 \Rightarrow \theta_2 = 53.13°$$

Resolving forces in vertical direction

$$\sum F_y = 0 \Rightarrow T\cos\theta_1 + T\cos\theta_2 + P\cos\theta_2 - 100(9.81) = 0$$

or $$T(\cos 69.44 + \cos 53.13) + P\cos 53.13 = 981$$

or $$0.9511T + 0.6P = 981 \qquad \text{(i)}$$

Resolving forces in horizontal direction

$$\sum F_x = 0 \Rightarrow T\sin\theta_2 + P\sin\theta_2 - T\sin\theta_1 = 0$$

or $$T\sin 53.13 + P\sin 53.13 - T\sin 69.44 = 0$$

or $$T\sin 53.13 + P\sin 53.13 - T\sin 69.44 = 0 \qquad \text{(ii)}$$

or $$0.799T + 0.799P - 0.636T = 0$$

$$0.1373T = 0.799P$$

$$T = 5.81P \qquad \text{(iii)}$$

Solving for T and P from Equations (i) and (ii), we get

$$P = 160.13\text{N}, T = 930.4\text{N}$$

PROBLEM 1.27

A smooth circular cylinder of weight W and radius r is supported in a horizontal position against a smooth vertical wall by a hinged bar AB as shown in Figure 1.27(a). The AB is supported by a horizontal cable BC. Find the value of the angle θ that AB should make with the wall to attain a minimum tension T in cable BD.

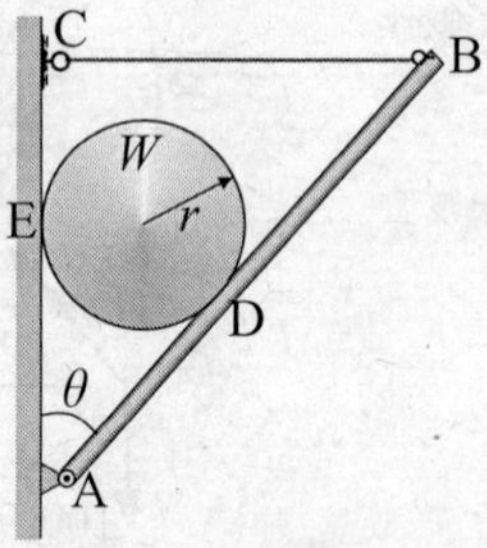

FIGURE 1.27(a) |

Solution The free body diagram of the cylinder is shown in Figure 1.27(b). Since the three concurrent forces R_E, W and R_D are in equilibrium, the algebraic sum of their projections on the y-axis must be zero. From this condition we have

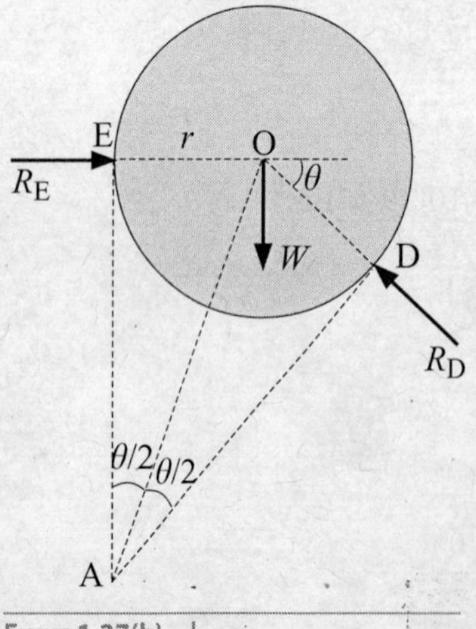

FIGURE 1.27(b) |

$$\sum F_y = 0 \Rightarrow R_D \sin\theta - W = 0$$

From which

$$R_D = \frac{W}{\sin\theta} \qquad \text{(i)}$$

Free body diagram of lever AB is shown in Figure 1.27(c). Summing moment about A, we get

$$\sum M_A = 0 \Rightarrow T \times \text{AC} - R_D \times \text{AD} = 0$$

or $$T = R_D \frac{\text{AD}}{\text{AC}} \qquad \text{(ii)}$$

Substituting $R_D = W / \sin\theta$, we have

$$T = \frac{W}{\sin\theta} \times \frac{\text{AD}}{\text{AC}}$$

From geometry AD = $r \cot \theta/2$ and AC = $l \cos\theta$,

$$T = \frac{W}{\sin\theta} \times \frac{r \cot\frac{\theta}{2}}{l\cos\theta} = \frac{Wr}{l} \times \frac{\cos\frac{\theta}{2}}{\sin\theta\cos\theta}$$

or

$$T = \frac{Wr}{2l\cos\theta\sin^2\frac{\theta}{2}} \qquad \text{(iii)}$$

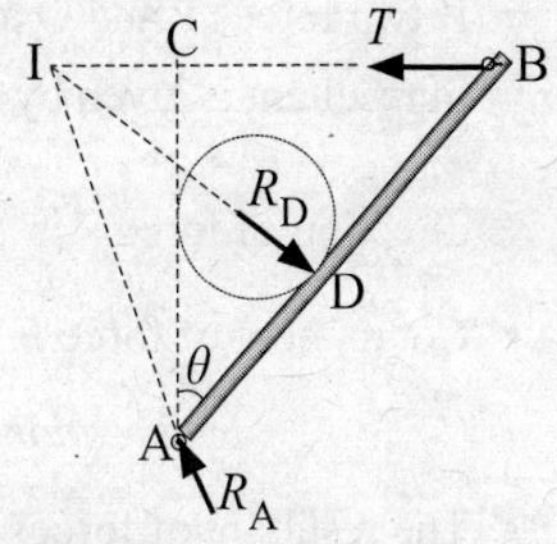

FIGURE 1.27(c) |

Equation (iii) shows that for $\theta = 0$ and for $\theta = 90°$, the tensile force T becomes infinitely large and that for all values of θ between these two extremes it will be finite. It follows then that there must be some value of θ between 0 and 90° for which the value of T is a minimum [Fig. 1.27(d)]. To determine this particular value of θ, we differentiate T with respect to θ and set the derivation $dT/d\theta = 0$, which gives

$$\frac{dT}{d\theta} = 0 \Rightarrow -\sin\theta\sin^2\frac{\theta}{2} + \cos\theta \times 2\sin\frac{\theta}{2}\cos\frac{\theta}{2} \times \frac{1}{2} = 0$$

or

$$\sin\theta\left[4\sin^2\frac{\theta}{2} - 1\right] = 0$$

Possible solution of above equation is

$$\theta = 2\sin^{-1}\left(\frac{1}{2}\right) = 60°$$

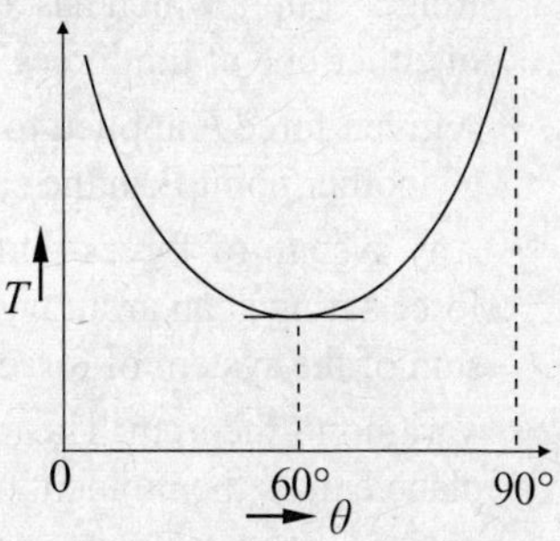

FIGURE 1.27(d) |

Substituting $\theta = 60°$ into Equation (iii), we have

$$T_{\min} = \frac{Wr}{2l\cos 60 \sin^2(60/2)} = \frac{4Wr}{l}$$

$$T_{\min} = \frac{4Wr}{l}$$

CHAPTER HIGHLIGHTS

- The principle of transmissibility allow the point of application of a force to be displaced along its line of action without altering the state the equilibrium of rigid body.
- When two or more forces act on a body, the net effect of the forces is equivalent to the action of a single force, known as the resultant.
- Polygon law of forces states that if a number of coplanar forces are acting at a point such that they can be represented in magnitude and direction by the sides of a polygon taken in the same order, then their resultant is represented in magnitude and direction by the closing side of the polygon taken in the opposite order.
- Law of parallelogram of forces states that "If two forces, acting at a point be represented in magnitude and direction by the two adjacent sides of a parallelogram, then their resultant is represented in magnitude and direction by the diagonal of the parallelogram passing through that point.

- If two forces P and Q act at a point and the angle between the two force be θ, then the magnitude of resultant is given by $R = \sqrt{P^2 + Q^2 + 2PQ\cos\theta}$ and the angle made by the resultant with the direction of force P is $\theta = \tan^{-1}\left(\frac{Q\sin\theta}{P + Q\cos\theta}\right)$
- Component of force **F** in *x, y and z directions are*

$$x\text{–}direction\ F_x = F\cos\theta_x, y\text{–}direction\ F_y = F\cos\theta_y\ and\ z\text{–}direction\ F_z = \cos\theta_z$$

- The resultant of forces is

$$\mathbf{R} = \left(\sum F_x\right)i + \left(\sum F_y\right)j + \left(\sum F_x\right)k$$

- If a force **F** acts at a point P, its moment M about another point O is given by the vector product **M** = **r** × **F**, where **r** is the vector from point O to point P.
- When two equal and opposite parallel forces act on a body at some distance apart, the two forces from a couple which has a tendency to rotate the body. The moment of this couple is the product of either one of the forces and perpendicular distance between the forces.
- A given force *F* applied to a body at any pint A can always be replaced by an equal force applied at another point B in the same direction together with a couple.
- Any system of forces that acts on a rigid body is dynamically equivalent to a single resultant force acting at an arbitrary point O and a single resultant couple. The resultant force is the vector sum of the system of forces, and the resultant couple is the moment of the forces about point O.
- Varignon's theorem: The moment of several concurrent coplanar forces about any point O in their plane equals the moment of their resultant about point O.
- A free-body diagram is a sketch of a body that shows the body free from other parts of the system and shows all forces acting on it.
- A body is in static equilibrium if, and only if, the vector sum of the external forces is zero and the moment of the external forces about any point is zero.

$$\sum \mathbf{F} = 0 \qquad \sum F_x = 0, \sum F_y = 0 \quad and \quad \sum F_z = 0$$

or

$$\sum \mathbf{M} = 0 \qquad \sum M_x = 0, \sum M_y = 0 \quad and \quad \sum M_z = 0$$

- "Lami's theorem," If three coplanar forces acting at a point are equilibrium, each force will be proportional to the sine of the angle between the other two forces."

$$\frac{P}{\sin\alpha} = \frac{Q}{\sin\beta} = \frac{R}{\sin\gamma}$$

REVIEW QUESTIONS

Q1. What are concurrent forces, coplanar forces, and collinear forces?

Q2. State the law of parallelogram of forces. Find the magnitude and direction of the resultant of two forces *P* and *Q* acting on a particle making an angle θ with each other.

Q3. State the law of triangle of forces and Lami's theorem.

Q4. Define moment of force about an axis.

Q5. State Varignon's theorem.

Q6. State the general algebraic conditions of equilibrium for concurrent and coplanar forces in terms of (x, y) projections.

Q7. State the three general algebraic conditions of equilibrium for concurrent and non-coplanar forces in terms of (*x*, *y*, *z*) projections.

Q8. State the equilibrium criterion for a rigid body that is subjected to forces with concurrent line of action.

Q9. What is a free body diagram? Explain with a suitable example.

Q10. A 100 N uniform ladder of length *L* shown in Figure 1.Q10 rests against a smooth wall at A and a smooth horizontal floor at B. The reactions of the wall and the floor on the ladder are normal to the wall and the floor at A and B. The cord holds the ladder in equilibrium. Determine the forces exerted on the ladder at A and B [Ans; $R_A = 28.87$ N ←, $R_B = 100$ N↑].

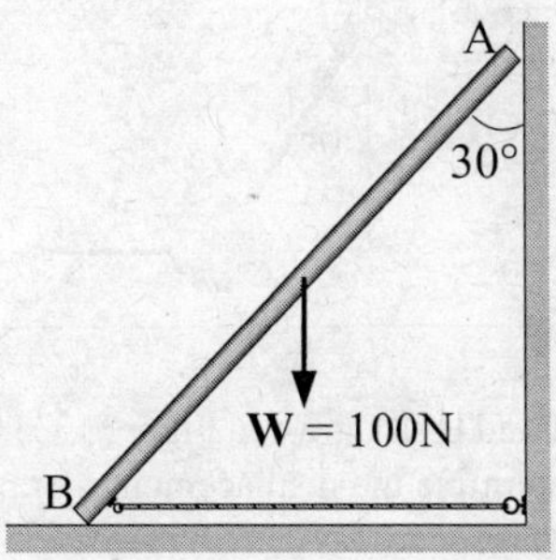

FIGURE 1.Q10 |

Q11. A block of weight 900 N as shown in Figure 1.Q11. The force *P* is just sufficient to lift the block off the floor. Determine the angle θ

[Ans; $\theta = 77.8$].

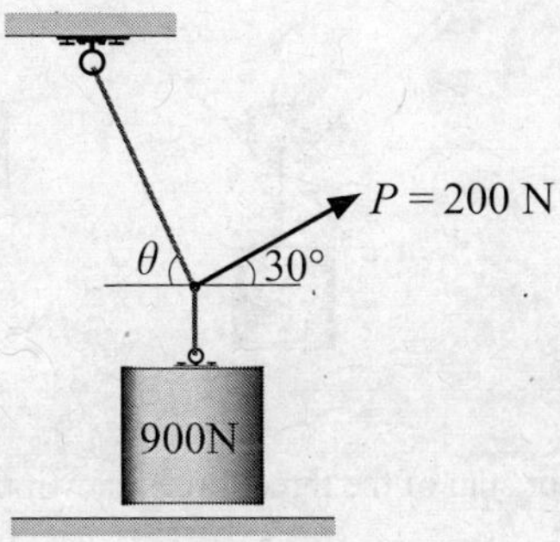

FIGURE 1.Q11 |

Q12. A cylinder of weight 50 kN rests on two small rollers as shown in Figure 1.Q12. Determine the reactions of the rollers on the cylinder. Assume that these forces act normal to the cylinder

[Ans; $R_A = 26.6$ kN, $R_B = 34.2$ kN].

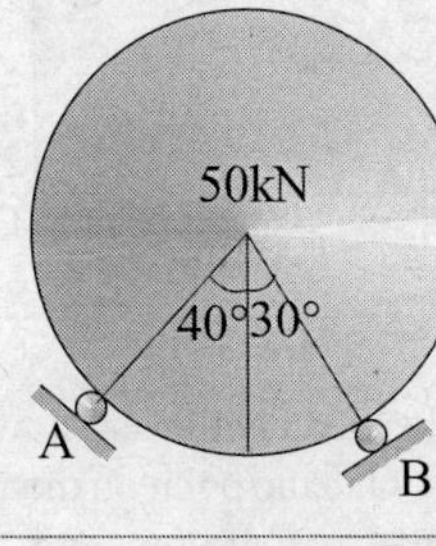

FIGURE 1.Q12 |

Q13. A double pendulum is held in equilibrium in the position shown in Figure 1.Q13 by a horizontal force F. The balls B and C each weigh 15 N. Determine the force F, the tensions in the strings AB and BC, and the angle θ [Ans; $F = 25.98$ N, $T_{AB} = 39.69$ N, $T_{AB} = 30$ N, $\theta = 40.9°$].

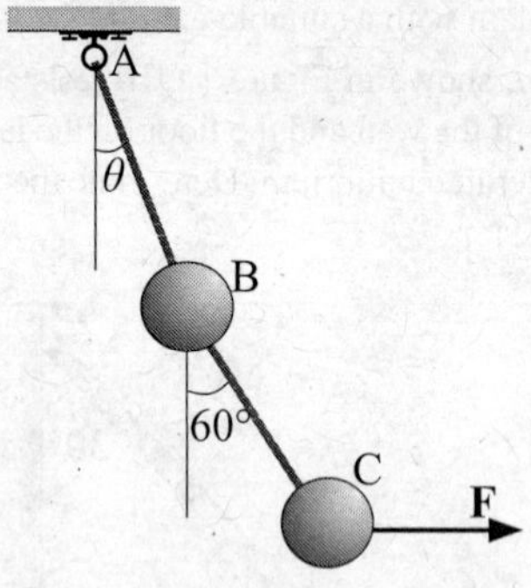

FIGURE 1.Q13 |

Q14. A cable is used to support blocks A and B as shown in Figure 1.Q14. Block A is supported by a small wheel that is free to roll on the cable. Determine the displacement y of block A for equilibrium if the weights of block A and B are 50 kg and 75 kg, respectively
[Ans; $y = 1.76$ m].

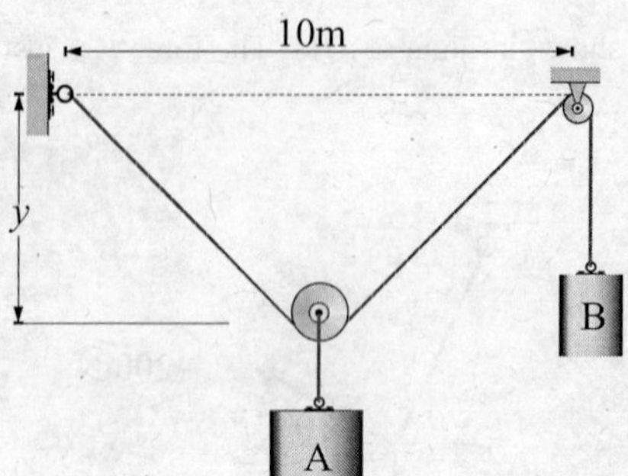

FIGURE 1.Q14 |

Q15. The angle θ is such that the vector sum of the three forces shown in Figure 1.Q15 is zero. Determine the moment of the three forces about P
[Ans; $M = 5000$N mm (clockwise)].

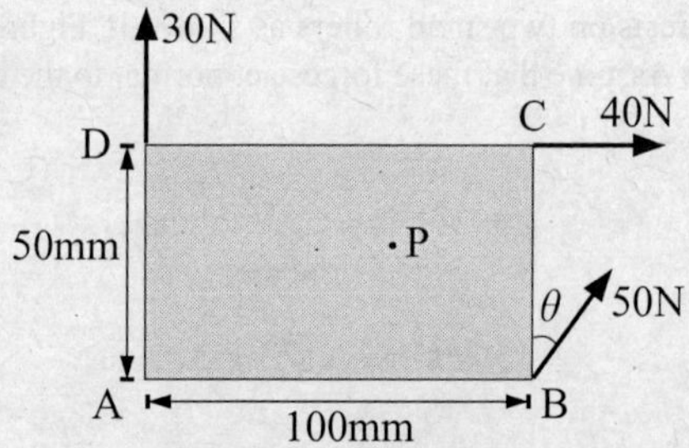

FIGURE 1.Q15 |

Q16. Replace the three forces acting on the bended cantilever shown in Figure 1.Q16 by a single equivalent force R. Specify the distance x from point O to the point on the x-axis through which the line of action of R passes [Ans; $R = -50i + 20j$, $x = 650$ mm].

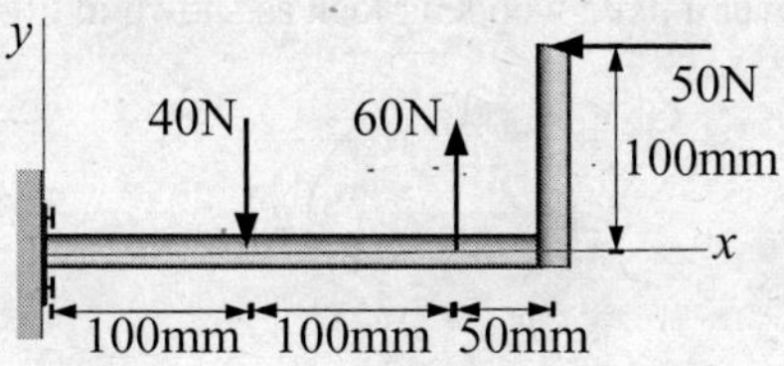

FIGURE 1.Q16 |

Q17. Reduce the force system in Figure 1.Q17 to a single force that acts at point O and a couple [Ans; $R = 236.6i - 36.6j$, $M_0 = -846.4$ Nm].

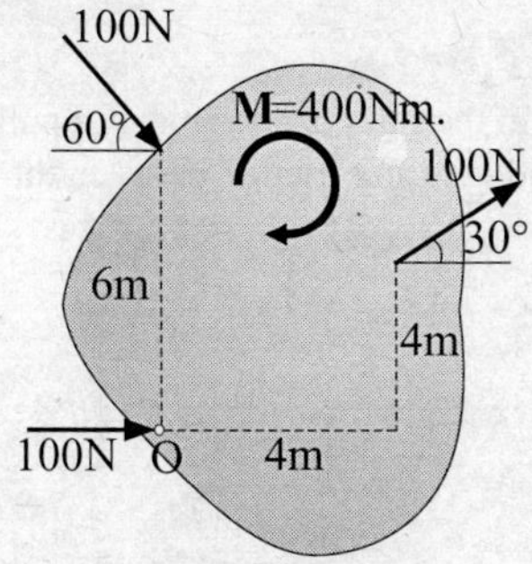

FIGURE 1.Q17 |

Q18. (a) Determine the moment of a 90 N force about point C for the condition $\theta = 15°$. Also determine the value of θ for which the moment about C is (b) zero and (c) a maximum (see Figure 1.Q18) [Ans; (a) $M_c = 33.5$ Nm, (b) 36.9° (or 21.7°), (c) 126.9° (or 307°)].

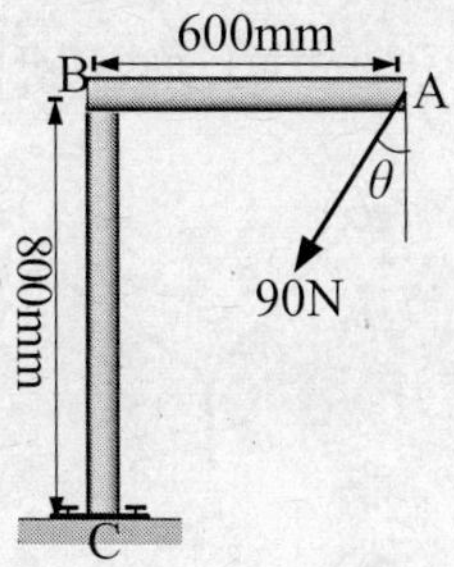

FIGURE 1.Q18 |

MULTIPLE CHOICE QUESTIONS

Q1. The forces, which meet at one point, but their lines of action do not lie in a plane, are called

(A) Coplanar non-concurrent forces (B) Non-coplanar concurrent forces
(C) Non-coplanar non-concurrent forces (D) Intersecting forces.

[Ans: B]

Q2. According to principle of transmissibility of forces, the effect of a force upon a body is

(A) maximum when it acts at the centre of gravity of a body
(B) different at different points in its line of action
(C) the same at every point in its line of action
(D) minimum when its acts at the centre of gravity of the body.

[Ans: C]

Q3. A roller of weight W is rolled over a fixed wooden block as shown in Figure 1.MQ3. The pull F required to just cause the said motion is

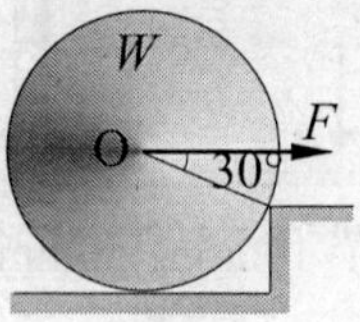

FIGURE 1.MQ3 |

(A) $\frac{W}{2}$ (B) W (C) $\sqrt{3}W$ (D) $2W$

[Ans: C]

Q4. A spring scale indicates a tension T in the right hand cable of a pulley system, as shown in Figure 1.MQ4. Neglecting the mass of the pulley and ignoring friction between the cable and pulley, the mass M is

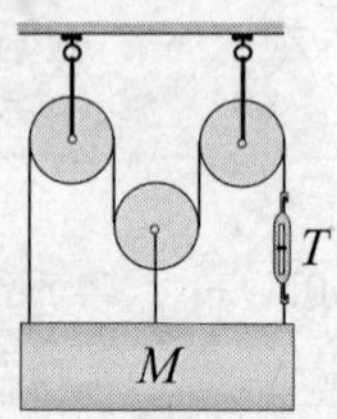

FIGURE 1.MQ4 |

(A) $\frac{2T}{g}$ (B) $T(1+e^{4\pi})g$ (C) $\frac{4T}{g}$ (D) None of these

[Ans: C]

Q5. A weight W is supported by two cables as shown in Figure 1.MQ5. The tension in the cable making an angle will be the minimum when the value of θ is

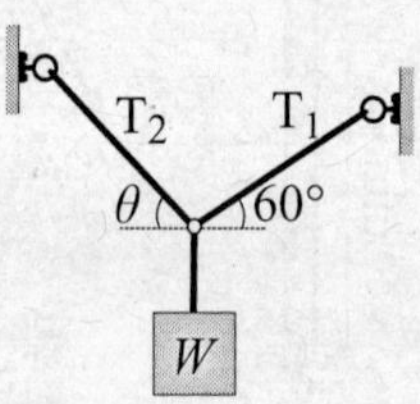

FIGURE 1.MQ5 |

(A) 0 (B) 30° (C) 45° (D) 60°

[Ans: B]

Q6. A ladder AB of weight W is supported as shown in Figure 1.MQ6. The reaction at A and B will be equal to

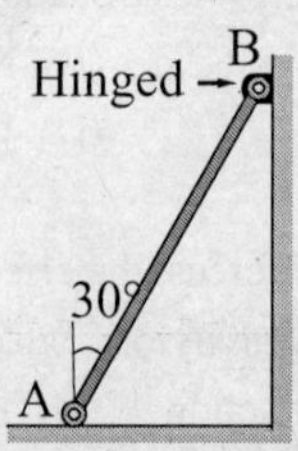

FIGURE 1.MQ6 |

(A) W, W (B) $W, 0$ (C) $\frac{W}{2}, \frac{W}{2}$ (D) $\frac{W}{2}, 0$

[Ans: C]

Q7. A string ABC has its extremities tied to two fixed points A and B in the same horizontal line. If a weight W is knotted at a given point C, then the tension in the portion CA is (where a, b, c are the sides and Δ is the area of triangle ABC)

(A) $\frac{Wb}{4c\Delta}(a^2 + b^2 + c^2)$ (B) $\frac{Wb}{4c\Delta}(b^2 + c^2 - a^2)$

(C) $\frac{Wb}{4c\Delta}(c^2 + a^2 - b^2)$ (D) $\frac{Wb}{4c\Delta}(a^2 + b^2 - c^2)$

[Ans: C]

Q8. Two like parallel forces P and Q ($P > Q$) act on a rigid body. If the force P is displaced parallel to itself through a distance d, then the resultant of the forces P and Q would be shifted by a distance

(A) $\frac{P}{P+Q}d$ (B) $\frac{P}{P-Q}d$ (C) $\frac{Q}{P-Q}d$ (D) $\frac{Q}{Q-P}d$

[Ans: A]

Q9. A weight W hangs by a string. It is pushed aside by a horizontal force until the string makes an angle of 30° with the vertical. The tension in the string is

(A) W (B) $\frac{2}{\sqrt{3}}W$ (C) 2W (D)3W

[Ans: B]

Q10. The resultant of the forces P and Q is R. If Q is doubled then R gets doubled in magnitude. R is again doubled if Q is reversed. Then $P2$, $Q2$ and $R2$ are in the ratio

(A) 2:2:3 (B) 3:2:2 (C) 2:3:2 (D) 2:3:3

[Ans: C]

Q11. P and Q are two like parallel forces. If P is moved parallel to itself through a distance of 4 cm, then the resultant of P and Q moves through a distance of

(A) $2cm$ (B) $4cm$ (C) $\frac{4Q}{P+Q}cm$ (D) $\frac{4P}{P+Q}cm$

[Ans: D]

Q12. A particle acted upon by three forces whose magnitudes are proportional to (i) 3, 5, 8; (ii) 5, 7, 10 and (iii) 5, 7, 14 can be arranged at rest

(A) only in case (ii) (B) only in case (i)

(C) only in cases (i) and (ii) (D) in all the three cases.

[Ans: C]

Q13. A uniform bar AB of weight 100 N is hinged at A to the vertical wall and held in horizontal position by a cord BC as shown in Figure 1.MQ13. The tension in the cord BC will be

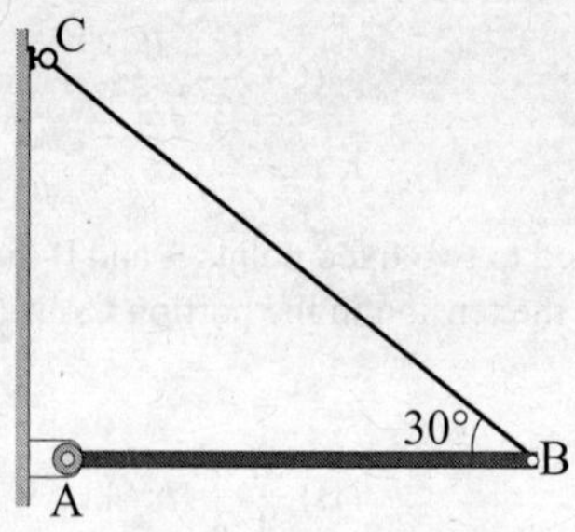

FIGURE 1.MQ13 |

(A) 100 N (B) 50 N (C) 200 N (D) 150 N

[Ans: A]

Q14. Suppose R is the resultant of a system of several coplanar forces $P_1, P_2, \ldots, P_n$ acting simultaneously at a point, then

$$R\cos\theta = \sum_{i=1}^{n} P_i \cos a_i = \sum_{i=1}^{n} X_i$$

and

$$R\sin\theta = \sum_{i=1}^{n} P_i \sin a_i = \sum_{i=1}^{n} Y_i$$

where $a_1, a_2 \ldots, a_n, \theta$ are the angles which the system of forces $P_1, P_2, \ldots, P_n$ and R makes with the horizontal line. If the system is in equilibrium, then which one of the following relations is correct?

(A) $\sum X_i + \sum Y_i = 0$ (B) $\sum X_i - \sum Y_i = 0$

(C) $\sum X_i = \sum Y_i = 0$ (D) None of these

[Ans: C]

Q15. A uniform heavy rod AB of length L and weight W is hinged at A and tied to a weight W_1 by a string at B. The massless string passes over a frictionless pulley (of negligible dimensions) at C as shown in Figure 1.MQ15. If the rod is in equilibrium at horizontal configuration, then

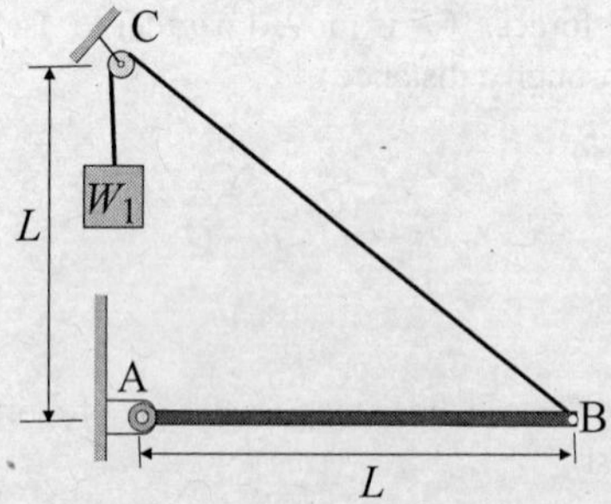

FIGURE 1.MQ15 |

(A) $W_1 = W$ (B) $W_1 = \dfrac{W}{2}$ (C) $W_1 = \sqrt{2}W$ (D) $W_1 = \dfrac{W}{\sqrt{2}}$

[Ans: D]

2 Plane Trusses and Frames

INTRODUCTION

A structure refers to a system of connected parts used to support a load. Important examples related to civil engineering include buildings, bridges and towers. In other branches of engineering, ship and aircraft frames, tanks, pressure vessels, mechanical systems and electrical supporting structures are important.

STRUCTURAL ELEMENTS

Tie Rods

Structure members subjected to a tensile force are often referred to as tie rods or bracing struts. Due to the nature of load, these members are rather slender and are often chosen from rods, bars, angles or channels.

Beams

Beams are usually straight horizontal members used to carry vertical loads. Depending upon the type of supports, beams are classified as follows:

(1) **Cantilever.** A cantilever is a beam whose one end is fixed and the other end is free. Figure 2.1 shows a cantilever with one end (A) rigidly fixed into its supports and the other end (B) free. The length between A and B is known as the length of cantilever.

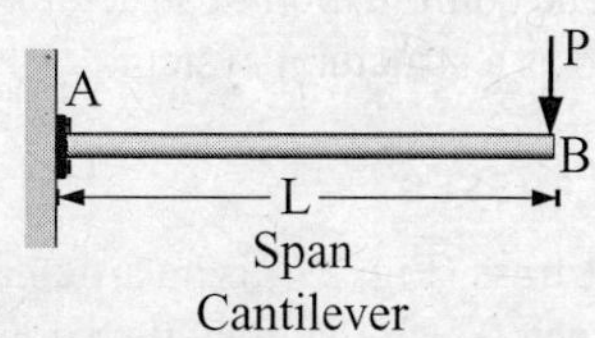

FIGURE 2.1

(2) **Simply supported beam.** A simply supported beam is one whose ends freely rest on walls (or columns or knife edges) as shown in Figure 2.2.

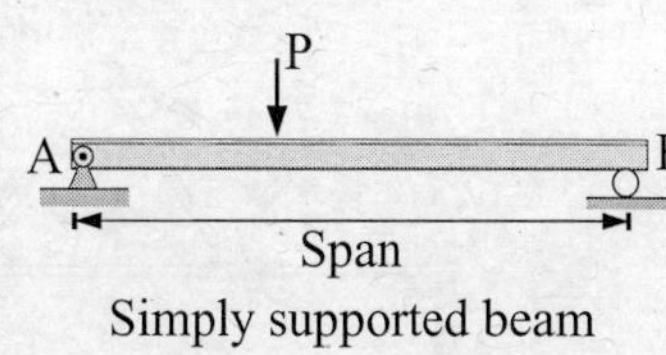

FIGURE 2.2

(3) **Overhanging beam.** An overhanging beam is one in which the supports are situated at the ends, i.e., one or both the ends project beyond the supports. In Figure 2.3, C and D are two supports and both the ends A and B of the beam are overhanging beyond the supports C and D, respectively.

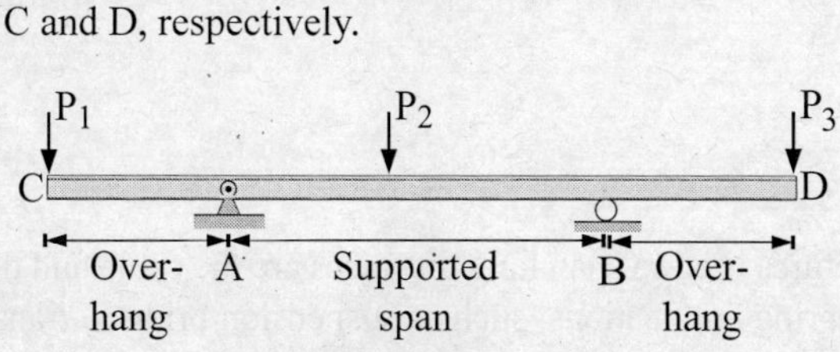

FIGURE 2.3

(4) **Fixed Beam.** A fixed beam is one whose both ends are rigidly fixed or built in into its supporting walls or columns (Fig. 2.4).

(5) **Continuous Beam.** A continuous beam is one which has more than two supports (Fig. 2.5). The supports at the extreme left and right are called the end supports and all the other supports, except the extreme, are called intermediate.

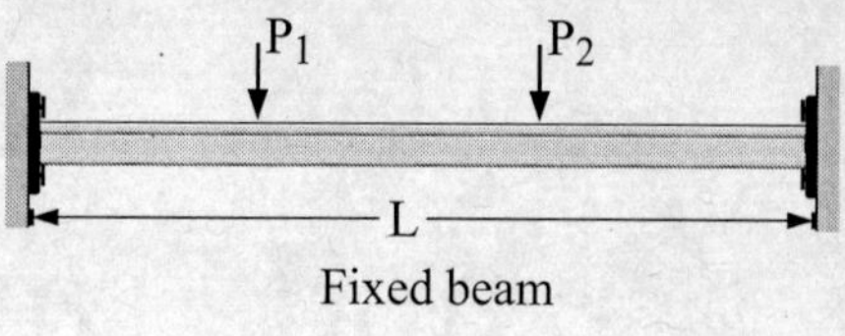

FIGURE 2.4 |

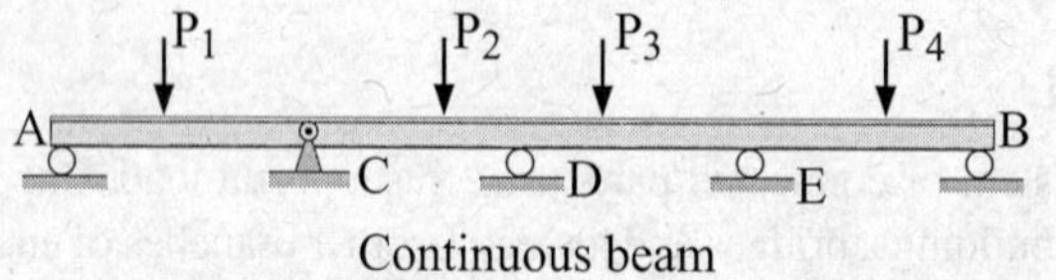

FIGURE 2.5 |

COLUMNS

Members that are generally vertical and resist axial compressive loads are referred to as columns. Tubes and wide-flange cross-section are often used for metal columns, and circular and square cross-sections with reinforcing rods are used for those made of concrete.

TYPES OF STRUCTURES

The combination of structural elements and the material from which they are composed of is referred to as a structural system.

TRUSSES

A truss (Fig. 2.6) is an articulated structure composed of straight members arranged and connected in such a way that they transmit primarily axial forces. If all the members lie in one plane, it is called a *plane truss*. A three-dimensional truss is called a *space truss*.

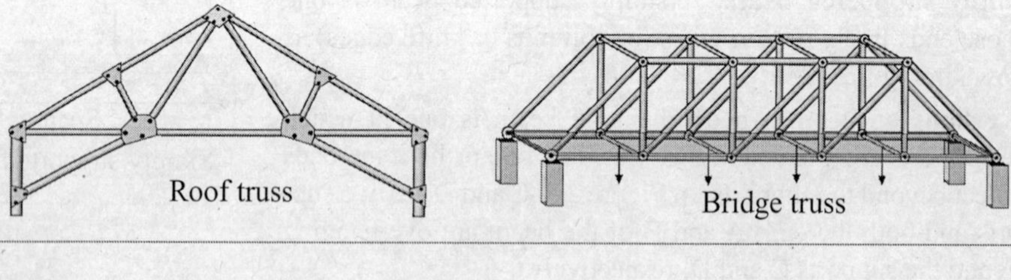

FIGURE 2.6 |

CABLES AND ARCHES

Two other forms of structures used to span long distances are the cable and the arch (Fig. 2.7). *Cables* are used in many engineering applications, such as suspension bridges, transmission line, messenger cables for supporting heavy trolley and many other applications. *Arches* are frequently used in bridge structures, dome roofs and the openings in masonry walls.

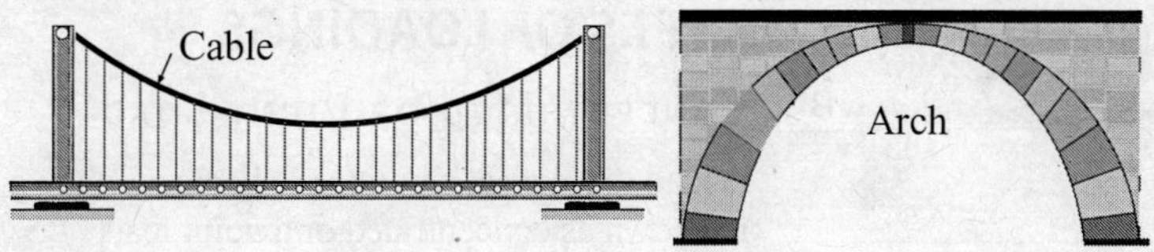

FIGURE 2.7 |

FRAMES

These are structures composed of jointed members, any one of which has more than two forces acting on it. Frames are designed to support applied loads and are usually fixed in a position (Fig. 2.8).

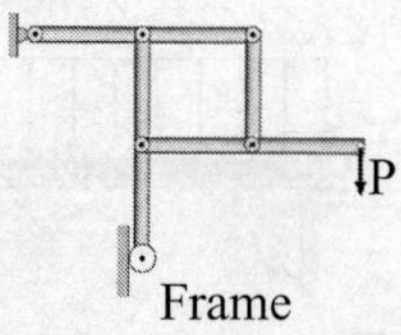

FIGURE 2.8 |

MACHINES

Structures are composed of jointed members, any one of which has more than two forces acting on it. Structures that contain moving parts and that are designed to transmit forces or couples from inputs values to output values are classified as machines (Fig. 2.9).

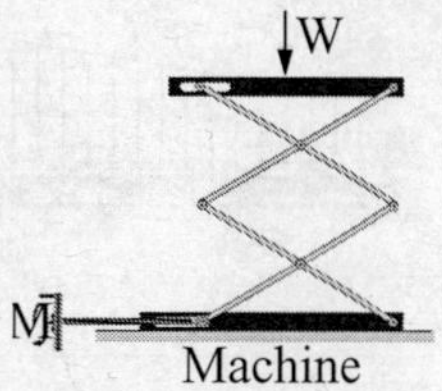

FIGURE 2.9 |

SUPPORT CONDITIONS

The nature of the reactive forces developed on a loaded body depends on the exact way in which the body is either supported by or connected to other bodies. Several basic types of support conditions are indicated; others are possible. Of primary importance are pinned connections, roller connections and fixed connections (Fig. 2.10).

In *pinned connections*, the joint allows an attached member to rotate freely but does not allow translations to occur in any direction. Consequently, the joint cannot provide moment resistance but can provide force resistance in any direction.

A *roller connection* allows rotations to occur freely. It resists translations, however, only in the direction perpendicular to the surface of the support. It does not have any force resistance parallel to the surface of support.

A *fixed joint* completely restrains rotations and translation in any direction. Consequently, it can provide moment resistance and force resistance in any direction.

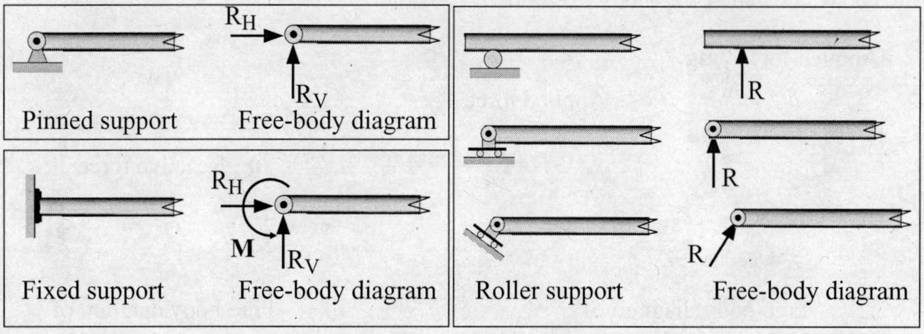

FIGURE 2.10 |

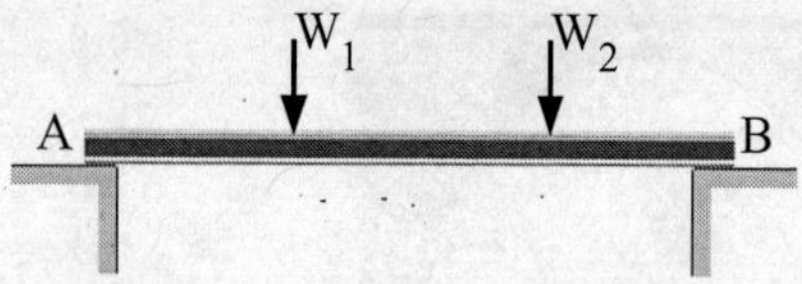

FIGURE 2.11 |

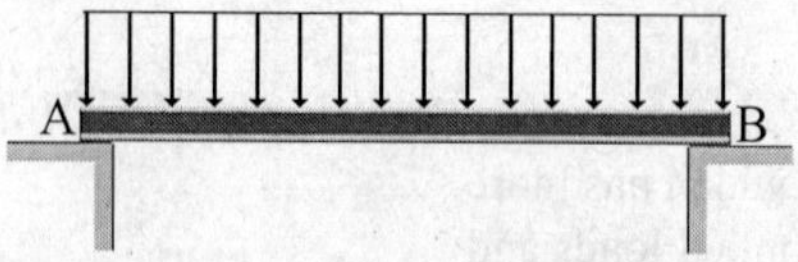

FIGURE 2.12 |

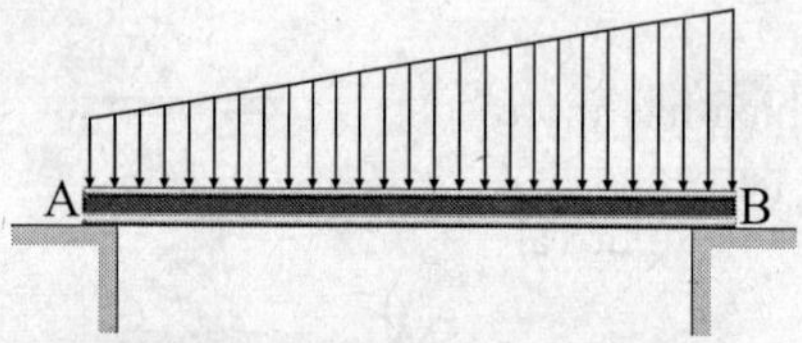

FIGURE 2.13 |

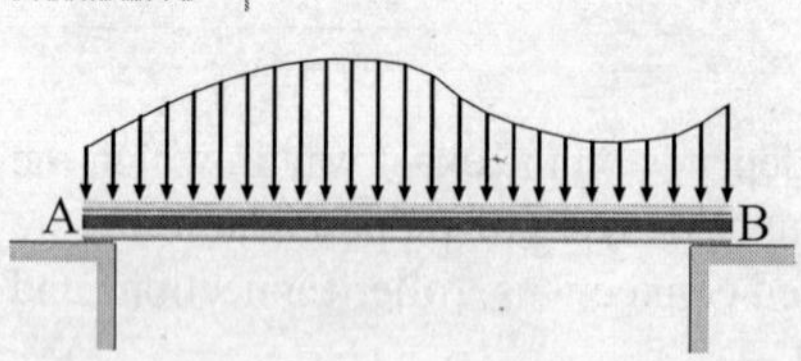

FIGURE 2.14 |

TYPES OF LOADINGS

CONCENTRATED OR POINT LOAD

A load that acts at a point on a beam, as shown in Figure 2.11, is known as concentrated or a point load.

UNIFORMLY DISTRIBUTED LOAD

If a beam is loaded in such a way that each unit length of the beam carries same intensity of the load, then that type of load is known as uniformly distributed load (written as u.d.l.). Figure 2.12 shows a beam AB, which carries a uniformly distributed load.

UNIFORMLY VARYING LOAD

Figure 2.13 shows a beam AB, which carries load in such a way that the rate of loading on each unit length of the beam varies uniformly. This type of load is known as uniformly varying load.

VARYING LOAD

Figure 2.14 shows a varying load. Here the ordinate represents the intensity of loading and abscissa represents the position of the load on the beam.

APPLIED AND REACTIVE FORCES

Forces and moments that act on a rigid body can be divided into two primary types: *applied* and *reactive*. In common engineering usage, applied forces are those that act directly on a structure.

Reactive forces are those generated by the action of one body on another and hence typically occur at connections or supports. The existence of reactive forces follows from Newton's Third Law, which generally states that to every action there is an equal and opposite reaction.

The diagrams in Figure 2.15, which show the complete system of applied and reactive forces active on a body, are called free-body diagrams.

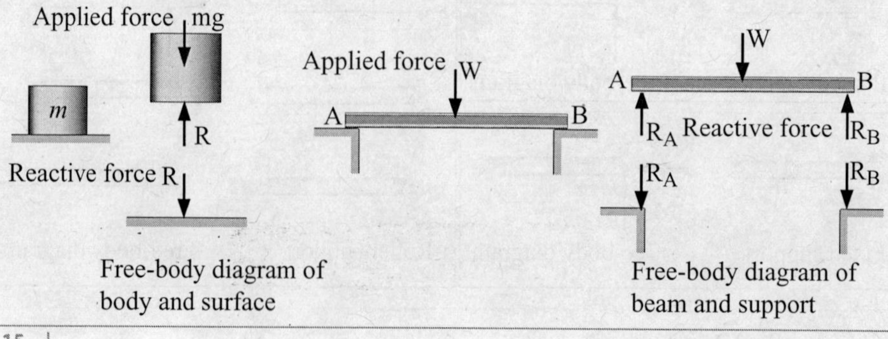

FIGURE 2.15 |

If a body is indeed in a state of equilibrium, it is clear that the general conditions of equilibrium for a rigid body must be satisfied. The whole system of applied and reactive forces acting on a body (as represented by the free body diagram) must be in a state of equilibrium. Free body diagrams are consequently often called equilibrium diagrams. Drawing equilibrium diagrams and finding reactions for loaded structural members is a common first step in a complete structural analysis.

EQUATIONS OF EQUILIBRIUM

A structure or one of its members is in equilibrium when it maintains a balance of force and moment. In general, this requires that the force and moment equations of equilibrium be satisfied along x-, y- and z-axis, namely,

$$\sum F_x = 0 \qquad \sum F_y = 0 \qquad \sum F_z = 0$$

$$\sum M_x = 0 \qquad \sum M_y = 0 \qquad \sum M_z = 0$$

Whenever these equations are applied, it is first necessary to draw a free body diagram of the truss or its members. If a member is selected, it must be isolated from its supports and surroundings, and its outlined shapes must be drawn. All the forces and couple moments that act on the member must be shown.

DETERMINACY OF BEAMS

The equilibrium equations provide both the necessary and sufficient conditions for equilibrium. When all the forces in a beam can be determined strictly from these equations, the beam is referred to as statically determinate. Truss having more unknown forces than available equilibrium equations is called statically indeterminate (Fig. 2.16). For a beam/truss, there are at most three equilibrium equations for each part, so that if there is a total of m parts and r forces and moment reaction components, we have

$$\begin{aligned} &r = 3m, \text{ statically determinate} \\ &r > 3m, \text{ statically indeterminate} \\ &(r - 3m) = \text{degree of indeterminacy} \end{aligned}$$

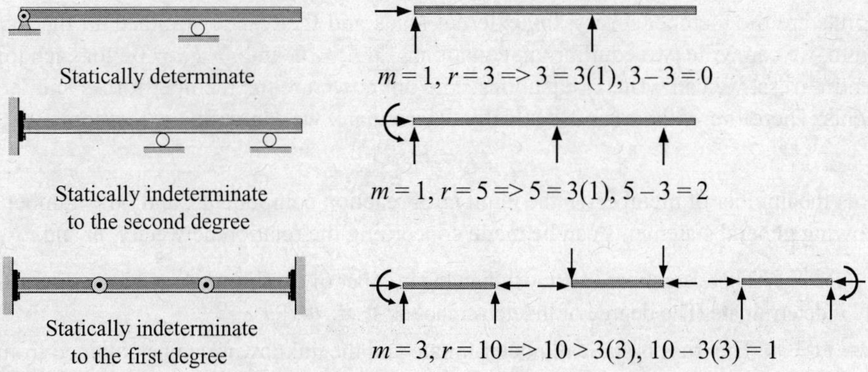

FIGURE 2.16 |

CLASSIFICATION OF TRUSS

EFFICIENT OR PERFECT TRUSS

A truss is said to be perfect when the number of members in the truss is just sufficient to prevent distortion of its shape when loaded with an external load.

A perfect truss has to satisfy the following equation:

$$m = 2j - 3$$

where m is the number of members.

IMPERFECT TRUSS

An imperfect truss is that which does not satisfy the equation, $m = 2j - 3$. Or in other words, a truss in which the number of members is more or less than $2j - 3$. The imperfect truss may be further classified into the following two types:

(1) deficient truss
(2) redundant truss.

Deficient Truss

A deficient truss is an imperfect truss, in which the number of members is less than $2j - 3$ (Fig. 2.17).

$$m < 2j - 3$$

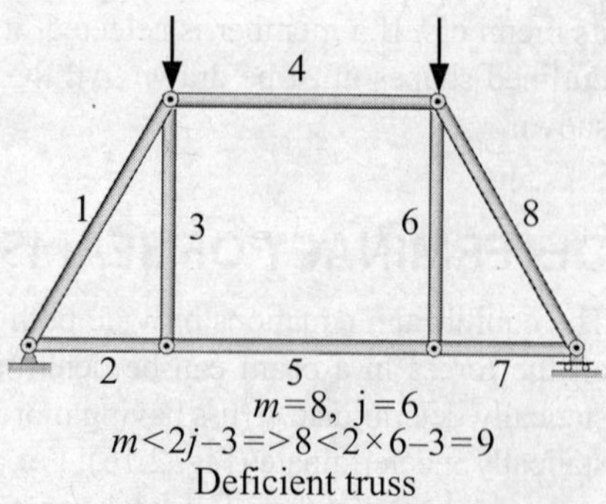

FIGURE 2.17

Redundant Truss

A redundant truss is an imperfect truss, in which the number of members is more than $2j - 3$.

$$m > 2j - 3$$

In this chapter, we shall discuss only perfect trusses.

DETERMINACY OF TRUSSES

A truss which possesses just sufficient numbers or bars to maintain its stability and equilibrium under any system of forces applied at joints is called a statically determinate and stable truss.

A planar truss may be thought of as a truss device having 'j' joints in a plane. The forces that act on the joints are the member forces, the external loads and the reactions. Since all the joints are in equilibrium, we can write two equilibrium equations, $\sum F_x = 0$ and $\sum F_y = 0$, for each joint. Thus, for the entire truss, we can write $2j$ equations. The unknown are the member forces and the reaction components. Therefore, if the truss is statically determinate, we can write the relation,

$$2j = m + r$$

where m is the number of members, r the number of reaction components, and j the number of joints. The following general statements can be made concerning the relation between j, m and r.

- $2j < m + r$. There are more unknown than the number of equilibrium equations. The truss is statically indeterminate. The degree of indeterminacy is $n = m + r - 2j$.
- $2j = m + r$. The truss is statically determinate and the unknowns can be obtained from $2j$ equations. The degree of indeterminacy $n = 0$.
- $2j > m + r$. There are not enough unknowns. The truss is a mechanism and always unstable.

NATURE OF FORCES

The members of a truss are subjected to either tensile or compressive force. A truss ABC loaded at joint A is shown in Figure 2.18(a).

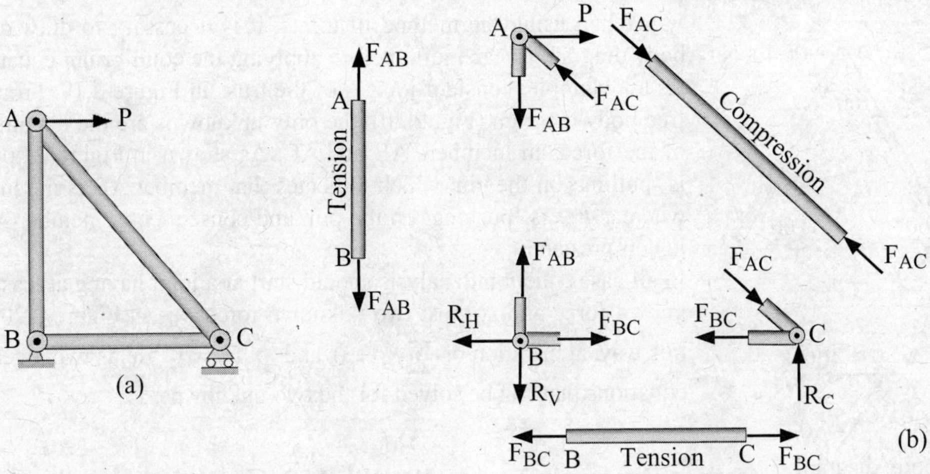

FIGURE 2.18 |

- The member AB is subjected to tensile force as shown in Figure 2.18(b)
- The member AC is subjected to compressive force as shown in Figure 2.18(b).
- The member BC is subjected to tensile force as shown in Figure 2.18(b).

It may be noted that tensile force in a member is represented in Figure 2.18(b). by two arrows going away from each other and a compressive force by two arrows coming towards each other. This is quite logical considering the fact that the markings on the members represent the internal reactive forces developed, which are opposite in direction to the applied forces.

BASIC ASSUMPTIONS OF TRUSS ANALYSIS

A plane truss can be defined as a structure composed of a number of bars, all lying in one plane and being hinged together at their ends in such a manner as to form a rigid configuration.

1. Members are connected at their ends by smooth frictionless pins.
2. Loads and reactions are applied to the truss at joints only.
3. The centroidal axis of each member is straight and coincides with the line connecting the joint centres at the end of the member.
4. All members function as two-force members, i.e., they are subjected to either tension or compression.

It is physically impossible for all these conditions to be satisfied exactly in an actual truss, and therefore a truss in which these idealized conditions are assumed to exist is called an ideal truss.

METHODS OF TRUSS ANALYSIS

Two analytical methods are usually utilized for truss analysis:

1. method of joints
2. method of sections.

Trusses can also be solved by graphical methods. Graphical method, however, is not explained in this chapter.

METHOD OF JOINTS

If a truss is in equilibrium, then each of its joints must also be in equilibrium. Hence, the method of joints consists of satisfying the equilibrium conditions $\sum F_x = 0$ and $\sum F_y = 0$ for the forces exerted on the pin at each joint of the truss.

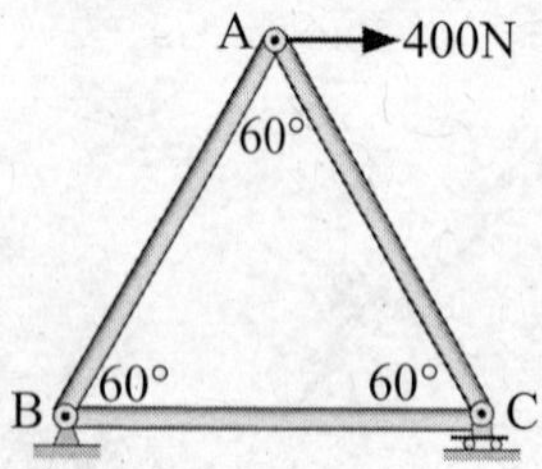

FIGURE 2.19 |

When using the method of joints, it is necessary to draw a free body diagram of each joint before applying the equilibrium equations. As an example, consider joint A of the truss in Figure 2.19. From the free body diagram (Fig. 2.20), the only unknowns are the magnitudes of the forces in members AB and AC. As shown in Figure 2.20, F_{AB} is 'pulling' on the pin, which indicates that member AB is in tension, whereas F_{AC} is 'pushing' on the pin, and consequently member AC is in compression.

In all cases, the joint analysis should start at a joint having at least one known force and at most two unknown forces, as in Figure 2.20b. In this way, application of $\sum F_x = 0$ and $\sum F_y = 0$ yields two algebraic equations that can be solved for the two unknowns.

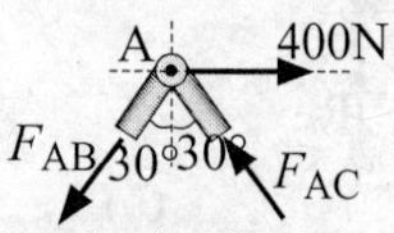

Free-body diagram of joint A

FIGURE 2.20 |

Steps for Analysis of Truss by Method of Joints

- Draw the free body diagram of a joint having at least one known force and at most two unknown forces. (If this joint is at one of the supports, it may be necessary to know the external reactions at the support.)
- Apply the two force equilibrium equations, $\sum F_x = 0$ and $\sum F_y = 0$ solve for the two unknown member forces, and verify their correct directional sense.
- Continue to analyze each of the other joints.

METHOD OF SECTIONS

The method of joints is most effective when the forces in all the members of a truss are to be determined. If, however, the force in only one member or the forces in a very few members are to be found then the method of sections is more efficient. This method is very quick as it does not involve the solution of other joints of the truss.

In this method, a section line is passed through the members, in which forces are to be determined. The section line should be drawn in such a way that it does not cut more than three members in which the forces are unknown. The part of the truss, on any one side of the section line, is treated as a free body in equilibrium under the action of external forces on that part and forces in the members cut by the section line. The unknown forces in the members are then determined by using equations of equilibrium, $\sum F_x = 0$, $\sum F_y = 0$ and $\sum M = 0$. If the magnitude of the forces in the member cut by a section line is positive, then the assumed direction is correct. If the magnitude of a force is negative, then reverse the direction of that force.

In this method, the equilibrium of a portion of the truss is considered, which is obtained by cutting the truss by an imaginary section. The following procedure is adopted to solve the problem by the method of sections.

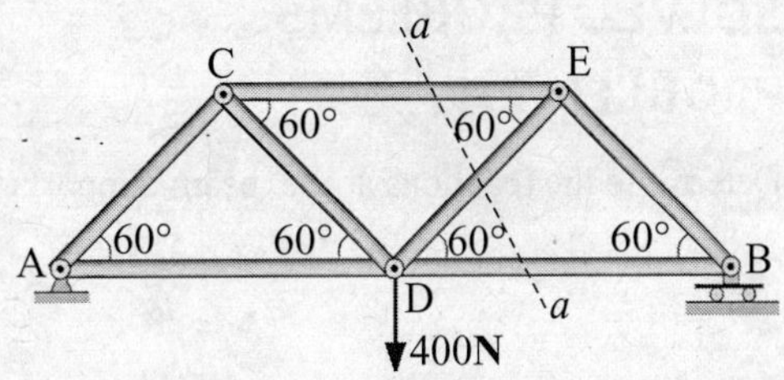

FIGURE 2.21 |

- Consider a truss as shown in Figure 2.21. Cut the truss through the member where forces are to be determined. In the figure, the section *aa* cuts the members CE, DE and DB.
- Before separating the two portions of the truss, determine the external reaction of load acting on the entire truss by applying equilibrium equations.
- Draw the free body diagram of the separate truss portions showing all forces (external and internal). The internal forces in the members cut by plane *aa* become the external forces acting on the two portions of the truss as shown in Figure 2.22.
- For analysis of free body diagram, choose the portion which involves the least number of forces.
- The equilibrium of the entire truss implies that every portion of the truss would also be in equilibrium. Using equations of equilibrium $\sum F_x = 0$, $\sum F_y = 0$ and $\sum M = 0$ for any portion of the truss, determine the unknown forces on members of the truss.

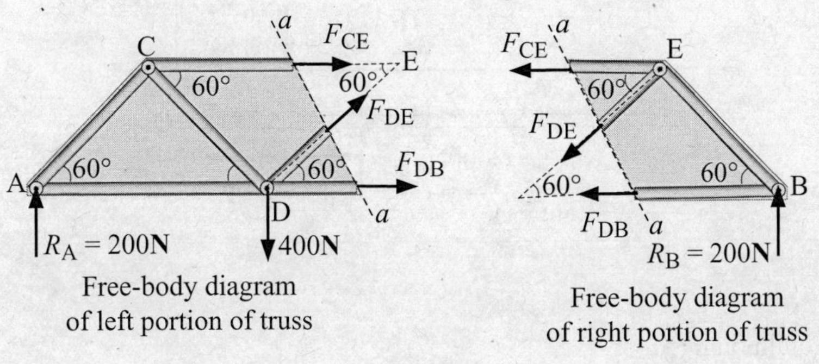

FIGURE 2.22 |

Steps for Analysis of Truss by Method of Sections

- Make a decision as to how to cut or section the truss through the members where forces are to be determined.
- Before isolating the appropriate section, it may first be necessary to determine the truss's external reactions, so that the three equilibrium equations are used only to solve for member forces at the cut section.
- Draw the free body diagram of that part of the sectioned truss which has the least number of forces on it.
- Use one of the two methods described above for establishing the science of an unknown force.

SOLVED PROBLEMS

PROBLEM 2.1

Determine the reactions at the beam supports for the given loading condition [Fig. 2.1(a)].

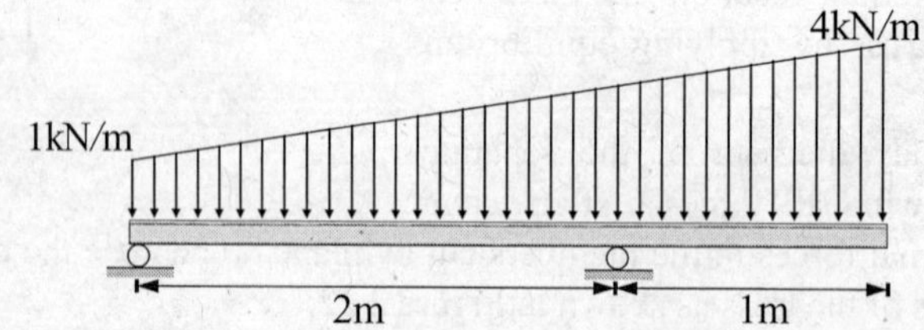

FIGURE 2.1(a) |

Solution. As shown in Figure 2.1(b), the trapezoidal distributed loading is segmented into a triangular and uniform load. The areas under the triangle and rectangle represent the resultant forces. These forces act through the centroid of their corresponding areas.

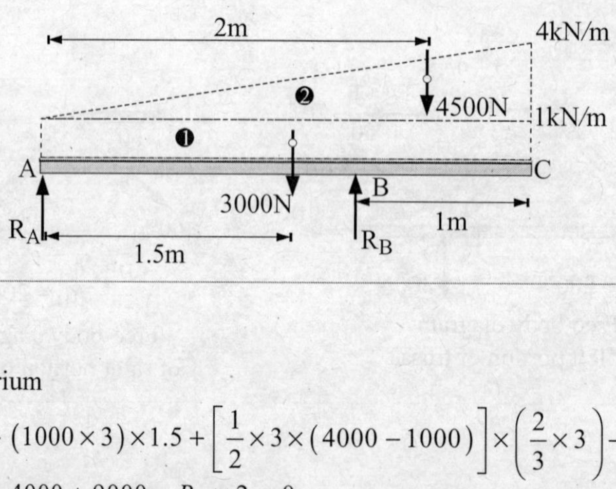

FIGURE 2.1(b) |

Equations of Equilibrium

$$\sum M_A = 0 \Rightarrow (1000\times 3)\times 1.5 + \left[\frac{1}{2}\times 3\times(4000-1000)\right]\times\left(\frac{2}{3}\times 3\right) - R_B\times 2 = 0$$

$$4000 + 9000 - R_B \times 2 = 0$$

$$\boldsymbol{R_B = 6750\text{N}}$$

$$\sum F_y = 0 \Rightarrow R_A + R_B - 1000\times 3 - \frac{1}{2}\times 3\times(4000-1000) = 0$$

$$R_A + 6750 - 3000 - 4500 = 0$$

$$\boldsymbol{R_A = 750\text{N}}$$

PROBLEM 2.2

A cantilever beam [Fig. 2.2(a)] is subjected to the load intensity which varies as. $\omega = \omega_0 \sin(\pi x / l)$. Determine the reactions at the beam supports.

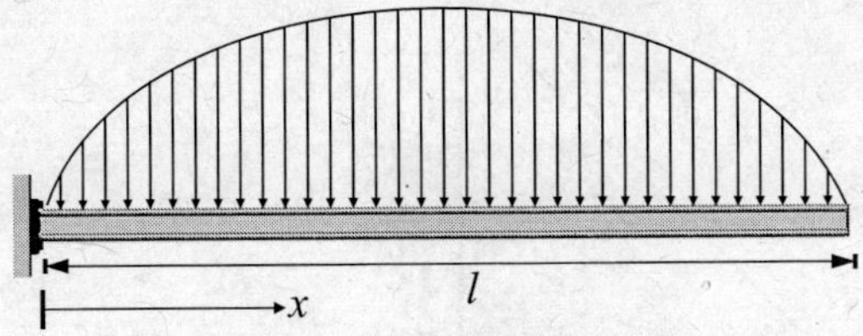

FIGURE 2.2(a) |

Solution. As shown in Figure 2.2(b), the fixed support at A exerts a normal force A_y and moment M_A. Equilibrium equations are

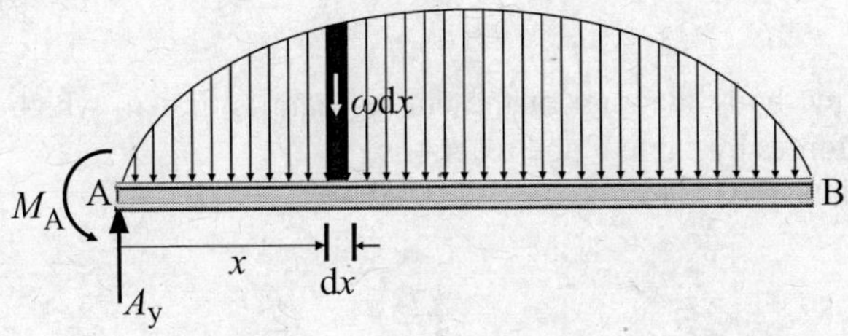

FIGURE 2.2(b) |

$$\sum F_y = 0 \Rightarrow A_y - \int \omega \mathrm{d}x = 0$$

or

$$A_y = \int_0^l \omega_0 \sin \frac{\pi x}{l} \mathrm{d}x$$

or

$$A_y = -\frac{\omega_0 l}{\pi}\left[\cos \frac{\pi x}{l}\right]_0^l = \frac{-\omega_0 l}{\pi}[-1-1]$$

$$A_y = \frac{2\omega_0 l}{\pi}$$

$$\sum M_A = 0 \Rightarrow -M_A + \int_0^l (\omega \mathrm{d}x)x = 0$$

or

$$M_A = \int_0^l \left(\omega_0 \sin \frac{\pi x}{l}\right) x\mathrm{d}x$$

or

$$M_A = \omega_0 \int_0^l x \sin \frac{\pi x}{l} \mathrm{d}x$$

$$M_A = \frac{\omega_0 l^2}{\pi}$$

PROBLEM 2.3

Determine the tension in the cable and the horizontal and vertical reaction at C [Fig. 2.3(a)].

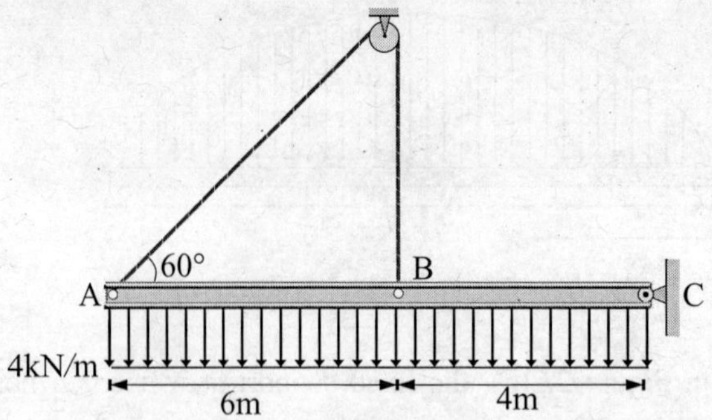

FIGURE 2.3(a)

Solution. The free body diagram of the entire beam is drawn, which shows load 40 kN at centroid of u.d.l. load, forces by cable T and reactions.

Equilibrium equations are

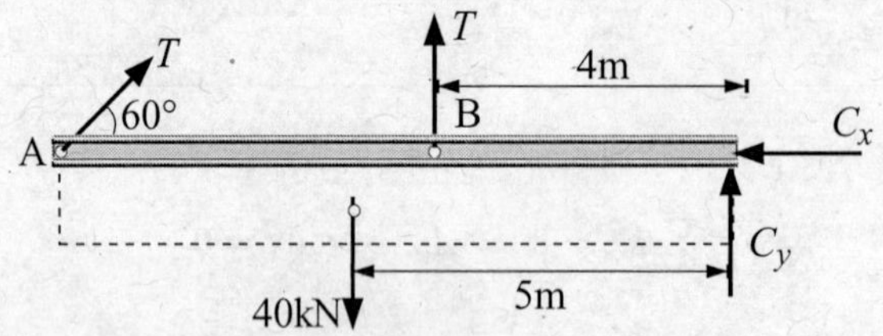

FIGURE 2.3(b)

$$\sum M_C = 0 \Rightarrow T \sin 60 \times 10 + T \times 4 - 40 \times 5 = 0$$

or $$\frac{\sqrt{3}T}{2} \times 10 + 4T = 200$$

$$T = 15.79\text{kN}$$

$$\sum F_y = 0 \Rightarrow T \sin 60 + T - 40 + C_y = 0$$

or $$C_y = 40 - T - T \sin 60$$

or $$C_y = 40 - 15.79 - 15.79 \times \frac{\sqrt{3}}{2}$$

$$C_y = 10.53\text{kN}$$

$$\sum F_x = 0 \Rightarrow T \cos 60 - C_x = 0$$

or $$C_x = 15.79 \cos 60$$

$$C_x = 7.895 \text{ kN}$$

PROBLEM 2.4

Determine the reactions at the supports [Fig. 2.4(a)].

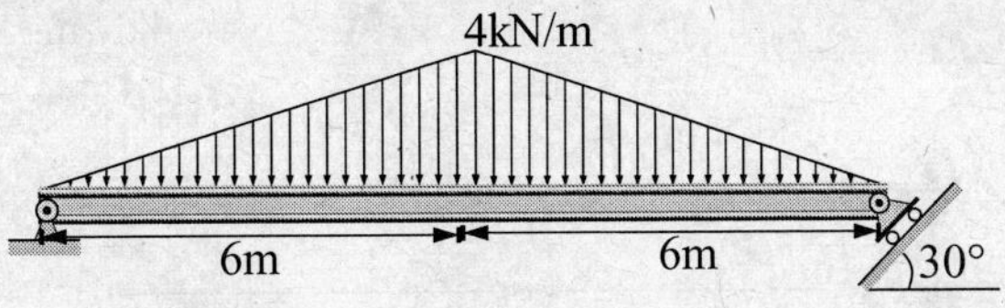

FIGURE 2.4(a) |

Solution. As shown in Figure 2.4(b), the triangular loading is segmented into two parts. The area of the triangle represents the resultant forces. These forces act through the centroid of their corresponding areas.

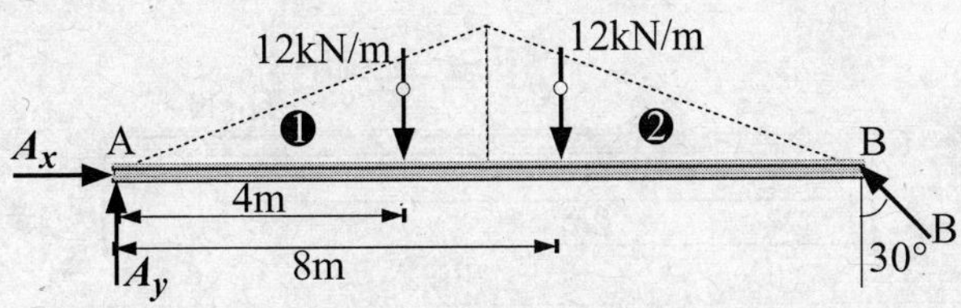

FIGURE 2.4(b) |

Equations of Equilibrium

$$\sum M_A = 0 \Rightarrow 12 \times 4 + 12 \times 8 - \text{B}\cos 30 \times 12 = 0$$

or
$$48 + 96 - \frac{\sqrt{3}}{2}\text{B} \times 2 = 0$$

$$B = 13.85 \text{ kN}$$

$$\sum F_x = 0 \Rightarrow \text{A}_x - \text{B}\sin 30 = 0$$

or
$$\text{A}_x = \frac{\text{B}}{2}$$

or
$$\text{A}_x = \frac{13.85}{2} = 6.92$$

$$\text{A}_x = 6.92\text{kN}$$

$$\sum F_y = 0 \Rightarrow \text{A}_y - \text{B}\sin 30 - 12 - 12 = 0$$

or
$$\text{A}_y + \frac{\sqrt{3}}{2}\text{B} = 24$$

or
$$\text{A}_y + \frac{\sqrt{3}}{.2}\text{B} = 24$$

or
$$\text{A}_y = 24 - \frac{\sqrt{3}}{2} \times 13.85 = 12.0$$

$$\text{A}_y = 12.0\text{kN}$$

PROBLEM 2.5

Find the reaction at the support A, B and C. There is a pin at B as shown in Figure 2.5(a).

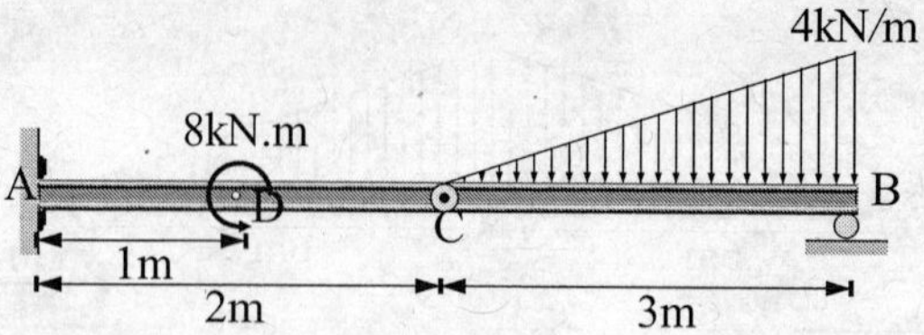

FIGURE 2.5(a) |

Solution. The free body diagram of each segment is shown in Figure 2.5(b) and 2.5(c).
Consider the beam BC Fig 2.5(b)

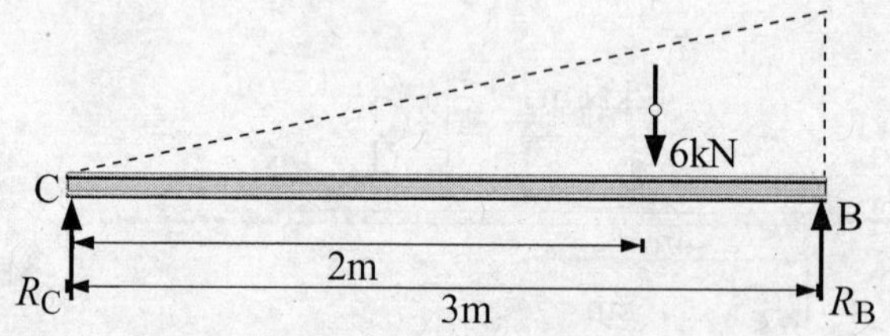

FIGURE 2.5(b) |

Equilibrium equations are

$$\sum M_C = 0 \Rightarrow 6 \times 2 - R_B \times 3 = 0$$

$$R_B = 4\text{kN}$$

$$\sum F_y = 0 \Rightarrow R_C + R_B - 6 = 0$$

$$R_C = -4 + 6 = 2$$

or

$$R_C = 2\text{kN}$$

Consider the beam AC. [Fig. 2.5(c)]

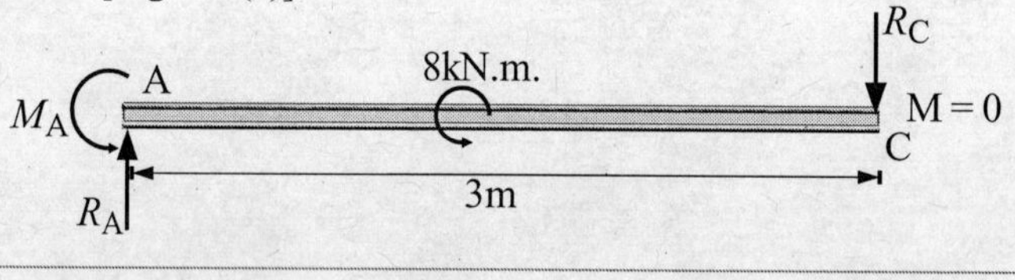

FIGURE 2.5(c) |

Equilibrium equations are

$$\sum F_y = 0 \Rightarrow R_A - R_C = 0$$

$$R_A = 2\text{kN}$$

$$\sum M_A = 0 \Rightarrow R_C \times 3 - 8 - M_A = 0$$

$$M_A = 2 \times 3 - 4$$

$$M_A = 2\text{kN.m}$$

PROBLEM 2.6

Figure 2.6(a) shows the cross-section of a gate 4 m high and 8 m long (perpendicular to the paper) which blocks water channel. The gate is hinged about a horizontal axis along its upper edge through A and is restrained from opening by the fixed ridge B which bears horizontally against the lower edge of the plate. Find the force B exerted on the plate by the ridge.

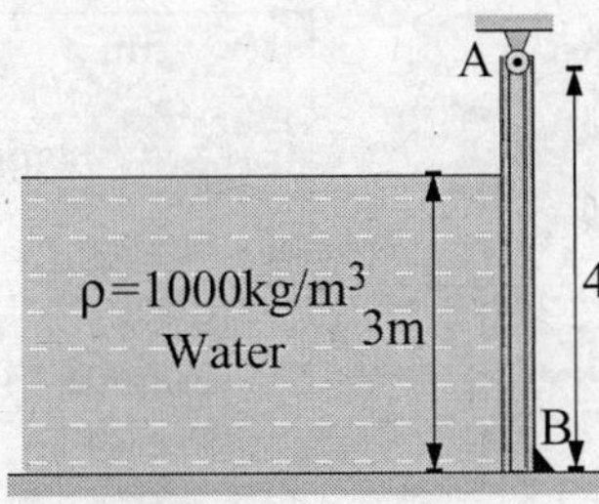

FIGURE 2.6(a)

Solution. The free body diagram of the plate is shown in Figure 2.6(b). Force per unit area due to water is given by (hydrostatic law)

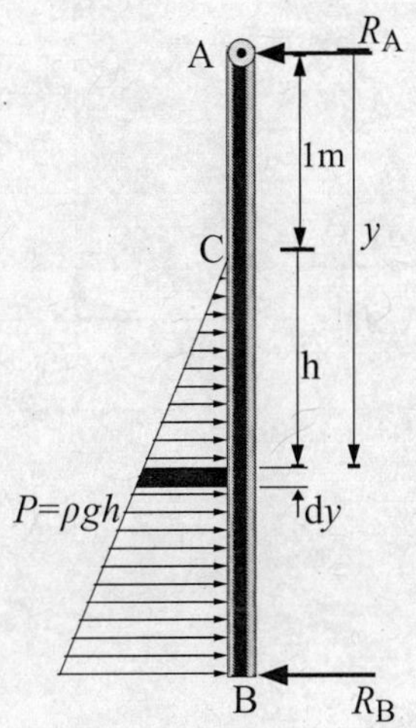

FIGURE 2.6(b)

$$P = \rho g h$$

or $$P = 1000 \times 9.81 \times (y - 1)\text{Nm}^{-2}$$

or $$P = 9810(y - 1)\text{Nm}^{-2}; \qquad 1 < y < 4$$

Taking moment of forces about A

$$\sum M_A = 0 \Rightarrow R_B \times 4 - \int dF \times y = 0$$

or $$R_B = \frac{\int dF \times y}{4} = \frac{\int P da \times y}{4}$$

$$R_B = \frac{\int_1^4 9810\,(y-1) \times (5dy)\,y}{4}$$

or $$= 122625 \int_1^4 (y^2 - y)\,dy$$

$$R_B = 165.54\text{kN}$$

PROBLEM 2.7

Determine the horizontal and vertical components of forces that the pins at A, B, C, E and F shown in Fig. 2.7(a) exert on their connecting members.

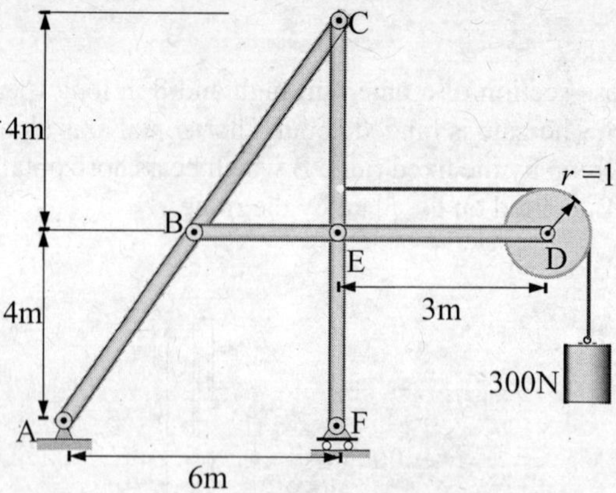

FIGURE 2.7(a) |

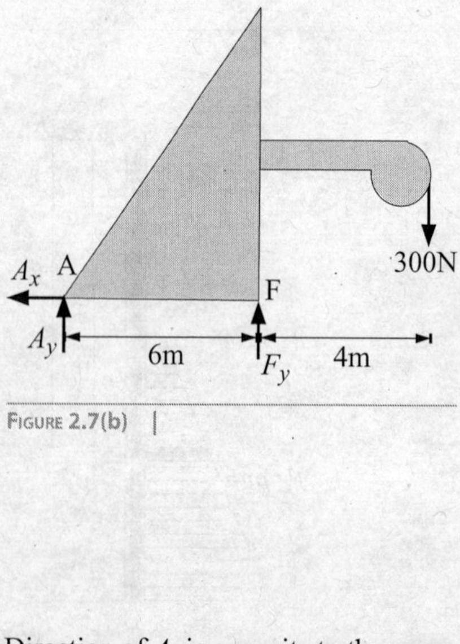

FIGURE 2.7(b) |

Solution. From the free body diagram of the entire frame as shown in Fig. 2.7(b), we determine the external reactions. Thus

$$\sum M_A = 0 \Rightarrow F_y \times 6 - 300\,(6+3+1) = 0$$

$$F_y = \frac{300\,(6+3+1)}{6} = 500$$

$$F_y = 500\text{N}$$

$$\sum F_x = 0 \Rightarrow A_x = 0$$

$$A_x = 0$$

$$\sum M_F = 0 \Rightarrow 300\,(3+1) + A_y \times 6 = 0$$

$$A_y = -200\text{N}$$

Direction of A_y is opposite to the assumed direction.
Next, we dismember the frame and draw separate free body diagrams for each member.

Consider Pulley

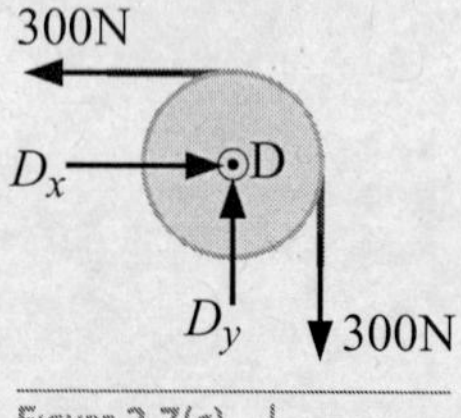

FIGURE 2.7(c) |

Free body diagram of the pulley is shown in Figure 2.7(c).

$$\left(\sum F_x\right)_D = 0 \Rightarrow D_x - 300 = 0$$

$$D_x = 300\text{N}$$

$$\left(\sum F_y\right)_D = 0 \Rightarrow D_y - 300 = 0$$

$$D_y = 300\text{N}$$

Consider Member BD

Free body diagram of member BD is shown in Figure 2.7(d)

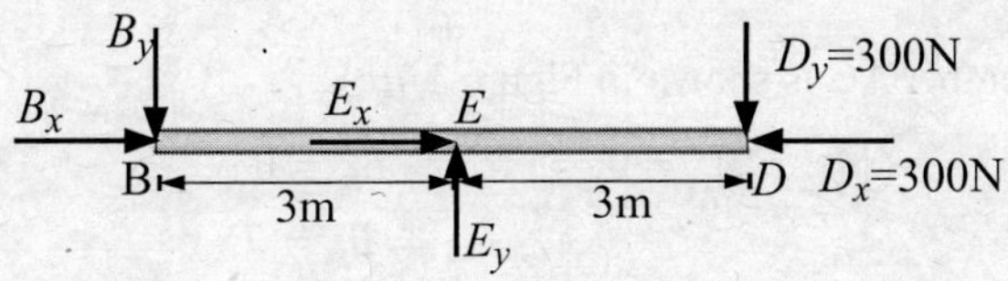

FIGURE 2.7(d)

$$\left(\sum M_E\right)_{BD} = 0 \Rightarrow 300 \times 3 - B_y \times 3 = 0$$

$$B_y = 300\text{N}$$

$$\left(\sum F_y\right)_{BD} = 0 \Rightarrow E_y - B_y - D_y = 0$$

or

$$E_y = 300 + 300$$

$$E_y = 600\text{N}$$

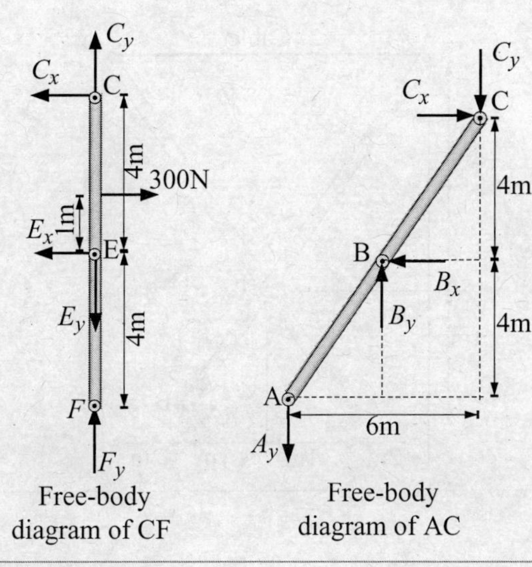

FIGURE 2.7 (e)

$$\left(\sum F_x\right)_{BD} = 0 \Rightarrow B_x + E_x - D_x = 0$$

$$B_x + E_x = 300 \quad \text{(i)}$$

Consider Member CF

Free body diagram of member CF is shown in Figure 2.7(e).

$$\left(\sum F_y\right)_{CF} = 0 \Rightarrow F_y - E_y + C_y = 0$$

or

$$500 - 600 + C_y = 0$$

$$C_y = 100\text{N}$$

$$\left(\sum M_E\right)_{CF} = 0 \Rightarrow 300 \times 1 - C_x \times 4 = 0$$

$$C_x = 75\text{N}$$

Consider Member AC

Free body diagram of member AC is shown in Figure 2.7(e).

$$\left(\sum F_x\right)_{AC} = 0 \Rightarrow \quad C_x - B_x = 0$$

$$C_x = B_x = 75$$

$$B_x = 75\text{N}$$

From Equation (i), we have

$$B_x + E_x = 300 \Rightarrow E_x = 300 - 75$$

$$E_x = 225\text{N}$$

PROBLEM 2.8

A two-member frame is pin connected at E, C and B as shown in Figure 2.8(a). The cable is attached to A, passes over the smooth peg at D, and is attached to a load 500 N. Determine the horizontal and vertical components of reactions at pin E, C and B.

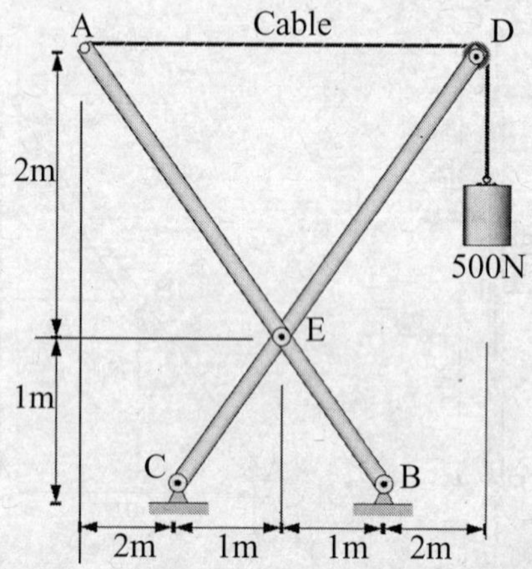

FIGURE 2.8(a) |

Solution. From the free body diagram of the entire frame, we determine the external reactions. Figure 2.8(b) Thus

$$\sum M_C = 0 \Rightarrow 500 \times 4 - B_y \times 2 = 0$$

$$B_y = 1000\text{N}$$

$$\sum F_y = 0 \Rightarrow C_y + B_y = 500$$

$$C_y = 500 - B_y = 500 - 1000$$

$$C_y = -500\text{N}$$

Negative sign indicate that the direction of C_y is opposite to assumed direction

$$\sum F_x = 0 \Rightarrow \quad B_x - C_x = 0$$

$$B_x = C_x$$

Next, we dismember the frame and draw a separate free body diagram of each member.

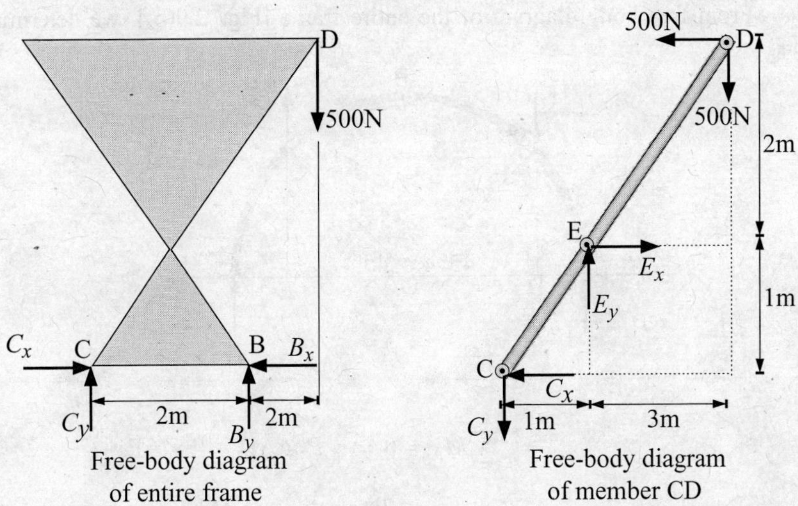

FIGURE 2.8(b) |

Consider Member CD

$$\left(\sum F_y\right)_{CD} = 0 \Rightarrow \quad E_y - C_y - 500 = 0$$

or

$$E_y = C_y + 500 = 500 + 500$$

$$E_y = 1000\text{N}$$

$$\left(\sum M_E\right)_{CD} = 0 \Rightarrow \quad 500 \times 3 - 500 \times 2 - C_y \times 1 + C_x \times 1 = 0$$

or

$$1500 - 1000 - 500 + C_x = 0$$

$$C_x = 0$$

From above equation $B_x = C_x$,

$$B_x = 0$$

$$\sum F_x = 0 \Rightarrow \quad E_x - 500 - C_x = 0 \text{ or } E_x = 500 + 0$$

$$E_x = 500\text{N}$$

PROBLEM 2.9

Two curved beams AB and BC are hinged at B and supported at A and C, as shown in Figure 2.9(a). Determine the horizontal and vertical components of all reactive forces acting on pins A, B and C.

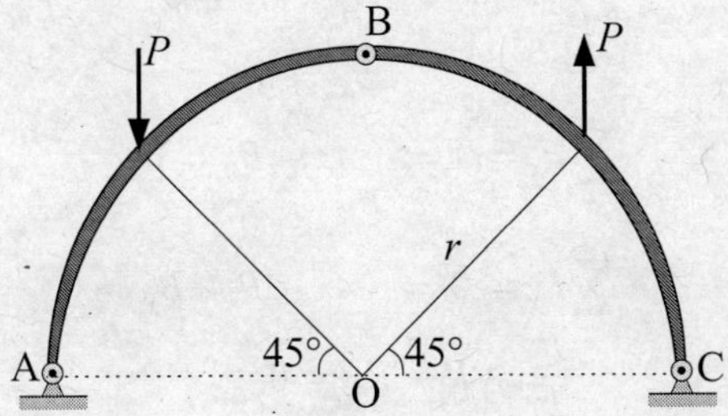

FIGURE 2.9(a) |

Solution. From free body diagram of the entire frame [Figs 2.9(b)], we determine external reactions. Thus

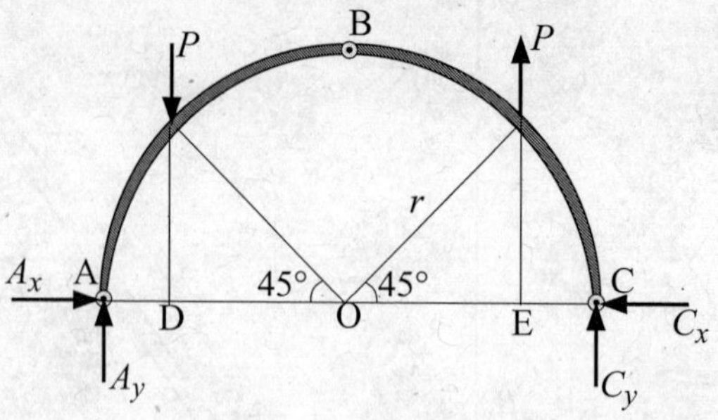

FIGURE 2.9(b) |

$$\sum M_A = 0 \Rightarrow \quad P \times AD - P \times AE - C_y \times 2r = 0$$

or
$$C_y = \frac{P \times (r - r\cos 45) - P \times (r + r\cos 45)}{2r}$$

$$C_y = -\frac{P}{\sqrt{2}}$$

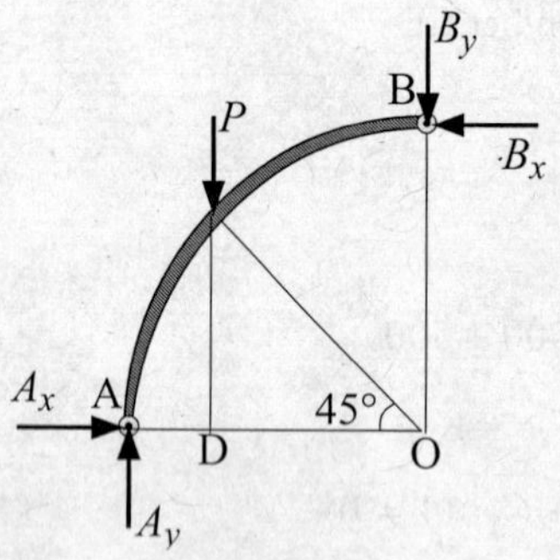

FIGURE 2.9(c) |

Negative sign indicates that the direction of C_y is opposite to the assumed direction

$$\sum F_y = 0 \Rightarrow \quad A_y + C_y - P + P = 0 \text{ or } A_y = -C_y$$

$$A_y = \frac{P}{\sqrt{2}}$$

$$\sum F_x = 0 \Rightarrow \quad A_x = C_x$$

Consider AB

$$\sum M_B = 0 \Rightarrow \quad -P \times r\cos 45 + A_y \times r - A_x \times r = 0$$

or
$$A_x = A_y - P\cos 45$$

or
$$A_x = \frac{P}{\sqrt{2}} - \frac{P}{\sqrt{2}} = 0$$

$$A_x = 0$$

$$\sum F_x = 0 \Rightarrow \quad A_x - B_x = 0$$

$$B_x = 0$$

$$\sum F_x = 0 \Rightarrow \quad A_x - B_x = 0$$

$$B_x = 0$$

$$\sum F_y = 0 \Rightarrow \quad A_y - P - B_y = 0$$

or
$$\frac{P}{\sqrt{2}} - P - B_y = 0$$

$$B_y = -0.2928P$$

PROBLEM 2.10

In the radial bar OA of the hinged regular octagon, a tensile force P is produced by tightening a turnbuckle T shown in Figure 2.10(a). Determine the forces produced in the other bars of the system.

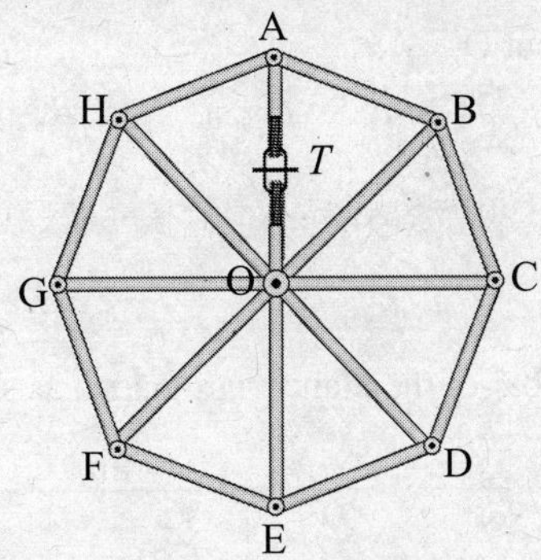

FIGURE 2.10(a)

Solution.

Joint A

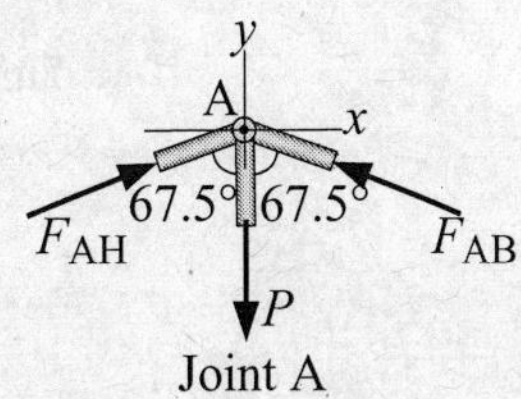

FIGURE 2.10(b)

Free body diagram of joint A is shown in Figure 2.10(b). Equilibrium in the horizontal direction:

$$\left(\sum F_x\right)_A = 0 \Rightarrow F_{AH}\sin 67.5 - F_{AB}\sin 67.5 = 0$$

or

$$F_{AH} = F_{AB}$$

Equilibrium in the vertical direction:

$$\left(\sum F_y\right)_A = 0 \Rightarrow F_{AH}\cos 67.5 + F_{AB}\cos 67.5 - P = 0$$

or

$$F_{AH} + F_{AB} = \frac{P}{\cos\ 67.5}$$

or

$$F_{AH} + F_{AB} = \frac{P}{2\cos\ 67.5} = 1.306P$$

$$F_{AH} = F_{AB} = 1.306P \text{ (compression)}$$

Member AH and AB are in compression as assumed, since the sign is positive.

Joint B

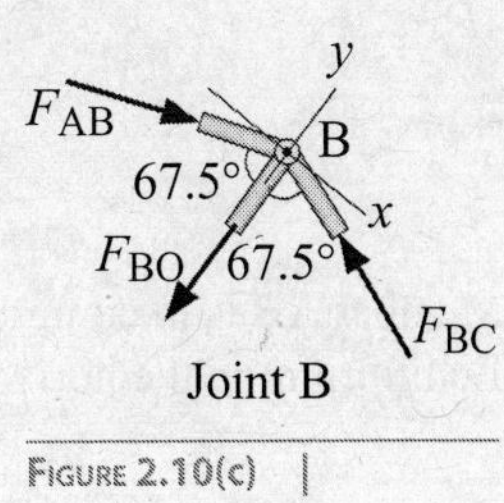

FIGURE 2.10(c)

Free body diagram of joint B is shown in Figure *2.10(c)*

Equilibrium in the horizontal direction:

$$\left(\sum F_x\right)_B = 0 \Rightarrow F_{AH}\sin 67.5 - F_{BC}\sin 67.5 = 0$$

$$F_{AB} = F_{BC}$$

Equilibrium in the vertical direction:

or

$$\left(\sum F_y\right)_B = 0 \Rightarrow F_{AB}\cos 67.5 + F_{BC}\cos 67.5 - F_{BO} = 0$$

$$F_{BO} = F_{AB}\cos 67.5 + F_{BC}\cos 67.5 = 0$$

substituting $F_{AB} = F_{BC}$ and $F_{AB} = 1.306P$, we have

$$F_{BO} = 2 \times 1.306P \times \cos 67.5 = P$$

or

$$F_{BO} = P \text{ (tension)}$$

Since the frame is symmetrical about O,

$$F_{AB} = F_{BC} = F_{CD} = F_{DE} = F_{EF} = F_{GH} = F_{HA} = 1.306P \text{ (compression)}$$

$$F_{BO} = F_{CO} = F_{DO} = F_{EO} = F_{FO} = F_{GO} = F_{HO} = F_{OA} = P \text{ (tension)}$$

PROBLEM 2.11

Determine the axial forces in each bar of the plane truss loaded as shown in Figure 2.11(a).

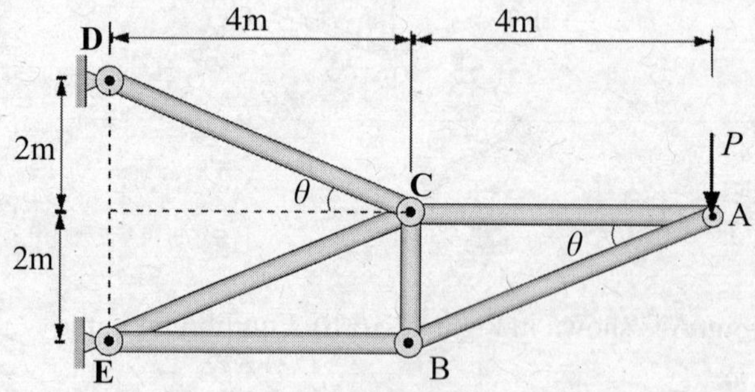

FIGURE 2.11(a) |

Solution. From the geometry of the figure,

$$\tan\theta = \frac{2}{4} \Rightarrow \theta = 26.56°$$

or

$$\sin\theta = \frac{1}{\sqrt{5}} \text{ and } \cos\theta = \frac{2}{\sqrt{5}}$$

Consider Joint A

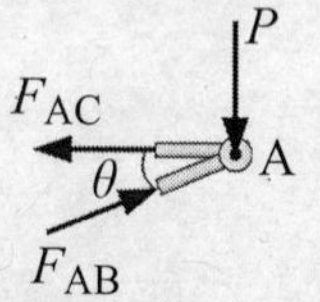

FIGURE 2.11(b) |

Free body diagram of joint A is shown in Figure 2.11(b).
Equilibrium in the vertical direction:

$$\left(\sum F_y\right)_A = 0 \Rightarrow -P + F_{AB}\sin\theta = 0$$

or

$$F_{AB} = \frac{P}{\sin\theta} = \sqrt{5}P$$

$$F_{AB} = \sqrt{5}P \text{ (compression)}$$

Member AB is in compression as assumed, since the sign is positive.
Equilibrium in the horizontal direction:

$$\left(\sum F_x\right)_A = 0 \Rightarrow -F_{AC} + F_{AB}\cos\theta$$

or

$$F_{AC} = F_{AB}\cos\theta = \sqrt{5}P \times \frac{2}{\sqrt{5}}$$

$$F_{AC} = 2P \text{ (tension)}$$

Member AC is in tension as assumed, since the sign is positive.

Consider Joint B

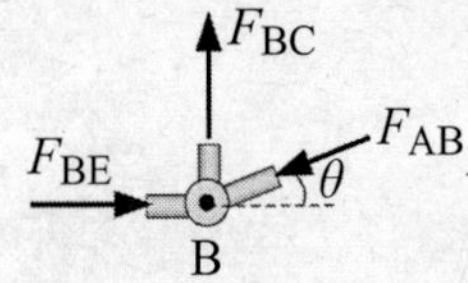

FIGURE 2.11(c)

Free body diagram of joint B is shown in Figure 2.11(c).
Equilibrium in the horizontal direction:

$$\left(\sum F_x\right)_B = 0 \Rightarrow \quad F_{BE} - F_{AB}\cos\theta = 0$$

or

$$F_{BE} = F_{AB}\cos\theta$$

or

$$F_{BE} = \sqrt{5}P \times \frac{2}{\sqrt{5}}$$

or

$$F_{BE} = 2P\,(\text{compression})$$

Member BE is in compression as assumed, since the sign is positive.
Equilibrium in the vertical direction:

$$\left(\sum F_x\right)_B = 0 \Rightarrow \quad F_{BC} - F_{AB}\sin\theta = 0$$

or

$$F_{BC} = \sqrt{5}P \times \frac{1}{\sqrt{5}} = P$$

$$F_{BC} = P\,(\text{tension})$$

Member BC is in tension as assumed, since the sign is positive.

Consider Joint C

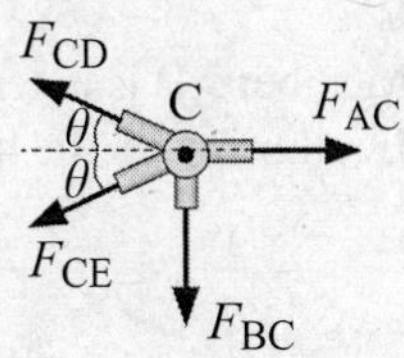

FIGURE 2.11(d)

Free body diagram of joint C is shown in Figure 2.11(d).
Equilibrium in the horizontal direction:

$$\left(\sum F_x\right)_C = 0 \Rightarrow \quad F_{AC} - F_{CD}\cos\theta - F_{CE}\cos\theta = 0$$

or

$$F_{CD} + F_{CE} = \frac{F_{AC}}{\cos\theta} = \frac{2P}{\frac{2}{\sqrt{5}}}$$

$$F_{CD} + F_{CE} = \sqrt{5}P \tag{i}$$

Equilibrium in the vertical direction:

$$\left(\sum F_y\right)_C = 0 \Rightarrow \quad F_{CD}\sin\theta - F_{CE}\sin\theta - F_{BC} = 0$$

or

$$F_{CD} - F_{CE} = \frac{F_{BC}}{\sin\theta} = \frac{P}{\frac{1}{\sqrt{5}}}$$

$$F_{CD} - F_{CE} = \sqrt{5}P \tag{ii}$$

From Equations (i) and (ii), we obtain

$$F_{CD} = 2\sqrt{5}P\,(\text{tension}) \text{ and } F_{CE} = 0$$

PROBLEM 2.12

A three-panel cantilever-type truss as shown in Figure 2.12(a) is pinned at E and tied to a vertical wall by a member DF. Determine the forces in the bars due to the vertical load 2 kN applied at A.

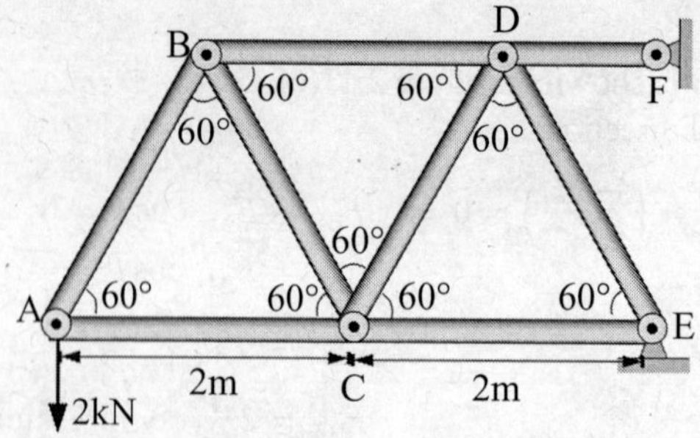

FIGURE 2.12(a) |

Solution.

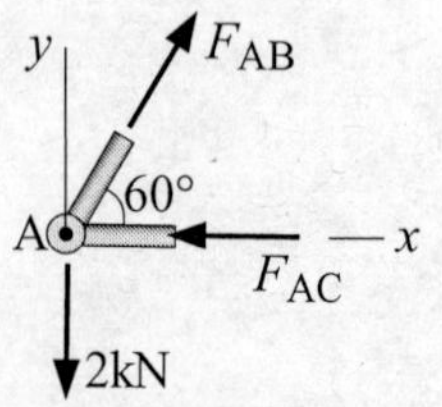

FIGURE 2.12(b) |

Joint A

Free body diagram of joint A is shown in Figure 2.12(b).
Equilibrium in the vertical direction:

$$\left(\sum F_y\right)_A = 0 \Rightarrow \quad F_{AB} \sin 60 - 2 = 0 \text{ or } F_{AB} = \frac{2}{\sin 60}$$

$$F_{AB} = \frac{4}{\sqrt{3}} \text{ kN (tension)}$$

Member AB is in tension as assumed, since the sign is positive.
Equilibrium in the horizontal direction:

$$\left(\sum F_x\right)_A = 0 \Rightarrow \quad F_{AB} \cos 60 - F_{AC} = 0$$

or

$$F_{AC} = F_{AB} \cos 60 = \frac{4}{\sqrt{3}} \times \frac{1}{2}$$

$$F_{AC} = \frac{2}{\sqrt{3}} \text{ kN (compression)}$$

Member AC is in compression as assumed, since the sign is positive.

Joint B

Free body diagram of joint B is shown in Figure 2.12(c).
Equilibrium in the vertical direction:

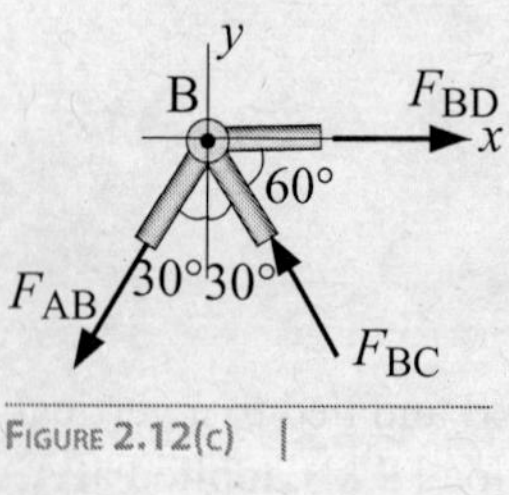

FIGURE 2.12(c) |

$$\left(\sum F_y\right)_B = 0 \Rightarrow \quad F_{BC} \cos 30 - F_{AB} \cos 30 = 0$$

or

$$F_{BC} = F_{AB} = \frac{4}{\sqrt{3}} \text{ kN}$$

$$F_{BC} = \frac{4}{\sqrt{3}} \text{ kN (compression)}$$

Member BC is in compression as assumed, since the sign is positive.
Equilibrium in the horizontal direction:

$$\left(\sum F_x\right)_B = 0 \Rightarrow \quad F_{BD} - F_{BC}\sin 30 - F_{AB}\sin 30 = 0$$

or

$$F_{BD} = (F_{BC} + F_{AB})\sin 30$$

$$F_{BD} = \left(\frac{4}{\sqrt{3}} + \frac{4}{\sqrt{3}}\right)\frac{1}{2} = \frac{4}{\sqrt{3}}$$

$$F_{BD} = \frac{4}{\sqrt{3}}\,\text{kN (tension)}$$

Member BD is in tension as assumed, since the sign is positive.

Joint C

Free body diagram of joint C is shown in Figure 2.12(d).
Equilibrium in the vertical direction:

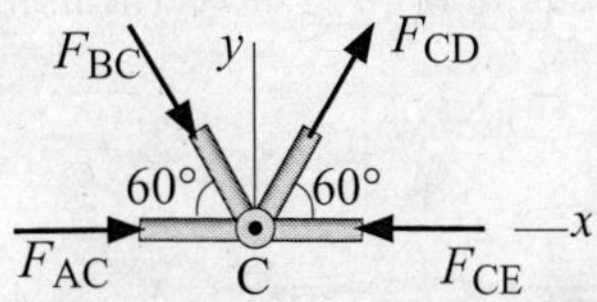

FIGURE 2.12(d)

$$\left(\sum F_y\right)_C = 0 \Rightarrow \quad F_{CD}\sin 60 - F_{BC}\sin 60 = 0$$

or

$$F_{CD} = F_{BC} = \frac{4}{\sqrt{3}}$$

$$F_{CD} = \frac{4}{\sqrt{3}}\,\text{kN (tension)}$$

Member CD is in tension as assumed, since the sign is positive.
Equilibrium in the vertical direction:

$$\left(\sum F_x\right)_C = 0 \Rightarrow \quad F_{AC} + F_{BC}\cos 60 + F_{CD}\cos 60 - F_{CE} = 0$$

$$\frac{2}{\sqrt{3}} + \frac{4}{\sqrt{3}} \times \frac{1}{2} + \frac{4}{\sqrt{3}} \times \frac{1}{2} - F_{CE} = 0$$

or

$$F_{CE} = 2\sqrt{3}\text{kN (compression)}$$

Member CE is in compression as assumed, since the sign is positive.

Joint D

Free body diagram of joint D is shown in Figure 2.12(e).
Equilibrium in the vertical direction:

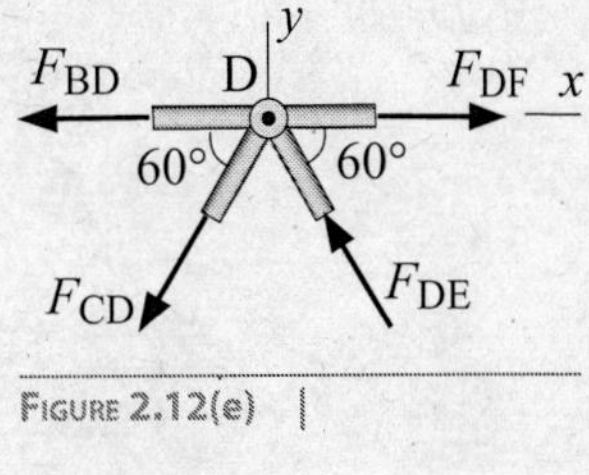

FIGURE 2.12(e)

$$\left(\sum F_y\right)_D = 0 \Rightarrow \quad -F_{CD}\sin 60 + F_{DE}\sin 60 = 0$$

or

$$F_{DE} = F_{CD} = \frac{4}{\sqrt{3}}$$

$$F_{DE} = \frac{4}{\sqrt{3}}\,\text{kN (compression)}$$

Member DE is in compression as assumed, since the sign is positive.
Equilibrium in the horizontal direction:

$$\left(\sum F_x\right)_D = 0 \Rightarrow \quad F_{DF} - F_{DE}\cos 60 - F_{CD}\cos 60 - F_{BD} = 0$$

or

$$F_{DF} - \frac{4}{\sqrt{3}} \times \frac{1}{2} - \frac{4}{\sqrt{3}} \times \frac{1}{2} - \frac{4}{\sqrt{3}} = 0$$

$$F_{DF} = \frac{8}{\sqrt{3}}\ \text{kN (tension)}$$

Member DF is in tension as assumed, since the sign is positive.

Joint E

Free body diagram of joint E is shown in Figure 2.12(f).
Equilibrium in the vertical direction:

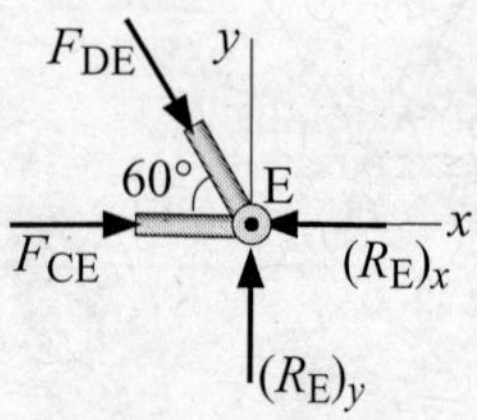

FIGURE 2.12(f) |

$$\left(\sum F_y\right)_E = 0 \Rightarrow \quad (R_E)_y - F_{DE}\sin 60 = 0$$

or

$$(R_E)_y - \frac{4}{\sqrt{3}} \times \frac{\sqrt{3}}{2} = 2$$

$$(R_E)_y = 2\text{kN}$$

Equilibrium in the horizontal direction [Fig. 2.12(g)]:

$$\left(\sum F_x\right)_E = 0 \Rightarrow \quad F_{DE}\cos 60 + F_{CE} - (R_E)_x = 0$$

or

$$\frac{4}{\sqrt{3}} \times \frac{1}{2} + 2\sqrt{3} - (R_E)_x = 0$$

$$(R_E)_x = \frac{8}{\sqrt{3}}\ \text{kN}$$

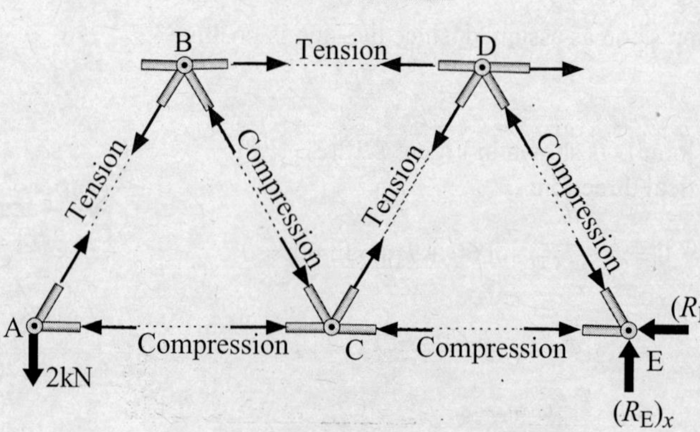

FIGURE 2.12(g) |

PROBLEM 2.13

Determine the forces in members BD and CD of the truss in Figure 2.13(a). All triangles are equilateral.

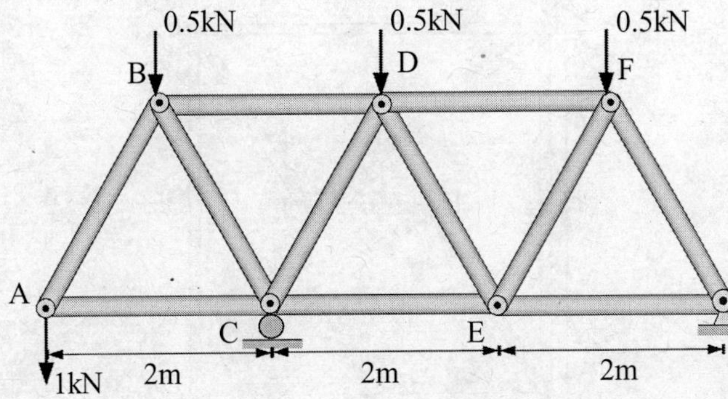

FIGURE 2.13(a)

Solution. Using the entire truss of Figure 2.13(b) as a free body and writing moments about G,

$$\sum M_G = 0$$

$$\Rightarrow R_C \times 4 - 1 \times 6 - 0.5 \times 5 - 0.5 \times 3 - 0.5 \times 1 = 0$$

$$R_C = \frac{10.5}{4} = 2.625\text{kN}$$

$$R_C = 2.625\text{kN}$$

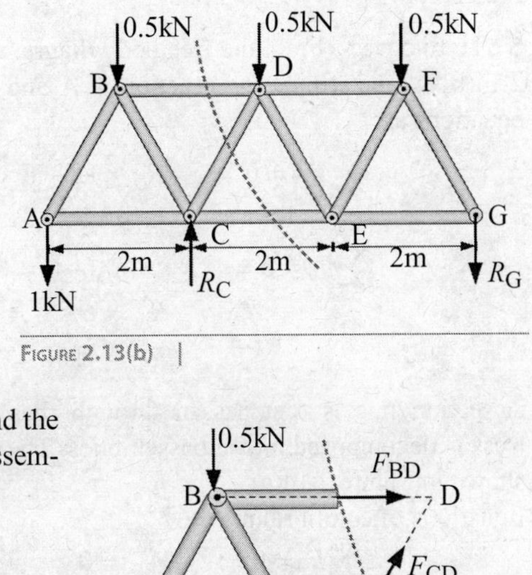

FIGURE 2.13(b)

A section line is passed through the member. BD, CD and the truss are decomposed into subassemblies. The left subassembly is shown in Figure 2.13 (c).

Equilibrium in the vertical direction:

$$\sum F_y = 0 \Rightarrow \quad F_{CD} \sin 60 + R_C - 0.5 - 1 = 0$$

$$F_{CD} \times \frac{\sqrt{3}}{2} + 2.625 - 0.5 - 1 = 0$$

$$F_{CD} = -1.3\text{kN}$$

$$F_{CD} = 1.3\text{kN (compression)}$$

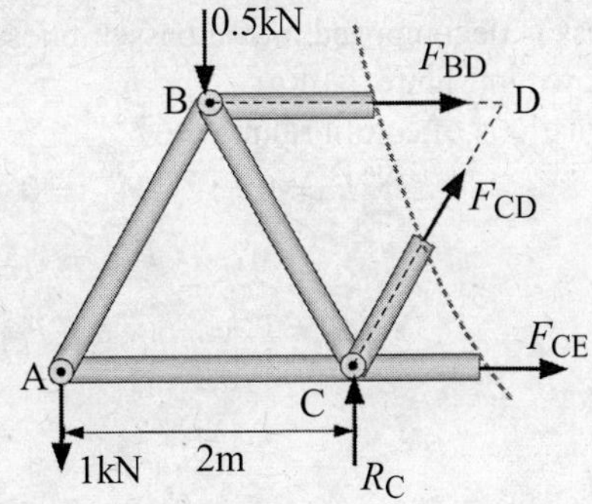

FIGURE 2.13(c)

Direction of force in member CD is opposite to the assumed direction.

$$\sum M_C = 0 \Rightarrow \quad F_{BD} \times 2 \sin 60 - 0.5 \times 1 - 1 \times 2 = 0$$

or

$$F_{BD} = \frac{2.5}{2 \sin 60} = 1.443$$

$$F_{BD} = 1.443\text{kN (tension)}$$

PROBLEM 2.14

A square truss consists of nine bars arranged as shown in Figure 2.14(a) and is subject to a 6 kN vertical load at E. Determine the forces in bar CD, AB and EF by method of sections.

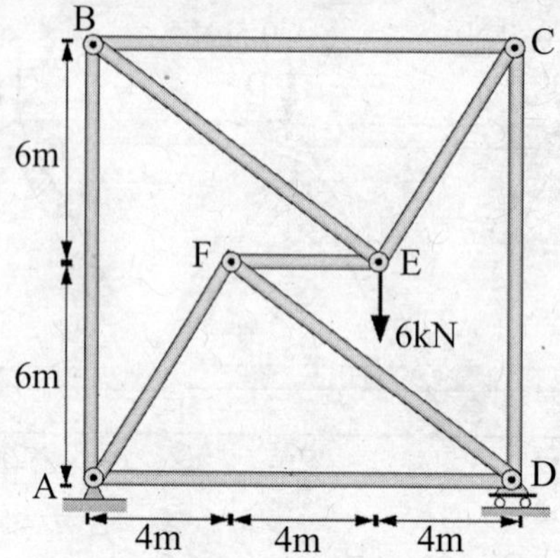

FIGURE 2.14(a)

Solution. From the free body diagram of entire truss Figure 2.14(b) we determine the reaction at A and D. Thus Equilibrium equations are

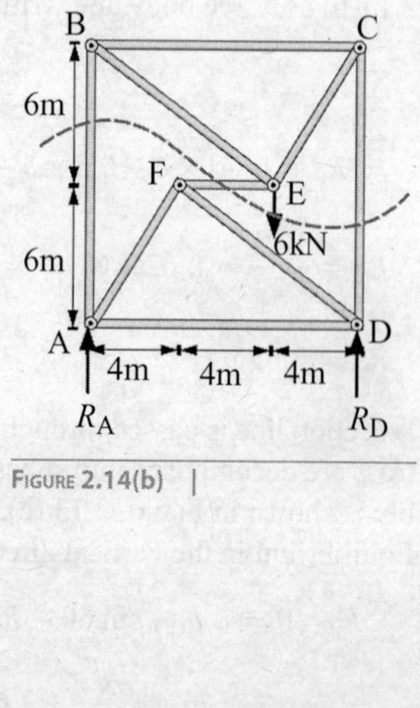

FIGURE 2.14(b)

$$\sum M_A = 0 \Rightarrow \quad -R_D \times 12 + 6 \times 8 = 0$$

$$R_D = 4\text{kN}$$

$$\sum F_y = 0 \Rightarrow \quad R_A \times R_D - 6 = 0$$

$$R_A = 6 - 4 = 2$$

$$R_A = 2\text{kN}$$

A section line is first passed through these three members and truss is decomposed into subassemblies. The upper subassembly is shown in Figure 2.14(c).

Equations of equilibrium:

$$\sum F_x = 0 \Rightarrow \quad -F_{EF} = 0$$

$$F_{EF} = 0$$

$$\sum M_B = 0 \Rightarrow \quad 6 \times 8 + F_{CD} \times 12 = 0$$

or

$$F_{CD} = -4$$

$$F_{CD} = 4\text{kN (compression)}$$

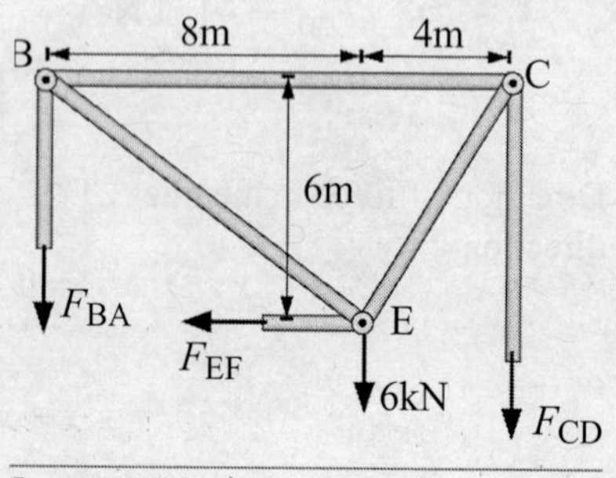

FIGURE 2.14(c)

$$\sum M_C = 0 \Rightarrow \quad F_{EF} \times 6 - F_{BA} \times 12 - 6 \times 4 = 0$$

or

$$0 \times 6 - F_{BA} \times 12 - 24 = 0$$

$$F_{BA} = -2\text{kN}$$

or

$$F_{BA} = 2\text{kN (compression)}$$

PROBLEM 2.15

Determine the forces in members FH, HG and IG of the truss shown in Figure 2.15(a). All triangles are equilateral with a side of 4 m (solve by method of sections).

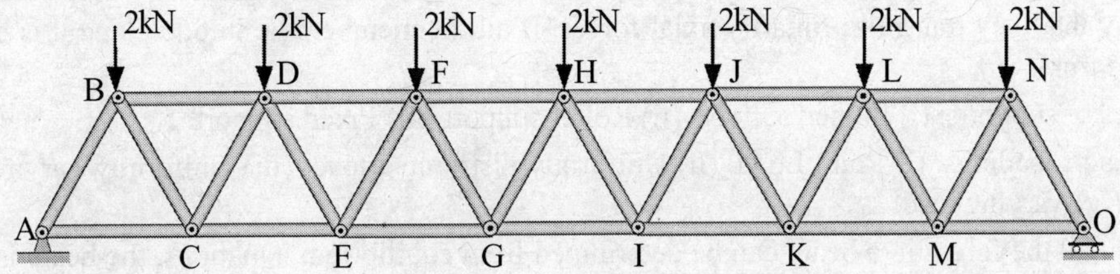

FIGURE 2.15(a) |

Solution. From the symmetry, the reactions at A and O are

$$R_A = R_O = \frac{7 \times 2}{2} = 7\text{kN}$$

To find forces in FH, HG and IG, it is more convenient to decompose the structure as shown in Figure 2.15(b), in which left subassembly is shown.

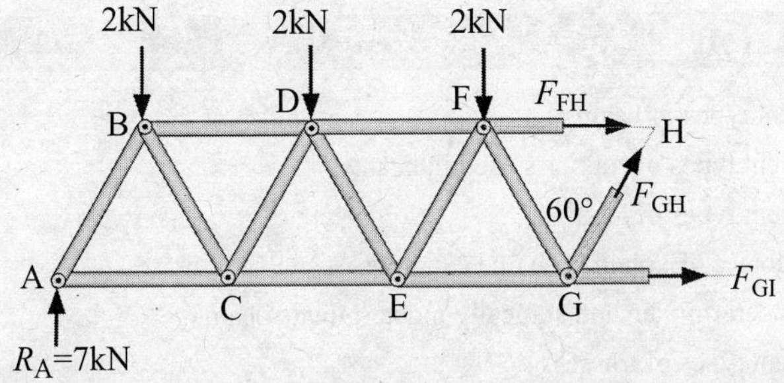

FIGURE 2.15(b) |

Equilibrium equations are

$$\sum F_y = 0 \Rightarrow 7 - 2 - 2 - 2 - F_{GH} \sin 60 = 0$$

or

$$F_{GH} = \frac{1}{\sin 60} = \frac{2}{\sqrt{3}}$$

$$F_{GH} = 1.15\text{kN (tension)}$$

$$\sum M_H = 0 \Rightarrow R_A \times 14 - 2 \times 12 - 2 \times 8 - 2 \times 4 - F_{GI} \times 4 \sin 60 = 0$$

or

$$F_{GI} = 14.43$$

$$F_{GI} = 14.43\text{kN (tension)}$$

$$\sum M_G = 0 \Rightarrow F_{FH} \times 4 \sin 60 + R_A \times 12 - 2 \times 10 - 2 \times 6 - 2 \times 2 = 0$$

or

$$F_{FH} \times \frac{4\sqrt{3}}{2} + 7 \times 12 - 20 - 12 - 4 = 0$$

$$F_{FH} = 13.85\text{kN (compression)}$$

CHAPTER HIGHLIGHTS

- Types of beams (i) Cantilever, (ii) Simply supported beam, (iii) Overhanging beam, (iv) Fixed beam, and (v) Continuous beam.
- A truss is an articulated structure composed of straight members arranged and connected in such a way that they transmit primarily axial forces. If all the members lie in one plane it is called a plane truss.
- Types of supports: (i) Pinned support (ii) Roller support (iii) Fixed support.
- Types of loadings: (i) Point Load, (ii) Uniformly distributed load, (iii) Uniformly varying load, and (iv) Varying load.
- When all the forces in a beam can be determined from equilibrium equations, the beam is called statically determinate. [*Condition for statically determinate beam* $r = 3m$].
- When all the forces in a beam cannot be determined from equilibrium equations, the beam is called statically indeterminate. [*Condition for statically indeterminate beam* $r > 3m$].
- Truss is perfect truss if $m = 2j - 3$
- Truss is imperfect truss if $m \neq 2j - 3$. If $m \leq 2j - 3$ is called deficient truss, and if $m > 2j - 3$ is called redundant truss.
- Two analytical methods are used for truss analysis (i) Method of joints, and (ii) Method of sections.

REVIEW QUESTIONS

Q1. Explain the different types of beams.

Q2. Describe the different types of supports used in beams.

Q3. Describe the different types of loadings.

Q4. What are reactive forces? Explain with one example.

Q5. What are statically determinate and statically indeterminate beams?

Q6. Classify the different types of trusses.

Q7. What are the assumptions made in the analysis of a simple truss?

Q8. What are the different methods of analyzing a perfect frame? Which one is used, where and why?

Q9. Explain the method of joints for finding forces in the member of a perfect truss.

Q10. Explain the method of sections for finding forces in the member of a perfect truss.

Q11. An overhang beam is loaded as shown in Figure 2.Q11. Determine the reactions at supports A and B [Ans; $R_A = 1119$ N ↑, $R_B = 19$ N ↓].

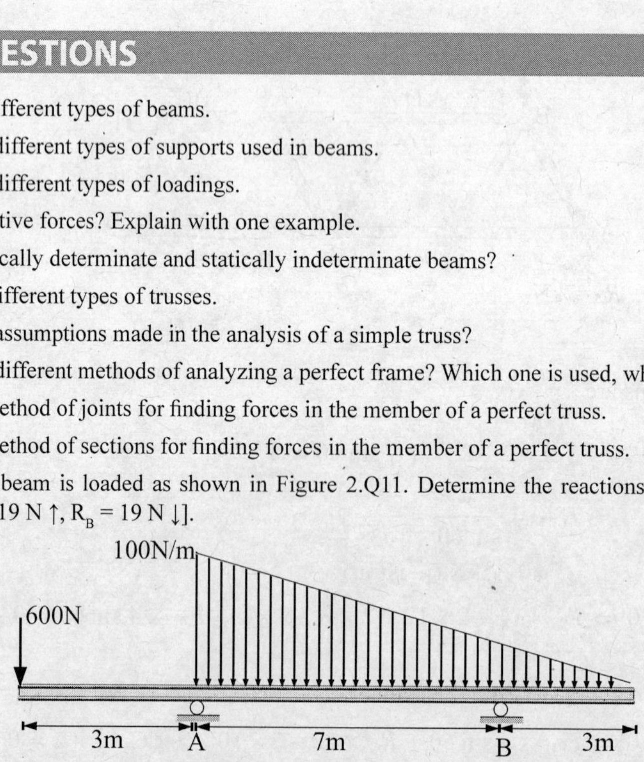

FIGURE 2.Q11 |

Q12. A frame supports a 1000 N load in the manner shown in Figure 2.Q12. Neglect the weights of the members compared with the forces induced by the load and calculate the compressive force in the bar AB and the shear force on the pin D. The pulley at E has a radius $r = 1$ m [Ans; $F_{AB} = 1250$ N, $R_D = 2016$ N].

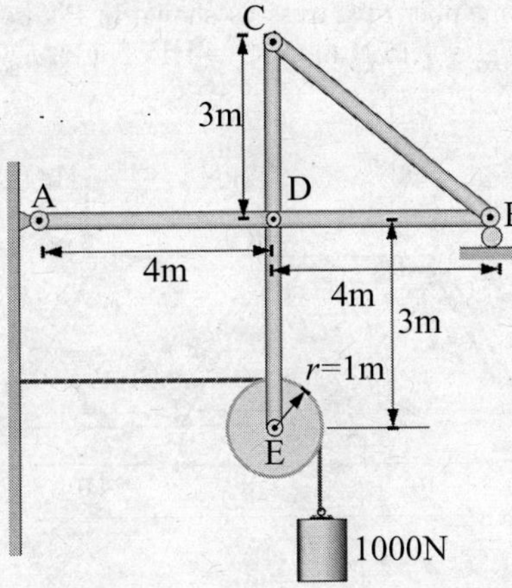

FIGURE 2.Q12

Q13. Neglecting the weight of frame, calculate the shear force on the pins A, B, C and E of the plane frame supported and loaded as shown in Figure 2.Q13. Assume that the horizontal floor is smooth [Ans; $R_A = 100$ N, $R_B = 300$ N, $R_C = 200$ N, $R_E = 400$ N].

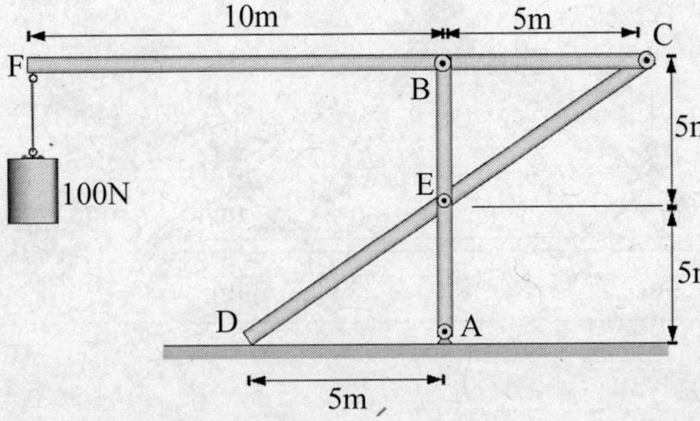

FIGURE 2.Q13

Q14. In a loaded ladder as shown in Figure 2.Q14, determine the tensile force in the string. Assume that the horizontal floor is smooth [Ans; 175 N].

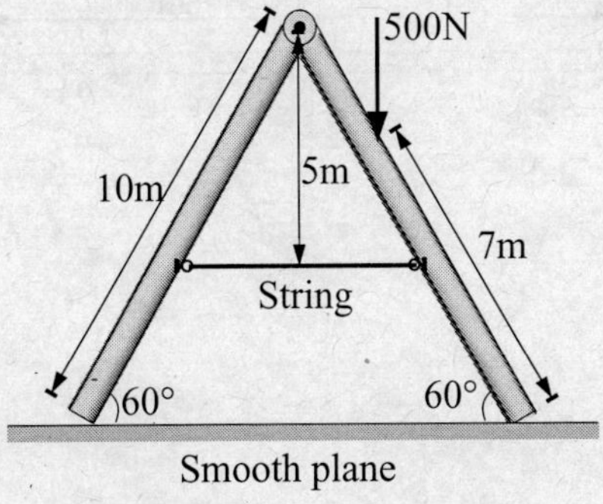

FIGURE 2.Q14

Q15. Determine the forces in each member of a truss as shown in Figure 2.Q15. All triangles are equilateral [Ans; F_{AB} = 2.31 N (C), F_{AG} = 1.15 N (T), F_{BG} = 1.15 N (T), F_{BC} = 1.73 N (C), F_{GF} = 2.31 N (T), F_{GC} = 1.55 N (C)].

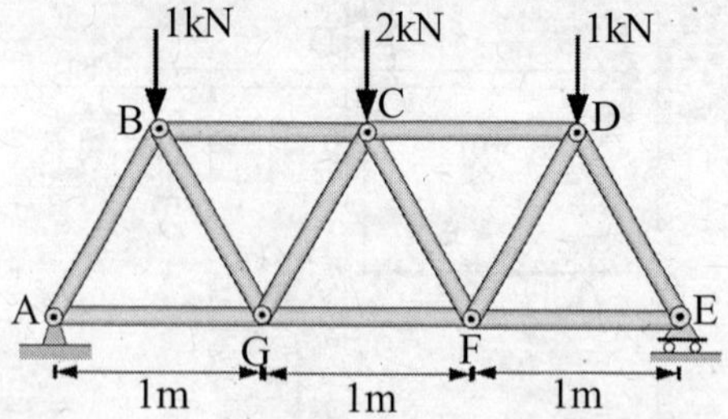

FIGURE 2.Q15 |

Q16. A triangular truss as shown in Figure 2.Q16 is subjected to three vertical loads. Determine the forces in the bars BD, CD and CE by (a) the method of sections and (b) method of joints [Ans; F_{BD} = 2.5 kN (C), F_{CD} = 0.866 kN (T), F_{CE} = 1.73 kN (T)].

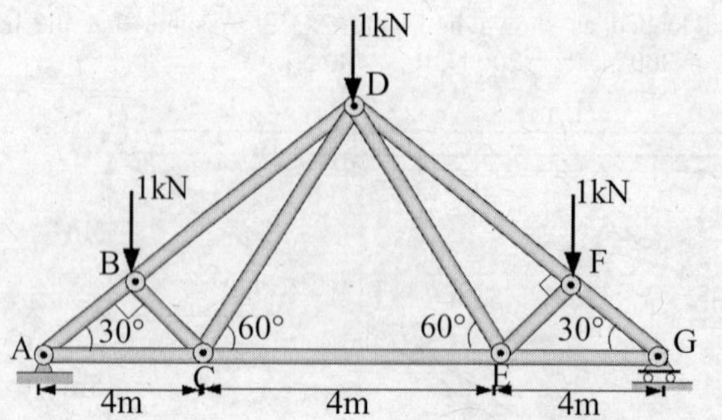

FIGURE 2.Q16 |

Q17. Find out the forces in the members AB, BD and CD of a pin jointed truss as shown in Figure 2.Q17 by (a) method of joints and (b) method of sections [Ans; F_{AB} = (80/3) kN (T), F_{BD} = 20 kN (C), F_{CD} = (100/3) kN (C)].

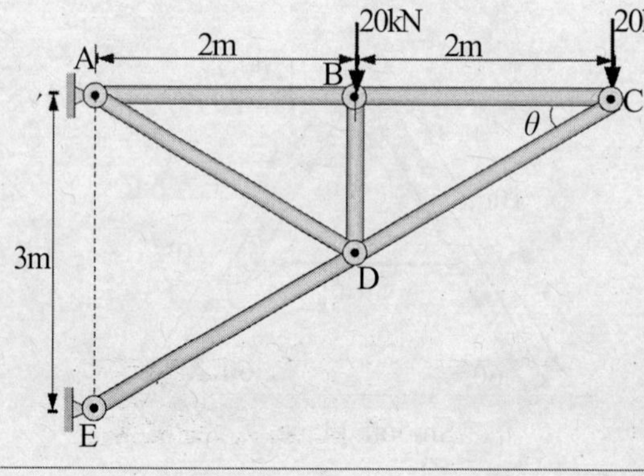

FIGURE 2.Q17 |

MULTIPLE CHOICE QUESTIONS

Q1. To solve a truss by the method of joints, the number of unknowns at a joint should not be

(A) less than 2 (B) more than 2 (C) less than 3 (D) more than 3

[Ans: B]

Q2. To solve a truss by the method of sections, the number of unknowns at a section should not be

(A) less than 2 (B) more than 2 (C) less than 3 (D) more than 3

[Ans: D]

Q3. In a perfect truss, the number of members should be

(A) more than $(2j - 3)$ (B) less than $(2j - 3)$ (C) equal to $(2j - 3)$ (D) equal to $2j$

[Ans: C]

Q4. Method of sections is more suitable when

(A) only reactions at the supports are desired

(B) only forces in few of the members are desired

(C) only forces in few of the members away from the supports are desired

(D) forces in all the members are desired

[Ans: C]

Q5. In a redundant truss, the number of members is

(A) more than $(2j - 3)$ (B) less than $(2j - 3)$ (C) equal to $(2j - 3)$ (D) more than to $2j$

[Ans: B]

Q6. A frame in which the number of members is just sufficient to keep it in equilibrium is known as

(A) theoretical frame (B) perfect frame (C) deficient frame (D) redundant frame

[Ans: B]

Q7. An overhang simply supported beam is shown in Figure 2.MQ7. The reactions at supports C and B are

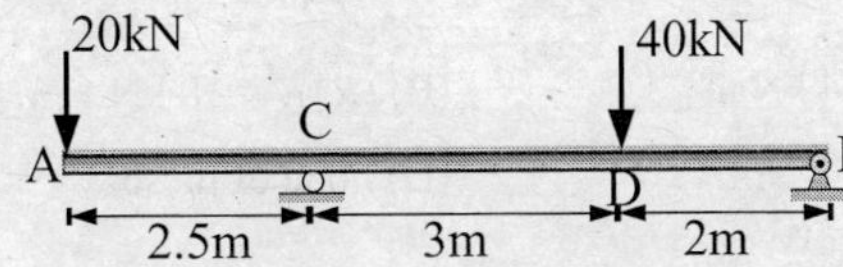

FIGURE 2.MQ7 |

(A) $R_C = 42$ kN, $R_B = 18$ kN (B) $R_C = 46$ kN, $R_B = 14$ kN

(C) $R_C = 40$ kN, $R_B = 20$ kN (D) none of these

[Ans: B]

Q8. A simply supported beam is subjected to a couple of moment 4 kN m at centre as shown in Figure 2.MQ8. The reaction at support B is

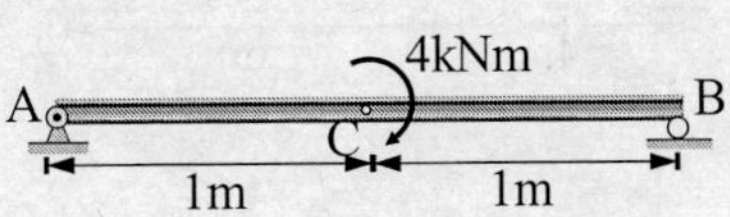

FIGURE 2.MQ8 |

(A) $R_B = 2$ kN (B) $R_B = 4$ kN (C) $R_B = 1$ kN (D) none of these.

[Ans: A]

Q9. Two beams AB and BC are jointed at B and loaded as shown in Figure 2.MQ9. The reaction at joint B is

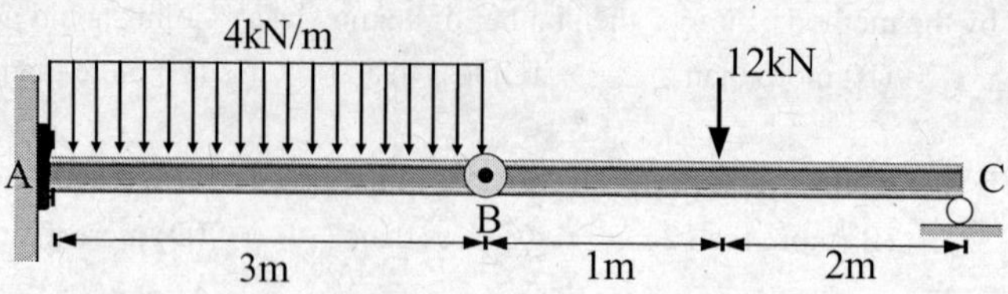

FIGURE 2.MQ9 |

(A) $R_B = 8$ kN (B) $R_B = 12$ kN (C) $R_B = 4$ kN (D) $R_B = 1$ kN

[Ans: A]

Q10. A cantilever beam AB is supported to triangular loading of intensity from 0 to 12 kN/m, as shown in Figure 2.MQ10. the reaction at fixed support is

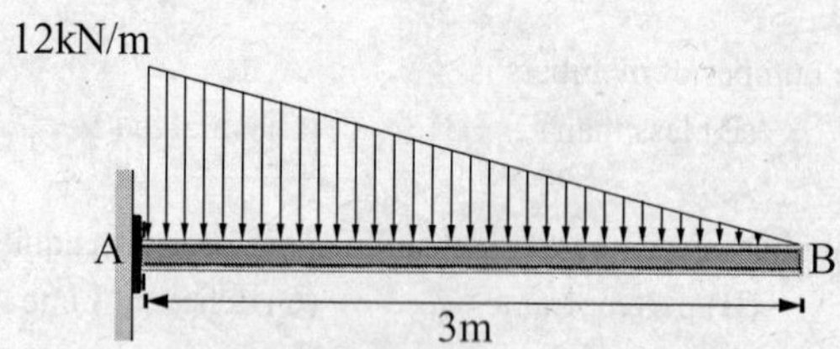

FIGURE 2.MQ10 |

(A) $(R_A)y = 18$ kN ↑, $M_A = 24$ kN m (B) $(R_A)y = 18$ kN ↑, $M_A = 12$ kN *m*

(C) $(R_A)y = 18$ kN ↑, $M_A = 6$ k Nm (D) none of these

[Ans: D]

Q11. A simply supported beam is shown in Figure 2.MQ11. The reactions at supports A and B are

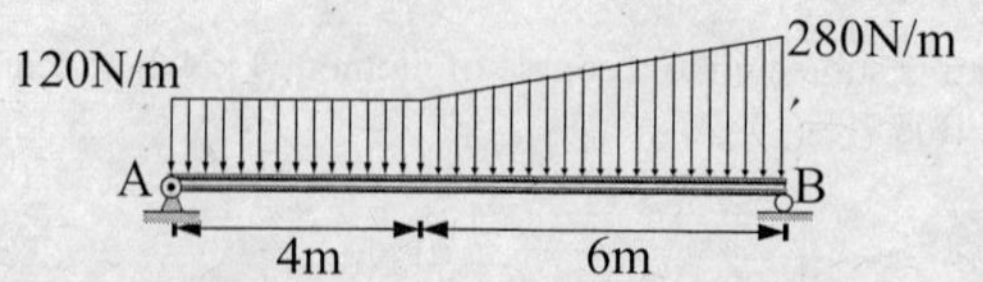

(A) $R_A = 690$ N, $R_B = 990$ N (B) $R_A = 696$ N, $R_B = 984$ N

(C) $R_A = 694$ N, $R_B = 986$ N (D) none of these

[Ans: B]

Q12. The force in member OC of the truss shown in Figure 2.MQ12 is

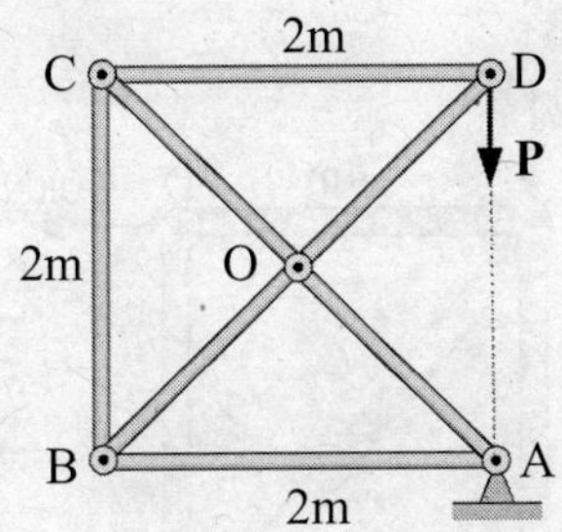

FIGURE 2.MQ12 |

(A) P (B) $P\sqrt{2}$ (C) $P/\sqrt{2}$ (D) none of these

[Ans: B]

Q13. The force in member AD of the truss shown in Figure 2.MQ13 is

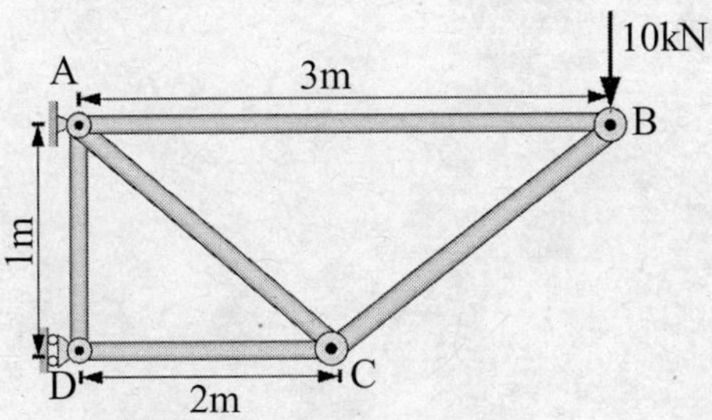

FIGURE 2.MQ13 |

(A) zero (B) 8. 73 kN (C) 3. 25 kN (D) none of these.

[Ans: A]

Q14. The force in member AC of the loaded frame shown in Figure 2.MQ14 is

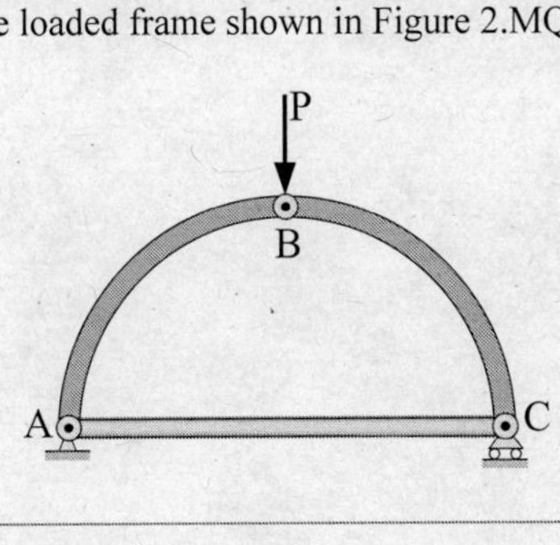

FIGURE 2.MQ14 |

(A) P (compression) (B) $2P$ (tension) (C) $P/2$ (tension) (D) none of these.

[Ans: C]

Q15. The force in member DB of the truss shown in Figure 2.MQ15 is

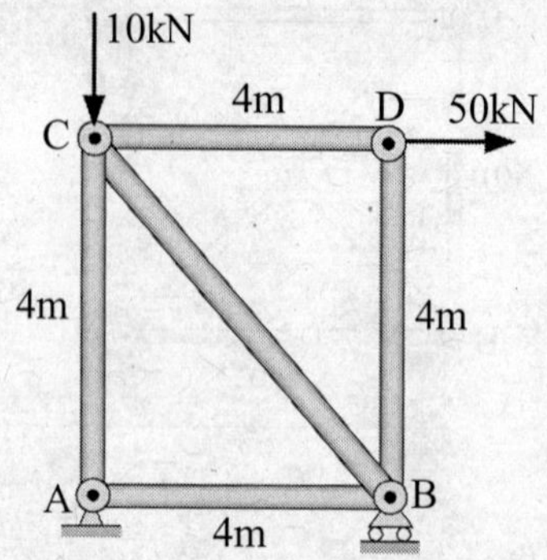

FIGURE 2.MQ15 |

(A) zero (B) $50\sqrt{2}$kN (compression) (C) $(50/\sqrt{2})$kN (tension) (D) none of these

[Ans: A]

3 Shear Force and Bending Moment

SHEAR FORCES AND BENDING MOMENTS

Shear force. The shear force S at any cross-section of a horizontal beam is the algebraic sum of the vertical projection of all forces (including reactions) to the *left* of the given cross-section (where, in the sum, the projection of a force is positive if its direction is upward and negative if its direction is downward).

Bending moment. The bending moment M at any cross-section of a horizontal beam is the sum of the moments about the given cross-section of all the forces (including support reactions) to the *left* of the cross-section (where, in the sum, a clockwise moment is positive and an anticlockwise moment is negative).

Consider a simply supported beam AB subjected to various concentrated and distributed forces as shown in Figure 3.1. We propose to find the shear force and bending moment at any section $m-n$ of the beam.

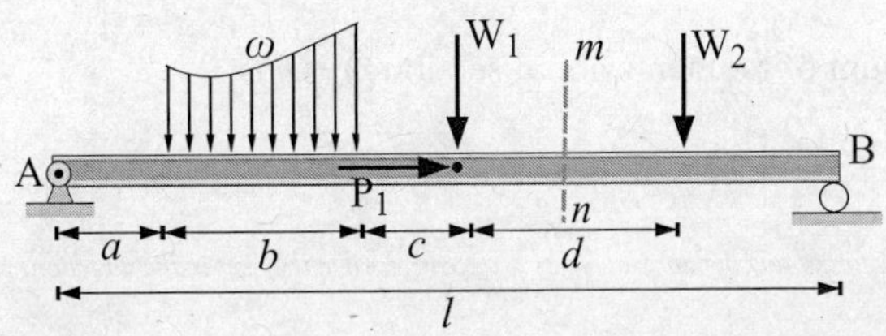

FIGURE 3.1 |

First, we determine the reactions at supports A and B by considering the entire beam as a free body diagram as shown in Figure 3.2; applying equilibrium equations $\Sigma M = 0$ and $\Sigma F = 0$, we obtain reactions at supports.

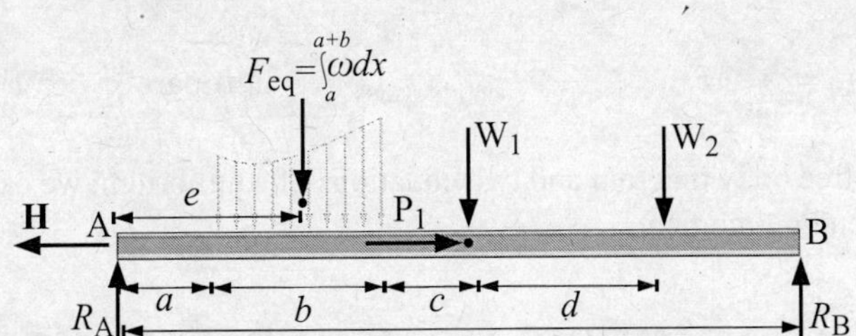

FIGURE 3.2 |

To determine the shear force S and the bending moment M at a section $m - n$, a distance x from the left support, we cut the beam at section $m - n$ and draw the free body diagram of the left portion of the beam (Fig. 3.3). The free body is held in equilibrium by the external forces, support reaction, the internal forces (transverse force S and axial force A) and couple (moment M) that act over the cut cross-section.

The force S is the resultant of the transverse forces that are distributed over the free body. It is called the shear force S acting on the beam at section $m - n$. The couple with moment M is the resultant of the moment of various forces and reactions. It is called the bending moment M.

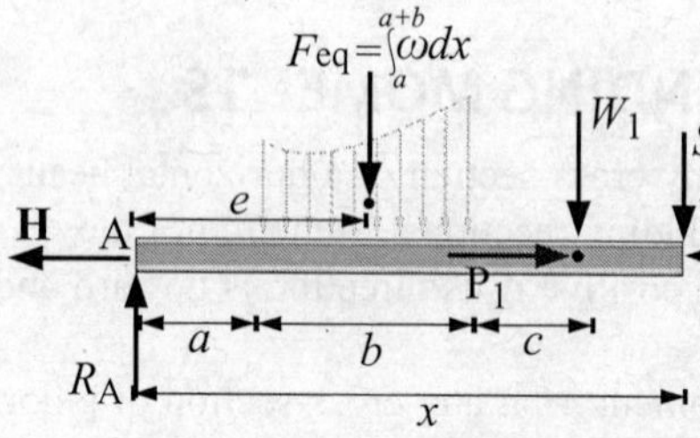

FIGURE 3.3 |

The shear force S and the bending moment M may be determined by the conditions of equilibrium for the isolated part (free body diagram) of the beam. The equation of equilibrium of forces is

$$\sum F_y = 0 \Rightarrow R_A - F_{eq} - W_1 - S = 0$$

The equation of equilibrium of moments about section $m - n$ is

$$\sum M = 0 \Rightarrow M + W_1(x - a - b - c) - F_{eq}(x - e) - R_A x' = 0$$

It is customary to assume that shear forces and bending moments are positive when they act in the directions shown in Figure 3.3.

From which,

$$\text{shear force, } S = R_A - F_{eq} - W_1,$$

$$\text{bending moment, } M = R_A x - W_1(x - a - b - c) - F_{eq}(x - e)$$

In general,

$$S = \sum F\uparrow - \sum F\downarrow \quad \text{left part of beam,}$$

$$M = \sum M_{\text{clockwise}} - M_{\text{anti-clockwise}} \quad \text{left part of beam}$$

Thus, by using a free body diagram and two equations of equilibrium, we can calculate easily the shear force and the bending moment.

SHEAR FORCE AND BENDING MOMENT DIAGRAMS

The graphical representation of shear force in which y-axis represents the shear force and x-axis represents the position of the section is called *shear force diagram* (SFD).

The graph in which y-axis represents the bending moment and x-axis represents the position of the section is called *bending moment diagram* (BMD).

The points (other than the ends) in a beam at which the bending moment is zero are called points of contraflexure or points of inflexion.

RELATIONSHIP BETWEEN SHEAR FORCE AND BENDING MOMENT

Let us consider a beam AB, as shown in Figure 3.4, subjected to a load whose intensity per unit length ω varies as a continuous function of x.

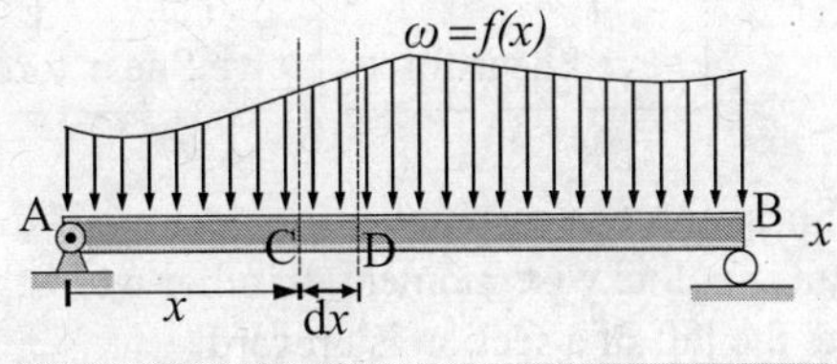

FIGURE 3.4 |

Let CD be a slice of the beam of infinitesimal length dx, located at a distance x from one end. As dx is very small, ω over this length may be taken as constant (Fig. 3.5).

The equilibrium equation for forces acting on the slice of length dx in the y-direction is

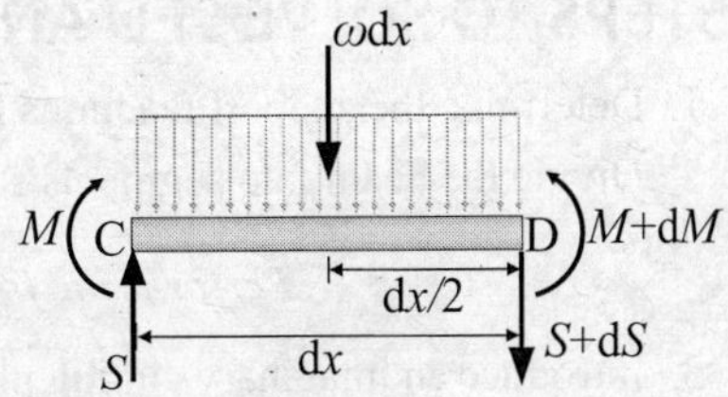

FIGURE 3.5 |

$$\sum F_y = 0;\; S - \omega \mathrm{d}x - (S + \mathrm{d}S) = 0$$

or

$$\omega \mathrm{d}x = \mathrm{d}S$$

Dividing both sides of equation by dx, we obtain

$$\omega = -\frac{\mathrm{d}S}{\mathrm{d}x}$$

That is, the rate of shear force is equal to the intensity of loading.

Now, summing moment about D, we find

$$\sum M = 0;\; M + S\mathrm{d}x - \omega\mathrm{d}x\left(\frac{\mathrm{d}x}{2}\right) - (M + \mathrm{d}M) = 0$$

or

$$S\mathrm{d}x - \omega\frac{(\mathrm{d}x)^2}{2} - \mathrm{d}M = 0$$

Neglecting small quantity $[(dx)^2/2]$

$$S\mathrm{d}x = \mathrm{d}M$$

$$S = \frac{\mathrm{d}M}{\mathrm{d}x}$$

That is, the rate of change of the bending moment with respect to x is equal to the shear force. Whenever the bending moment is maximum or minimum, the shear force is zero.

Taking he derivative again, we get

$$\omega = -\frac{\mathrm{d}S}{\mathrm{d}x} = -\frac{\mathrm{d}^2 M}{\mathrm{d}x^2}$$

SIGN CONVENTION OF SHEAR FORCE AND BENDING MOMENT

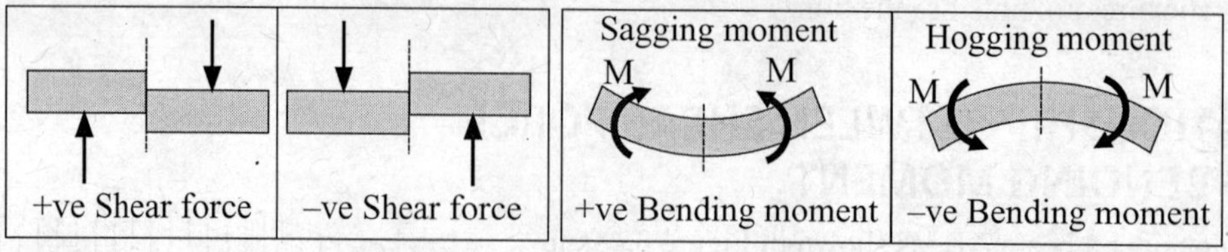

Shear force is assumed to be positive if it produces a clockwise moment and negative if it produces an anticlockwise moment. In other words, the shear force is positive if the net resultant external force to the left of a section is upward.

Positive bending moment produces compression on the top fibre of the beam and negative bending moment produces tension on the top fibre of the beam.

STEPS TO SOLVE SFD AND BMD PROBLEMS

1. Determine the support reactions from the free body diagram of entire beam.
2. Divide the beam into segments so that the loading within each segment is continuous.

Perform the following steps for each segment of the beam:

3. Introduce an imaginary cutting plane within the segment, located at a distance x from the left end of the beam, that cuts the beam into two parts.
4. Draw free body diagram for the segment of the beam. Show S and M at the cut section.
5. Write the equilibrium equations, obtainable from the free body diagram.
6. Solve the equilibrium equations for the shear force S and the bending moment M.
7. Plot the expressions for S and M for the segment. It is desirable to draw the shear force diagram below the entire beam and then draw the bending moment diagram below the shear force diagram.

Note: Perform the last five steps (steps 3–7) for each segment of the beam.

SOLVED PROBLEMS

PROBLEM 3.1

Draw shear force and bending moment diagrams for a cantilever beam AB carrying a concentrated load W [Fig. 3.1(a)].

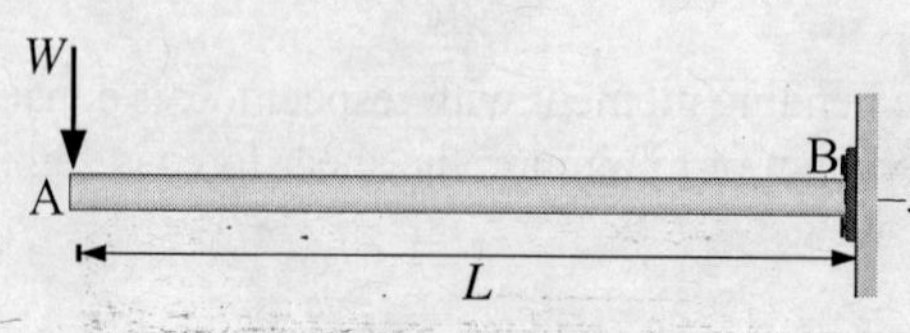

FIGURE 3.1 (a) |

Solution. We cut through the beam at a distance x from the free end A. Then we draw a free body diagram of the left-hand part of the beam [Fig. 3.1(b)]. From the equations of equilibrium for this free body, we obtain the shear force S and bending moment M at a distance x from the free end:

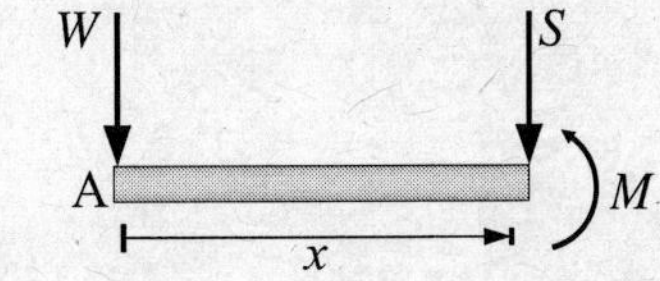

FIGURE 3.1 (b)

$$\sum F_y = 0; -W - S = 0,$$

$$S = -W$$

$$\sum M = 0; Wx + M = 0,$$

$$M = -Wx$$

$$M_{x=L} = -WL$$

The shear force and bending moment diagrams in Figure 3.1(c) are the plots of the expressions for S and M derived in above equations.

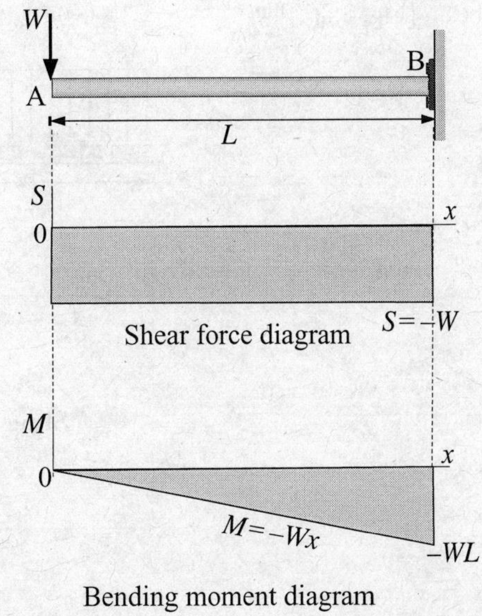

FIGURE 3.1 (c)

PROBLEM 3.2

A cantilever beam supporting a uniform load of constant intensity ω is shown in Figure 3.2(a).

Draw the shear force and bending moment diagrams for this beam.

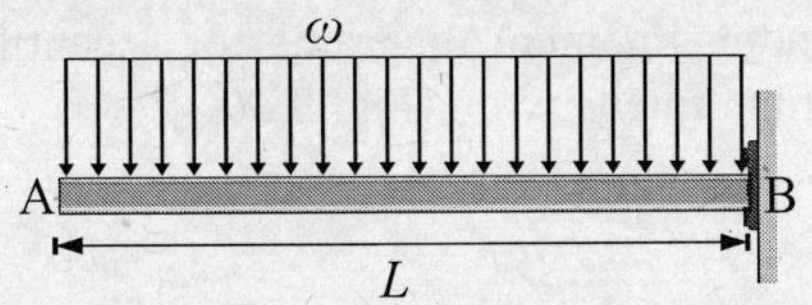

FIGURE 3.2 (a)

Solution. We cut through the beam at a distance x from free end A and draw free body diagram of left-hand part of the beam [Fig. 3.2(b)]. For purpose of calculating the shear force and the bending moment, the uniform distributed load is replaced by its resultant force ωx which acts at centroid. Applying equilibrium equations, we have

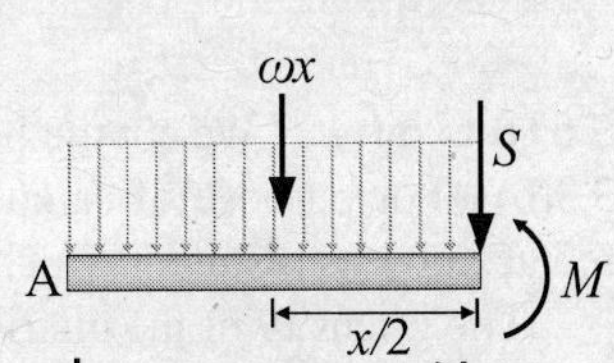

FIGURE 3.2 (b)

$$\sum F_y = 0 \Rightarrow -\omega x - S = 0,$$

$$S = -\omega x,$$

$$S_{max} = -\omega L,$$

$$\sum M = 0 \Rightarrow \omega x \times \frac{x}{2} + M = 0,$$

$$M = -\frac{\omega x^2}{2}, \quad M_{max} = -\frac{\omega L^2}{2}$$

The equations for the shear force and the bending moment are plotted below the sketch of the beam [Fig. 3.2(c)].

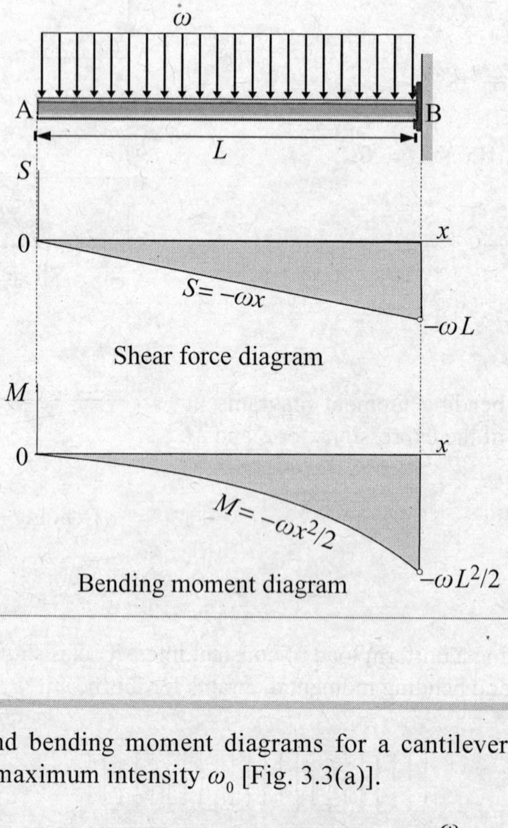

FIGURE 3.2 (c) |

PROBLEM 3.3

Draw the shear force and bending moment diagrams for a cantilever beam AB supporting a linearly varying load of maximum intensity ω_0 [Fig. 3.3(a)].

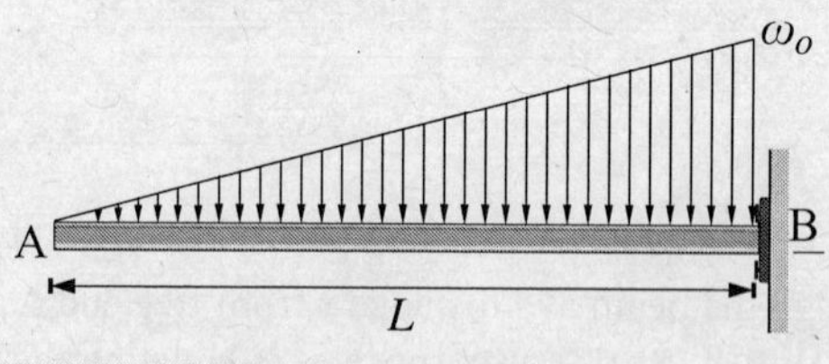

FIGURE 3.3 (a) |

Solution. We cut the beam at a distance x from free end and draw the free body diagram [Fig. 3.3(b)]. For purpose of calculating the shear force and the bending moment, the triangular load is replaced by its resultant which acts at centroid.

The intensity of the distributed load at a distance x from the end is

$$\omega = \frac{\omega_0 x}{L}$$

Therefore, the total downward load on the free body, equal to the area of the triangular loading diagram, is

$$\frac{1}{2}\omega x = \frac{1}{2}\frac{\omega_0 x}{L}x = \frac{\omega_0 x^2}{2L}$$

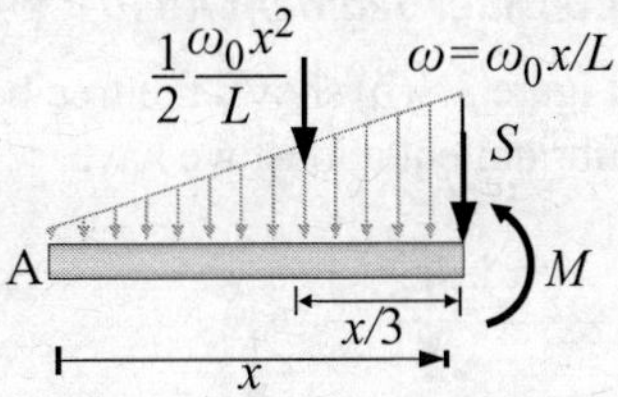

FIGURE 3.3 (b) |

Applying equations of equilibrium, we have

$$\sum F_y = 0 \Rightarrow -\frac{\omega_0 x^2}{2L} - S = 0,$$

$$S = -\frac{\omega_0 x^2}{2L}$$

$$S_{max} = -\frac{\omega_0 L}{2},$$

$$\sum M = 0 \Rightarrow \frac{\omega_0 x^2}{2L} \times \frac{x}{3} + M = 0,$$

$$M = -\frac{\omega_0 x^3}{6L},$$

$$M_{max} = -\frac{\omega_0 L^2}{6}$$

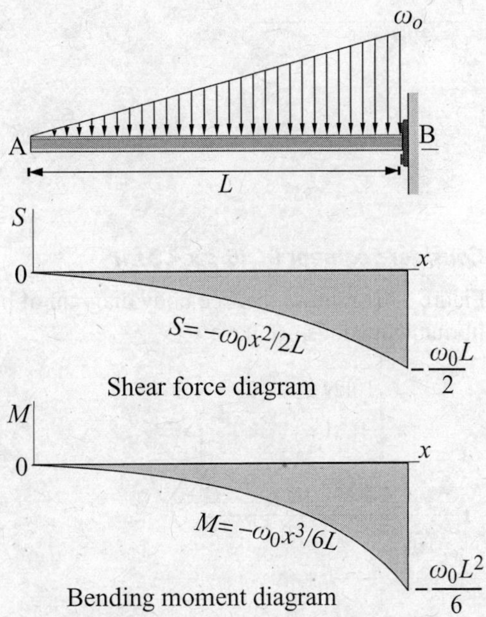

FIGURE 3.3 (c) |

The shear force and bending moment diagrams in Figure 3.3(c) are the plots of the expressions for S and M derived in above equations.

PROBLEM 3.4

Draw the shear force and bending moment diagrams for a cantilever beam carrying uniform load of intensity 5 kN/m over 2 m and concentrated load 10 kN as shown in Figure 3.4(a).

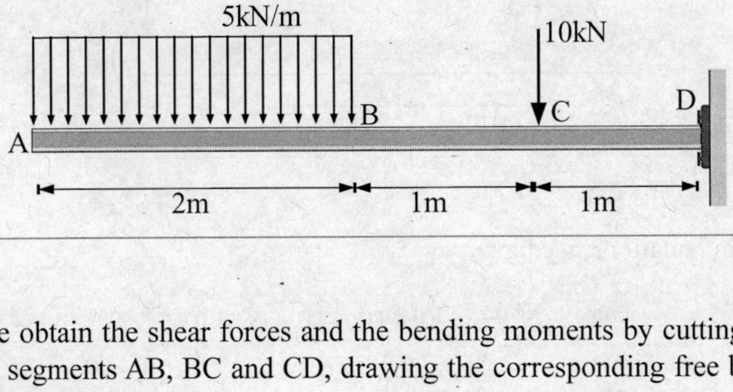

FIGURE 3.4 (a) |

Solution. We obtain the shear forces and the bending moments by cutting through the beam in each of the three segments AB, BC and CD, drawing the corresponding free body diagrams, and solving the equations of equilibrium.

Consider Segment AB (0 < x < 2 m)

Figure 3.4(b) shows the free body diagram of beam segment of length $x(0 < x < 2$ m). Applying equilibrium equations, we have

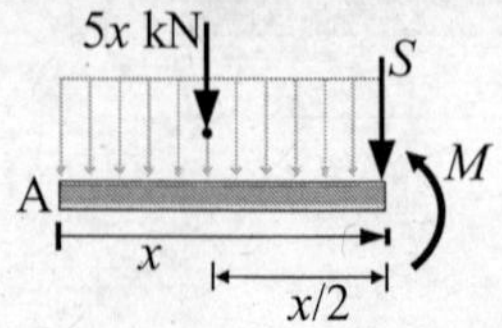

FIGURE 3.4 (b) |

$$\sum F_y = 0; -5x - S = 0,$$

$$S = -5x,$$

$$S_{x=2} = -10\text{N},$$

$$\sum M = 0; 5x \times \frac{x}{2} + M = 0,$$

$$M = -\frac{5}{2}x^2$$

$$M_{x=2} = -10\text{Nm}$$

Consider Segment BC (0 < x < 3 m)

Figure 3.4(c) shows the free body diagram of beam segment of length x $(2 < x < 3m)$. Applying equilibrium equations, we have

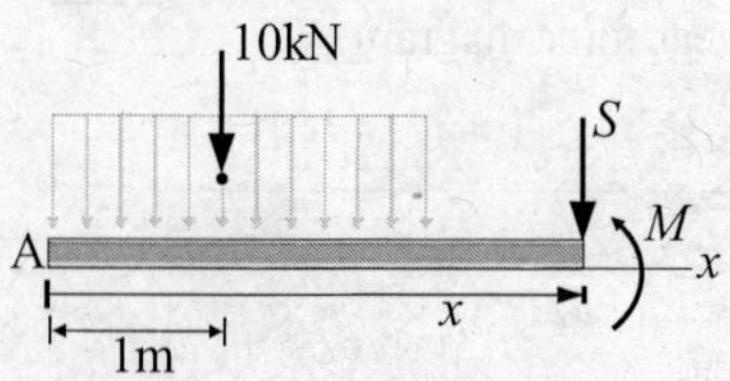

FIGURE 3.4 (c) |

$$\sum F_y = 0; -10 - S = 0,$$

$$S = -10,$$

$$\sum M = 0; 10(x-1) + M = 0,$$

$$M = 10(1 - x),$$

$$M_{x=3} = -20\text{Nm}$$

Consider Segment CD (0 < x < 4 m)

Figure 3.4(d) shows the free body diagram of beam segment of length x $(3 < x < 4$ m).

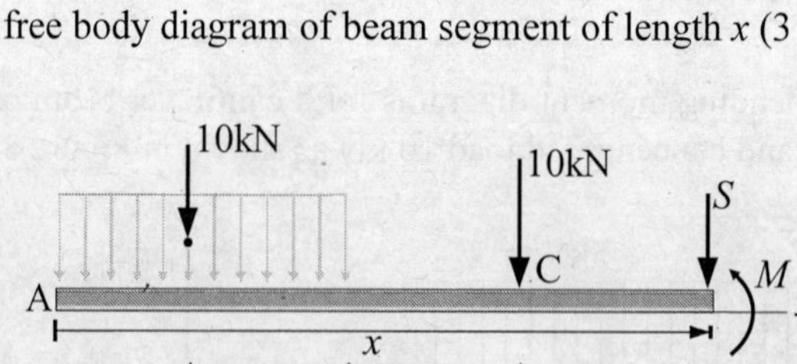

FIGURE 3.4 (d) |

Applying equilibrium equations, we have

$$\sum F_y = 0; -10 - 10 - S = 0$$

$$S = -20$$

$$\sum M = 0; 10(x - 1) + 10(x - 3) + M = 0$$

$$M = 40 - 20x$$

$$M_{x=4\text{m}} = -40\text{kNm}$$

The shear force and bending moment diagrams [Fig. 3.4(e)] are obtained by plotting the above-derived equations.

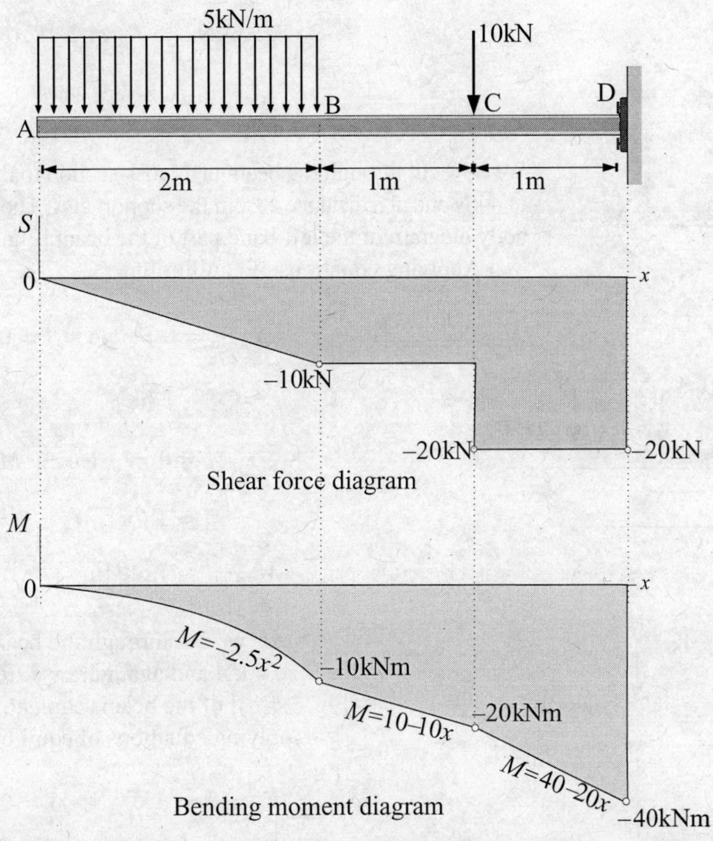

FIGURE 3.4 (e) |

PROBLEM 3.5

Draw the shear force and bending moment diagrams for the simply supported beam AB [Figure 3.5(a)].

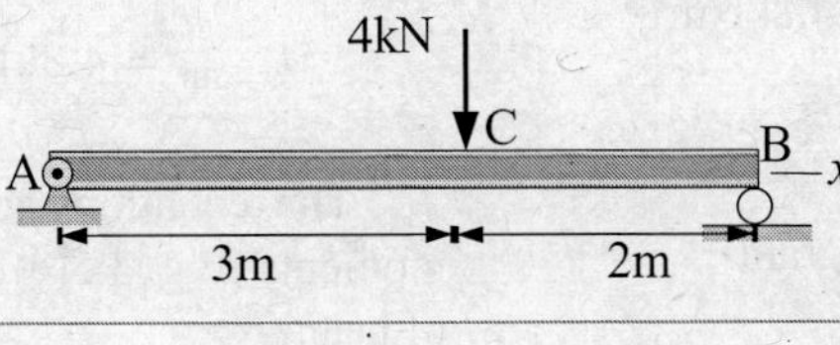

FIGURE 3.5 (a) |

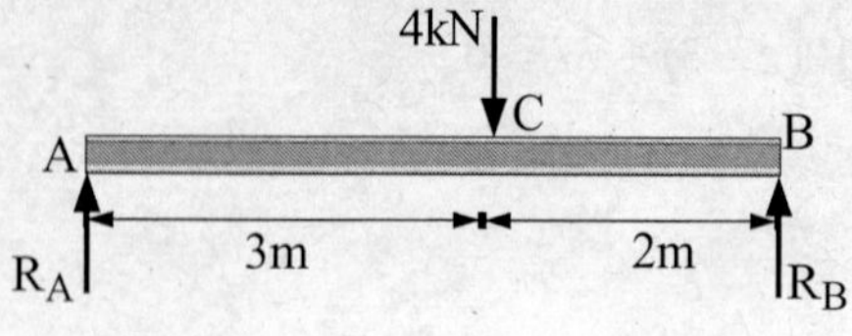

FIGURE 3.5 (b) |

Solution. From the free body diagram [Fig. 3.5(b)] of the entire beam, we find the support reactions by applying the static equilibrium equations:

$$\sum M_A = 0 \Rightarrow 4 \times 3 - R_B \times 5 = 0$$

$$\sum F_y = 0 \Rightarrow R_A + R_B - 4 = 0$$

From which

$$R_A = 1.6\text{kN}, \qquad R_B = 2.4\text{kN}$$

Segment AC (0 < x < 3 m)

We now cut through the beam at a cross-section to the left of the load of 4 kN and at a distance *x* from the support at A. Then, we draw a free body diagram of the left-hand part of the beam [Fig. 3.5(c)].

Applying equations of equilibrium,

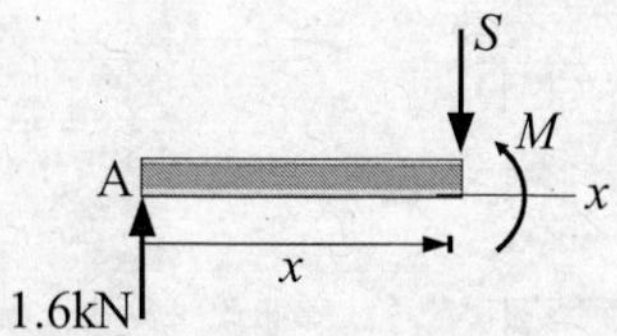

FIGURE 3.5 (c) |

$$\sum F_y = 0 \Rightarrow 1.6 - S = 0$$

$$S = 1.6\text{kN},$$

$$\sum M = 0 \Rightarrow -1.6x + M = 0$$

$$M = 1.6x$$

$$M_{x=3\text{m}} = 4.8\text{kNm}$$

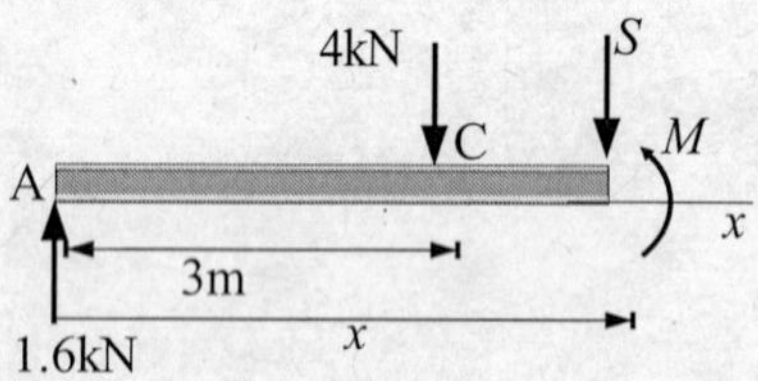

FIGURE 3.5 (d) |

Next, we cut through the beam to the right of the load 4 kN and again draw a free body diagram [Fig. 3.5(d)] of the beam segment.

Applying equations of equilibrium,

$$\sum F_y = 0 \Rightarrow 1.6 - 4 - S = 0$$

$$S = -2.4$$

$$\sum M = 0 \Rightarrow -1.6 \times x + 4 \times (x - 3) + M = 0$$

$$M = 12 - 2'4x$$

$$M_{x=3\text{m}} = 4.8\text{kNm}$$

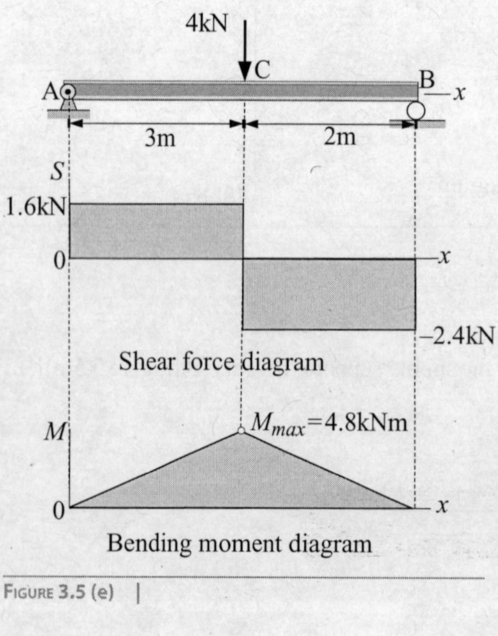

FIGURE 3.5 (e) |

The equations for shear forces and bending moments [Fig. 3.5(e)] are plotted below the sketch of beam.

PROBLEM 3.6

Draw the shear force and bending moment diagrams for the simply supported beam and loading shown in Figure 3.6(a).

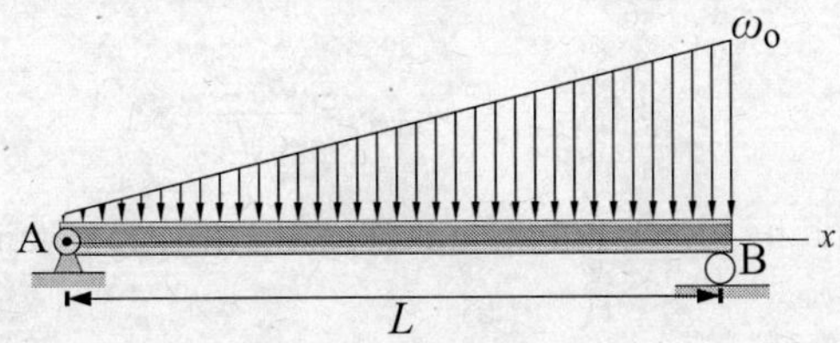

FIGURE 3.6 (a) |

Solution. We begin the analysis by determining the reactions of the beam from a free body diagram of the entire beam [Fig. 3.6(b)].

$$\sum M_A = 0 \Rightarrow \frac{1}{2}\omega_0 L \times \frac{2L}{3} - R_B \times L = 0$$

$$\sum F_y = 0 \Rightarrow R_A + R_B = \frac{1}{2}\omega_0 L$$

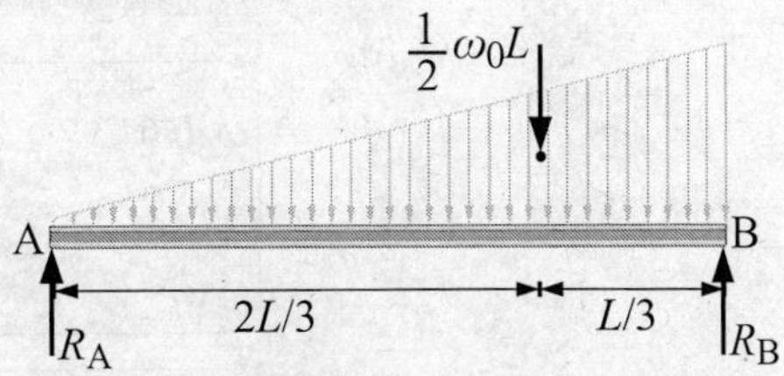

FIGURE 3.6 (b) |

From which,

$$R_B = \frac{\omega_0 L}{3} \text{ and } R_A = \frac{\omega_0 L}{6}$$

We now cut through the beam at a cross-section at a distance x from the support A. Then we draw a free body diagram of the left-hand part of the beam [Fig. 3.6(c)]. From the equations of equilibrium for this free body, we obtain the shear force S and the bending moment M at a distance x from the support:

$$\sum F_y = 0 \Rightarrow \frac{\omega_0 L}{6} - \frac{\omega_0 x^2}{2L} - S = 0$$

$$S = \frac{\omega_0 L}{6} - \frac{\omega_0 x^2}{2L},$$

$$\sum M = 0 \Rightarrow -\frac{\omega_0 L}{6}x + \frac{\omega_0 x^2}{2L} \times \frac{x}{3} + M = 0$$

or

$$M = \frac{\omega_0 Lx}{6} - \frac{\omega_0 x^{3\prime}}{6L}$$

$$M = \frac{\omega_0 L^2}{6}\left[\left(\frac{x}{L}\right) - \left(\frac{x}{L}\right)^3\right]$$

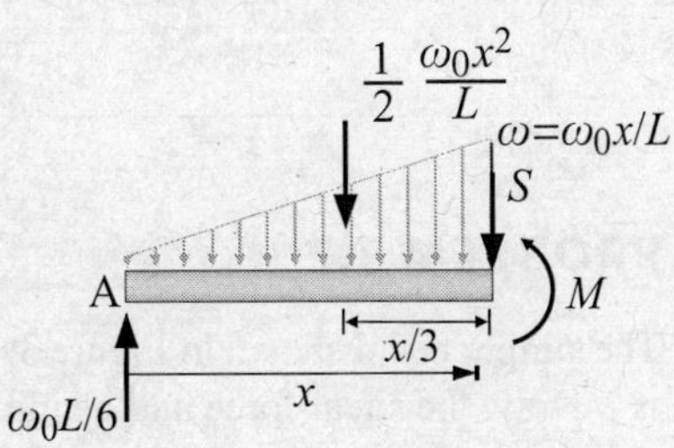

FIGURE 3.6 (c) |

The maximum bending moment occurs where the slope of the bending moment curve is zero: that is, where

$$\frac{d}{dx}M = 0$$

Differentiating the expression for M, we obtain

$$\frac{d}{dx} M = 0 \Rightarrow \frac{1}{L} - 3\frac{x^2}{L^3},$$

which yields

$$x = \frac{L}{\sqrt{3}} \qquad M_{max} = \frac{\omega_0 L^2}{9\sqrt{3}}$$

The shear force and bending moment diagrams [Fig. 3.6(d)] are obtained by plotting the above-derived equations.

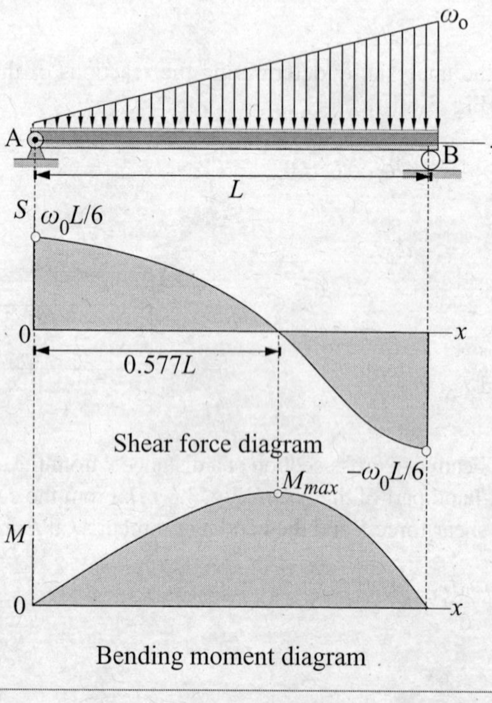

FIGURE 3.6 (d) |

PROBLEM 3.7

The simple beam shown in Figure 3.7(a) supports a concentrated load and a segment of uniform load. Draw the shear force and bending moment diagrams for this beam.

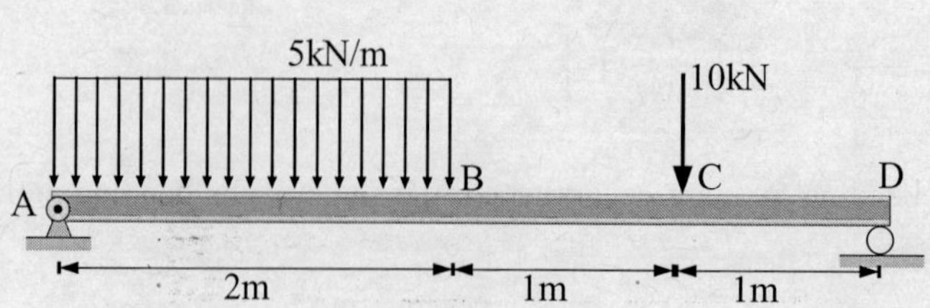

FIGURE 3.7 (a) |

Solution. From the free body diagram of the entire beam, we find the support reactions by applying the static equilibrium equations:

$$\sum F_y = 0 \Rightarrow R_A - 10 - 10 + R_D = 0$$

$$R_A + R_D = 20 \qquad \text{(i)}$$

$$\sum M_A = 0 \Rightarrow 10 \times 1 + 10 \times 3 - R_D \times 4 = 0$$

$$R_D = 10\text{kN} \qquad \text{(ii)}$$

From Equations (i) and (ii), we have

$$R_A = R_D = 10\text{kN}$$

We obtain shear forces and bending moments by cutting through the beam in each of the three segments, drawing the corresponding free body diagrams and solving the equations of equilibrium.

From A to B (0 < x < 2)

Considering the portion of the beam to the left of section B [Fig. 3.7(b)] and replacing the distributed load by its resultant, we obtain

$$\sum F_y = 0;\ 10 - 5x - S = 0$$

$$S = 10 - 5x$$

$$\sum M = 0;\ -10x + 5x \times \frac{x}{2} + M = 0$$

$$M = 10x - \frac{5}{2}x^2$$

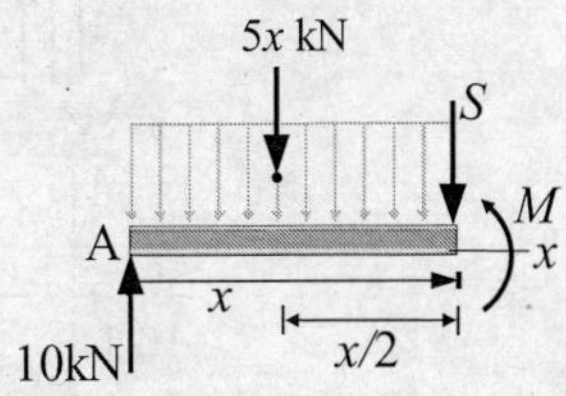

FIGURE 3.7 (b)

From B to C (2 < x < 3)

We consider as free body the portion of the beam to the left of section C [Fig. 3.7(c)] and write equilibrium equations:

$$\sum F_y = 0; 10 - 10 - S = 0$$

$$\sum M = 0; -10x + 10 \times (x - 1) + M = 0$$

From which,

$$S = 0 \text{ and } M = 10\text{kNm}$$

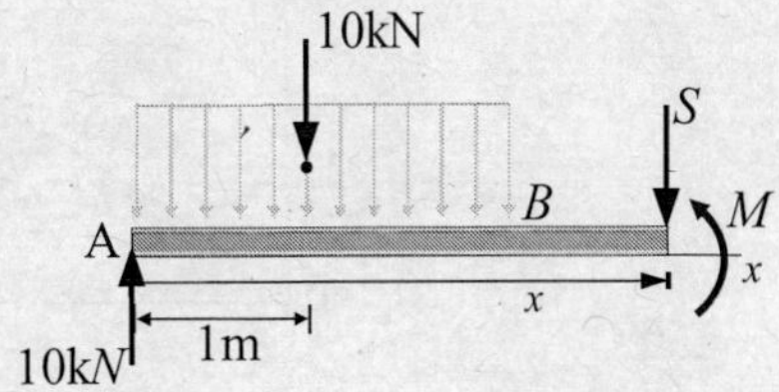

FIGURE 3.7 (c)

From C to D (3 < x < 4)

We consider as free body the portion of the beam to the left of section C [Fig. 3.7(d)] and write equilibrium equations:

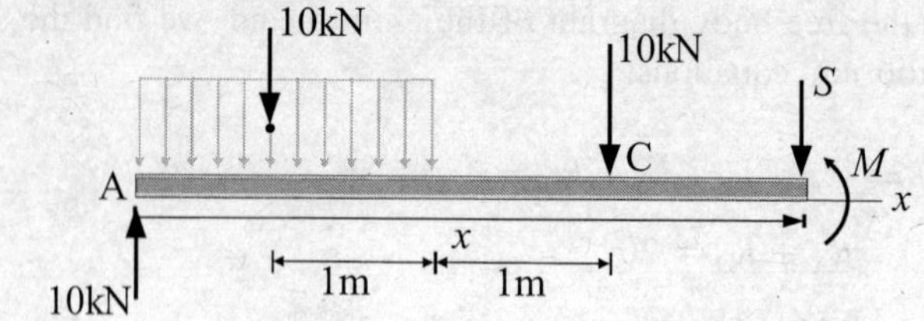

FIGURE 3.7 (d) |

$$\sum F_y = 0; 10 - 10 - 10 - S = 0$$

$$S = -10\text{kNm}$$

$$\sum M = 0; -10x + 10(x-1) + 10(x-3) + M = 0$$

$$M = 40 - 10x$$

The shear forces and bending moments for the entire beam may now be plotted [Fig. 3.7(e)].

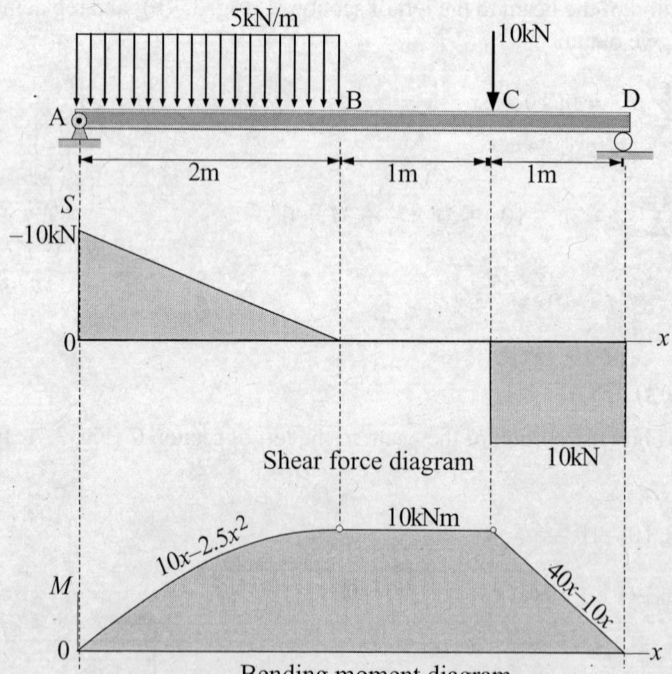

FIGURE 3.7 (e) |

PROBLEM 3.8

Draw the shear-force and bending moment diagrams for a simply supported beam AB supporting a linearly varying load of maximum intensity 3 kN/m [Fig. 3.8(a)].

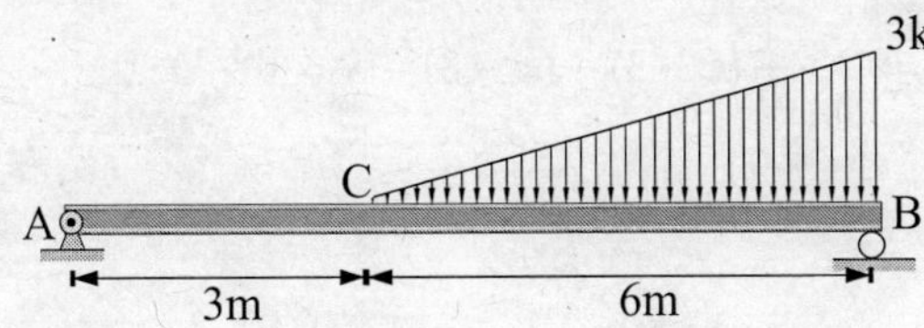

FIGURE 3.8 (a) |

Solution. From the free body diagram [Fig. 3.8(b)] of the entire beam, we find the reaction at supports by applying equation of equilibrium; thus

$$\sum M_A = 0 \Rightarrow 9 \times 7 - R_B \times 9 = 0$$

$$\sum F_y = 0 \Rightarrow R_A + R_B - 9 = 0$$

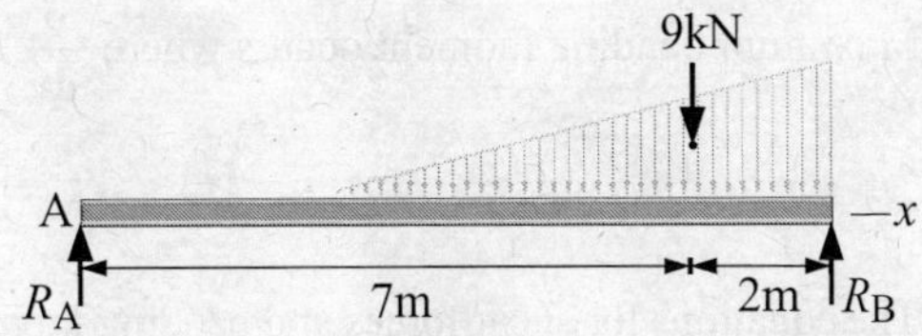

FIGURE 3.8 (b) |

From which

$$R_A = 2\text{kN},\ R_B = 7\text{kN}$$

Consider Segment AC (0 < x < 3 m)

From the free body diagram [Fig. 3.8(c)] of the section for which ($0 < x < 3$ m), equilibrium in the vertical direction and a moment sum about the cut section give

$$\sum F_y = 0 \Rightarrow 2 - S = 0$$

$$S = 2\text{kN},$$

$$\sum M = 0 \Rightarrow -2x + M = 0$$

$$M = 2x$$

$$M_{x=3\text{m}} = 6\text{kNm}$$

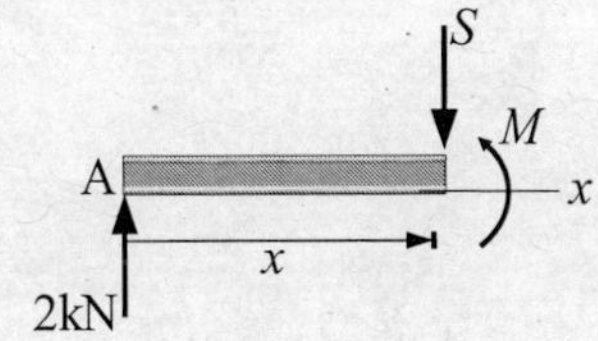

FIGURE 3.8 (c) |

Consider Segment CB (3 < x < 9)

From the free body diagram [Fig. 3.8(d)] of the section for which ($3 < x < 9$), equilibrium in the vertical direction and a moment sum about the cut section give

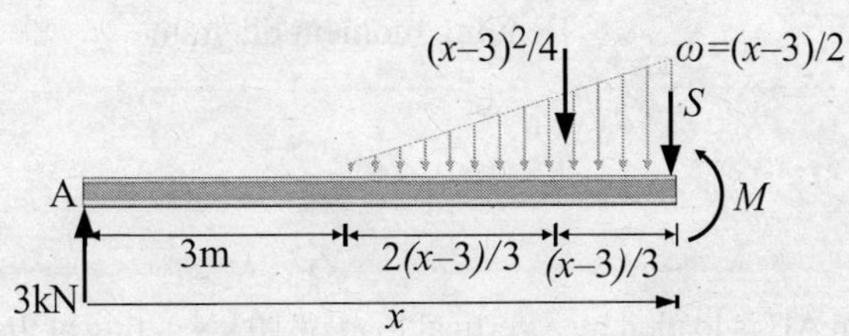

FIGURE 3.8 (d) |

$$\sum F_y = 0 \Rightarrow 2 - \frac{1}{2}(x-3)\frac{1}{2}(x-3) - S = 0$$

$$S = 2 - \frac{(x-3)^2}{4}$$

$$\sum M = 0 \Rightarrow -2x + \frac{1}{2}(x-3)\frac{1}{2}(x-3) \times \left(\frac{x-3}{3}\right) + M = 0$$

$$M = 2x - \frac{(x-3)^3}{12}$$

Maximum bending moment occurs where $\frac{d}{dx}M = 0 \Rightarrow 2 - \frac{(x-3)^2}{4} = 0 \Rightarrow x = 5.82\text{m},$

$$M_{x=5.82\text{m}} = M_{\max} = 9.77\text{kNm}$$

The equations for shear forces and bending moments are plotted in the sketch of the beam [Fig. 3.8(e)].

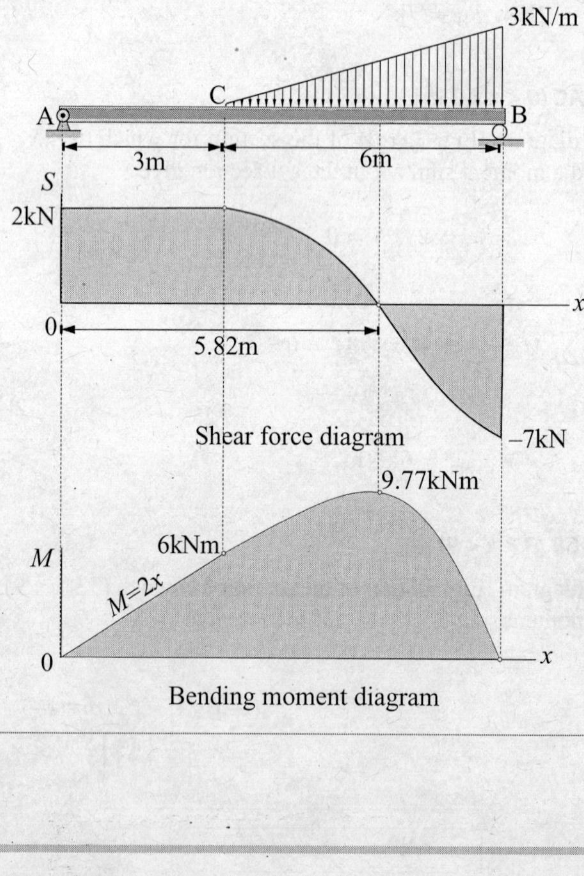

FIGURE 3.8 (e) |

PROBLEM 3.9

A simply supported beam AB is loaded by a vertical load of 10 kN acting at the end of a bracket shown in Figure 3.9(a).

Draw the shear force and bending moment diagrams for the beam.

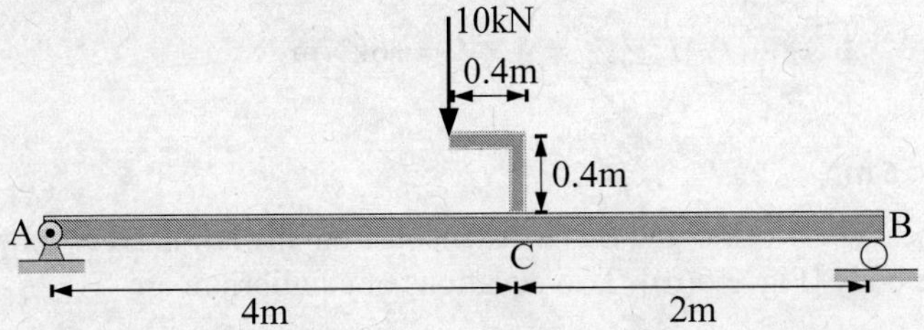

FIGURE 3.9 (a) |

Solution. By transferring a load 10 kN at C by a force 10 kN and a moment 4 kN m, we draw free body diagram of the beam [Fig. 3.9(b)]. and applying equations of equilibrium, we have

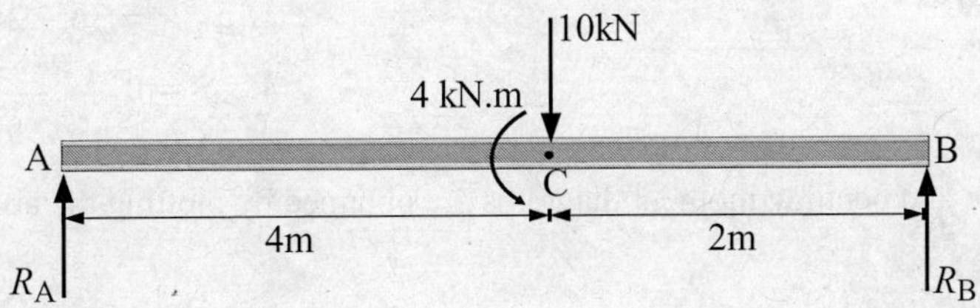

FIGURE 3.9 (b) |

$$\sum M_A = 0 \Rightarrow 10 \times 4 - 4 - R_B \times 6 = 0$$

$$\sum F_y = 0 \Rightarrow R_A + R_B - 10 = 0$$

From which,

$$R_A = 4\text{kN} \text{ and } R_B = 6\text{kN}$$

Segment AC (0 < x < 4 m)

We cut through the beam between A and C at a distance x from the left-hand support A and isolate part of the beam as a free body [Fig. 3.9(c)].

Two equations of equilibrium are

$$\sum F_y = 0 \Rightarrow 4 - S = 0$$

$$\sum M = 0 \Rightarrow -4x + M = 0$$

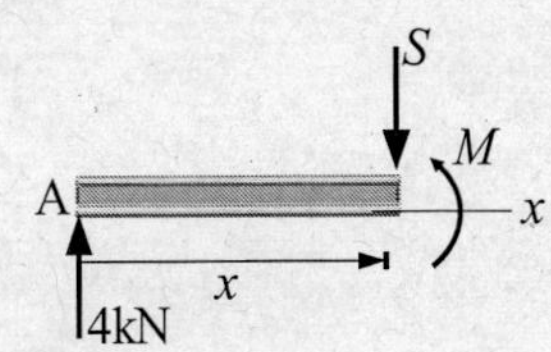

FIGURE 3.9 (c) |

From which

$$S = 4\text{kN and } M = 4x$$

$$M_{x=4-0} = 4 \times 4 = 16\text{kNm}$$

Segment CB (4 < x < 6 m)

We cut through the beam between C and B at a distance x from the left-hand support A and isolate part of the beam as a free body [Fig. 3.9(d)]. Two equations of equilibrium are

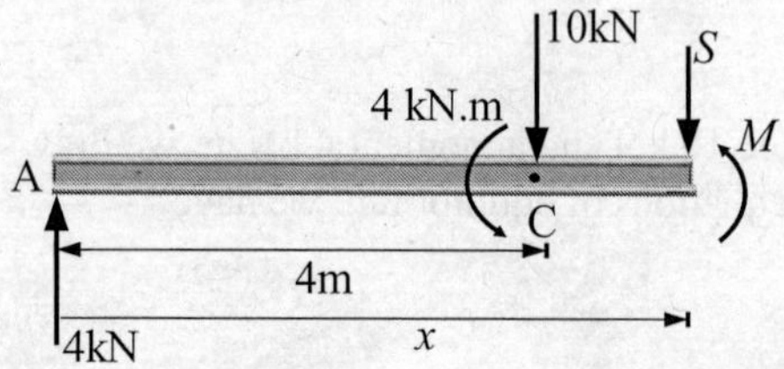

FIGURE 3.9 (d) |

$$\sum F_y = 0 \Rightarrow 4 - 10 - S = 0$$

$$S = -6\text{kN}$$

$$\sum M = 0 \Rightarrow -4x + 10(x-4) + 4 + M = 0$$

$$M = 36 - 6x$$

$$M_{x=4+0} = 36 - 24 = 12\text{kNm}$$

$$M_{x=6} = 0$$

The shear force and bending moment diagrams are obtained by plotting the above-derived equations [Fig. 3.9(e)].

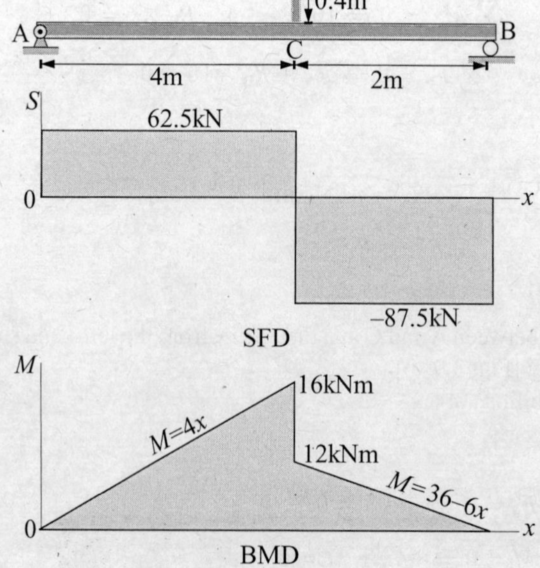

FIGURE 3.9 (e) |

PROBLEM 3.10

A beam OAB with an overhang at the left-hand is shown in Figure 3.10(a). Draw the shear force and bending moment diagrams for this beam.

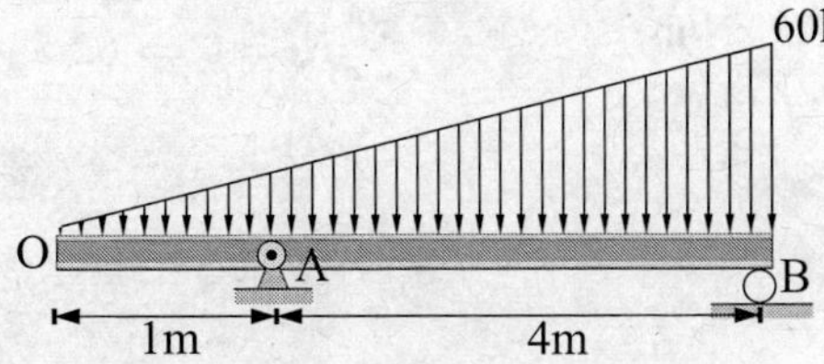

FIGURE 3.10 (a)

Solution. From the free body diagram [Fig. 3.10(b)] of the entire beam, we find the reactions, R_A and R_B:

$$\sum M_B = 0 \Rightarrow R_A \times 4 - 150 \times \frac{5}{3} = 0$$

$$\sum F_y = 0 \Rightarrow R_A + R_B - 150 = 0$$

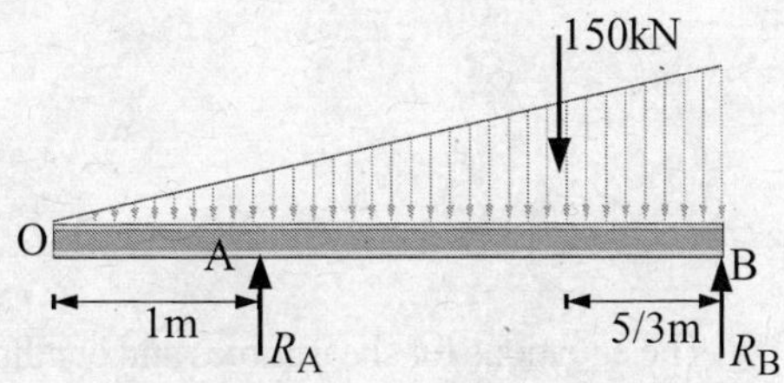

FIGURE 3.10 (b)

From which

$$R_A = 62.5\text{kN and } R_B = 87.5\text{kN}$$

Segment OA (0 < x < 1 m)

We now cut through the overhang beam at a cross-section between O and A, and at a distance x from O. Then we draw a free body diagram of the left-hand part of the beam [Fig. 3.10(c)]. Applying the equations of equilibrium to the segment, we have

$$\sum F_y = 0 \Rightarrow -6x^2 - S = 0,$$

$$S = -6x^2,$$

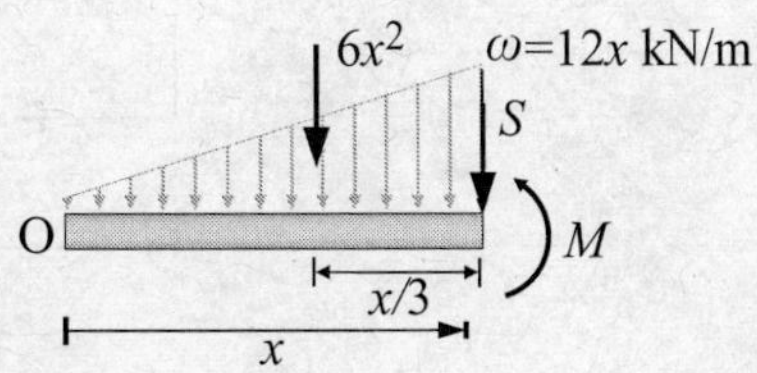

FIGURE 3.10 (c)

$$S_{x=1\text{m}} = -6 \times 1^2 = -6\text{kNm},$$

$$\sum M = 0 \Rightarrow M + 6x^2 \times \frac{x}{3} = 0,$$

$$M = -2x^3,$$

$$M_{x=1} = -2\text{kNm}$$

Segment AB (1< x < 5 m)

We now cut through the overhang beam at a cross-section between A and B, and at a distance x from O. Then we draw a free body diagram of the left-hand part of the beam [Fig. 3.10(d)]. Applying the equations of equilibrium to the segment, we have

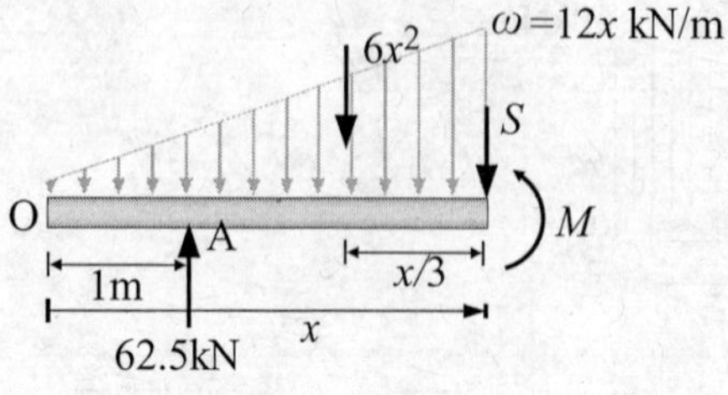

FIGURE 3.10 (d) |

$$\sum F_y = 0 \Rightarrow 62.5 - 6x^2 - S = 0,$$

$$S = 62.5 - 6x^2$$

$$S_{x=1+0\text{m}} = 56.5\text{kN},$$

$$S_{x=5-0\text{m}} = -87.5\text{kN},$$

$$\sum M = 0 \Rightarrow M + 6x^2 \times \frac{x}{3} - 62.5(x-1) = 0,$$

$$M = 62.5(x-1) - 2x^3$$

$$M_{x=1+0} = -2\text{kNm},$$

$$M_{x=5-0} = 0$$

The equations for shear forces and bending moments are plotted in the sketch of beam [Fig. 3.10(e)].

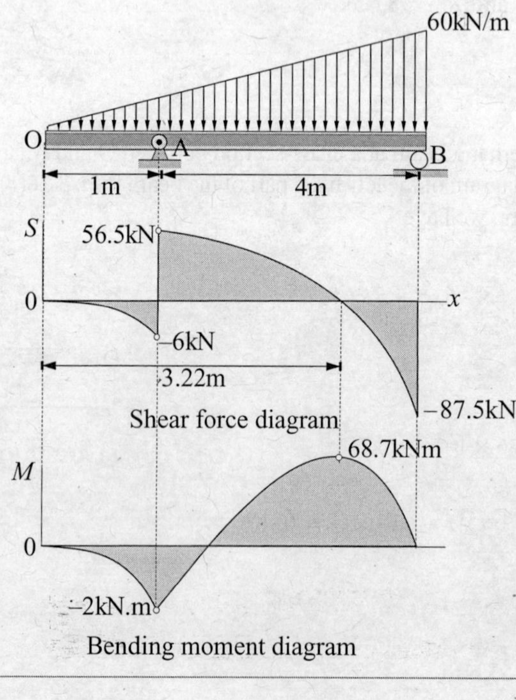

FIGURE 3.10 (e) |

PROBLEM 3.11

Beam CABD is simply supported at A and B and has overhangs at each end [Fig. 3.11(a)]. Draw the shear force and bending moment diagrams for this beam.

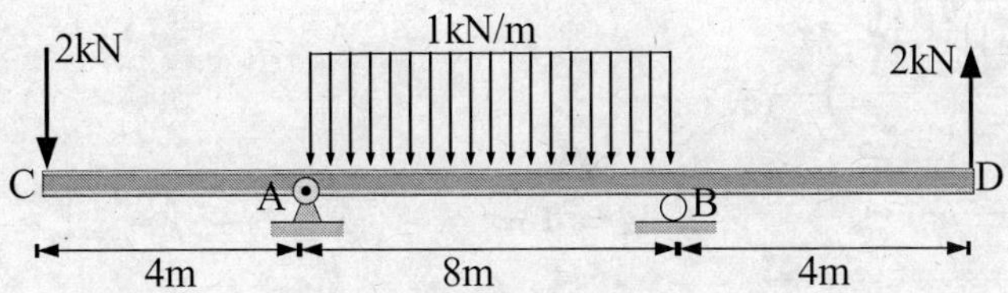

FIGURE 3.11 (a)

Solution. We begin by calculating the reactions R_A and R_B from equations of equilibrium for the beam considered as a free body [Fig. 3.11(b)].

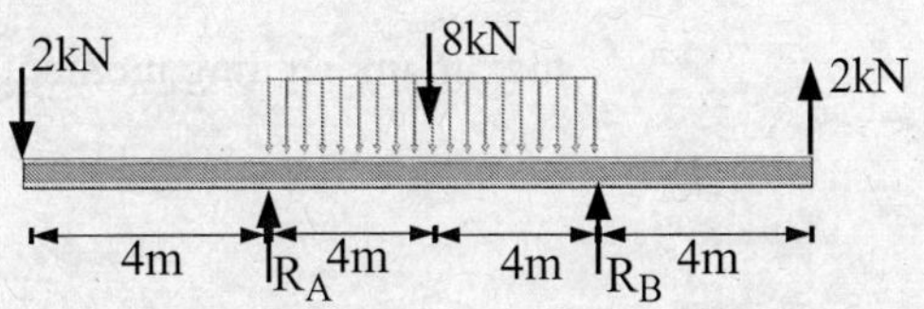

FIGURE 3.11 (b)

$$\sum M_A = 0 \Rightarrow 8 \times 4 - R_B \times 8 - 2 \times 12 - 2 \times 4 = 0,$$

$$\sum F_y = 0 \Rightarrow R_A + R_B - 2 - 8 + 2 = 0$$

From which

$$R_B = 0 \text{ and } R_A = 8\text{kN}$$

Segment CA (0 < x < 4 m)

We now cut through the beam at a cross-section between C and A, and at a distance x from C. Then we draw a free body diagram of the left-hand part of the beam [Fig. 3.11(c)]. Applying the equations of equilibrium to the segment, we have

$$\sum F_y = 0 \Rightarrow -2 - S = 0,$$

$$\sum M = 0 \Rightarrow M + 2x = 0$$

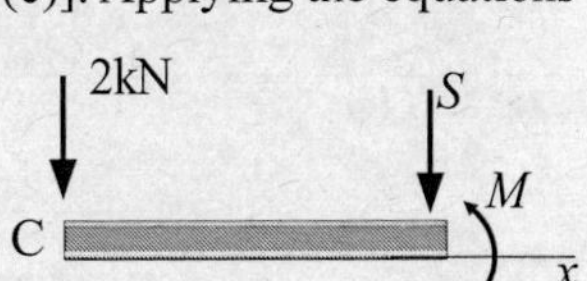

FIGURE 3.11 (c)

From which

$$S = -2\text{kN and } M = -2x$$

$$M_{x=4} = -8\text{kNm}$$

Segment AB (4 < x < 12 m)

We now cut through the beam at a cross-section between A and B, and at a distance x from C. Then we draw a free body diagram of the left-hand part of the beam [Fig. 3.11(d)]. Applying the equations of equilibrium to the segment, we have

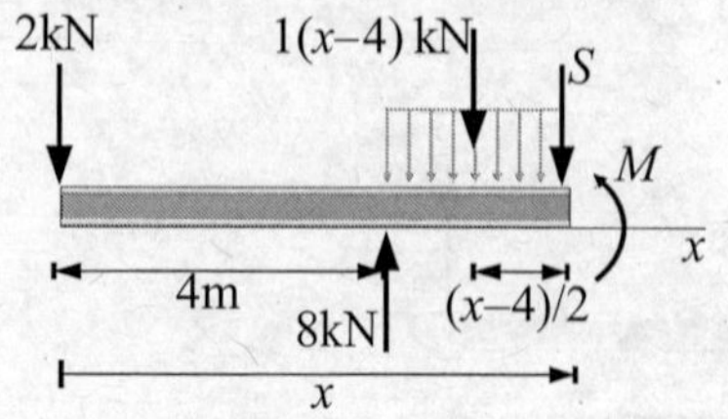

FIGURE 3.11 (d) |

$$\sum F_y = 0 \Rightarrow -2 + 8 - (x - 4) - S = 0,$$

$$S = -x + 10\text{kN}$$

$$\sum M = 0 \Rightarrow 2x - 8(x - 4) + (x - 4)\left(\frac{x - 4}{2}\right) + M = 0,$$

$$M = 6x - 32 - \frac{(x - 4)^2}{2}$$

$$\text{maximum bending moment, } M_{\max} = 6 \times 10 - 32 - \frac{(10 - 4)^2}{2} = 10\text{kNm}$$

Segment BD (12 < x < 16 m)

We now cut through the beam at a cross-section between B and D, and at a distance x from C. Then we draw a free body diagram of the left-hand part of the beam [Fig. 3.11(e)]. Applying the equations of equilibrium to the segment, we have

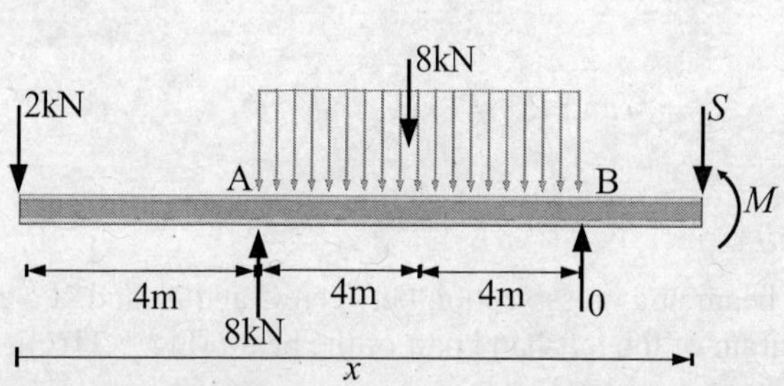

FIGURE 3.11 (e) |

$$\sum F_y = 0 \Rightarrow -2 + 8 - 8 - S = 0$$

$$\sum M = 0 \Rightarrow 2x - 8(x - 4) + 8(x - 8) + M = 0$$

$$S = 2\text{kN}$$

$$M = 32 - 2x$$

The equations for shear forces and bending moments are plotted in the sketch of the beam [Fig. 3.11(f)].

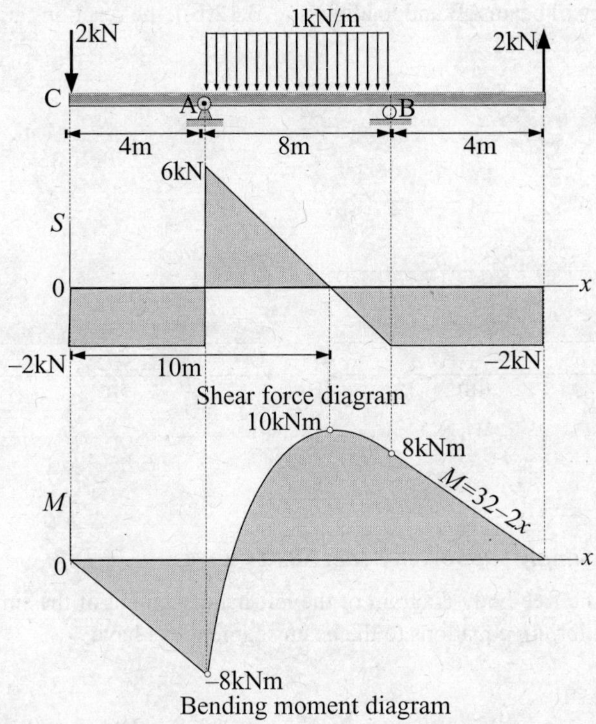

FIGURE 3.11 (f) |

PROBLEM 3.12

The compound beam ABC shown in Figure 3.12(a) consists of two beam (AB and BC) joined by a hinged connection at B. A uniform load of intensity 2 kN/m acts downward on beam AB. Draw the shear force and bending moment diagrams for this compound beam.

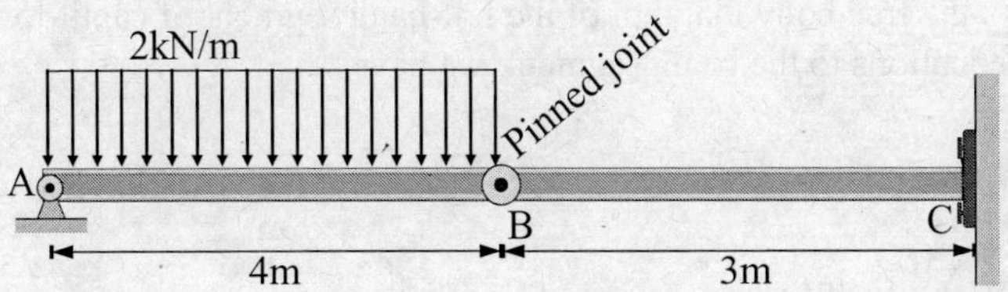

FIGURE 3.12 (a) |

Solution. A hinged or pinned joint is capable of transmitting only horizontal and vertical forces. No moment can be transmitted at the pinned joint. Therefore, the point where a hinge or pinned joint occurs is a particularly convenient location for separation of the structure into parts for the purpose of calculating the reactions.

From the symmetry of beam AB and loading [Fig. 3.12(b)], the reactions at support A and pinned joint B are

$$R_A = R_B = \frac{4 \times 2}{2} = 4\text{kN}$$

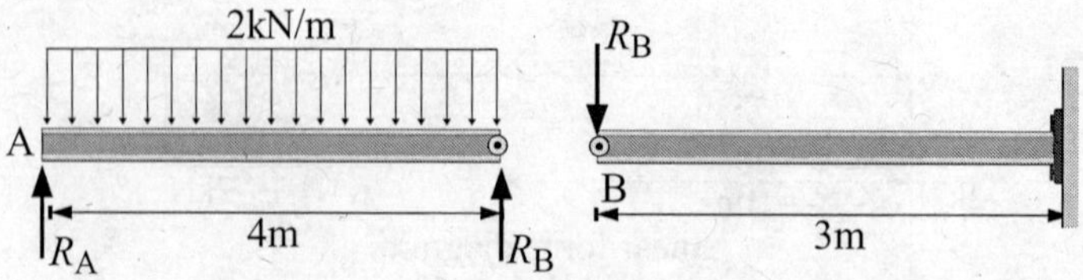

FIGURE 3.12 (b) |

Consider Segment of Simply Supported Beam AB (0 < x < 4 m)

Figure 3.12(c) shows the free body diagram of the left-hand segment of the simply supported beam, AB. Applying the equilibrium equations to the beam segment, we have

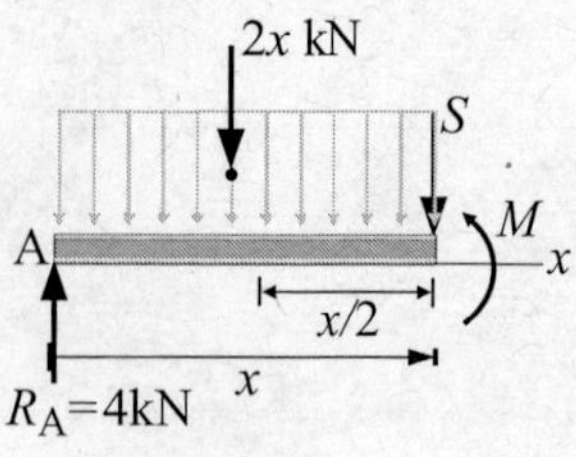

FIGURE 3.12 (c) |

$$\sum F_y = 0 \Rightarrow -S - 2x + 4 = 0,$$

$$S = 4 - 2x,$$

$$\sum M = 0 \Rightarrow M + 2x \times \frac{x}{2} - 4x = 0$$

$$M = 4x - x^2$$

Consider Segment of Cantilever Beam BC (4 < x < 7 m)

Figure 3.12(d) shows the free body diagram of the left-hand segment of cantilever beam BC. Applying the equilibrium equations to the beam segment, we have

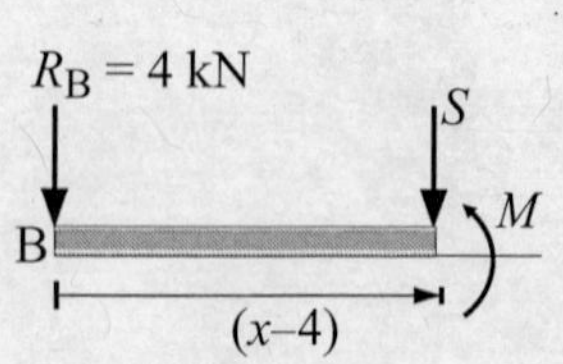

FIGURE 3.12 (d) |

$$\sum F_y = 0 \Rightarrow -S - 4 = 0$$

$$S = -4 \text{ kN}$$

$$\sum M = 0 \Rightarrow M + 4(x - 4) = 0$$

$$M = -4(x - 4)$$

The shear force and bending moment diagrams are plotted as shown in Figure 3.12(e). Problem

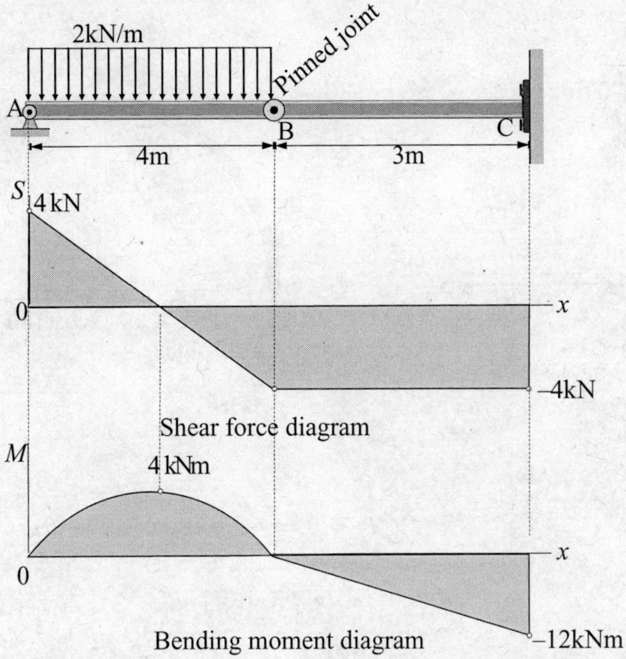

FIGURE 3.12 (e) |

PROBLEM 3.13

Draw the shear force and bending moment diagrams for a curved beam supported on A and B as shown in Figure 3.13(a).

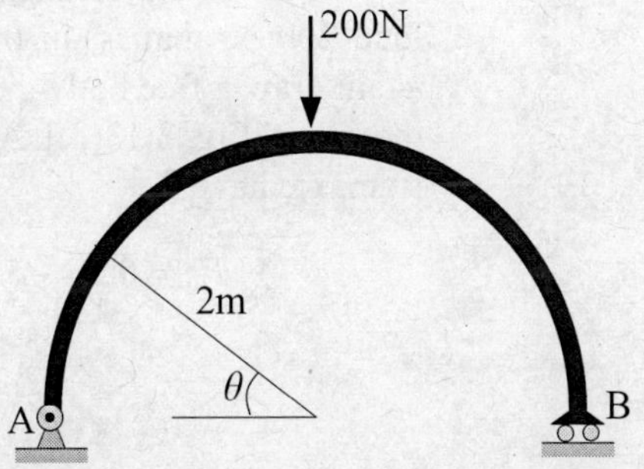

FIGURE 3.13 (a) |

Solution. The first step is to find the reactions R_A and R_B at the supports. Applying equations of equilibrium, $\sum F_r = 0$ and $\sum M = 0$, from which we find, respectively,

$$R_A = 100\text{N and } R_B = 100\text{N}$$

We now cut through the beam at a cross-section to the left of the load 200 N and at an angle θ from the support at A. Then we draw a free body diagram of the left-hand part of the beam [Fig. 3.13(b)]. Applying equations of equilibrium, we have

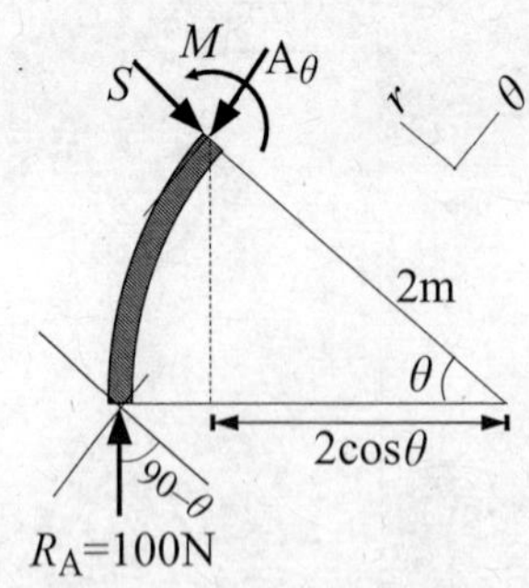

FIGURE 3.13 (b) |

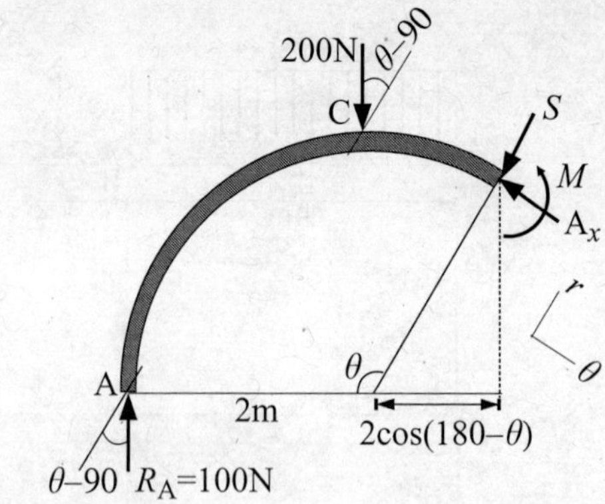

FIGURE 3.13 (c) |

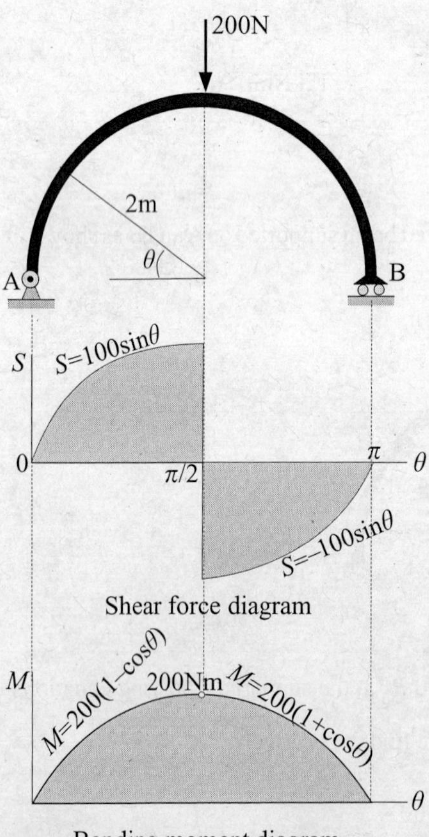

FIGURE 3.13 (d) |

$$\sum F_r = 0 \Rightarrow -S + 100\cos(90 - \theta) = 0,$$

$$S = 100 \sin \theta,$$

$$\sum M = 0 \Rightarrow M - 100(2 - 2\cos\theta) = 0,$$

$$M = 200(1 - \cos\theta)$$

Next, we cut through the beam to the right of the load 200 N, that is, in the region $90 < \theta < 180$, and again draw a free body diagram of the left-hand part of the beam [Fig. 3.13(c)]. Applying equations of equilibrium, we have

$$\sum F_r = 0 \Rightarrow -S - 200\cos(\theta - 90) + 100\cos(\theta - 90) = 0,$$

$$S = -100 \sin \theta$$

$$\sum M = 0 \Rightarrow M + 200 \times 2\cos(180 - \theta) - 100 \times [2 + 2\cos(180 - \theta)] = 0$$

$$M = 200(1 + \cos\theta)$$

The shear force and bending moment diagrams are plotted as shown in Figure 3.13(d) Problem

PROBLEM 3.14

Draw the shear force and bending moment diagrams for the bent beam shown in Figure 3.14(a).

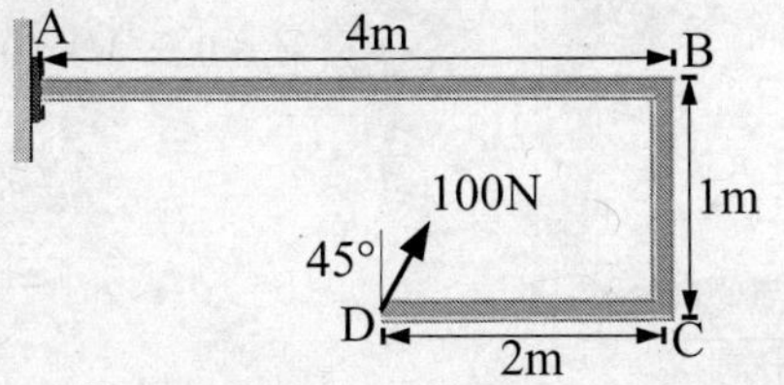

FIGURE 3.14 (a) |

Solution. The determination of the expressions for the shear force S and the bending moment M for each of three segments (AB, BC and DC) is explained as follows.

Segment DC (0 < x < 2 m)

Figure 3.14(b) shows the free body diagram for one part of the beam, located within segment DC. Applying the equations of equilibrium to the beam segment, we have

$$\sum F_y = 0 \Rightarrow 100 \cos 45 - S = 0,$$

$$S = 70.71\text{N}$$

$$\sum M = 0 \Rightarrow M - 100 \cos 45 \times x = 0,$$

$$M = 70.1x$$

$$M_{x=2} = 141.4\text{Nm}$$

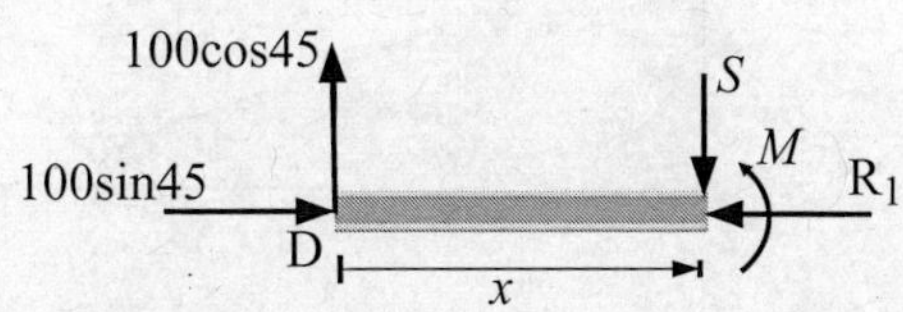

FIGURE 3.14 (b) |

Segment BC (0 < y < 1 m)

Figure 3.14(c) shows the free body diagram for the second part of the beam, located within segment BC. Applying the equations of equilibrium to the beam segment, we have

$$\sum F_x = 0 \Rightarrow 100 \sin 45 + S = 0,$$

$$S = -70.71\text{N},$$

$$\sum M = 0 \Rightarrow M - 100 \cos 45 \times 2 + 100 \sin 45 \times y = 0,$$

$$M = 141.42 - 70.71y$$

$$M_{y=0} = 141.4\text{Nm and } M_{y=1} = 70.71\text{Nm}$$

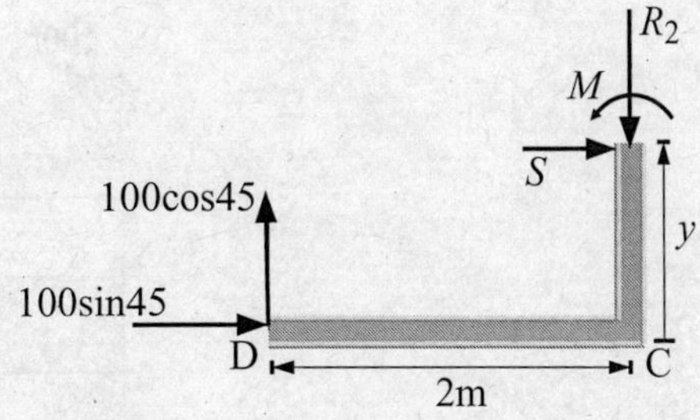

FIGURE 3.14 (c) |

Segment AB (2 m < z < 2 m)

Figure 3.14(d) shows the free body diagram for the third part of the beam, located within segment AB. Applying the equations of equilibrium to the beam segment, we have

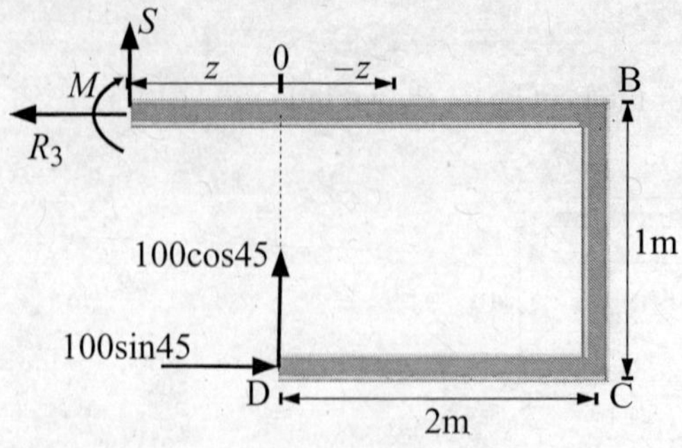

FIGURE 3.14 (d) |

$$\sum F_y = 0 \Rightarrow 100\cos 45 + S = 0,$$

$$S = -70.71\text{N}$$

$$\sum M = 0 \Rightarrow M - 100\cos 45 \times z - 100\sin 45 \times 1 = 0,$$

$$M = 70.71 + 70.71z,$$

$$M_{z=2\text{m}} = 212.13\text{Nm}$$

$$M_{z=0} = 70.71\text{Nm}$$

$$M_{z=-2} = 70.71\text{Nm}$$

We now construct the shear force and bending moment diagrams for every segment [Fig. 3.14(e)].

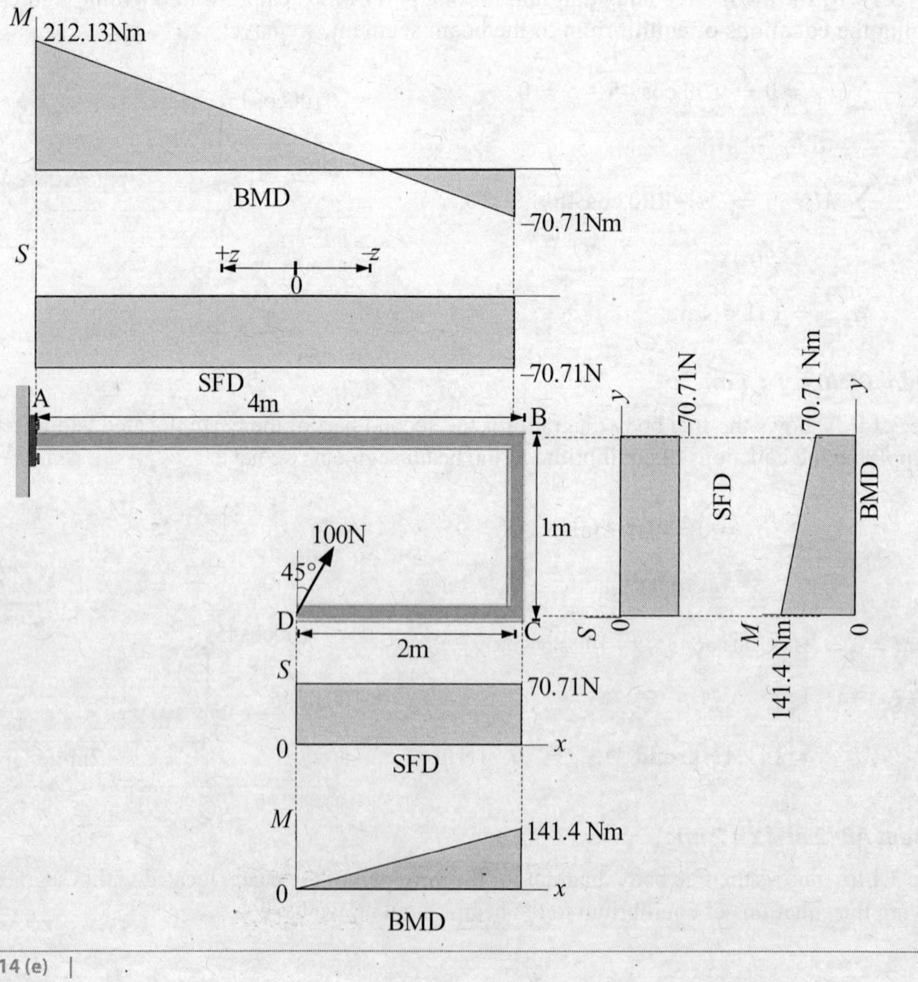

FIGURE 3.14 (e) |

CHAPTER HIGHLIGHTS

- The resultant force S of the distributed transverse forces that act on a beam's cross-section is called the shear force at the section. Shear force is internal to the beam.
- The resultant couple, with moment M, of the distributed forces that act on a beam's cross-section is called the bending moment at the section. Bending moment is internal to the beam.
- Shear force is assumed to be positive if it produces a clockwise moment and negative if it produces an anticlockwise moment.
- The diagram that shows the variation of the shear force along the length of the beam is called SFD. Similarly, the diagram that shows the variation of the bending moment along the length of a beam is known as BMD.
- The points (other than ends) in a beam at which the bending moment is zero are called points of contraflexure or points of inflexion.
- Relation between shear force and bending moment is $\omega = -\dfrac{dS}{dx} = -\dfrac{d^2M}{dx^2}$.

REVIEW QUESTIONS

Q1. Define shear force and bending moment.

Q2. Derive the relationship between shear force and moment.

Q3. What is the point of contraflexure?

Q4. What are sagging and hogging moments?

Q5. Draw the shear force diagram (SFD) and the bending moment diagram (BMD) for a loaded beam as shown in Figure 3.Q5.

$$\left[\begin{array}{lll} \text{Ans; } S = 14.5, & M = 14.5x & \textit{for } 0 < x < 3. \\ S = 1.5 - 5(x-3), & M = 14.5x - 15 - 2.5(x-3)^2 & \textit{for } 3 < x < 7 \\ S = -15.5, & M = 14.5x - 15 - 2.5(x-3)^2 - 10(x-7) & \textit{for } 7 < x < 10 \end{array}\right]$$

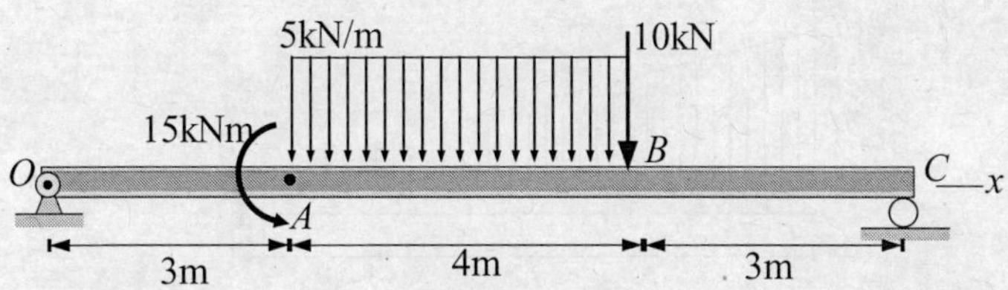

FIGURE 3.Q5 |

Q6. Draw the shear force and bending moment diagrams for the loaded beam as shown in Figure 3.Q6. [Ans; $S = 9 - 0.75x^2$, $M = 9x - 0.25x^3$ kN m]

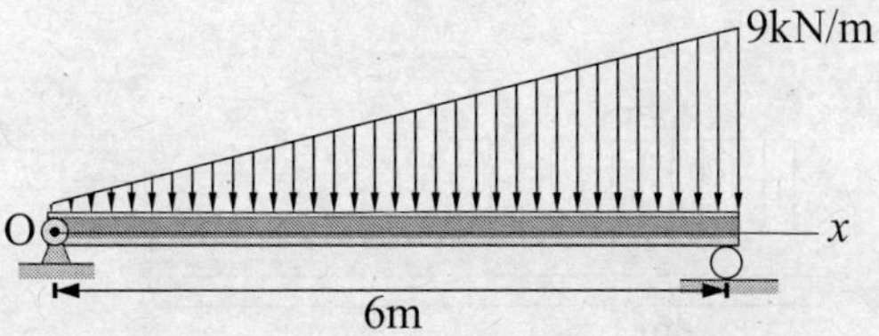

FIGURE 3.Q6 |

Q7. Draw the shear force and bending moment diagrams for the loaded beam as shown in Figure 3.Q7.

$$\begin{bmatrix} \text{Ans; } S = 18, & M = 18x & \text{for } 0 < x < 2m \\ S = 14, & M = 4x + 28 & \text{for } 2m < x < 5m \\ S = -24, & M = -24x + 168 & \text{for } 5m < x < 7m \end{bmatrix}$$

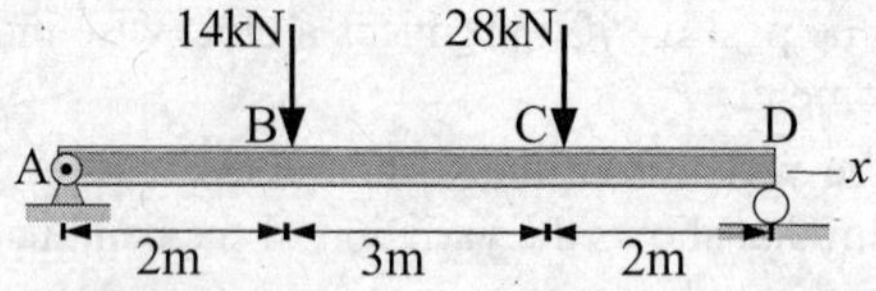

FIGURE 3.Q7 |

Q8. Derive formulas for the shear force and the bending moment of semicircular beam at the section located by the angle θ in Figure 3.Q8.
[Ans; $S = -2 \cos \theta$ kN, $M = \sin \theta$ kN m].

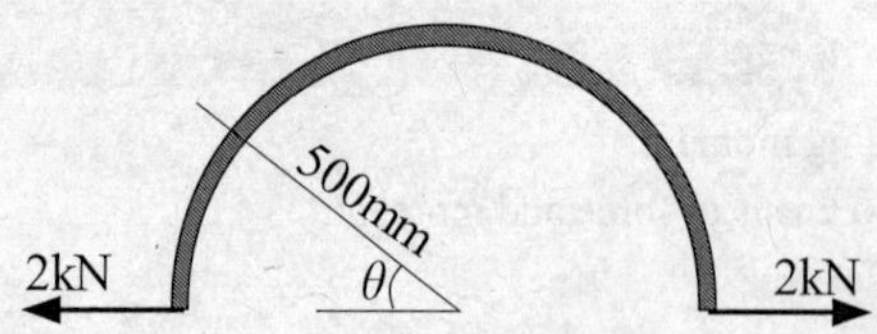

FIGURE 3.Q8 |

Q9. Draw the shear force and bending moment diagrams for a loaded beam as shown in Figure 3.Q9.
[Ans; $S = -\omega_0 x + \omega_0 x^2/2L$, $M = -\omega_0 x^2/2 + \omega_0 x^3/6L$]

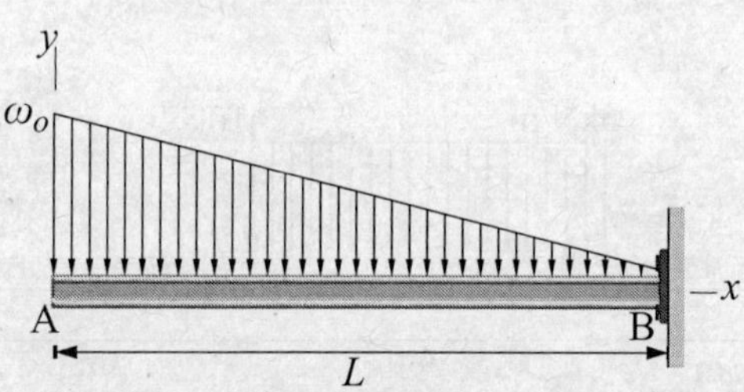

FIGURE 3.Q9 |

Q10. Draw the shear force and bending moment diagrams for the loaded beam as shown in Figure 3.Q10 .
[Ans; for AB $S = -120x$ N, $M = -60x^2$ N m.
for BC $S = -960$ N, $M = -960x + 3840$ N m].

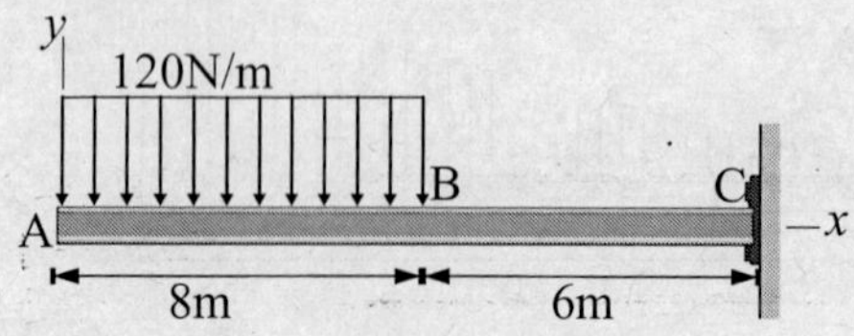

FIGURE 3.Q10 |

Q11. Derive the shear force and the bending moment as a function of the angle θ ?for the arch shown in Figure 3.Q11

$$\left[\begin{array}{l} \text{Ans; for AB } S = \dfrac{P}{2}\sin\theta,\ M = \dfrac{PR}{2}(1-\cos\theta). \\[2ex] \text{for BC } S = -\dfrac{P}{2}\sin\theta,\ M = \dfrac{PR}{2}(1+\cos\theta) \end{array}\right]$$

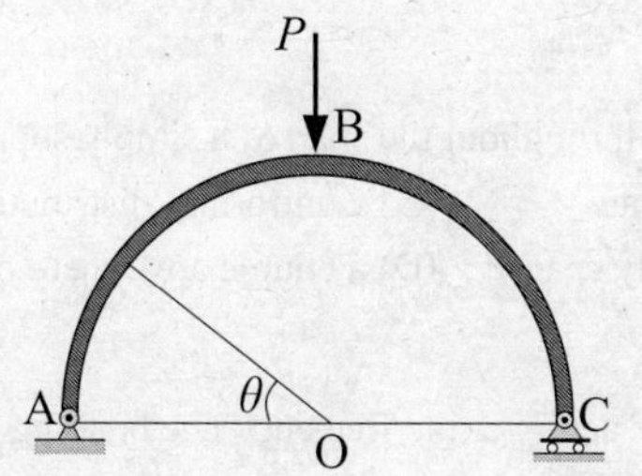

FIGURE 3.Q11 |

Q12. Draw the load and the bending moment diagram that correspond to the given shear force diagram in Figure 3.Q12. Assume that no couples are applied to the beam.

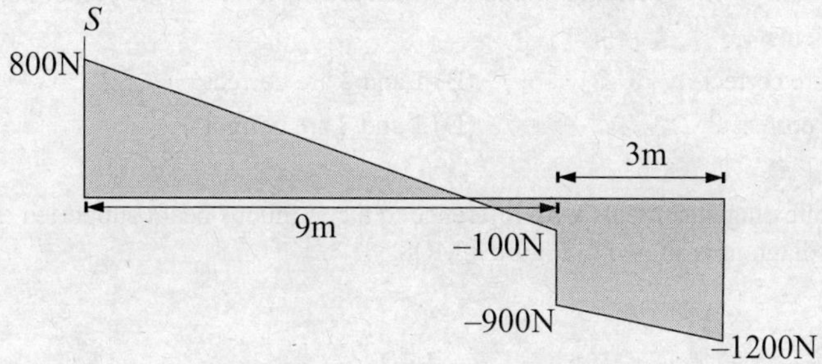

FIGURE 3.Q12 |

MULTIPLE CHOICE QUESTIONS

Q1. The bending moment diagram for a cantilever beam subjected to bending moment at the free end of the beam will be

(A) rectangle (B) triangle (C) parabola (D) cubic parabola

[Ans: A]

Q2. The rate of change of bending moment is equal to

(A) shear force at that section
(B) deflection at that section
(C) loading at that section.
(D) intensity of loading at that section

[Ans: A]

Q3. If a simply supported beam at its ends is loaded by a central concentrated load, then maximum moment is M. If the same weight be equally distributed over the beam, then its maximum moment will be

(A) M (B) $\frac{M}{2}$ (C) $\frac{M}{3}$ (D) $\frac{M}{4}$

[Ans: B]

Q4. The maximum bending moment in a cantilever of span l carrying a uniformly distributed load of intensity ω per unit length is

(A ωl (B) $2\omega l^2$ (C) $\frac{\omega l^2}{2}$ (D) $\frac{\omega l^2}{3}$

[Ans: C]

Q5. For the shear force to be uniform throughout the span of a simply supported beam, it should carry

(A) a concentrated load at mid-span (B) a uniformly distributed load over its entire span
(C) two concentrated loads equally spaced (D) a couple any where on its span

[Ans: D]

Q6. A cantilever of span l has a load P acting at the free end. The bending moment at the support end will be

(A) Zero (B) Pl (C) $\frac{Pl}{4}$ (D) $\frac{Pl}{2}$

[Ans: B]

Q7. Consider the following statements:

A simply supported beam is subjected to a couple somewhere in the span. It would produce

1. a rectangular shear force diagram
2. parabolic bending moment diagram
3. both positive and positive bending moments which are maximum at the point of application of the couple

Of these statements

(A) 1, 2 and 3 are correct (B) 1 and 2 are correct
(C) 2 and 3 are correct (D) 1 and 3 are correct

[Ans: D]

Q8. Consider the following statements with reference to a continuous beam supported at A, C and E for which the shear force diagram is shown in Figure 3.MQ8.

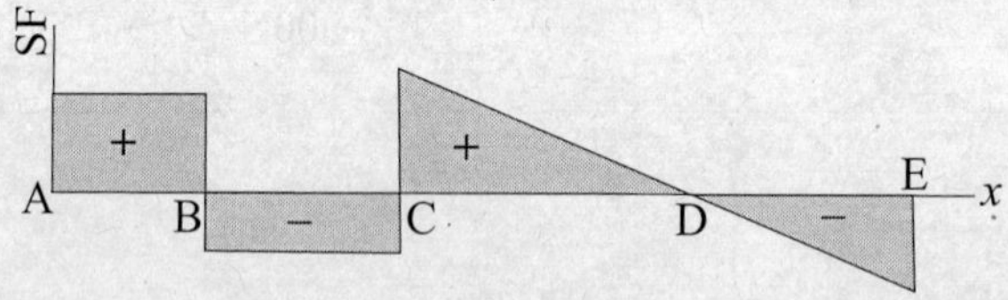

FIGURE 3.MQ8 |

There is

1. a concentrated load acting at point B
2. a concentrated load acting at point D
3. a uniformly distributed load acting on the portion CE

Of these statements,

(A) 1, 2 and 3 are correct (B) 1 and 2 are correct
(C) 2 and 3 are correct (D) 1 and 3 are correct

[Ans: D]

Q9. The bending moment diagram for an overhanging beam is shown in Figure 3.MQ9. The points of contraflexure would include

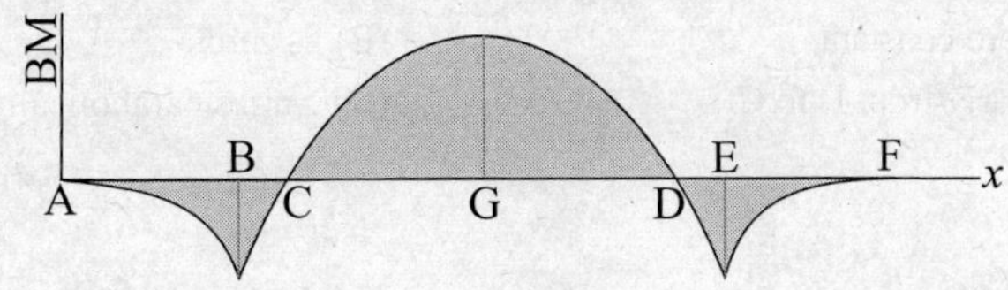

FIGURE 3.MQ9 |

(A) A and F (B) B and E (C) C and D (D) A and D

[Ans: C]

Q10. A simply supported beam is loaded as shown in Figure 3.MQ10. The maximum shear force in the beam will be

(A) zero (B) W (C) $2W$ (D) $4W$

[Ans: C]

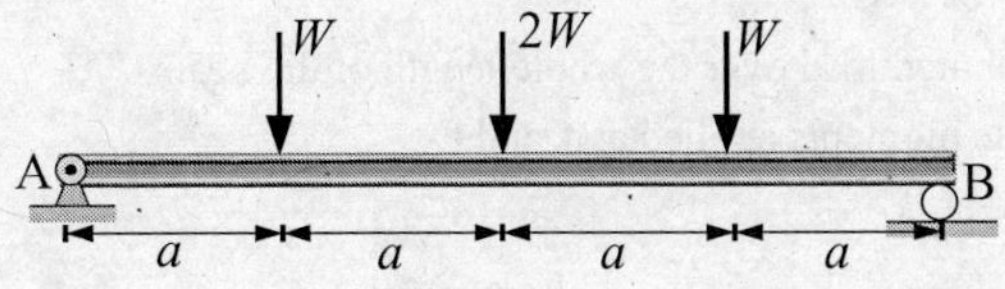

FIGURE 3.MQ10 |

Q11. A 2 m long beam BC carries a uniformly distributed load and is simply supported at its ends by two cantilever AB 2 m long and CD 2 m long as shown in Figure 3.MQ11. The shear force at end A of the cantilever AB will be

(A) zero (B) 2 kN. (C) 5 kN (D) 10 kN

[Ans: C]

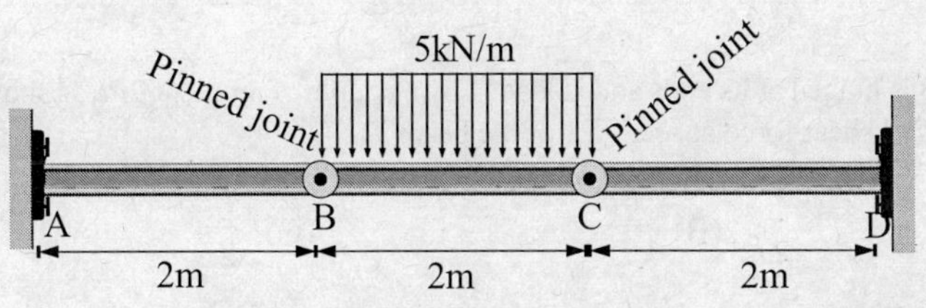

FIGURE 3.MQ11 |

Q12. A beam subjected to a load 2 kN is shown in Figure 3.MQ12. The bending moment at the support A of the beam will be

(A) 1 kN m (B) 3 kN m (C) 4 kN m (D) zero

[Ans: A]

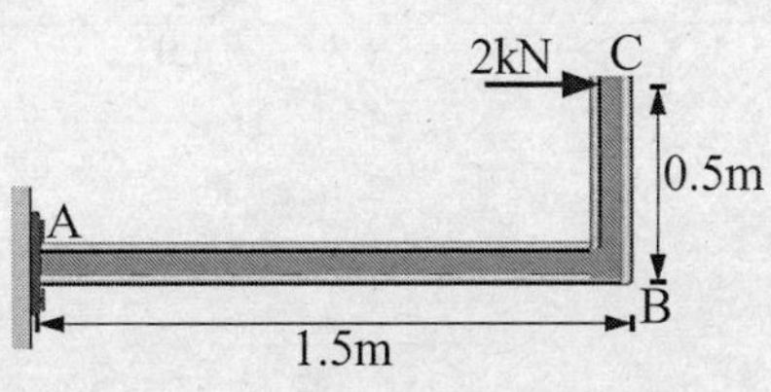

FIGURE 3.MQ12 |

Q13. Figure 3.MQ13 shows the shear force diagram for the beam ABCD. The bending moment in the portion BC of the beam

(A) is a non-zero constant (B) is zero

(C) varies linearly from B to C (D) varies parabolically from B to C

[Ans: A]

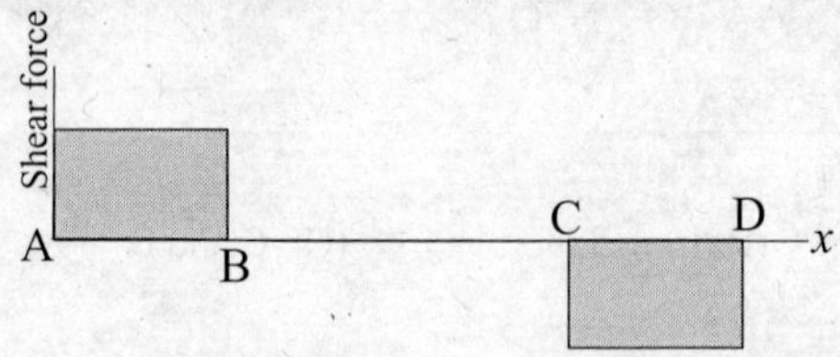

FIGURE 3.MQ13 |

Q14. If the bending moment diagram for a simply supported beam is of the form given in Figure 3.MQ14, the load acting on the beam is

(A) a concentrated force at B

(B) a uniformly distributed load over the whole length of the beam

(C) equal and opposite moments applied at A and C

(D) a moment applied at B

[Ans: D]

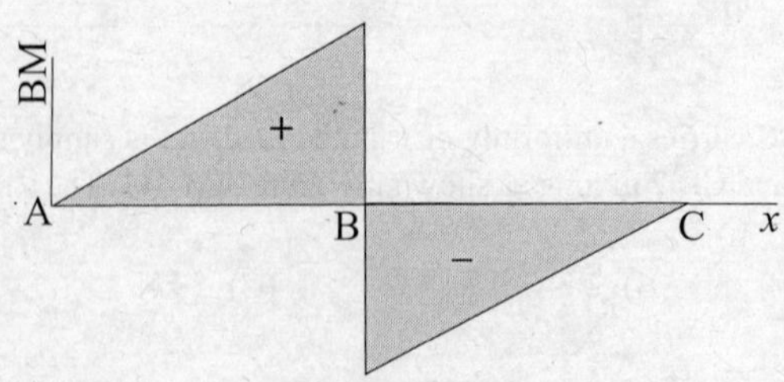

FIGURE 3.MQ14 |

Q15 A beam AB is hinged at its ends and is loaded by a couple of moment $2Pc$ as shown in Figure 3.MQ15. The magnitudes of shear force at section x of the beam is

(A) 0 (B) $\frac{Pc}{L}$ (C) $\frac{Pc}{2L}$ (D) $\frac{Pc}{4L}$

[Ans: B]

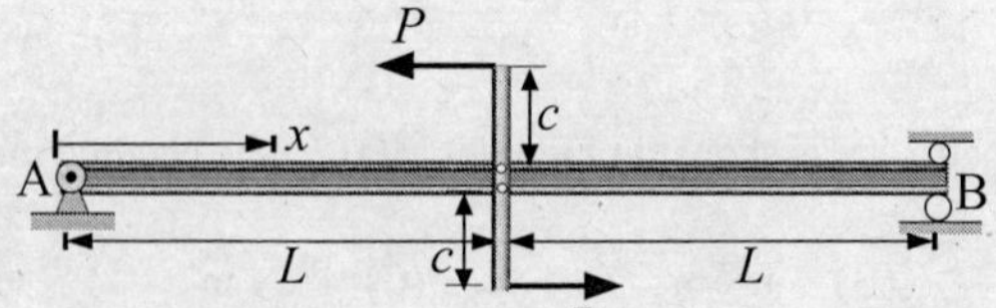

FIGURE 3.MQ15 |

4 Friction

INTRODUCTION

When a body slides over another body, a force is exerted at the surface of contact by the stationary body on the moving body. This resisting force is called the *force of friction* and acts in a direction opposite to the direction of motion. The property of the bodies by virtue of which a force is exerted by a stationary body on a moving body to resist the motion of the moving body is called *friction*. Friction acts parallel to the surface of contact and depends upon the nature of surface of contact.

TYPES OF FRICTION

Dry Friction

Dry friction is said to occur when there is relative motion between two completely unlubricated surfaces. It is further divided into two types:

(a) **Solid friction:** When the two surfaces have a sliding motion relative to each other, it is called solid friction.

(b) **Rolling friction**: Friction due to rolling of one body over another is called rolling friction.

Film Friction

When two surfaces in contact are completely separated by lubrication, friction will occur due to the shearing of different layers of the lubricant. This is known as film friction or viscous friction.

Boundary (Or Skin) Friction

When two surfaces in contact have a minute thin layer of lubricant between them, it is known as boundary or skin friction.

STATIC AND KINETIC FRICTION

Consider a block of weight W, placed on a horizontal plane surface, as shown in Figure 4.1. If we apply an external horizontal force P to the block, acting to the right, the block will remain stationary, if P is not too large. The force that keeps the block from moving acts to the left and is called the friction force, F. As long as the block is in equilibrium, $F = P$. Since the block is stationary, we call this frictional force is *force of static friction*, F_s. Experiments show that frictional force arises from the roughness of the two surfaces.

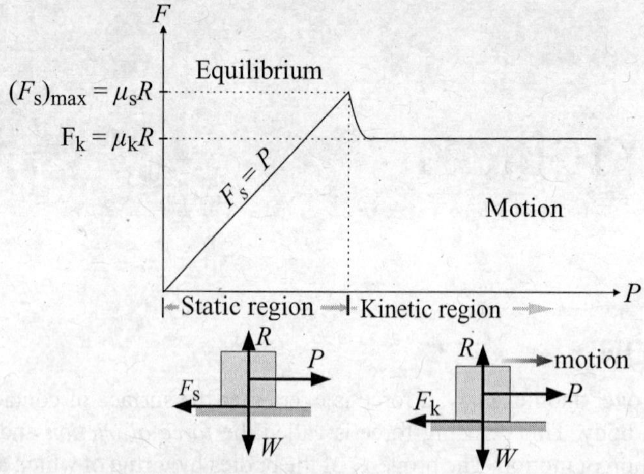

FIGURE 4.1 |

If we increase the magnitude of P, the block will eventually slip. When the block is on the verge of slipping, F_s is at maximum. When P exceeds $(F_s)_{max}$, the block moves and accelerates to the right. When the block is in motion, the retarding frictional force becomes less than $(F_s)_{max}$. When the block is in motion, we call the retarding force the *force of kinetic friction*, F_k. The unbalanced force $(P - F_k)$ produces acceleration. If $P = F_k$, the block moves with a constant speed. If the applied force is removed, then the frictional force acting to left decelerates the block.

Experimental evidence shows that $(F_s)_{max}$ and F_k are proportional to normal force acting on the block. We have

- The force of static friction between any two surfaces in contact is opposite the applied force and can have values given by

$$F_s \leq \mu_s R,$$

where μ_s is called coefficient of static friction.
When the block is on the verge of slipping, force of static friction is given by

$$F_s = (F_s)_{max} = \mu_s R$$

- The force of kinetic friction acting on an object is opposite to the direction of motion of the object and is given by

$$F_k = \mu_k R\,,$$

where μ_k is a constant called coefficient of kinetic friction.

- The values of coefficients of friction μ_s and μ_k do not depend upon the area of the surfaces in contact. Both the coefficients, however, depend strongly on the nature of the surface in contact, but μ_k is generally less than μ_s. Approximate values of coefficients of static and kinetic friction are given in Table 1.

TABLE 4.1 | Approximate values of coefficient of friction (static & kinetic) for surfaces

	μ_s	μ_k
Aluminum on steel	0.61	0.47
Steel on steel	0.74	0.57
Copper on steel	0.53	0.36
Wood on wood	0.25–0.5	0.2
Glass on glass	0.94	0.4
Teflon on Teflon	0.04	0.04
Ice on ice	0.1	0.03
Metal on metal (lubricated)	0.15	0.06
Rubber on concrete	1.0	0.8

LAWS OF SOLID FRICTION (LAWS OF COULOMB FRICTION)

Based upon experimental evidence, the following laws of friction have been established for dry contact surfaces:

- The force of friction acts in the opposite direction in which surface is having a tendency to move.
- The force of friction is equal to the force applied to the surface, so long as the surface is at rest.
- When the surface is on the point of motion, the force of friction is at maximum and this maximum frictional force is called the limiting frictional force.
- The limiting frictional force bears a constant ratio to the normal reaction between two surfaces.
- The limiting frictional force does not depend up on the shape and areas of the surfaces in contact.
- The ratio between limiting friction and normal reaction is slightly less when the two surfaces are in motion.
- The force of friction is independent of the velocity of sliding.

ANGLE OF FRICTION

It is the angle ϕ which resultant S subtends with the normal to the plane, when the body just starts sliding over the horizontal plane. This angle is also known as limiting angle of reaction or friction.

It is sometimes found convenient to replace the normal force R and the friction force F_s by their resultant S. From the geometry of Figure 4.2(a), we note that

$$\tan\phi = \frac{F_s}{R} = \frac{\mu R}{R}$$

or

$$\tan\phi = \mu$$

$$\text{limiting angle of friction } \phi = \tan^{-1}\mu$$

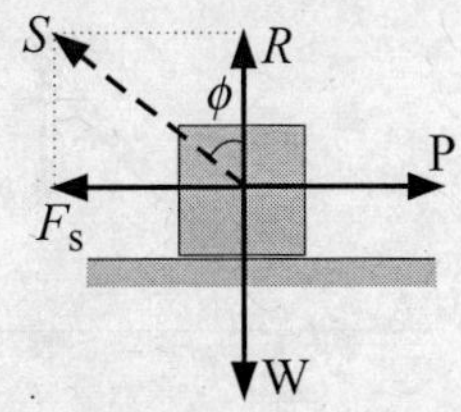

FIGURE 4.2(a) |

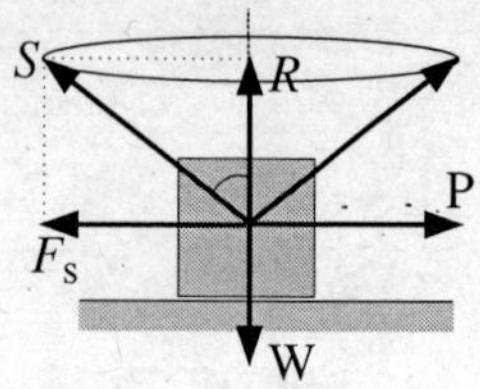

FIGURE 4.2(b) |

This indicates that the friction coefficient is tan of angle of friction. If we describe a right circular cone of apex angle 2ϕ about the line of action R as its axis, then this cone is called a *cone of friction* (Fig. 4.2(b)).

ANGLE OF REPOSE

Consider a block of weight W resting on an inclined plane making an angle θ which is increased gradually till the block is just at the point of sliding. The block is then in an equilibrium state under the action of following forces (Fig. 4.3):

(i) Weight W of the block acting vertically downward.
(ii) Normal reaction R acting perpendicular to the inclined plane.
(iii) Limiting frictional force $F = \mu R$ acting up parallel to the inclined plane.

Resolving these forces along and perpendicular to the plane,

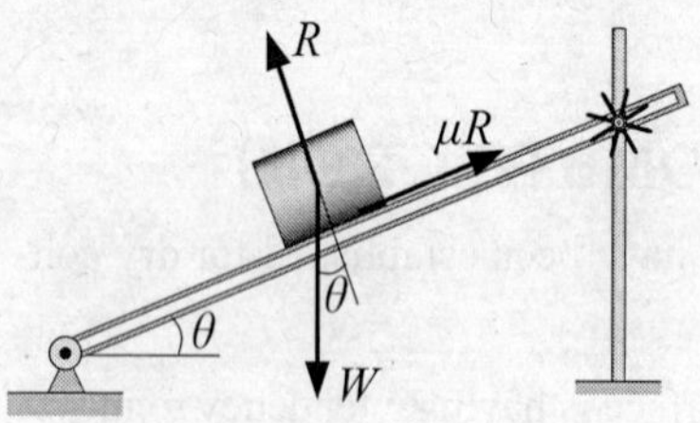

FIGURE 4.3 |

$$W \sin\theta - \mu R = 0 \quad \text{(i)}$$

$$R - W\cos\theta = 0 \quad \text{(ii)}$$

Equations (i) and (ii) give

$$\mu = \tan\theta$$

In terms of the angle of friction ϕ ($\mu = \tan\phi$)

$$\tan\phi = \tan\theta,$$

or

$$\theta = \phi$$

The maximum angle θ of the inclined plane, at which a block resting on it is about to slide down is called the *angle of repose*, and it is equal to the angle of friction between the block and the inclined plane.

LADDER

A ladder is an arrangement used for climbing on the walls or roofs. It essentially consists of two long uprights of wood or iron which are connected by a number of cross bars. These cross bars are called rungs and provide steps for climbing.

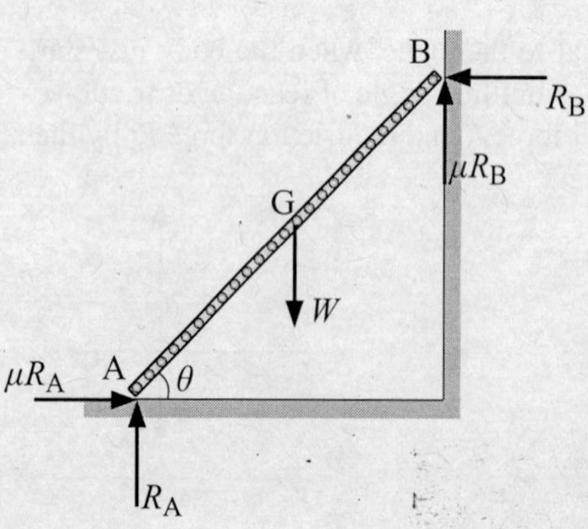

FIGURE 4.4 |

Figure 4.4 shows a ladder AB with its end A resting on the ground and end B leaning against a wall. The ladder is acted upon by the following set of forces:

- Weight, W, acting downward at its centre of mass.
- Normal force R_B and friction force $F_B = \mu R_B$ at the end B leaning against the wall. Since the ladder has a tendency to slip downwards, the friction force will be acting upwards. If the wall is smooth ($\mu = 0$), the friction force will be zero.
- Normal force R_A and friction force $F_A = \mu R_A$ at the end A resting on the floor. Since the ladder, upon slipping, tends to move away from the wall, the direction of friction force will be towards the wall. For the

equilibrium of the system, the algebraic sum of the horizontal and vertical components of the forces must be zero. In addition, the moments of all the forces about any point must be zero.

WEDGE

A wedge is used to produce small adjustments in the position of a body or raise the heavy loads. These loads may be raised by applying to the wedge a force usually considerably smaller than the weight of the load. It consists of three sliding pairs, as shown in Figure 4.5, formed by the frame, wedge and slider. When a force F is applied to the wedge, the slider is raised in the guides raising the load.

Mechanical efficiency of the wedge is defined as the ratio of the load raised when friction is considered to the load raised when friction is neglected, while the force applied is the same. Free body diagrams of wedge and slider are shown in Figure 4.5. We neglect the mass of the wedge. From the free body diagram, we write the equilibrium force equations by equating to zero the sum of the force vectors acting on each body.

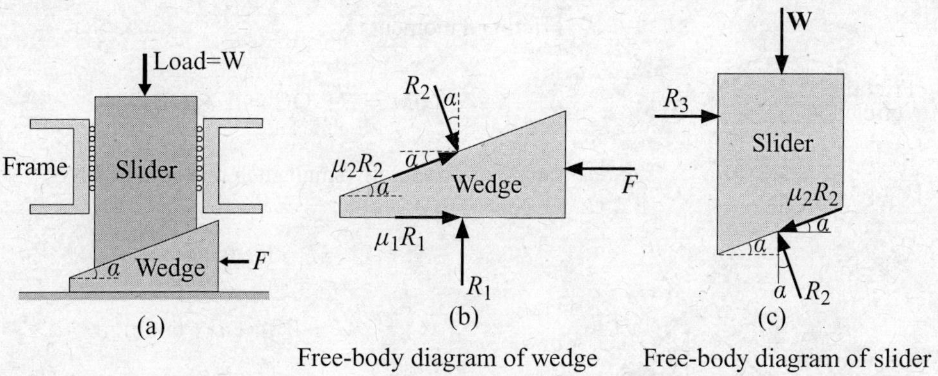

FIGURE 4.5 |

EQUILIBRIUM FORCE EQUATIONS OF WEDGE (Part B of Fig. 4.5)

$$\sum F_y = 0 \Rightarrow R_1 - R_2 \cos a + \mu_2 R_2 \sin a = 0 \tag{i}$$

$$\sum F_x = 0 \Rightarrow \mu_1 R_1 + \mu_2 R_2 \cos a + R_2 \sin a - F = 0 \tag{ii}$$

EQUILIBRIUM FORCE EQUATIONS OF SLIDER (Part C of Fig. 4.5)

$$\sum F_y = 0 \Rightarrow R_2 \cos a - \mu_2 R_2 \sin \alpha - W = 0 \tag{iii}$$

$$\sum F_x = 0 \Rightarrow R_3 - \mu_2 R_2 \cos a - R_2 \sin \alpha = 0 \tag{iv}$$

By solving above equations, we can find the unknown values.

JOURNAL BEARING

Journal bearings are used to provide lateral support to rotating shafts and axles. When a shaft revolves in a bearing, some power is lost due to friction between the two surfaces. Figure 4.6 shows the position of the shaft in a bearing, when it is stationary. At the instant the journal rests at the centre of the bearing, the film of lubricant is being crushed.

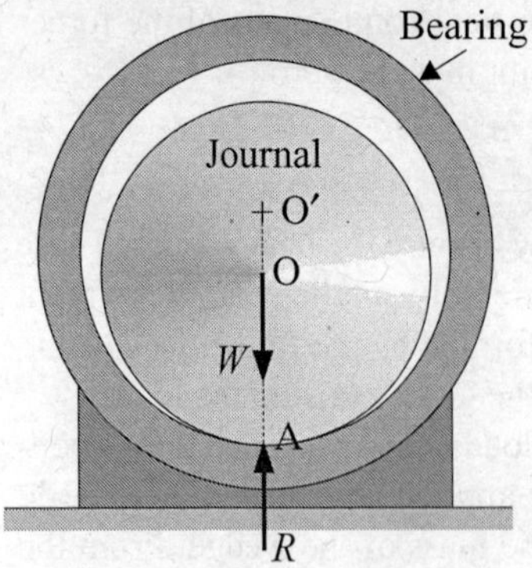

FIGURE 4.6 |

Consider a journal bearing, as shown in Figure 4.6. When the journal is at standstill, then the point of contact is at A. The load W is balanced by the reaction R. When the journal starts rotating in the clockwise direction, then the point of contact shifts from A to B. The friction force $F = \mu R$ will act at right angle to R, with the resultant of F and R making an angle ϕ with R. The resultant reaction S acts in vertical direction opposite to the direction of W. Let S intersect the horizontal centre line of the shaft at C. Then a small circle drawn with radius OC is known as friction circle.

For the equilibrium of the journal, we have

$$\sum F_y = 0 \Rightarrow W = S$$

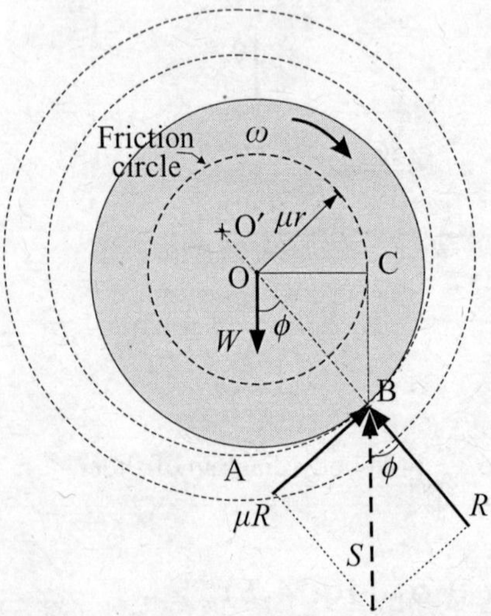

FIGURE 4.7 |

Frictional moment,

$$M = S \times \text{OC} = W \times \text{OB}\sin\phi,$$

(angle ϕ is generally small such that $\sin\phi \approx \tan\phi$)

$$M \approx Wr\tan\phi$$

or

$$M = W\mu r \ (\mu = \tan\phi)$$

Hence, the radius of friction circle is μr.

Thus, the effect of friction is equivalent to displacing the reaction through a distance equal to $r\sin\phi$ or such that it is tangential to the friction circle (Fig. 4.7).

DISC FRICTION (THRUST BEARING)

Force acting along the axis of a shaft is known as thrust. This produces lateral motion along the axis, which is prevented by providing one or more bearing surfaces, at right angles to the axis of rotation (Fig. 4.8). Thrust bearing is provided at the end of the shaft. It may be of flat or conical type.

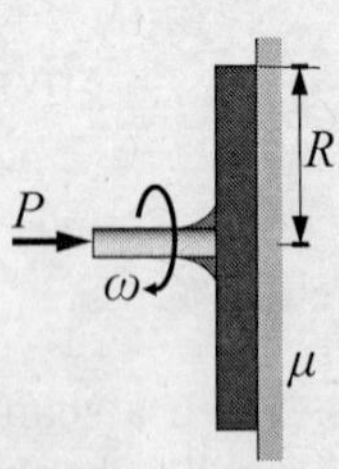

FIGURE 4.8 |

Friction torque of a thrust bearing is calculated, usually, on the basis of two assumptions. Each assumption leads to a different value of torque. In one case, it is assumed that the intensity of pressure on the bearing surface is constant, whereas in the second case, it is the uniform wearing of the bearing surface.

Consider an element of width dr at radius r (Fig. 4.9). The normal force on the area of element is then

$$\text{d}R = p\text{d}A = pr\text{d}\theta\text{d}r,$$

where p is the normal pressure between contact surfaces.

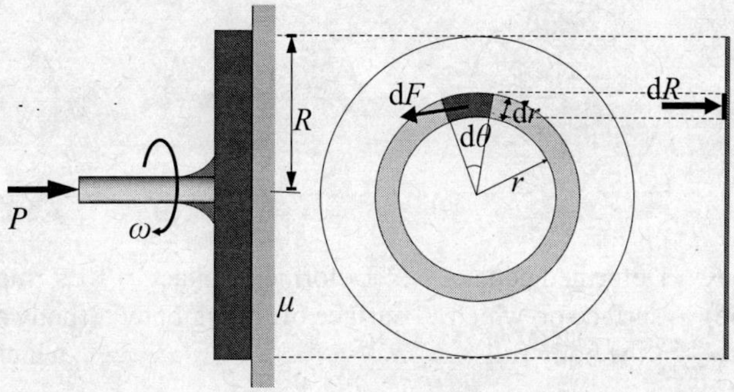

FIGURE 4.9 |

The friction force associated with this force during motion is

$$dF = \mu dR = \mu pr d\theta \, dr$$

The moment of this elemental friction force about the shaft is

$$dM = r \times dF = \mu pr^2 d\theta dr$$

Total moment, $$M = \int dM = \iint \mu pr^2 \, dr d\theta$$

Assuming that pressure p is uniformly distributed over the entire surface. So that

$$p = \frac{P}{\pi R^2}$$

Substituting the constant value of p in the expression for M gives

$$M = \int_0^{2\pi}\int_0^{R} \mu pr^2 \, dr d\theta = \mu\left(\frac{W}{\pi R^2}\right)\int_0^{2\pi}\int_0^{R} r^2 \, dr d\theta$$

or $$M = \mu\left(\frac{P}{\pi R^2}\right) \times \frac{R^3}{3} \times 2\pi$$

$$M = \frac{2}{3}\mu PR$$

ROLLING FRICTION

When cylinder rolls over a flat surface, the contact is theoretically along a line parallel to the axis. Similarly, when a ball rolls over a flat surface, the contact is theoretically a point. However, the ball or cylinder possesses weight, and due to the pressure of the same, deformation of the flat surfaces or of the rolling body or of both takes place. The amount of deformation depends upon the elasticity of the materials in contact and the pressure.

To describe rolling friction, we consider a rolling body (cylinder or sphere), as shown in Figure 4.10. The action of a load W on the axle and a force F applied at its centre produces rolling. The deformations of the rolling body, supporting surface and both surfaces are shown in Figure 4.10. We find

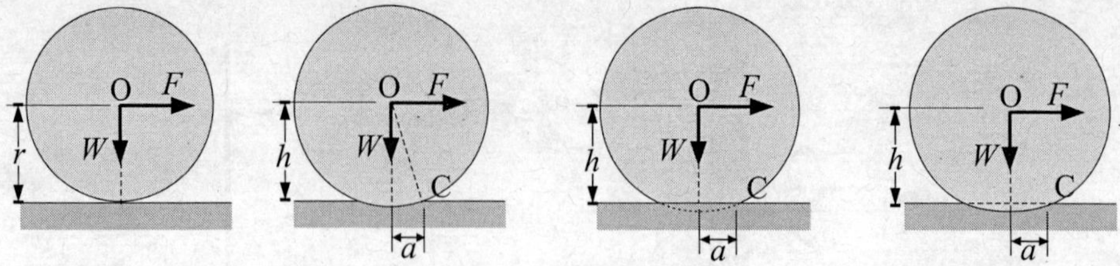

Pure rolling of body (cylinder or sphere) (a)

Deformed contact surface on which the body rolls. (b)

Deformed contact surface of rolling body (c)

Deformed both surfaces (body and surface on which body rolls) (d)

FIGURE 4.10 |

force F necessary to maintain rolling at contact speed by equating the moments of all forces about C to zero. This gives us

$$Fh = Wa \text{ or } F = \frac{a}{h}W = \mu_r W$$

The ratio $\mu_r = a / h$ is called the coefficient of rolling resistance.

The body rolls without sliding, if $\mu > \frac{a}{h}$,

the body will slide, if $\mu < \frac{a}{h}$

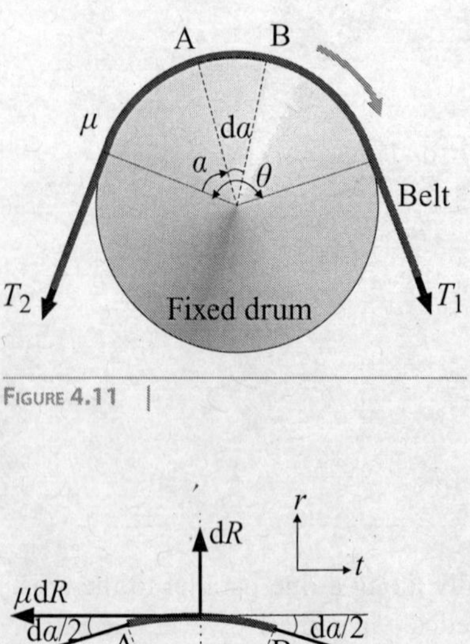

FIGURE 4.11 |

BELT FRICTION

Consider a flat belt passing over a fixed cylindrical drum. Let T_1 and T_2 be the belt tensions on the tight and slack sides, respectively, when the belt is just about to slide towards the right.

Consider a small element AB of the belt. Figure 4.11 shows the free body diagram of the element of the belt The tension increases from T at the angle θ to $T + \mathrm{d}T$ at the angle $\theta + \mathrm{d}\theta$. The friction force $\mu\, \mathrm{d}R$ will act tangentially to the surface resisting the slipping of the elementary belt on the drum. We write the equation of equilibrium for the element AB of the belt.

Resolving the forces in the tangential direction

$$\sum F_t = 0 \Rightarrow \mu \mathrm{d}R + T \cos\frac{\mathrm{d}\alpha}{2} - (T + \mathrm{d}T)\cos\frac{\mathrm{d}\alpha}{2} = 0$$

For small angle of $\mathrm{d}\alpha$, $\cos(\mathrm{d}\alpha / 2) \approx 1$.

Therefore,

$$\mu dR + T - (T + \mathrm{d}T) = 0$$

or

$$\mathrm{d}T = \mu \mathrm{d}R \qquad \text{(i)}$$

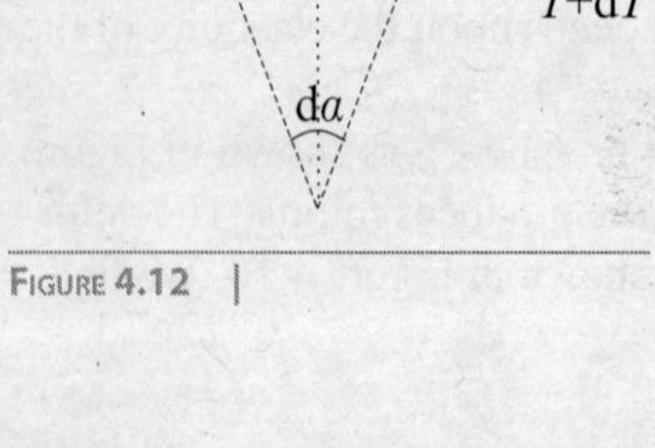

FIGURE 4.12 |

Resolving the forces in the radial direction,

$$\sum F_r = 0 \Rightarrow dR - T\sin\frac{d\alpha}{2} - (T + dT)\sin\frac{d\alpha}{2} = 0$$

$$\left[\text{as } d\alpha \text{ is small; } \sin\frac{d\alpha}{2} \approx \frac{d\alpha}{2}\right]$$

or $$dR - T\frac{d\alpha}{2} - T\frac{d\alpha}{2} - dT\frac{d\alpha}{2} = 0$$

or $$dR - T\,d\alpha - \frac{dT \times d\alpha}{2} = 0$$

Neglecting product of two small quantities,

$$dR = T\,d\alpha \tag{ii}$$

Substituting the value of dR from equation (i), we have

$$\frac{dT}{\mu} = T\,d\alpha \text{ or } \frac{dT}{T} = \mu d\alpha$$

Integrating between corresponding limits yields

$$\int_{T_2}^{T_1}\frac{dT}{T} = \int_0^{\theta}\mu d\alpha$$

or $$\log\frac{T_1}{T_2} = \mu\theta$$

or $$\frac{T_1}{T_2} = e^{\mu\theta}$$

$$\boxed{\frac{T_1}{T_2} = e^{\mu\theta}}$$

It is to be noted that the above relation is valid only when the belt is on the point of slipping on the drum (Fig. 4.12).

SOLVED PROBLEMS

PROBLEM 4.1

Determine the couple M required to spin the cylinder of mass m in its position against the vertical wall [Fig. 4.1(a)]. The coefficient of friction for each pair of contacting surfaces is μ.

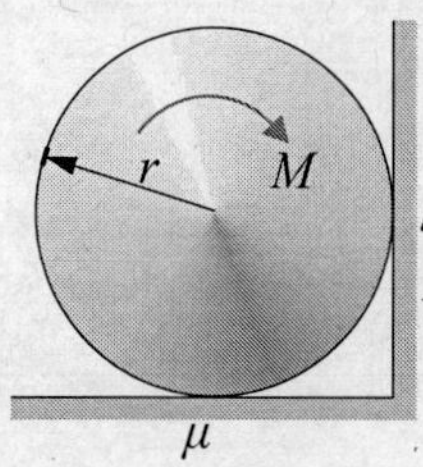

FIGURE 4.1(a) |

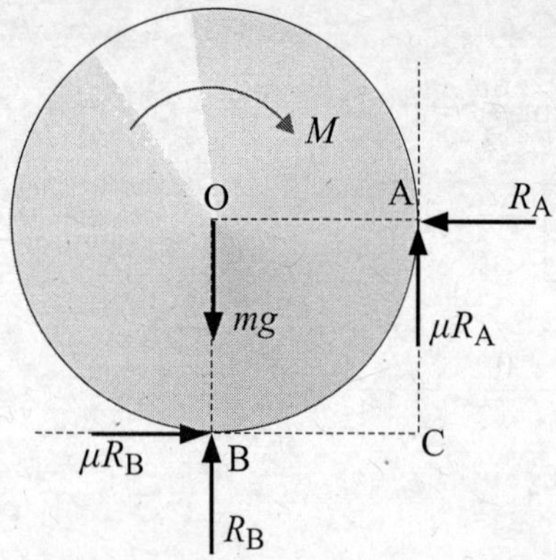

FIGURE 4.1(b) |

Solution. A free body diagram of cylinder is drawn in Figure 4.1(b). When the couple M is applied clockwise, the cylinder is just about to slip and the friction forces at contact point have reached their maximum. The equilibrium equations are

$$\sum F_y = 0 \Rightarrow \mu R_A + R_B - mg = 0$$

or

$$\mu R_A + R_B = mg \qquad \text{(i)}$$

$$\sum F_x = 0 \Rightarrow \mu R_B - R_A = 0$$

or

$$\mu R_B = R_A \qquad \text{(ii)}$$

$$\sum M_o = 0 \Rightarrow M - \mu R_A \times r - \mu R_B \times r = 0$$

or

$$M = \mu r(R_A + R_B) \qquad \text{(iii)}$$

Solving for R_A and R_B from Equations (i) and (ii), we obtain

$$R_B = \left(\frac{1}{1+\mu^2}\right) mg \text{ and } R_A = \left(\frac{\mu}{1+\mu^2}\right) mg$$

Substitute the value of R_A and R_B in Equation (iii), we obtain

$$M = \mu r\left[\left(\frac{\mu}{1+\mu^2}\right) mg + \left(\frac{1}{1+\mu^2}\right) mg\right]$$

On simplifying, we get

$$M = mgr\left(\frac{\mu + \mu^2}{1+\mu^2}\right)$$

PROBLEM 4.2

A semi-circular disc of weight W rests in a vertical plane with its curved edge on a rough horizontal plane and an equally rough vertical plane as shown in Figure 4.2(a). Find the greatest angle θ that the diameter can make with the horizontal plane.

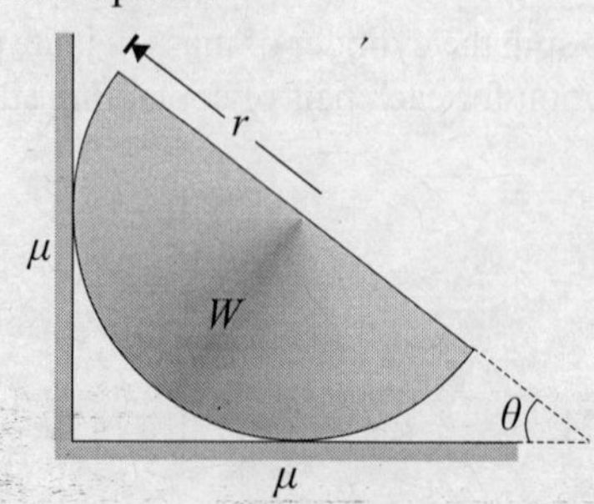

FIGURE 4.2(a) |

Solution. Figure 4.2(b) shows a free body diagram of semi-circular disc. The condition of equilibrium is at the instant before sliding occurs. Thus, the sum of the horizontal forces, vertical forces and moments is zero, which gives

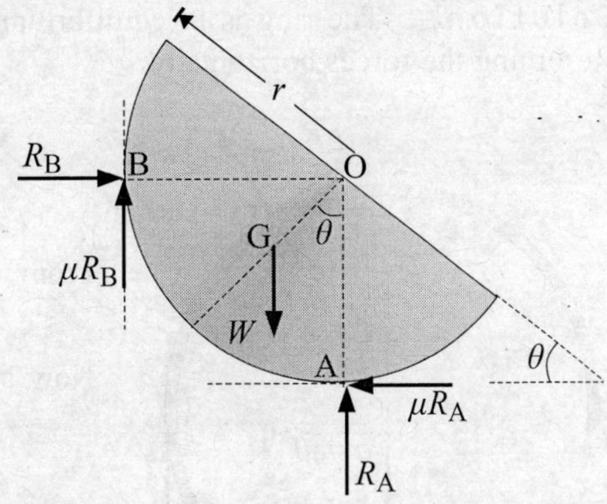

FIGURE 4.2(b)

$$\sum F_x = 0 \Rightarrow R_B - \mu R_A = 0 \tag{i}$$

$$\sum F_y = 0 \Rightarrow R_A + \mu R_B - W = 0 \tag{ii}$$

$$\sum M_A = 0 \Rightarrow R_B \times r + \mu R_B \times r - W \times \text{OG} \sin\theta = 0 \qquad \left[\text{OG} = \frac{4r}{3\pi}\right]$$

or $$(1+\mu)R_B \times r = W\frac{4r}{3\pi}\sin\theta \tag{iii}$$

Solving for R_B from Equations (i) and (ii), we have

$$R_B = \frac{\mu}{1+\mu^2}W$$

Substituting the value of R_B in the Equation (iii), we have

$$\frac{\mu(1+\mu)}{1+\mu^2} = \frac{4}{3\pi}\sin\theta$$

or $$\sin\theta = \frac{3\pi}{4} \times \left(\frac{\mu+\mu^2}{1+\mu^2}\right)$$

From which,

$$\theta = \sin^{-1}\left(\frac{3\pi}{4} \times \frac{\mu+\mu^2}{1+\mu^2}\right)$$

PROBLEM 4.3

A thin ring of negligible thickness has a mass m and a radius r is hanged from a support at O [Fig. 4.3(a)]. If the coefficient of friction between the ring and the support is μ, determine the vertical force F applied to the periphery of the ring so as to slip the ring on the support. Also determine the angle θ at which slipping occurs.

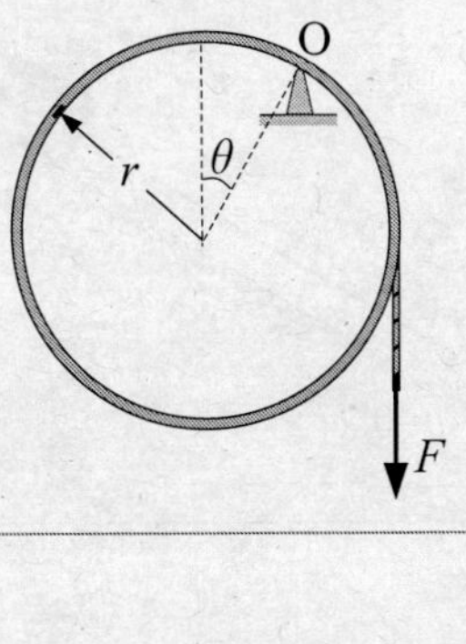

FIGURE 4.3(a)

Solution. The ring is in equilibrium under the action of forces shown in Figure 4.3(b). Resolving the forces horizontally,

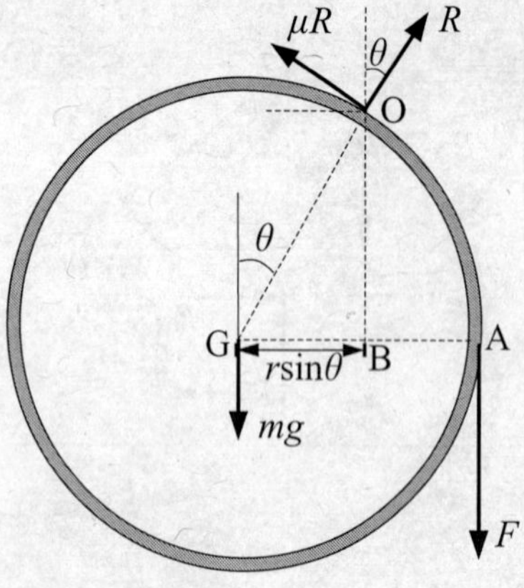

FIGURE 4.3(b) |

$$\sum F_x = 0 \Rightarrow R\sin\theta - \mu R\cos\theta = 0$$

$$\tan\theta = \mu \tag{i}$$

From which,

$$\theta = \tan^{-1}\mu$$

Now, summing moment about O

$$\sum M_O = 0 \Rightarrow F(r - r\sin\theta) - mgr\sin\theta = 0$$

On simplifying, we get

$$F = \frac{mg\sin\theta}{1-\sin\theta}$$

$$= \frac{mg\dfrac{\tan\theta}{\sqrt{1+\tan^2\theta}}}{1-\dfrac{\tan\theta}{\sqrt{1+\tan^2\theta}}}$$

$$= \frac{mg\tan\theta}{\sqrt{1+\tan^2\theta}-\tan\theta}$$

Substituting the value of $\tan\theta = \mu$ from Equation (i), we have,

$$F = mg\frac{\mu}{\sqrt{1+\mu^2}-\mu}$$

PROBLEM 4.4

Two blocks A and B of weight 100 N and 50 N are tied together by a rope parallel to the plane as shown in Figure 4.4(a). The coefficient of friction between block A and the plane is 0.2, and that between block B and the plane is 0.5. Determine the value of the angle θ at which sliding will occur. What is the tension in the rope?

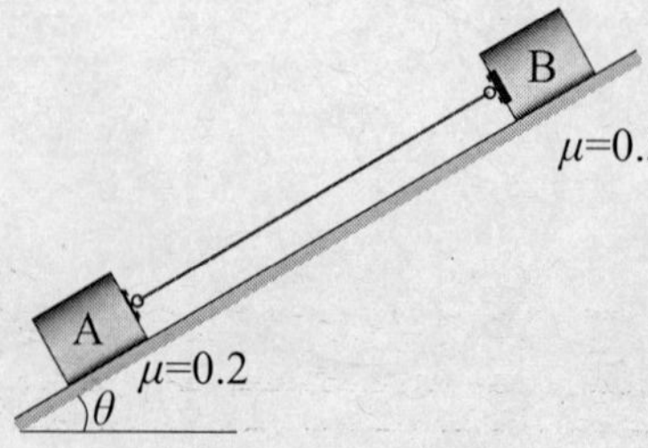

FIGURE 4.4(a) |

Solution. Let the greatest inclination of the plane be θ, then the blocks would be on the point of sliding down the plane, hence the forces of friction $0.2R_A$ and $0.5R_B$ will be up the plane, where R_A and R_B are the normal reactions of the plane [Fig. 4.4(b)].

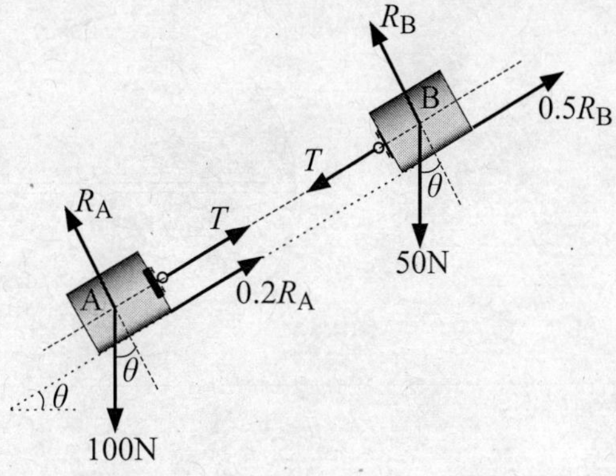

FIGURE 4.4(b)

Consider Block A

Resolving forces acting on block A along and perpendicular to the inclined plane, we have

$$0.2R_A + T - 100 \sin \theta = 0 \quad \text{(i)}$$

$$R_A - 100 \cos \theta = 0 \quad \text{(ii)}$$

Consider Block B

Resolving forces along and perpendicular to the plane, we have

$$0.5R_B - T - 50 \sin \theta = 0 \quad \text{(iii)}$$

$$R_B - 50 \cos \theta = 0 \quad \text{(iv)}$$

From Equations (i) and (ii), we have

$$T = 100 \sin \theta - 0.2(100 \cos \theta) \quad \text{(v)}$$

From Equations (iii) and (iv), we have

$$T = 0.5(50 \cos \theta) - 50 \sin \theta \quad \text{(vi)}$$

Equating the value of T from Equations (v) and (vi), we obtain

$$100 \sin \theta - 0.2(100 \cos \theta) = 0.5(50 \cos \theta) - 50 \sin \theta$$

or

$$150 \sin \theta = 45 \cos \theta \Rightarrow \theta = \tan^{-1}\left(\frac{45}{150}\right)$$

From which,

$$\theta = 16.7^\circ$$

Substitute the value of $\theta = 16.7^\circ$ into Equation (v), we obtain

$$T = 100 \sin 16.7^\circ - 0.2(100 \cos 16.7^\circ) = 9.57\text{N}$$

$$T = 9.57\text{N}$$

PROBLEM 4.5

A uniform thin rod of length $(4r + 2a)$ and mass m is supported by the pegs A and B, which are fixed in the wheel [Fig. 4.5(a)]. If the coefficient of friction between the rod and the pegs is μ, determine the angle θ through which the wheel may turn about its horizontal axis through O before the rod begins to slip.

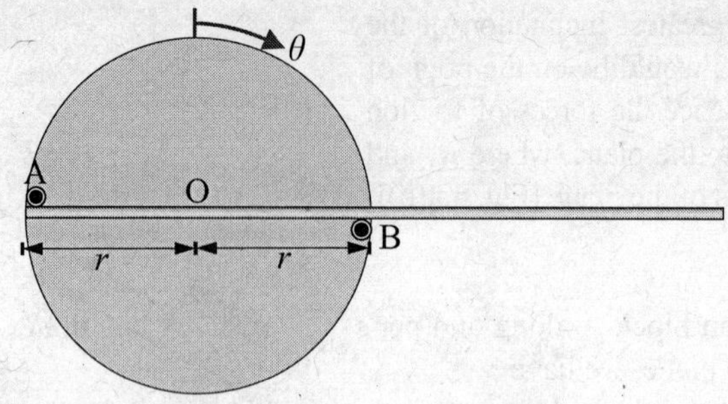

FIGURE 4.5(a) |

Solution. A free body diagram of the rod when it just starts moving down is shown in Figure 4.5(b). Resolving the forces in y-direction, we get

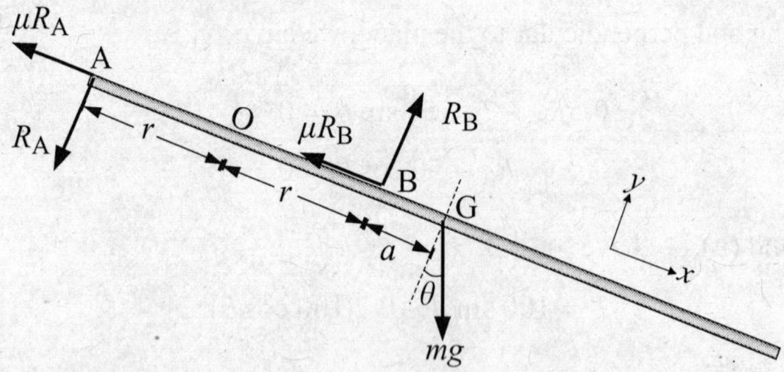

FIGURE 4.5(b) |

$$\sum F_y = 0 \Rightarrow R_B - R_A - mg\cos\theta = 0$$

or
$$R_B - R_A = mg\cos\theta \tag{i}$$

Resolving the forces in x-direction, we get

$$\sum F_x = 0 \Rightarrow \mu R_A + \mu R_B - mg\sin\theta = 0$$

or
$$R_A + R_B = \frac{mg\cos\theta}{\mu} \tag{ii}$$

Solving for R_B from Equations (i) and (ii), we obtain

$$R_B = \frac{mg}{2}\left(\cos\theta + \frac{\sin\theta}{\mu}\right) \tag{iii}$$

Now summing moment about A

$$\sum M_A = 0 \Rightarrow R_B \times 2r - mg\cos\theta(2r + a) = 0$$

$$R_B = mg\cos\theta\left(1 + \frac{a}{2r}\right) \tag{iv}$$

Equating Equations (iii) and (iv),

$$\frac{mg}{2}\left(\cos\theta + \frac{\sin\theta}{\mu}\right) = mg\cos\theta\left(1 + \frac{a}{2r}\right)$$

On simplifying, we get

$$\tan\theta = \mu\left(1 + \frac{a}{r}\right)$$

$$\theta = \tan^{-1}\mu\left(1 + \frac{a}{r}\right)$$

PROBLEM 4.6

Two large rollers of rolling mill rotate in opposite direction as shown in Figure 4.6(a). Determine the maximum thickness t which the slab may have and still enter the rolls by means of the friction between the slab and the rolls. The coefficient of friction between the rolls and the slab is μ.

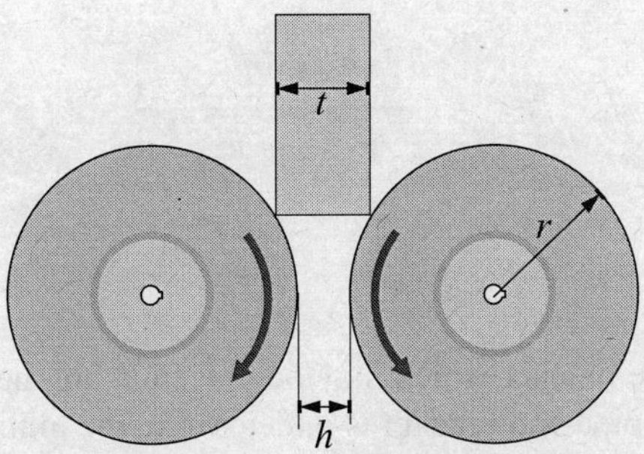

FIGURE 4.6(a) |

Solution. From the geometry [Fig. 4.6(b)],

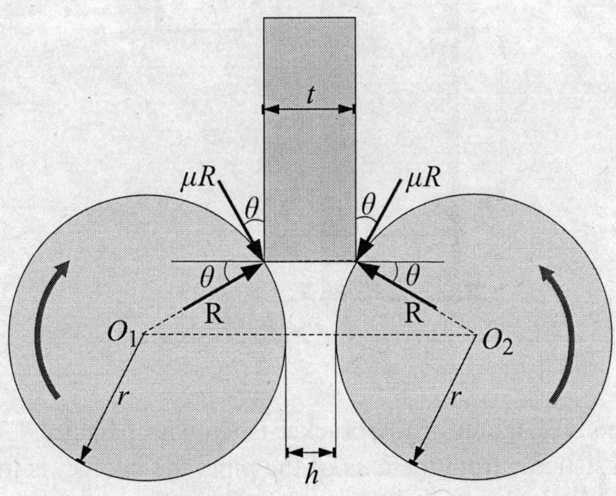

FIGURE 4.6(b) |

$$O_1O_2 = 2r + h = 2r\cos\theta + t$$

or
$$t = h + 2r(1 - \cos\theta) \qquad \text{(i)}$$

$$\sum F_y = 0 \Rightarrow 2\mu R\cos\theta - 2R\sin\theta = 0$$

or
$$\tan\theta = \mu$$

or
$$\frac{\sin\theta}{\cos\theta} = \mu \Rightarrow \frac{1-\cos^2\theta}{\cos^2\theta} = \mu$$

or
$$\cos\theta = \frac{1}{\sqrt{1+\mu^2}}$$

Substituting $\cos\theta = 1/\sqrt{1+\mu^2}$ into the Equation (i), we obtain

$$t = h + 2r\left[1 - \frac{1}{\sqrt{1+\mu^2}}\right]$$

$$t = h + \frac{2r\left(\sqrt{1+\mu^2} - 1\right)}{\sqrt{1+\mu^2}}$$

PROBLEM 4.7

Figure 4.7(a) shows a movable bracket which may be placed at any height on the pipe. If the coefficient of friction between the pipe and bracket is μ, determine the minimum distance x at which the load W can be supported. Neglect the weight of the bracket.

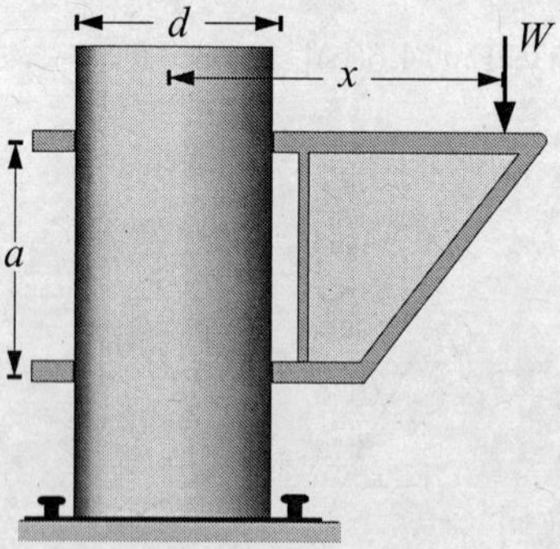

FIGURE 4.7(a) |

Solution. A free body diagram of the bracket is shown in Figure 4.7(b). When a load W is placed at the minimum distance x from the axis of the pipe, the bracket is just about to slip and the friction forces at contact point have reached their maximum values.

The equation of equilibrium gives

$$\sum F_y = 0 \Rightarrow \mu R_A + \mu R_B - W = 0,$$

or

$$R_A + R_B = \frac{W}{\mu},$$

$$\sum F_x = 0 \Rightarrow R_B - R_A = 0$$

or

$$R_B = R_A$$

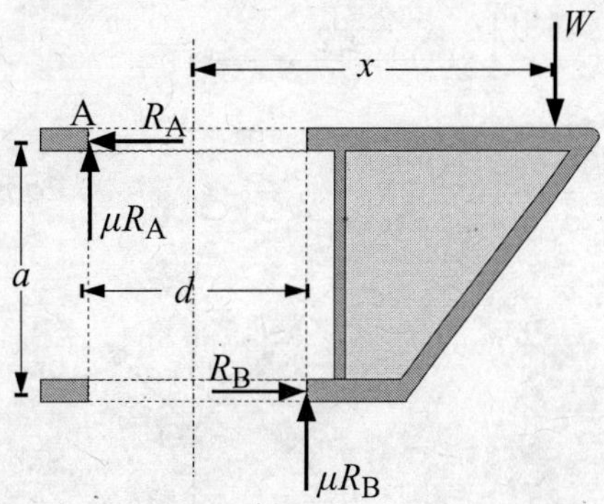

FIGURE 4.7(b)

Solving for R_B and R_A, we obtain

$$R_B = R_A = \frac{W}{2\mu}$$

$$\sum M_A = 0 \Rightarrow W\left(x + \frac{d}{2}\right) - \mu R_B \times d - R_B \times a = 0$$

Substituting the value of R_B, we get

$$W\left(x + \frac{d}{2}\right) - \mu \times \frac{W}{2\mu} \times d - \frac{W}{2\mu} \times a = 0$$

On simplifying, we get

$$x = \frac{a}{2\mu}$$

PROBLEM 4.8

Two blocks of equal weight W can slide on a horizontal surface. A string of length L is suspended between the blocks and carries a weight M at its midpoint as shown in Figure 4.8(a).

The coefficient of friction between the blocks and the surface is μ. How far will the blocks be in equilibrium.

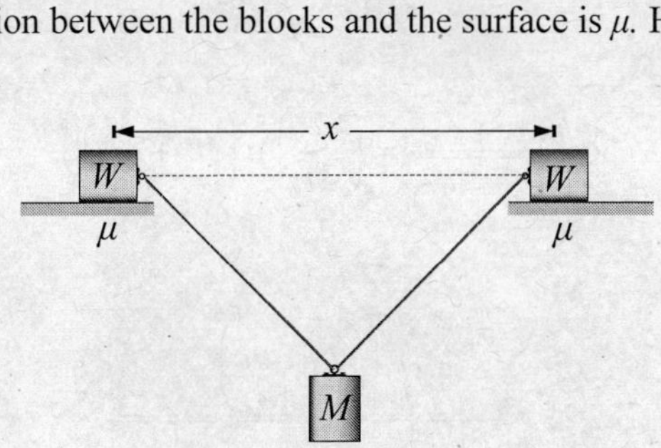

FIGURE 4.8(a)

Solution. The free body diagrams of the two blocks in their relative positions are shown in Figure 4.8(b).

Consider Block M

$$\sum F_y = 0 \Rightarrow T\cos\theta + T\cos\theta - M = 0$$

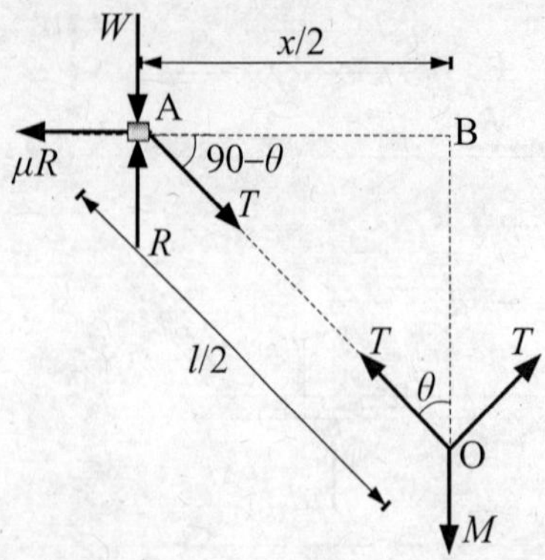

FIGURE 4.8(b) |

or $$T = \frac{M}{2\cos\theta} \quad \text{(i)}$$

Consider Block W

$$\sum F_y = 0 \Rightarrow R - W - T\sin(90 - \theta) = 0$$

or $$R = W + T\cos\theta \quad \text{(ii)}$$

Substituting the value of $T = M/2\cos\theta$ in Equation (ii) we get

$$R = W + \frac{M}{2} \quad \text{(iii)}$$

$$\sum F_x = 0 \Rightarrow T\cos(90 - \theta) - \mu R = 0$$

$$T\sin\theta = \mu R \quad \text{(iv)}$$

Substituting $T = M/2\cos\theta$ and $R = W + M/2$, we get

$$\frac{M}{2\cos\theta} \times \sin\theta = \mu\left(W + \frac{M}{2}\right)$$

or $$\tan\theta = \left(\frac{2}{M}\right)\mu\left(W + \frac{M}{2}\right)$$

From geometry, $\tan\theta = \dfrac{AB}{OB} = \dfrac{AB}{\sqrt{OA^2 - AB^2}} = \dfrac{x}{\sqrt{l^2 - x^2}}$

or $$\frac{x}{\sqrt{l^2 - x^2}} = \left(\frac{2}{M}\right)\mu\left(W + \frac{M}{2}\right)$$

On simplifying, we get

$$x = \frac{l\mu\left(W + \frac{M}{2}\right)}{\sqrt{\left(\frac{M}{2}\right)^2 + \mu^2\left(W + \frac{M}{2}\right)^2}}$$

PROBLEM 4.9

A thick semi-circular disc of radius r and weight W rests on a horizontal surface and is pulled at right angle to its geometric axis by a horizontal force P, applied at the middle of the front edge [Fig. 4.9(a)]. Find the angle θ that the flat face will make with the horizontal plane just before sliding begins, if the coefficient of friction between disc and surface is μ.

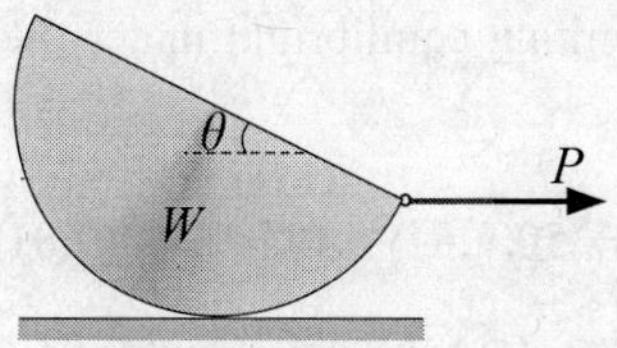

FIGURE 4.9(a)

Solution. Figure 4.9(b) shows the free body diagram of the semi-circular disc at the time of impending motion. The three equilibrium equations are

$$\sum F_x = 0 \Rightarrow P - \mu R = 0 \quad \text{(i)}$$

$$\sum F_y = 0 \Rightarrow R - W = 0 \quad \text{(ii)}$$

$$\sum M_C = 0 \Rightarrow P(r - r\sin\theta) - W\left(\frac{4r}{3\pi}\sin\theta\right) = 0 \quad \text{(iii)}$$

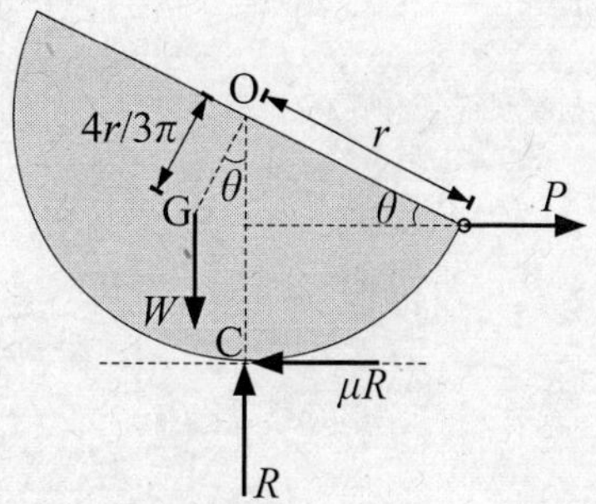

FIGURE 4.9(b)

From Equations (i) and (ii), we obtain $P = \mu W$

Substituting $P = \mu W$ in Equation (iii),

$$\mu W(r - r\sin\theta) - W\left(\frac{4r}{3\pi}\sin\theta\right) = 0$$

From which

$$\sin\theta = \frac{3\pi\mu}{3\pi\mu + 4},$$

$$\theta = \sin^{-1}\left(\frac{3\pi\mu}{3\pi\mu + 4}\right)$$

PROBLEM 4.10

A uniform 200 kg wheel rolls on its hub up the circular incline under the action of a 50 kg mass attached to the cable around the rim [Fig. 4.10(a)]. Assuming the friction is sufficient to prevent slippage, determine the angle θ at which the wheel comes to rest. Also find the minimum coefficient of friction that will permit this position to be reached with no slipping.

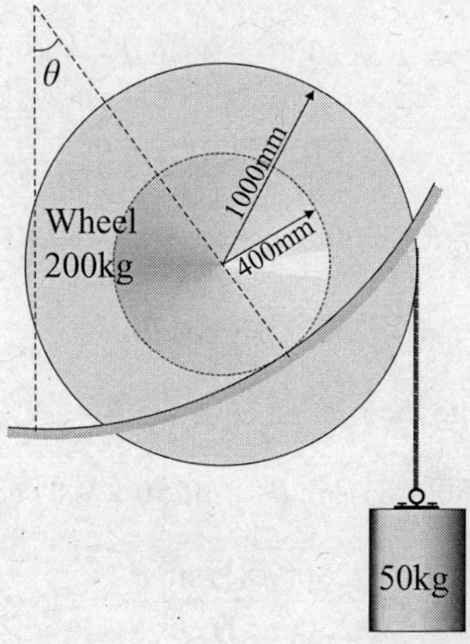

FIGURE 4.10(a)

Solution. The wheel is in limiting equilibrium under the action of forces shown in Figure 4.10(b). Taking moment about C,

$$\sum M_c = 0 \Rightarrow 50(9.81) \times BC - 200(9.81) \times CD = 0$$

or
$$50(9.81) \times (OA - CD) - 200(9.81) \times CD = 0$$

or
$$50(9.81) \times (1000 - 400 \sin \theta) - 200(9.81) \times 400 \sin \theta = 0$$

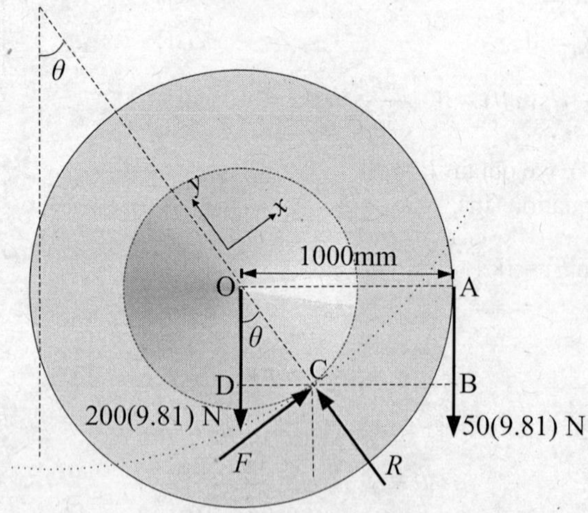

FIGURE 4.10(b) |

or
$$\sin \theta = \frac{1000}{5 \times 400} = \frac{1}{2}$$

$$\theta = 30°$$

$$\sum F_x = 0 \Rightarrow F - 200(9.81) \sin \theta - 50(9.81) \sin \theta$$

or
$$F = 250(9.81) \sin \theta$$

$$\sum F_y = 0 \Rightarrow R - 200(9.81) \cos \theta - 50(9.81) \cos \theta = 0$$

or
$$R = 250(9.81) \cos \theta$$

Substituting the value of F and R into the friction equation $F = \mu R$,

or
$$250(9.81) \sin \theta = \mu(250 \times 9.81 \cos \theta)$$

or
$$\mu = \frac{250(9.81) \sin \theta}{250(9.81) \cos \theta} = \tan \theta$$

Substituting $\theta = 30°$, we get

$$\mu = \tan 30 = \frac{1}{\sqrt{3}}$$

$$\mu = 0.577$$

PROBLEM 4.11

Three blocks, of masses 120 kg, 160 kg, and 200 kg, are arranged on top of a rough, rigid surface as shown in Figure 4.11(a). The top block is retained against horizontal motion by a surrounding cap. Find the maximum value that the horizontal force P may have before motion begins.

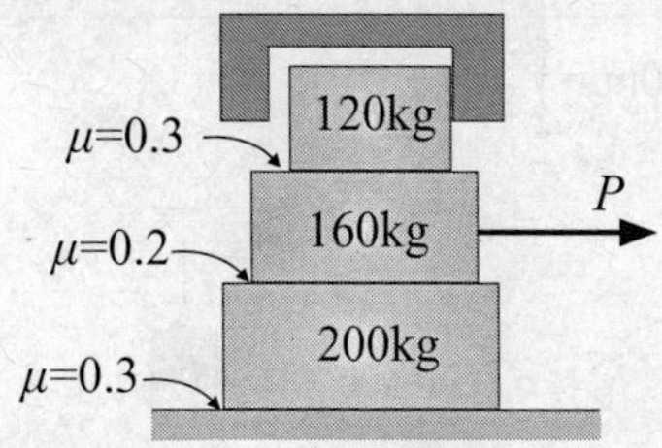

FIGURE 4.11(a)

Solution. When motion is impending, two possibilities exist. The first is that the middle block along will begin to slip to the right, and the second is that the middle and lowest blocks will begin to move together (as one body) to the right.

For the first possibility, the free body diagram of the middle block is shown in Figure 4.11(b). Then, for equilibrium,

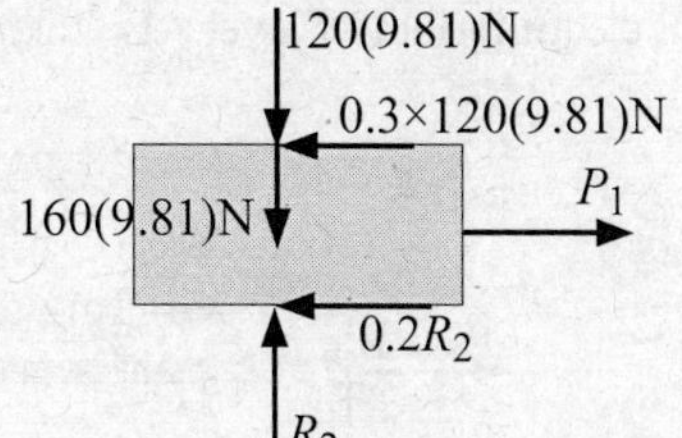

FIGURE 4.11(b)

$$\sum F_y = 0 \Rightarrow R_2 - 120(9.81) - 160(9.81) = 0$$

or $$R_2 = 2476.8\text{N}$$

$$\sum F_x = 0 \Rightarrow P_1 - 0.3 \times 120(9.81) - 0.2R_2 = 0$$

or $$P_1 = 902.5\text{N}$$

For the second possibility, the free body diagram of the middle and the lowest blocks (as one unit), again when motion is impending, is shown in Figure 4.11(c). Now, for equilibrium

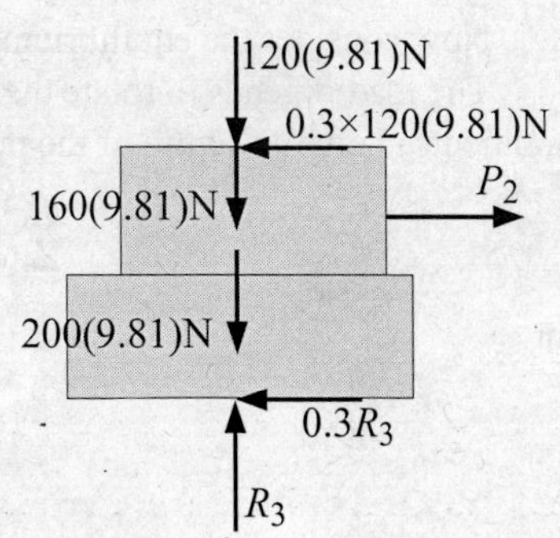

FIGURE 4.11(c)

$$\sum F_y = 0 \Rightarrow R_3 - 120(9.81) - 160(9.81) - 200(9.81) = 0$$

or $$R_3 = 4708.8\text{N}$$

$$\sum F_x = 0 \Rightarrow P_2 - 0.3 \times 120(9.81) - 0.2R_3 = 0$$

or $$P_2 = 1765.8\text{N}$$

Since $P_1 = 902.5$ N is smaller than $P_2 = 1765.8$ N, that is, the force at which motion first impends

$$P_1 = 902.5\text{N}$$

PROBLEM 4.12

What is the maximum load W that a force will hold up, if the coefficient of friction between lever and pulley is 0.2 in the arrangement shown in Figure 4.12(a). Neglect the weight of lever.

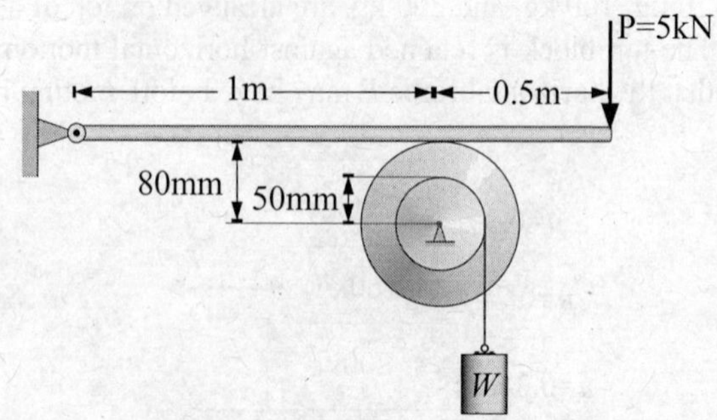

FIGURE 4.12(a) |

A free body diagram of lever and pulley is drawn as shown in Figure 4.12(b). First of all, consider the equilibrium of lever AB. Taking moment about B

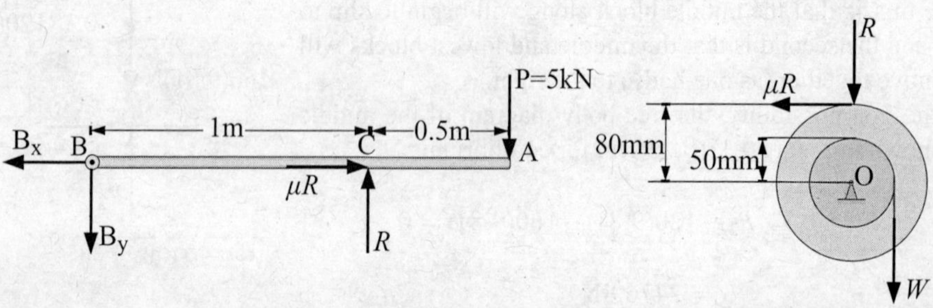

FIGURE 4.12(b) |

$$\sum M_B = 0 \Rightarrow 5 \times (1 + 0.5) - R \times 1 = 0$$

$$R = 7.5\text{kN}$$

Now consider the equilibrium of pulley.

The load W tends to rotate the pulley and frictional force acts to prevent the rotation. Now taking moments about the centre of the pulley,

$$\sum M_O = 0 \Rightarrow W \times 50 - \mu R \times 80 = 0$$

or

$$W \times 50 = 0.2 \times 7.5 \times 80$$

or

$$W = \frac{0.2 \times 7.5 \times 80}{50} = 2.4\text{kN}$$

$$W = 2.4\text{kN}$$

PROBLEM 4.13

What should be the value of the angle θ so that motion of the 60 kg block impends down the plane? The coefficient of friction μ for all surfaces is 0.25 [Fig. 4.13(a)].

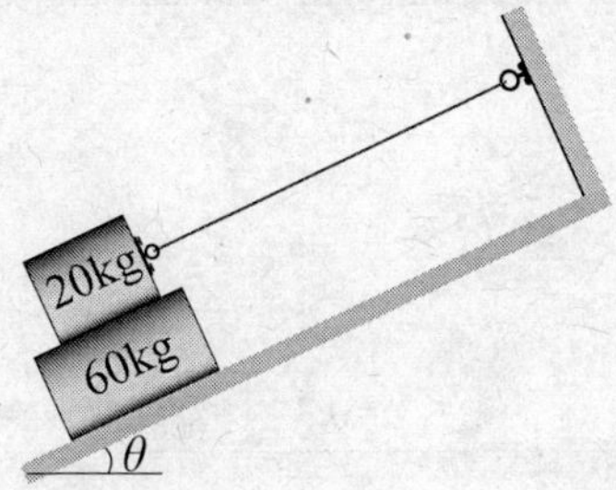

FIGURE 4.13(a) |

Solution. Consider the body of mass 20 kg is in equilibrium under the action of forces shown in Figure 4.13(b).

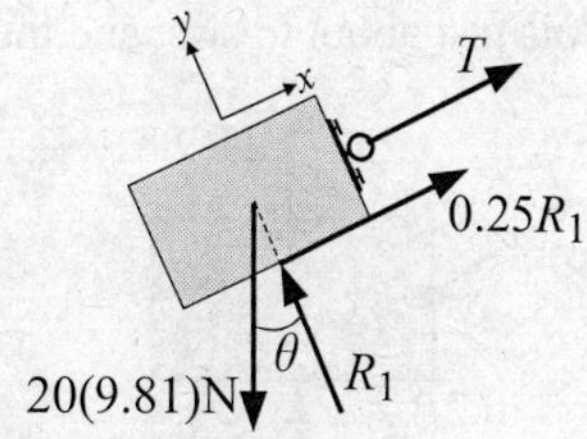

FIGURE 4.13(b) |

Resolving the forces perpendicular to the plane,

$$\sum F_y = 0 \Rightarrow R_1 - 20(9.81)\cos\theta = 0$$

or $$R_1 = 196.2\cos\theta \quad \text{(i)}$$

Now consider the body of mass 60 kg under the action of forces shown in Figure 4.13(c). Resolving the forces perpendicular to the plane,

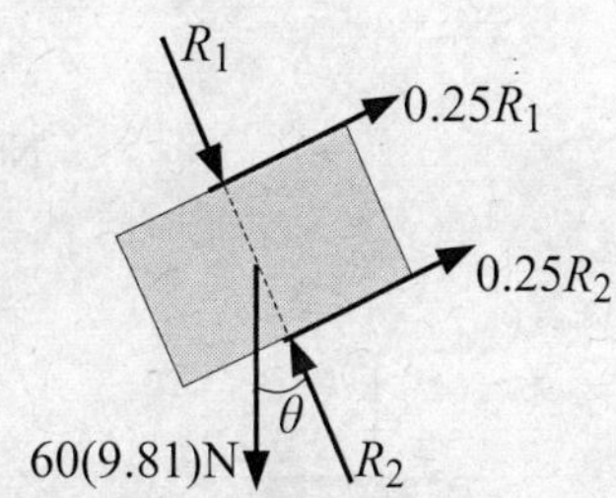

FIGURE 4.13(c) |

$$\sum F_y = 0 \Rightarrow R_2 - R_1 - 60(9.81)\cos\theta = 0$$

or $$R_2 = R_1 + 60(9.81)\cos\theta \quad \text{(ii)}$$

Substituting the value of R_1 in Equation (ii), we get

$$R_2 = 196.2\cos\theta + 60(9.81)\cos\theta$$

or $$R_2 = 784.8\cos\theta \quad \text{(iii)}$$

Resolving the forces parallel to the plane,

$$\sum F_x = 0 \Rightarrow 0.25R_1 + 0.25R_2 - 60(9.81)\sin\theta = 0$$

Substituting the value of R_1 and R_2, we get,

$$0.25(196.2\cos\theta + 784.8\cos\theta) = 588.6\sin\theta$$

or $$245.25\cos\theta = 588.6\sin\theta \Rightarrow \tan\theta = 0.416$$

$$\theta = 22.6^\circ$$

PROBLEM 4.14

Block A of mass 100 kg rests on a horizontal surface and supports another block B of mass 25 kg on top of it. Block B is attached to a vertical wall by an inclined string as shown in Figure 4.14(a). Determine the force P applied to the lower block that will be necessary to cause the slipping to impend.

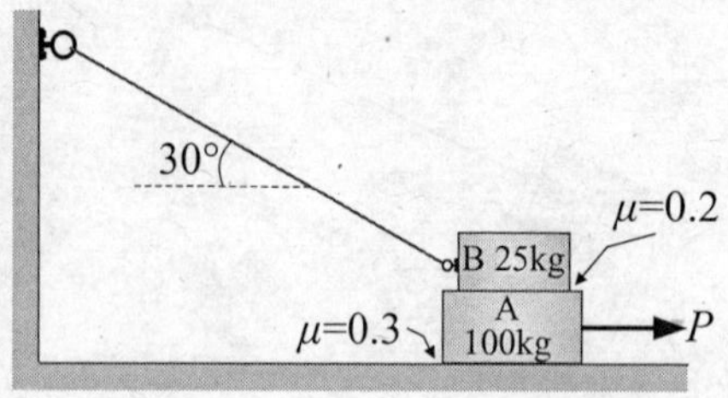

FIGURE 4.14(a) |

Solution. We draw the free body diagrams of blocks A and B. When force P is increased, block A is just about to slip, and the forces of friction at A and B have reached their maximum value.

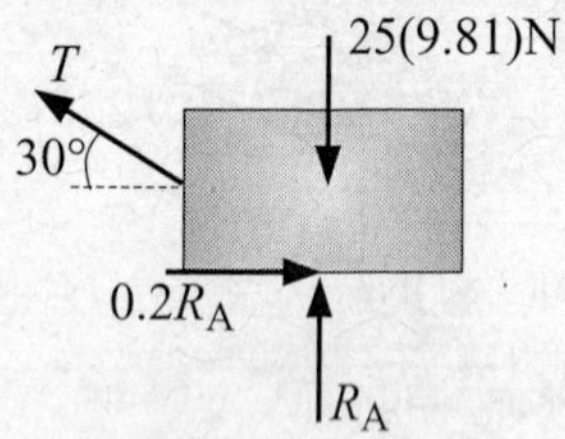

FIGURE 4.14(b) |

Consider block B [Fig. 4.14(b)]. We have

$$\sum F_x = 0 \Rightarrow 0.2R_A - T\cos 30 = 0 \quad \text{(i)}$$

$$\sum F_y = 0 \Rightarrow R_A + T\sin 30 - 25(9.81) = 0 \quad \text{(ii)}$$

Solving for R_A from Equations (i) and (ii), we obtain

$$R_A = 219.86\text{N}$$

Now consider block A [Fig. 4.14(c)]. We have

$$\sum F_y = 0 \Rightarrow R_B - R_A - 100(9.81) = 0$$

$$R_B = R_A + 100(9.81)$$

$$R_B = 219.86 + 981 = 1200.86\text{N}$$

$$\sum F_x = 0 \Rightarrow P - 0.2R_A - 0.3R_B = 0$$

$$P = 0.2 \times 219.86 + 0.3 \times 1200.86$$

$$P = 404.23\text{N}$$

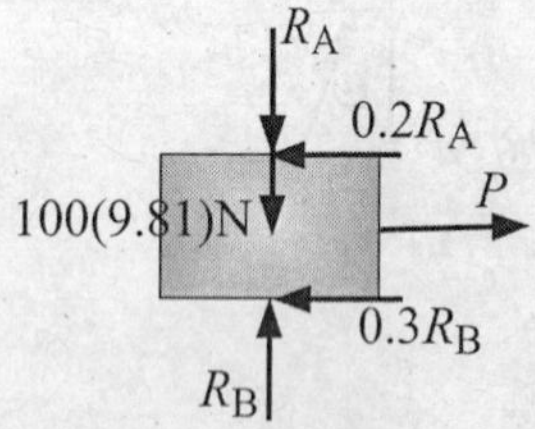

FIGURE 4.14(c) |

PROBLEM 4.15

Two blocks A and B are connected by a cable as shown in Figure 4.15(a). Determine the magnitude P of the smallest force required to move the blocks. The coefficient of friction at all surfaces of contact is 0.25.

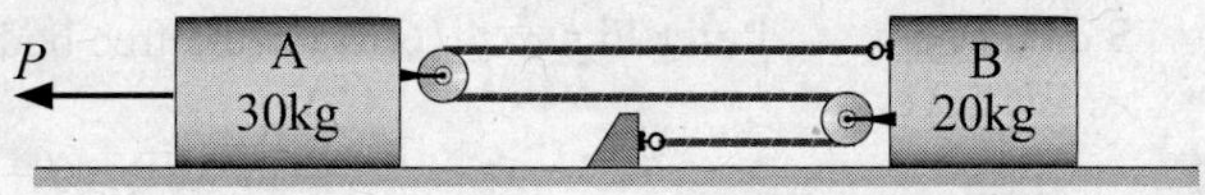

FIGURE 4.15(a)

Solution. Considering block B [Fig. 4.15(b)],

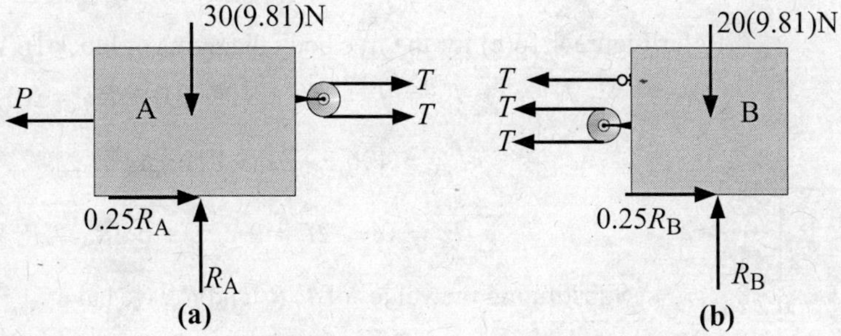

FIGURE 4.15(b)

$$\sum F_y = 0 \Rightarrow R_B - 20(9.81) = 0$$

or
$$R_B = 196.2\text{N}$$

$$\sum F_x = 0 \Rightarrow 0.25R_B - 3T = 0$$

or
$$T = \frac{0.25 \times 196.2}{3} = 16.35\text{N}$$

Considering block A [Fig. 4.15(b)],

$$\sum F_y = 0 \Rightarrow R_A - 30(9.81) = 0$$

or
$$R_A = 294.3\text{N}$$

$$\sum F_x = 0 \Rightarrow T + T + 0.25R_A - P = 0$$

or
$$P = 2 \times 16.35 + 0.25 \times 294.3 = 106.2\text{N}$$

$$P = 106.2\text{N}$$

PROBLEM 4.16

Determine the largest load W that can be suspended without moving blocks A and B. Coefficient of friction at all surfaces is 0.2 [Fig. 4.16(a)].

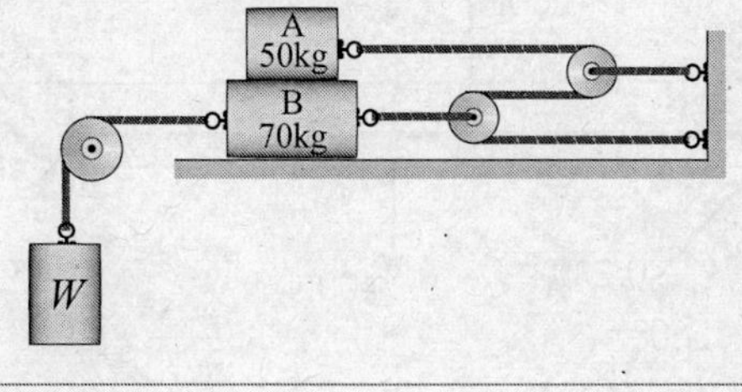

FIGURE 4.16(a)

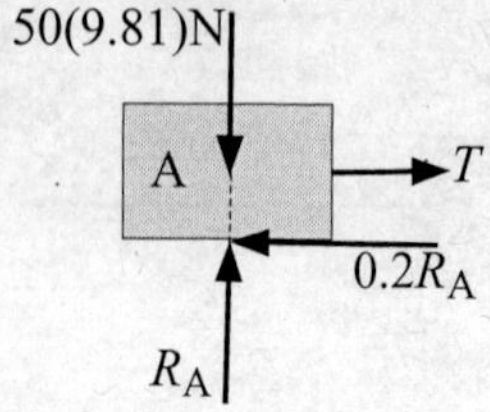

FIGURE 4.16(b) |

Solution. Refer Figure 4.16(b) for the free body diagrams of block A. We have.

$$\sum F_y = 0 \Rightarrow R_A - 50(9.81) = 0$$

$$R_A = 294.3\text{N} \qquad \text{(i)}$$

$$\sum F_y = 0 \Rightarrow T - 0.2R_A = 0$$

$$T = 0.2 \times 490.5 = 98.1\text{N}$$

Refer Figure 4.16(c) for the free body diagrams of block B. We have

$$\sum F_y = 0 \Rightarrow R_A + 70(9.81) - R_B = 0$$

$$R_B = 490.5 + 70(9.81) = 1177.2\text{N} \qquad \text{(ii)}$$

$$\sum F_x = 0 \Rightarrow 2T + 0.2R_B + 0.2R_A - W = 0$$

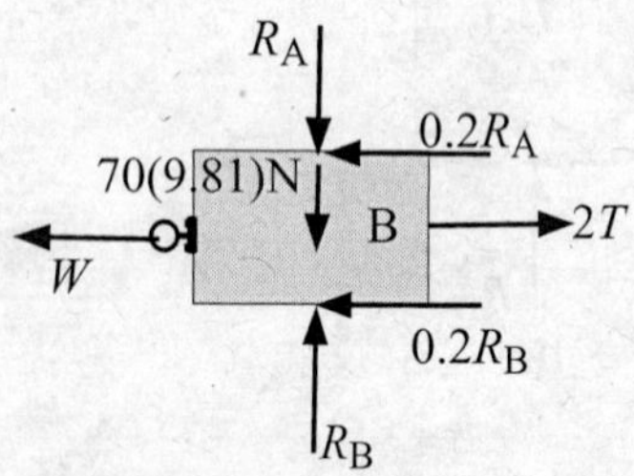

FIGURE 4.16(c) |

Substituting the values of T, R_A and R_B, we have

$$W = 2 \times 98.1 + 0.2 \times 1177.2 + 0.2 \times 490.5$$

$$W = 529.74\text{N}$$

$$W = 529.74\text{N (or 54kg)}$$

PROBLEM 4.17

A horizontal bar AB of length 4 m and weight 400 N is lying on inclined planes as shown in Figure 4.17(a). Find how close to the ends A and B a load of 500 N can be placed safely, if coefficient of friction between the bar and supports is 0.2.

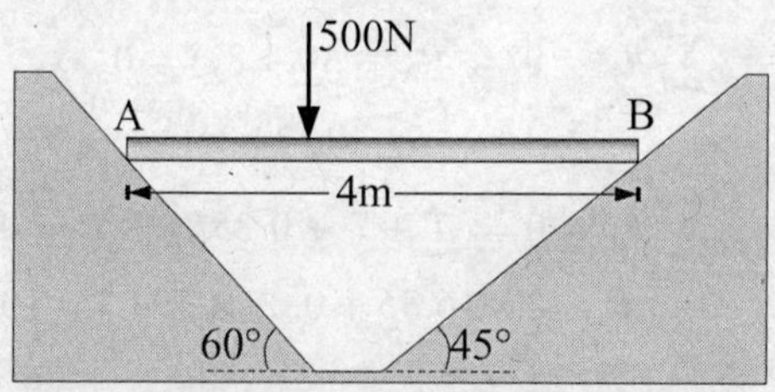

FIGURE 4.17(a) |

Solution. When the load is close to A, the end A will slip down. For this position, the free body diagram is shown in Figure 4.17(b). The equilibrium equations are

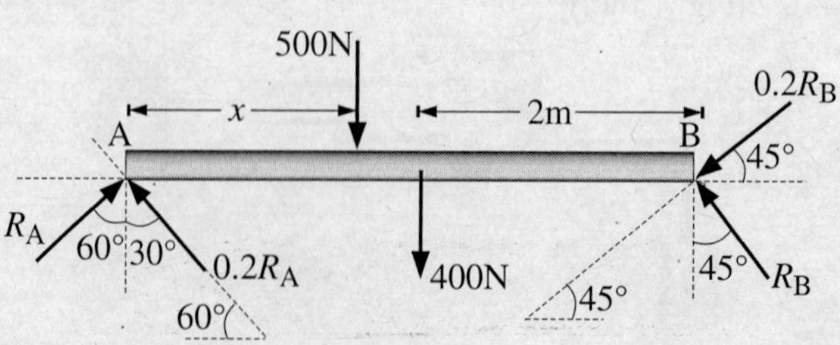

FIGURE 4.17(b) |

$$\sum F_y = 0 \Rightarrow R_A \cos 60 + 0.2R_A \cos 30 + R_B \cos 45 - 0.2R_B \sin 45 - 500 - 400 = 0$$

or $$0.67R_A + 0.56R_B = 900$$

$$\sum F_x = 0 \Rightarrow R_A \sin 60 - 0.2R_A \sin 30 - R_B \sin 45 - 0.2R_B \cos 45 = 0$$

or $$R_A = 1.107R_B \qquad \text{(ii)}$$

Solving for R_A and R_B, we have

$$R_B = 691.14\text{N} \qquad R_A = 765.09\text{N}$$

Taking moment about A,

$$\sum M_A = 0 \Rightarrow 500 \times x + 400 \times 2 - R_B \cos 45 \times 4 - 0.2R_B \sin 45 \times 4 = 0$$

Substituting the value of R_B = 691.14 N, we obtain

$$x = 3.09\text{m}$$

When the load is close to end B, end B may slide down and end A may slide up. For this position, the free body diagram is shown in Figure 4.17(c).

$$\sum F_y = 0 \Rightarrow R_A \sin 30 - 0.2R_A \sin 60 + 0.2R_B \cos 45 + R_B \cos 45 - 500 - 400 = 0$$

or $$0.326R_A + 0.84R_B = 900 \qquad \text{(i)}$$

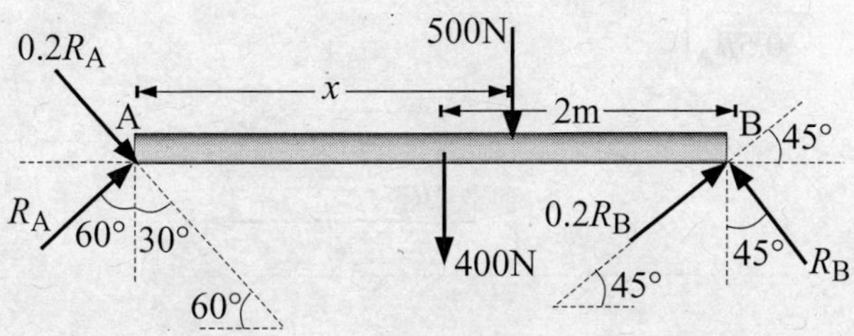

FIGURE 4.17(c) |

$$\sum F_x = 0 \Rightarrow R_A \cos 30 + 0.2R_A \cos 60 + 0.2R_B \sin 45 - R_B \sin 45 = 0$$

or $$R_B = 1.707R_A \qquad \text{(ii)}$$

Solving for R_A and R_B, we have

$$R_A = 511.23\text{N} \qquad R_B = 872.68\text{N}$$

$$\sum M_A = 0 \Rightarrow R_B \cos 45 \times 4 + 0.2R_B \cos 45 \times 4 - 400 \times 2 - 500 \times x = 0$$

Substituting R_B = 872.68 N, we get

$$x = 4.32\text{m}$$

It means that the motion will be impending when the load is at 4.32 m from end A, which is not a possible case of loading. Thus, 500 N load can be placed anywhere between 3.09 m and 4.32 m from end A.

PROBLEM 4.18

Determine the force F required to rotate a 100 kg reel of cable that rests on its hubs and bears against a vertical wall [Fig. 4.18(a)]. The coefficient of friction for each pair of contacting surfaces is 0.5.

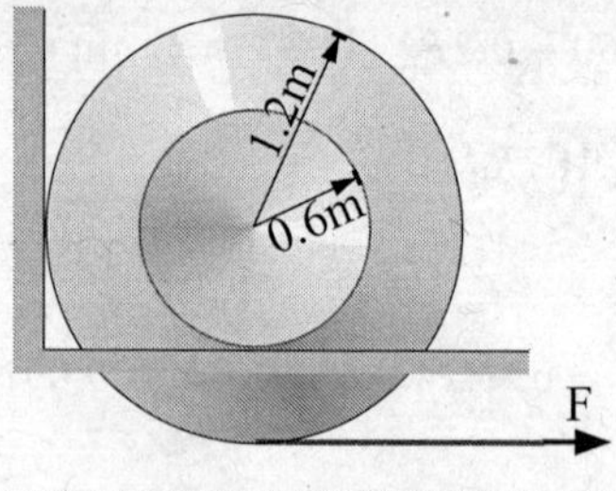

FIGURE 4.18(a) |

Solution. We draw the free body diagram [Fig. 4.18(b)] of the reel when it is just about to slip, and the forces of friction at A and B have reached their maximum values. The equilibrium equations are

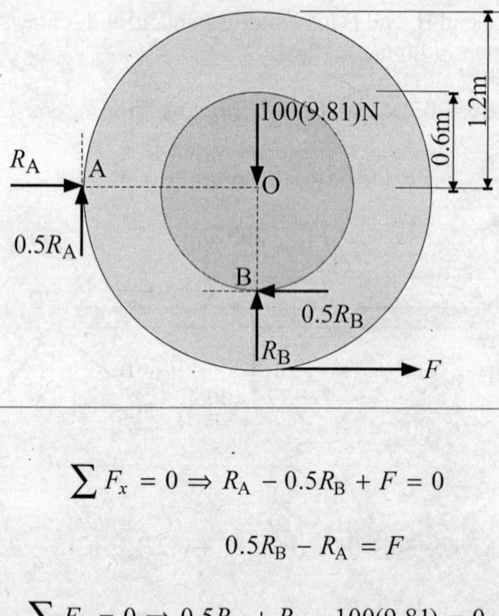

FIGURE 4.18(b) |

$$\sum F_x = 0 \Rightarrow R_A - 0.5R_B + F = 0$$

or

$$0.5R_B - R_A = F \qquad \text{(i)}$$

$$\sum F_y = 0 \Rightarrow 0.5R_A + R_B - 100(9.81) = 0$$

or

$$0.5R_A + R_B = 981 \qquad \text{(ii)}$$

$$\sum M_O = 0 \Rightarrow 0.5R_A \times 1.2 + 0.5R_B \times 0.6 - F \times 1.2 = 0$$

or

$$F = 0.5R_A + 0.25R_B \qquad \text{(iii)}$$

Solving for R_A and R_B in terms of F from Equations (i) and (ii), we obtain

$$R_A = 392.4 - \frac{4}{5}F \text{ and } R_B = 784.8 + \frac{F}{2.5}$$

Substituting R_A and R_B in Equation (iii), we have

$$F = 0.5\left(392.4 - \frac{4}{5}F\right) + 0.25\left(784.8 + \frac{F}{2.5}\right)$$

From which, we obtain

$$F = 301.8\text{N}$$

PROBLEM 4.19

A smooth circular cylinder A rests on top of two half-cylinders B and C, both having the same radius r. The weight of cylinder A is W and that of B and C is $W/2$ each [Fig. 4.19(a)]. Assume that the coefficient of friction between the flat surface of the half-cylinders and the horizontal table top is μ. Determine the maximum distance d between the centres of the half-cylinders to maintain equilibrium.

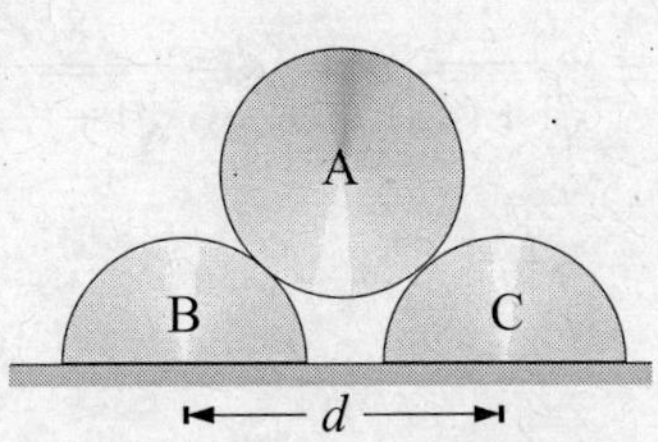

FIGURE 4.19(a) |

Solution. Part (a) of Figure 4.19(b) shows the free body diagram of combined cylinders. We have

$$\sum F_y = 0 \Rightarrow R + R - W - \frac{W}{2} - \frac{W}{2} = 0$$

$$R = W \tag{i}$$

Part (b) of Figure 4.19(b) shows the free body diagram of cylinder A. We have

$$\sum F_y = 0 \Rightarrow 2N\cos(90 - \theta) - W = 0$$

$$N = \frac{W}{2\sin\theta} \tag{ii}$$

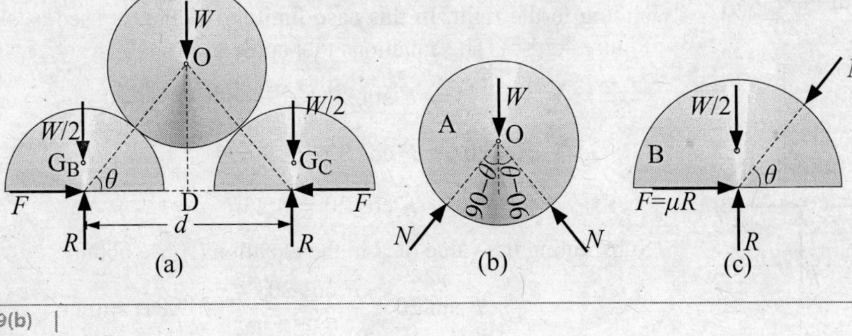

FIGURE 4.19(b) |

Part (c) of Figure 4.19(b) shows the free body diagram of semi-cylinder B. We have

$$\sum F_x = 0 \Rightarrow N \cos \theta - \mu R = \theta \qquad \text{(iii)}$$

Substituting the value of N and R from Equations (i) and (ii), we have

$$\left(\frac{W}{2 \sin \theta}\right) \cos \theta - \mu W = 0$$

or

$$\tan \theta = \frac{1}{2\mu} \Rightarrow \cos \theta = \frac{2\mu}{\sqrt{1 + (2\mu)^2}}$$

From part (a) of Figure 4.19(b),

$$\cos \theta = \frac{\frac{d}{2}}{r + r} = \frac{d}{4r}$$

$$\frac{d}{4r} = \frac{2\mu}{\sqrt{1 + (2\mu)^2}} \Rightarrow d = \frac{8\mu r}{\sqrt{1 + (2\mu)^2}}$$

$$d = \frac{8\mu r}{\sqrt{1 + 4\mu^2}}$$

PROBLEM 4.20

Determine the value of F that will cause a 100 kg block in Figure 4.20(a) to move. The coefficient of static friction between the block and surface is 0.25.

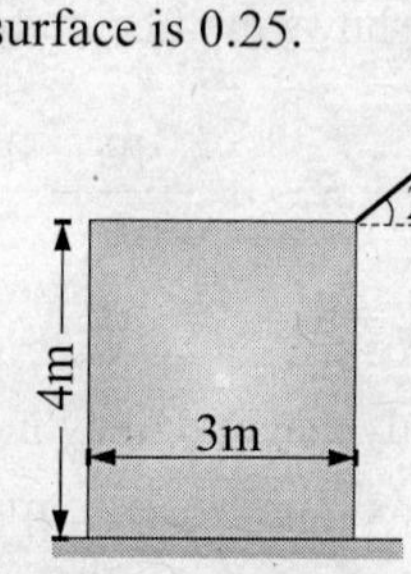

FIGURE 4.20(a) |

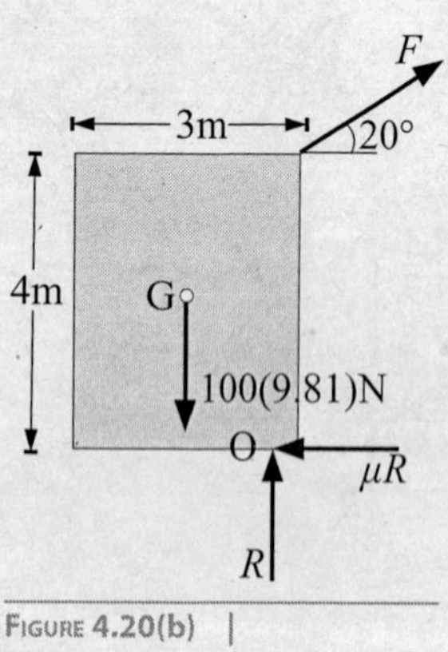

FIGURE 4.20(b) |

Solution. The block may slide to the right or it may tip about the forward edge O. First, determine the value of F needed to cause sliding to the right. In this case limiting friction is used as shown in Figure 4.20(b). The equations that apply are

$$\sum F_y = 0 \Rightarrow F \sin 20 + R - 100(9.81) = 0 \qquad \text{(i)}$$

$$\sum F_x = 0 \Rightarrow F \cos 20 - \mu R = 0$$

$$F \cos 20 = 0.25R \qquad \text{(ii)}$$

Substituting the value of R in the equation (i), we obtain

$$F \sin 20 + \frac{F \cos 20}{0.25} - 100(9.81) = 0$$

$$F = 239.2\text{N}$$

Refer to Figure 4.20(c). Next assume that the block will tip about the forward edge at O. The equation of equilibrium is

$$\sum M_O = 0 \Rightarrow F \cos 20 \times 4 - 100(9.81) \times \frac{3}{2} = 0$$

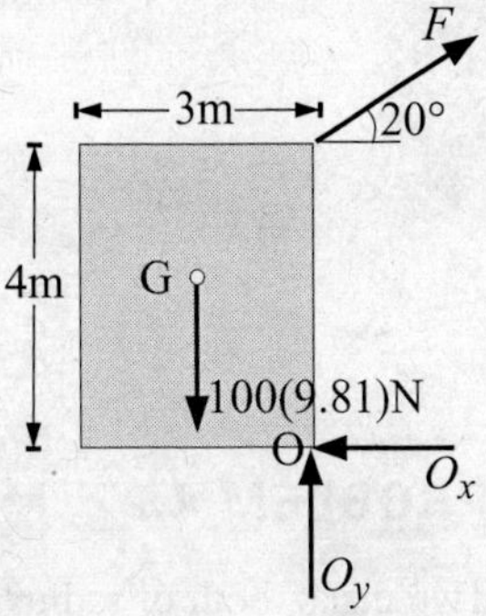

FIGURE 4.20(c) |

or

$$F = 391.4\text{N}$$

$$\sum F_x = 0 \Rightarrow F \cos 20 - O_x = 0$$

or

$$O_x = 391.4 \cos 20 = 367.8\text{N}$$

To cause sliding, F must equal 239.2 N, whereas to cause tipping F must go up to 391.4 N. It is thus seen that sliding will be the first to occur as F increases steadily from zero to a maximum.

$$F = 239.2\text{N}$$

PROBLEM 4.21

A homogeneous cylinder of diameter 500 mm and mass 50 kg rests against the vertical and inclined surfaces as shown in Figure 4.21(a). If the coefficient of friction between cylinder and the surfaces is 0.2, calculate the applied clockwise couple M which would cause the cylinder to slip.

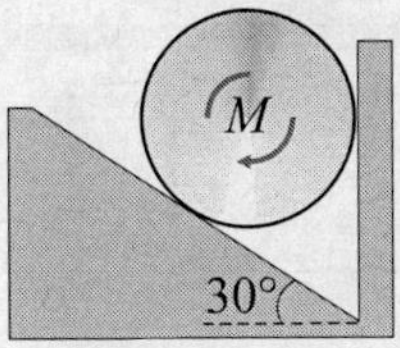

FIGURE 4.21(a) |

Solution. Figure 4.21(b) shows the free body diagram of the cylinder when it is just about to slip, and forces of friction at A and B have reached their maximum values. The equilibrium equations are

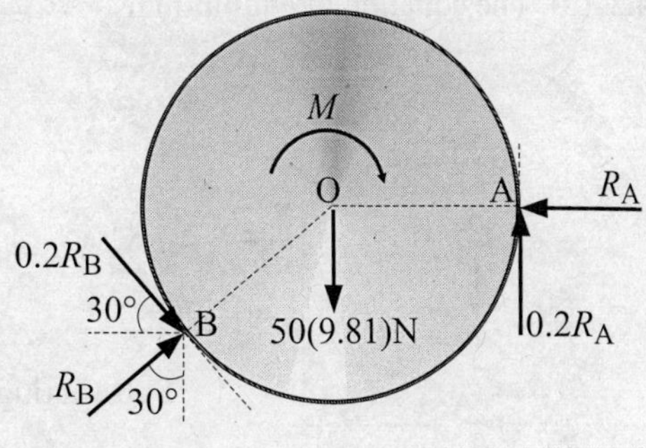

FIGURE 4.21(b) |

$$\sum F_x = 0 \Rightarrow R_B \sin 30 + 0.2R_B \cos 30 - R_A = 0$$

$$R_A = 0.67R_B$$

$$\sum F_y = 0 \Rightarrow 0.2R_A + R_B \cos 30 - 0.2R_B \sin 30R_A$$

$$- 50(9.81) = 0$$

$$R_A + 3.83R_B = 2452.5$$

On simplifying

$$R_A = 365.15\text{N}, R_B = 545\text{N}$$

$$\sum M_o = 0 \Rightarrow M - 0.2R_A \times \frac{250}{1000} - 0.2R_B \times \frac{250}{1000} = 0$$

or
$$M = 0.2(R_A + R_B) \times \frac{250}{1000}$$

$$M = 0.2(365.15 + 545) \times \frac{250}{1000} = 45.5$$

$$M = 45.5\text{Nm}$$

PROBLEM 4.22

Two disks, both of radius 30 cm, are joined by a short axle of radius 15 cm, as shown side view in Figure 4.22(a). The weight of the assembly is 500 N. The coefficient of friction between the disks and the horizontal plane is 0.20, and the vertical wall is smooth. A cord is wound the axle, and a weight *W* is suspended from the cord. Determine the value of *W* for impending motion.

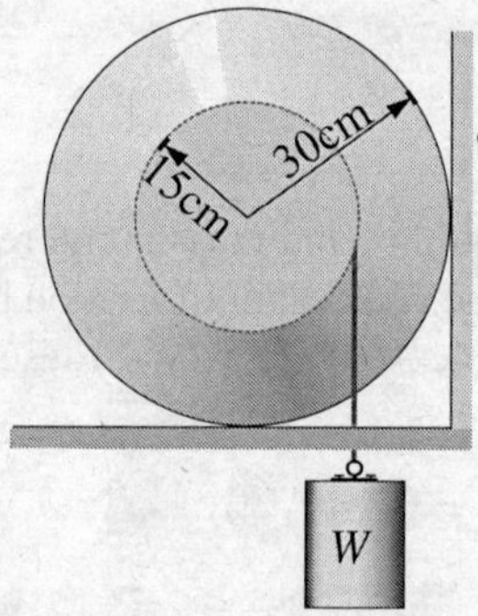

FIGURE 4.22(a) |

Solution. The system of two disks is in equilibrium under the action of forces shown in Figure 4.22(b). The equation of equilibrium

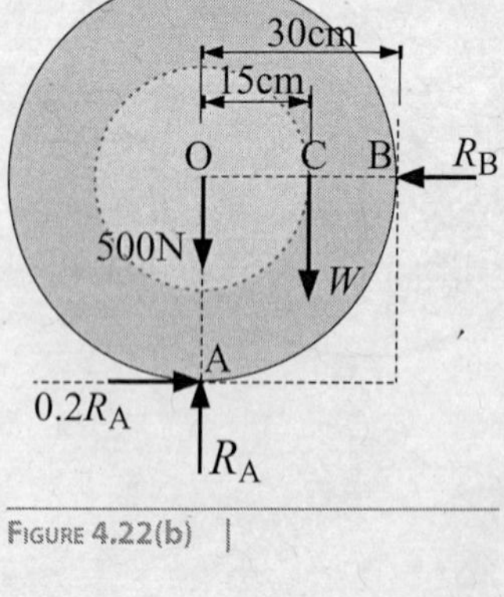

FIGURE 4.22(b) |

$$\sum F_y = 0 \Rightarrow R_A - 500 - W = 0$$

$$R_A - 500 + W \qquad \text{(i)}$$

$$\sum M_o = 0 \Rightarrow W \times 15 - 0.2R_A \times 30 = 0$$

$$W = 0.4R_A \qquad \text{(ii)}$$

Substituting $R_A = 500 + W$, we obtain

$$W = 0.4(500 + W)$$

From which

$$W = 333.3\text{N}$$

PROBLEM 4.23

A tong is designed to handle hot cylinders when they are heat treated in a furnace [Fig. 4.23(a)]. For a 30° jaw opening, what is the minimum coefficient of friction between the jaws and the cylinder which will enable the tong to grip the cylinder without slipping.

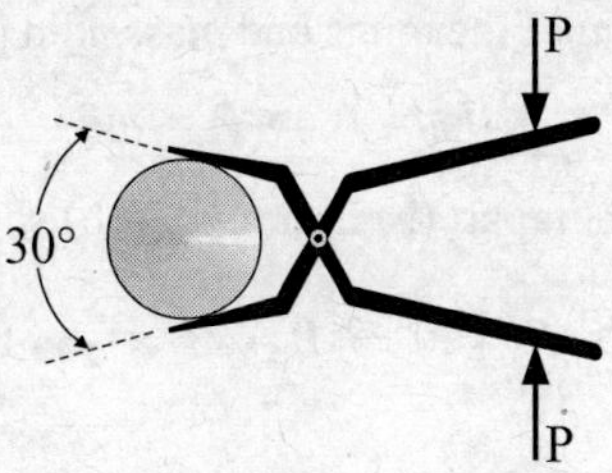

FIGURE 4.23(a) |

Solution. The free body diagram of cylinder is shown in Figure 4.23(b).

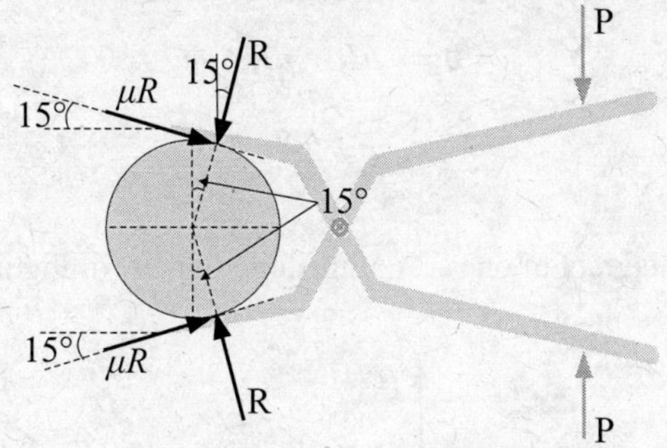

FIGURE 4.23(b)

$$\sum F_x = 0 \Rightarrow \mu R \cos 15 + \mu R \cos 15 - R \sin 15 - R \sin 15 = 0$$

or
$$2\mu R \cos 15 - 2R \sin 15 = 0$$

or
$$\mu = \tan 15$$

$$\mu = 0.267$$

PROBLEM 4.24

What must be the coefficient of friction between the gripping surfaces of a tong and a block of mass m to prevent slipping [Fig. 4.24(a)].

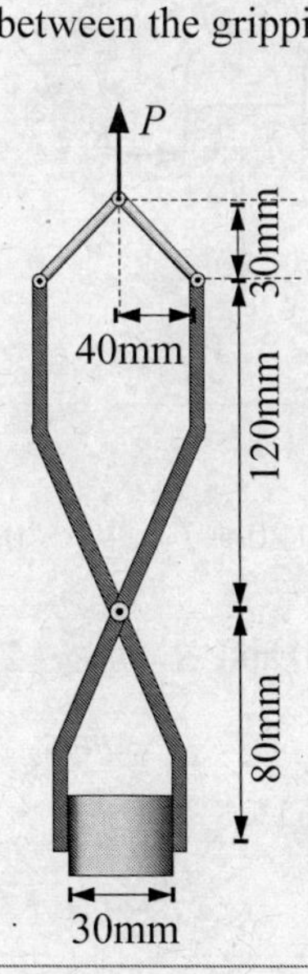

FIGURE 4.24(a) |

Solution. A free body diagram of the tongs and mass m in part (a) of Figure 4.24(b) shows that

$$\sum F_y = 0 \Rightarrow P = mg$$

The free body diagram of the top pin in part (b) Figure 4.24(b) shows that

$$\sum F_y = 0 \Rightarrow P - 2T \times \frac{3}{5} = 0$$

or

$$T = \frac{5}{6}P = \frac{5}{6}mg \qquad \text{(i)}$$

The free body diagram of the mass m in part (c) of Figure 4.24(b) shows that

$$\sum F_y = 0 \Rightarrow \mu R + \mu R - mg = 0$$

$$\mu R = \frac{mg}{2} \qquad \text{(ii)}$$

Now draw a free body diagram of one arm of the tong in part (d) Figure 4.24(b). Taking moment about O

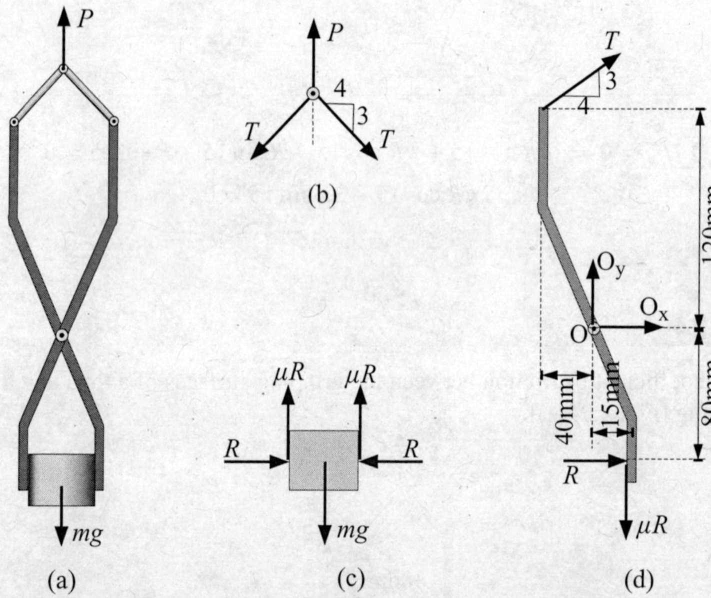

FIGURE 4.24(b) |

$$\sum M_o = 0 \Rightarrow T \times \frac{4}{5} \times 120 + T \times \frac{3}{5} \times 40 - R \times 80 + \mu R \times 15 = 0 \qquad \text{(iii)}$$

Substitute $T = (5/6)mg$ from Equation (i) and $R = mg/2\mu$ from Equation (ii):

$$\frac{5}{6}mg \times \frac{4}{5} \times 120 + \frac{5}{6}mg \times \frac{3}{5} \times 40 - \frac{mg}{2\mu} \times 80 + \mu \times \frac{mg}{2\mu} \times 15 = 0$$

On simplifying, we get

$$\mu = 0.37$$

PROBLEM 4.25

Determine the maximum angle θ with the horizontal at which the incline can be tilted before the semi-cylinder slips [Fig. 4.25(a)]. For this condition, what is the corresponding angle α?

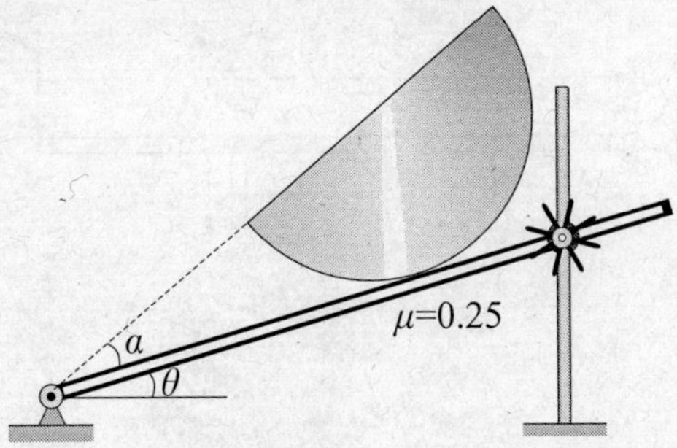

FIGURE 4.25(a)

Solution. The free body diagram of semi-cylinder is shown in Figure 4.25(b). We have

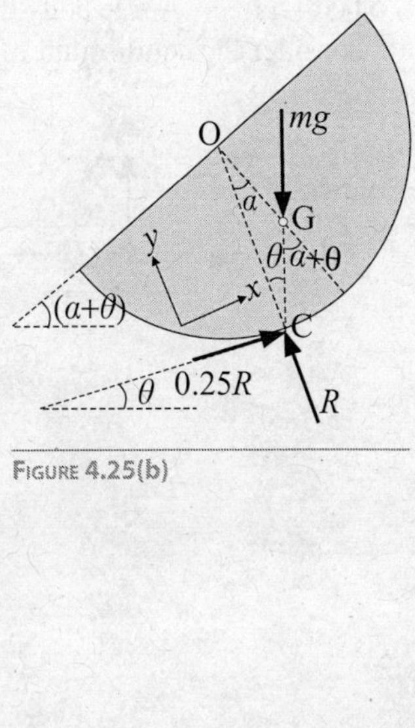

FIGURE 4.25(b)

$$\sum F_y = 0 \Rightarrow mg\cos\theta = R$$

$$\sum F_x = 0 \Rightarrow mg\sin\theta = 0.25R$$

Solving the above equations, we obtain

$$\tan\theta = 0.25$$

From which

$$\theta = 14.03°$$

In ΔOGC,

$$\frac{OG}{\sin\theta} = \frac{OC}{\sin[180 - (\alpha+\theta)]} = \frac{GC}{\sin\alpha}$$

or

$$\frac{\frac{4r}{3\pi}}{\sin\theta} = \frac{r}{\sin(\alpha+\theta)} \qquad \left[OG = \frac{4r}{3\pi}\right]$$

or

$$\sin(\alpha+\theta) = \frac{3\pi}{4}\sin\theta$$

or

$$\alpha = \sin^{-1}\left[\frac{3\pi}{4}\sin\theta\right] - \theta$$

Substituting $\theta = 14.03°$ in the above equation, we have

$$\alpha = \sin^{-1}\left[\frac{3\pi}{4}\sin 14.03°\right] - 14.03° = 20.8°$$

$$\alpha = 20.8°$$

PROBLEM 4.26

A single-lever block brake is shown in Figure 4.26(a). Find the force P required to prevent rotation if the coefficient of friction is μ.

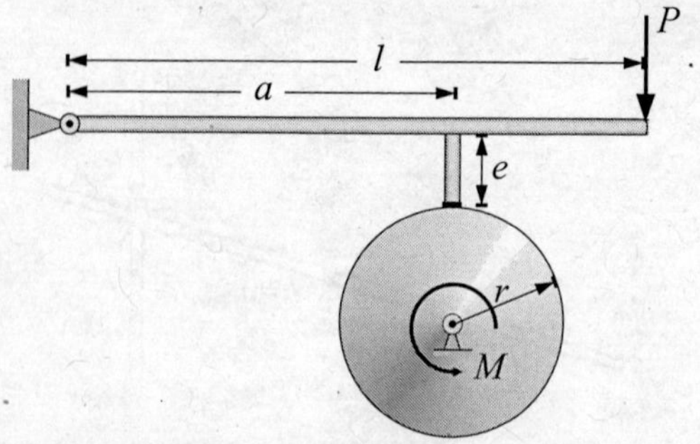

FIGURE 4.26(a)

Solution. A free body diagram of lever and brake is drawn as shown in Figure 4.26(b). First of all, consider the equilibrium of lever AB. Taking the moment about A,

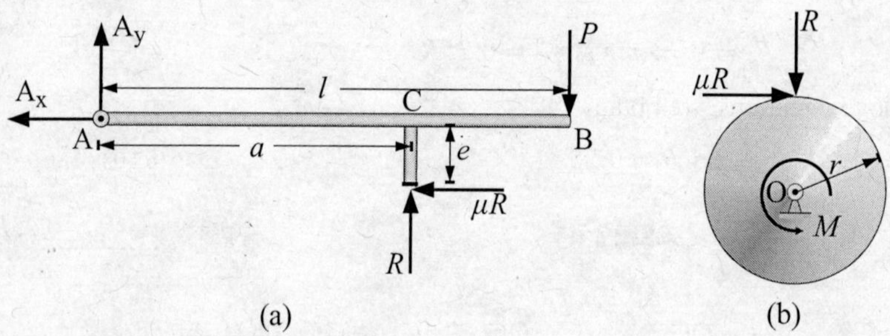

FIGURE 4.26(b)

$$\sum M_A = 0 \Rightarrow P \times l + \mu R \times e - R \times a = 0$$

$$R = \frac{Pl}{a - \mu e} \tag{i}$$

$$\sum M_o = 0 \Rightarrow \mu R \times r - M = 0$$

$$M = \mu r R \tag{ii}$$

Substituting the value of R in Equation (ii), we obtain

$$M = \mu r \frac{Pl}{a - \mu e}$$

$$\boxed{M = \frac{Plr\mu}{a - \mu e}}$$

PROBLEM 4.27

Figure 4.27(a) shows a solid shaft of diameter d that rotates at a constant angular velocity ω in a bearing of slightly larger diameter D. The coefficient of friction between the shaft and the bearing surface is μ. Determine the location of the centre of the shaft with respect to the centre O of the bearing.

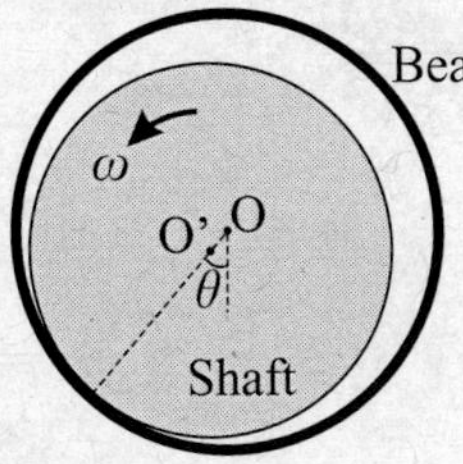

FIGURE 4.27(a) |

Solution. In the drawing the free body diagram of the rotating shaft [Fig. 4.27(b)], the normal force R at the contact of the shaft and bearing acts along a line that passes through the centres of both.

Equilibrium of the shaft is represented by the closed force triangle in Figure 4.27(b). Since OO " = $(D - d)/2$, from the force triangle,

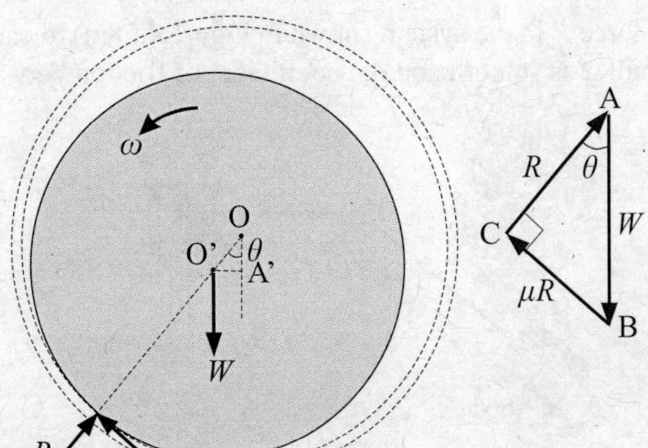

FIGURE 4.27(b) |

$$AB^2 = BC^2 + AC^2$$

or

$$W^2 = (\mu R)^2 + R^2$$

or

$$W = R\sqrt{1 + \mu^2}$$

and

$$\cos\theta = \frac{R}{W} = \frac{R}{R\sqrt{1+\mu^2}} = \frac{1}{\sqrt{1+\mu^2}}$$

$$\sin\theta = \frac{\mu R}{W} = \frac{\mu R}{R\sqrt{1+\mu^2}} = \frac{\mu}{\sqrt{1+\mu^2}}$$

The horizontal displacement of the centre of the shaft from the centre of the bearing is

$$O'A = OO' \sin\theta = \frac{D-d}{2} \times \frac{\mu}{\sqrt{1+\mu^2}}$$

$$O'A = \frac{(D-d)\mu}{2\sqrt{1+\mu^2}}$$

and its vertical displacement is

$$OA = OO' \cos\theta = \frac{D-d}{2} \times \frac{1}{\sqrt{1+\mu^2}}$$

$$OA = \frac{D-d}{2\sqrt{1+\mu^2}}$$

PROBLEM 4.28

What is the value of the force P in the system shown in Figure 4.28(a) to cause the motion to impend rightward. Assume the pulley is smooth and the coefficient of friction between the other contact surfaces is 0.20.

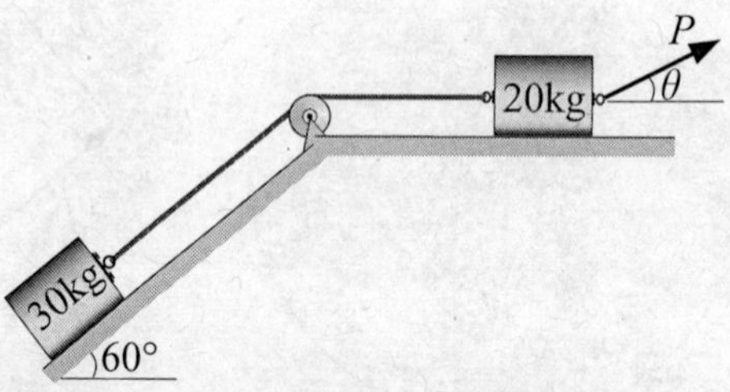

FIGURE 4.28(a) |

Solution. The equation of equilibrium for the 30 kg block are [Fig. 4.28(b)]

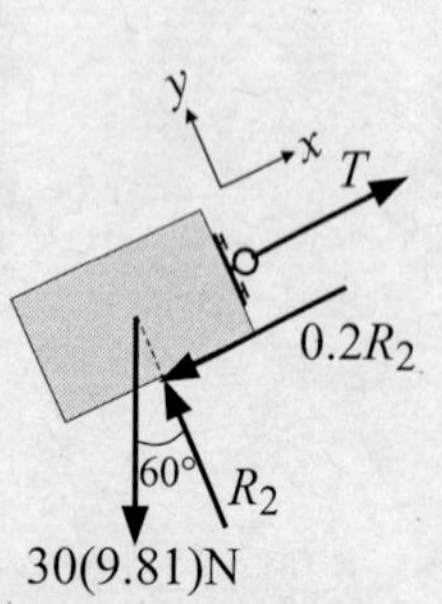

FIGURE 4.28(b) |

$$\sum F_y = 0 \Rightarrow R_2 - 30(9.81)\cos 60 = 0$$

or $$R_2 = 147.15\text{N} \quad \text{(i)}$$

$$\sum F_x = 0 \Rightarrow T - 0.2R_2 - 30(9.81)\sin 60 = 0$$

or $$T = 0.2 \times 147.15 + 30 \times 9.81 \times \frac{\sqrt{3}}{2}$$

or $$T = 284.3\text{N} \quad \text{(ii)}$$

The equations of equilibrium for the 20 kg block are [Fig. 4.28(c)]

$$\sum F_x = 0 \Rightarrow P\cos\theta - T - 0.2R_1 = 0 \quad \text{(iii)}$$

$$\sum F_y = 0 \Rightarrow P \sin\theta + R_1 - 20(9.81) = 0 \qquad \text{(iv)}$$

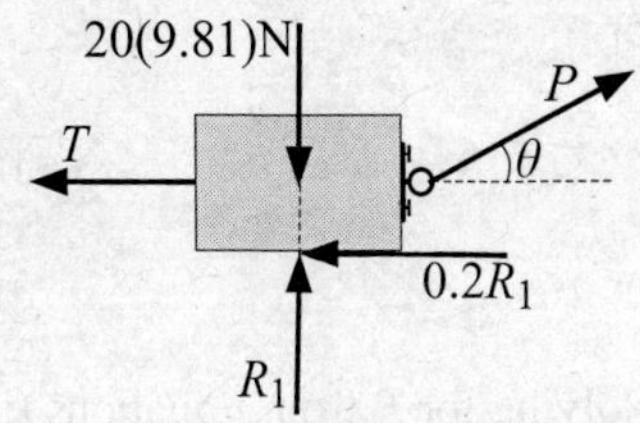

FIGURE 4.28(c) |

Eliminating R_1 from Equations (iii) and (iv), we obtain

$$P = \frac{323.54}{\cos\theta + 0.2 \sin\theta}$$

The value of P will be a minimum when the denominator $(\cos\theta + 0.2 \sin\theta)$ is a maximum. Take the derivative of the denominator with respect to θ and set this equal to zero to determine the value of θ that will make P a minimum

$$\frac{d}{d\theta}(\cos\theta + 0.2 \sin\theta) = 0$$

or

$$-\sin\theta + 0.2 \cos\theta = 0 \Rightarrow \tan\theta = 0.2$$

$$\theta = 11.3°$$

Hence, the minimum value of P is,

$$P = 317.28\text{N}$$

PROBLEM 4.29

A solid cylinder of mass 30 kg rests in a trough formed by two inclined planes as shown in Figure 4.29(a). Find the minimum vertical force P needed to cause impending rotary motion of the bar. The coefficient of friction is 0.2 between all pairs of surfaces.

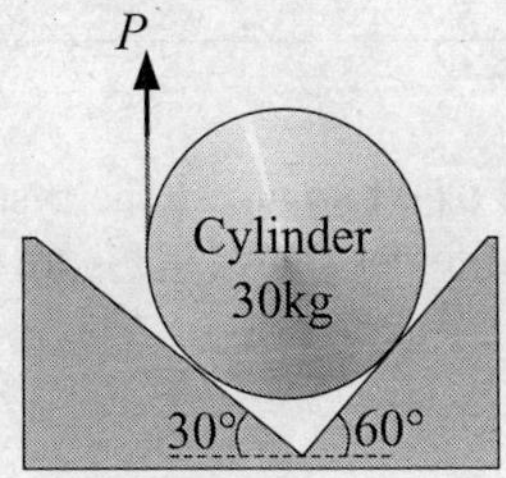

FIGURE 4.29(a) |

Solution. A free body diagram is shown in Figure 4.29(b). We have

$$\sum F_x = 0 \Rightarrow R_A \cos 60 + 0.2R_A \cos 30 + 0.2R_B \sin 30 - R_B \sin 60 = 0$$

or

$$R_A = 1.137R_B \qquad \text{(i)}$$

$$\sum F_y = 0 \Rightarrow R_A \sin 60 - 0.2R_A \sin 30 + 0.2R_B \cos 30 + R_B \cos 60 - 30(9.81) + P = 0$$

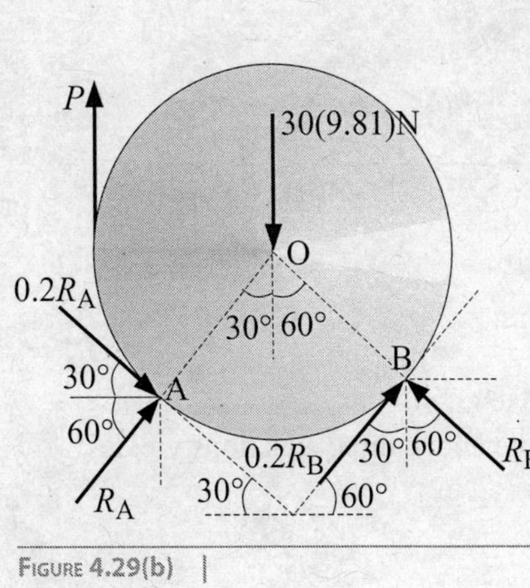

FIGURE 4.29(b) |

or
$$1.137R_A + R_B = 437.1 - P \quad \text{(ii)}$$

$$\sum M_o = 0 \Rightarrow P \times r - 0.2R_A \times r - 0.2R_B \times r = 0$$

or
$$P = 0.2(R_A + R_B) \quad \text{(iii)}$$

Solving for P from Equations (i), (ii) and (iii), we have

$$P = 68.69\text{N}$$

PROBLEM 4.30

Two disks, both of radius $2r$, are attached to the ends of a short axle of radius r. The system is placed on an inclined plane and held in equilibrium by horizontal cable as shown in Figure 4.30(a). The total weight of the system (two discs and axle) is W. Find the tension in the cable and minimum value of the coefficient of friction between the discs and the plane.

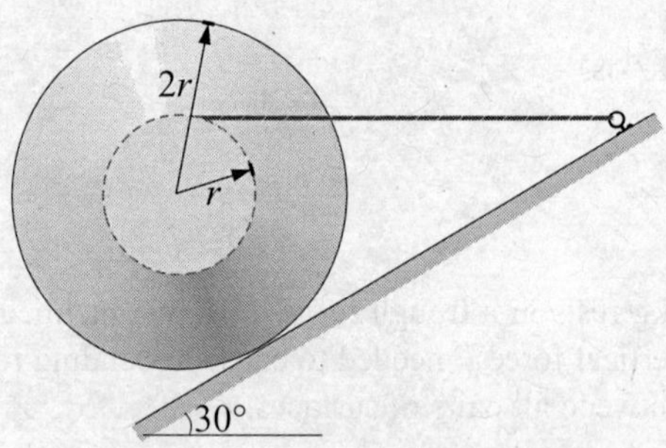

FIGURE 4.30(a) |

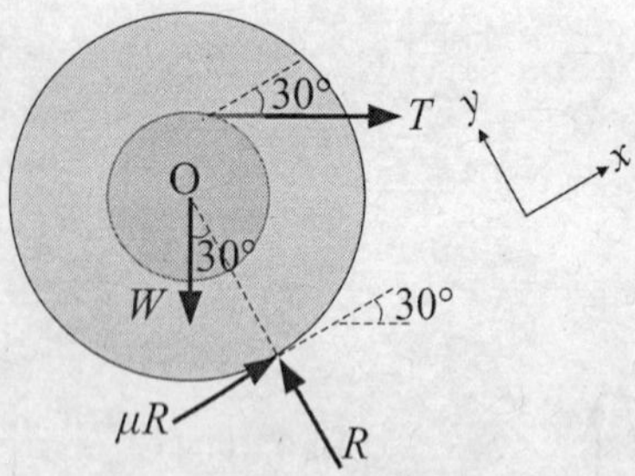

FIGURE 4.30(b) |

Solution. The system is in equilibrium under the action of forces shown in Figure 4.30(b). The equilibrium equations are

$$\sum F_x = 0 \Rightarrow T\cos 30 + \mu R - W\sin 30 = 0$$

or
$$\sqrt{3}T + 2\mu R = W \quad \text{(i)}$$

$$\sum F_y = 0 \Rightarrow T\sin 30 + W\cos 30 - R = 0$$

or
$$R = \frac{T + \sqrt{3}W}{2} \quad \text{(ii)}$$

$$\sum M_o = 0 \Rightarrow -\mu R \times 2r + T \times r = 0$$

$$T = 2\mu R \quad \text{(iii)}$$

The above three equations yield

$$T = \frac{W}{\sqrt{3}+1} = 0.366W$$

$$\text{tension of cable, } T = 0.366W$$

and
$$\mu = \frac{1}{4+\sqrt{3}} = 0.174$$

$$\text{coefficient of friction } \mu = 0.174$$

PROBLEM 4.31

A cabinet of weight 2 kN is supported on the horizontal surfaces by a skid at A and a roller at B as shown in Figure 4.31(a). If a force F of 500 N is required to initiate motion of the cabinet, determine the coefficient of friction at B.

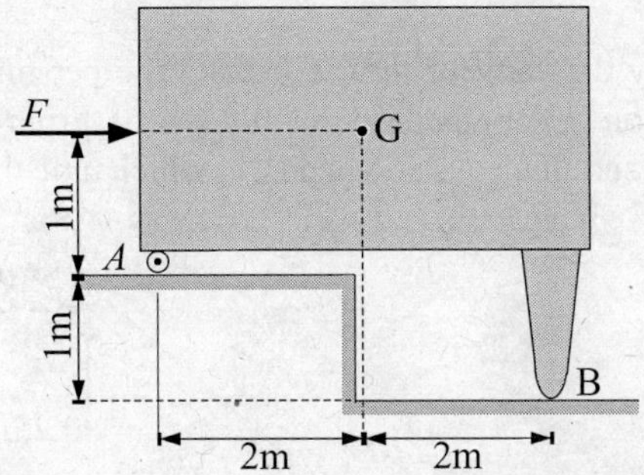

FIGURE 4.31(a)

Solution. Figure 4.31(b) shows the free body diagram of the cabinet when it is just about to slip and force of friction at B has reached its maximum value. The equilibrium equations are

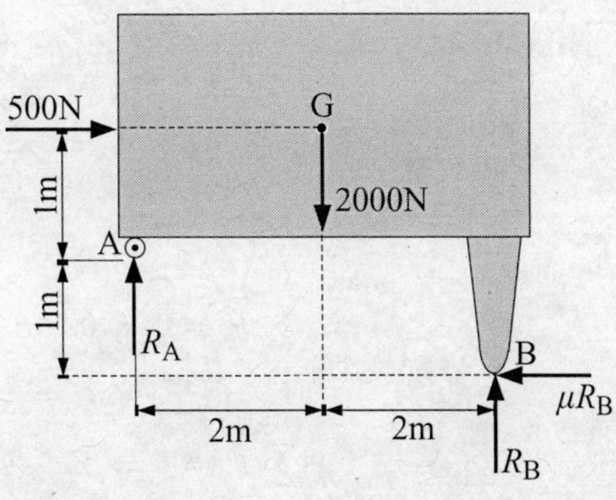

FIGURE 4.31(b)

$$\sum M_B = 0 \Rightarrow R_A \times 4 + 500 \times 2 - 2000 \times 2 = 0$$

or
$$R_A = \frac{2000 \times 2 - 500 \times 2}{4} = 750\text{N} \qquad \text{(i)}$$

$$\sum F_y = 0 \Rightarrow R_A + R_B - 2000 = 0$$

or
$$R_B = 2000 - R_A \qquad \text{(ii)}$$

Substituting the value of R_A

or
$$R_B = 2000 - 750 = 1250\text{N}$$

$$\sum F_x = 0 \Rightarrow 500 - \mu R_B = 0 \qquad \text{(iii)}$$

Substituting the value of R_B

or
$$\mu = \frac{500}{R_B} = \frac{500}{1250} = 0.4$$

$$\mu = 0.4$$

PROBLEM 4.32

A ladder of mass 35 kg and length 10 m rests against a vertical wall and is inclined at 60° to the horizontal [Fig. 4.32(a)]. The coefficient of friction between the ladder and the wall as well as between the ladder and the ground is 0.25. How far up the ladder can a 72 kg person climb before the ladder begins to slip?

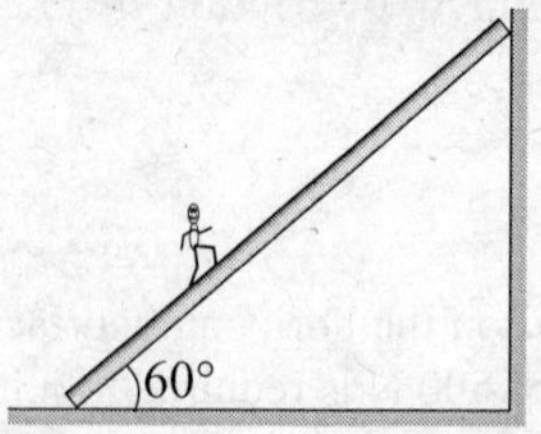

FIGURE 4.32(a) |

Solution. The free body diagram for the ladder and the person is shown in Figure 4.32(b), where x is the distance the person has climbed, so that $0 < x > 10$. From the geometry of the problem, it is clear that slipping impends simultaneous at A and B, at which time the friction forces are as shown in the figure. For equilibrium, we have

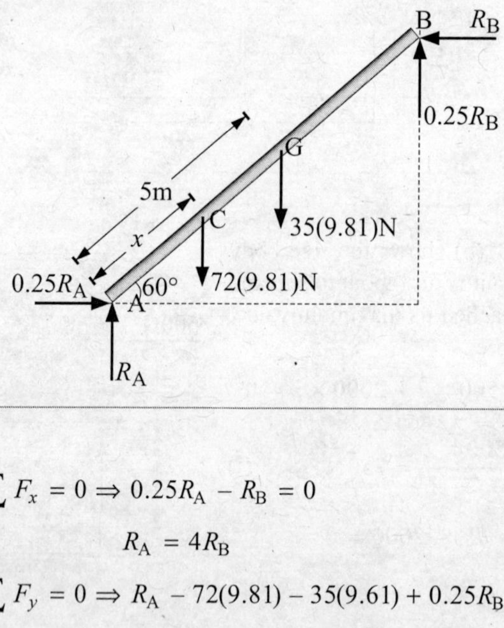

FIGURE 4.32(b) |

$$\sum F_x = 0 \Rightarrow 0.25R_A - R_B = 0$$

$$R_A = 4R_B \tag{i}$$

$$\sum F_y = 0 \Rightarrow R_A - 72(9.81) - 35(9.61) + 0.25R_B = 0$$

$$R_A + 0.25R_B = 1049.6 \tag{ii}$$

Solving for R_A and R_B from Equations (i) and (ii), we have

$$R_B = 246.9\text{N and } R_A = 987.8\text{N}$$

$$\sum M_A = 0 \Rightarrow 72(9.81) \times x \cos 60 + 35(9.61) \times 5 \cos 60 - 0.25R_B$$

$$\times 10 \cos 60 - R_B \times 10 \sin 60 = 0$$

From which we find that

$$x = 4.5\text{m}$$

PROBLEM 4.33

Two identical blocks of weight W are supported by a link inclined at 45° with the horizontal as shown in Figure 4.33(a). If both the blocks are in limiting equilibrium, find the coefficient of friction, assuming it to be the same at floor as well as wall.

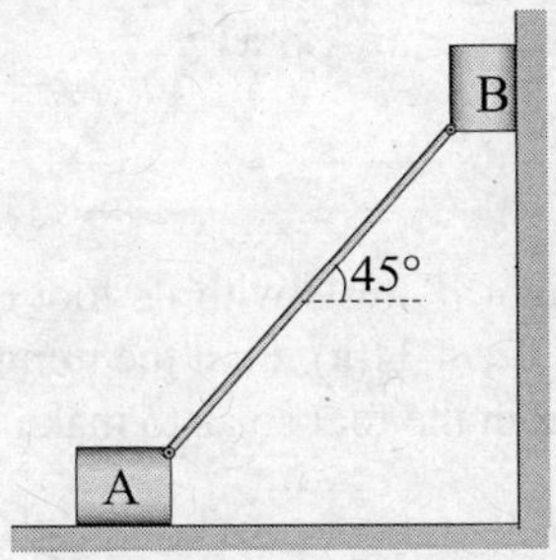

FIGURE 4.33(a) |

Solution. The free body diagram of the ladder is shown in Figure 4.33(b).

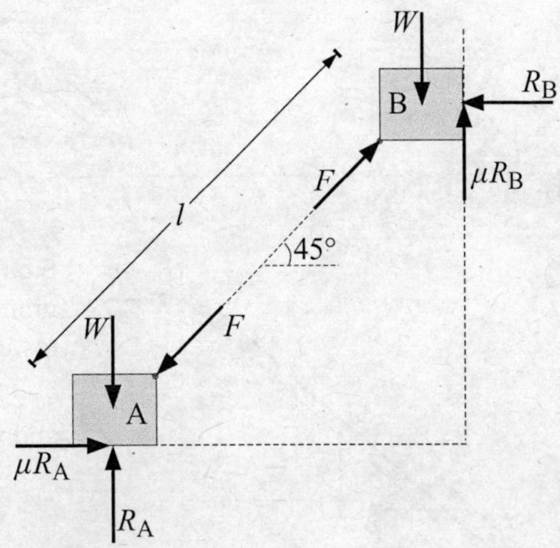

FIGURE 4.33(b) |

Summing forces vertically,

$$\sum F_y = 0 \Rightarrow \mu R_B + R_A - W - W = 0$$

$$\mu R_B + R_A = 2W$$

Summing forces horizontally,

$$\sum F_x = 0 \Rightarrow \mu R_A - R_B = 0$$

$$\mu R_A = R_B$$

$$R_B = 2W\left(\frac{\mu}{1+\mu^2}\right)$$

Taking moment about block A,

$$\sum M_A = 0 \Rightarrow W \times l\cos 45 - R_B \times l\sin 45 - \mu R_B \times l\cos 45 = 0$$

$$W = R_B + \mu R_B$$

Substitute the value of R_B in the above equation,

$$W = 2W\left(\frac{\mu}{1+\mu^2}\right) + \mu \times 2W\left(\frac{\mu}{1+\mu^2}\right)$$

or $$\mu^2 + 2\mu - 1 = 0$$

Solving the quadratic equation, we have

$$\mu = \frac{-2 \pm \sqrt{4+4}}{2} = 0.412, -2.412$$

Therefore,

$$\mu = 0.412$$

PROBLEM 4.34

A uniform ladder, of length l and weight W, rests with its foot on a rough ground and its upper end against a smooth wall as shown in Figure 4.34(a). Find the minimum force P, which is applied horizontally to the ladder at a distance c from the foot so as to make the foot approach the wall.

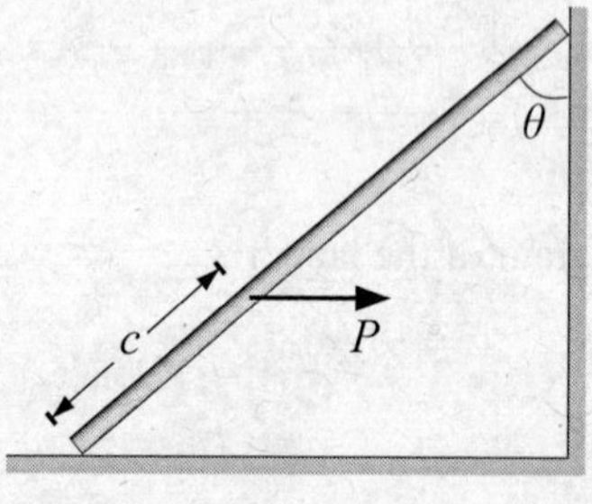

FIGURE 4.34(a) |

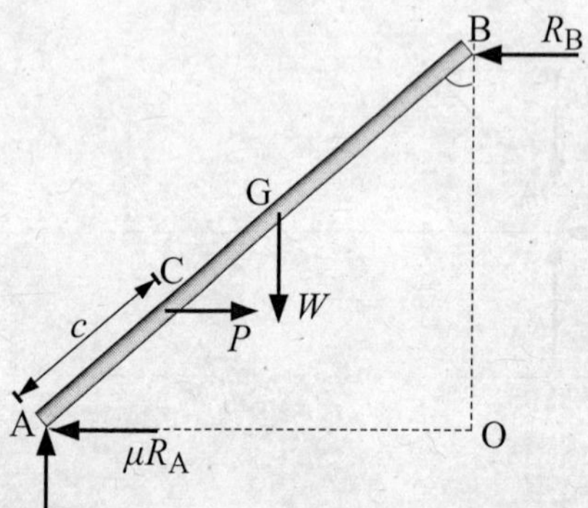

FIGURE 4.34(b) |

Solution. Figure 4.34(b) shows the free body diagram of the ladder when its lower end is just about to slip towards the wall.

Resolving the forces acting on the ladder horizontally and vertically, we get

$$\sum F_x = 0 \Rightarrow P - \mu R_A - R_B = 0 \quad \text{(i)}$$

$$\sum F_y = 0 \Rightarrow R_A - W = 0 \quad \text{(ii)}$$

Eliminating R_A from Equations (i) and (ii), we have

$$R_B = P - \mu W \quad \text{(iii)}$$

Now taking moment about A, we have

$$\sum M_A = 0 \Rightarrow W(\text{AG} \sin\theta) - R_B(\text{AB} \cos\theta) + Pc\cos\theta = 0$$

Substituting R_B from Equation (iii), we get

$$W(\text{AG} \sin\theta) - (P - \mu W)\text{AB}\cos\theta + Pc\cos\theta = 0$$

or

$$W\frac{l}{2}\sin\theta - (P - \mu W)l\cos\theta + Pc\cos\theta = 0$$

or $$P(l - c)\cos\theta = Wl\left(\mu\cos\theta + \frac{1}{2}\sin\theta\right)$$

$$P = W\frac{l}{l - c}\left(\mu + \frac{1}{2}\tan\theta\right)$$

PROBLEM 4.35

A ladder rests in limiting equilibrium with its lower end on a rough horizontal plane and other end against a rough vertical wall as shown in Figure 4.35(a). If the ladder is on the point of slipping at both ends, find the inclination of the ladder to the horizontal.

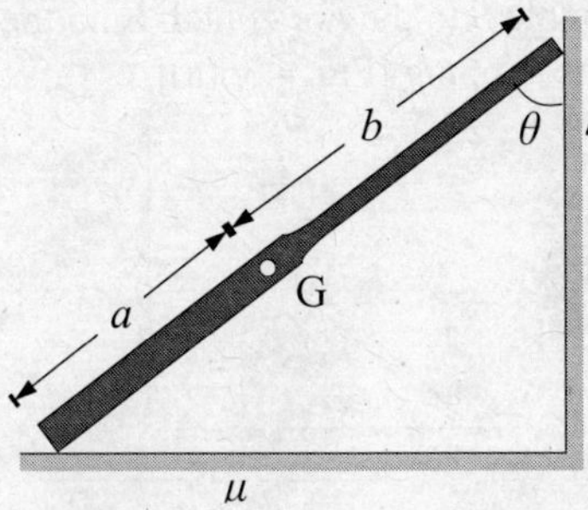

FIGURE 4.35(a) |

Solution. The ladder is in limiting equilibrium under the action of the forces shown in Figure 4.35(b). The equation of equilibrium gives

$$\sum F_y = 0 \Rightarrow R_A + \mu' R_B - W = 0$$

or $$R_A + \mu' R_B = W \qquad \text{(i)}$$

$$\sum F_x = 0 \Rightarrow \mu R_A - R_B = 0$$

or $$R_B = \mu R_A \qquad \text{(ii)}$$

$$\sum M_A = 0 \Rightarrow Wa\sin\theta - R_B(a + b)\cos\theta$$
$$- \mu' R_B\,(a + b)\sin\theta = 0 \qquad \text{(iii)}$$

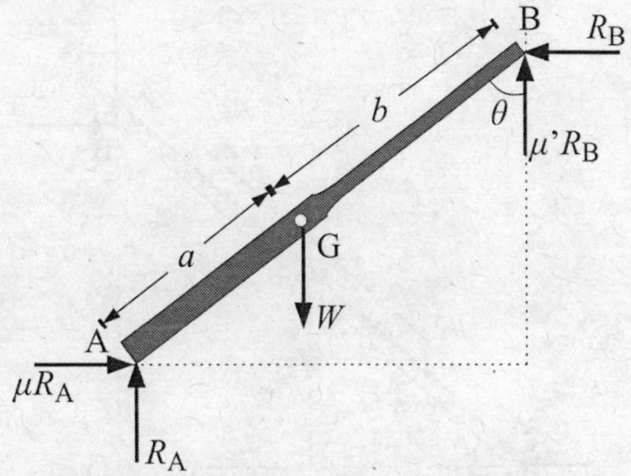

FIGURE 4.35(b) |

Solving R_B from Equations (i) and (ii), we get

$$R_B = \frac{\mu}{1 + \mu\mu'}W$$

Substituting R_B in Equation (iii), we have

$$Wa\sin\theta - \frac{\mu W}{1 + \mu\mu'}(a + b)\cos\theta - \frac{\mu'\mu W}{1 + \mu\mu'}(a + b)\sin\theta = 0$$

or $$\left[Wa - \frac{\mu\mu' W(a+b)}{1+\mu\mu'}\right]\sin\theta - \frac{\mu W}{1+\mu\mu'}(a+b)\cos\theta$$

or $$\tan\theta = \frac{\mu(a+b)}{a - \mu\mu' b}$$

$$\theta = \tan^{-1}\frac{\mu(a+b)}{a - \mu\mu' b}$$

PROBLEM 4.36

A uniform ladder of weight W is resting against a smooth vertical wall and on a rough horizontal plane. The ladder is inclined at an angle θ to the horizontal. Find the weight of a man who can climb up to the top of the ladder without its slipping [Fig. 4.36(a)].

FIGURE 4.36(a) |

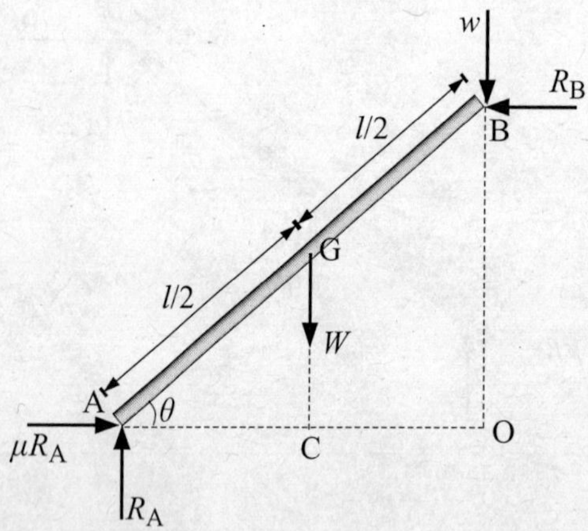

FIGURE 4.36(b) |

Solution. Let the man of weight w be at the top of the ladder, then it is on the point of slipping [Fig. 4.36(b)]. Resolving the forces acting on the ladder horizontally and vertically, we get

$$\sum F_y = 0 \Rightarrow R_A - W - w = 0$$

or $$R_A = (W + w) \quad \text{(i)}$$

$$\sum F_x = 0 \Rightarrow \mu R_A - R_B = 0 \quad \text{(ii)}$$

Taking the moment of the forces about A, we get

$$\sum M_A = 0 \Rightarrow w \times AO - R_B \times OB + W \times AC = 0$$

or $$wl\cos\theta - R_B \times l\sin\theta + W \times \frac{l}{2}\cos\theta = 0$$

or $$R_B = \left(w + \frac{W}{2}\right)\cot\theta \quad \text{(iii)}$$

The man can climb up to the top of the ladder without its slipping provided the force μR_A, which prevents slipping, is greater than the horizontal reaction R_B of the wall

i.e., if $$\mu R_A > R_B$$

Substituting the values of R_A and R_B from Equations (i) and (iii), we have

$$\mu(w + W) > \left(w + \frac{W}{2}\right) \cot\theta$$

or

$$2\mu w + 2\mu W > 2w \cot\theta + W \cot\theta$$

On simplifying, we obtain

$$w > \frac{W}{2}\left(\frac{\cot\theta - 2\mu}{\mu - \cot\theta}\right)$$

weight of man smallar than $\dfrac{W}{2}\left(\dfrac{\cot\theta - 2\mu}{\mu - \cot\theta}\right)$

PROBLEM 4.37

A uniform beam of length $2h$ can rest with one end on a rough horizontal plane and against the top of a wall of height h, in vertical plane perpendicular to the wall and at any inclination to the wall as shown in Figure 4.37(a). Assuming the coefficients of friction between the beam and both wall and ground are same, find the minimum coefficient of friction.

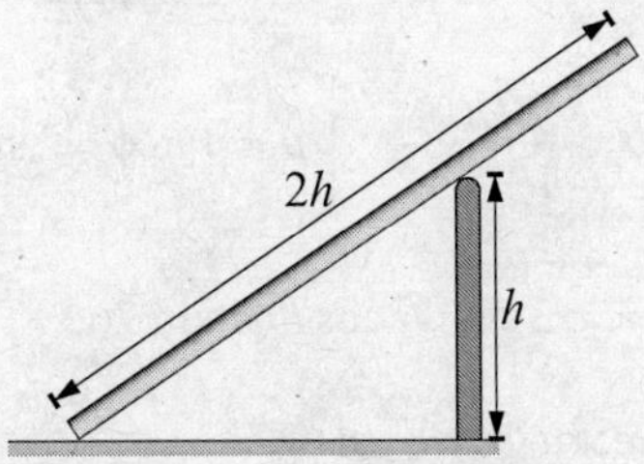

FIGURE 4.37(a) |

Solution. The beam is in limiting equilibrium under the action of forces shown in Figure 4.37(b).

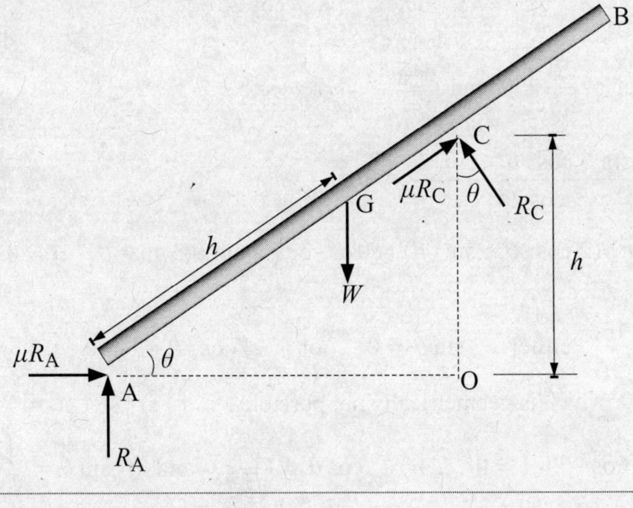

FIGURE 4.37(b) |

$$\sum F_y = 0 \Rightarrow R_C \cos\theta + \mu R_C \sin\theta - W + R_A = 0$$

or
$$R_C(\cos\theta + \mu\sin\theta) + R_A = W \tag{i}$$

$$\sum F_x = 0 \Rightarrow \mu R_A - R_C \sin\theta + \mu R_C \cos\theta = 0$$

or
$$R_C(\mu\cos\theta - \sin\theta) + \mu R_A = 0 \tag{ii}$$

Solving for R_C from Equations (i) and (ii), we get

$$R_C = \frac{\mu W}{(1+\mu^2)\sin\theta}$$

Now taking moment about A,

$$\sum M_A = 0 \Rightarrow W(\text{AG}\cos\theta) - R_C \times \text{AC} = 0 \tag{iii}$$

or
$$\text{W}h\cos\theta = \frac{\mu \text{W}}{(1+\mu^2)\sin\theta} \times \frac{h}{\sin\theta}$$

or
$$2\sin^2\theta\cos\theta = \frac{2\mu}{1+\mu^2} \qquad \left[\mu = \tan\phi \Rightarrow \sin^2\phi = \frac{2\mu}{1+\mu^2}\right]$$

or
$$2\sin^2\theta\cos\theta = \sin 2\phi \tag{iv}$$

Differentiating both sides with respect to θ, we get

$$2\sin^2\theta(-\sin\theta) + (2\sin\theta\cos\theta)\cos\theta = 2\cos 2\phi \times \frac{d\phi}{d\theta}$$

$$\frac{d\phi}{d\theta} = \frac{2\sin\theta\cos^2\theta - \sin^3\theta}{\cos 2\phi}$$

For minimum value of ϕ, $\dfrac{d\phi}{d\theta} = 0$ and $\dfrac{d^2\phi}{d\theta^2} = +ve\phi$

$$\frac{2\sin\theta\cos^2\theta - \sin^3\theta}{\cos 2\phi} = 0$$

$$\sin\theta(2\cos^2\theta - \sin^2\theta) = 0 \qquad \text{or} \qquad \sin\theta(3\cos^2\theta - 1) = 0$$

Therefore

$$\text{either} \quad \sin\theta = 0 \quad \text{or} \quad 3\cos^2\theta - 1 = 0$$

If $\sin\theta = 0$, then $\theta = 0$, which is geometrically not possible,

$$\therefore\ 3\cos^2\theta - 1 = 0 \qquad \text{or} \qquad \cos\theta = \frac{1}{\sqrt{3}} \qquad \text{and} \qquad \sin\theta = \sqrt{\frac{2}{3}}$$

Substituting the values of cos θ and sin θ in Equation (iv), we obtain

$$2\left(\sqrt{\frac{2}{3}}\right)^2 \times \frac{1}{\sqrt{3}} = \sin 2\phi$$

or
$$\sin 2\phi = \frac{4}{3\sqrt{3}} \Rightarrow \phi = \frac{1}{2}\sin^{-1}\frac{4}{3\sqrt{3}} = 25.16°$$

Minimum value of the coefficient of friction,

$$\mu = \tan\phi = \tan 25.16 = 0.469$$

$$\boxed{\mu = 0.469}$$

PROBLEM 4.38

Determine the horizontal force P on the wedges A and B that is necessary to raise the weight of 1000 kg resting on C as shown in Figure 4.38(a). The coefficient of friction for all contact surfaces is 0.25.

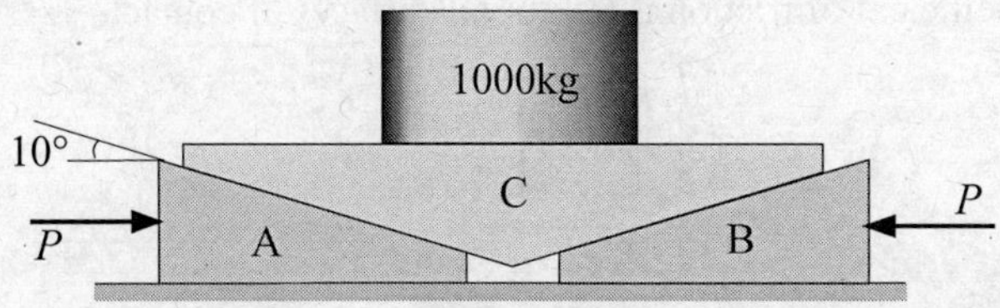

FIGURE 4.38(a) |

Solution. Figure 4.38(b) shows the free body diagram of block C. We get

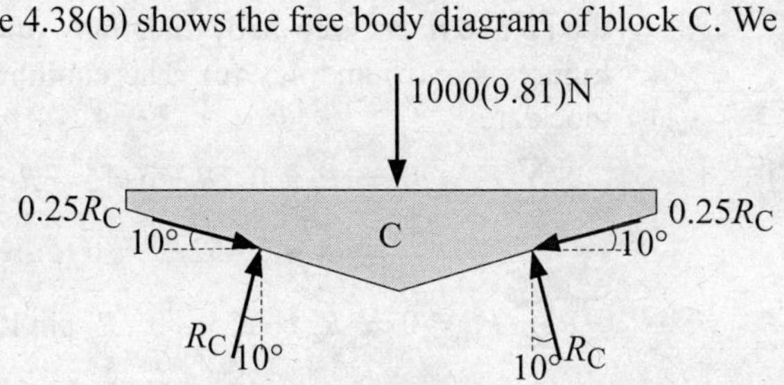

FIGURE 4.38(b) |

$$\sum F_y = 0 \Rightarrow 2R_C\cos 10 - 1000(9.81) - 2\times 0.25R_C\sin 10 = 0$$

$$R_C = \frac{1000(9.81)}{2(\cos 10 - 0.25\sin 10)} = 5210.3\text{N}$$

Figure 4.38(c) shows the free body diagram of wedge A. We get

$$\sum F_y = 0 \Rightarrow R_A - R_C\cos 10 + 0.25R_C\sin 10 = 0$$

$$R_A = R_C(\cos 10 - 0.25\sin 10)$$

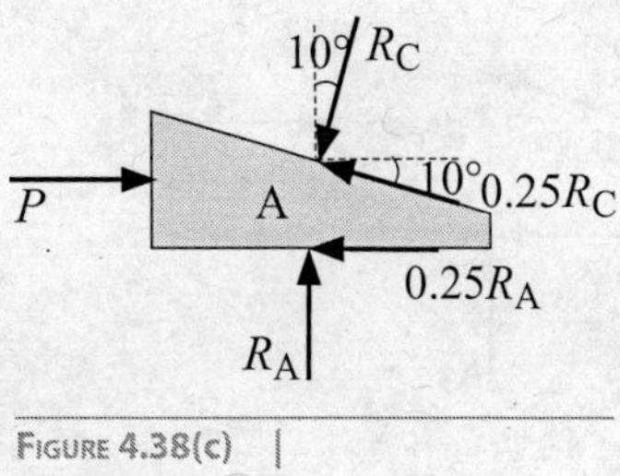

FIGURE 4.38(c) |

Substituting the value of R_C, we obtain

$$R_A = 5210.3(\cos 10 - 0.25 \sin 10) = 4905\text{N}$$

or

$$R_A = 4905\text{N}$$

$$\sum F_x = 0 \Rightarrow P - 0.25R_A - 0.25R_C \cos 10 - R_C \sin 10 = 0$$

$$P = 0.25R_A + R_C(0.25 \cos 10 + \sin 10)$$

Substituting the value of R_A and R_C, we have

$$P = 0.25 \times 4905 + 5210.3(0.25 \cos 10 + \sin 10)$$

$$P = 3413.79\text{N}$$

PROBLEM 4.39

Determine the horizontal force F required for wedge B to raise block A of weight 4000 N as shown in Figure 4.39(a). The coefficient of friction is 0.2 for all surfaces of contact.

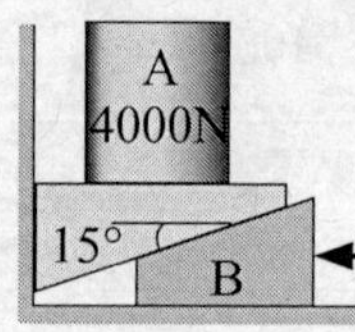

FIGURE 4.39(a)

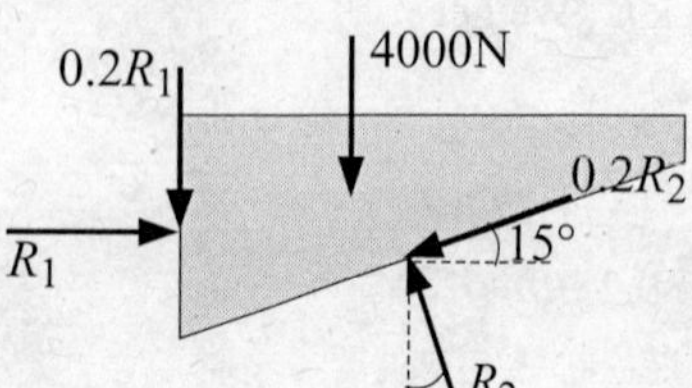

FIGURE 4.39(b)

Solution. Free body diagrams of wedges are shown in Figures 4.39(b) and 4.39(c). The equilibrium equations for the block are

$$\sum F_x = 0 \Rightarrow R_1 - 0.2R_2 \cos 15 - R_2 \sin 15 = 0$$

$$R_1 - 0.452R_2 = 0 \qquad \text{(i)}$$

$$\sum F_y = 0 \Rightarrow R_2 \cos 15 - 0.2R_2 \sin 15 - 0.2R_1 - 4000 = 0$$

$$0.914R_2 - 0.2R_1 = 4000 \qquad \text{(ii)}$$

Solving R_2 from Equation (i) and (ii), we have

$$R_2 = 4856.7\text{N}$$

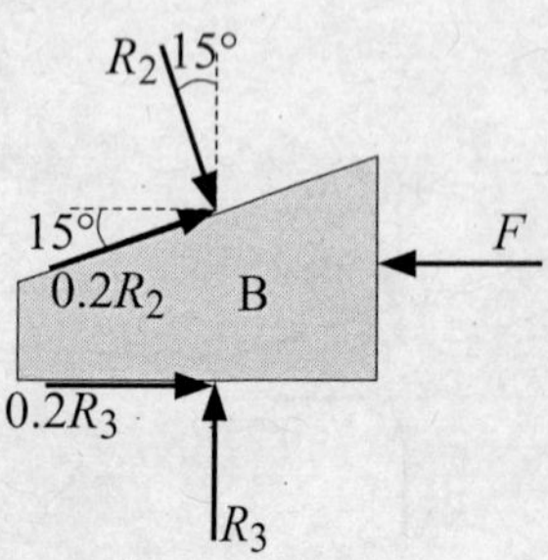

FIGURE 4.39(c)

The equilibrium equations for wedge B are

$$\sum F_y = 0 \Rightarrow R_2 \cos 15 - 0.2R_2 \sin 15 - R_3 = 0$$

$$R_3 = 4856.7(\cos 15 - 0.2 \times \sin 15)$$

$$R_3 = 4439.8\text{N}$$

$$\sum F_x = 0 \Rightarrow 0.2R_2 \cos 15 + R_2 \sin 15 + 0.2R_3 - F = 0$$

$$F = R_2(0.2 \cos 15 + \sin 15) + 0.2R_3$$

Substituting the value of R_2 and R_3, we obtain

$$F = 4856.7 \times (0.2 \cos 15 + \sin 15) + 0.2 \times 4439.8$$

$$F = 3083.2\text{N}$$

PROBLEM 4.40

The horizontal position of the 200 kg rectangular block is adjusted by a 6° wedge under the action of a force *P* [Fig. 4.40(a)]. If the coefficient of friction for both the pairs of wedge surface is 0.25 and if the coefficient of friction between the block and the horizontal surface is 0.5, determine the least force *P* required to move the block.

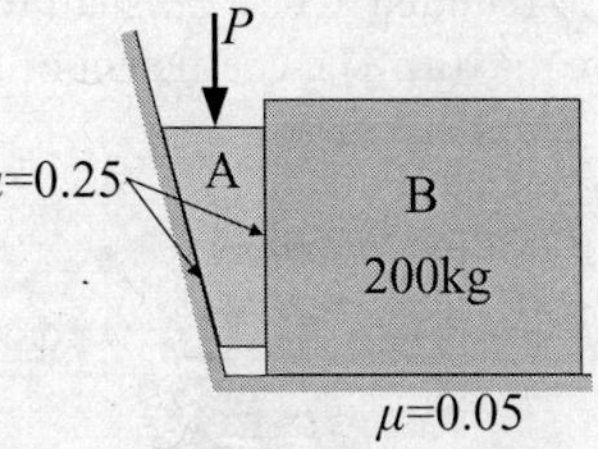

FIGURE 4.40(a) |

Solution. Figure 4.40(b) shows the free body diagram of block B. We have

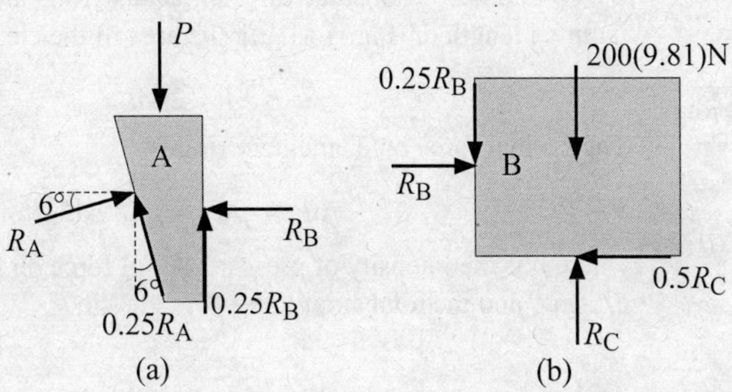

FIGURE 4.40(b) |

$$\sum F_x = 0 \Rightarrow R_B - 0.5R_C = 0$$

$$R_C = 2R_B$$

$$\sum F_y = 0 \Rightarrow R_C - 0.25R_B - 200(9.81) = 0$$

Substituting $R_C = 2R_B$, we get

$$R_B = 1121.1\text{N}$$

Part (a) of Figure 4.40(b) shows the free body diagram of wedge A. We have

$$\sum F_x = 0 \Rightarrow R_A \cos 6 - 0.25R_A \sin - R_B = 0$$

$$R_A = \frac{R_B}{\cos 6 - 0.25 \sin 6} = \frac{1121.1}{0.994 - 0.25 \times 0.1045}$$

or

$$R_A = 1157.6\text{N}$$

$$\sum F_y = 0 \Rightarrow R_A \sin 6 + 0.25 R_A \cos 6 + 0.25 R_B - P = 0$$

$$P = 1157.6 \times \sin 6 + 0.25 \times 1157.6 \cos 0.25 \times 1121.1$$

$$P = 688.9\text{N}$$

PROBLEM 4.41

Derive the expression for the torque M required to turn the shaft whose thrust W is supported by a conical pivot bearing, as shown in Figure 4.41(a). The coefficient of friction is μ and the bearing pressure is constant.

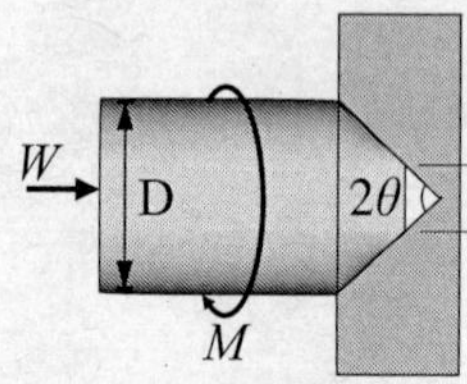

FIGURE 4.41(a) |

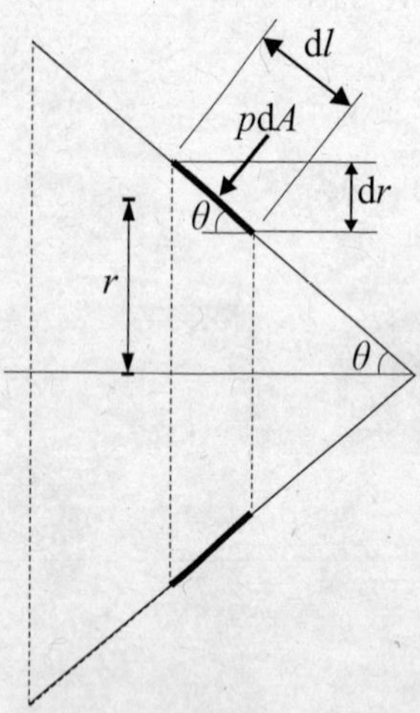

FIGURE 4.41(b) |

Solution. Consider an elementary ring at a radius r and an inclined length dl [Fig. 4.41(b)], the area of the element is

$$dA = 2\pi dl$$

The normal force on elementary ring is

$$dP = p dA = p2\pi r dl,$$

where p is the intensity of pressure. Axial force on the elemental ring is d$F \sin\theta$ and the total axial force W is $\int dP \sin\theta$.

$$W = \int_{d/2}^{D/2} p2\pi r dl \sin\theta$$

or

$$W = \int_{d/2}^{D/2} p2\pi r \frac{dr}{\sin\theta} \sin\theta$$

From the geometry, dr = d$l \sin\theta$.

$$W = p\frac{\pi}{4}(D^2 - d^2) \quad \text{or} \quad p = \frac{4W}{\pi(D^2 - d^2)}$$

Frictional force on the bearing surface of the elementary ring

$$dF = \mu dP = \mu p 2\pi r dl$$

Frictional moment about the axis of rotation,

$$dM = dF \times r = \mu p 2\pi r dl \times r = \mu p 2\pi r \frac{dr}{\sin\theta} \times r$$

Therefore, total frictional moment working against the rotation of the shaft,

$$M = \int dM = \int_{d/2}^{D/2} \mu p 2\pi r \frac{dr}{\sin\theta} \times r$$

or

$$M = \frac{\mu p 2\pi}{\sin\theta} \int_{d/2}^{D/2} r^2 \, dr = \frac{2\pi\mu p}{\sin\theta} \times \frac{D^3 - d^3}{3 \times 8}$$

Substituting $p = \dfrac{4W}{\pi(D^2 - d^2)}$

$$M = \frac{2\pi\mu}{\sin\theta} \times \frac{4W}{\pi(D^2 - d^2)} \times \frac{D^3 - d^3}{3 \times 8}$$

$$M = \frac{\mu W}{3\sin\theta} \times \frac{(D^3 - d^3)}{(D^2 - d^2)}$$

PROBLEM 4.42

A rope holding a block of mass 150 kg passes over a pulley and is attached to a frame as shown in Figure 4.42(a). Find the minimum coefficient of friction between the rope and the pulley for which equilibrium is possible.

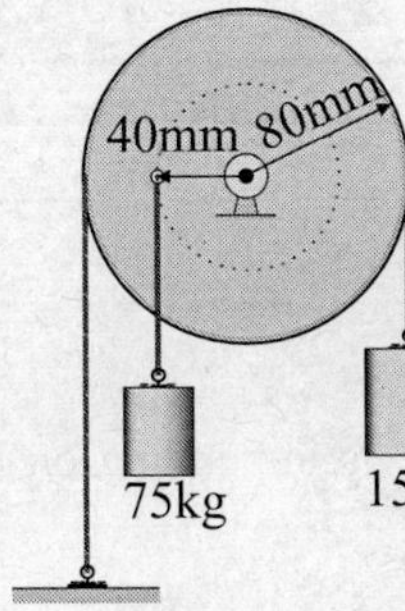

FIGURE 4.42(a)

Solution. Figure 4.42(b) shows the free body diagram of the pulley when it is just about to rotate counter-clockwise. Taking moment about O,

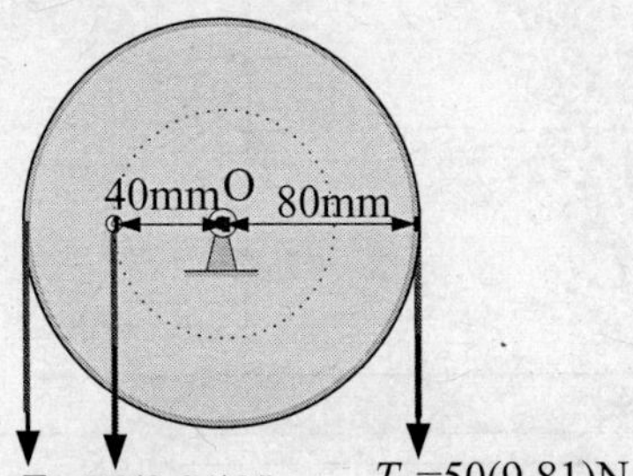

FIGURE 4.42(b)

$$\sum M_o = 0 \Rightarrow T_1 \times 80 - T_2 \times 80 - 75(9.81) \times 40 = 0$$

or

$$150(9.81) - T_2 = \frac{75(9.81) \times 40}{80}$$

$$T_2 = 1103.62\text{N}$$

Tension ratio of rope, $\dfrac{T_1}{T_2} = e^{\mu\theta}$

or $$\frac{150 \times 9.81}{1103.62} = e^{\mu \times \pi}$$

or $$e^{\pi\mu} = 1.33$$

or $$\mu = \frac{k \log_e 1.33}{\pi} \Rightarrow \mu = 0.0907$$

$$\mu = 0.09$$

PROBLEM 4.43

A uniform beam of 5 m length is suspended by a rope that passes over a large pulley [Fig. 4.43(a)]. A locking pin C prevents rotation of the pulley. The coefficient of friction between the rope and the pulley is 0.20. Determine the minimum value of *x* for which the rope will not slip on the pulley.

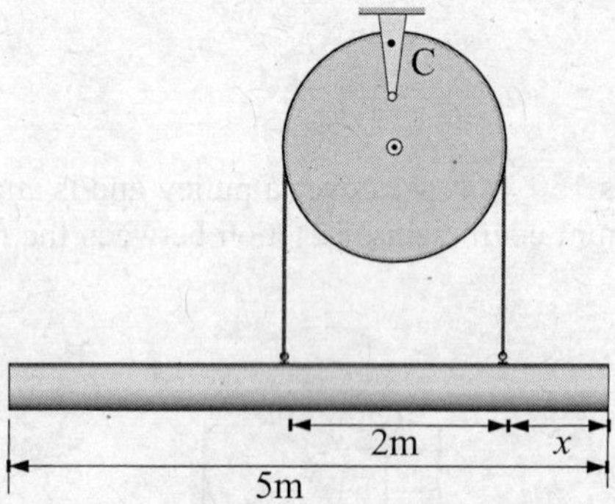

FIGURE 4.43(a) |

Solution. Since slipping of the rope is impending on the pulley in counter-clockwise [Fig. 4.43(b)], we use $T_1 / T_2 = e^{\mu\theta}$

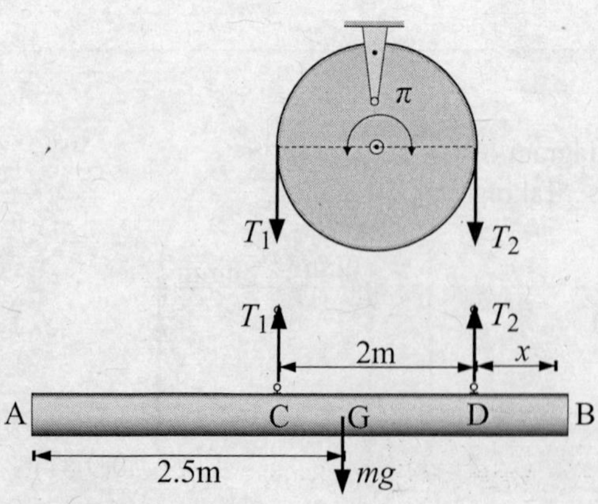

FIGURE 4.43(b) |

$$\frac{T_1}{T_2} = e^{0.2\pi} \Rightarrow T_1 = 1.874T_2 \qquad \text{(i)}$$

Now consider the beam. The equilibrium equations are

$$\sum F_y = 0 \Rightarrow T_1 + T_2 - mg = 0 \qquad \text{(i)}$$

$$\sum M_c = 0 \Rightarrow mg(2 + x - 2.5) - T_2 \times 2 = 0 \qquad \text{(ii)}$$

From Equations (i) and (ii), we obtain

$$T_2 = \frac{mg}{2.874}$$

Substituting the value of T_2 in Equation (iii), we have

$$mg(2 + x - 2.5) - \left[\frac{mg}{2.874}\right] \times 2 = 0$$

On simplifying, we obtain

$$x = 1.195\text{m}$$

PROBLEM 4.44

A rope is placed around three fixed cylinders, each of diameter 20 cm, located in the same horizontal plane [Fig. 4.44(a)]. Knowing that the coefficient of friction between fixed cylinders and rope is 0.2, determine the largest value of the mass m for which equilibrium is possible.

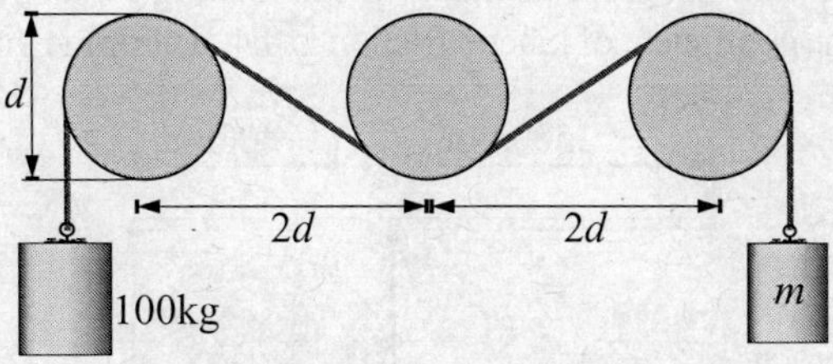

FIGURE 4.44(a)

Solution. Since slipping of the rope is impending [Fig. 4.44(b)], we use the tension ratio formula.

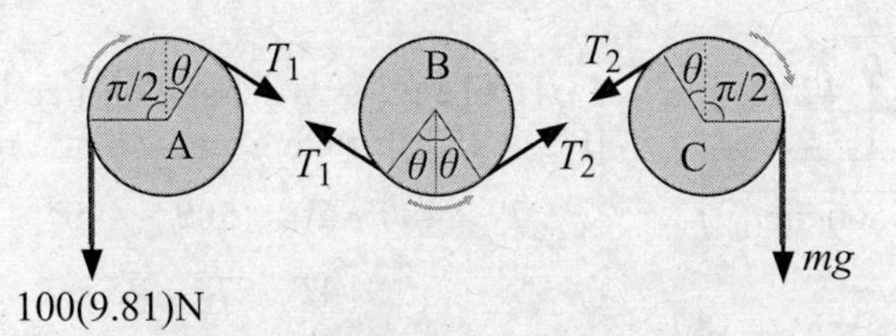

FIGURE 4.44(b)

For the fixed cylinder A,

$$\frac{T_1}{100(9.81)} = e^{0.2\left[\left(\frac{\pi}{2}\right)+\theta\right]} \tag{i}$$

For the fixed cylinder B,

$$\frac{T_2}{T_1} = e^{0.2(\theta+\theta)} \tag{ii}$$

For the fixed cylinder C

$$\frac{mg}{T_2} = e^{0.2\left[\left(\frac{\pi}{2}\right)+\theta\right]} \tag{iii}$$

Multiplying Equations (i), (ii) and (iii), we obtain

$$\frac{T_1}{100(9.81)} \times \frac{T_2}{T_1} \times \frac{mg}{T_2} = e^{0.2\times\left[\left(\frac{\pi}{2}\right)+\theta+\theta+\theta+\theta+\left(\frac{\pi}{2}\right)+\theta\right]}$$

or $$\frac{mg}{100(9.81)} = e^{0.2(\pi+4\theta)} \quad \text{or} \quad \frac{m}{100} = e^{0.2(\pi+4\theta)}$$

From the geometry, $\sin\theta = \dfrac{d}{2d} \Rightarrow \theta = \dfrac{\pi}{6}$

$$m = 100e^{0.2\left(\pi+\frac{4\pi}{6}\right)} = 284.96\text{kg}$$

$$m = 284.96\text{kg}$$

PROBLEM 4.45

A band brake drum is used to control the speed of a flywheel, rotating counter-clockwise as shown in Figure 4.45(a). Knowing that the coefficient of kinetic friction is 0.4, determine the frictional torque about O.

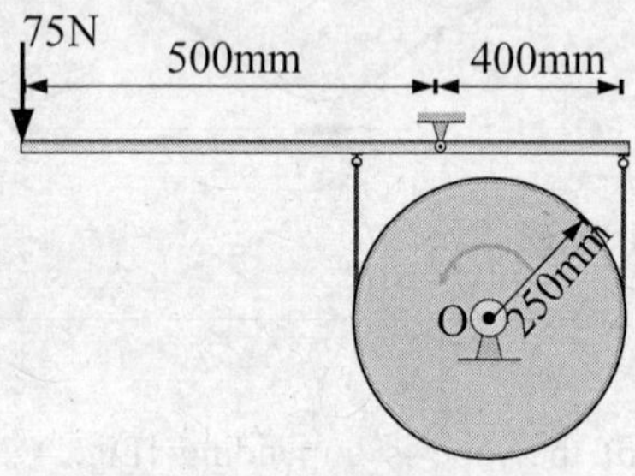

FIGURE 4.45(a) |

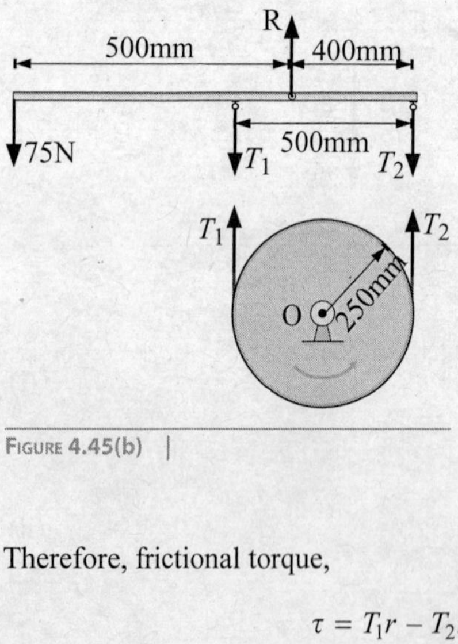

FIGURE 4.45(b) |

Solution. We draw the free body diagrams of lever [Fig. 4.45(b)]. Taking moment of forces about the hinge,

$$T_2 \times 400 - T_1 \times (500 - 400) - 75 \times 500 = 0$$

or $$4T_2 - T_1 = 75 \times 5 \qquad \text{(i)}$$

Tension ratio of belt

$$\frac{T_1}{T_2} = e^{0.4\pi}$$

or $$T_1 = 3.51T_2 \qquad \text{(ii)}$$

Solving Equations (i) and (ii), we obtain

$$T_2 = 770.94\text{N}, \; T_1 = 2706.0\text{N}$$

Therefore, frictional torque,

$$\tau = T_1 r - T_2 r = (2706.0 - 770.94) \times \frac{250}{1000}\text{ Nm}$$

$$\tau = 483.76\text{Nm}$$

PROBLEM 4.46

A force $F = 6$ mg is required to raise the load of mass m with the rope as shown in Figure 4.46(a). Calculate the coefficient of friction μ between the rope and the fixed pegs.

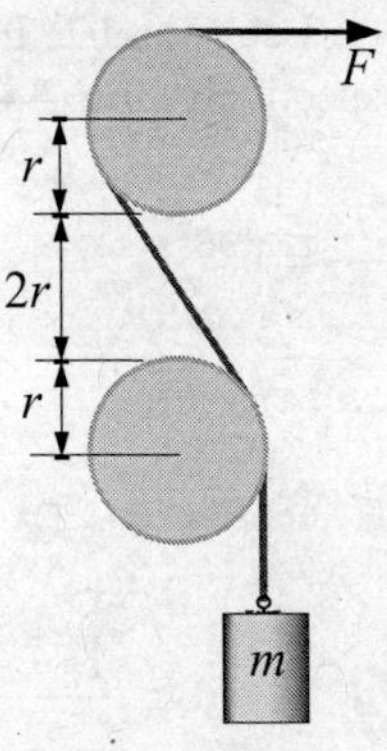

FIGURE 4.46(a) |

Solution. Since slipping of the rope is impending upward, we apply the formula of tension ratio [Fig. 4.46(b)]. Tension ratios of belt for upper and lower pegs are given by

$$\frac{F}{T} = e^{\mu\left[\left(\frac{\pi}{2}\right)+\theta\right]} \quad \text{(i)}$$

and

$$\frac{T}{mg} = e^{\mu\theta} \quad \text{(ii)}$$

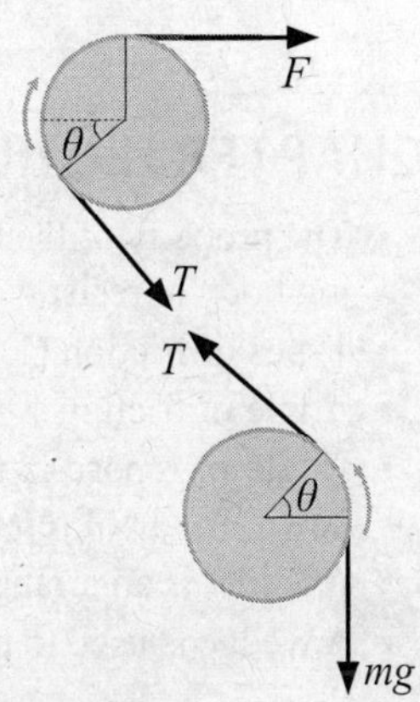

FIGURE 4.46(b) |

Multiplying Equations (i) and (ii), we obtain

$$\frac{F}{T} \times \frac{T}{mg} = e^{\mu\left[\left(\frac{\pi}{2}\right)+\theta\right]+\mu\theta} \quad \text{or} \quad \frac{6mg}{mg} = e^{\mu\left[\left(\frac{\pi}{2}\right)+2\theta\right]}$$

From the geometry $\sin\theta = \frac{2r}{4r} \Rightarrow \theta = \frac{\pi}{6}$

$$6 = e^{\mu\left[\left(\frac{\pi}{2}\right)+\left(\frac{2\pi}{6}\right)\right]} \quad \text{or} \quad 6 = e^{\pi\times 2.61}$$

or

$$\mu \frac{\log_e 6}{2.61} = 0.68$$

$$\mu = 0.68$$

PROBLEM 4.47

Figure 4.47(a) shows a rope brake type dynamometer, which is used to measure the power output of a small IC engine. The coefficient of kinetic friction between rope and brake drum is 0.2. When the flywheel is at rest, the reading of each spring scale is 50 N. What will be the reading of each scale when the flywheel is rotating clockwise? Assume that the belt is of constant length.

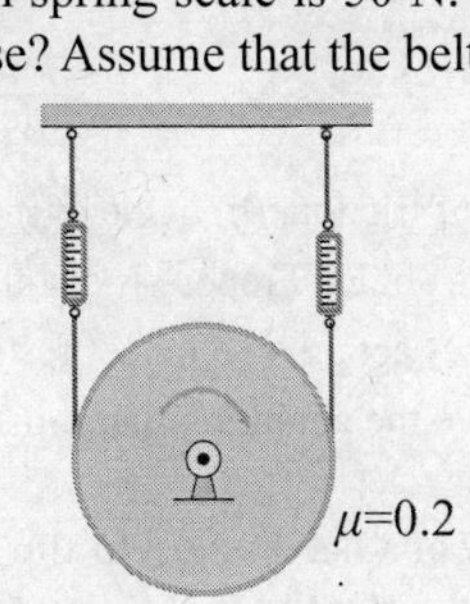

FIGURE 4.47(a) |

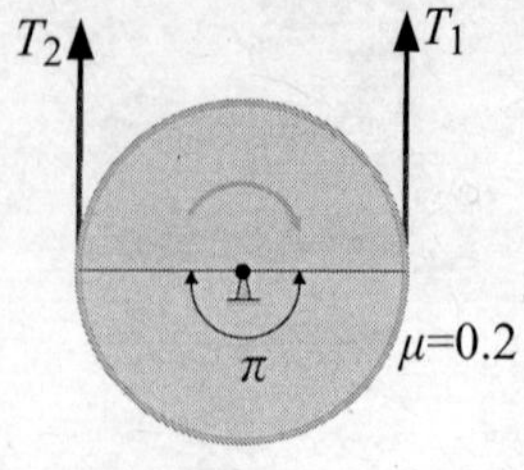

FIGURE 4.47(b) |

Solution. Let T_1 and T_2 be the tension of the belt on the tight and stock sides, respectively [Fig. 4.47(b)].

Tension ratio $\dfrac{T_1}{T_2} = e^{\mu\theta}$

or $$\frac{T_1}{T_2} = e^{0.2\pi} = 1.87$$

or $$T_1 = 1.87T_2 \qquad \text{(i)}$$

Given $$\frac{T_1 + T_1}{2} = 50$$

or $$T_1 + T_2 = 100 \qquad \text{(ii)}$$

Solving Equations (i) and (ii), we get

$$T_2 = 34.78\text{N}, \; T_1 = 65.0\text{N}$$

Readings of scale are 65.05 N and 34.78 N

CHAPTER HIGHLIGHTS

- The property of the bodies by virtue of which a force is exerted by a stationary body on the moving body to resist the motion of the moving body is called friction.
- Types of friction (1) Dry Friction, (2) Film Friction and (3) Boundary (or skin) Friction
- Angle of friction $\phi = \tan{-1}\ \mu$
- Angle of repose is the angle of inclined plane, at which a block resting on it is about to slide down. Angle of repose $\theta = \phi$
- A ladder is an arrangement used for climbing on the walls or roofs.
- A wedge is used to produce small adjustments in the position of body or raise the heavy loads.
- Moment require to rotate the circular disc is $\text{M} = \dfrac{2}{3}\mu\text{PR}$
- Coefficient of rolling resistance $\mu_r = \dfrac{a}{h}$.
- Tension ratio of belt (when the belt is on the point of slipping on the drum) is $\dfrac{T_1}{T_2} = e^{\mu}$.

REVIEW QUESTIONS

Q1. Define friction and give its types.

Q2. Explain static and kinetic friction.

Q3. State the laws of solid friction.

Q4. Define the following terms: (a) angle of friction, (b) angle of repose, and (c) coefficient of friction.

Q5. What is angle of repose? Show that the angle of repose is equal to angle of static friction.

Q6. What is ladder friction? Show the forces acting on a ladder resting against a wall.

Q7. What is wedge? Where is it used? Give the condition that must be satisfied in the state of equilibrium of wedge and slider.

Q8. Derive an expression for a moment M of a flat disc just to slip with a thrust P (given: radius of disc = R, coefficient of friction between disc and surface = μ).

Q9. What is rolling resistance? How is coefficient of rolling resistance evaluated?

Q10. Derive an expression for the ratio of tensions for a flat belt passing over a pulley, when it is just at the point of slipping.

Q11. Derive an expression for the least force P required to move the body on an inclined plane in (a) upward direction, (b) downward direction as shown in Figure 4.Q11 (friction angle $\phi = \tan_{-1} \mu$).

[Ans; (a) $P = W \tan(\theta + \phi)$; (b) $P = W \tan(\theta - \phi)$]

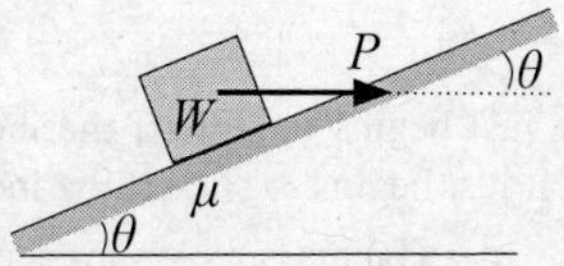

FIGURE 4.Q11 |

Q12. Derive an expression for the least force P required to move the body on an inclined plane in (a) upward direction, (b) downward direction as shown in Figure 4.Q12 (friction angle $\phi = \tan^{-1} \mu$).

$$\left[\text{Ans; (a) } P = W\frac{\sin(\theta+\phi)}{\cos(a-\phi)}\text{; (b) } P = W\frac{\sin(\theta-\phi)}{\cos(a+\phi)}\right]$$

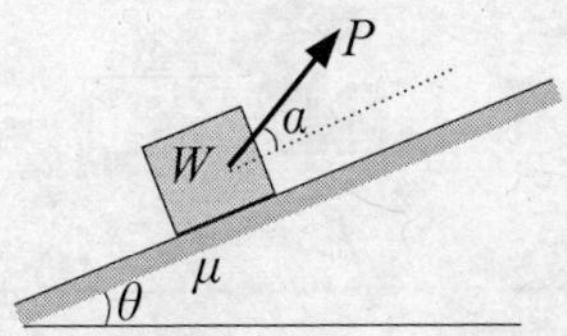

FIGURE 4.Q12 |

Q13. Derive an expression for the least force required to drag a body on a rough horizontal plane.

Q14. A ladder of length 6 m and weight 500 N rests on a rough wall and smooth ground making an angle of 60° with the ground. A force of 200 N is applied to the end of the ladder on the ground opposite to x-direction. If the ladder is in equilibrium condition, find the friction force acting on the ladder resting on the wall.
[Ans; 96.4 N]

Q15. Find out the height of a particle of weight W from the ground which rests inside a hollow sphere of radius r, assuming $\mu = 1/\sqrt{3}$.
[Ans; 0.134r]

MULTIPLE CHOICE QUESTIONS

Q1. The ratio of limiting friction and normal reaction is known as

(A) coefficient of friction
(B) angle of friction
(C) angle of repose
(D) friction resistance

[Ans: A]

Q2. Limiting force of friction is the

(A) tangent of angle between normal reaction and the resultant of normal reaction and limiting friction
(B) ratio of limiting friction and normal reaction
(C) the friction force acting when the body is just about to move
(D) the friction force acting when the body is in motion

[Ans: C]

Q3. A 13 m ladder is placed against a smooth vertical wall with its lower end 5 m from the wall. What should be the coefficient of friction between ladder and the floor so that it remains in equilibrium?

(A) 0.1 (B) 0.15 (C) 0.25 (D) None of these

[Ans: D]

Q4. A body of weight W on an inclined plane of angle α being pulled up by a horizontal force P will be on the point of motion up the plane when P is equal to

(A) W (B) $W\sin(\alpha + \phi)$ (C) $W\tan(\alpha + \phi)$ (D) $W\tan\alpha$

[Ans: C]

Q5. Block A kept on an inclined surface just begins to slide if the inclination is 30°. Block A is replaced by another block, B, and it is found that it just begins to slide if the inclination is 40°. Then

(A) mass of A > mass of B (B) mass of A < mass of B

(C) mass of A = mass of B (D) all the three are possible

[Ans: D]

Q6. A block of mass m rests on a horizontal surface. The coefficient of friction between the mass and the surface is μ. If the mass is pulled by a force F as shown in Figure 4.MQ6, the limiting friction between the block and the surface will be

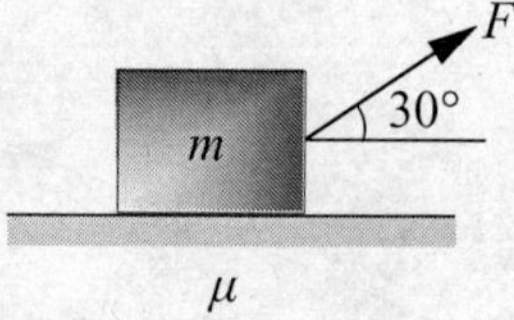

FIGURE 4.MQ6 |

(A) μmg (B) $\mu[mg - (\sqrt{3}/2)F]$ (C) $\mu\,[mg - (1/2)F]$ (D) $\mu[mg + (1/2)F]$

[Ans: C]

Q7. A body of mass 2 kg rests on a rough inclined plane making an angle of 30° with the horizontal. The coefficient of static friction between the block and the plane is 0.7. The frictional force on the block is

(A) 9.81 N (B) 11.88 N (C) 16.9 N (D) 6.86 N

[Ans: A]

Q8. A cubical block of side l rests on a rough horizontal surface with a coefficient of friction μ. A horizontal force F is applied on the block as shown in Figure 4.MQ8. If the coefficient of friction is sufficiently high so that the block does not slide before toppling, the minimum force required to topple the block is

(A) Infinite (B) $mg/4$ (C) $mg/2$ (D) $mg(1 - \mu)$

[Ans: C]

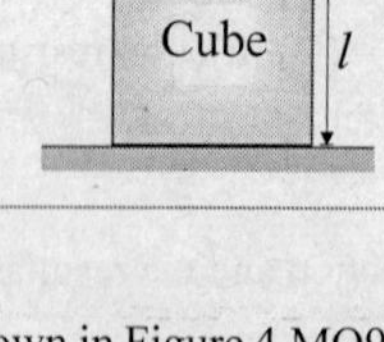

FIGURE 4.MQ8 |

Q9. Block A of mass M rests on a table as shown in Figure 4.MQ9. The coefficient of friction between block and the table is μ. The maximum weight of block B that would keep the system in equilibrium is

(A) $\mu Mg\tan\theta$ (B) Mg (C) μMg (D) $\mu Mg\cos\theta$

[Ans: A]

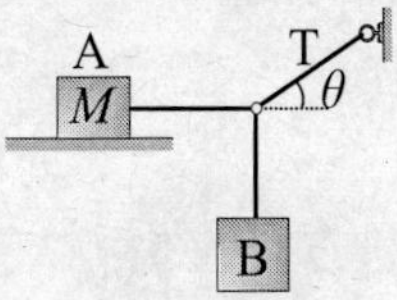

FIGURE 4.MQ9 |

Q10. The block shown in Figure 4.MQ10 is kept in equilibrium and prevented from sliding down by applying a force of 500 N. The coefficient of friction is $\frac{\sqrt{3}}{5}$. The weight of the block would be

(A) 4000 N (B) 2500 N (C) 1000 N (D) 500 N

[Ans: B]

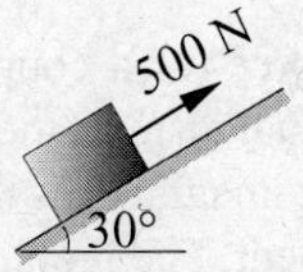

FIGURE 4.MQ10 |

Q11. A block of weight W is resting on an inclined plane set at an angle θ to the horizontal. If the coefficient of dry friction between the surfaces is μ, the frictional force acting on the block is equal to

(A) $\mu W \cos\theta$ subject to a maximum of μW (B) $\mu W \sin\theta$ subject to a maximum of μW

(C) $W \sin\theta$ subject to a maximum of $\mu W \cos\theta$ (D) $\mu W \tan\theta$ subject to a maximum limit

[Ans: C]

Q12. A particle inside a hollow sphere of radius r, having a coefficient of friction $1/\sqrt{3}$ can rest up to a height of

(A) $r/2$ (B) $r/4$ (C) $r/8$ (D) $0.134r$

[Ans: D]

Q13. An inverted "V" is made up of two uniform boards smoothly joined and each weighing 200 N. Each side has the same length and makes an angle of 30° with the vertical as shown in Figure 4.MQ13. The magnitude of static frictional force that acts on each of the lower end of the V is

(A) $100\sqrt{3}$ N (B) $100/\sqrt{3}$ N (C) $200\sqrt{3}$ N (D) $200/\sqrt{3}$ N

[Ans: B]

FIGURE 4.MQ13 |

Q14. A ladder 5 m long rests on a rough horizontal plane and against a smooth vertical wall. The ladder is about to slip on the rough plane when its lower end is at a distance 4 m from the wall. The coefficient of friction is

(A) 2/3 (B) 1/2 (C) 3/5 (D) 4/5

[Ans: A]

Q15. A light ladder of length 10 m is supported on a rough floor having a coefficient of friction $\sqrt{3}/4$ and leans against a smooth wall. If the ladder makes an angle of 30° with the wall and a man can climb up the ladder without slipping taking place up to a distance D along the ladder from the foot of the ladder, then D is equal to

(A) 9.5 m (B) 7.5 m (C) 6.5 m (D) 5.5 m

[Ans: B]

5 Virtual Work

INTRODUCTION

The virtual work is the work done by a force (or a couple) on a body due to a small virtual (i.e., imaginary) displacement of the body. A virtual displacement is any arbitrary very small change in the position of a body. The displacement is only imagined, as the body is not actually displaced, for which the displacement is called virtual displacement. The product of the force and the virtual displacement in the direction of the force is called virtual work.

WORK OF FORCE

Consider a body that moves from a point A to B under the force **F**. If r denotes the position vector corresponding to point A, the small vector **dr** is called the displacement of the particle. The work done by the force **F** during the displacement dr is defined as

$$dU = \mathbf{F}\mathbf{dr}$$

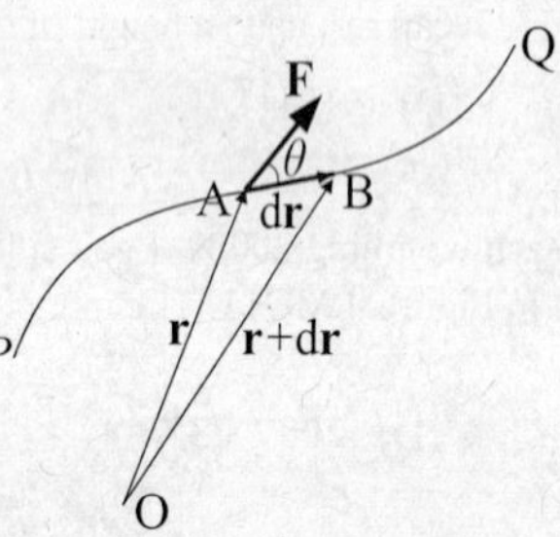

FIGURE 5.1 |

If F denotes the magnitude of the force **F**, ds denotes the magnitude of the differential displacement dr, and θ denotes the angle between F and ds (Fig. 5.1). By the definition of the dot product, we write

$$dU = F\,ds\cos\theta$$

The work dU is positive if the angle θ is acute and negative if θ is obtuse.

We may write

$$dU = (F\cos\theta)ds$$

$$= (\text{force component } F\cos\theta \text{ in the direction of displacement}) \times \text{displacement}$$

$$dU = F(ds\cos\theta)$$

$$= \text{Force } F \times \text{displacement component in the direction of force } F$$

If we express **F** and **dr** in terms of their rectangular components, we have

$$dU = (F_x i + F_y j + F_z k)\cdot(dx i + dy j + dz k)$$

$$dU = (F_x dx + F_y dy + F_z dz)$$

To obtain the total work U done by the force **F** during a finite movement of point A from P to Q. We integrate dU between these positions. Thus,

$$U = \int \mathbf{F} \cdot dr$$

or

$$U = \int F \cos\theta \, ds$$

or

$$U = \int (F_x dx + F_y dy + F_z dz)$$

WORK OF COUPLE

Consider a rigid body acted upon by a couple M as shown in Figure 5.2 whose angular movement is represented by an angle $d\theta$. The work done by the couple M during the angular displacement $d\theta$ is defined as

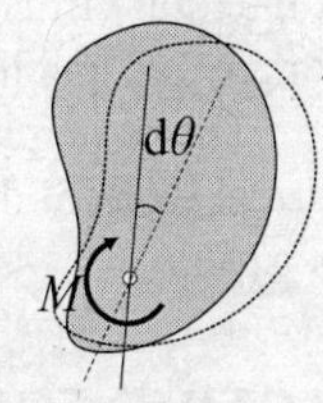

FIGURE 5.2 |

$$dU = M \cdot d\theta$$

Total work done by the couple M during finite rotation is

$$U = \int M \cdot d\theta$$

The work of the couple is positive if M has same sense as $d\theta$ and negative if M has a sense opposite to that of the rotation.

VIRTUAL WORK

When a system or a body is in equilibrium under the action of a number of forces, there is no displacement and accordingly no work is done. However, if the system is imagined to undergo any arbitrary but small displacement, some work can be imagined to have been done.

The imaginary small displacement given to the system is called virtual displacement. The product of the net resultant force and the virtual displacement in the direction of the force is called virtual work.

The concept of virtual work is used in solving problems related to static equilibrium.

PRINCIPLE OF VIRTUAL WORK

Consider a particle at A, subjected to several forces $\mathbf{F}_1, \mathbf{F}_2, \mathbf{F}_3, \ldots, \mathbf{F}_n$ as shown in Figure 5.3. We assume that the particle undergoes a small virtual displacement $\delta\mathbf{r}$ from A to A. The work of each of the force $\mathbf{F}_1, \mathbf{F}_2, \mathbf{F}_3, \ldots, \mathbf{F}_n$ during the virtual displacement $\delta\mathbf{r}$ is called virtual work. The virtual work of all the forces acting on the particle is

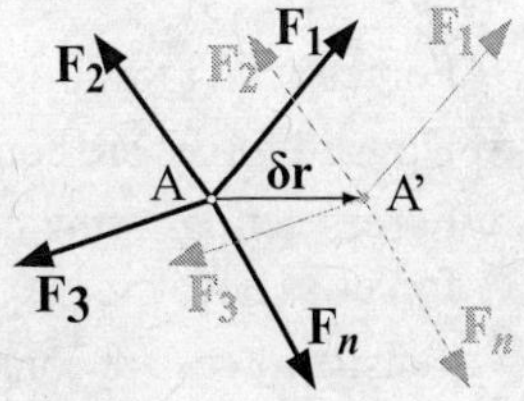

FIGURE 5.3 |

$$dU = \mathbf{F}_1 \cdot \delta\mathbf{r} + \mathbf{F}_2 \cdot \delta\mathbf{r} + \mathbf{F}_3 \cdot \delta\mathbf{r} + \cdots + \mathbf{F}_n \cdot \delta\mathbf{r}$$

$$= (\mathbf{F}_1 + \mathbf{F}_2 + \mathbf{F}_3 + \cdots + \mathbf{F}_n) \cdot \delta\mathbf{r}$$

or

$$dU = (\Sigma\, \mathbf{F}) \cdot \delta\mathbf{r}, \quad \text{(i)}$$

where $\Sigma\, \mathbf{F}$ is the resultant of the given forces.

If we express $\Sigma\, \mathbf{F}$ and $\delta\mathbf{r}$ in terms of their rectangular components, we have

$$dU = \left(i \sum F_x + j \sum F_y + k \sum F_z\right) \cdot \left(i\delta x + j\delta y + k\delta z\right)$$

$$dU = \sum F_x \delta x + \sum F_y \delta y + \sum F_z \delta z$$

The principle of virtual work for a *particle* states that "if a particle is in an equilibrium, the total virtual work of the forces acting on the particle is zero for any virtual displacement of the particle". This condition is necessary; if the particle is in equilibrium, the resultant $\sum \mathbf{F}$ of the forces is zero, and it follows from Equation (i) that the total virtual work dU is zero. The condition is also sufficient: if the total virtual work dU is zero for any virtual displacement, the scalar product $(\sum \mathbf{F}) \cdot \delta \mathbf{r}$ is zero for any $\delta \mathbf{r}$, and the resultant $\sum \mathbf{F}$ must be zero.

The principle of virtual work for a *rigid body* states that "if a rigid body is in equilibrium the total virtual work of the external forces acting on the rigid body is zero for any virtual displacement of the body."

ACTIVE FORCES AND ACTIVE FORCE DIAGRAM

Active forces are external forces capable of doing virtual work during possible virtual displacement. In Figure 5.4, forces P and Q are active forces because they would do work as the links move.

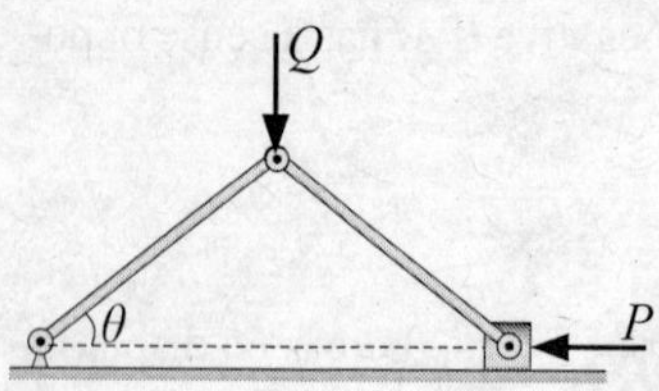

Active force diagram

FIGURE 5.4 |

When using the method of virtual work, we draw a diagram that isolates the system under consideration. Unlike the free body diagram, where all forces are shown, the diagram for the method of virtual work needs to show only the active forces, since the reactive forces do not enter into the application of $\delta U = 0$. Such a drawing will be termed an active force diagram. Figure 5.4 is an active force diagram for the system shown.

SIGN CONVENTIONS

There are different sign conventions for finding out the virtual work done in different books; we shall use the following sign conventions.

- Forces acting towards right are considered as positive, whereas those towards left as negative.
- Upward forces are considered as positive, whereas the downward forces as negative.
- Forces acting in the clockwise direction are considered as positive, whereas the anticlockwise forces as negative.
- Tensile forces are considered as positive whereas the compressive forces as negative.

STABILITY OF EQUILIBRIUM

STABLE EQUILIBRIUM

Stable equilibrium occurs if the potential energy V is a minimum. If, as shown in Figure 5.5, a uniform rod is hinged at one end and placed in the vertically downward position, intuition indicates that this is a position of stable equilibrium with the potential energy of the rod at minimum because any disturbance will be followed by return to the original position. Using the x-axis as a standard (datum), the potential energy of the rod at an angle θ is

$$V = -mgy$$

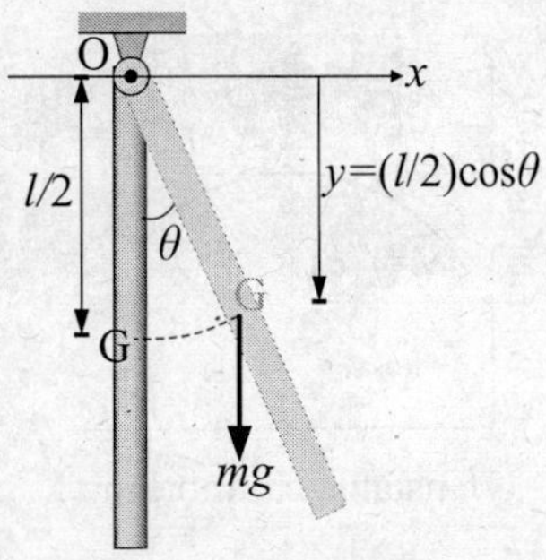

FIGURE 5.5 |

The vertical position of centre of mass G is $y = (l\,/\,2)\cos\theta$,

$$V = -mg\frac{l}{2}\cos\theta$$

Set $\frac{dV}{d\theta}$ equal to zero to find the equilibrium position:

$$\frac{dV}{d\theta} = mg\frac{l}{2}\sin\theta = 0$$

Hence, the solution needed here is $\theta = 0$ (the rod is then vertically downwards). To determine the type of equilibrium, it is necessary to evaluate $\frac{d^2V}{d\theta^2}$ at the equilibrium position. Thus,

$$\frac{d^2V}{d\theta^2} = mg\frac{l}{2}\cos\theta$$

and at $\theta = 0$,

$$\frac{d^2V}{d\theta^2} = mg\frac{l}{2}\,\text{(positive)},$$

showing a stable equilibrium (Fig. 5.6).

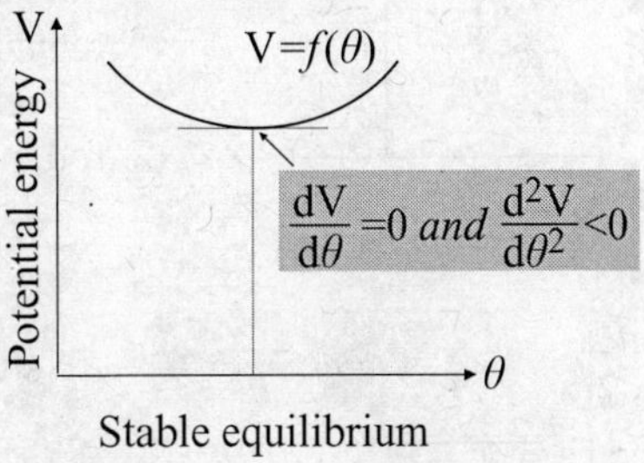

FIGURE 5.6 |

UNSTABLE EQUILIBRIUM

Unstable equilibrium occurs if the potential energy V is a maximum. If, as shown in Figure 5.7, the rod is hinged at one end and placed in the vertically upward position, intuition indicates that this is a position of unstable equilibrium, with the potential energy of the rod a maximum. Using the *x*-axis as a standard (datum), the potential energy of the rod at any position θ above the *x*-axis is

$$V = mgy$$

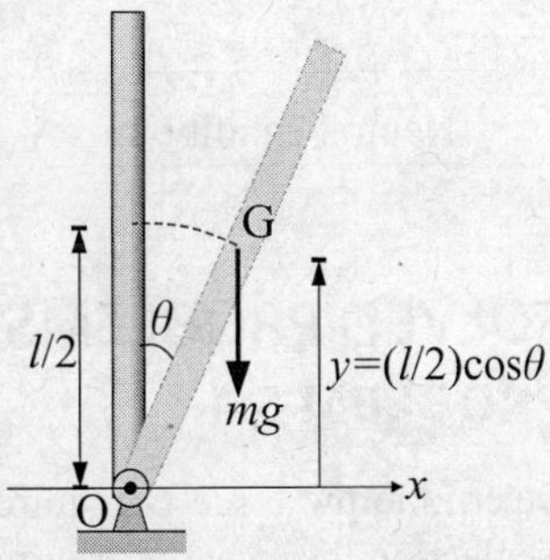

FIGURE 5.7 |

The vertical position of centre of mass G is $y = (l\,/\,2)\cos\theta$,

$$V = mg\frac{l}{2}\cos\theta$$

Set $\frac{dV}{d\theta}$ equal to zero to find the equilibrium position:

$$\frac{dV}{d\theta} = -mg\frac{l}{2}\sin\theta = 0$$

Hence, the solution needed here is $\theta = 0$ (the rod is then vertically upwards). To determine the type of equilibrium, it is necessary to evaluate $\frac{d^2V}{d\theta^2}$ at the equilibrium position. Thus,

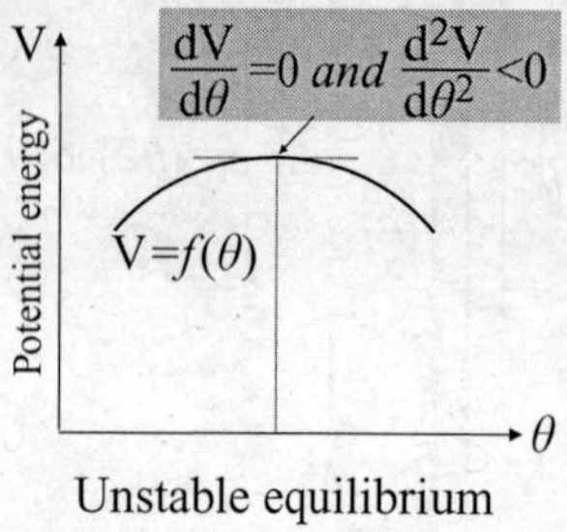

FIGURE 5.8 |

$$\frac{d^2V}{d\theta^2} = -mg\,\frac{l}{2}\cos\theta$$

and at $\theta = 0$,

$$\frac{d^2V}{d\theta^2} = -mg\,\frac{l}{2}\,\text{(negative)},$$

showing unstable equilibrium (Fig. 5.8).

NEUTRAL EQUILIBRIUM

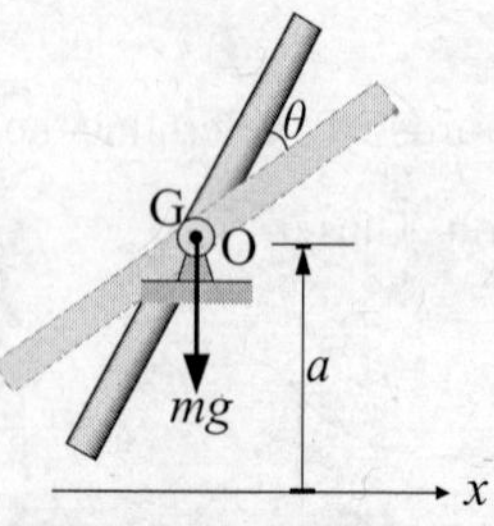

FIGURE 5.9 |

Neutral equilibrium exists if a system remains in any position in which it is placed. Neutral equilibrium occurs if the potential energy V remains constant. For example, a rod, which is hinged about the centre of mass as shown in Figure 5.9 can be located any position (θ) and will remain there.

Using the x-axis as a standard (datum), the potential energy of the rod in any position θ is

$$V = mga$$

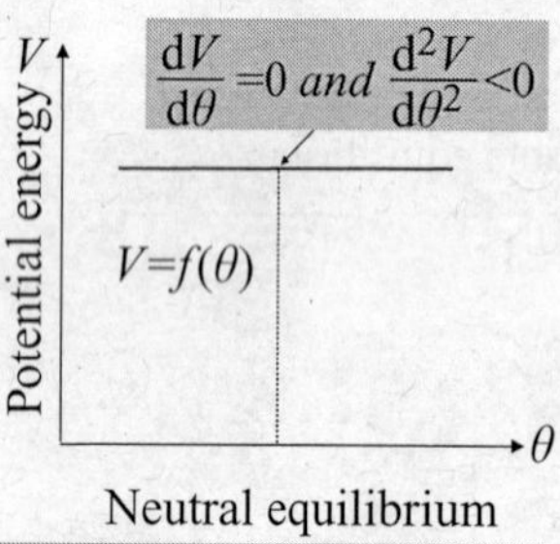

FIGURE 5.10 |

To study equilibrium, set $\dfrac{dV}{d\theta} = 0$

$$\frac{dV}{d\theta} = \frac{d}{d\theta}(mga) = 0,$$

where θ may be any angle and $\dfrac{d^2V}{d\theta^2} = 0$ independent of θ (Fig. 5.10).

SOLVED PROBLEMS

PROBLEM 5.1

Determine the force Q required to maintain the equilibrium of the linkage shown in Figure 5.1(a). Neglecting the effect of friction.

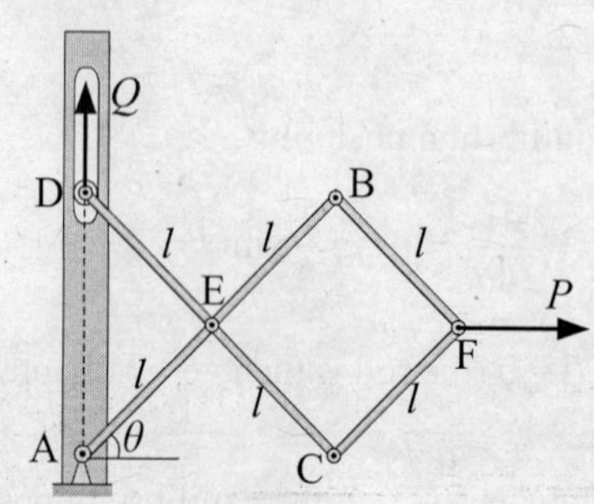

FIGURE 5.1(a) |

Solution. The active force diagram shows the two forces P and Q. There are no other external forces that do work on the linkage during a change in the angle θ. The principle of virtual work requires that the total work of all external forces be zero for any virtual displacement:

$$\delta u = 0 \Rightarrow P\delta x + Q\delta y = 0$$

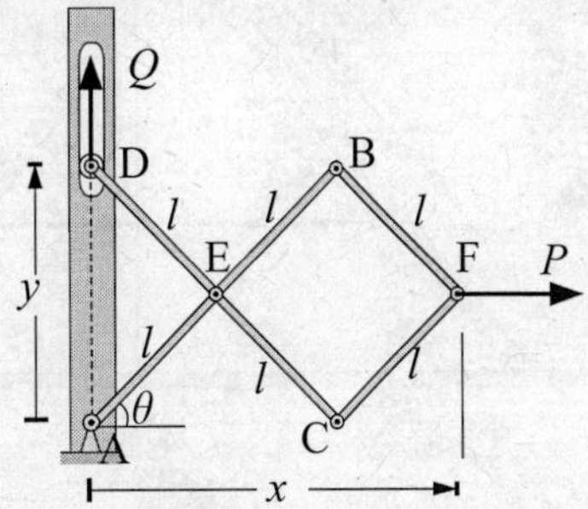

FIGURE 5.1(b) |

From Figure 5.1(b)

$$x = 3l\cos\theta \quad \text{or} \quad \delta x = -3l\sin\theta\delta\theta$$

$$y = 2l\sin\theta \quad \text{or} \quad \delta y = 2l\cos\theta\delta\theta$$

Substituting the values of δx and δy into the virtual work gives us

$$P(-3l\sin\theta\delta\theta) + Q(2l\cos\theta\delta\theta) = 0$$

or

$$3P\sin\theta = 2Q\cos\theta$$

From which, we obtain

$$Q = \frac{3}{2}P\tan\theta$$

PROBLEM 5.2

Two pin-joined bars are held on a frictionless plane by a rope as shown in Figure 5.2(a). Find the tension T in the cable in terms of the load W and the angle θ.

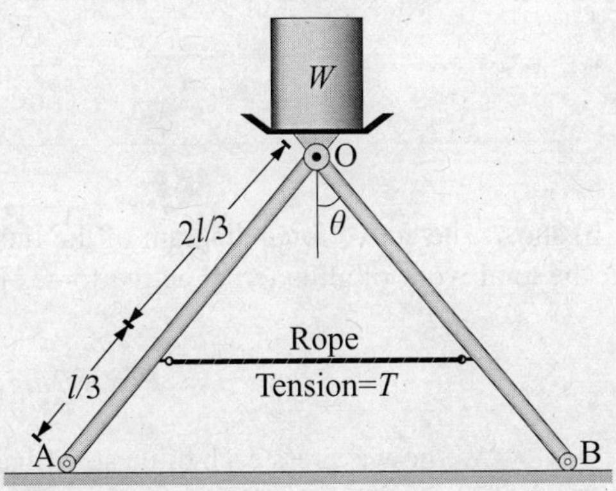

FIGURE 5.2(a) |

Solution. Figure 5.2(b) shows the active force diagram. The reactive forces do not do any work during a virtual displacement and are therefore not shown.

The virtual work done by W is

$$-W\delta y = -W\delta(l\cos\theta)$$

$$= Wl\sin\theta\delta\theta$$

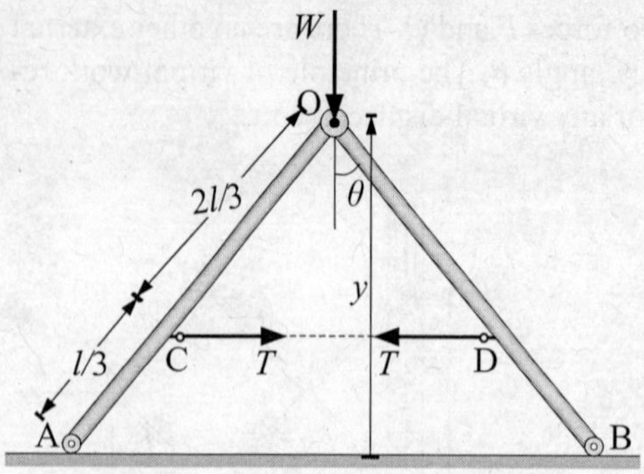

FIGURE 5.2(b) |

The virtual work done by the tension T is

$$-T\delta(\text{CD}) = -T\delta\left(2 \times \frac{2l}{3}\sin\theta\right)$$

$$= -\frac{4Tl}{3}\cos\theta\delta\theta$$

The total virtual work δu is zero; hence

$$\delta u = 0 \Rightarrow -W\delta y - T\delta(\text{CD}) = 0$$

or $$Wl\sin\theta\delta\theta - \frac{4Tl}{3}\cos\theta\delta\theta = 0$$

From which, we obtain

$$T = \frac{3}{4}W\tan\theta$$

PROBLEM 5.3

Determine the magnitude of the force P required to maintain the equilibrium of the two bar linkages [Fig. 5.3(a)]. The mass and the length of each link are m and l, respectively.

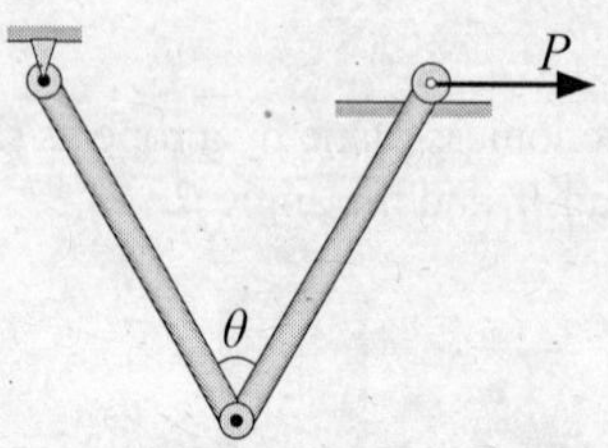

FIGURE 5.3(a) |

Solution. Figure 5.3(b) shows the active force diagram of the linkage system. The principle of virtual work requires that the total work of all external active forces be zero for any virtual displacement. Thus,

$$\delta u = 0 \Rightarrow P\delta x + 2mg\delta y = 0$$

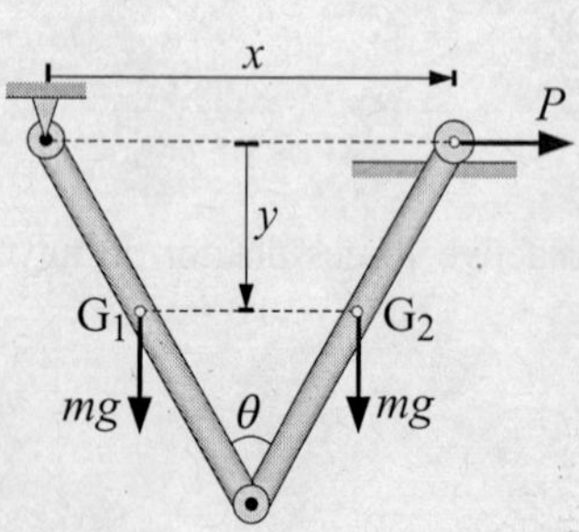

FIGURE 5.3(b) |

We now express each of these virtual displacements in terms of the variable θ. From the geometry

$$y = \frac{l}{2}\cos\frac{\theta}{2} \quad \text{and} \quad x = 2l\sin\frac{\theta}{2},$$

then $$\delta x = l\cos\frac{\theta}{2}\delta\theta \quad \text{and} \quad \delta y = -\frac{l}{4}\sin\frac{\theta}{2}\delta\theta$$

Substituting the values of δx and δy into the equation of virtual work gives us

$$Pl\cos\frac{\theta}{2}\delta\theta - 2mg\frac{l}{4}\sin\frac{\theta}{2}\delta\theta = 0$$

From which, we have

$$\tan\frac{\theta}{2} = \frac{2P}{mg}$$

or

$$P = \frac{mg}{2}\tan\frac{\theta}{2}$$

PROBLEM 5.4

Determine the force P acting on a sliding collar required to maintain the equilibrium of the linkage shown in Figure 5.4(a). Neglect the mass and friction of the moving part.

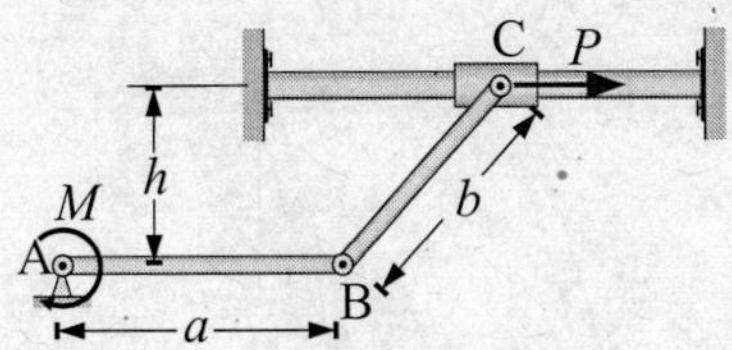

FIGURE 5.4(a) |

Solution. Let the slider be given a small virtual displacement δx, then the virtual work done by the force P is $P\,\delta x$ and virtual work done by couple M is $M\,\delta\theta$. The horizontal displacement δx of slider is expressed in terms of the angular displacement $\delta\theta$ of the link AB [Fig. 5.4(b)] as follows.

$$\delta y = a\delta\theta,$$

where $\delta\theta$ is in radians.

From the right triangle for which link BC is the constant hypotenuse, we may write

$$b^2 = x^2 + y^2$$

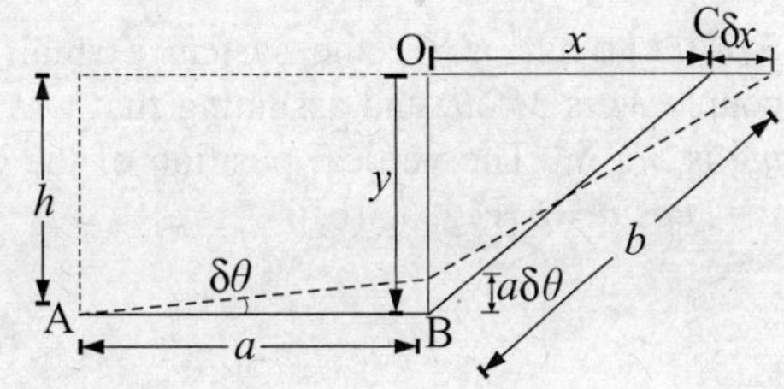

FIGURE 5.4(b) |

We now take the differential of the equation and get

$$0 = 2x\delta x + 2y\delta y \text{ or } \delta x = -\frac{y}{x}\delta y$$

Thus

$$\delta x = -\frac{y}{x}a\delta\theta$$

and the virtual work equation becomes

$$\delta u = 0 \Rightarrow M\delta\theta + P\delta x = 0$$

Substituting the value of δx, we have

$$M\delta\theta + P\left(-\frac{y}{x}a\delta\theta\right) = 0$$

or

$$P = \frac{Mx}{ya} = \frac{Mx}{ha}$$

$$P = \frac{Mx}{ha}$$

PROBLEM 5.5

Derive an expression for the magnitude of the couple M required to maintain the equilibrium of the linkage in the position shown in Figure 5.5(a).

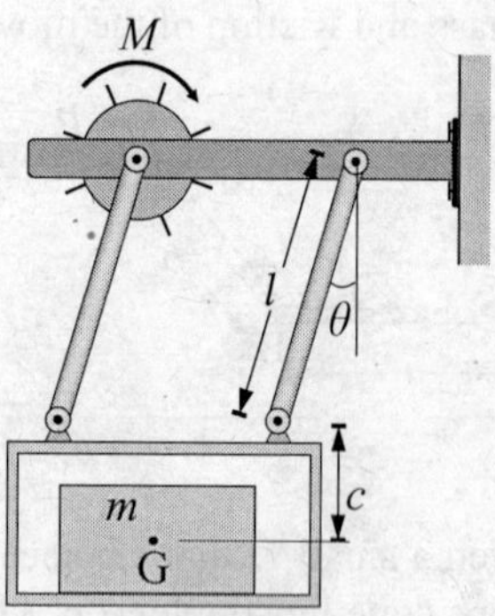

FIGURE 5.5(a) |

Solution. Give the system a small virtual displacement $\delta\theta$ then the work done by the couple M is $M\,\delta\theta$, and assuming that y is positive to downwards then the virtual work done by mg is $mg\,\delta y$. The vertical position of the centre of mass G from the horizontal reference line is $y = l\cos\theta + c$ [Fig. 5.5(b)].

$$mg\delta y = mg\delta(l\cos\theta + c)$$
$$= -mgl\sin\theta\delta\theta$$

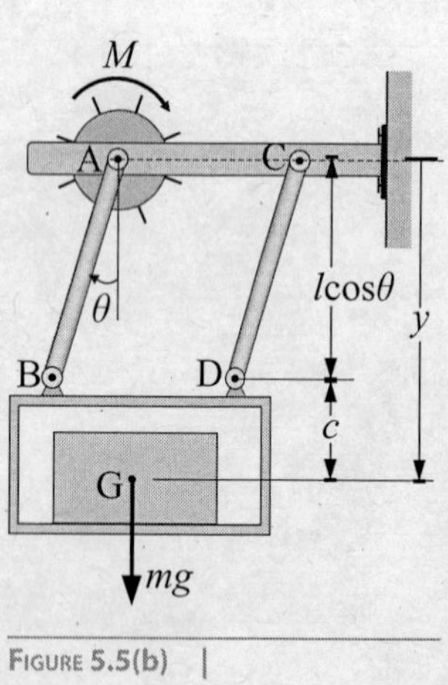

FIGURE 5.5(b) |

The total virtual work done by mg and the couple M must be zero:

$$\delta u = 0 \Rightarrow M\delta\theta + mg\delta y = 0$$

Substituting value of δy into the virtual work equation gives us

$$M\delta\theta - mgl\sin\theta\delta\theta = 0$$

or

$$M = mgl\sin\theta$$

From which we obtain

$$\theta = \sin^{-1}\frac{M}{mgl}$$

PROBLEM 5.6

Figure 5.6(a) shows the slider-crank mechanism. Derive a relation between the moment M applied to the crank and the force F applied to the crosshead required to maintain equilibrium.

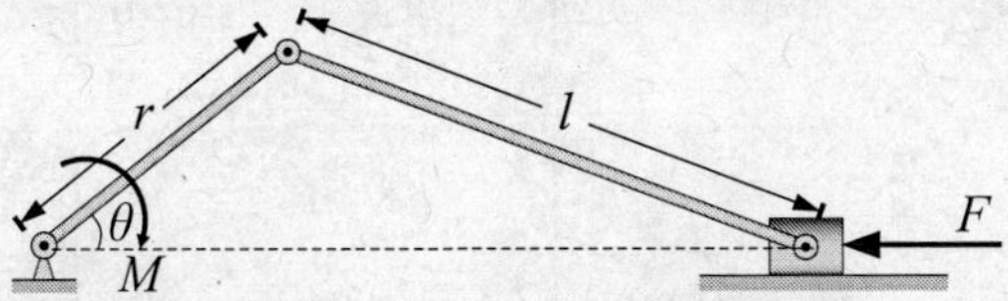

FIGURE 5.6(a) |

Solution. Figure 5.6(b) shows the active force diagram of the slider-crank mechanism. Assuming that x is positive to right, the virtual work done by F is $-F\,\delta x$ and virtual work done by the couple M is $-M\,\delta\theta$. The total virtual work done by the force F and the couple M must be zero. Hence

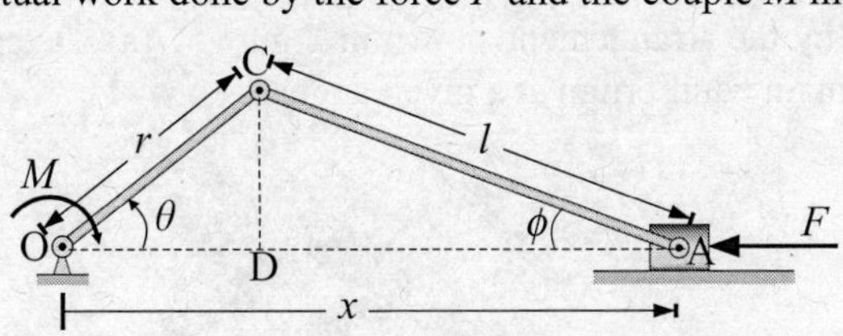

FIGURE 5.6(b) |

$$\delta u = 0 \Rightarrow -M\delta\theta - F\delta x = 0$$

From the geometry of the figure,

$$x = r\cos\theta + l\cos\phi$$

and

$$CD = r\sin\theta = l\sin\phi$$

or

$$\sin\phi = \frac{r}{l}\sin\theta$$

$$\cos\phi = \sqrt{1-\sin^2\phi} \Rightarrow \cos\phi = \sqrt{1-\left(\frac{r}{l}\right)^2\sin^2\theta}, \text{ then}$$

$$x = r\cos\theta + l\sqrt{1-\left(\frac{r}{l}\right)^2\sin\theta},$$

or

$$\delta x = -r\sin\theta\delta\theta - \frac{r^2}{l}\frac{\sin\theta\cos\theta}{\sqrt{1-\left(\frac{r}{l}\right)^2\sin^2\theta}}\delta\theta$$

Substituting δx into the virtual work equation

$$-M\delta\theta - F\left(-r\sin\theta\delta\theta - \frac{r^2}{l}\frac{\sin\theta\cos\theta}{\sqrt{1-\left(\frac{r}{l}\right)^2\sin^2\theta}}\delta\theta\right) = 0$$

or

$$M = Fr\sin\theta\left(1 + \frac{\cos\theta}{\sqrt{\left(\frac{l}{r}\right)^2 - \sin^2\theta}}\right)$$

$$M = Fr\sin\theta\left(1 + \frac{\cos\theta}{\sqrt{\left(\frac{l}{r}\right)^2 - \sin^2\theta}}\right)$$

PROBLEM 5.7

A box of mass m is tilted by the arrangement shown in Figure 5.7(a). Determine the force F_C in the cylinder necessary to maintain equilibrium at a given angle θ.

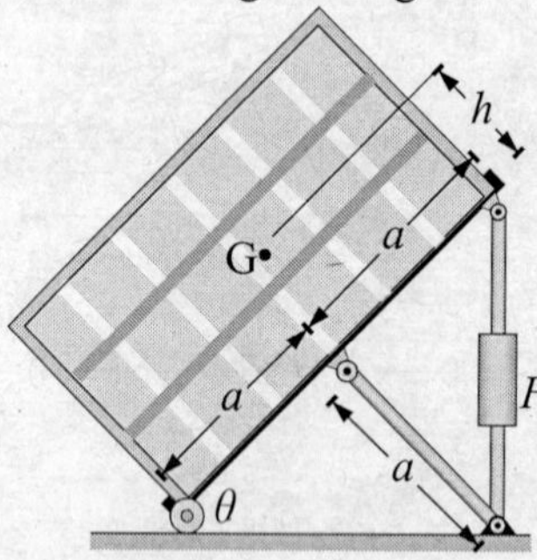

FIGURE 5.7(a) |

Solution. We first observe that the reactions at O and C do not do any work. We denote by y the elevation of the centre of mass G above the base and by s the length of the cylinder and piston assembly. The total virtual work by mg and F_C must be zero:

$$\delta u = 0 \Rightarrow -mg\,\delta y + F_C\,\delta s = 0$$

From Figure 5.7(b),

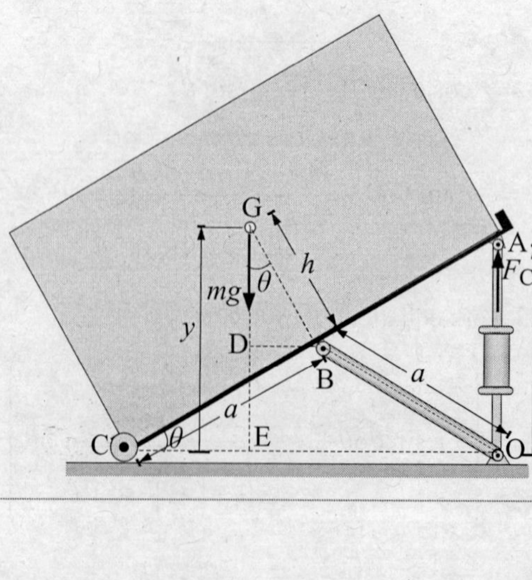

FIGURE 5.7(b) |

$$s = 2a \sin \theta, \text{ and}$$

$$y = GD + DE = h \cos \theta + a \sin \theta, \text{ then}$$

$$\delta s = 2a \cos \delta\theta$$

$$\delta y = -h \sin \theta\delta\theta + a \cos \theta\delta\theta$$

Substituting the values of δs and δy into the virtual work equation, we have

$$-mg\left(-h \sin \theta\delta\theta + a \cos \theta\delta\theta\right) + F_C\left(2a \cos \delta\theta\right) = 0$$

From which, we get

$$F_C = \frac{mg}{2}\left(1 - \frac{h}{a} \tan \theta\right)$$

$$\boxed{F_C = \frac{mg}{2}\left(1 - \frac{h}{a} \tan \theta\right)}$$

PROBLEM 5.8

A portable hydraulic-lift is used to raise a load W, as shown in Figure 5.8(a). Determine the compressive force F_C in the piston rod of the cylinder to support the load W at a height h.

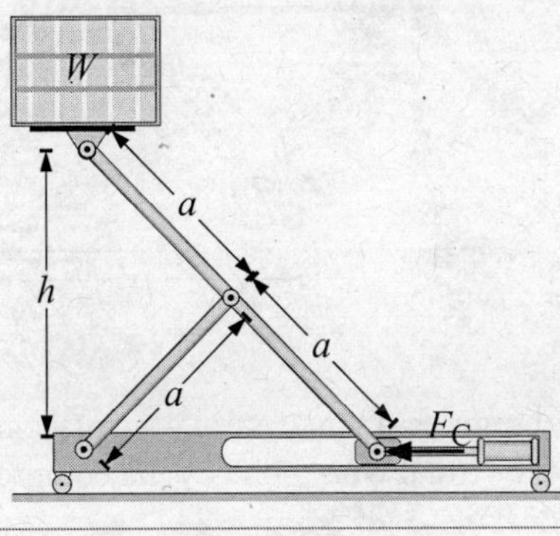

FIGURE 5.8(a) |

Solution. Choosing a coordinate system with origin at O [Fig. 5.8(b)], we write

$$x = 2a \cos \theta \quad \text{or} \quad \delta x = -2a \sin \theta\delta\theta$$

$$y = 2a \sin \theta \quad \text{or} \quad \delta y = -2a \cos \theta\delta\theta$$

The total virtual work done by F_C and W must be zero:

$$\delta u = 0 \Rightarrow -W\,\delta y - F_C\,\delta x = 0$$

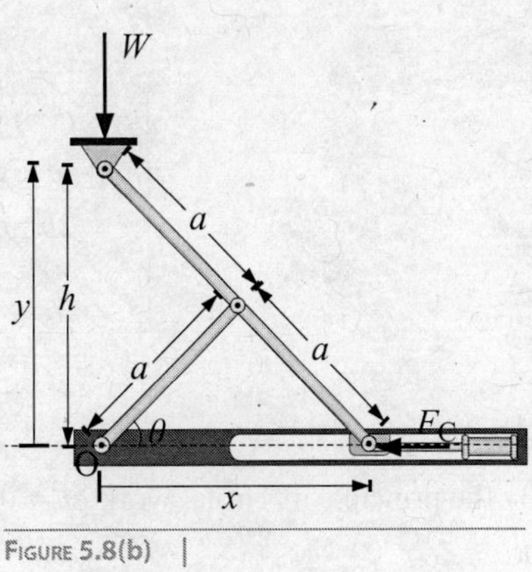

FIGURE 5.8(b) |

Substituting the values of δx and δy into the above equation,

$$-W(2a\cos\theta\delta\theta) - F_C\ (-2a\sin\theta\delta\theta) = 0$$

or

$$2aW\cos\theta = 2aF_C\sin\theta$$

or

$$F_C = W\cot\theta = W\frac{\sqrt{a^2 - \left(\frac{h}{2}\right)^2}}{\frac{h}{2}}$$

$$F_C = W\sqrt{\left(\frac{2a}{h}\right)^2 - 1}$$

PROBLEM 5.9

A hydraulic lift is used to raise a bus of weight W, as shown in Figure 5.9(a). Determine the compressive force F_C exerted by the hydraulic cylinder in terms of the angle θ.

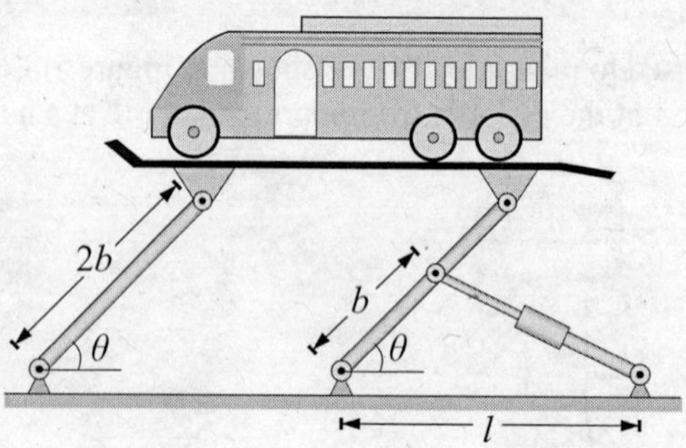

FIGURE 5.9(a) |

Solution. Let the link OA be given a small virtual displacement [Fig. 5.9(b)] such that the length s changes into $s + \delta s$, then the virtual work done by the compressive force F_C is $F_C\ \delta s$ and virtual work done by the weight W is $-W\delta y$.

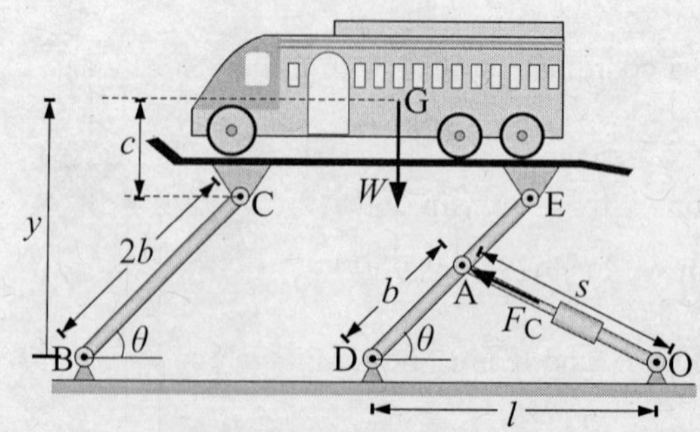

FIGURE 5.9(b) |

By the principle of virtual work $\delta u = 0$.

or

$$-W\,\delta y + F_C\delta s = 0$$

The vertical displacement δy of the platform is expressed in terms of the angular displacement $\delta\theta$ of DE as follows

$$y = \text{DE} \sin\theta + c = 2b \sin\theta + c$$

$$\delta y = 2b \cos\theta\delta\theta$$

To express δs similarly in terms of $\delta\theta$, we first note that by the law of cosines,

$$s^2 = b^2 + l^2 - 2lb\cos\theta$$

Differentiating,

$$2s\delta s = 2lb \sin\theta\delta\theta$$

$$\delta s = \frac{lb\sin\theta}{s}\delta\theta$$

Substituting for δy and δs into the virtual work equation,

$$-W(2b\cos\theta\delta\theta) + F_C\left(\frac{lb\sin\theta}{s}\delta\theta\right) = 0$$

or
$$F_C = 2W\frac{s}{l}\cot\theta$$

Substituting $s = \sqrt{b^2 + l^2 - 2lb\cos\theta}$, we have

$$F_C = 2W\frac{\sqrt{b^2 + l^2 - 2lb\cos\theta}}{l}\cot\theta$$

$$F_C = 2W\cot\theta\sqrt{1 + \left(\frac{b}{l}\right)^2 - 2\left(\frac{b}{l}\right)\cos\theta}$$

PROBLEM 5.10

A smooth cone of weight W stands inverted in a circular hole with its axis vertical as shown in Figure 5.10(a). A string is wrapped twice round the cone just above the hole and pulled tight. What must be the tension in the string so that it will just raise the cone?

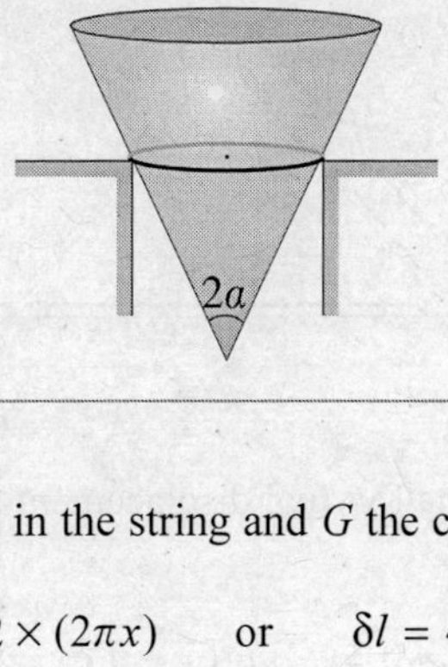

FIGURE 5.10(a) |

Solution. Let T be the tension in the string and G the centre of mass of the cone. The length of string

$$l = 2\times(2\pi x) \quad \text{or} \quad \delta l = 4\pi\delta x$$

Assuming that y is positive to the up direction from the centre of hole O, the virtual work done by the weight W is $-W\,\delta y$.

Virtual work done by string tension is $-T \times \delta l$.

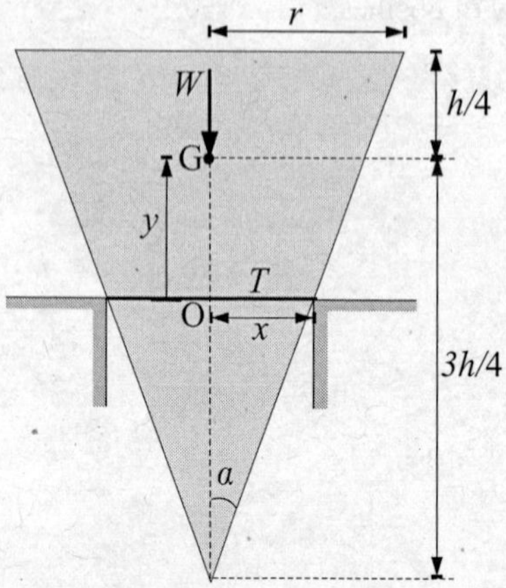

FIGURE 5.10(b) |

By the principle of virtual work done; $\delta u = 0$,

$$\text{or} \qquad -W\,\delta y - T \times \delta l = 0$$

From the geometry of Figure 5.10(b),

$$\frac{r}{h} = \frac{x}{(3h/4) - y} \Rightarrow x = \frac{3r}{4} - \frac{r}{h}y \qquad \text{or} \qquad \delta y = -\frac{h}{r}\,\delta x$$

Substituting the values of δy and δl into the virtual work equation, we have

$$-W\left(-\frac{h}{r}\,\delta x\right) - T \times 4\pi\delta x = 0$$

$$\text{or} \qquad T = \frac{Wh}{4\pi r} = \frac{W}{4\pi}\cot a$$

$$T = \frac{W}{4\pi}\cot a$$

PROBLEM 5.11

Two light rods AD and BC are smoothly hinged at O, and ends of rods A and B are connected by a string as shown in Figure 5.11(a). The rods rest in a vertical plane with ends A and B on a smooth horizontal table. A smooth circular disc of radius r and weight W is placed on the rods above O with its plane vertical so that the rods are tangents to the disc. Find the tension of the string.

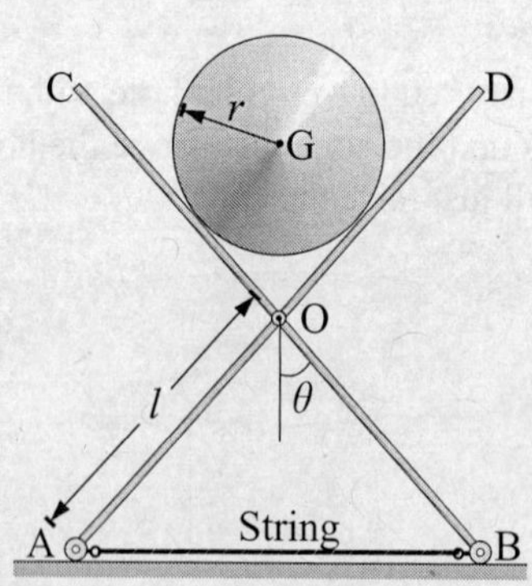

FIGURE 5.11(a) |

Solution. Give the system a small virtual displacement such that θ changes into $\theta + \delta\theta$, then the equation of virtual work is

$$\delta u = 0 \Rightarrow -mg\delta y - T\delta x = 0$$

From the geometry of Figure 5.11(b),

$$y = GO + OE = r\operatorname{cosec}\theta + l\cos\theta$$

or
$$\delta y = -r\operatorname{cosec}\theta\cot\theta\,\delta\theta - l\sin\theta\,\delta\theta$$

Length of the string

$$x = 2l\sin\theta \qquad \text{or} \qquad \delta x = 2l\cos\theta\,\delta\theta$$

Substituting the values of δy and δl into the virtual work equation, we have

$$-mg(-r\operatorname{cosec}\theta\cot\theta\,\delta\theta - l\sin\theta\,\delta\theta) - T\times 2l\cos\theta\,\delta\theta = 0$$

From which, we obtain

$$T = \frac{mg}{2}\left(\frac{r}{l}\operatorname{cosec}^2\theta + \tan\theta\right)$$

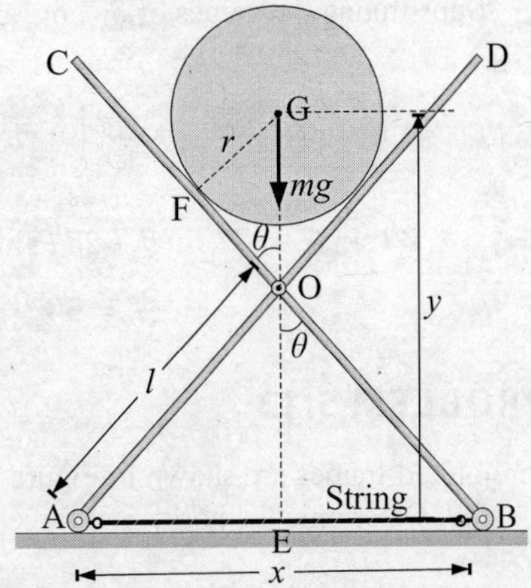

FIGURE 5.11(b) |

PROBLEM 5.12

Six equal rods each of weight W are freely jointed at their extremities so as to form a hexagon as shown in Figure 5.12(a); the rod AF is fixed in a horizontal position and the middle points of AF and CD are jointed by a rope. Find the tension of rope.

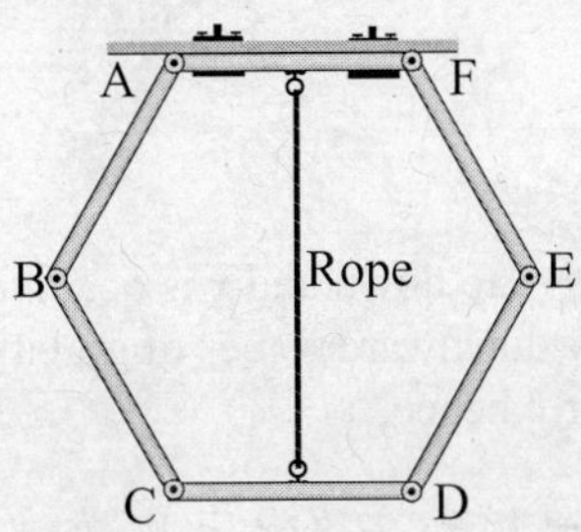

FIGURE 5.12(a) |

Solution. Let W and l be the weight and length of each rod and T be the tension in the rope. Give the system a small virtual displacement, then the principle of virtual work

$$\delta u = 0 \Rightarrow 2W\,\delta y_1 + 2W\,\delta y_2 + W\,\delta y_3 - T\,\delta y_3 = 0$$

From Figure 5.12(b),

$$y_1 = \frac{l}{2}\cos\theta,\ y_2 = \frac{3l}{2}\cos\theta,\ y_3 = 2l\cos\theta, \text{ then}$$

$$\delta y_1 = -\frac{l}{2}\sin\theta\,\delta\theta,\ \delta y_2 = -\frac{3l}{2}\cos\theta\,\delta\theta \text{ and}$$

$$\delta y_3 = -2l\sin\theta\,\delta\theta$$

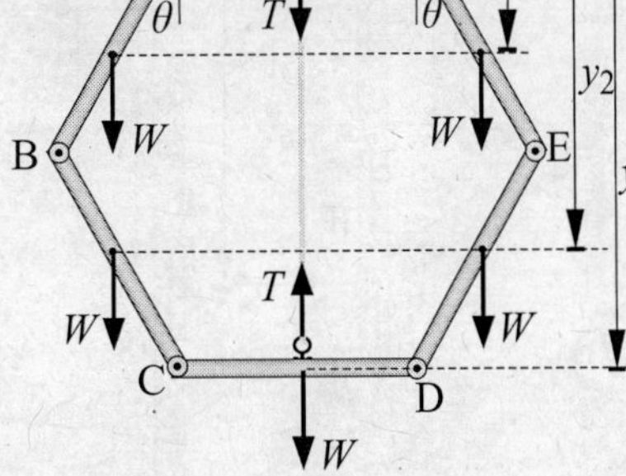

FIGURE 5.12(b) |

Substituting the values of δy_1, δy_2 and δy_3 into the virtual work equation,

$$2W\left(-\frac{l}{2}\sin\theta\delta\theta\right)+2W\left(-\frac{3l}{2}\sin\theta\delta\theta\right)+W\left(-2l\sin\theta\delta\theta\right)-T\left(-2l\sin\theta\delta\theta\right)=0$$

or $$Wl\sin\theta+3Wl\sin\theta+2Wl\sin\theta-2Tl\sin\theta=0$$

$$T=3W$$

PROBLEM 5.13

Pin-jointed frames are shown in Figure 5.13(a). Each of the links of the frame has a weight W and a length l. Determine the horizontal force P required to hold them in place in the vertical plane shown.

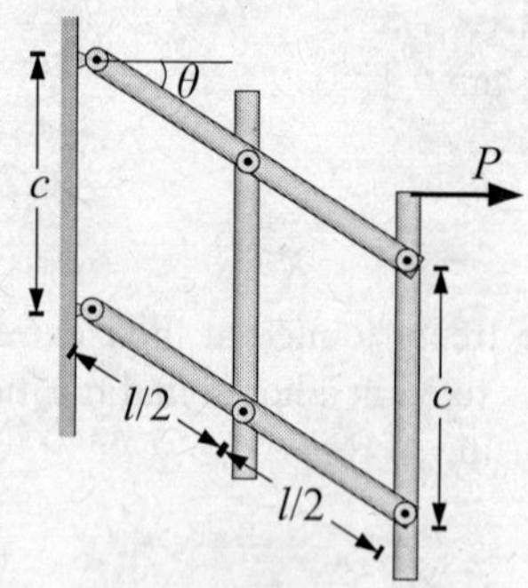

FIGURE 5.13(a) |

Solution. Assuming y is positive to down and x is positive to the right, the virtual work done by P is $P\,\delta x$ and virtual work done by downward forces of gravity is $W\delta y_1+W\delta y_2+W\delta y_3+W\delta y_4$.

The total virtual work done is zero: hence,

$$\delta u=0\Rightarrow W\,\delta y_1+W\,\delta y_2+W\,\delta y_3+W\,\delta y_4+P\delta x=0$$

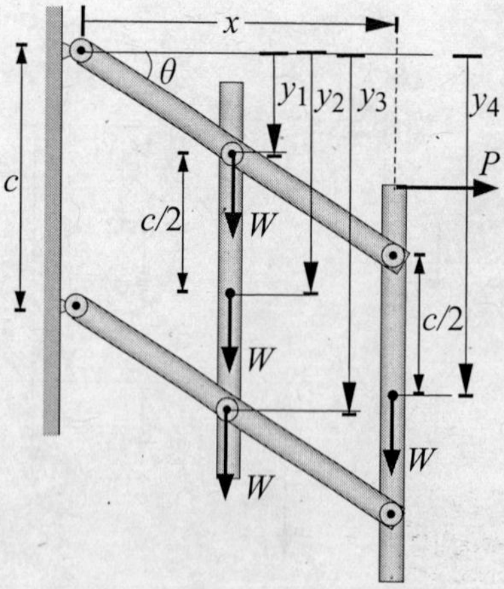

FIGURE 5.13(b) |

From Figure 5.13(b)

$$y_1=\frac{l}{2}\sin\theta,\ y_2=\frac{l}{2}\sin\theta+\frac{c}{2},\ y_3=\frac{l}{2}\sin\theta+c,$$

$$y_4=l\sin\theta+\frac{c}{2},\ x=l\cos\theta,\text{ then}$$

$$\delta y_1=\frac{l}{2}\cos\theta\delta\theta,\ \delta y_2=\frac{l}{2}\cos\theta\delta\theta,\ \delta y_3=\frac{l}{2}\cos\theta\delta\theta,$$

$$\delta y_4=l\cos\theta\delta\theta,\ \delta x=-l\sin\theta\delta\theta$$

Substituting the values of $\delta y_1,\delta y_2,\delta y_3,\delta y_4$ and δx into the virtual work equation, we get

$$W\left(\frac{l}{2}\cos\theta\delta\theta\right)+W\left(\frac{l}{2}\cos\theta\delta\theta\right)+W\left(\frac{l}{2}\cos\theta\delta\theta\right)+W(l\cos\theta\delta\theta)+P(-l\sin\theta\delta\theta)=0$$

On simplifying, we obtain

$$P=\frac{5}{2}Wl\cot\theta$$

PROBLEM 5.14

An endless chain of mass m rests in the form of a circular band round a smooth vertical cone, which has its vertex upwards as shown in Figure 5.14(a). Find the tension in the chain due to its weight.

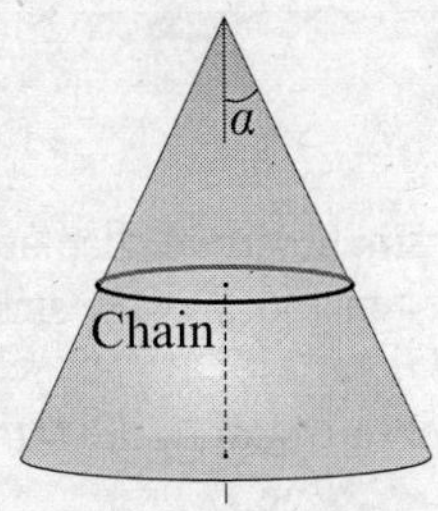

FIGURE 5.14(a) |

Solution. Let the chain be given a small virtual displacement such that radius x changes into $x + \delta x$, then the virtual work done by tension T in the string is $-T\delta l$ and virtual work done by the weight mg of the string is $-mg\ \delta y$.

Therefore, the equation of virtual work is

$$\delta u = 0 \Rightarrow -mg\delta y - T\delta l = 0$$

From Figure 5.14(b)

$$y = h - \frac{h}{r}x \quad \text{or} \quad \delta y = -\frac{h}{r}\delta x$$

$$l = 2\pi x \quad \text{or} \quad \delta l = 2\pi\delta x$$

Substituting the values of δl and δx into the virtual work equation,

$$-mg\left(-\frac{h}{r}\delta x\right) - T(2\pi\delta x) = 0$$

or

$$T = \frac{mgh}{2\pi r} = \frac{mg}{2\pi}\left(\frac{h}{r}\right) = \frac{mg}{2\pi}\cot\alpha$$

$$T = \frac{mg}{2\pi}\cot\alpha$$

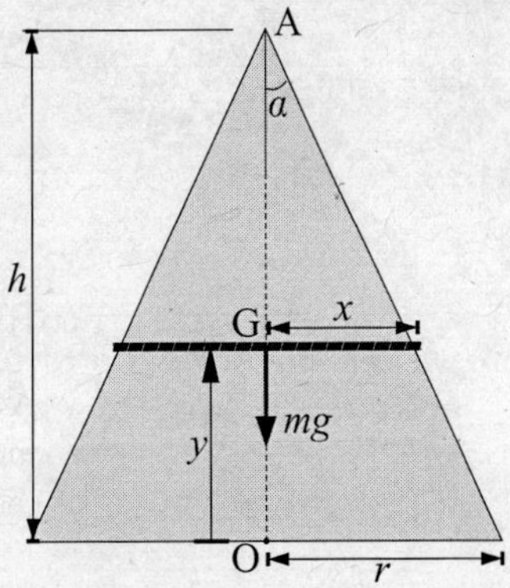

FIGURE 5.14(b) |

PROBLEM 5.15

A uniform bar of mass m with end rollers is held in the equilibrium position by the horizontal cord CD as shown in Figure 5.15(a). Determine the tension T in the cord CD.

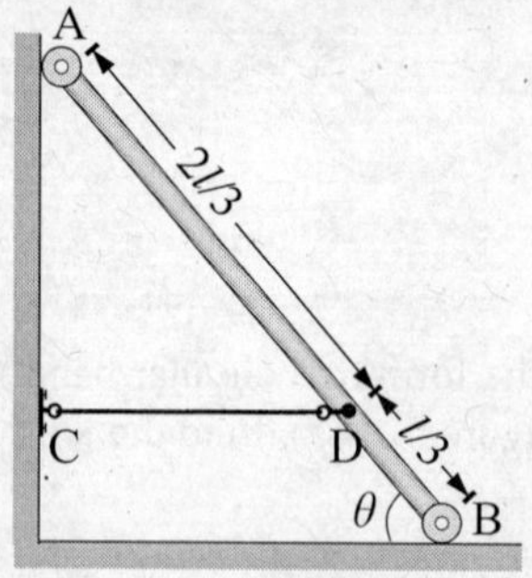

FIGURE 5.15(a) |

Solution. Let the bar be given a small virtual displacement such that radius θ changes into $\theta - \delta\theta$, then the virtual work done by the tension T in the string is $-T\delta(\text{CD})$ and virtual work done by the weight mg of the string is $-mg\delta(\text{GF})$.

By the principle of virtual work, algebraic sum of the total virtual work done is zero:

$$\delta u = 0 \Rightarrow -T\delta(\text{CD}) - mg\delta(\text{GF}) = 0$$

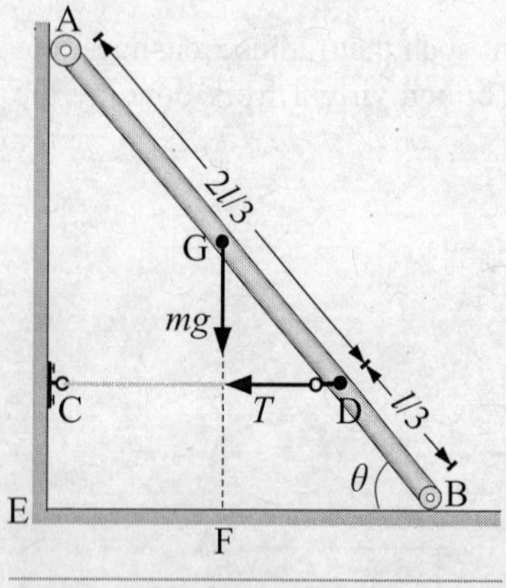

FIGURE 5.15(b) |

From Figure 5.15(b)

$$\text{CD} = \frac{2l}{3}\cos\theta \quad \text{or} \quad \delta(\text{CD}) = -\frac{2l}{3}\sin\theta\delta\theta$$

$$\text{GF} = \frac{l}{2}\sin\theta \quad \text{or} \quad \delta(\text{GF}) = \frac{l}{2}\cos\theta\delta\theta$$

Substituting the values of δ(GF) and δ(CD) into the virtual work equation

$$-T\left(-\frac{2l}{3}\sin\theta\delta\theta\right) - mg\left(\frac{l}{2}\cos\theta\delta\theta\right) = 0$$

$$T = \frac{3}{4}mg\cot\theta$$

PROBLEM 5.16

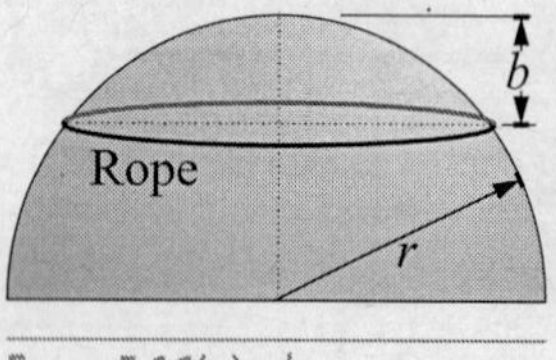

FIGURE 5.16(a) |

A heavy rope of weight W is placed on a smooth hemisphere in a symmetrical position as shown in Figure 5.16(a). Determine the tension of the rope.

Solution. Let the rope be given a small virtual displacement such that x changes into $x + \delta x$, then the virtual work by the tension T is $-T\,\delta l$ and virtual work by the weight mg is $-mg\ \delta y$. By the principle of virtual work,

$$\delta u = 0 \Rightarrow -mg\delta y - T\delta l = 0$$

The length of the rope $l = 2\pi x$ or $\delta l = 2\pi\ \delta x$.
From Fig. 5.16(b)

$$x^2 + y^2 = r^2 \quad \text{or} \quad 2x\delta x + 2y\delta y = 0$$

$$\delta y = -\frac{x}{y}\delta x$$

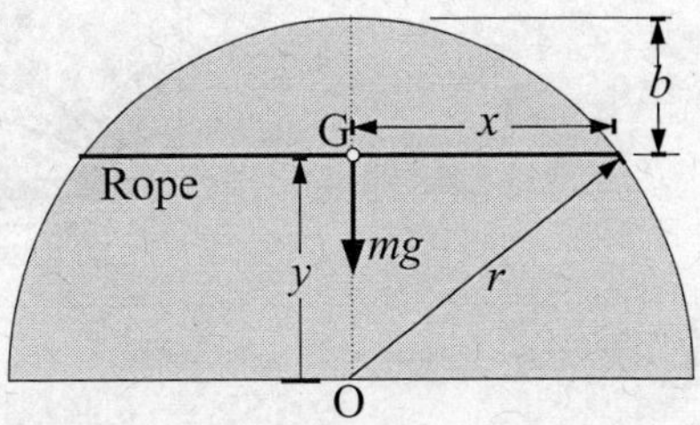

FIGURE 5.16(b)

Substituting the values of δl and δy into the virtual work equation,

$$-mg\left(-\frac{x}{y}\delta x\right) - T(2\pi\delta x) = 0$$

or

$$T = \frac{mgx}{2\pi y} = \frac{mg\sqrt{r^2 - y^2}}{2\pi y}$$

Substituting $y = r - b$, we have

$$T = \frac{mg\sqrt{r^2 - (r-b)^2}}{2\pi(r-b)} = \frac{mg\sqrt{2rb - b^2}}{2\pi(r-b)}$$

$$T = \frac{mg\sqrt{2rb - b^2}}{2\pi(r-b)}$$

PROBLEM 5.17

A structure, with a pin-jointed member, is shown in Figure 5.17(a). Determine the tension T in the turnbuckle.

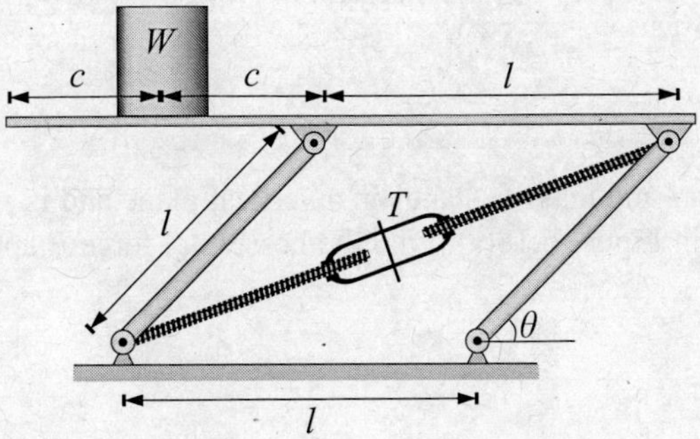

FIGURE 5.17(a)

Solution. Let T be the tension in the turnbuckle. Give the system a small virtual displacement such that θ becomes $\theta + \delta\theta$, then the equation of virtual work is

$$\delta u = 0 \Rightarrow -W\delta y - T\delta s = 0$$

From Figure 5.17(b)

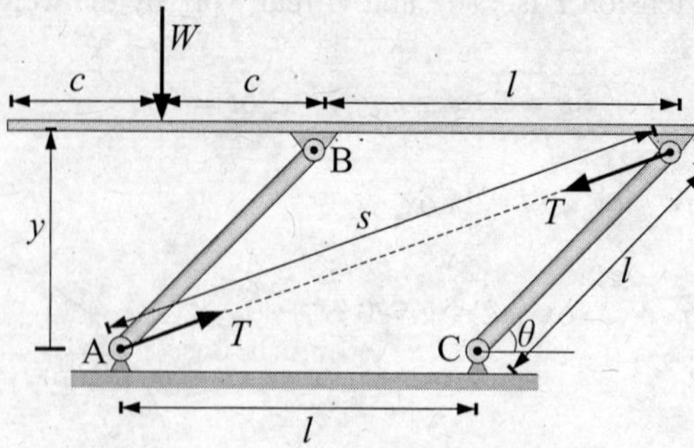

FIGURE 5.17(b) |

$$y = l \sin \theta \quad \text{or} \quad \delta y = l \cos \theta \delta\theta$$

$$AD = s = \sqrt{l^2 + l^2 + 2l^2 \cos \theta}$$

$$= 2l \cos \frac{\theta}{2} \quad \text{or} \quad \delta s = -l \sin \frac{\theta}{2} \delta\theta$$

Substituting the values of δs and δy into the virtual equation,

$$-W(l \cos \theta \delta\theta) - T\left(-l \sin\left(\frac{\theta}{2}\right)\delta\theta\right) = 0$$

or

$$Wl \cos \theta = Tl \sin \frac{\theta}{2}$$

or

$$T = W \frac{\cos \theta}{\sin\left(\frac{\theta}{2}\right)}$$

$$T = W\left(\operatorname{cosec} \frac{\theta}{2} - 2 \sin \frac{\theta}{2}\right)$$

PROBLEM 5.18

A uniform bar AB of length l and mass m stands in a vertical plane and is supported by smooth surfaces at A and B as shown in Figure 5.18(a). Find the horizontal force P applied at B if the bar is in equilibrium.

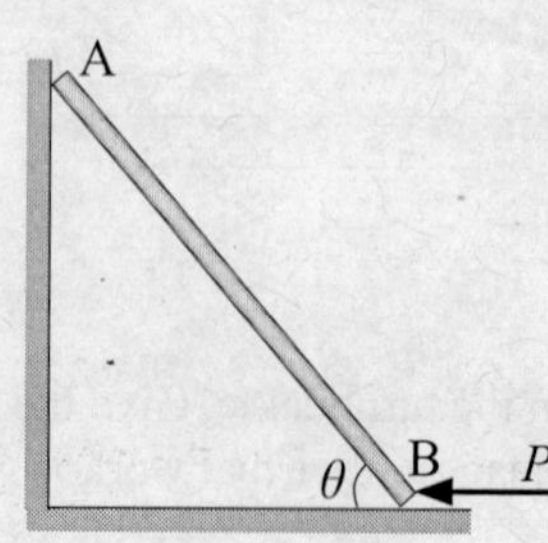

FIGURE 5.18(a) |

Solution. Let the end B of the rod be given a small virtual displacement δx. Then the work done by the force P is $-P\,\delta x$ and virtual work done by the weight mg is $-mg\,\delta y$.

By the principle of virtual work for equilibrium,

$$\delta u = 0 \Rightarrow -P\delta x - mg\delta y = 0$$

From Figure 5.18(b),

$$x = l\cos\theta \quad \text{or} \quad \delta x = -l\sin\theta\delta\theta$$

$$y = \frac{l}{2}\sin\theta \quad \text{or} \quad \delta y = \frac{l}{2}\cos\theta\delta\theta$$

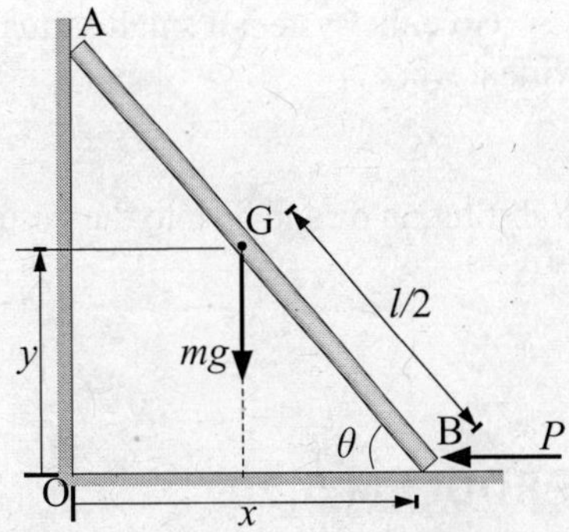

Figure 5.18(b)

Substituting the values of δx and δy into the virtual work equation,

$$-P(-l\sin\theta\delta\theta) - mg\left(\frac{l}{2}\cos\theta\delta\theta\right) = 0$$

or

$$\left(Pl\sin\theta - mg\frac{l}{2}\cos\theta\right)\delta\theta = 0$$

From which

$$P = \frac{mg}{2}\cot\theta$$

PROBLEM 5.19

Determine the relation between the forces Q and P for equilibrium of the system of hinged bars shown in Figure 5.19(a).

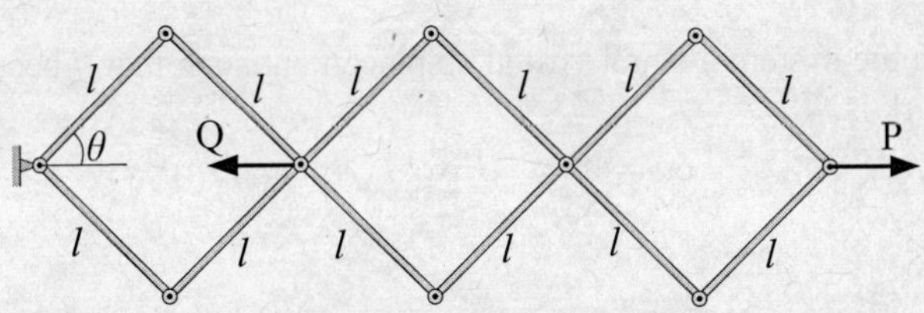

Figure 5.19(a)

Solution. The horizontal distance of joint I from the hinged support O [Fig. 5.19(b)],

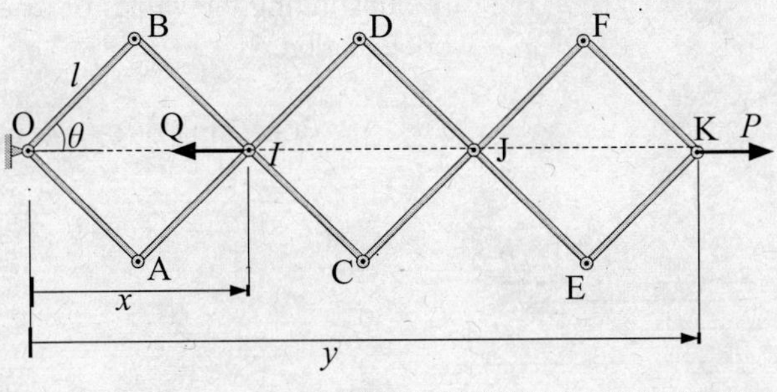

Figure 5.19(b)

$$x = 2l\cos\theta \quad \text{or} \quad \delta x = -2l\sin\theta\delta\theta,$$

and the horizontal distance of L from the hinged support O,

$$y = 6l\cos\theta \quad \text{or} \quad \delta y = -6l\sin\theta\delta\theta$$

Give the system a small virtual displacement such that θ becomes $\theta + \delta\theta$, then the equation of virtual work is

$$\delta u = 0 \Rightarrow -Q\delta x + P\delta y = 0$$

Substituting δx and δy into the virtual work equation

$$-Q(-2l\sin\theta\delta\theta) + P(-6l\sin\theta\delta\theta) = 0$$

$$Q = 3P$$

PROBLEM 5.20

A simple linkage consisting of two equal mass-less bars and a slider B is subjected to a set of two forces P and Q as shown in Figure 5.20(a). Find the angle θ as a function of P and Q for equilibrium.

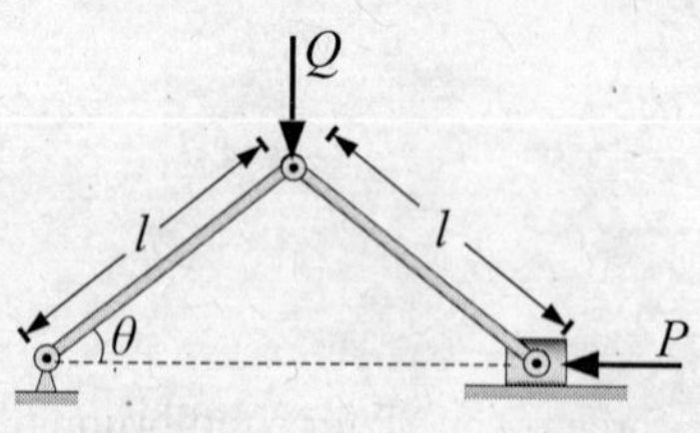

FIGURE 5.20(a) |

Solution. Give the system a small virtual displacement such that θ becomes $\theta + \delta\theta$, then the equation of virtual work is.

$$\delta u = 0 \Rightarrow -P\delta x - Q\delta y = 0$$

From Figure 5.20(b),

$$x = 2l\cos\theta \quad \text{or} \quad \delta x = -2l\sin\theta\delta\theta$$

$$y = l\sin\theta \quad \text{or} \quad \delta y = l\cos\theta\delta\theta$$

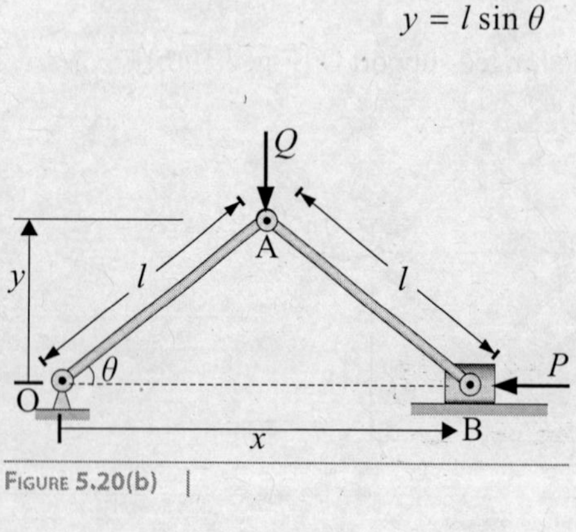

FIGURE 5.20(b) |

Substituting the values of δx and δy into the virtual work equation,

$$-P(-2l\sin\theta\delta\theta) - Q(l\cos\theta\delta\theta) = 0$$

or

$$2P\sin\theta = Q\cos\theta$$

$$\theta = \tan^{-1}\frac{Q}{2P}$$

PROBLEM 5.21

Three equal rods, each of length 2 m and weight 50 N, are freely jointed at their ends as shown in Figure 5.21(a). A 200 N force is applied horizontally at the lower end to keep it in equilibrium. Determine the angle θ_1, θ_2 and θ_3.

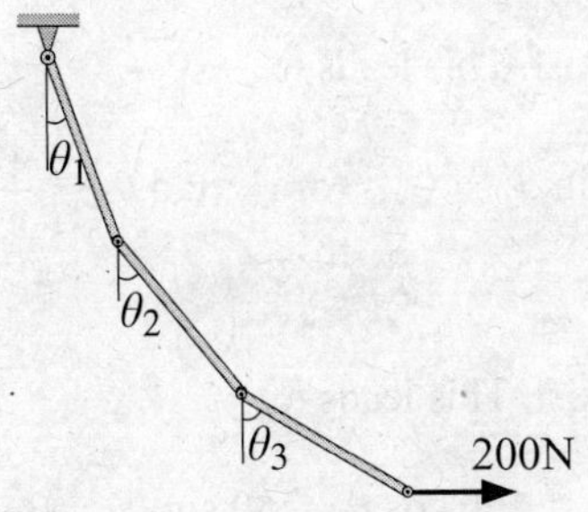

FIGURE 5.21(a) |

Solution. Assuming that x is positive to right, the virtual work done by 200 N for an increase δx is $200\delta x$.

Assuming that y is positive to down, the virtual work done by downward forces of gravity is

$$50\delta y_1 + 50\delta y_2 + 50\delta y_3$$

The total virtual work δu is zero; hence,

$$\delta u = 0 \Rightarrow 200\delta x + 50\delta y_1 + 50\delta y_2 + 50\delta y_3$$

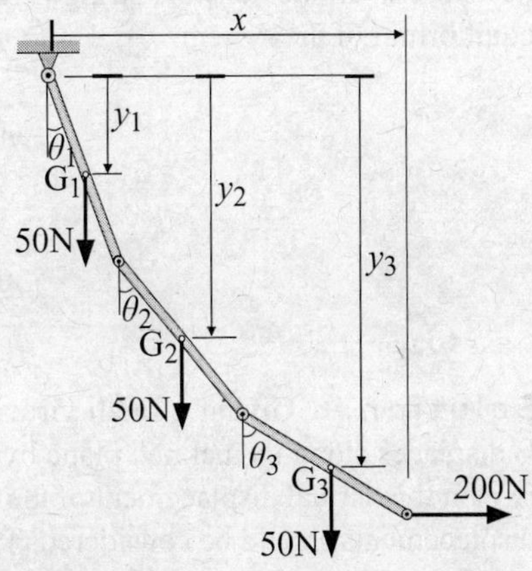

FIGURE 5.21(b) |

From Figure 5.21(b),

$$x = 2\sin\theta_1 + 2\sin\theta_2 + 2\sin\theta_3$$
$$y_1 = 1\cos\theta_1,$$
$$y_2 = 2\cos\theta_1 + 1\cos\theta_2,$$
$$y_3 = 2\cos\theta_1 + 2\cos\theta_2 + 1\cos\theta_3$$

Hence,

$$\delta x = 2\cos\theta_1\,\delta\theta_1 + 2\cos\theta_2\,\delta\theta_2 + 2\cos\theta_3\,\delta\theta_3$$
$$\delta y_1 = -\sin\theta_1\,\delta\theta_1,$$
$$\delta y_2 = -2\sin\theta_1\,\delta\theta_1 - \sin\theta_2\,\delta\theta_2,$$
$$\delta y_3 = -2\sin\theta_1\,\delta\theta_1 - 2\sin\theta_2\,\delta\theta_2 - \sin\theta_3\,\delta\theta_3$$

Substituting the values of δx, δy_1, δy_2 and δy_3 into the virtual work, we have

$$200(2\cos\theta_1\,\delta\theta_1 + 2\cos\theta_2\,\delta\theta_2 + 2\cos\theta_3\,\delta\theta_3) + 50(-\sin\theta_1\delta\theta_1)$$
$$+ 50(-2\sin\theta_1\,\delta\theta_1 - \sin\theta_2\,\delta\theta_2) + 50(-2\sin\theta_1\,\delta\theta_1 - 2\sin\theta_2\,\delta\theta_2 - \sin\theta_3\,\delta\theta_3) = 0$$

or

$$(400\cos\theta_1 - 50\sin\theta_1 - 100\sin\theta_1 - 100\sin\theta_1)\delta\theta_1 + (400\cos\theta_2 - 50\sin\theta_2$$
$$- 100\sin\theta_2)\delta\theta_2 + (400\cos\theta_3 - 50\sin\theta_3)\delta\theta_3 = 0$$

To solve for θ_1, allow only $\delta\theta_1$ to exist. This leads to

$$400\cos\theta_1 - 50\sin\theta_1 - 100\sin\theta_1 - 100\sin\theta_1 = 0 \quad \text{or} \quad \tan\theta_1 = \frac{400}{250}$$

$$\theta_1 = 57.9^\circ$$

To solve for θ_2, allow only $\delta\theta_2$ to exist. This leads to

$$400\cos\theta_2 - 150\sin\theta_2 = 0 \quad \text{or} \quad \tan\theta_2 = \frac{400}{150} \Rightarrow \theta_2 = 69.4^\circ$$

$$\theta_2 = 69.4^\circ$$

To solve for θ_3, allow only $\delta\theta_3$ to exist. This leads to

$$400\cos\theta_3 - 50\sin\theta_3 = 0$$

or

$$\tan\theta_3 = \frac{400}{150} \Rightarrow \theta_3 = 82.8^\circ$$

$$\theta_3 = 82.8^\circ$$

PROBLEM 5.22

A frictionless double-incline with angles 30° and 60° carries a set of sliding masses m_1 and m_2 connected with an inextensible string [Fig. 5.22(a)]. Find the value of angle θ defining configuration of equilibrium of the system.

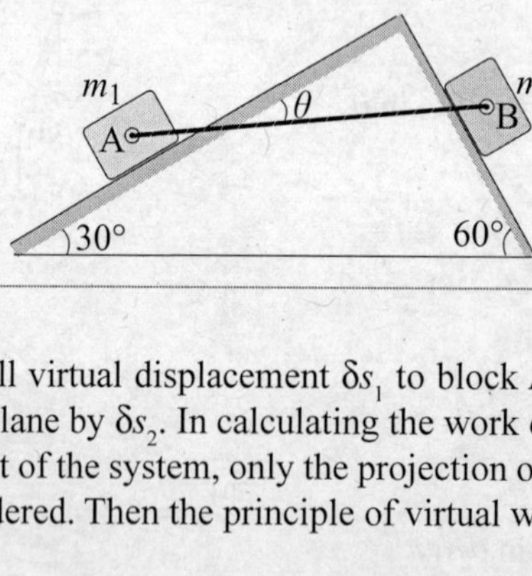

FIGURE 5.22(a) |

Solution. Giving a small virtual displacement δs_1 to block A along the inclined plane, block B displaces along its inclined plane by δs_2. In calculating the work done by the active forces m_1g and m_2g on this virtual displacement of the system, only the projection of each force on the corresponding displacements should be considered. Then the principle of virtual work gives

From Figure 5.22(b)

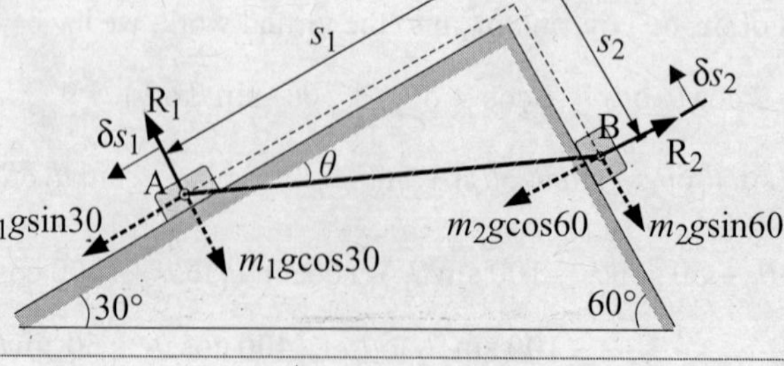

FIGURE 5.22(b) |

$$\delta u = 0 \Rightarrow m_1 g \sin 30\delta s_1 + m_2 g \sin 60\delta s_2 = 0$$

$$s_1 = l \cos\theta \quad \text{or} \quad \delta s_1 = -l \sin\theta\delta\theta$$

$$s_2 = l \sin\theta \quad \text{or} \quad \delta s_2 = l \cos\theta\delta\theta$$

where l is the length of the string.

Substituting the values of δs_1 and δs_2 into the virtual work equation, we obtain

$$m_1 g \sin 30(-l \sin\ \theta\delta\theta) + m_2 g \sin 60\left(l \cos\theta\delta\theta\right) = 0$$

or
$$\tan\theta = \frac{m_2 \sin 60}{m_1 \sin 30}$$

From which

$$\theta = \tan^{-1}\left(\frac{\sqrt{3}m_2}{m_1}\right)$$

PROBLEM 5.23

A screw-press shown in Figure 5.23(a) has a single square thread with pitch p. Neglecting friction in the screw, find the relation between the total compressive force Q on the body and the force P applied to the handle. The force P is acting at right angle to the handle in a plane perpendicular to the axis of the screw.

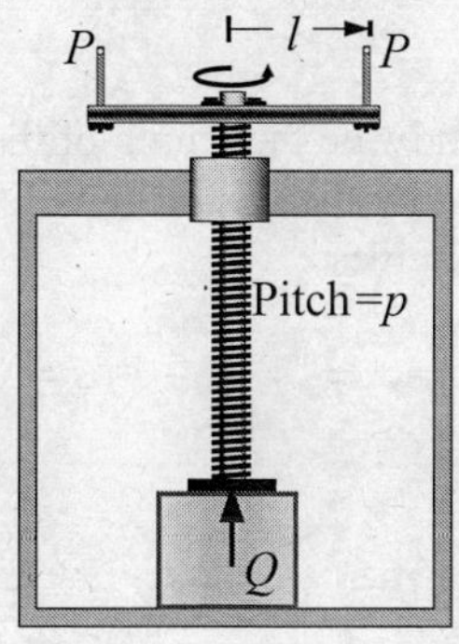

FIGURE 5.23(a)

Solution. Give the screw a small virtual angular displacement by an angle $\delta\theta$ [Fig. 5.23(b)]. Then the work done by a couple of the moment M (=2Pl) is $2Pl\delta\theta$.

The virtual displacement of the end of the screw is $\delta y = (p/2\pi)\delta\theta$ and the work done on this displacement by the force Q is $-Q\delta y$. Then the principle of virtual work gives,

$$\delta u = 0 \Rightarrow M\delta\theta - Q\delta y = 0$$

Substituting $M = 2Pl$ and $\delta y = (p / 2\pi)\delta\theta$, we have

or
$$2Pl\delta\theta - Q\frac{p}{2\pi}\delta\theta = 0$$

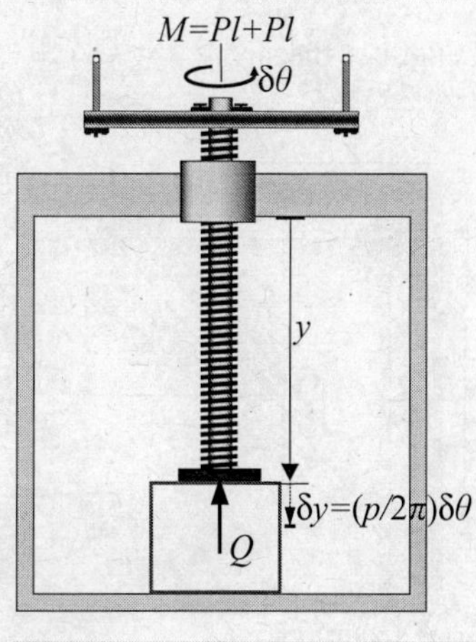

FIGURE 5.23(b)

$$2Pl = Q\frac{p}{2\pi}$$

$$Q = \frac{4\pi Pl}{p}$$

PROBLEM 5.24

A uniform rod AB of weight W is attached to blocks A and B that move freely in the frictionless guides as shown in Figure 5.24(a). The blocks are connected by an inextensible string that passes over a pulley at O. Find the tension in the string.

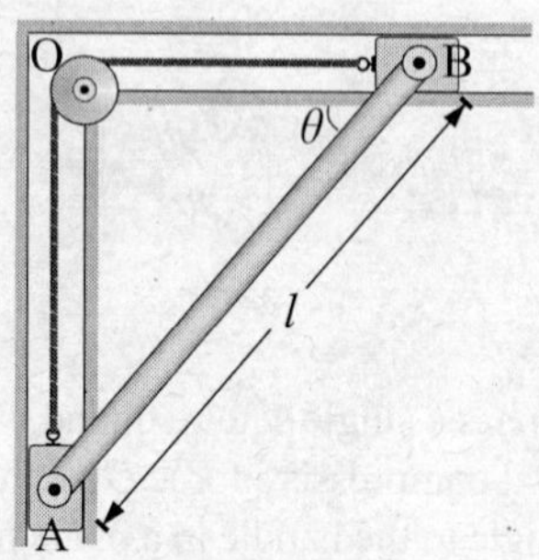

FIGURE 5.24(a) |

Solution. Let T be the tension and s be the length of the string. Let the rod AB gives a small virtual displacement $\delta\theta$, then work done by the weight W is $W\delta y$ and work done by the tension T is $-T\delta s$. Then the principle of virtual work gives

$$\delta u = 0 \Rightarrow W\delta y - T\delta s = 0$$

From Figure 5.24(b)

$$y = \frac{l}{2}\sin\theta \quad \text{or} \quad \delta y = \frac{l}{2}\cos\theta\delta\theta$$

Length of the string $s = \text{OA} + \text{OB} = l(\cos\theta + \sin\theta)$

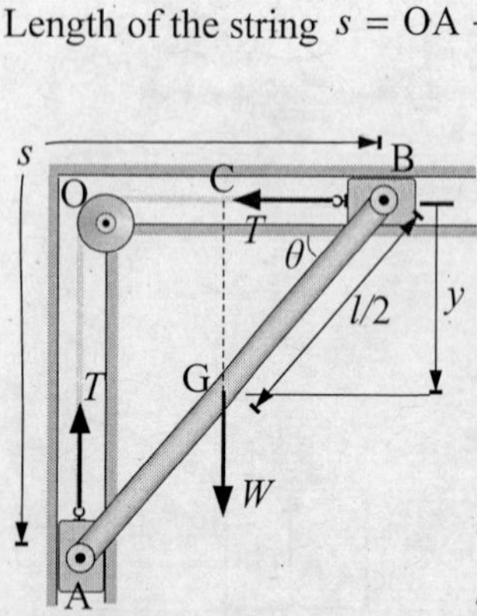

FIGURE 5.24(b) |

Substituting the value δs and δy into the virtual work equation,

$$W\left(\frac{l}{2}\cos\theta\delta\theta\right) - T \times l(-\sin\theta + \cos\theta)\delta\theta = 0$$

or

$$T = W\frac{\cos\theta}{2(\cos\theta - \sin\theta)}$$

$$T = \frac{W}{2(1 - \tan\theta)}$$

PROBLEM 5.25

Derive an expression for the relation between P and Q to maintain the equilibrium in the mechanism shown in Figure 5.25(a).

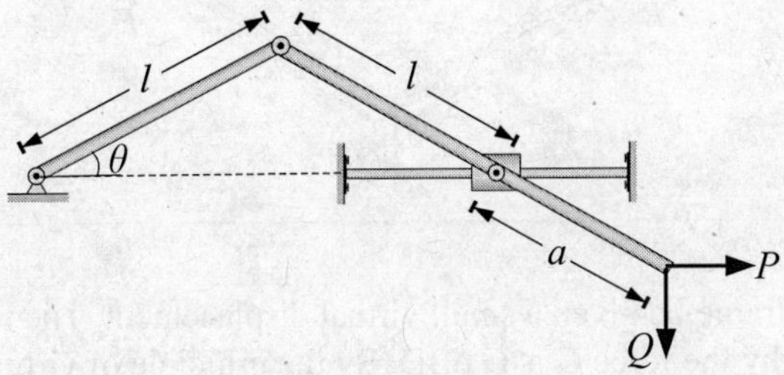

FIGURE 5.25(a) |

Solution. Choosing the coordinate system with origin at O [Fig. 5.25(b)], we write

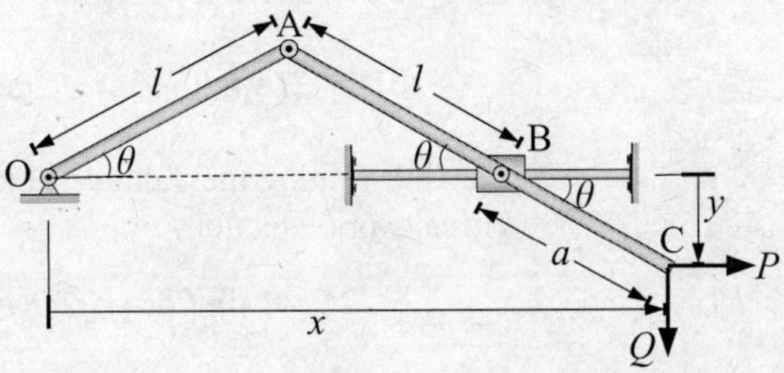

FIGURE 5.25(b) |

$$x = 2l\cos\theta + a\cos\theta \quad \text{or} \quad \delta x = -(2l + a)\sin\theta\delta\theta$$

$$y = a\sin\theta \quad \text{or} \quad \delta y = a\cos\theta\delta\theta$$

Total virtual work done by the forces P and Q must be zero. Therefore,

$$\delta u = 0 \Rightarrow P\delta x + Q\delta y = 0$$

Substituting the values of δx and δy into the virtual work equation,

$$P[-(2l + a)\sin\theta\delta\theta] + Q(a\cos\theta\delta\theta) = 0$$

or

$$Qa\cos\theta = P(2l + a)\sin\theta$$

$$Q = P\left(1 + \frac{2l}{a}\right)\tan\theta$$

PROBLEM 5.26

Find the relation between the forces P and Q for equilibrium of the linkage shown in Figure 5.26(a).

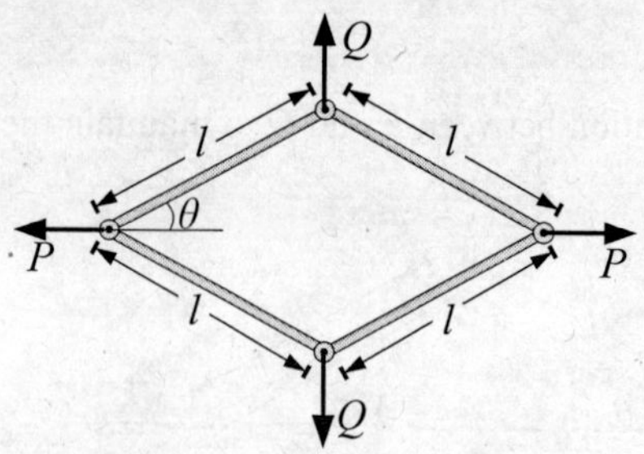

FIGURE 5.26(a) |

Solution. Let the frame be given a small virtual displacement. Then the virtual work done by the force P is $P\,\delta$AC and by the force Q is $Q\,\delta$BD. By the principle of virtual work $\delta u = 0$,

$$2P\delta(\text{AC}) + 2Q\delta(\text{BD}) = 0$$

From Figure 5.26(b)

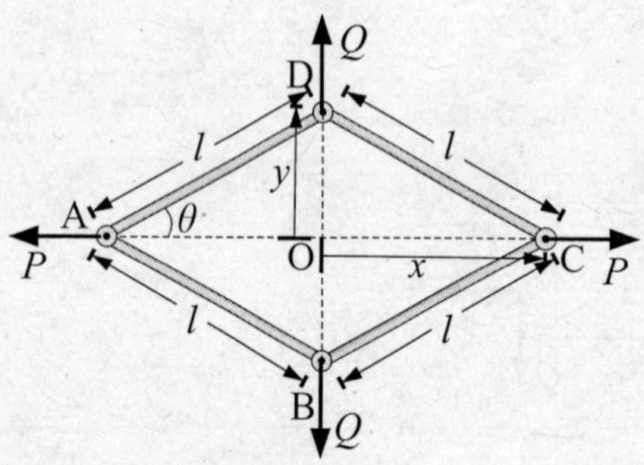

FIGURE 5.26(b) |

$$\text{AC} = 2l\cos\theta \quad \text{or} \quad \delta(\text{AC}) = -2l\sin\theta\delta\theta$$

$$\text{BD} = 2l\sin\theta \quad \text{or} \quad \delta(\text{BD}) = l\cos\theta\delta\theta$$

Substituting the values of δ(AC) and δ(BD) into the virtual work equation,

$$2P(-l\sin\theta\delta\theta) + 2Q(l\cos\theta\delta\theta) = 0$$

From which

$$\frac{Q}{P} = \tan\theta$$

PROBLEM 5.27

A vertical load W is applied to a linkage at B as shown in Figure 5.27(a). The unstretched length of spring is L and the spring stiffness is k. Determine an expression for θ and for the tension in the spring that corresponds to the equilibrium position of the linkage.

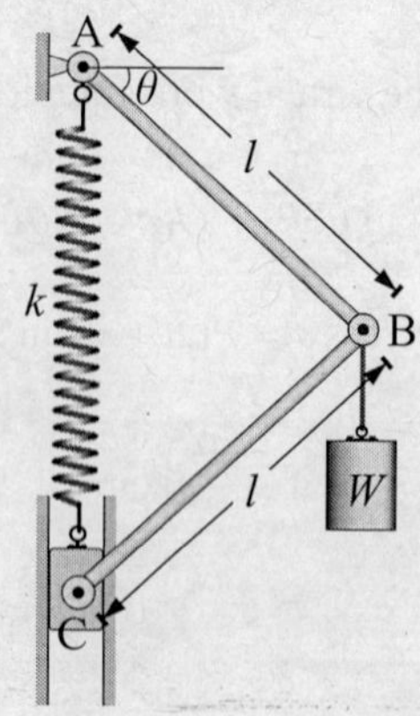

FIGURE 5.27(a) |

Solution. The active force diagram of the system is shown in Figure 5.27(b). The force exerted at C by the spring is

$$F_s = k\Delta$$

From the figure,

$$\Delta = 2l \sin\theta - L$$

$$y_B = l\sin\theta \quad \text{and} \quad y_C = 2l\sin\theta, \text{then}$$

$$\delta y_B = l\cos\theta\delta\theta \quad \text{and} \quad \delta y_C = 2l\cos\theta\delta\theta$$

The virtual work done by the load W is $W\,\delta y_B$ and virtual work done by the spring force is $-F_s\,\delta y_C$.

By the principle of virtual work

$$\delta u = 0 \Rightarrow W\,\delta y_B - F_s\,\delta y_C = 0$$

Substituting the values of F_s, δy_B and δy_C, we have

$$W(l\cos\theta\delta\theta) - k(2l\sin\theta - L)(2l\cos\theta\delta\theta) = 0$$

From which

$$\sin\theta = \frac{W + 2kL}{4kl}$$

$$\theta = \sin^{-1}\left(\frac{W + 2kL}{4kl}\right)$$

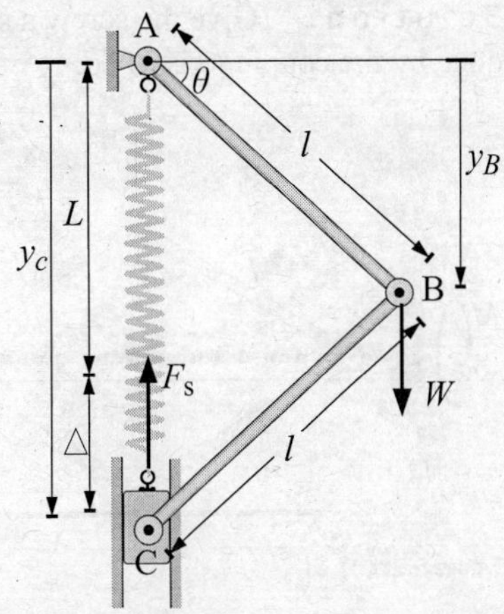

FIGURE 5.27(b)

Substituting $\sin\theta = (W + 2kL) / 4kl$ in $F_s = k\Delta = k(2l\sin\theta - L)$, we obtain

$$F_s = k\left[2l\left(\frac{W + 2kL}{4kl}\right) - L\right] = \frac{W}{2}$$

From which

$$F_s = \frac{W}{2}$$

PROBLEM 5.28

The vertical position of a load W is controlled by an adjusting screw that connects joints A and B, as shown in Figure 5.28(a). The change in the distance between A and B for one revolution of the screw equals the lead p of the screw. Determine an expression for the couple M applied to the adjusting screw, necessary to raise the load.

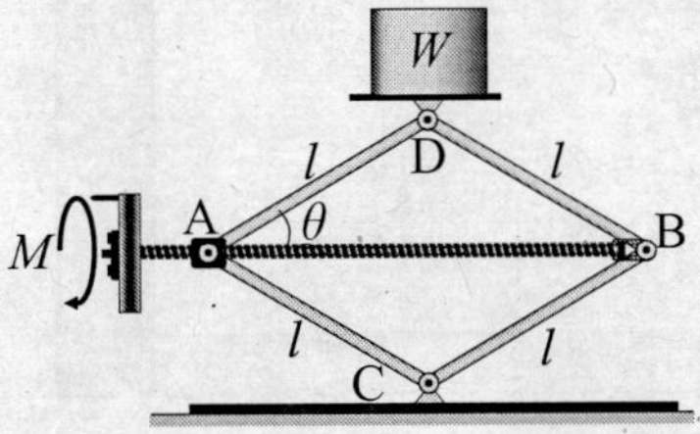

FIGURE 5.28(a)

Solution. Give the screw a small virtual angular displacement by an angle $\delta\alpha$. Then the work done by the couple M is $M\,\delta\alpha$.

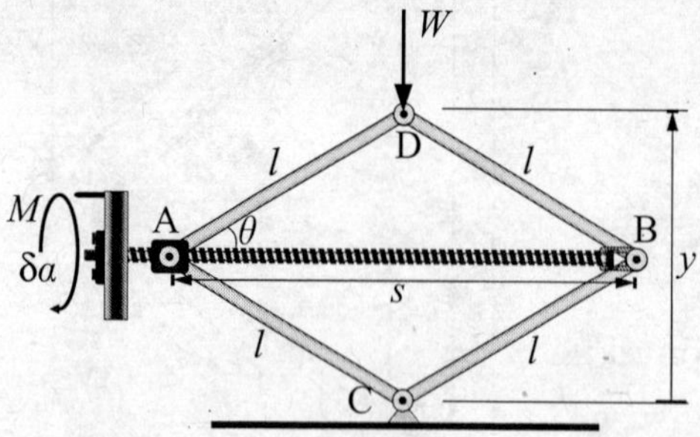

FIGURE 5.28(b) |

From Figure 5.28(b)

$$y = 2l\sin\theta \quad \text{or} \quad \delta y = 2l\cos\theta\delta\theta$$

$$s = 2l\cos\theta \quad \text{or} \quad \delta s = -2l\sin\theta\delta\theta$$

The relation between angular movement and axial movement of the screw is

$$\delta a = \frac{2\pi}{p}\delta s = \frac{2\pi}{p}\times(-2l\sin\theta\delta\theta)$$

The work done by the load W is $-W\,\delta y$.

By the principle of virtual work

$$\delta u = 0 \Rightarrow M\delta a - W\delta y = 0$$

Substituting the values of $\delta\alpha$ and δy, we obtain

$$M\left(\frac{2\pi}{p}\times(-2l\sin\theta\delta\theta)\right) - W\times 2l\cos\theta\delta\theta = 0$$

From which

$$M = W\frac{p}{2\pi}\cot\theta$$

PROBLEM 5.29

Figure 5.29(a) shows a rack and pinion type press. Determine the force F exerted on the body B, if the pitch circle radius of the pinon is 60 mm and friction in all moving parts is neglected.

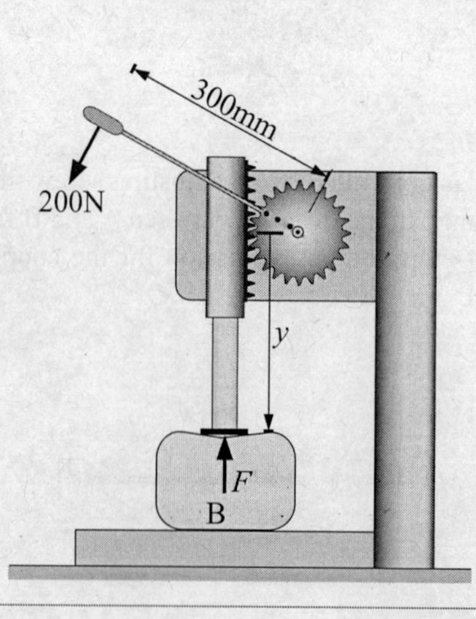

FIGURE 5.29(a) |

Solution. Give the lever a small virtual displacement by an angle $\delta\theta$. Then the work done by 200 N is $200 \times 300\ \delta\theta$.

The virtual displacement of the rack is $\delta y = 60\ \delta\theta$ and the virtual work done against the compressive force F is $-F\ \delta y$. Then the principle of virtual work gives

$$\delta u = 0 \Rightarrow 200 \times 300\delta\theta - F\delta y = 0$$

Substituting $\delta y = 60\ \delta\theta$, we get

$$200 \times 300\delta\theta - F \times 60\delta\theta = 0$$

From which

$$\boxed{F = 1000\text{N}}$$

PROBLEM 5.30

A toggle press is shown in Figure 5.30(a). The upper jaw of the toggle press slides with a smooth vertical column. Determine the compressive force F exerted on the body B.

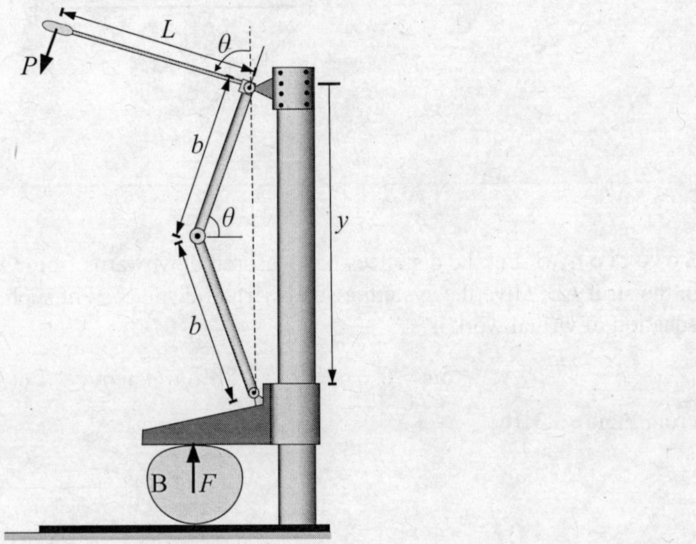

FIGURE 5.30(a) |

Solution. The active force diagram shows the reactive compressive force F and the applied force P.

There is no other external active force or moment that does work on the system.

The vertical position of the body from the fixed reference is $y = 2b \sin \theta$. The virtual work done by F during moment δy in the opposite direction of F is $-F\ \delta y$. With θ having measured positive in the anticlockwise direction, the work done by the force P is PL $\delta\theta$. Then the principle of virtual work gives

$$\delta u = 0 \Rightarrow PL\delta\theta - F\delta y = 0$$

or

$$PL\delta\theta - F\delta(2b \sin \theta) = 0$$

or

$$PL\delta\theta - 2Fb \cos \theta\delta\theta = 0$$

From which, we obtain

$$F = \frac{PL}{2b} \sec \theta$$

PROBLEM 5.31

Four equal uniform bars, each of weight W and length l are jointed together so as to form a rhombus. This is suspended vertically from one of the joints, and a disc of weight w and radius r is placed on bars AC and BC as shown in Figure 5.31(a). The rhombus is kept in shape by a slight strut AB. Determine the compressive force in the strut AB.

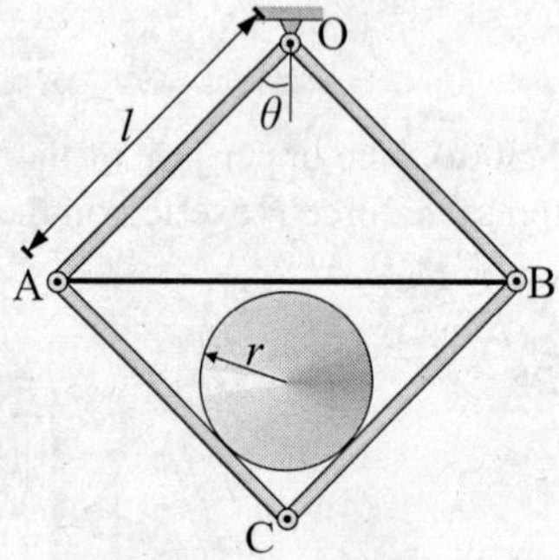

FIGURE 5.31(a) |

Solution. Let the distances be measured downwards from O as O is fixed. Let T be the thrust in the strut AB. Give the system a small virtual displacement such that θ becomes $\theta + \delta\,\theta$, then the equation of virtual work is

$$\delta u = 0 \Rightarrow 2W\,\delta y_1 + 2W\,\delta y_2 + w\delta y_3 + T\,\delta(\text{AB}) = 0$$

From Figure 5.31(b)

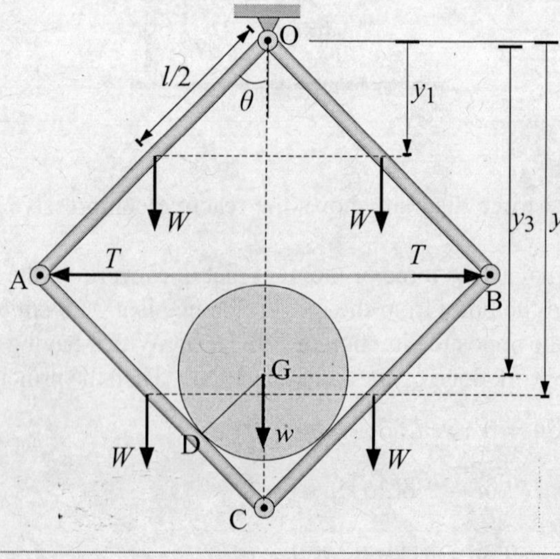

FIGURE 5.31(b) |

$$AB = 2l \sin\theta \quad \text{or} \quad \delta(AB) = 2l\cos\theta\delta\theta$$

$$y_1 = \frac{l}{2}\cos\theta \quad \text{or} \quad \delta y_1 = -\frac{l}{2}\sin\theta\delta\theta$$

$$y_2 = \frac{3l}{2}\cos\theta \quad \text{or} \quad \delta y_2 = -\frac{3l}{2}\sin\theta\delta\theta$$

$$y_3 = OC - GC = 2l\cos\theta - r\text{cosec}\theta$$

or
$$\delta y_3 = -(2l\sin\theta - r\text{cosec}\theta\cot\theta)\delta\theta$$

Substituting the values of δy_1, δy_2, δy_3 and $\delta(AB)$ into the virtual work equation,

$$2W\left(-\frac{l}{2}\sin\theta\delta\theta\right) + 2W\left(-\frac{3l}{2}\sin\theta\delta\theta\right) - w(2l\sin\theta - r\text{cosec}\theta\cot\theta)\delta\theta_1 + T\times 2l\cos\theta\delta\theta = 0$$

or
$$4Wl\sin\theta + w(2l\sin\theta - r\text{cosec}\theta\cot\theta) = 2Tl\cos\theta$$

or
$$T = 4W\tan\theta + w\left[\tan\theta - \frac{r}{2l}\text{cosec}^2\theta\right]$$

$$T = (4W + w)\tan\theta - \frac{wr}{2l}\text{cosec}^2\theta$$

PROBLEM 5.32

A uniform rod of length $2l$ rests in equilibrium against a smooth vertical wall upon a smooth peg P at a distance a from the wall [Fig. 5.32(a)]. Neglecting friction, determine the position of equilibrium of the rod as defined by the angle θ that it makes with the face of the wall.

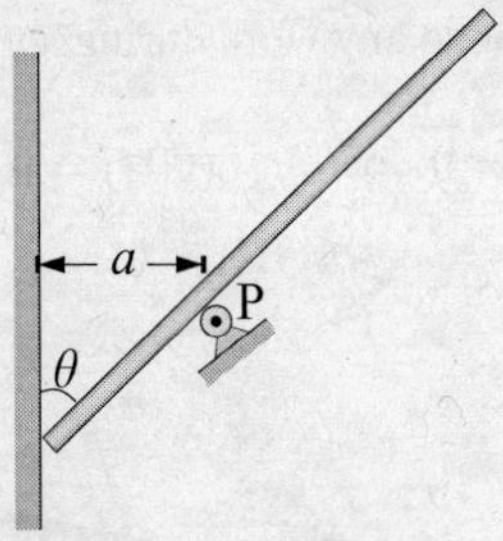

FIGURE 5.32(a) |

Solution. Let P be the peg and G be the centre of gravity of the rod. The weight mg of the rod acts vertically downwards of its centre of gravity.

The vertical height of G above the peg is given by

$$y = l\cos\theta - a\cot\theta$$

Give the system a small virtual displacement such that θ changes into $\theta + \delta\theta$ [Fig. 5.32(b)], then equation of virtual work is

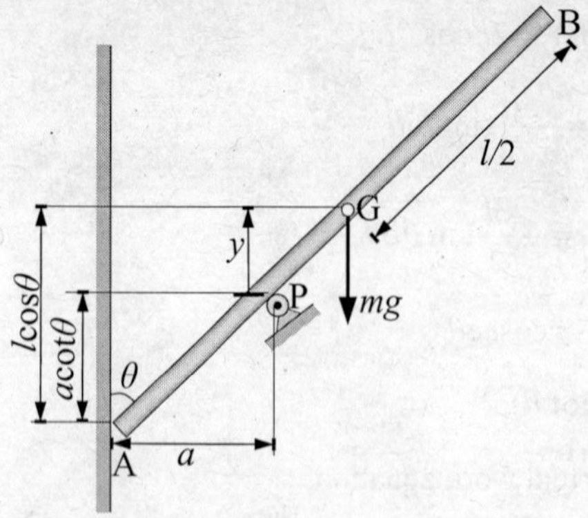

FIGURE 5.32(b) |

$$\delta u = 0 \Rightarrow -mg\delta y = 0$$

or
$$-mg\delta(l\cos\theta - a\cot\theta) = 0$$

or
$$-l\sin\theta + a\operatorname{cosec}^2\theta = 0$$

On simplifying, we obtain

$$\sin^3\theta = \frac{a}{l}$$

$$\theta = \sin^{-1}\left(\frac{a}{l}\right)^{1/3}$$

PROBLEM 5.33

A heavy uniform rod of length $2l$ rests with its ends in contact with two smooth inclined planes of inclination α and β to the horizontal as shown in Figure 5.33(a). Find the angle θ for the equilibrium.

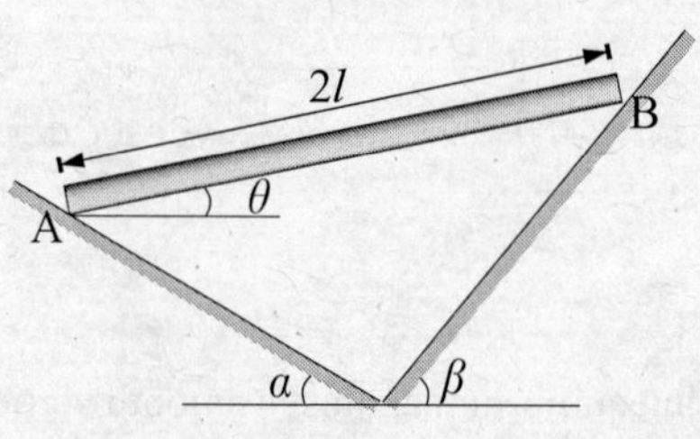

FIGURE 5.33(a) |

Solution. The active force diagram shows the weight acting through the centre of mass G. Reactions at the ends of the rod that do not do any work during a virtual displacement $\delta\theta$ are not shown.

The principle of virtual work is

$$\delta u = 0 \Rightarrow -mg\delta(\text{GH}) = 0$$

or
$$\delta(\text{GH}) = 0$$

From Figure 5.33(b),

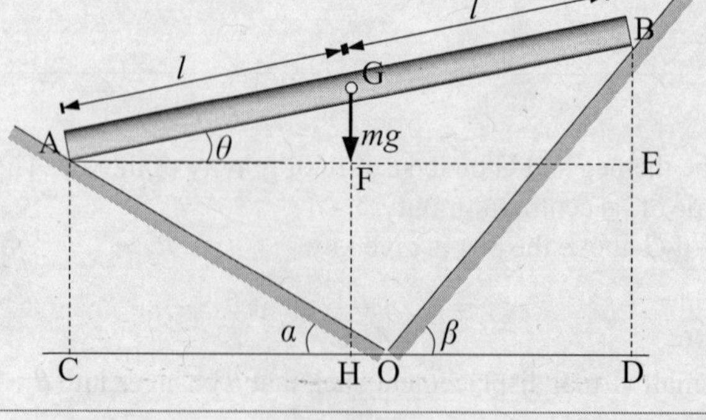

FIGURE 5.33(b) |

$$GH = \frac{1}{2}(AC + BD)$$

$$= \frac{1}{2}(OA\sin\alpha + OB\sin\beta)$$

$$\frac{OA}{\sin(\beta-\theta)} = \frac{OB}{\sin(\alpha+\theta)} = \frac{AB}{\sin[\pi-(\alpha+\beta)]} \Rightarrow OA = \frac{2l\sin(\beta-\theta)}{\sin(\alpha+\beta)} \quad \text{and} \quad OB = \frac{2l\sin(\alpha+\theta)}{\sin(\alpha+\beta)}$$

$$GH = \frac{1}{2}\left[\frac{2l\sin(\beta-\theta)}{\sin(\alpha+\beta)} \times \sin\alpha + \frac{2l\sin(\alpha+\theta)}{\sin(\alpha+\beta)} \times \sin\beta\right]$$

$$GH = \frac{l}{\sin(\alpha+\beta)}[\sin(\beta-\theta)\times\sin\alpha + \sin(\alpha+\theta)\times\sin\beta]$$

The virtual work equation

$$\delta(GH) = 0 \Rightarrow \delta\left[\frac{l}{\sin(\alpha+\beta)}[\sin(\beta-\theta)\times\sin\alpha + \sin(\alpha+\theta)\times\sin\beta]\right] = 0$$

or $$-\cos(\beta-\theta)\times\sin\alpha + \cos(\alpha+\theta)\times\sin\beta = 0$$

or $$\cos\beta\cos\theta\sin\alpha + \sin\beta\sin\theta\sin\alpha = \cos\alpha\cos\theta\sin\beta - \sin\alpha\sin\theta\sin\beta$$

or $$\tan\theta = \frac{\sin(\beta-\alpha)}{2\sin\beta\sin\alpha}$$

$$\theta = \tan^{-1}\left[\frac{\sin(\beta-\alpha)}{2\sin\beta\sin\alpha}\right]$$

PROBLEM 5.34

A uniform rectangular block of height h and mass m is centred on a horizontal position on the fixed circular surface of radius r. Determine the maximum height of h for which the block will be stable in the position shown in Figure 5.34(a).

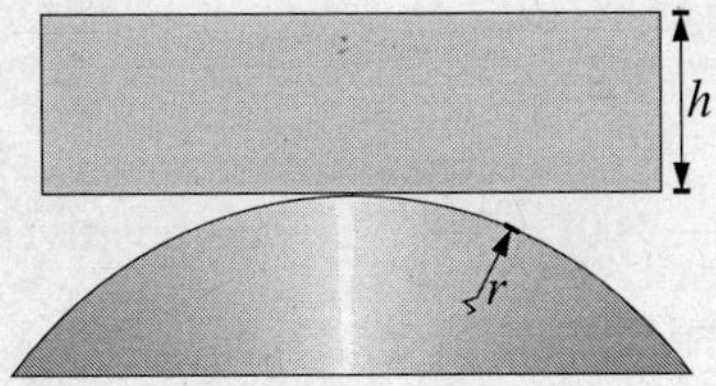

FIGURE 5.34(a) |

Solution. If the block is stable, its potential energy must be a minimum; i.e., the centre of mass G must be raised slightly for any infinitesimal virtual displacement defined by the angle $\delta\theta$, as shown in Figure 5.34(b). During this displacement, the block can roll on the circular surface without slipping. Hence, A′C is equal to the arc AC = $r\,\delta\theta$ and the elevation of the point G above the centre O of the circular surface is

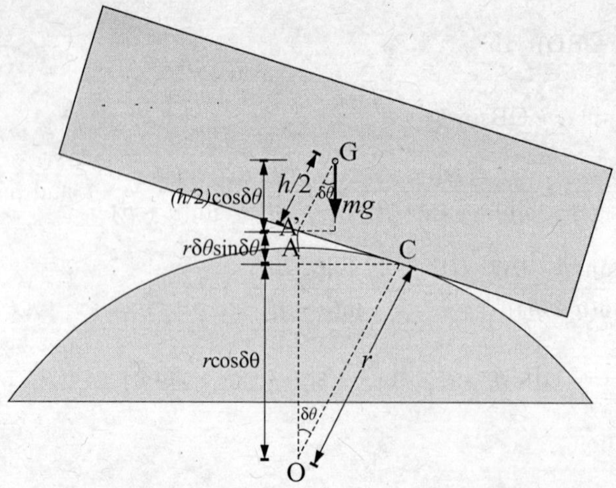

FIGURE 5.34(b) |

$$r\cos\delta\theta + r\delta\theta\sin\delta\theta + \frac{h}{2}\cos\delta\theta$$

In the configuration of equilibrium [Fig. 5.34(a)], the elevation of G above O is $r + (h/2)$. Hence, for stability, we must have

$$r\cos\delta\theta + r\delta\theta\sin\delta\theta + \frac{h}{2}\cos\delta\theta > r + \frac{h}{2}$$

Since $\delta\theta$ is very small, we can take

$$\cos\delta\theta = 1 - \frac{(\delta\theta)^2}{2}, \text{ and } \sin\delta\theta \approx \delta\theta$$

Then

$$r\left(1 - \frac{(\delta\theta)^2}{2}\right) + r\delta\theta \times \delta\theta + \frac{h}{2}\left(1 - \frac{(\delta\theta)^2}{2}\right) > r + \frac{h}{2}$$

or

$$-r\frac{(\delta\theta)^2}{2} + r\delta\theta \times \delta\theta - \frac{h}{2} \times \frac{(\delta\theta)^2}{2} > 0$$

or

$$\left(r - \frac{h}{2}\right) \times \frac{(\delta\theta)^2}{2} > 0 \Rightarrow \frac{(\delta\theta)^2}{2} \text{ always positive, then } \left(r - \frac{h}{2}\right) > 0$$

or

$$r > \frac{h}{2}$$

$$h < 2r$$

PROBLEM 5.35

Two identical blocks are attached to gears as shown in Figure 5.35(a). Determine the values of angle θ for equilibrium and identify the type of equilibrium for each position.

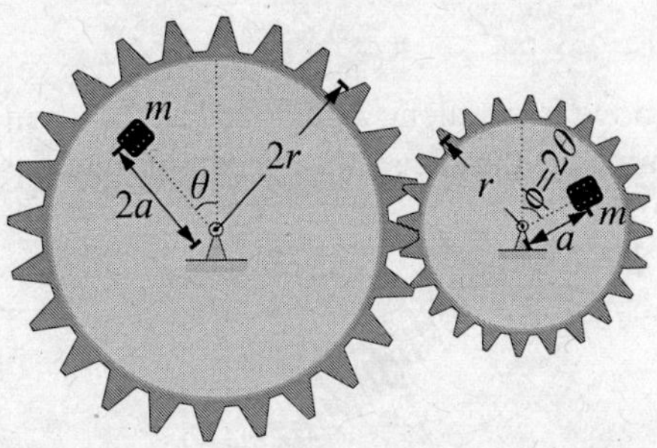

FIGURE 5.35(a) |

Solution. If $\theta = 0°$ is selected as the standard or reference configuration, then, at any angle θ the total potential energy of system is

$$V_g = mg(2a\cos\theta) + mg(a\cos\phi)$$

Since $2r\theta = r\phi \Rightarrow \phi = 2\theta$,

$$V_g = 2mga\cos\theta + mga\cos 2\theta$$

Equilibrium occurs for $\frac{d}{d\theta}(V_g) = 0$, so that

$$\frac{d}{d\theta}(2mga\cos\theta + mga\cos 2\theta) = -2mga(\sin\theta + \sin 2\theta) = 0$$

or

$$\sin\theta(1 + 2\cos\theta) = 0$$

Solution to this equation are given by

$$\sin\theta = 0; (\theta = 0°, 180°)$$

$$\cos\theta = -\frac{1}{2}; (\theta = 120°, 240°)$$

We now determine the stability by examining the sign of $\frac{d^2}{d\theta^2}(V_g)$,

$$\frac{d^2}{dx^2}(V_g) = -2mga(\cos\theta + 2\cos 2\theta)$$

At $\theta = 0°$	$\frac{d^2}{d\theta^2}(V_g) = -6mga$	which is negative	[Unstable]
At $\theta = 120°$	$\frac{d^2}{d\theta^2}(V_g) = 3mga$	which is positive	[Stable]
At $\theta = 180°$	$\frac{d^2}{d\theta^2}(V_g) = -2mga$	which is negative	[Unstable]
At $\theta = 240°$	$\frac{d^2}{d\theta^2}(V_g) = 3mga$	which is positive	[Stable]

PROBLEM 5.36

A collar C, of weight W, may move freely along a guide rod as shown in Figure 5.36(a). The spring is unstreched when $y = 0$. Determine the distance y for the equilibrium position of collar.

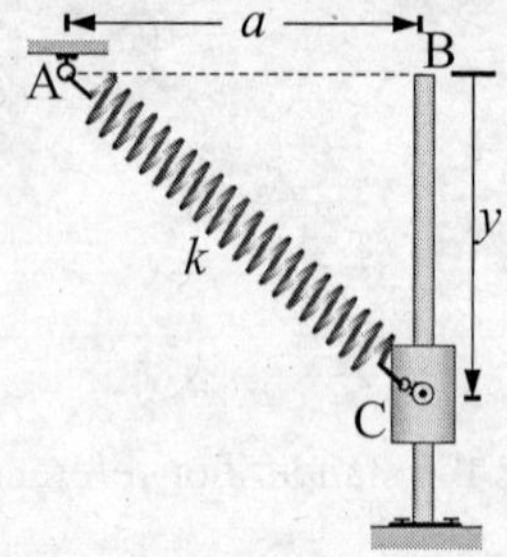

FIGURE 5.36(a)

Solution. If $y = 0$ is used as the standard or reference configuration and the spring is unstreched, then for any value of y, the potential energy of the collar is $-Wy$ and the elastic energy of spring is $(1/2)k\Delta^2$.

The total potential energy of the system is

$$V = -Wy + \frac{1}{2}k\Delta^2$$

From the geometry, extension of the spring, $\Delta = \sqrt{a^2 + y^2} - a$, hence,

$$V = -Wy + \frac{1}{2}k\left(\sqrt{a^2 + y^2} - a\right)^2$$

To study equilibrium, set $\frac{d}{dy}(V) = 0$,

$$\frac{d}{dy}\left(-Wy + \frac{1}{2}k(\sqrt{a^2 + y^2} - a)^2\right) = 0$$

or

$$-W + k\left(\sqrt{a^2 + y^2} - a\right)\left(\frac{2y}{\sqrt{a^2 + y^2}}\right) = 0$$

On simplifying, we obtain

$$y = \frac{a}{\sqrt{\left(\frac{W}{k}\right)^2 + 2\left(\frac{W}{k}\right)a}}$$

PROBLEM 5.37

Two bars EF and BC are connected at P by a smooth pin that is attached to BC and may slide freely in a slot cut in EF as shown in Figure 5.37(a). Determine the largest mass m_2 for which the equilibrium of the mechanism is stable in the position shown.

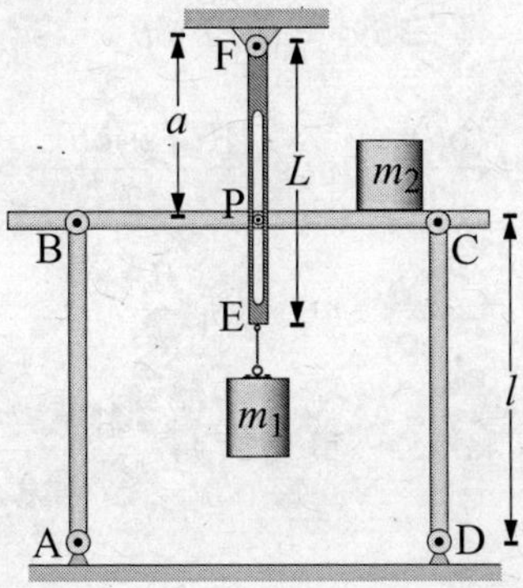

FIGURE 5.37(a) |

Solution. If the system is stable, its gravitational potential energy must be a minimum; i.e., the sum of potential energy of blocks m_1 and m_2 must rise slightly for any small virtual displacement by an angle $\delta\theta$, as shown in Figure 5.37(b). In the displaced position, total potential energy is

$$m_2 gl \cos \delta\theta + m_1 g(l + a - L \cos \delta\phi)$$

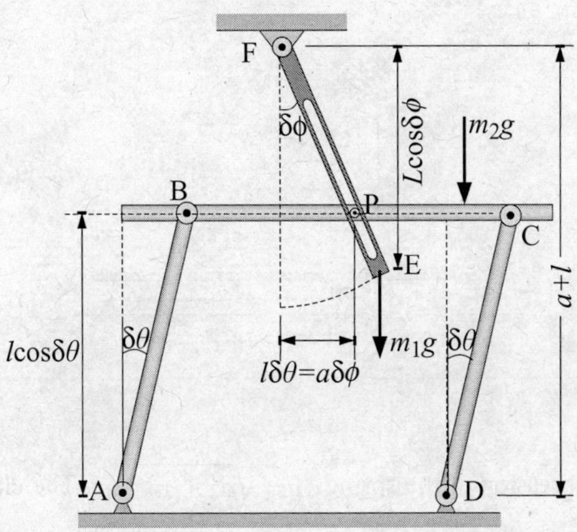

FIGURE 5.37(b) |

In the configuration of equilibrium [Fig. 5.37(a)], total potential energy is

$$m_2 gl + m_1 g(l + a - L)$$

Hence, for stability, we must have

$$m_2 gl \cos \delta\theta + m_1 g(l + a - L \cos \delta\phi) > m_2 gl + m_1 g(l + a - L)$$

or

$$m_2 gl(\cos \delta\theta - l) - m_1 gL(\cos \delta\phi - 1) > 0$$

Since $\delta\theta$ and $\delta\phi$ are very small, then $\cos \delta\theta = 1 - [(\delta\theta)^2 / 2]$ and $\delta\phi = 1 - [(\delta\phi)^2 / 2]$:

$$-m_2 gl \frac{(\delta\theta)^2}{2} + m_1 gL \frac{(\delta\phi)^2}{2} > 0$$

The relation between θ and ϕ is $l\theta = a\phi \Rightarrow \delta\phi = (l / a)\delta\theta$:

$$-m_2 gl \frac{(\delta\theta)^2}{2} + m_1 gL \frac{[(l / a)\delta\theta]^2}{2} > 0$$

or

$$m_2 < m_1 \frac{lL}{a^2}$$

The largest value of $m_2 = m_1 \frac{lL}{a^2}$

PROBLEM 5.38

In the mechanism shown in Figure 5.38(a), a spring of stiffness k is uncompressed when $x = 0$. As x increases, the rod slides through the pivoted collar at B and compresses the spring between the collar and the end of the rod. Determine the force F required to produce a given displacement x. Neglect the mass of the links.

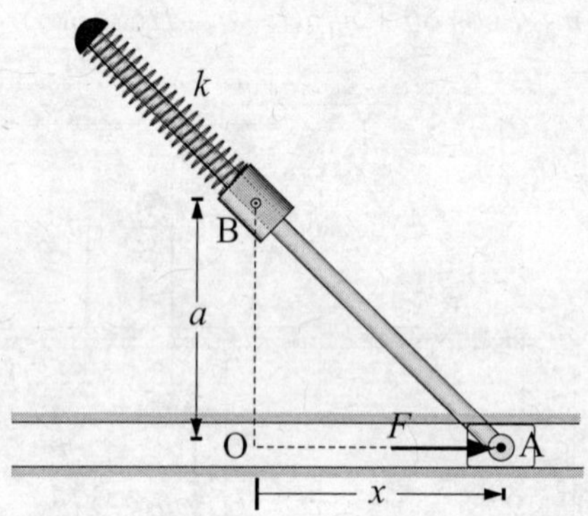

FIGURE 5.38(a) |

Solution. The compression of the spring $\Delta = \sqrt{a^2 + x^2} - a$. The elastic potential energy of the spring is

$$V_e = \frac{1}{2} k \left(\sqrt{a^2 + x^2} - a \right)^2 \qquad \left[V_e = \frac{1}{2} k\Delta^2 \right]$$

The distance between O and A is x, so that the virtual work done by F is

$$\delta u = F\delta x$$

The virtual work equation now gives

$$\delta u = \delta V_e$$

or

$$F\delta x = \delta \left[\frac{1}{2} k(\sqrt{a^2 + x^2} - a)^2 \right]$$

or

$$F\,\delta x = \frac{kx(\sqrt{a^2 + x^2} - a)}{\sqrt{a^2 + x^2}}\delta x$$

$$F = kx\left(1 - \frac{a}{\sqrt{a^2 + x^2}}\right)$$

PROBLEM 5.39

For the mechanism shown in Figure 5.39(a), the spring is uncompressed when $\theta = 0$. Determine the stiffness k of the spring, which will establish an equilibrium position θ in the vertical plane. The mass of the links is negligible.

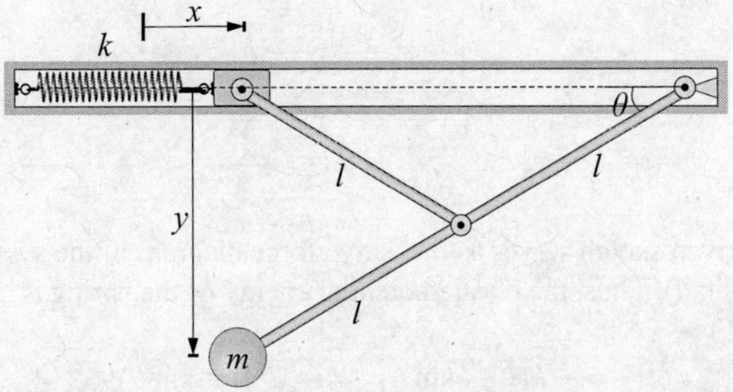

FIGURE 5.39(a) |

Solution. The system consists of the spring and the mass. We will take the x-axis as the datum for zero gravitational potential energy. In the displaced position, the elastic and gravitational potential energies are

$$V_e = \frac{1}{2}kx^2 = \frac{1}{2}k(2l - 2l\cos\theta)^2 = 2kl^2(1 - \cos\theta)^2$$

and

$$V_g = -mg(2l\sin\theta)$$

The total potential energy is then

$$V = V_e + V_g = 2kl^2(1 - \cos\theta)^2 - mg(2l\sin\theta)$$

Equilibrium occurs; $\dfrac{dV}{d\theta} = 0$, so that

$$\frac{dV}{d\theta} = 4kl^2\sin\theta(1 - \cos\theta) - 2mgl\cos\theta = 0$$

or

$$4kl^2\sin\theta(1 - \cos\theta) = 2mgl\cos\theta$$

$$k = \frac{mg\cot\theta}{2l(1 - \cos\theta)}$$

PROBLEM 5.40

When $x = 0$, the spring of stiffness k is uncompressed [Fig. 5.40(a)]. Neglecting the mass of the rod, determine the force F required to produce a given displacement x.

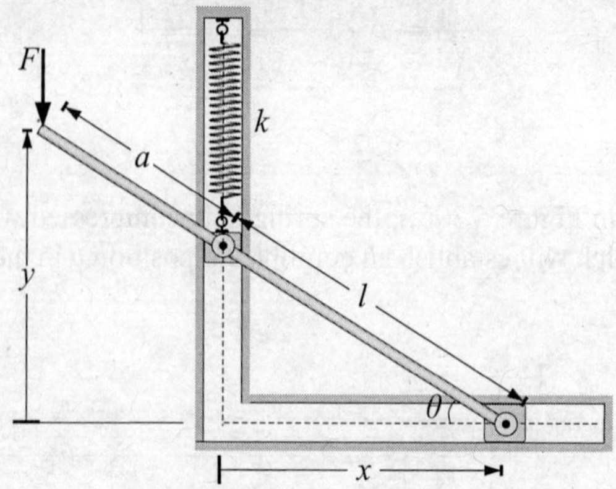

FIGURE 5.40(a) |

Solution. The given sketch serves as the active force diagram of the system. The extension of the spring is $\Delta = (l - l\sin\theta)$. Thus, the elastic potential energy of the spring is

$$V_e = \frac{1}{2}k(l - l\sin\theta)^2 = \frac{1}{2}kl^2(1 - \sin\theta)^2$$

The virtual work by F is

$$\delta u = -F\delta y = -F\delta(l + a)\sin\theta$$

The virtual work equation now gives

$$\delta u = \delta V_e$$

or

$$-F\delta(l + a)\sin\theta = \delta\left(\frac{1}{2}kl^2(1 - \sin\theta)^2\right)$$

or

$$-F(l + a)\cos\theta\delta\theta = -kl^2\cos\theta(1 - \sin\theta)\delta\theta$$

$$F = \frac{kl^2}{(l + a)}(1 - \sin\theta),$$

or

$$\left[\text{from the figure, } \cos\theta = \frac{x}{l} \Rightarrow \sin\theta = \sqrt{1 - \left(\frac{x}{l}\right)^2}\right]$$

Substituting $\sin\theta = \sqrt{1 - (x/l)^2}$, we obtain

$$F = \frac{kl^2}{l + a}\left[1 - \sqrt{1 - \left(\frac{x}{l}\right)^2}\right]$$

PROBLEM 5.41

A spring of stiffness k is un-deformed when $\theta = 0$ [Fig. 5.41(a)]. Neglecting the mass of the links, determine the force F required to produce equilibrium of the two hinged links at an angle θ:

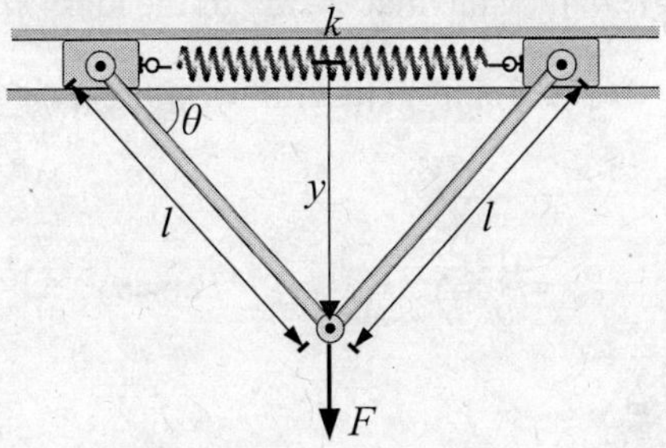

FIGURE 5.41(a) |

Solution. The compression of the spring is

$$x = (2l - 2l\cos\theta)$$

Thus, the elastic potential energy of the spring is

$$V_e = \frac{1}{2}kx^2 = \frac{1}{2}k(2l - 2l\cos\theta)^2 = 2kl^2(1-\cos\theta)^2$$

Virtual work done by F is

$$\delta u = F\delta y = F\delta(l\sin\theta)$$

The virtual work equation,

$$\delta u = \delta V_e$$

or

$$F\delta(l\sin\theta) = \delta[2kl^2(1-\cos\theta)^2]$$

or

$$Fl\cos\theta = 4kl^2\sin\theta(1-\cos\theta)$$

$$F = 4kl\tan\theta(1-\cos\theta)$$

PROBLEM 5.42

The constant of a spring is k and the spring is unstreched when $\theta = 0°$. The bar ABC is attached to collars B and C, which may move freely on the guide rod as shown in Figure 5.42(a).

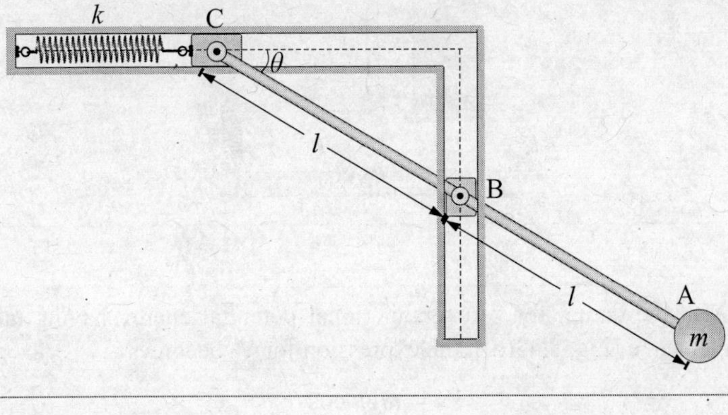

FIGURE 5.42(a) |

Neglecting the weight of the bar ABC, derive an equation in m, θ, k and l that must be satisfied when the bar ABC is in equilibrium.

Solution. If $\theta = 0$ is used as the standard or reference configuration and the spring is unstreched, then for any other angle θ, the potential energy of the mass m is

$$V_g = -mg(2l \sin \theta) = -2mgl \sin \theta$$

And elastic energy of spring

$$V_e = \frac{1}{2} k\Delta^2 = \frac{1}{2} k(l - l \cos \theta)^2 = \frac{1}{2} kl^2(1 - \cos \theta)^2$$

Thus, the total potential energy is

$$V = V_g + V_e = -2mgl \sin \theta + \frac{1}{2} kl^2(1 - \cos \theta)^2$$

Equilibrium occurs for $\dfrac{dV}{d\theta} = 0$ so that

$$\frac{d}{d\theta}\left(-2mgl \sin \theta + \frac{1}{2} kl^2(1 - \cos \theta)^2\right) = 0$$

or

$$-2mgl \cos \theta + kl^2 \sin \theta(1 - \cos \theta) = 0$$

$$\tan \theta - \sin \theta - \frac{2mg}{kl} = 0$$

PROBLEM 5.43

Figure 5.43(a) shows an inverted pendulum. Determine the maximum height h of the mass m for which the inverted pendulum will be stable in the vertical position as shown in the figure. Each of the springs has a stiffness k, and they have equal pre-compression in this position. Neglect the weight of the frame.

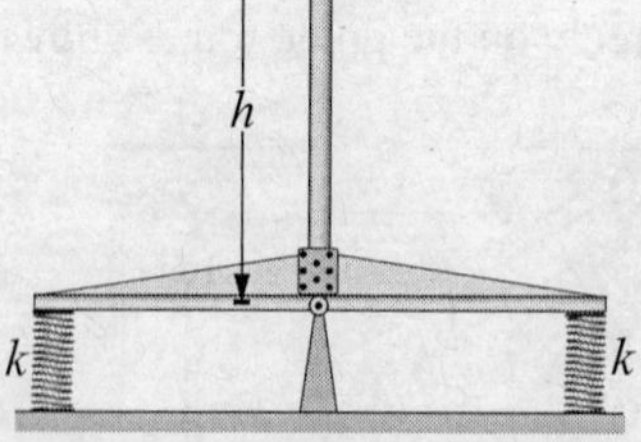

FIGURE 5.43(a) |

Solution. With the datum for zero gravitational potential energy having taken through the hinge at O for convenience [Fig. 5.43(b)], the expression for V_g becomes

$$V_g = mgh \cos \theta$$

The elastic potential energy of the spring is

$$V_e = 2\left(\frac{1}{2}k\Delta^2\right) = k(a\sin\theta)^2$$

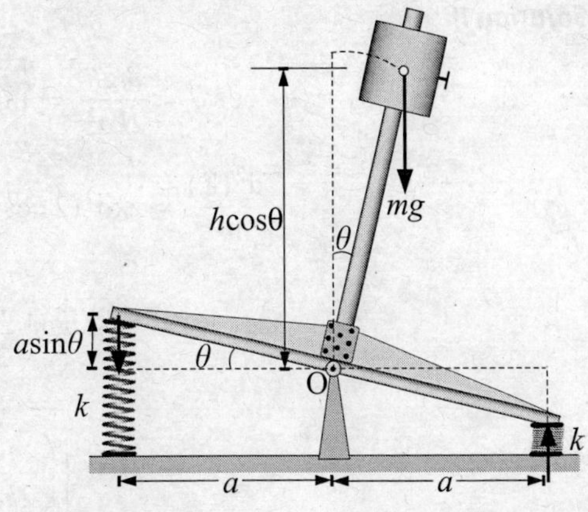

FIGURE 5.43(b)

The total potential energy is then

$$V = V_g + V_e = ka^2\sin^2\theta + mgh\cos\theta$$

Equilibrium occurs for $\dfrac{dV}{d\theta} = 0$ so that

$$\frac{dV}{d\theta} = 2ka^2\sin\theta\cos\theta - mgh\sin\theta = 0$$

or $$\sin\theta(2ka^2\cos\theta - mgh) = 0$$

The two solutions to this equation are given by

$$\sin\theta = 0 \Rightarrow [\theta = 0°]$$

and $$\cos\theta = \frac{mgh}{2ka^2} \Rightarrow \left[\theta = \cos^{-1}\left(\frac{mgh}{2ka^2}\right)\right]$$

We now determine the stability by examining the sign of the second derivative of V for each of the two equilibrium positions. The second derivative is

$$\frac{d^2V}{d\theta^2} = 2ka^2\cos 2\theta - mgh\cos\theta$$

$$= 2ka^2(2\cos^2\theta - 1) - mgh\cos\theta$$

Solution I

$$\sin\theta = 0 \Rightarrow [\theta = 0°]$$

$$\frac{d^2(V)}{d\theta^2} = 2ka^2(2\cos^2\theta - 1) - mgh\cos\theta$$

$$= 2ka^2 - mgh = 2ka^2\left(1 - \frac{mgh}{2ka^2}\right)$$

$$= \text{positive (stable) if } h < \frac{2ka^2}{mg}$$

$$= \text{negative (unstable) if } h > \frac{2ka^2}{mg}$$

Thus, if the height h is less than $2ka^2 / mg$, the mass m will return to the vertical position.

Solution II

$$\cos\theta = \frac{mgh}{2ka^2} \Rightarrow \left[\theta = \cos^{-1}\left(\frac{mgh}{2ka^2}\right)\right]$$

$$\frac{d^2(V)}{d\theta^2} = 2ka^2(2\cos^2\theta - 1) - mgh\cos\theta$$

$$= 2ka^2\left\{2\left(\frac{mgh}{2ka^2}\right)^2 - 1\right\} - mgh\left(\frac{mgh}{2ka^2}\right)$$

$$= 2ka^2\left\{\left(\frac{mgh}{2ka^2}\right)^2 - 1\right\}$$

Since the cosine must be less than unity, we see that this solution is limited to the case where $h < (2ka^2 / mg)$, which makes the second derivative of V negative. Thus, equilibrium for solution II is ever stable. Therefore

$$h_{\max} = \frac{2ka^2}{mg}$$

PROBLEM 5.44

In the mechanism shown in Figure 5.44(a), a spring of stiffness k is uncompressed when $\theta = 0$. Determine the angle θ for the equilibrium position and specify the minimum spring stiffness k that will limit θ to 30°. A collar of mass m slides freely on the fixed vertical shaft.

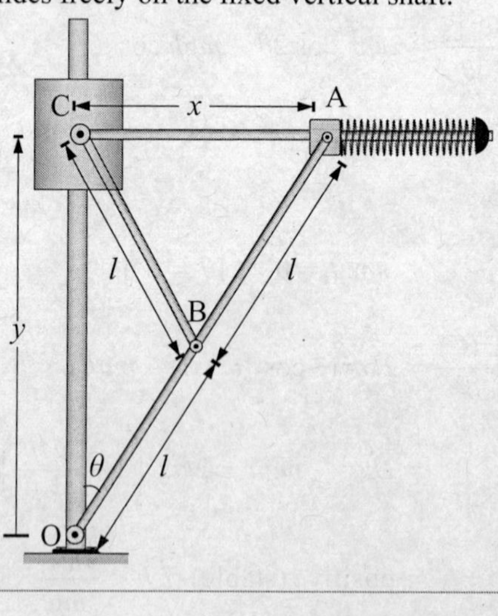

FIGURE 5.44(a)

Solution. The compression x of the spring is the distance that A has moved away from C, which is $x = 2l\sin\theta$. Thus the elastic potential energy of the spring is

$$\left[V_e = \frac{1}{2}kx^2\right] \qquad V_e = \frac{1}{2}k(2l\sin\theta)^2 = 2kl^2\sin^2\theta$$

With the datum for zero gravitational potential energy having taken through the support O for convenience, the expression for V_g becomes

$$[V_g = mgy] \qquad V_g = mgy = 2mgl\cos\theta$$

The total potential energy taken is then

$$V = V_e + V_g = 2kl^2\sin^2\theta + 2mgl\cos\theta$$

Equilibrium occurs for $\dfrac{dV}{d\theta} = 0$ so that

$$\frac{dV}{d\theta} = 4kl^2\sin\theta\cos\theta - 2mgl\sin\theta = 0$$

or

$$\sin\theta\left(\cos\theta - \frac{mg}{2kl}\right) = 0$$

The solution of equation $\sin\theta = 0$ $(\theta = 0°)$, and $\cos\theta = \dfrac{mg}{2kl}\left[\theta = \cos^{-1}\left(\dfrac{mg}{2kl}\right)\right]$

Given that the system is a stable equilibrium at $\theta = 30°$; therefore,

$$\cos\theta = \frac{mg}{2kl} \Rightarrow \cos 30 = \frac{mg}{2kl}$$

or

$$\frac{\sqrt{3}}{2} = \frac{mg}{2kl} \Rightarrow k = \frac{mg}{\sqrt{3}l}$$

$$k_{\min} = \frac{mg}{\sqrt{3}l}$$

PROBLEM 5.45

Determine the minimum stiffness k of each spring that will ensure stability of the system in the position shown in Figure 5.45(a). Neglect the mass of links, spring and platform. Each spring has a tensile preset deflection Δ.

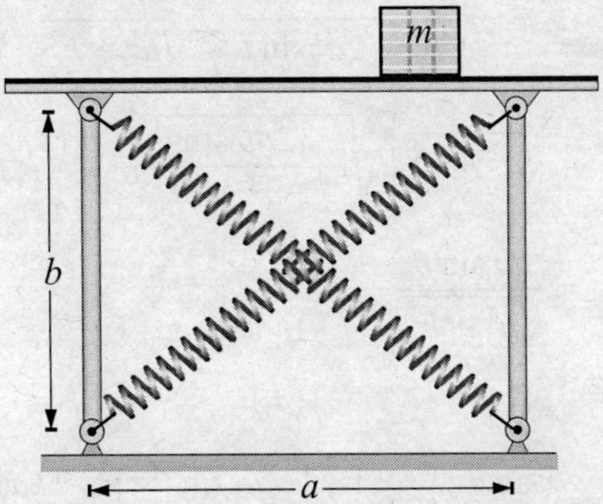

FIGURE 5.45(a) |

Solution. The system consists of the springs and platform. Since there are no external active forces, the given sketch in Figure 5.45(b) serves as the active force diagram. We shall take AC as the datum for zero gravitational potential energy. In the displaced position, the elastic and gravitational energies are

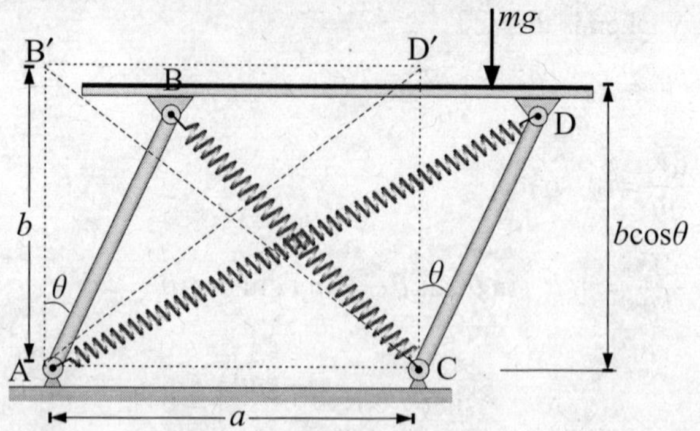

FIGURE 5.45(b) |

$$V_e = \frac{1}{2}k\Delta_1{}^2 + \frac{1}{2}k\Delta_2{}^2 \text{ and } V_g = mgb\cos\theta,$$

where

$$\Delta_1 = \text{compression of spring BC} = \text{B}'\text{C} - \text{BC}$$

$$\Delta_1 = \sqrt{a^2 + b^2} - \sqrt{a^2 + b^2 - 2ab\sin\theta}$$

$$= \sqrt{a^2 + b^2}\left[1 - \sqrt{1 - \frac{2ab\sin\theta}{a^2 + b^2}}\right]$$

or

$$\Delta_1 = \sqrt{a^2 + b^2}\left[1 - 1 + \frac{2ab\sin\theta}{2(a^2 + b^2)}\right] = \frac{ab\sin\theta}{\sqrt{a^2 + b^2}}$$

and

$$\Delta_2 = \text{extension of spring AD} = \text{AD} - \text{AD}'$$

$$\Delta_2 = \sqrt{a^2 + b^2 + 2ab\sin\theta} - \sqrt{a^2 + b^2}$$

$$= \sqrt{a^2 + b^2}\left[\sqrt{1 + \frac{2ab\sin\theta}{a^2 + b^2}} - 1\right]$$

or

$$\Delta_2 = \frac{ab\sin\theta}{\sqrt{a^2 + b^2}}$$

Thus, elastic potential of spring is

$$V_e = \frac{1}{2}k\left(\frac{ab\sin\theta}{\sqrt{a^2 + b^2}}\right)^2 + \frac{1}{2}k\left(\frac{ab\sin\theta}{\sqrt{a^2 + b^2}}\right)^2 = \frac{ka^2b^2\sin^2\theta}{a^2 + b^2}$$

The total potential energy is, then

$$V = V_e + V_g = \frac{ka^2b^2 \sin^2\theta}{a^2 + b^2} + mgb\cos\theta$$

Equilibrium occurs for $\frac{dV}{d\theta} = 0$ so that

$$\frac{dV}{d\theta} = \frac{2ka^2b^2 \sin\theta\cos\theta}{a^2 + b^2} - mgb\sin\theta = 0$$

$$\sin\theta\left(\cos\theta - \frac{mg}{2ka^2b}(a^2 + b^2)\right) = 0$$

The two solutions to this equation are given by

$$\sin\theta = 0 \Rightarrow [\theta = 0°]$$

and $$\cos\theta = \frac{mg}{2ka^2b}\left(a^2 + b^2\right) \qquad \left[\theta = \cos^{-1}\left(\frac{mg}{2ka^2b}(a^2 + b^2)\right)\right]$$

We now determine the stability by examining the sign of the second derivative of V for the given position $[\theta = 0°]$. The second derivative is

$$\frac{dV}{d\theta} = \frac{2ka^2b^2\cos 2\theta}{a^2 + b^2} - mgb\cos\theta$$

$$= \frac{2ka^2b^2}{a^2 + b^2}(2\cos^2\theta - 1) - mgb\cos\theta$$

Solution I

$$\sin\theta = 0 \Rightarrow [\theta = 0°]$$

$$\frac{d^2V}{d\theta^2} = \frac{2ka^2b^2}{a^2 + b^2}(2 - 1) - mgb$$

$$= \frac{2ka^2b^2}{a^2 + b^2} - mgb = \frac{2ka^2b^2}{a^2 + b^2}\left(1 - \frac{mgh(a^2 + b^2)}{2ka^2b}\right)$$

$$= \text{positive (stable) if } k > \frac{mg(a^2 + b^2)}{2a^2b}$$

$$= \text{negative (unstable) if } k < \frac{mg(a^2 + b^2)}{2a^2b}$$

Thus, if the stiffness of each spring is greater than $mg(a^2 + b^2)\,/\,2a^2b$, the platform will return to the initial position.

$$k_{min} = \frac{mg(a^2 + b^2)}{2a^2b}$$

PROBLEM 5.46

Two beams AC and BC are hinged at point C and loaded with a load W as shown in Figure 5.46(a). Using the principle of virtual work, find the reaction at D.

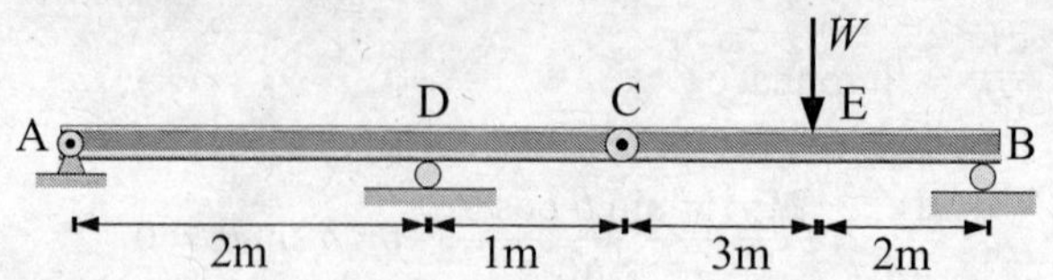

FIGURE 5.46(a) |

Solution. Assume that a vertical displacement at the hinged point C in upward direction is y_c. From the geometry of Figure 5.46(b), virtual displacement at points D and E may be calculated as follows:

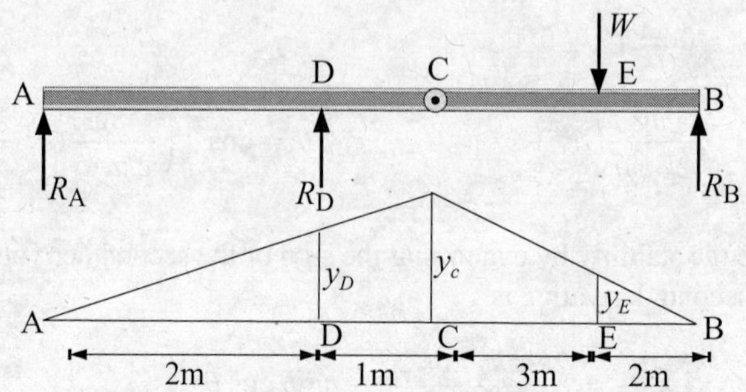

FIGURE 5.46(b) |

$$\frac{y_c}{3} = \frac{y_D}{2} \Rightarrow y_D = \frac{2}{3} y_c$$

and

$$\frac{y_c}{5} = \frac{y_E}{2} \Rightarrow y_E = \frac{2}{5} y_c$$

As per the principle of virtual work, the algebraic sum of the total virtual work done is zero. Therefore,

$$R_D \times y_D - W \times y_E = 0$$

or

$$R_D \times \frac{2}{3} y_c - W \times \frac{2}{5} y_c = 0$$

$$R_D = \frac{3}{5} W$$

CHAPTER HIGHLIGHTS

- An imaginary small displacement given to the system is called *virtual displacement*.
- The product of the force and the virtual displacement in the direction of the force is called *virtual work*.
- Active forces are external forces capable of doing virtual work during possible virtual displacement.
- Active force diagram is the sketch of system which shows the active forces.

- The principle of virtual work for a ***particle (or system)*** states that "*If a particle (or system) is in a equilibrium, the total virtual work of the forces acting on the particle (or system) is zero for any virtual displacement*".
- Stable equilibrium occurs if the potential energy V of system is a minimum.

$$\text{Condition for stable equilibrium } \frac{dV}{d\theta} = 0 \text{ and } \frac{d^2V}{d\theta^2} > 0 \text{ } (+ve \text{ } value)$$

- Unstable equilibrium occurs if the potential energy V of system is a maximum.

$$\text{Condition for unstable equilibrium } \frac{dV}{d\theta} = 0 \text{ and } \frac{d^2V}{d\theta^2} < 0 \text{ } (-ve \text{ } value)$$

- Neutral equilibrium exists if a system remains in any position in which it is placed.

$$\text{Condition for neutral equilibrium } \frac{dV}{d\theta} = 0 \text{ and } \frac{d^2V}{d\theta^2} = 0$$

REVIEW QUESTIONS

Q1. Define virtual displacement and virtual work.

Q2. Explain the principle of virtual work.

Q3. What are the active force and the active force diagram?

Q4. Explain stable equilibrium, unstable equilibrium and neutral equilibrium.

Q5. A uniform rod of length l and weight W is held in equilibrium by a string as shown in Figure 5.Q5. Find the tension in string in terms of the weight and the angle θ.

$\left[\text{Ans; T} = \frac{W}{2}\tan\theta\right]$

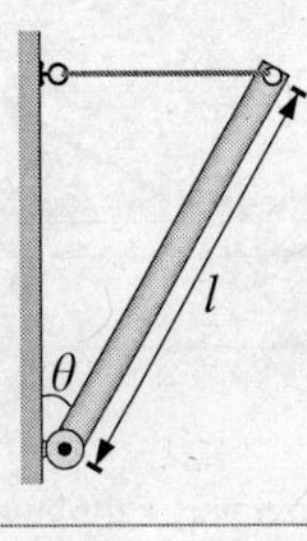

FIGURE 5.Q5 |

Q6. A beam supported on four rollers, each of radius r, is moved up an inclined plane by a force F that acts parallel to the plane (Fig. 5.Q6). Determine the magnitude of the force F necessary to maintain equilibrium if there is no slipping between either rollers and the inclined plane or the rollers and the beam. The weight of the beam is W and the weight of each roller is w.
[Ans; $F = (W + 2w)\sin\theta$].

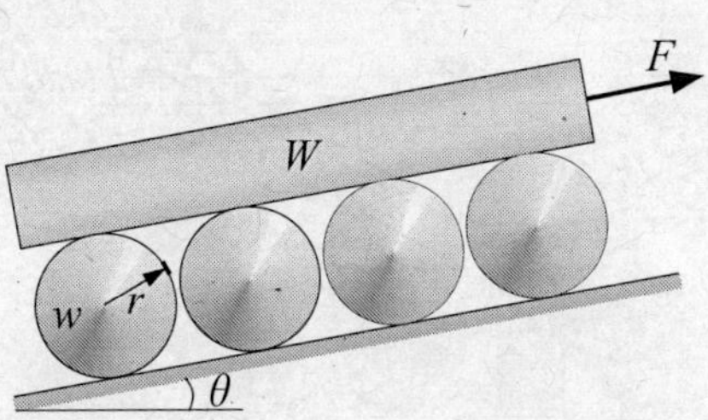

FIGURE 5.Q6 |

Q7. The elevation of a platform of mass m is controlled by adjusting a screw with torque M as shown in Figure 5.Q7. The single-thread screw with square threads has a pitch p. Neglecting all friction and weight of links, find the torque M, applied to the adjusting screw, necessary to raise the load.

$\left[\text{Ans; } M = (mgp / \pi)\cot\theta\right]$.

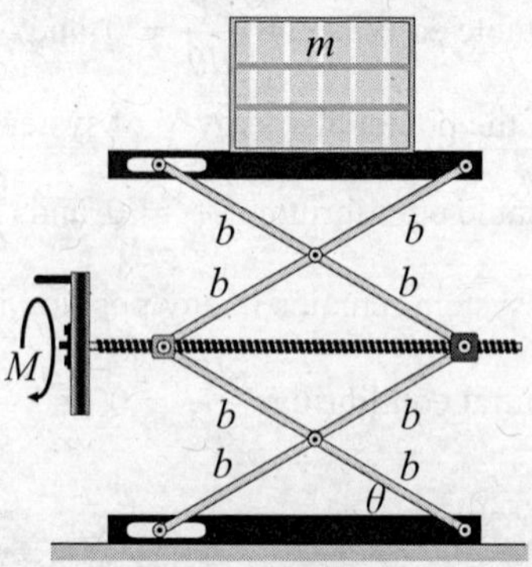

FIGURE 5.Q7 |

Q8. A beam AB of length 1.5 m and negligible weight is hinged at end A and carries a 200 N weight at its free end. The beam is supported at its mid point C by a wire CD as shown in Figure 5.Q8. Use the method of virtual work to determine the tension induced in the wire.

[Ans; $T = 800$ N].

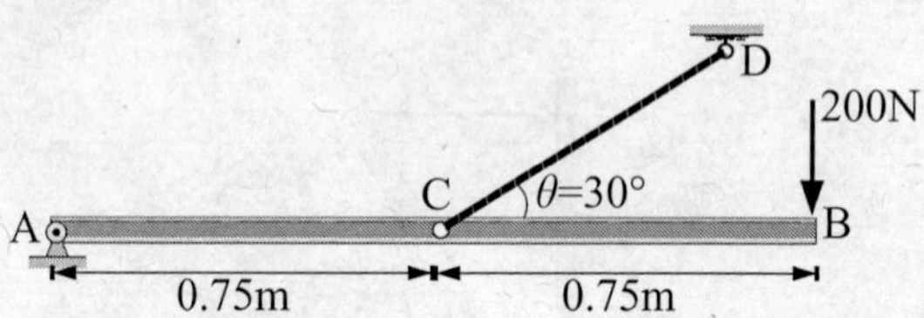

FIGURE 5.Q8 |

Q9. Two uniform links, each of weight w, are in a vertical plane and are connected by a spring as shown in Figure 5.Q9. Specify the stiffness k of the spring that will establish an equilibrium position θ. The spring has a stiffness k and is unstretched when $\theta = 0°$.

$$\left[\text{Ans; } k = \frac{(W + w)\cot\theta}{4l(1 - \cos\theta)}\right].$$

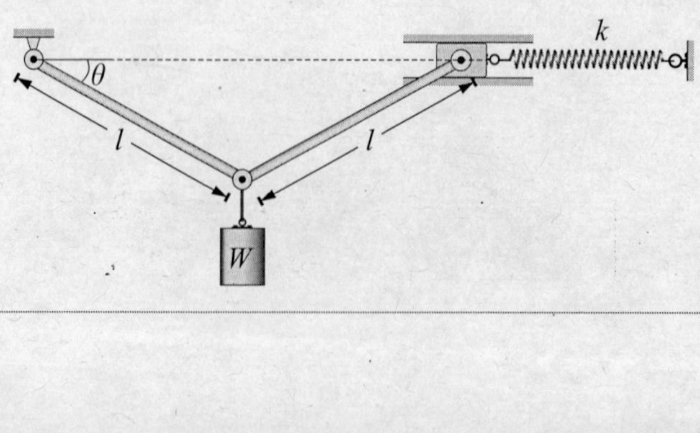

FIGURE 5.Q9 |

Q10. A homogenous cone has a conical cavity cut into it as shown in Figure 5.Q10. Determine the depth of h of the cavity in terms of H so that the cone balances on the pivot and remains in neutral equilibrium. [(Ans; $h = H/3$)].

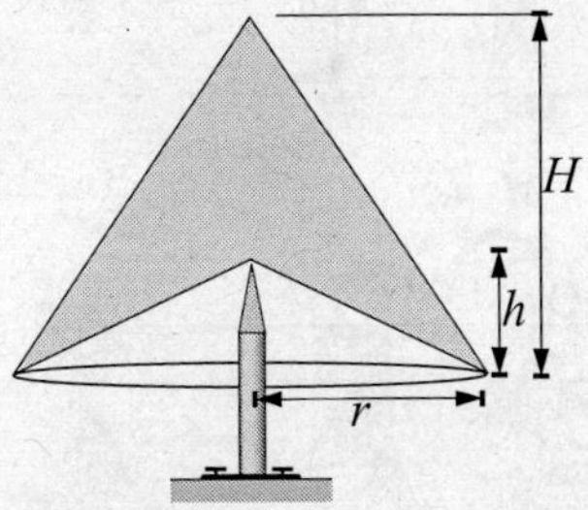

FIGURE 5.Q10 |

Q11. For the mechanism shown in Figure 5.Q11, the spring is uncompressed when $\theta = 0°$. The uniform link AB has mass m and its end A travels freely in horizontal slot. End B is attached to the vertical plunger. Determine the angle θ for equilibrium. Neglect the mass of the remainder of the mechanism.

$$\left[\text{Ans; } \theta = \sin^{-1}\left(\frac{mg}{2kl}\right);\ k > \frac{mg}{2l}\right].$$

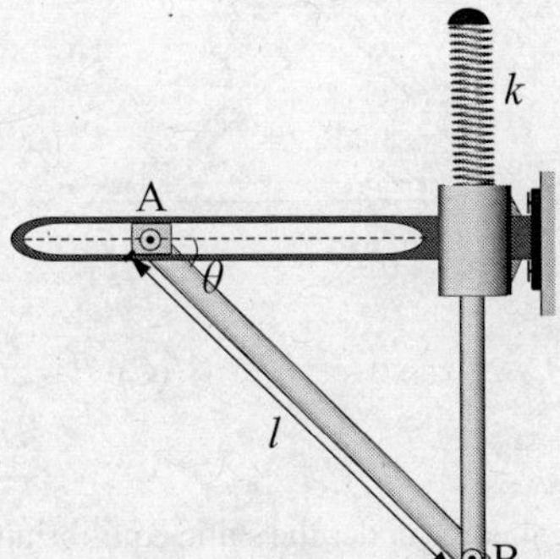

FIGURE 5.Q11

MULTIPLE CHOICE QUESTIONS

Q1. A sphere of weight W and radius R is resting on a circular smooth belt of radius $r = R/2$ as shown in Figure 5.MQ1. The tension of the rope is

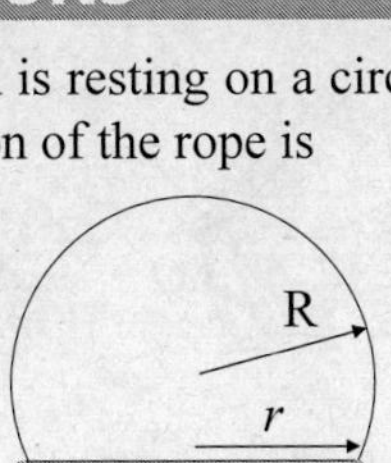

FIGURE 5.MQ1

(A) $\dfrac{W}{\pi\sqrt{6}}$ (B) $\dfrac{W}{\pi 2\sqrt{3}}$ (C) $\dfrac{W}{\pi\sqrt{10}}$ (D) $\dfrac{W}{3\pi}$

[Ans: B]

Q2. Figure 5.MQ2 shows the system of linkage under the static equilibrium. Which one of the following options is true?

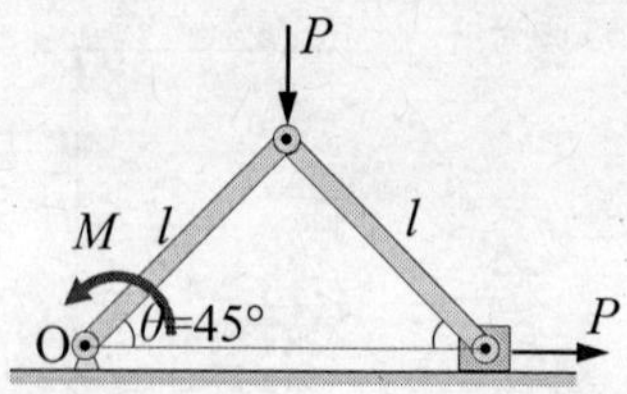

FIGURE 5.MQ2

(A) $M = \frac{3}{\sqrt{2}} Pl$ (B) $M = 2Pl$ (C) $M = \frac{1}{\sqrt{2}} Pl$ (D) None of these

[Ans: A]

Q3. Figure 5.MQ3 shows the system of linkage under the static equilibrium. Which one of the following options is true?

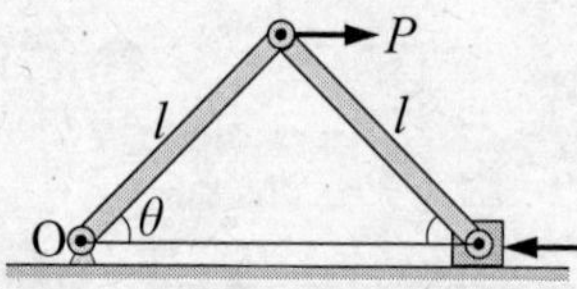

FIGURE 5.MQ3

(A) $P = 2Q \cot\theta$ (B) $P = 2Q \tan\theta$ (C) $P = 2Q$ (D) None of these

[Ans: C]

Q4. Figure 5.MQ4 shows the system of linkage under the static equilibrium. Which one of the following options is true?

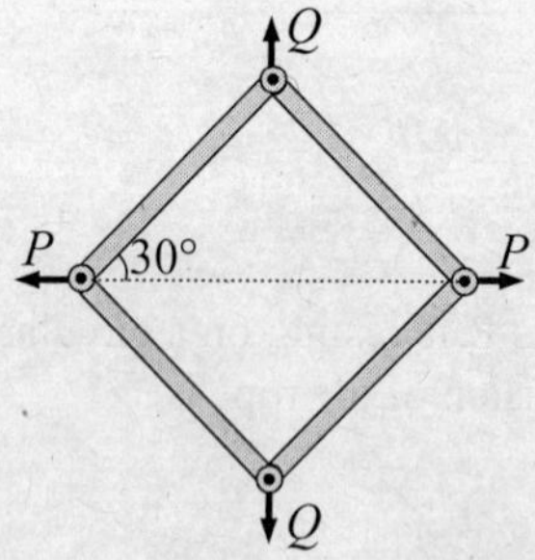

FIGURE 5.MQ4

(A) $Q = \sqrt{3}P$ (B) $Q = \frac{P}{\sqrt{3}}$ (C) $Q = \frac{\sqrt{3}}{2} P$ (D) None of these

[Ans: B]

Q5. A uniform ladder of length 10 m and weight 500 N rests on a smooth floor at A and against a smooth wall at B as shown in Figure 5.MQ5. A horizontal rope PQ prevents the ladder from sliding. The tension of the rope is

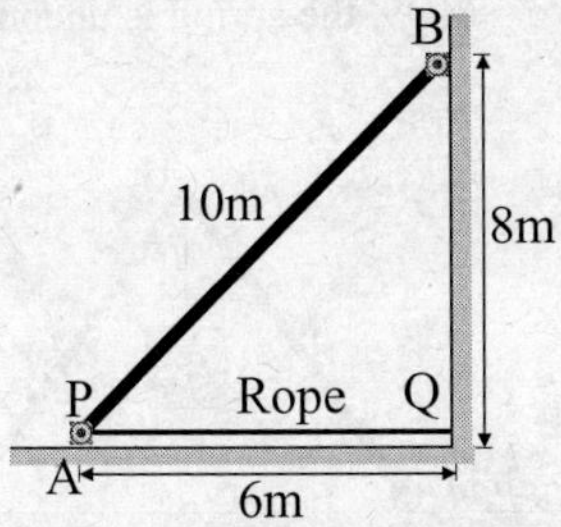

FIGURE 5.MQ5

(A) 2000 N (B) 1000 N (C) $\frac{500}{3}$ N (D) $\frac{1000}{3}$ N

[Ans: D]

Q6. The potential energy of a mechanical system is given by $V = 8x^3 + 6x^2 - 7x$, where x is in meters. The system is under static equilibrium at

(A) $x = 0.345$ m (B) $x = 2.145$ m (C) $x = 0.145$ m (D) None of these

[Ans: A]

Q7. A heavy block of weight 10 kN is to be raised vertically by a horizontal force P applied to a smooth wedge as shown in Figure 5.MQ7. The magnitude of force P is

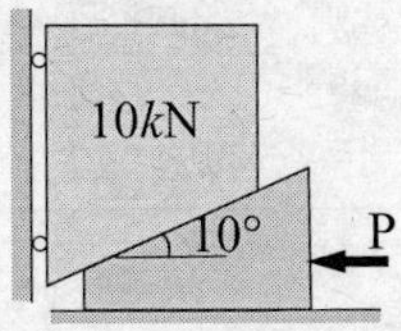

FIGURE 5.MQ7

(A) 500 N (B) 1763.2 N (C) 1235.2 N (D) None of these

[Ans: B]

Q8. Figure 5.MQ8 shows an inverted pendulum. The maximum height of collar C for which the system remains stable equilibrium, is

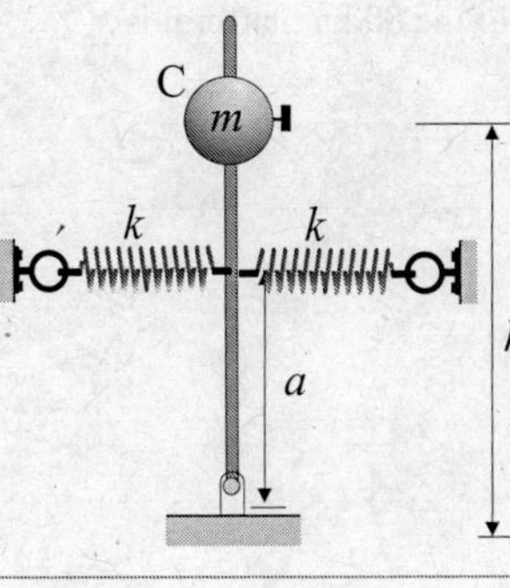

FIGURE 5.MQ8

(A) $h_{max} = \frac{2ka^2}{mg}$ (B) $h_{max} = \frac{ka^2}{mg}$ (C) $h_{max} = \frac{4ka^2}{3mg}$ (D) None of these

[Ans: A]

Q9. For the mechanism shown in Figure 5.MQ9, the spring is uncompressed when $\theta = 0$. The angle θfor the equilibrium position is

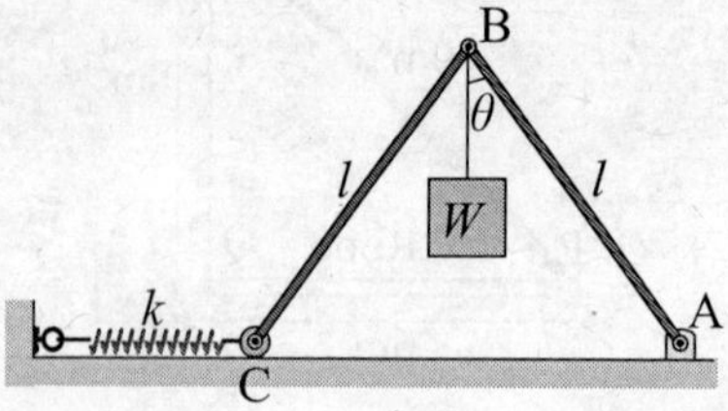

FIGURE 5.MQ9 |

(A) 30° (B) $\theta = \cos^{-1}\left(\frac{W}{kl}\right)$ (C) $\theta = \cos^{-1}\left(\frac{W}{2kl}\right)$ (D) None of these

[Ans: C]

Q10. The elevation of the load is controlled by the adjusting screw that connects joints A and C (Fig. 5.MQ10). The tensile force in adjusting screw AC is

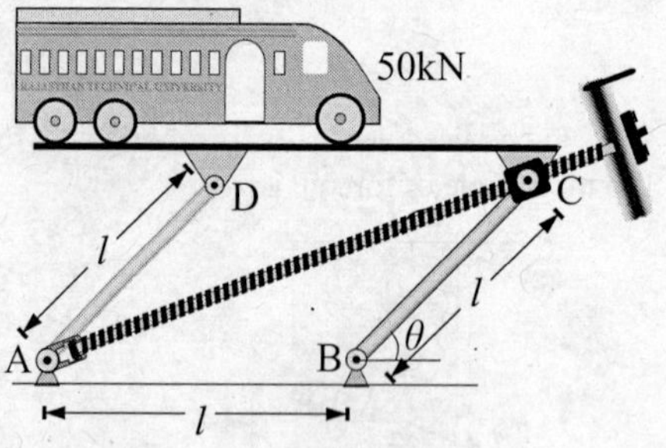

FIGURE 5.MQ10

(A) $\frac{100\cos\theta}{\sin(\theta/2)}$ kN (B) 50 cos θ kN (C) $\frac{50\cos\theta}{\sin(\theta/2)}$ kN (D) None of these

[Ans: C]

Q11. Each of the three uniform links has a mass of 10 kg as shown in Figure 5. MQ11. The horizontal force F necessary to maintain equilibrium of the 80 kg platform is

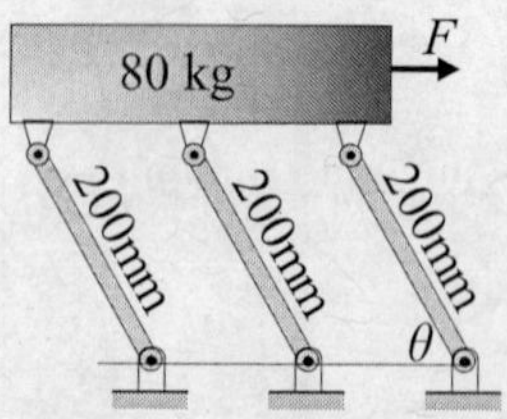

FIGURE 5.MQ11 |

(A) 931.95 cot θ N (B) 931.95 tan θ N (C) 784.8 cot θ N (D) None of these

[Ans: B]

Q12. A light rod BC is hinged at C and supported in horizontal position by a string AB as shown in Figure 5.MQ12. The tension of string is

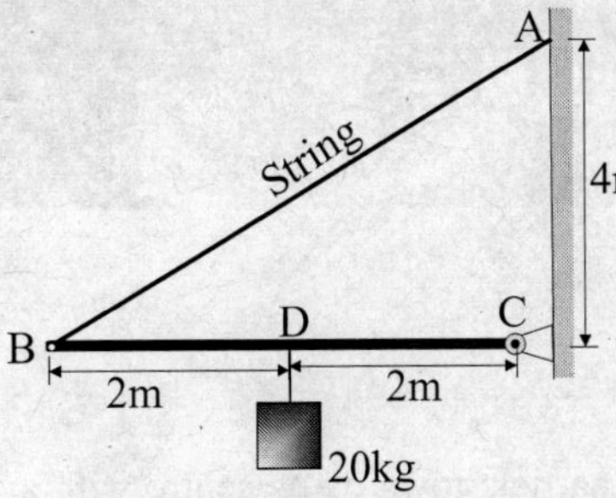

FIGURE 5.MQ12 |

(A) 196.2 N (B) 138.73 N (C) 98.1 N (D) None of these

[Ans: B]

6 Centroid and Moment of Inertia

CENTRE OF GRAVITY

We know that earth attracts every particle towards its centre with a force that is proportional to the mass of the particle. For different particles of a rigid body, these forces, which meet at the centre of the earth, may be considered parallel, as the distance to the centre is usually large in comparison to the size of the body. Thus, they form a system of parallel forces. The resultant of these forces is called the weight of the body.

The weight of a body acts through a definite point in the body, viz. the centre of the above-mentioned system of parallel forces. Such a point is called the centre of gravity of the body.

The centre of gravity of a body, or the system of particles rigidly connected together, is that point through which the line of action of the weight of the body always passes.

Figure 6.1 shows a body of mass M. This mass can be split into an infinite number of small parts each of mass dm. Consider one such elemental mass. Let this mass be at a distance x, y and z from the zy-plane, zx-plane, and xy-plane, respectively. The weight of the elemental mass acts through the elemental mass is

$$\text{weight of the element mass} = \mathrm{d}m \times g$$

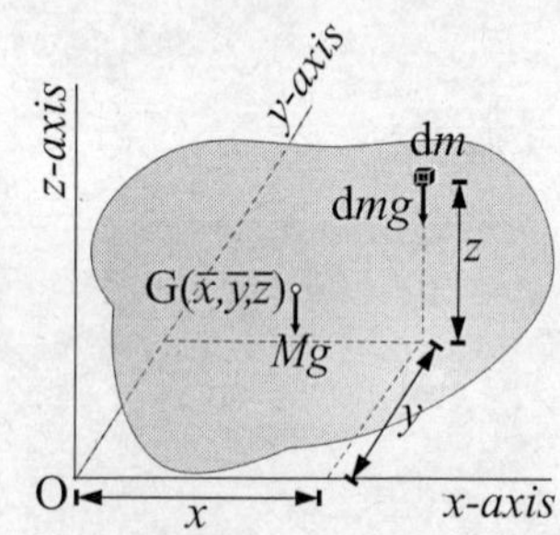

FIGURE 6.1

Moment of the weight of the elemental mass about zy-plane is $x(\mathrm{d}m \times g)$.

Therefore, moment of the weight of the whole body about zy-plane is $\int x\mathrm{d}mg$.

$$\text{Weight of total mass} = Mg$$

Let $\bar{x}$ be the distance of the centre of gravity of the body from zy-plane. It is at this distance that the resultant of the weights of all the elementary masses acts:

$$Mg\bar{x} = \int x\mathrm{d}mg$$

$$\bar{x} = \frac{\int x\mathrm{d}mg}{Mg} \quad \text{or} \quad \bar{x} = \frac{\int x\mathrm{d}m}{M}$$

Similarly, the distance of centre of gravity from zx-plane and xy-plane are $\bar{y} = (\int y\mathrm{d}m)/M$ and $\bar{z} = (\int z\mathrm{d}m)/M$. It gives

$$\bar{x} = \frac{\int x\mathrm{d}m}{\int \mathrm{d}m}, \quad \bar{y} = \frac{\int y\mathrm{d}m}{\int \mathrm{d}m}, \quad \bar{z} = \frac{\int z\mathrm{d}m}{\int \mathrm{d}m},$$

where $\int \mathrm{d}m = M$

CENTROID OF LINE

For a wire or a rod of length L, density ρ, and cross-sectional area A, as shown in Figure 6.2, the element has a mass $dm = \rho A\, dL$. The coordinates of the centre of mass may be written as

$$\bar{x} = \frac{\int x\,dm}{\int dm} = \frac{\int x\rho A dL}{\int \rho A dL}$$

$$\bar{y} = \frac{\int y\rho A dL}{\int \rho A dL} \quad \text{and} \quad \bar{z} = \frac{\int z\rho A dL}{\int \rho A dL}$$

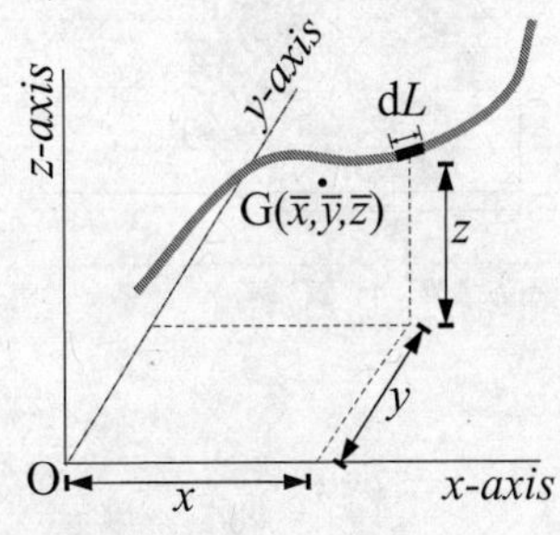

FIGURE 6.2 |

If the density ρ and the cross-sectional area A are constant over a length of the wire (or the rod), the coordinates of the centre of mass also become the coordinates of the centroid G of the line segment. In that case, the above equations may be written as

$$\bar{x} = \frac{\int x dL}{L}, \quad \bar{y} = \frac{\int y dL}{L} \quad \text{and} \quad \bar{z} = \frac{\int z dL}{L},$$

where $$\int dL = L$$

CENTROID OF AREA

For a thin plate of area A, thickness t and density ρ, as shown in Figure 6.3, the mass of an element is $dm = \rho t\, dA$. The coordinates of the centre of mass may be written as

$$\bar{x} = \frac{\int x\,dm}{\int dm} = \frac{\int x\rho t dA}{\int \rho t dA}$$

$$\bar{y} = \frac{\int y\rho t dA}{\int \rho t dA} \quad \text{and} \quad \bar{z} = \frac{\int z\rho t dA}{\int \rho t dA}$$

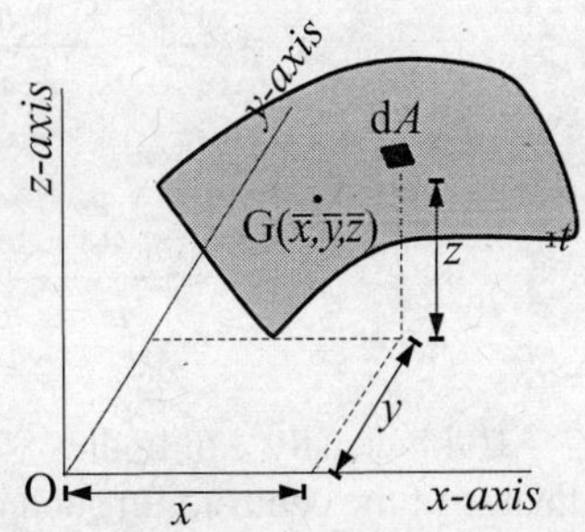

FIGURE 6.3 |

If the density ρ and the thickness of plate t are constant, the coordinates of the centre of mass also become the coordinate of the centroid of the line segment. In such case, the above equations may be written as

$$\bar{x} = \frac{\int x dA}{A}, \quad \bar{y} = \frac{\int y dA}{A} \quad \text{and} \quad \bar{z} = \frac{\int z dA}{A},$$

where $$\int dA = A$$

CENTROID OF VOLUME

For a body of volume V and density ρ, the element has mass $dm = \rho\, dV$. The coordinates of the centre of mass may be written as (Fig. 6.4),

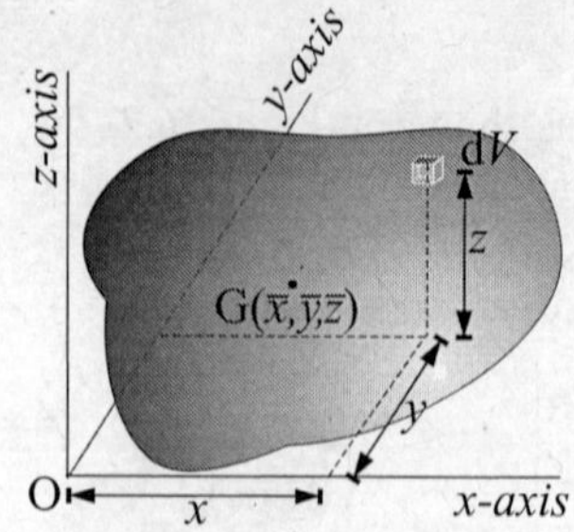

FIGURE 6.4 |

$$\bar{x} = \frac{\int x\,dm}{\int dm} = \frac{\int x\rho dV}{\int \rho dV}$$

$$\bar{y} = \frac{\int y\rho dV}{\int \rho dV} \quad \text{and} \quad \bar{z} = \frac{\int z\rho dV}{\int \rho dV}$$

If density ρ of a body is a constant, the coordinates of the centre of mass also become the coordinates of the centroid G of the line segment. In such cases, the above equations may be written as

$$\bar{x} = \frac{\int x dV}{V}, \quad \bar{y} = \frac{\int y dV}{V}, \quad \text{and} \quad \bar{z} = \frac{\int z dV}{V}, \qquad \text{where } \int dV = V$$

CENTROID OF COMPOSITE BODIES

To locate the centre of gravity (or centroid) of a composite body, the composite body will be split into a number of simple bodies of weight (or volume), and centre of gravity (or centroid) can be found using a standard expression. Then the centre of gravity (or centroid) of the composite body is

$$\bar{x} = \frac{x_1m_1 + x_2m_2 + x_3m_3 + \cdots + x_nm_n}{m_1 + m_2 + m_3 + \cdots + m_n} = \frac{\sum_{i=1}^{i=n} x_im_i}{\sum_{i=1}^{i=n} m_i} = \frac{\sum_{i=1}^{i=n} x_im_i}{M}$$

$$\bar{y} = \frac{y_1m_1 + y_2m_2 + y_3m_3 + \cdots + y_nm_n}{m_1 + m_2 + m_3 + \cdots + m_n} = \frac{\sum_{i=1}^{i=n} y_im_i}{\sum_{i=1}^{i=n} y_im_i} = \frac{\sum_{i=1}^{i=n} y_im_i}{M}$$

and

$$\bar{z} = \frac{z_1m_1 + z_2m_2 + z_3m_3 + \cdots + z_nm_n}{m_1 + m_2 + m_3 + \cdots + m_n} = \frac{\sum_{i=1}^{i=n} z_im_i}{\sum_{i=1}^{i=n} m_i} = \frac{\sum_{i=1}^{i=n} z_im_i}{M}$$

If the density ρ of bodies is constant, the coordinates of the centre of mass also become the coordinates of the centroid. In such cases, the above equations may be written as

$$\bar{x} = \frac{\sum_{i=1}^{i=n} x_iv_i}{V}, \quad \bar{y} = \frac{\sum_{i=1}^{i=n} y_iv_i}{V} \quad \text{and} \quad \bar{z} = \frac{\sum_{i=1}^{i=n} z_iv_i}{V}$$

where

$$\sum_{i=1}^{i=n} v_i = V$$

FOR COMPOSITE PLATES (CONSTANT THICKNESS)

$$\bar{x} = \frac{\sum_{i=1}^{i=n} x_i\rho_iA_i}{\sum_{i=1}^{i=n} \rho_iA_i}, \quad \bar{y} = \frac{\sum_{i=1}^{i=n} y_i\rho_iA_i}{\sum_{i=1}^{i=n} \rho_iA_i}, \quad \text{and} \quad \bar{z} = \frac{\sum_{i=1}^{i=n} z_i\rho_iA_i}{\sum_{i=1}^{i=n} \rho_iA_i}$$

FOR COMPOSITE RODS

$$\bar{x} = \frac{\sum_{i=1}^{i=n} x_i \rho_i a_i L_i}{\sum_{i=1}^{i=n} \rho_i a_i L_i}, \quad \bar{y} = \frac{\sum_{i=1}^{i=n} y_i \rho_i a_i L_i}{\sum_{i=1}^{i=n} \rho_i a_i L_i}, \quad \text{and} \quad \bar{z} = \frac{\sum_{i=1}^{i=n} z_i \rho_i a_i L_i}{\sum_{i=1}^{i=n} \rho_i a_i L_i}$$

CENTROID OF CIRCULAR ARC

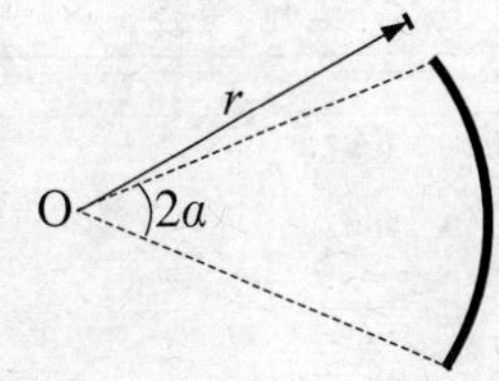

FIGURE 6.5 |

Figure 6.5 shows a circular arc of radius r, making an angle 2α at the centre. Since the circular arc AB is symmetrical about the x-axis, $\bar{y} = 0$. Consider an elementary arc PQ (Fig. 6.6) of length $dl = r\,d\theta$. Its mass $dm = \rho r\,d\theta$, where ρ is the mass per unit length of arc. The x-coordinate of the element is $r\cos\theta$. Thus,

$$\bar{x} = \frac{\int x\,dm}{\int dm}$$

or

$$\bar{x} = \frac{\int_{-\alpha}^{+\alpha} r\cos\theta \rho r\,d\theta}{\int_{-\alpha}^{+\alpha} \rho r\,d\theta}$$

or

$$\bar{x} = r\frac{\int_{-\alpha}^{+\alpha} \cos\theta\,d\theta}{\int_{-\alpha}^{+\alpha} d\theta} = r\frac{[\sin\theta]_{-\alpha}^{+\alpha}}{[\theta]_{-\alpha}^{+\alpha}}$$

or

$$\bar{x} = r\frac{\sin\alpha}{\alpha}$$

$$\boxed{\bar{x} = \frac{r\sin\alpha}{\alpha}}$$

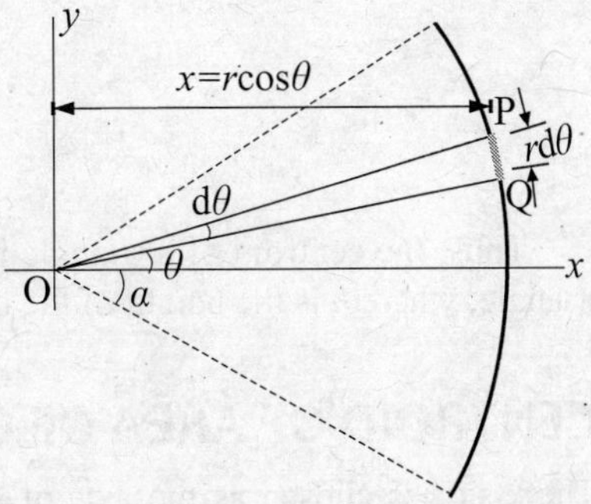

FIGURE 6.6 |

For a semicircular arc 2α = π (Fig. 6.7), which gives

$$\bar{x} = \frac{2r}{\pi}$$

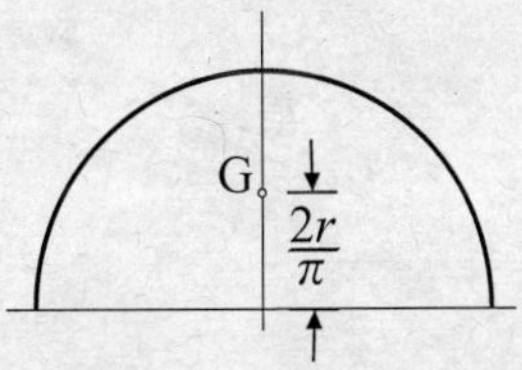

FIGURE 6.7 |

CENTROID OF AREA OF TRIANGLE

Figure 6.8 shows a triangle of height h and base b. Consider an elemental strip of thickness dy at a distance y from the base.

From the geometry of Figure 6.9,

$$\frac{x}{b} = \frac{h-y}{h}$$

or

$$x = \frac{b}{h}(h-y)$$

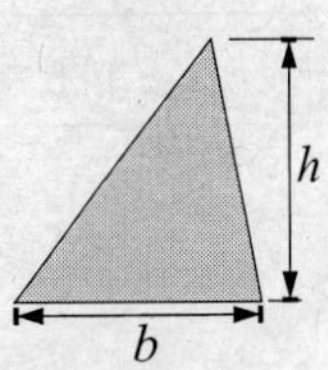

FIGURE 6.8 |

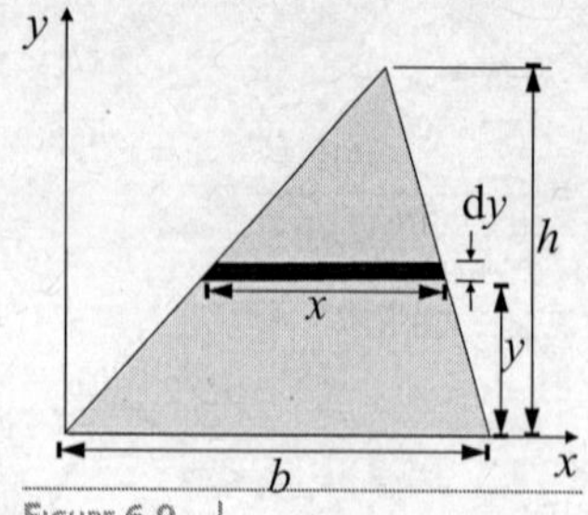

FIGURE 6.9 |

Area of the element $dA = x dy = \frac{b}{h}(h-y)dy$

The y-coordinate of the centroid of area is given by

$$\overline{y} = \frac{\int y_c \, dA}{\int dA} = \frac{\int y dA}{\int dA} \qquad [y_c = y]$$

or
$$\overline{y} = \frac{\int_0^h y \frac{b}{h}(h-y)dy}{\int_0^h \frac{b}{h}(h-y)dy}$$

or
$$\overline{y} = \frac{\int_0^h (hy - y^2)dy}{\int_0^h (h-y)dy} = \frac{h^3/6}{h^2/2}$$

$$\overline{y} = \frac{h}{3}$$

Thus, the centroid of a triangle is at a distance $\frac{h}{3}$ from the base (or $\frac{2h}{3}$ from the apex) of the triangle, where h is the height of the triangle.

CENTROID OF AREA OF CIRCULAR SECTOR

The x-axis is chosen as the axis of symmetry, and $\overline{y}$ is therefore zero. We choose the triangular elemental area (Fig. 6.10) about the vertex, which has an area $dA = \frac{1}{2}R^2 d\theta$. The centroid of the triangular element of area is $\frac{2}{3}R$ from its vertex, so that the x-coordinate of the centroid of the element is $x_c = \frac{2}{3}R\cos\theta$ (Fig. 6.11). Then

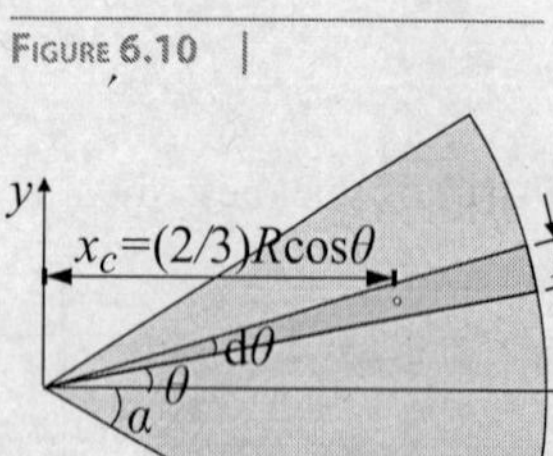

FIGURE 6.10 |

$$\overline{x} = \frac{\int x_c \, dA}{\int dA} = \frac{\int_{-\alpha}^{+\alpha} \left[\frac{2}{3}R\cos\theta\right]\frac{2}{3}R^2 d\theta}{\int_{-\alpha}^{+\alpha} \frac{1}{2}R^2 d\theta}$$

or
$$\overline{x} = \frac{2}{3}\frac{\int_{-\alpha}^{+\alpha} R^3 \cos\theta d\theta}{\int_{-\alpha}^{+\alpha} R^2 d\theta} = \frac{2}{3}R\frac{[\sin\theta]_{-\alpha}^{+\alpha}}{[\theta]_{-\alpha}^{+\alpha}} = \frac{2}{3}R\frac{\sin\alpha}{\alpha}$$

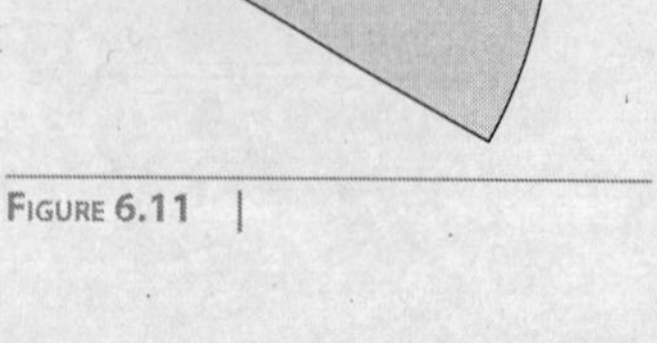

FIGURE 6.11 |

$$\overline{x} = \frac{2}{3}R\frac{\sin\alpha}{\alpha}$$

For a semicircular arc $2\alpha = \pi$, which gives

$$\overline{x} = \frac{4r}{\pi}$$

DOUBLE INTEGRATION METHOD

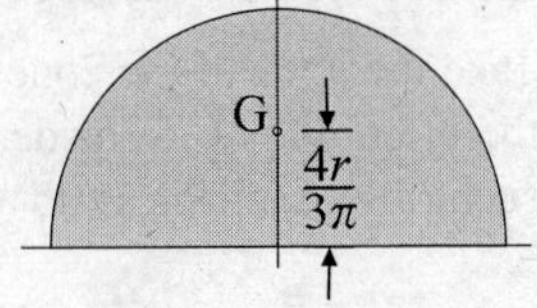

FIGURE 6.12

$$\overline{x} = \frac{\iint x_c \, dA}{\iint dA} = \frac{\int_{-a}^{+a}\int_0^R r\cos\theta r d\theta dr}{\int_{-a}^{+a}\int_0^R r d\theta dr}$$

or
$$\overline{x} = \frac{\int_{-a}^{+a}\int_0^R r^2\cos\theta d\theta dr}{\int_{-a}^{+a}\int_0^R r d\theta dr} = \frac{\int_{-a}^{+a}\frac{R^3}{3}\cos\theta d\theta}{\int_{-a}^{+a}\frac{R^2}{2}d\theta}$$

or
$$\overline{x} = \frac{2}{3}R\frac{[\sin\theta]_{-a}^{+a}}{[\theta]_{-a}^{+a}}$$

or
$$\overline{x} = \frac{2}{3}R\frac{\sin a}{a}$$

$$\overline{x} = \frac{2R\sin a}{3a}$$

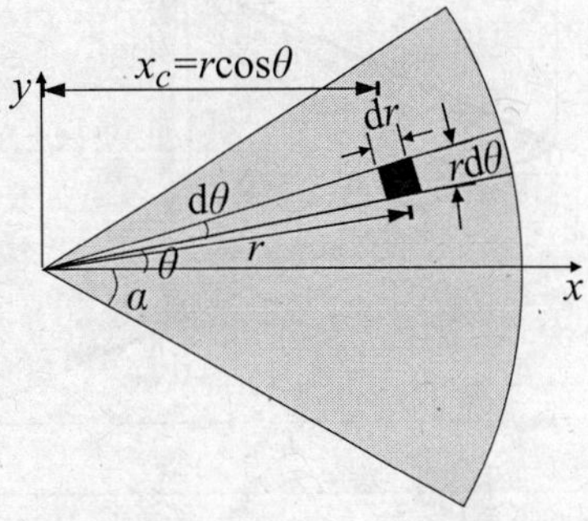

FIGURE 6.13

CENTRE OF GRAVITY OF RIGHT CIRCULAR SOLID CONE

Take the apex O of the cone shown in Figure 6.14 as the origin. The cone is symmetric about x-axis, therefore $\overline{y} = 0$ Consider the cone as split into an infinite number of horizontal discs. Consider one such disc of thickness dx at x.

$$\text{Radius of disc } y = \frac{r}{h}x$$

$$\text{Mass of disc, } dm = \rho dV = \rho\left(\pi y^2 dx\right)$$

where ρ is the density,

or
$$dm = \rho\pi\left(\frac{r}{h}x\right)^2 dx$$

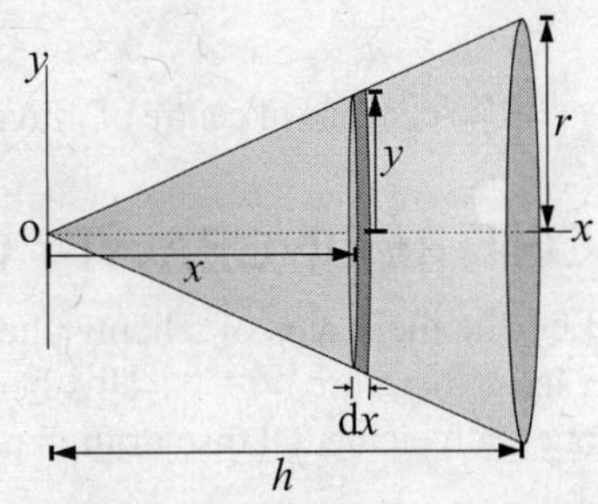

FIGURE 6.14

The x-coordinate of centre of gravity is given by

$$\overline{x} = \frac{\int x dm}{\int dm}$$

or
$$\overline{x} = \frac{\int_0^h x \times \rho\pi\left[\frac{r}{h}x\right]^2 dx}{\int_0^h \rho\pi\left[\frac{r}{h}x\right]^2 dx} = \frac{\int_0^h x^3 dx}{\int_0^h x^2 dx}$$

From which,

$$\overline{x} = \frac{3}{4}h$$

The distance of the centre of gravity of solid cone from the vertex is $\frac{3}{4}h$, and from the base is $\frac{h}{4}$.

CENTRE OF GRAVITY OF RIGHT CIRCULAR HOLLOW CONE

Take the apex of the cone shown in Figure 6.15 as the origin. The cone is symmetric about x -axis, therefore $\bar{y} = 0$. Consider the cone as split into an infinite number of horizontal rings. Consider one such ring of thickness do at x.

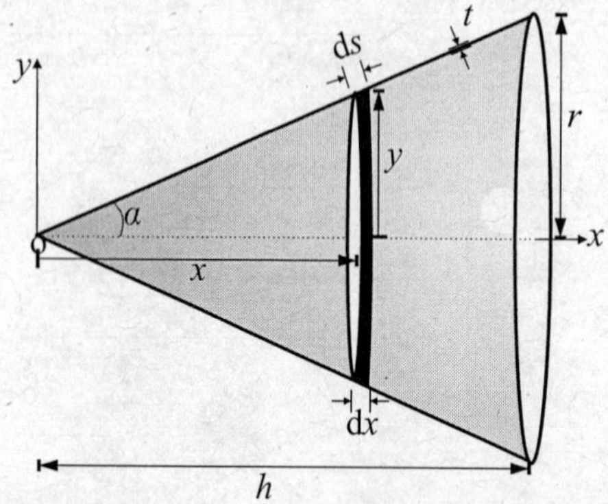

FIGURE 6.15 |

$$\text{Radius of ring, } y = \frac{r}{h}x$$

$$\text{Mass of ring, d}m = \rho \text{d}V = \rho\left(2\pi y \text{d}s\right)t$$

$$\text{d}m = \rho\left(2\pi \frac{r}{h} x\text{d}x\sec a\right)t,$$

where $\rho = \text{density}\ \frac{\text{d}x}{\text{d}s} = \cos a$

The x-coordinate of centre of gravity is given by

$$\bar{x} = \frac{\int x\text{d}m}{\int \text{d}m}$$

or

$$\bar{x} = \frac{\int_0^h x \times \rho[2\pi\left[\frac{r}{h}\right] x\text{d}x\sec a]t}{\int_0^h [2\pi\left[\frac{r}{h}\right] x\text{d}x\sec a]t} = \frac{\int_0^h x^2\text{d}x}{\int_0^h x\text{d}x}$$

$$\bar{x} = \frac{2}{3}h$$

The distance of centre of gravity of hollow cone from the vertex is $(2/3)h$ and from the base is $(1/3)h$.

CENTRE OF GRAVITY OF SOLID HEMISPHERE

Let r be the radius of a hemisphere. The hemisphere, as shown in Figure 6.16, is symmetrical about x-axis; therefore, $\bar{y} = 0$. This hemisphere can be split into infinite number of small discs. Consider one such elemental disc of thickness dx at a distance x from the centre of base O.

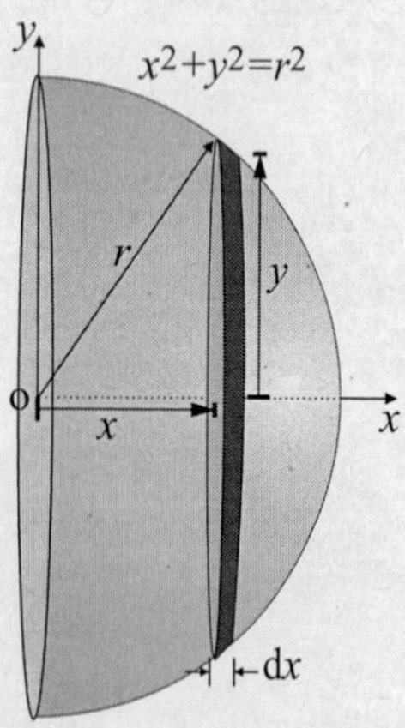

FIGURE 6.16 |

Mass of disc d$m = \rho$ d$V = \rho\pi y^2$ dx, where ρ is the density.

From the geometry of Figure 6.16,

$$\text{radius of elemental disc } y = \sqrt{r^2 - x^2}$$

$$\text{d}m = \rho\pi\left(\sqrt{r^2 - x^2}\right)^2 \text{d}x$$

The x-coordinate of centre of gravity is given by

$$\bar{x} = \frac{\int x dm}{\int dm}$$

or

$$\bar{x} = \frac{\int_0^r x \times \rho\pi(\sqrt{r^2 - x^2})^2 \text{d}x}{\int_0^r \rho\pi(\sqrt{r^2 - x^2})^2 \text{d}x}$$

or
$$\bar{x} = \frac{\int_0^r (xr^2 - x^3)\mathrm{d}x}{\int_0^r (r^2 - x^2)\mathrm{d}x} = \frac{3}{8}r$$

$$\bar{x} = \frac{3}{8}r$$

Distance of the centre of gravity of a solid hemisphere from the base is $\frac{3}{8}r$.

CENTRE OF GRAVITY OF THIN HOLLOW HEMISPHERE

Let r be the radius of a hemisphere. The hemisphere, as shown in Figure 6.17, is symmetrical about x-axis; therefore $\bar{y} = 0$. This hemisphere can be split into infinite number of small rings. Consider one such elemental ring of thickness t and width ds at a distance x from the centre of base O.

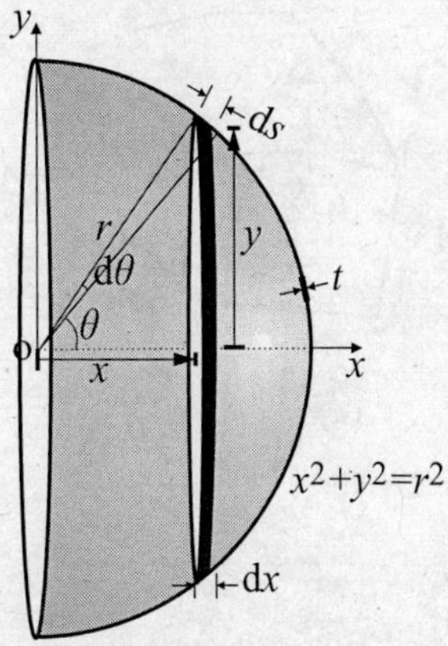

FIGURE 6.17 |

Mass of ring $\mathrm{d}m = \rho \mathrm{d}V = \rho 2\pi y \mathrm{d}st$,

where ρ is the density and t is the thickness.

The x-coordinate of centre of gravity is given by

$$\bar{x} = \frac{\int x\mathrm{d}m}{\int \mathrm{d}m}$$

or
$$\bar{x} = \frac{\int x \times \rho 2\pi y \mathrm{d}st}{\int \rho 2\pi y \mathrm{d}st}$$

or
$$\bar{x} = \frac{\int xy\mathrm{d}s}{\int y\mathrm{d}s}$$

From the geometry of Figure 6.17,

$$x = r\cos\theta, \qquad y = r\sin\theta, \qquad \text{and} \qquad \mathrm{d}s = r\mathrm{d}\theta$$

Substituting the values of x, y, and ds, we have

$$\bar{x} = \frac{\int r\cos\theta r\sin\theta r\mathrm{d}\theta}{\int r\sin\theta r\mathrm{d}\theta}$$

or
$$\bar{x} = r\frac{\int_0^{\pi/2}\sin\theta\cos\theta\mathrm{d}\theta}{\int_0^{\pi/2}\sin\theta\mathrm{d}\theta} = \frac{r}{2}\frac{\int_0^{\pi/2}\sin 2\theta\mathrm{d}\theta}{\int_0^{\pi/2}\sin\theta\mathrm{d}\theta}$$

From which

$$\bar{x} = \frac{r}{2}$$

Distance of centre of gravity of a hollow thin hemisphere from the base is $\frac{1}{2}r$.

THEOREM OF PAPPUS

In the third century, an Alexandrian Greek, named Pappus, discovered two formulas that relate centroid to surfaces and solid of revolution. The formulas provide shortcuts to a number of otherwise lengthy calculations.

THEOREM 1: PAPPUS'S THEOREM FOR VOLUME

The volume of a body of revolution is equal to the generating area times the distance travelled by centroid of the area while the body is being generated.

Proof

Consider plane area A in the xy-plane as shown in Figure 6.18. Let the area A be rotated by the x-axis through an angle θ. Now we consider an element dA of the area A. The volume dV generated by the element dA is equal to $(y\theta)$ dA. Thus, the entire volume generated by A is

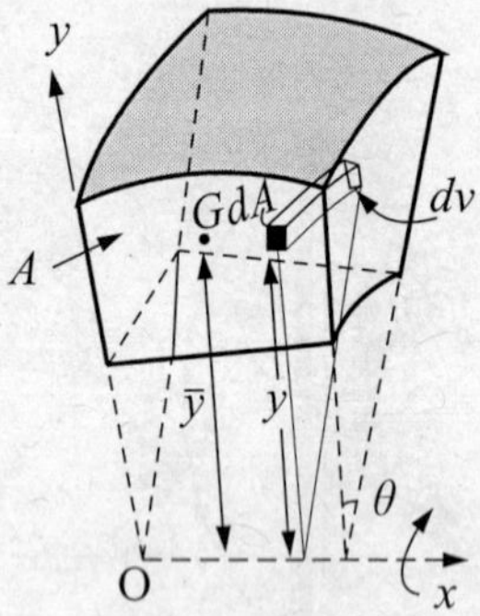

FIGURE 6.18 |

$$V = \int y\theta \mathrm{d}A, \text{ for } 0 \le \theta \le 2\pi$$

Since $\overline{y} = (\int y dA) / A$, we have

$$V = \theta \overline{y} A,$$

where $\overline{y}\theta$ is the distance travelled by a centroid of area A.

In particular, if $\theta = 2\pi$, the volume V is a complete volume of revolution given by

$$V = 2\pi \overline{y} A$$

THEOREM 2: PAPPUS'S THEOREM FOR SURFACE AREAS

The area of a surface of revolution is equal to the length of the generating curve times the distance travelled by the centroid of the curve while the surface is being generated.

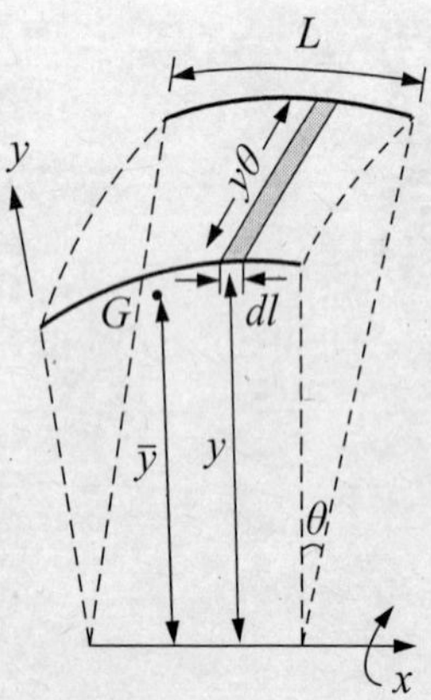

FIGURE 6.19 |

Proof

Consider an arc of length L in xy-plane as shown in Figure 6.19. Let the arc L be rotated by the x-axis through an angle θ. Now consider an element dL of the arc L. The surface area dA generated by the element dL is equal to $y\theta$ dL. Thus, the entire area generated by L is

$$A = \int y\theta \mathrm{d}L$$

Since $\overline{y} = (\int y dL) / L$, we have

$$A = \overline{y}\theta L, \qquad \text{for} \qquad 0 \le \theta \le 2\pi$$

In particular, if $\theta = 2\pi$, the surface is a complete surface of revolution with area

$$A = 2\pi \overline{y} L$$

AREA MOMENT OF INERTIA

The area moment of inertia of a beam's cross-sectional area measures the beam's ability to resist bending. The larger the moment of inertia, the less the beam will bend. The moment of inertia is a geometrical property of a beam and depends on a reference axis. The smallest moment of inertia about any axis passes through the centroid.

It is always considered with respect to a reference axis such as *x-x* or *y-y* (Fig. 6.20). It is a mathematical property of a section concerned with a surface area and how that area is distributed about the reference axis. The reference axis is usually a centroidal axis.

The moment of inertia is also known as the second moment of area and is expressed mathematically as

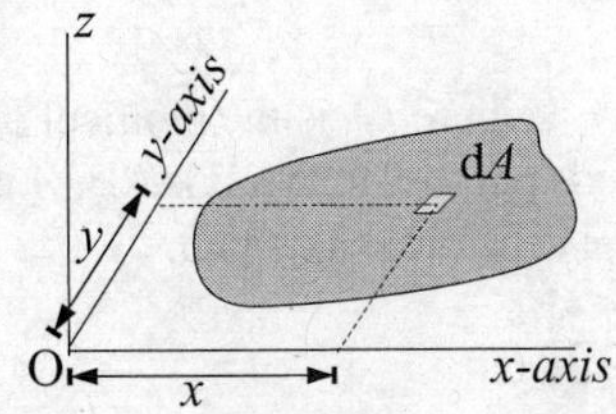

FIGURE 6.20

$$I_x = \int y^2 \, dA$$

$$I_y = \int x^2 \, dA,$$

where y is the distance from the x-axis to an infinitesimal area dA and x is the distance from the y-axis to an infinitesimal area dA.

POLAR MOMENT OF INERTIA

The polar area moment of inertia of a beam's cross-sectional area measures the beam's ability to resist torsion. The larger the polar moment of inertia, the less the beam will twist. The following are the mathematical equations to calculate the polar moment of inertia:

$$I_z = \int r^2 \, dA$$

or $$I_z = \int \left(x^2 + y^2\right) dA \quad \text{or} \quad I_z = \int x^2 \, dA + \int y^2 \, dA$$

or $$I_z = I_x + I_y,$$

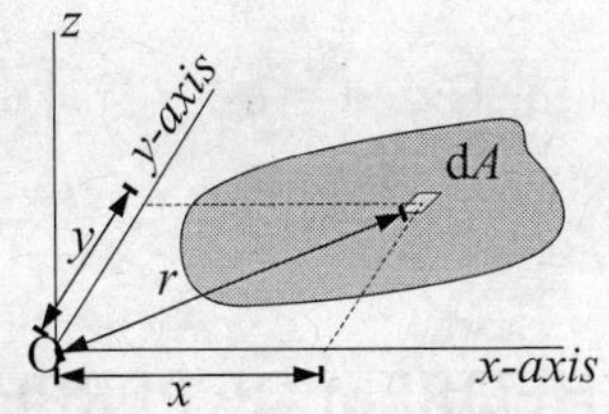

FIGURE 6.21

where x is the distance from the y-axis to an infinitesimal area dA, y is the distance from the x-axis to an infinitesimal area dA, I_x is the moment of inertia about x-axis and I_y is the moment of inertia about y-axis (Fig. 6.21).

RADIUS OF GYRATION

The radius of gyration k describes the way in which the area of a cross-section is distributed around its centroidal axis. If the area is concentrated far from the centroidal axis, it will have a greater value of k and a greater resistance to buckling (or bending). A cross-section can have more than one radius of gyration and most sections have at least two. If this is the case, the section tends to buckle around the axis with the smallest value. The radius of gyration is defined as:

$$k = \sqrt{\frac{I}{A}},$$

where k is the radius of gyration, I is the moment of inertia, and A is the area of cross-section.

THEOREM OF PARALLEL AXES

The moment of inertia of a lamina about any axis in the plane of the lamina equals the sum of the moment of inertia about a parallel centroidal axis in the plane of the lamina and the product of the area of the lamina and the square of the distance between the two axes.

$$I_{AB} = I_G + Ay_c^2$$

where I_{AB} is the moment of inertia about the axis AB, I_G is the moment of inertia about centroidal axis parallel to the axis AB, A is the area of plane, y_c is the distance between the axis AB and the parallel centroidal axis GG.

PROOF

Figure 6.22 shows a plane of area A and G is the centroid. Consider an elemental parallel strip of area dA at distance y from centroidal axis. The moment of inertia of this strip about the axis AB is

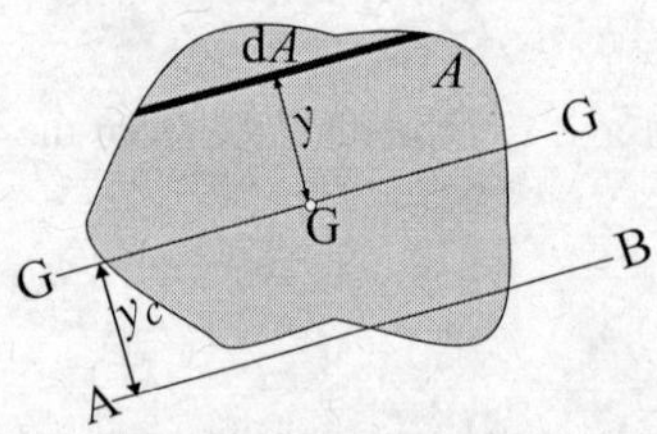

FIGURE 6.22

$$dI_{AB} = (y + y_c)^2\, dA$$

Total moment of inertia of the plane about AB is

$$I_{AB} = \int (y + y_c)^2 dA$$

or $$I_{AB} = \int y^2 dA + \int y_c^2 dA + \int 2yy_c dA$$

or $$I_{AB} = \int y^2 dA + \int y_c^2 dA + 2y_c \int y dA$$

or $$I_{AB} = I_G + y_c^2 A + 2y_c \times 0,$$

where $\int y^2\, dA = I_G$, $y_c^2 \int dA = Ay_c^2$, and $\int y dA = 0$,

$$I_{AB} = I_G + Ay_c^2$$

THEOREM OF PERPENDICULAR AXES

The moment of inertia of an area about an axis perpendicular to its plane (polar moment of inertia) at any point O is equal to the sum of moments of inertia about any two axes mutually perpendicular through the same point O and lying in the plane of the area.

$$I_z = I_x + I_y$$

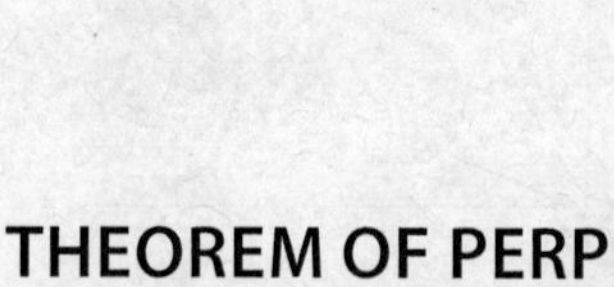

FIGURE 6.23

PROOF

Figure 6.23 shows a lamina of area A in xy plane. Consider an elemental area dA of coordinates (x, y) at a distance r from O. By the definition of moment of inertia,

$$dI_z = r^2\, dA$$

or $$I_z = \int (x^2 + y^2)\,dA, \quad \text{where} \quad r^2 = x^2 + y^2$$

or $$I_z = \int x^2\,dA + \int y^2\,dA$$

$$I_z = I_y + I_x$$

where
$$I_y = \int x^2\,dA \quad \text{and} \quad I_x = \int y^2\,dA$$

Moment of Inertia of Rectangle

Consider an elemental strip of thickness dy at a distance y from the x-axis (Fig. 6.24). The area of the element d$A = b$ dy. Moment of inertia of the elemental strip about the centroidal x-axis is

$$(dI_G)_x = y^2\,dA = y^2 b\,dy$$

or $$(dI_G)_x = y^2 b\,dy$$

The moment of inertia of the whole section about x-axis is

$$(I_G)_x = \int_{-d/2}^{+d/2} y^2 b\,dy = b\left[\frac{y^3}{3}\right]_{-d/2}^{+d/2} = \frac{bd^3}{12}$$

$$(I_G)_x = \frac{bd^3}{12}$$

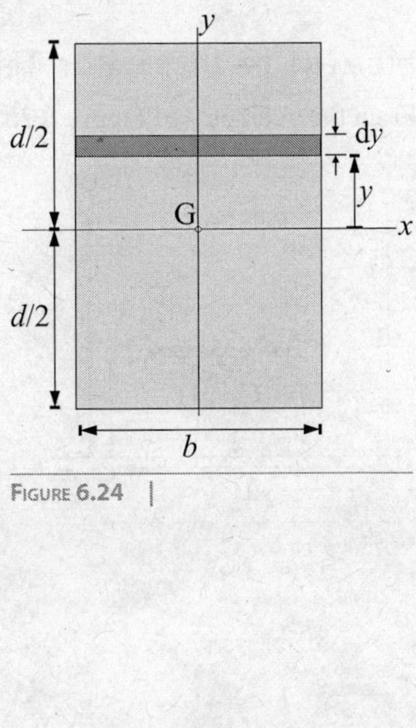

FIGURE 6.24 |

Similarly, the moment of inertia of the whole section about y-axis is

$$(I_G)_y = \int_{-b/2}^{+b/2} x^2\,dx\,d = \frac{db^3}{12}$$

$$(I_G)_y = \frac{db^3}{12}$$

Double Integration Method

The moment of inertia of the whole section about x-axis is

$$(I_G)_x = \iint y^2\,dA = \int_{-b/2}^{+b/2}\int_{-d/2}^{+d/2} y^2\,dx\,dy$$

or $$(I_G)_x = \int_{-b/2}^{+b/2} \frac{1}{3} \times 2\left(\frac{d}{2}\right)^3 dx$$

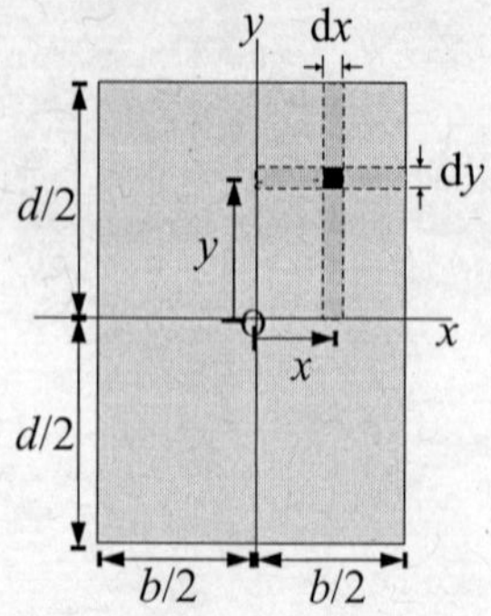

FIGURE 6.25 |

$$= \frac{1}{3} \times 2\left(\frac{d}{2}\right)^3 \times 2 \times \frac{b}{2} = \frac{bd^3}{12}$$

$$(I_G)_x = \frac{bd^3}{12}$$

The moment of inertia of the whole section about y-axis (Fig. 6.25) is

$$(I_G)_y = \iint x^2\, dA = \int_{-b/2}^{+b/2}\int_{-d/2}^{+d/2} x^2\, dx dy = \frac{db^3}{12}$$

$$(I_G)_y = \frac{db^3}{12}$$

MOMENT OF INERTIA OF TRIANGLE

From the geometry of Figure 6.26, we find that the two triangles ADE and ABC are similar. Therefore

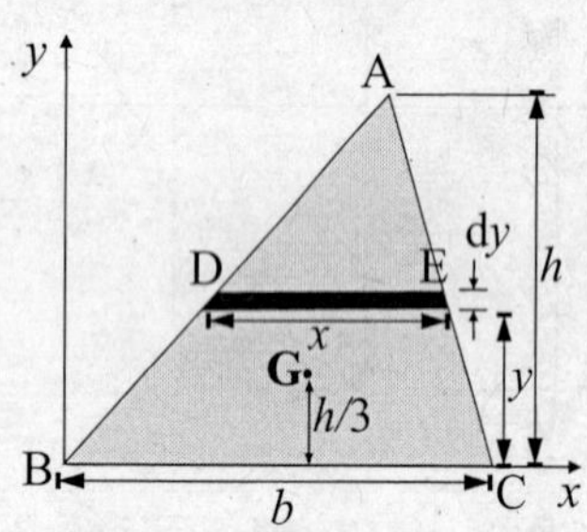

FIGURE 6.26 |

$$\frac{DE}{BC} = \frac{h-y}{h} \Rightarrow DE = \left(\frac{h-y}{h}\right)b$$

Area of the elemental strip DE is

$$\left(\frac{h-y}{h}\right)b \times dy$$

And moment of inertia of the strip about the base BC

$$dI_{BC} = y^2\, dA = y^2\left(\frac{h-y}{h}\right)b \times dy$$

or

$$dI_{BC} = \frac{b}{h}\left(hy^2 - y^3\right)dy$$

Moment of inertia of the triangle about base BC is

$$I_{BC} = \frac{b}{h}\int_0^h \left(hy^2 - y^3\right)dy$$

From which

$$I_{BC} = \frac{bh^3}{12}$$

We know that the distance between centre of gravity of the triangular section and base BC is

$$\bar{y} = \frac{h}{3},$$

Therefore, moment of inertia about an axis through its centre of gravity and parallel to x-axis is

$$I_G = I_{BC} - A\bar{y}^2$$

or
$$I_G = \frac{bh^3}{12} - \frac{bh}{2}\left(\frac{h}{2}\right)^2$$

$$I_G = \frac{bh^3}{36}$$

Moment of Inertia of Circular Area

Figure 6.27 shows a circular area of radius r. Consider an elemental strip of thickness dy at a distance y from the x-axis. The area of the element, $dA = b\,dy$. From geometry of the figure,

$$b = 2R\cos\theta,\ y = R\sin\theta \Rightarrow dy = R\cos\theta d\theta$$

Moment of inertia of the elemental strip about the centroidal axis x-x is

$$dI_x = y^2 dA = y^2 b dy$$

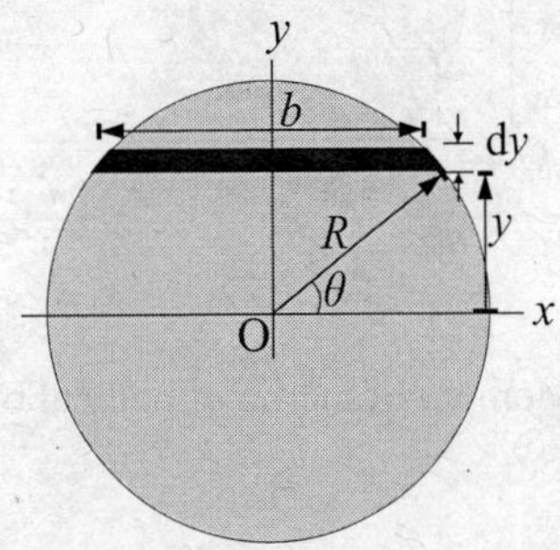

FIGURE 6.27 |

Total moment of inertia is,

$$I_x = \int y^2 dA = \int y^2 b dy$$

or
$$I_x = \int_{-\pi/2}^{+\pi/2} (R\sin\theta)^2\, 2R\cos\theta R\cos\theta d\theta$$

or
$$I_x = 4R^4 \int_0^{+\pi/2} \sin^2\theta\cos^2\theta d\theta$$

From which

$$I_x = \frac{\pi R^4}{4} \quad \text{or} \quad I_x = \frac{\pi d^4}{64}$$

Due to symmetry, $I_x = I_y = \pi d^4/64$, and applying the theorem of perpendicular axis

$$I_z = I_x + I_y \quad \text{or} \quad I_z = 2\times\frac{\pi d^4}{64} = \frac{\pi d^4}{32}$$

$$I_z = \frac{\pi d^4}{32}$$

MASS MOMENT OF INERTIA

The mass moment of inertia of a solid measures the solid's ability to resist changes in rotational speed about a specific axis. The larger the mass moment of inertia, the smaller is the angular acceleration about that axis for a given torque.

The mass moment of inertia depends on a reference axis and is usually specified with two subscripts. This helps to provide clarity during three-dimensional motion, where rotation can occur about multiple axes.

The moment of inertia of body about a given axis of rotation is the sum of the product of the mass of the various particles and square of their perpendicular distance from the axis of rotation.

Figure 6.28 shows a body that can be split into many elemental parts, each of mass dm. Consider an element at coordinates (x, y, z).

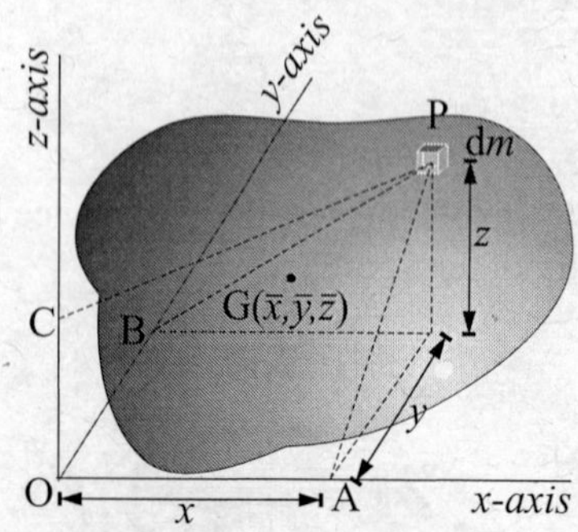

FIGURE 6.28 |

The moment of inertia of element about x-axis is

$$d(I_o)_x = (AP)^2 dm$$

Moment of inertia of body about x-axis

$$(I_o)_x = \int (y^2 + z^2) dm$$

Moment of inertia of body about y-axis

$$(I_o)_y = (BP)^2 dm$$

or

$$(I_o)_y = \int (x^2 + z^2) dm$$

Moment of inertia of body about z-axis

$$(I_o)_z = (CP)^2 dm$$

or

$$(I_o)_z = \int (x^2 + y^2) dm$$

THEOREM OF PARALLEL AXES

It states that *the moment of inertia of a body about any axis is equal to its moment of inertia about a parallel axis through its centre of mass plus the product of the mass of the body and the square of the perpendicular distance between the axes*. Thus, if I_G is the moment of inertia of a body about centroidal axis and M is the mass of the body and I_{AB} is the moment of inertia about parallel axis AB, which is at a distance y_c from centroidal axis, then

$$I_{AB} = I_G + My_c^2$$

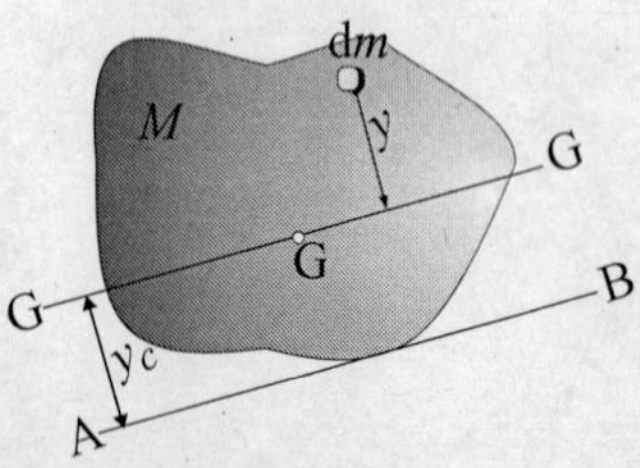

FIGURE 6.29 |

PROOF

Figure 6.29 shows a body of mass M. Consider an element dm at a distance y from centroidal axis GG. The moment of inertia of the element about the axis AB is

$$dI_{AB} = (y + y_c)^2 dm$$

or

$$dI_{AB} = y^2 dm + y_c^2 dm + 2y_c dm,$$

where y_c is the distance between AB and GG.

Moment of inertia of the whole body about AB is

$$I_{AB} = \int y^2\,dm + \int y_c^{\,2}\,dm + \int 2y_c y\,dm$$

or

$$I_{AB} = \int y^2\,dm + y_c^{\,2}\int dm + 2y_c\int y\,dm$$

Since $\int y^2\,dm = I_G$, $\int dm = M$, and $\int y\,dm = 0$,

$$I_{AB} = I_G + My_c^{\,2}$$

THEOREM OF PERPENDICULAR AXES

It states that *the moment of inertia of a plane lamina about an axis perpendicular to its plane is equal to the sum of the moment of inertia of the lamina about any two mutually perpendicular axes in its plane and intersecting each other at the point where the perpendicular axis passes through it.*

$$I_z = I_x + I_y$$

PROOF

Figure 6.30 shows a plane lamina of mass M in xy-plane. Consider an element of mass dm. The moment of inertia of the element about z-axis is

$$dI_z = r^2\,dm$$

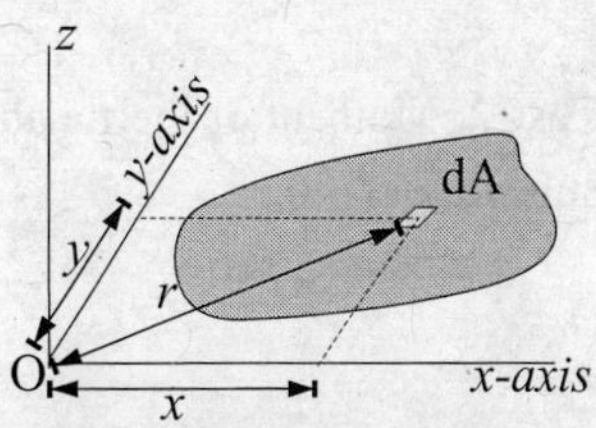

FIGURE 6.30 |

Since $r^2 = x^2 + y^2$, moment of inertia of body

$$I_z = \int \left(x^2 + y^2\right)dm$$

or

$$I_z = \int x^2\,dm + \int y^2\,dm$$

$$I_z = I_y + I_x$$

where $I_y = \int x^2 dm$ and $I_x = \int y^2\,dm$

MOMENT OF INERTIA OF UNIFORM ROD

Let ρ be the mass per unit length of a rod. Consider an elemental part of the rod at a distance x and of length dx (Fig. 6.31).

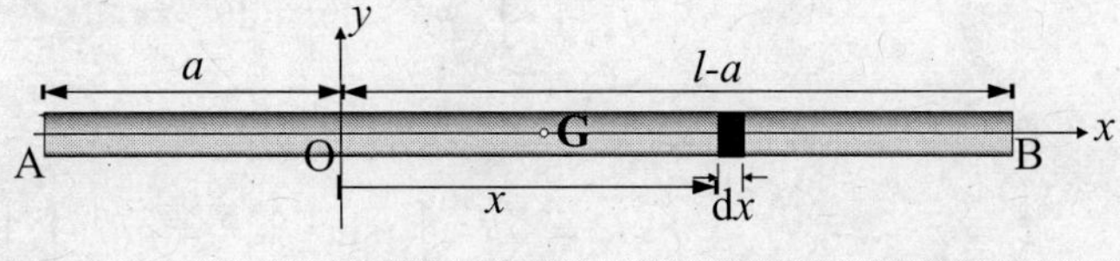

FIGURE 6.31 |

Mass of the element, d$m = \rho$ dx. Moment of inertia of the elemental part about the y-y axis is

$$\left(\mathrm{d}I_{\mathrm{O}}\right)_y = x^2\,\mathrm{d}m$$

Therefore, moment of inertia of the rod about the y-y axis is

$$\left(I_{\mathrm{O}}\right)_y = \int_{-a}^{l-a} x^2 \rho \mathrm{d}x$$

or

$$\left(I_{\mathrm{O}}\right)_y = \frac{\rho}{3}\left[(l-a)^3 + a^3\right]$$

or

$$\left(\mathrm{I}_{\mathrm{O}}\right)_y = \frac{m}{3l}\left[(l-a)^3 + a^3\right]$$

Case 1: Moment of inertia about centroidal axes

Substitute $l/2$ for $a = \dfrac{l}{2}$

$$\left(I_{\mathrm{G}}\right)_y = \frac{m}{3l}\left[\left(l-\frac{l}{2}\right)^3 + \left(\frac{l}{2}\right)^3\right]$$

$$I_{\mathrm{G}} = \frac{ml^2}{12}$$

Case 2: Moment of Inertia about one end and normal to rod

Substitute $a = 0$.

$$I_{\mathrm{A}} = \frac{ml^2}{3}$$

MOMENT OF INERTIA OF CIRCULAR RING

Figure 6.32 shows a ring of radius R and mass m. Mass per unit length of ring is $\rho = m / 2\pi R$. Consider an elemental length d$s = R$ dθ at angle θ to the x-axis. The mass of elemental length d$m = \rho$ ds = ρR dθ. Moment of inertia of elemental mass about x-axis is

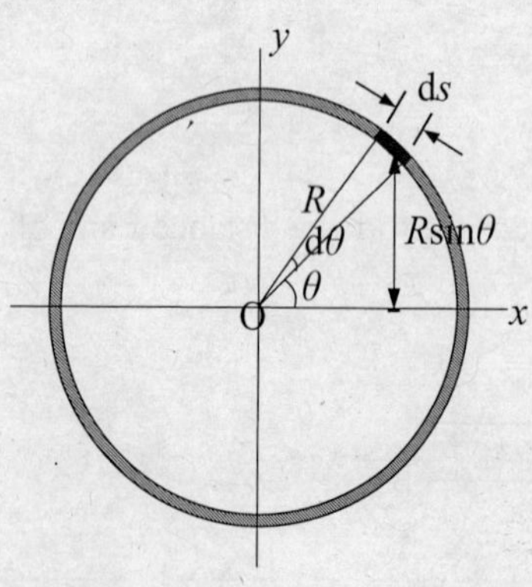

FIGURE 6.32 |

$$\mathrm{d}I_x = y^2\,\mathrm{d}m$$

or

$$\mathrm{d}I_x = (R\sin\theta)^2\ \rho R\mathrm{d}\theta$$

Total moment of inertia

$$I_x = \int_{-\pi/2}^{\pi/2} \rho R^3 \sin^2\theta\mathrm{d}\theta$$

or

$$I_x = \rho R^3 \int_{-\pi/2}^{\pi/2} \sin^2\theta\mathrm{d}\theta = \rho R^3 \int_{-\pi/2}^{\pi/2}\left(\frac{1-\cos 2\theta}{2}\right)\mathrm{d}\theta$$

or $$I_x = \rho \pi R^3$$

or $$I_x = \left(\frac{m}{2\pi R}\right)\pi R^3 = \frac{1}{2} mR^2$$

Moment of inertia of the ring about diametral axis is

$$I_x = \frac{1}{2} mR^2$$

Similarly,

$$I_y = \frac{1}{2} mR^2$$

Again, the moment of inertia of the element about z-axis (which is perpendicular to x and y axes)

$$dI_z = R^2 \, dm \quad \text{or} \quad dI_z = R^2 \rho R d\theta$$

Total moment of inertia

$$I_z = \int_0^{2\pi} \rho R^3 \, d\theta = \rho R^3 2\pi = \frac{m}{2\pi R} R^3 2\pi$$

$$I_z = mR^2$$

MOMENT OF INERTIA OF DISC

Figure 6.33 shows a disc of radius R, mass m and thickness t. Density of the disc material $\rho = m/\pi R^2 t$.

Consider an elemental area, $dA = r\,d\theta\,dr$ in polar coordinates. The mass of the elemental area $dm = \rho r\,d\theta\,dr\,t$. Moment of inertia of elemental mass about x-axis is

$$dI_x = y^2 \, dm$$

or $$dI_x = (r \sin\theta)^2 \, \rho r d\theta dr t$$

$$dI_x = \rho t r^3 \sin^2 \theta d\theta dr$$

Total moment of inertia

$$I_x = \rho t \int_0^R \int_0^{2\pi} r^3 \sin^2 \theta d\theta dr$$

$$I_x = \rho t \int_0^R \int_0^{2\pi} r^3 \sin^2 \theta d\theta dr = \frac{m}{\pi R^2 t} t \times \frac{R^4}{4} \times \frac{2\pi}{2}$$

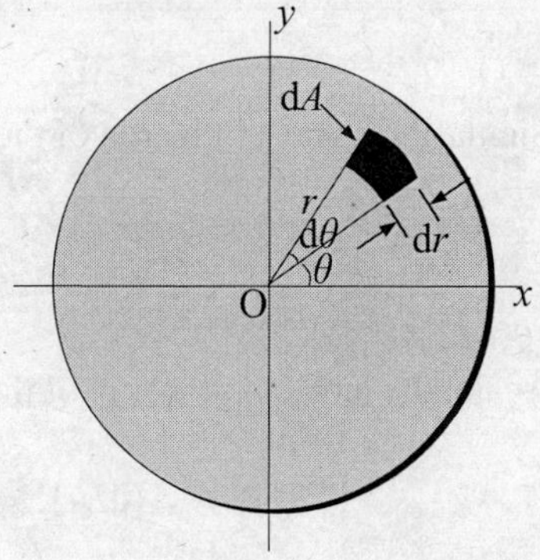

FIGURE 6.33 |

Moment of disc about diamentral axis,

$$I_x = \frac{1}{4} mR^2$$

Similarly,

$$I_y = \frac{1}{4} mR^2$$

Again, the moment of inertia of element about z-axis (which is perpendicular to x and y axes)

$$\mathrm{d}I_z = r^2 \mathrm{d}m \qquad \text{or} \qquad \mathrm{d}I_z = r^2 \rho r \mathrm{d}\theta \mathrm{d}r t$$

Total moment of inertia

$$I_z = \rho t \int_0^{2\pi} \int_0^R r^3 \mathrm{d}\theta \mathrm{d}r = \frac{m}{\pi R^2 t} t \times \frac{R^4}{4} \times 2\pi$$

$$I_z = \frac{1}{2} mR^2$$

MOMENT OF INERTIA OF PLATE

Figure 6.34 shows a plate of width b, depth d, thickness t and mass m. Density of the plate material, $\rho = m/\pi \mathrm{b}\mathrm{d}t$. Consider an elemental area $\mathrm{d}A = \mathrm{d}x\ \mathrm{d}y$. The mass of the elemental area, $\mathrm{d}m = \rho\ \mathrm{d}x\ \mathrm{d}y\ t$. Moment of inertia of the elemental mass about x-axis passing through centroid of plate is.

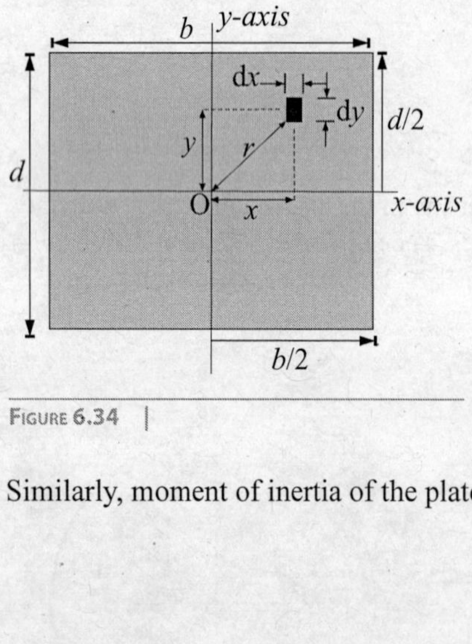

FIGURE 6.34 |

$$(\mathrm{d}I_G)_x = y^2 \rho \mathrm{d}x \mathrm{d}y t$$

Total moment of inertia

$$(I_G)_x = \int_{-b/2}^{+b/2} \int_{-d/2}^{+d/2} y^2 \rho \mathrm{d}x \mathrm{d}y t$$

or

$$(I_G)_x = \frac{m}{\pi b \mathrm{d}t} t \times \frac{1}{3} 2 \left(\frac{d}{2}\right)^3 \times 2 \left(\frac{b}{2}\right)$$

$$(I_G)_x = \frac{md^2}{12}$$

Similarly, moment of inertia of the plate about y-axis is

$$(I_G)_y = \frac{mb^2}{12}$$

Again, the moment of inertia of the element about z-axis (which is perpendicular to x and y axes)

$$(\mathrm{d}I_G)_z = r^2 \mathrm{d}m \qquad \text{or} \qquad (\mathrm{d}I_G)_z = (x^2 + y^2)\mathrm{d}m$$

Total moment of inertia

$$(I_G)_z = \int (x^2 + y^2)\mathrm{d}m = \int x^2 \mathrm{d}m + \int y^2 \mathrm{d}m$$

or

$$(I_G)_z = (I_G)_y + (I_G)_x = \frac{mb^2}{12} + \frac{md^2}{12}$$

Moment of inertia of the plate about z-axis (which is perpendicular to x and y axes) passing through centroid

$$(I_G)_z = m\left(\frac{b^2 + d^2}{12}\right)$$

Moment of Inertia of Solid Sphere

Figure 6.35 shows a sphere of radius R and mass m. Density of the sphere material is

$$\rho = \frac{m}{(4/3)\pi R^3} = \frac{3m}{4\pi R^3}$$

Consider an elemental disc of thickness dx at x from the centre of the sphere. Mass of the elemental disc d$m = \rho$ d$V = \rho\pi y^2$ dx. Moment of inertia of the elemental disc about x-axis is

$$(dI_G)_x = \frac{1}{2} y^2 \, dm$$

or

$$(dI_G)_x = \frac{1}{2} y^2 \rho\pi y^2 \, dx$$

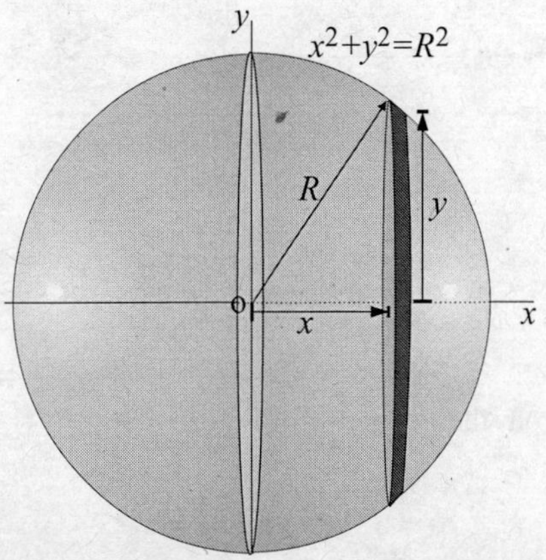

FIGURE 6.35

Moment of inertia of the sphere about its diameter is

$$(I_G)_x = \frac{1}{2}\rho\pi \int_{-R}^{+R} y^4 \, dx$$

or

$$(I_G)_x = \rho\pi \int_{0}^{+R} y^4 \, dx$$

From the geometry of Figure 6.35,

$$y^2 = R^2 - x^2$$

$$(I_G)_x = \rho\pi \int_{0}^{+R} (R^2 - x^2)^2 \, dx$$

From which

$$(I_G)_x = \rho\pi \frac{8}{15} R^5$$

Substituting , $\rho = 3m/4\pi R^3$ we have

$$(I_G)_x = \frac{3m}{4\pi R^3} \pi \frac{8}{15} R^5 = \frac{2}{5} mR^2$$

$$(I_G)_x = \frac{2}{5} mR^2$$

Moment of Inertia of Cone

Figure 6.36 shows a solid cone of radius R and mass m. Density of cone material

$$\rho = \frac{m}{\frac{1}{3}\pi r^2 h} = \frac{3m}{\pi r^2 h}$$

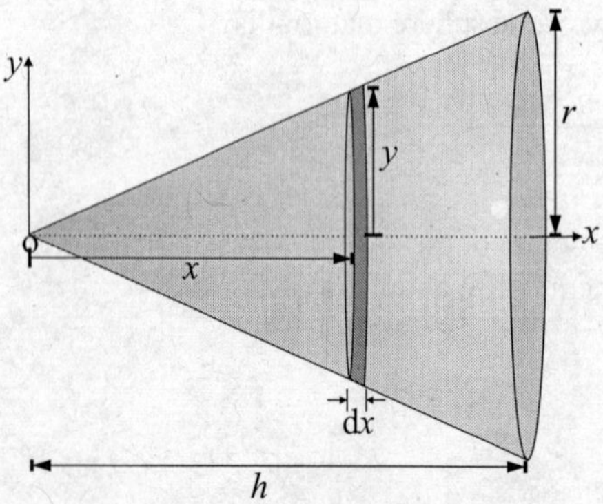

FIGURE 6.36 |

Consider an elemental disc of thickness dx at x from the apex of the cone. The mass of the elemental disc dm = ρ dV = $\rho\pi y^2$ dx. Moment of inertia of the elemental disc about x-axis is

$$(\mathrm{d}I_G)_x = \frac{1}{2} y^2 \,\mathrm{d}m$$

or

$$(\mathrm{d}I_G)_x = \frac{1}{2} y^2 \rho\pi y^2 \mathrm{d}x$$

Moment of inertia of the cone about x-axis is

$$(I_G)_x = \int \frac{1}{2} y^2 \rho\pi y^2 \,\mathrm{d}x$$

or

$$(I_G)_x = \frac{1}{2}\rho\pi \int_0^h y^4 \,\mathrm{d}x$$

From the geometry of Figure 6.36,

$$y = \frac{r}{h} x$$

$$(I_G)_x = \frac{1}{2}\rho\pi \int_0^{+R} \left(\frac{r}{h}x\right)^4 \mathrm{d}x$$

From which

$$(I_G)_x = \frac{1}{2}\rho\pi \left(\frac{r}{h}\right)^4 \frac{h^5}{5}$$

Substituting $\rho = 3m/\pi r^2 h$, we have

$$(I_G)_x = \frac{1}{2} \times \frac{3m}{\pi r^2 h} \pi \left(\frac{r}{h}\right)^4 \frac{h^5}{5}$$

$$(I_G)_x = \frac{3}{10} m r^2$$

Again, moment of inertia of elemental disc about y-axis is (by theorem of parallel axes)

$$(\mathrm{d}I_o)_y = \frac{1}{4}y^2\,\mathrm{d}m + x^2\,\mathrm{d}m$$

or
$$(I_o)_y = \int\left(\frac{1}{4}y^2 + x^2\right)\rho\pi y^2\,\mathrm{d}x$$

or
$$(I_o)_y = \int\left[\frac{1}{4}\left(\frac{r}{h}x\right)^2 + x^2\right]\rho\pi\left(\frac{r}{h}x\right)^2\mathrm{d}x$$

or
$$(I_o)_y = \int_0^h\left[\frac{1}{4}\left(\frac{r}{h}x\right)^2 + x^2\right]\rho\pi\left(\frac{r}{h}x\right)^2\mathrm{d}x$$

From which

$$(I_o)_y = \frac{3}{5}m\left[\frac{r^2}{4} + h^2\right]$$

PRODUCT OF INERTIA

The product of inertia of an area with respect to any two rectangular axes may be defined as the sum of the products obtained by multiplying each element of the area by the product of the two coordinates of the element with respect to two rectangular axes, i.e.,

$$I_{xy} = \int xy\,\mathrm{d}A,$$

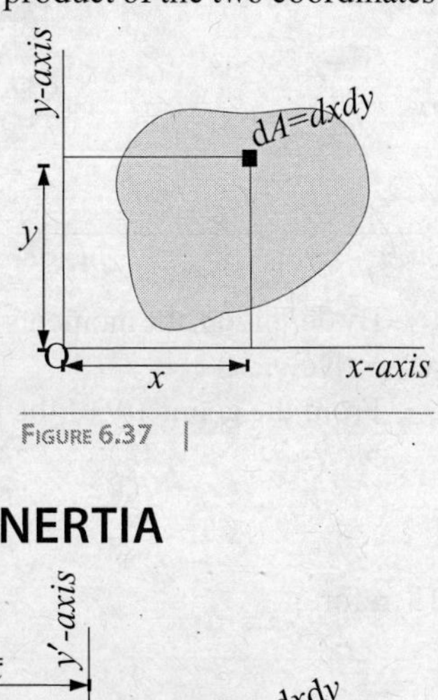

FIGURE 6.37

where $\mathrm{d}A$ is an element of the given area and x and y are the coordinates of the element with respect to the two rectangular axes (Fig. 6.37).

The product of inertia may be positive, negative or zero. The product of inertia of an area with respect to two rectangular axes is zero if either one of the axes is an axis of symmetry.

PARALLEL AXIS THEOREM FOR PRODUCT OF INERTIA

Let Gx' and Gy' be the axes through centroid G of the area, and Ox and Oy are parallel axes through the point O. By definition, the product of inertia of the area A in Figure 6.38 with respect to the x and y axes in terms of the coordinates $\bar{x}$, $\bar{y}$ to the centroidal axes is

$$I_{xy} = \int xy\,\mathrm{d}A = \int(x'+\bar{x})(y'+\bar{y})\,\mathrm{d}A$$

$$= \int x'y'\,\mathrm{d}A + \bar{x}\int y'\,\mathrm{d}A + \bar{y}\int x'\,\mathrm{d}A + \bar{x}\bar{y}\int \mathrm{d}A$$

$$= \bar{I}_{x'y'} + \bar{x}\times 0 + \bar{y}\times 0 + \bar{x}\bar{y}A$$

$$I_{xy} = \bar{I}_{x'y'} + \bar{x}\bar{y}A$$

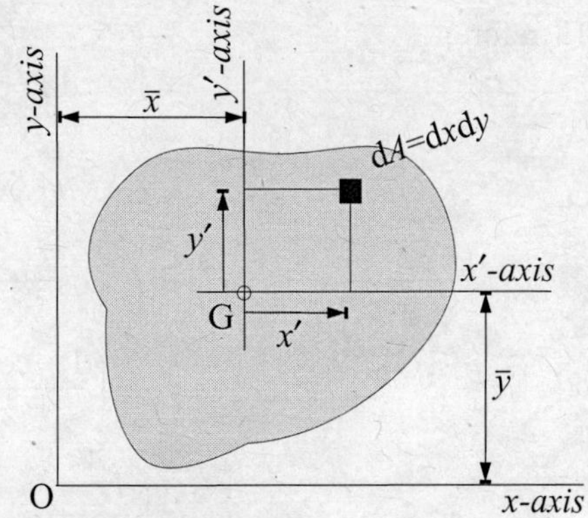

FIGURE 6.38

TRANSFORMATION EQUATIONS FOR MOMENTS AND PRODUCT OF INERTIA

The product of inertia is useful when we need to calculate the moment of inertia of an area about inclined axes. Let I_x, I_y and I_{xy} with respect to x–y coordinate system be known and we are interested to know $I_{x'}$, $I_{y'}$ and $I_{x'y'}$ with respect to x'–y' axes which have been obtained after rotation of the x–y axes through an angle θ in the anti-clockwise direction as shown in Figure 6.39.

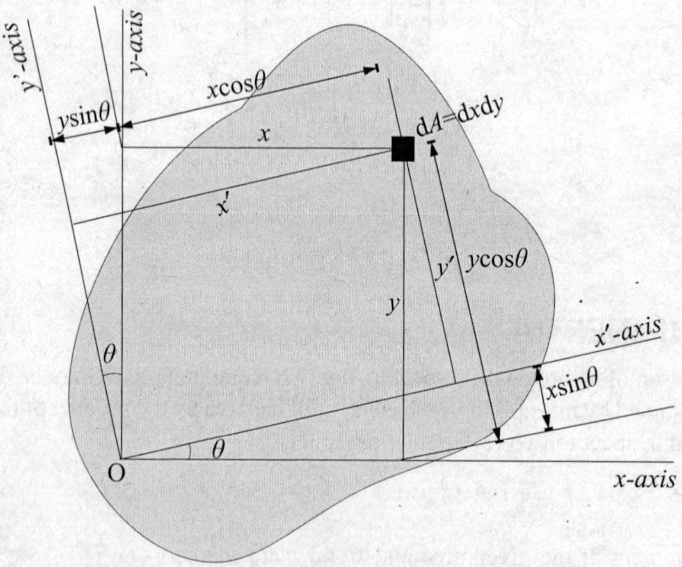

FIGURE 6.39 |

By definition, the moments of inertia about the x' and y' axes are $I_{y'} = \int x'^2\,dA$ and $I_{x'} = \int y'^2\,dA$, respectively.

From the geometry of the figure,

$$x' = y\sin\theta + x\cos\theta$$

$$y' = y\cos\theta - x\sin\theta$$

Therefore,

$$I_{x'} = \int y'^2\,dA = \int (y\cos\theta - x\sin\theta)^2\,dA$$

or

$$I_{x'} = \cos^2\theta \int y^2\,dA + \sin^2\theta \int x^2\,dA - 2\sin\theta\cos\theta \int xy\,dA$$

$$I_{x'} = I_x\cos^2\theta + I_y\sin^2\theta - I_{xy}\sin 2\theta$$

or

$$= I_x\left(\frac{1+\cos 2\theta}{2}\right) + I_y\left(\frac{1-\cos 2\theta}{2}\right) - I_{xy}\sin 2\theta$$

$$I_{x'} = \left(\frac{I_x + I_y}{2}\right) + \left(\frac{I_x - I_y}{2}\right)\cos 2\theta - I_{xy}\sin 2\theta$$

Now we have $I_{y'} = \int x'^2 \, dA$;

$$I_{y'} = \int (y \sin\theta + x\cos\theta)^2 \, dA$$

$$I_{y'} = \sin^2\theta \int y^2 \, dA + \cos^2\theta \int x^2 \, dA + 2\sin\theta\cos\theta \int xy \, dA$$

or

$$= \sin^2\theta I_x + \cos^2\theta I_y + \sin 2\theta I_{xy}$$

or

$$I_{y'} = \left(\frac{1-\cos 2\theta}{2}\right) I_x + \left(\frac{1+\cos 2\theta}{2}\right) I_y + \sin 2\theta I_{xy}$$

$$I_{y'} = \left(\frac{I_x + I_y}{2}\right) - \left(\frac{I_x - I_y}{2}\right) \cos 2\theta + I_{xy} \sin 2\theta$$

In a similar manner, we write the product of inertia about the inclined axes as

$$I_{x'y'} = \int x'y' \, dA = \int (y\sin\theta + x\cos\theta)(y\cos\theta - x\sin\theta) \, dA$$

or

$$I_{x'y'} = \left(\cos^2\theta - \sin^2\theta\right) \int xy \, dA + \sin\theta\cos\theta \int y^2 \, dA - \sin\theta\cos\theta \int x^2 \, dA$$

or

$$I_{x'y'} = \cos 2\theta I_{xy} + \sin 2\theta \frac{I_x}{2} - \sin 2\theta \frac{I_y}{2}$$

$$I_{x'y'} = \left(\frac{I_x - I_y}{2}\right) \sin 2\theta + I_{xy} \cos 2\theta$$

PRINCIPAL AXES

A principal axis of inertia of an area, for a given point in the area, is an axis about which the moment of inertia of the area is either maximum or minimum than for any other axes passing through the given point. There are two principal axes of inertia of an area for any point in the area. Axes for which the product of inertia is zero are called principal axes.

Now, we have

$$I_{x'} = \left(\frac{I_x + I_y}{2}\right) + \left(\frac{I_x - I_y}{2}\right) \cos 2\theta - I_{xy} \sin 2\theta$$

For $I_{x'}$, to be maximum or minimum

$$\frac{d}{d\theta} I_{x'} = 0 \Rightarrow -\left(I_x - I_y\right) \sin 2\theta - 2I_{xy} \cos 2\theta = 0$$

From which

$$\tan 2\theta = \frac{2I_{xy}}{I_y - I_x}$$

The above equation gives two values of 2θ, each differs by π from the other, since $\tan 2\theta = \tan(180 + 2\theta)$. Consequently, the two solutions for θ will differ by 90°. One value defines the axis of maximum moment of inertia, and the other value defines the axis of minimum moment of inertia. These two rectangular axes are called the principal axes of inertia:

$$I_{max} = \left(\frac{I_x + I_y}{2}\right) + \frac{1}{2}\sqrt{\left(I_x - I_y\right)^2 + 4I_{xy}}$$

$$I_{min} = \left(\frac{I_x + I_y}{2}\right) - \frac{1}{2}\sqrt{\left(I_x - I_y\right)^2 + 4I_{xy}}$$

SOLVED PROBLEMS

PROBLEM 6.1

Find the centre of gravity of a segment of a sphere of radius r as shown in Figure 6.1(a). Find the position of centre of gravity of the hemisphere from the centre.

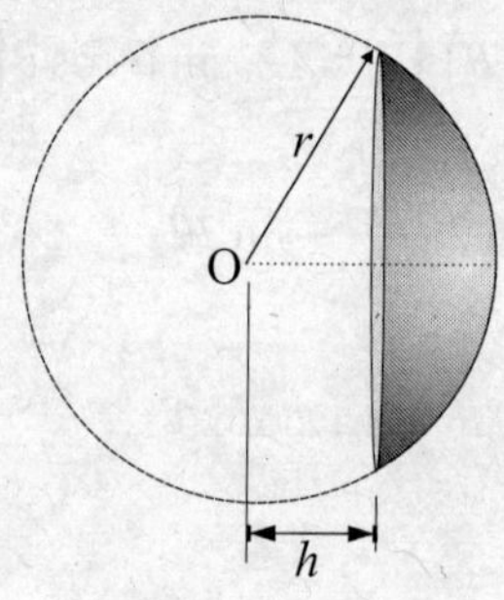

FIGURE 6.1(a)

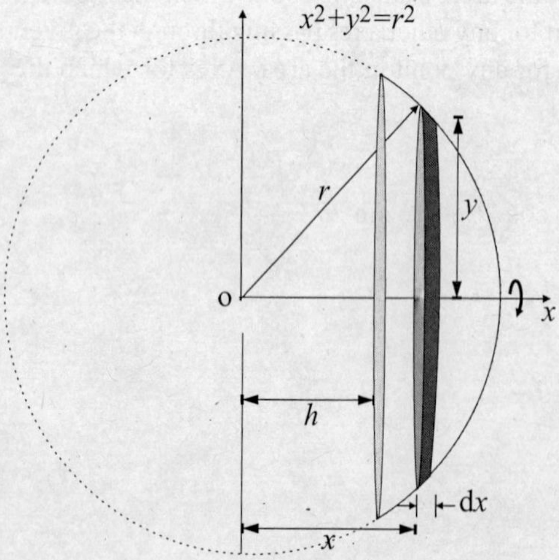

FIGURE 6.1(b)

Solution. Take the centre of the sphere as the origin [Fig. 6.1(b)]. The spherical segment is symmetrical about x-axis, therefore, $\bar{y} = 0$. Now consider an element of thickness dx at x. The volume of the element $\mathrm{d}V = \pi y^2\,\mathrm{d}x$. Hence, we have

$$\bar{x} = \frac{\int x\,\mathrm{d}m}{\int \mathrm{d}m} = \frac{\int x\rho\,\mathrm{d}V}{\int \rho\,\mathrm{d}V},$$

where ρ is the density of sphere,

or $$\bar{x} = \frac{\int_h^r x\,\mathrm{d}V}{\int_h^r \mathrm{d}V} = \frac{\int_h^r x\pi y^2\,\mathrm{d}x}{\int_h^r \pi y^2\,\mathrm{d}x} = \frac{\int_h^r xy^2\,\mathrm{d}x}{\int_h^r y^2\,\mathrm{d}x}$$

or
$$\bar{x} = \frac{\int_h^r xy^2\,dx}{\int_h^r y^2\,dx}$$

The sphere is generated by revolving the curve $x^2 + y^2 = r^2$ about x-axis. Hence, $y = \sqrt{r^2 - x^2}$

$$\bar{x} = \frac{\int_h^r x(r^2 - x^2)\,dx}{\int_h^r (r^2 - x^2)\,dx} = \frac{\left[\frac{r^2x^2}{2} - \frac{x^2}{2}\right]_h^r}{\left[r^2x - \frac{x^2}{3}\right]_h^r}$$

or
$$\bar{x} = \frac{\left(\frac{r^4 - 2r^2h^2 + h^4}{4}\right)}{\left(\frac{2r^3 - 3r^2h + h^3}{3}\right)} = \frac{3}{4}\left(\frac{r^4 - 2r^2h^2 + h^4}{2r^3 - 3r^2h + h^3}\right)$$

or
$$\bar{x} = \frac{3}{4}\frac{(r^2 - h^2)^2}{(r-h)^2(2r+h)} = \frac{3}{4}\frac{(r+h)^2}{(2r+h)}$$

$$\boxed{\bar{x} = \frac{3}{4}\frac{(r+h)^2}{(2r+h)}}$$

For a hemisphere, $h = 0$, which gives $\bar{x} = \frac{3r}{8}$.

PROBLEM 6.2

Find the centre of gravity of a plate bounded by the cardioid $r = 1 - \cos\theta$, if the surface density $\rho = \frac{k}{r}$.

Solution. The given plate is symmetrical about x-axis, therefore, $\bar{y} = 0$. The cardioid can be covered by radial strips for which r varies from 0 to $1 - \cos\theta$, and θ varies from 0 to 2π [Fig. 6.2(a)]. Hence, we have

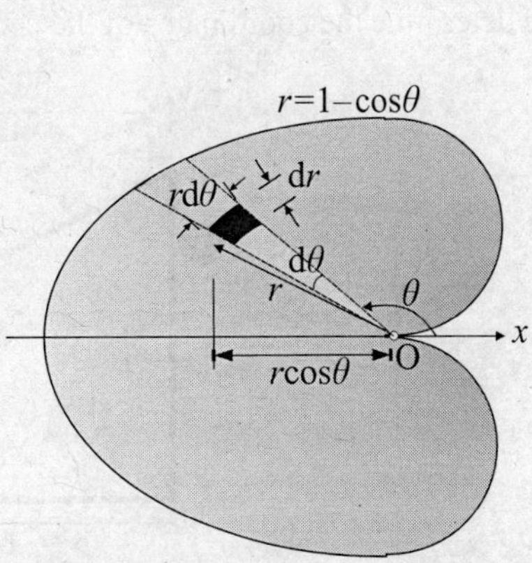

FIGURE 6.2(a)

$$\bar{x} = \frac{\iint \rho x\,dA}{\iint \rho\,dA} = \frac{\int_0^{2\pi}\int_0^{1-\cos\theta} \rho r\cos\theta\, r\,d\theta\,dr}{\int_0^{2\pi}\int_0^{1-\cos\theta} \rho r\,d\theta\,dr}$$

or
$$\bar{x} = \frac{\int_0^{2\pi}\int_0^{1-\cos\theta} \frac{k}{r}\, r\cos\theta\, r\,d\theta\,dr}{\int_0^{2\pi}\int_0^{1-\cos\theta} \frac{k}{r}\, r\,d\theta\,dr}$$

or
$$\overline{x} = \frac{\int_0^{2\pi}\int_0^{1-\cos\theta} r\cos\theta\, d\theta dr}{\int_0^{2\pi}\int_0^{1-\cos\theta} d\theta dr}.$$

or
$$\overline{x} = \frac{1}{2}\frac{\int_0^{2\pi}\cos\theta\,(1-\cos\theta)^2\, d\theta}{\int_0^{2\pi}(1-\cos\theta)d\theta}$$

Integrating gives

$$\overline{x} = -\frac{1}{2}$$

Hence, the centre of gravity is

$$\left(-\frac{1}{2}, 0\right)$$

PROBLEM 6.3

A thin homogeneous wire is bent as shown in Figure 6.3(a). Determine the coordinates of its centre of gravity.

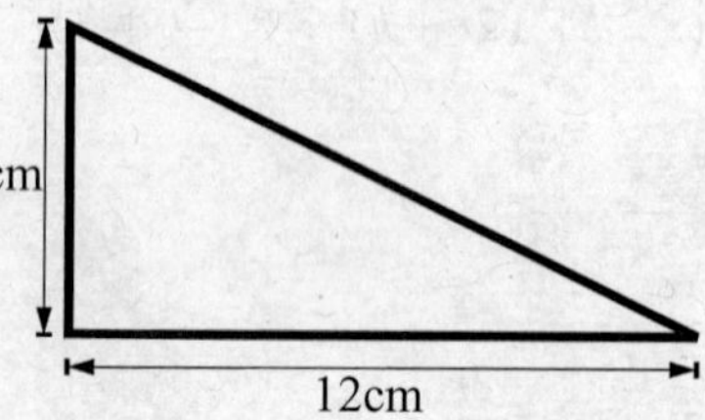

FIGURE 6.3(a) |

Solution. Let ρ be the mass per unit length of the wire. The body may be considered composed of three wires: OA, AB and OB. Choosing the coordinate axes shown [Fig. 6.3(b)], with origin at O, we determine the coordinates of the centroid and mass of each segment of wire.

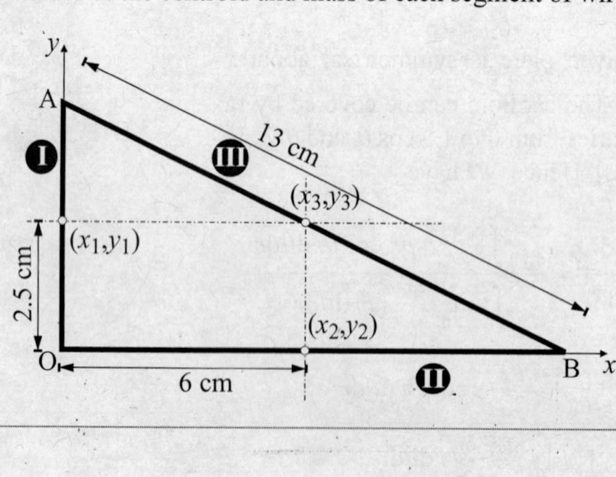

FIGURE 6.3(b) |

Part I

$$m_1 = 5\rho$$

$$x_1 = 0,\ y_1 = 2.5\text{cm}$$

Part II

$$m_2 = 12\rho$$

$$x_2 = 6\text{cm},\ y_2 = 6\text{cm}$$

Part III

$$m_3 = 13\rho$$

$$x_3 = 6\text{cm},\ y_3 = 2.5\text{cm}$$

x-coordinate of centre of gravity is given by

$$\bar{x} = \frac{\sum xm}{\sum m}$$

or $$\bar{x} = \frac{x_1m_1 + x_2m_2 + x_3m_3}{m_1 + m_2 + m_3}$$

or $$\bar{x} = \frac{0\times 5\rho + 6\times 12\rho + 6\times 13\rho}{5\rho + 12\rho + 13\rho} = \frac{150\rho}{30\rho} = 5$$

$$\boxed{\bar{x} = 5\text{cm}}$$

y-coordinate of centre of gravity is given by

$$\bar{y} = \frac{\sum ym}{\sum m}$$

or $$\bar{y} = \frac{y_1m_1 + y_2m_2 + y_3m_3}{m_1 + m_2 + m_3}$$

or $$\bar{y} = \frac{2.5\times 5\rho + 0\times 12\rho + 2.5\times 13\rho}{5\rho + 12\rho + 13\rho} = \frac{45\rho}{30\rho} = 1.5$$

$$\boxed{\bar{y} = 1.5\text{cm}}$$

Therefore, the coordinates of centre of gravity is (5 cm, 1.5 cm).

PROBLEM 6.4

From a semicircular lamina of radius r, a circular lamina of radius $(r/2)$ is removed as shown in Figure 6.4(a). Find the position of centre of gravity of the remainder.

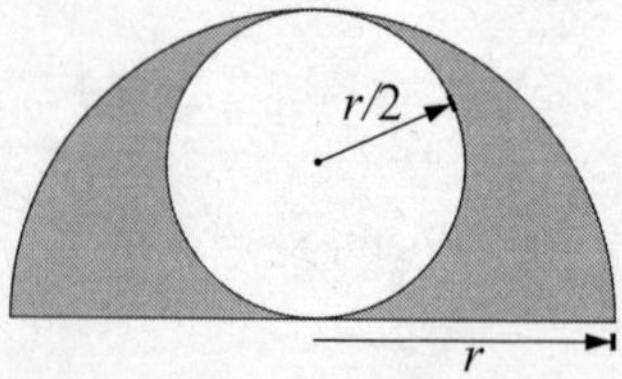

FIGURE 6.4(a) |

Solution. Let C and C_1 be the location of centre of gravity of the given circle and semicircle, respectively, as shown in Figure 6.4(b).

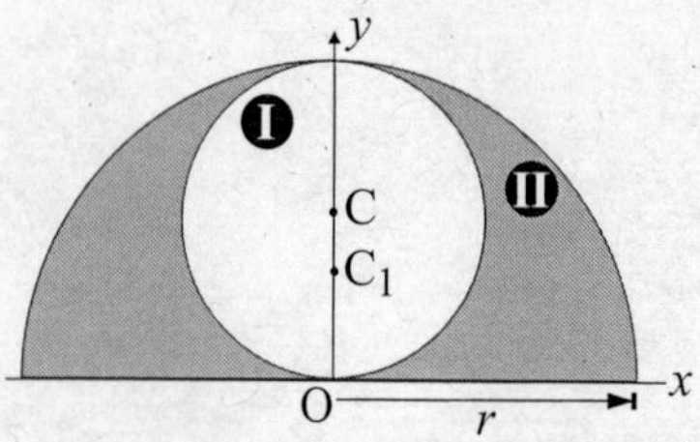

FIGURE 6.4(b) |

$$OC = \frac{r}{2} \quad \text{and} \quad OC_1 = \frac{4r}{3\pi}$$

Let ρ be the surface density, then mass of circle

$$m_1 = \pi\left(\frac{r}{2}\right)^2 \rho = \frac{1}{4}\pi r^2 \rho$$

and mass of semicircle

$$m_2 = \left(\frac{\pi r^2}{2}\right)\rho = \frac{1}{2}\pi r^2 \rho$$

The lamina has symmetry about y-axis, then y-coordinate of centre of gravity is given by

$$\bar{y} = \frac{\sum ym}{\sum m} = \frac{m_2 OC_1 - m_1 OC}{m_2 - m_1}$$

or

$$\bar{y} = \frac{\left(\frac{1}{2}\right)\pi r^2 \rho \times \left(\frac{4r}{3\pi}\right) - \left(\frac{1}{4}\right)\pi r^2 \rho \times \left(\frac{r}{2}\right)}{\left(\frac{1}{2}\right)\pi r^2 \rho - \left(\frac{1}{4}\right)\pi r^2 \rho} = \frac{(16-3\pi)r}{6\pi}$$

$$\bar{y} = \left(\frac{16-3\pi}{6\pi}\right)r$$

PROBLEM 6.5

A body consisting of a cone and a hemisphere of radius r fixed on the same base rests on a table, as shown in Figure 6.5(a). Find the greatest height h of the cone, so that the combined body may stand upright.

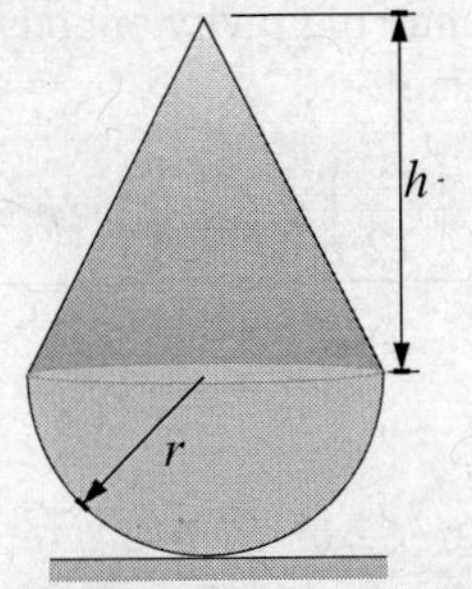

FIGURE 6.5(a) |

Solution. As the body is symmetrical about the vertical axis, its centre of gravity will lie on this axis. Consider two parts of the body, viz. hemisphere and cone. Let bottom of the hemisphere be the axis of reference [Fig. 6.5(b)].

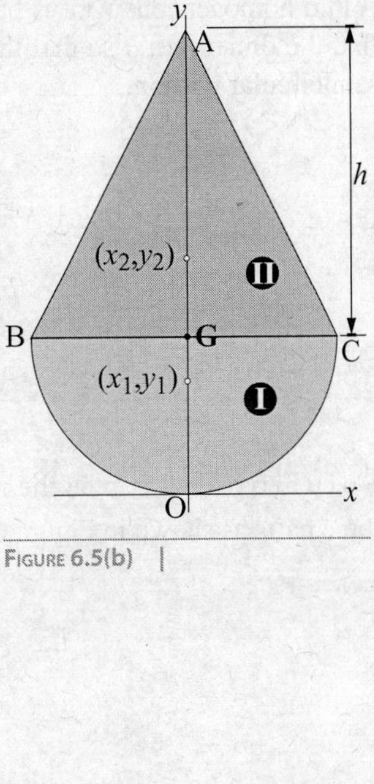

FIGURE 6.5(b) |

Part I: Hemisphere

y-Coordinate of centre of gravity of hemisphere.

$$y_1 = \frac{5}{8}r$$

Mass of hemisphere

$$m_1 = \rho\frac{2}{3}\pi r^3$$

Part II: Cone

y-Coordinate of centre of gravity of cone

$$y_2 = r + \frac{h}{4}$$

Mass of cone

$$m_2 = \rho\frac{1}{3}\pi r^2 h$$

Distance of centre of gravity of the combined body from O is

$$\bar{y} = \frac{m_1 y_1 + m_2 y_2}{m_1 + m_2}$$

$$= \frac{\rho\left(\frac{2}{3}\right)\pi r^3 \times \left(\frac{5}{8}\right)r + \rho\left(\frac{1}{3}\right)\pi r^2 h \times \left[r + \left(\frac{h}{4}\right)\right]}{\rho\left(\frac{2}{3}\right)\pi r^3 + \rho\left(\frac{1}{3}\right)\pi r^2 h}$$

$$\bar{y} = \frac{\left(\frac{5}{12}\right)r^4 + \left(\frac{1}{3}\right)r^3 h + \left(\frac{1}{12}\right)r^2 h^2}{\left(\frac{2}{3}\right)r^3 + \left(\frac{1}{3}\right)r^2 h}$$

Condition for stable equilibrium: centre of gravity of body should be below the common face BC or maximum it may coincide with it. Therefore,

$$\bar{y} = r \Rightarrow r = \frac{\left(\frac{5}{12}\right)r^4 + \left(\frac{1}{3}\right)r^3h + \left(\frac{1}{12}\right)r^2h^2}{\left(\frac{2}{3}\right)r^3 + \left(\frac{1}{3}\right)r^2h}$$

On simplifying, we obtain

$$3r^2 = h^2 \Rightarrow h = \sqrt{3}r$$

$$h = 1.73r$$

PROBLEM 6.6

A thin homogeneous wire is bent into the shape shown in Figure 6.6(a). If the dimension a is fixed, find the dimension b so that the centre of gravity of the wire will coincide with the centre C of the semicircular portion.

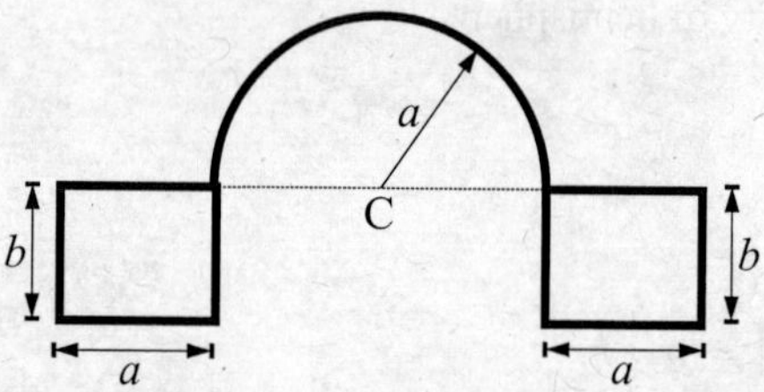

FIGURE 6.6(a) |

Solution. Let ρ be the mass per unit length of the wire. The wire frame can be divided into three parts as shown in Figure 6.6(b). The frame is symmetrical about y-axis, therefore, $\bar{x} = 0$.

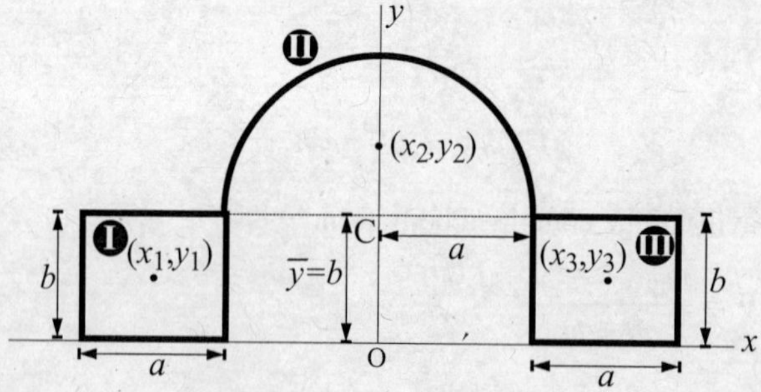

FIGURE 6.6(b) |

Part I

Mass of rectangular frame

$$m_1 = \rho(2b + 2a)$$

y-Coordinate of centre of gravity of rectangular frame

$$y_1 = \frac{1}{2}b$$

Part II: Semicircular Frame

Mass of cone

$$m_2 = \rho\pi a$$

y-Coordinate of centre of gravity of cone

$$y_2 = b + \frac{2a}{\pi}$$

Part III: Rectangular Frame

Mass of rectangular frame

$$m_3 = \rho(2b + 2a)$$

y-Coordinate of centre of mass of rectangular frame

$$y_3 = \frac{1}{2}b$$

y-Coordinate of centre of gravity is given by

$$\bar{y} = \frac{\sum ym}{\sum m} = \frac{y_1m_1 + y_2m_2 + y_3m_3}{m_1 + m_2 + m_3}$$

or
$$\bar{y} = \frac{\frac{1}{2}b \times \rho(2b + 2a) + \left[b + \frac{2a}{\pi}\right] \times \rho\pi a + \frac{1}{2}b \times \rho(2b + 2a)}{\rho(2\text{b} + 2a) + \rho\pi a + \rho(2b + 2a)}$$

Given
$$\bar{y} = b \Rightarrow b = \frac{\frac{1}{2}b \times \rho(2b + 2a) + \left[b + \frac{2a}{\pi}\right] \times \rho\pi a + \frac{1}{2}b \times \rho(2b + 2a)}{\rho(2b + 2a) + \rho\pi a + \rho(2b + 2a)}$$

Simplifying, we get

$$b^2 + ab - a^2 = 0$$

Possible solution of above quadratic equation is

$$b = \left(\frac{-1 + \sqrt{5}}{2}\right)a = 0.618a$$

$$b = 0.618a$$

PROBLEM 6.7

Find the position of centroid of a frustum of a solid right circular cone base having an axial hole of 0.25 m diameter as shown in Figure 6.7(a).

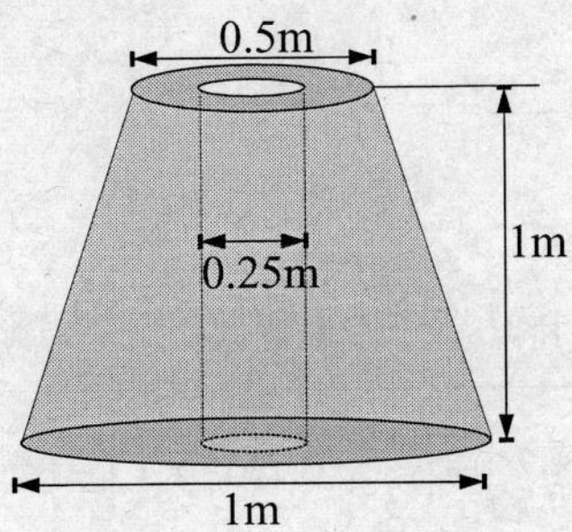

FIGURE 6.7(a) |

Solution. The body is obtained by subtracting cone CDE and cylinder PQRS from cone ABE as shown in Figure 6.7(b). Let base of cone be the axis of reference.

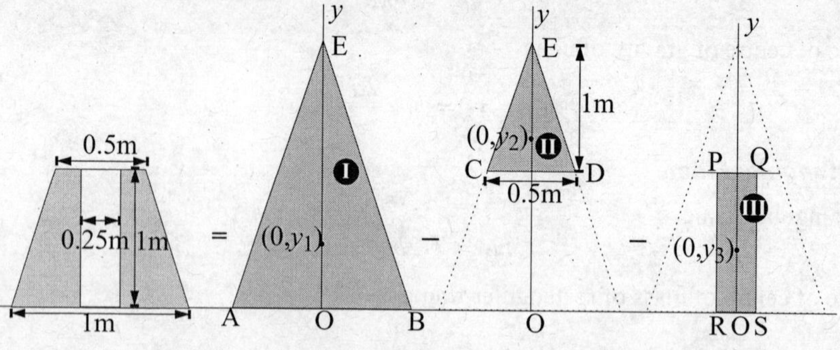

FIGURE 6.7(b) |

Part I: Right Circular Cone EAB

$$v_1 = \frac{1}{3}\pi r^2 h = \frac{1}{3}\pi(0.5)^2 \times 2$$

$$y_1 = \frac{2}{4}\text{m} = 0.5\text{m}$$

Part II: Right Circular Cone ECD (Removed)

$$v_2 = \frac{1}{3}\pi r^2 h = \frac{1}{3}\pi(0.25)^2 \times 1$$

$$y_2 = 1 + \frac{1}{4} = 1.25\text{m}$$

Part III: Cylinder PQRS (Removed)

$$v_3 = \pi r^2 h = \left(\frac{0.25}{2}\right)^2 \times 1$$

$$y_3 = \frac{1}{2} = 0.5\text{m}$$

Distance of centroid of body from O is

$$\bar{y} = \frac{\sum yv}{\sum v} = \frac{y_1 v_1 - y_2 v_2 - y_3 v_3}{v_1 - v_2 - v_3}$$

$$\bar{y} = \frac{0.5 \times \left(\frac{1}{3}\pi(0.5)^2 \times 2\right) - 1.25 \times \left(\frac{1}{3}\pi(0.25)^2 \times 1\right) - 0.5 \times \left[\pi\left(\frac{0.25}{2}\right)^2 \times 2\right]}{\left(\frac{1}{3}\pi(0.5)^2 \times 2\right) - \left(\frac{1}{3}\pi(0.25)^2 \times 1\right) - \left[\pi\left(\frac{0.25}{2}\right)^2 \times 2\right]}$$

Simplifying, we obtain

$$\overline{y} = 0.379\text{m}$$

PROBLEM 6.8

Locate the centre of gravity of the composite body as shown in Figure 6.8(a).

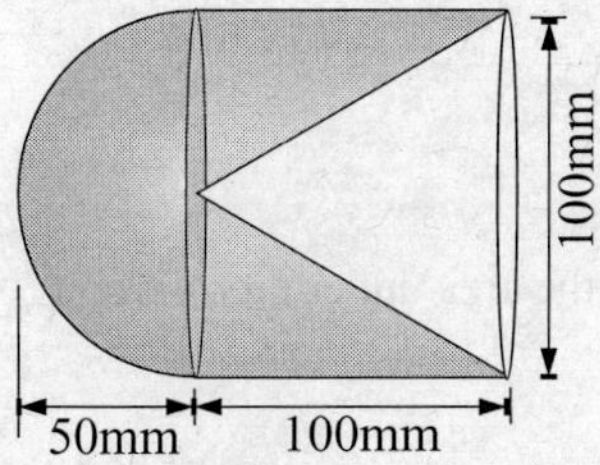

FIGURE 6.8(a) |

Solution. The body can be divided into three parts as shown in Figure 6.8(b). The body is symmetrical about x-axis, therefore $\overline{y} = 0$.

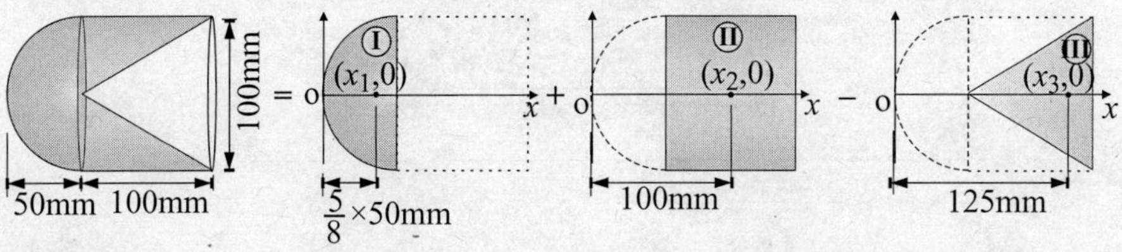

FIGURE 6.8(b) |

Part I: Hemisphere

$$v_1 = \frac{1}{2} \times \frac{4}{3}\pi(50)^3\,\text{mm}^3$$

$$x_1 = \frac{5}{8} \times 50\text{mm}$$

Part II: Cylinder

$$v_2 = \pi(50)^2 \times 100\text{mm}^3$$

$$x_2 = 50 + 50 = 100\text{mm}$$

Part III: Cone (Remove)

$$v_3 = \frac{1}{3}\pi(50)^2 \times 100\text{mm}^3$$

$$x_3 = 50 + \frac{3}{4} \times 100 = 125\text{mm}$$

Location of centroid of body is given by

$$\overline{x} = \frac{x_1 v_1 + x_2 v_2 - x_3 v_3}{v_1 + v_2 - v_3}$$

or $$\overline{x} = \frac{\left(\frac{5}{8}\times 50\right)\times\left(\frac{1}{2}\times\frac{4}{3}\pi(50)^3\right)+100\times(\pi(50)^2\times 100)-125\times\left[\frac{1}{3}\pi(50)^2\times 100\right]}{\left[\frac{1}{2}\times\frac{4}{3}\pi(50)^3+\pi(50)^2\times 100+\frac{1}{3}\pi(50)^2\times 100\right]}$$

On simplifying, we get

$$\overline{x} = 68.75\text{mm}$$

PROBLEM 6.9

Find the coordinate of the centroid of the area under the curve $y = b[1-(x^2/a^2)]$ from $x = 0$ to $x = a$ [Fig. 6.9(a)].

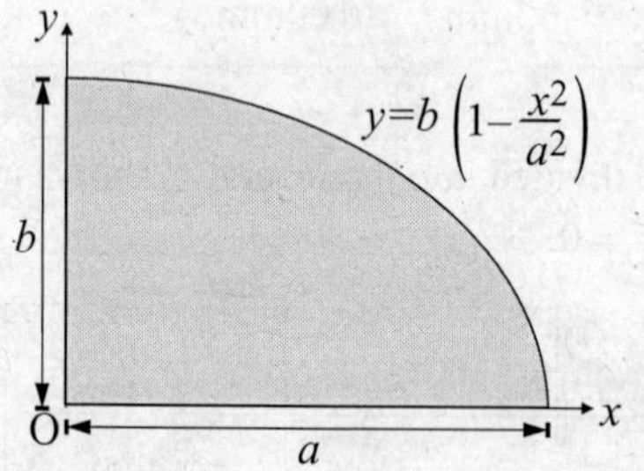

FIGURE 6.9(a) |

Solution. A vertical element of area $dA = y\,dx$ is chosen as shown in Figure 6.9(b). The x-coordinate of the centroid is given by

$$\overline{x} = \frac{\int x_c dA}{\int dA}$$

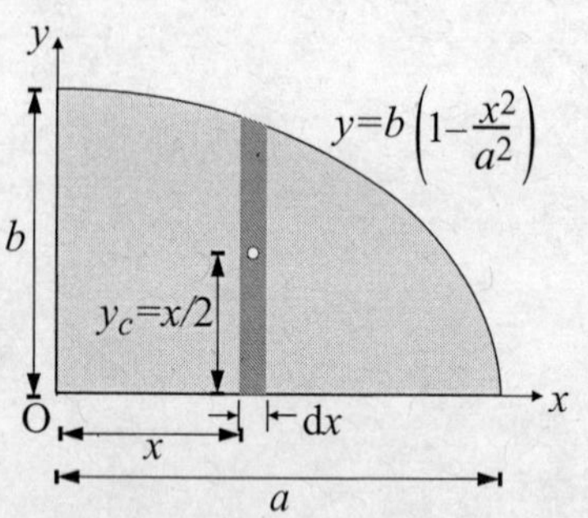

FIGURE 6.9(b) |

or $$\overline{x} = \frac{\int_0^a xy\,dx}{\int_0^a y\,dx} = \frac{\int_0^a xy\,dx}{\int_0^a y\,dx}$$

Substituting $y = b\left(1-\frac{x^2}{a^2}\right)$ we have

$$\overline{x} = \frac{\int_0^a xb\left(1-\frac{x^2}{a^2}\right)dx}{\int_0^a b\left(1-\frac{x^2}{a^2}\right)dx}$$

Integrating, we get

$$\overline{x} = \frac{3a}{8}$$

The y-coordinate of centroid of the rectangular element is $y_c = \frac{y}{2}$. Therefore,

$$\overline{y} = \frac{\int_0^a y_c dA}{\int_0^a dA} = \frac{\int_0^a \frac{y}{2} dA}{\int_0^a dA} = \frac{1}{2}\frac{\int_0^a y.y dx}{\int_0^a y dx}$$

or

$$\overline{y} = \frac{1}{2}\frac{\int_0^a y^2 dx}{\int_0^a y dx}$$

Substituting $y = b\left(1 - \dfrac{x^2}{a^2}\right)$, we have

$$\overline{y} = \frac{1}{2}\frac{\int_0^a b\left(1 - \frac{x^2}{a^2}\right)^2 dx}{\int_0^a b\left(1 - \frac{x^2}{a^2}\right) dx}$$

Integrating it gives,

$$\overline{y} = \frac{2b}{5}$$

Therefore, the coordinates of centroid are

$$\left(\frac{3a}{8}, \frac{2b}{5}\right)$$

Double Integration Method

A differential element of area dA = dx dy is chosen as shown in Figure 6.9(c). The x-coordinate of the centroid is given by

$$\overline{x} = \frac{\iint x dA}{\iint dA} = \frac{\int_0^a \int_0^{b\left(1-\frac{x^2}{a^2}\right)} x dx dy}{\int_0^a \int_0^{b\left(1-\frac{x^2}{a^2}\right)} dx dy}$$

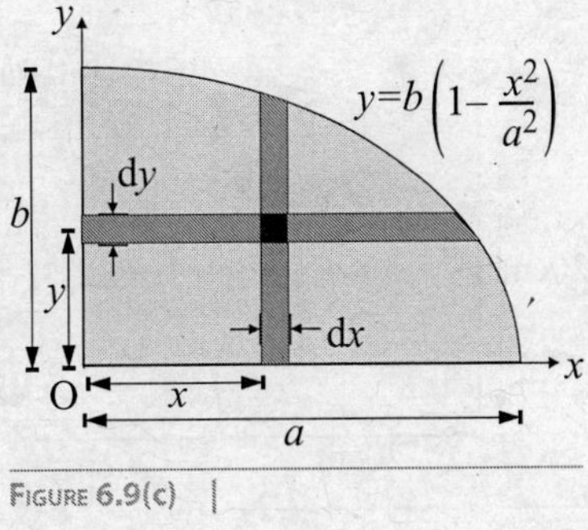

FIGURE 6.9(c) |

Integrating gives

$$\overline{x} = \frac{3a}{8}$$

The y-coordinate of centroid is given by

$$\overline{y} = \frac{\iint y dA}{\iint dA}$$

or

$$\bar{y} = \frac{\int_0^a \int_0^{b\left(1-\frac{x^2}{a^2}\right)} y\,dx\,dy}{\int_0^a \int_0^{b\left(1-\frac{x^2}{a^2}\right)} dx\,dy}$$

Integrating gives

$$\bar{y} = \frac{2b}{5}$$

PROBLEM 6.10

A semicircular area is removed from a trapezium as shown in Figure 6.10(a). Determine the centroid of the remaining area.

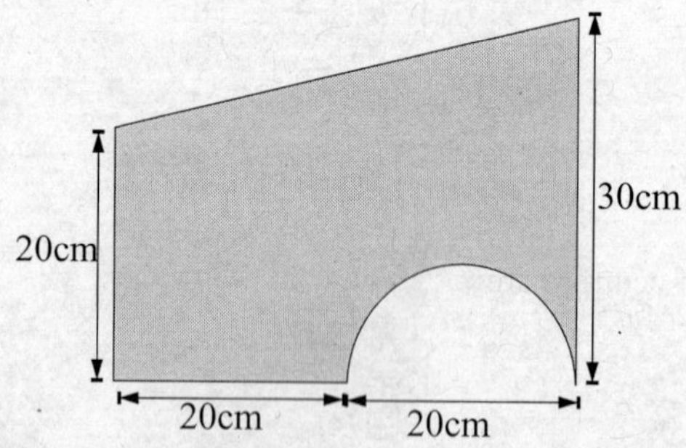

FIGURE 6.10(a) |

Solution. The given area can be divided into three parts as shown in Figure 6.10(b). Let left face and base of the trapezium be the axes of reference.

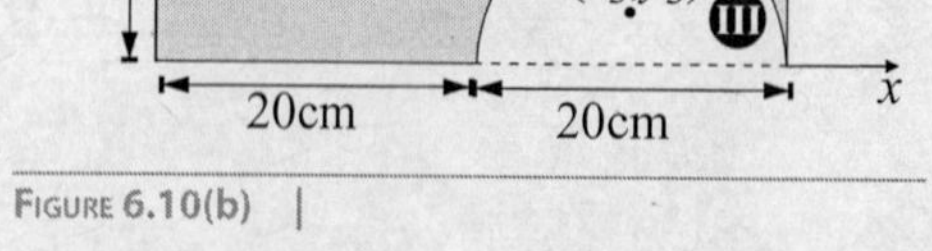

FIGURE 6.10(b) |

Part I: Rectangle

$$a_1 = 20 \times 40 = 800\text{cm}^2$$

$$x_1 = \frac{40}{2} = 20\text{cm}$$

$$y_1 = \frac{20}{2} = 10\text{cm}$$

Part II: Triangle

$$a_2 = \frac{1}{2} \times 40 \times 10 = 200\text{cm}^2$$

$$x_2 = \frac{2}{3} \times 40 = \frac{80}{3}\text{cm}$$

$$y_2 = 20 + \frac{10}{3} = \frac{70}{3}\text{cm}$$

Part III Semicircle (Removed)

$$a_3 = \frac{\pi r^2}{2} = \frac{\pi}{2} \times 10^2 = 50\pi \text{cm}^2$$

$$x_3 = 20 + 10 = 30\text{cm},$$

$$y_3 = \frac{4r}{3\pi} = \frac{4 \times 10}{3\pi} = \frac{40}{3\pi}\text{cm}^2$$

The x-coordinate of centroid

$$\overline{x} = \frac{\sum xa}{\sum a} = \frac{x_1a_1 + x_2a_2 - x_3a_3}{a_1 + a_2 - a_3}$$

or

$$\overline{x} = \frac{20 \times 800 + \frac{80}{3} \times 200 - 30 \times 50\pi}{800 + 200 - 50\pi}$$

or

$$\overline{x} = \frac{16620.9}{842.9} = 19.71\text{cm}$$

$$\boxed{\overline{x} = 19.71\text{cm}}$$

The y-coordinate of centroid

$$\overline{y} = \frac{\sum ya}{\sum a} = \frac{y_1a_1 + y_2a_2 - y_3a_3}{a_1 + a_2 - a_3}$$

or

$$\overline{y} = \frac{800 \times 10 + 200 \times \frac{70}{3} \times \frac{40}{3\pi} \times 50\pi}{800 + 200 - 50\pi}$$

or

$$\overline{y} = \frac{14333.3}{842.9} = 17.00\text{cm}$$

$$\boxed{\overline{y} = 17.00\text{cm}}$$

Therefore, the coordinates of centroid of area are (19.71, 17.00).

PROBLEM 6.11

Determine the coordinates of centroid of the shaded area between the two curves shown in Figure 6.11(a).

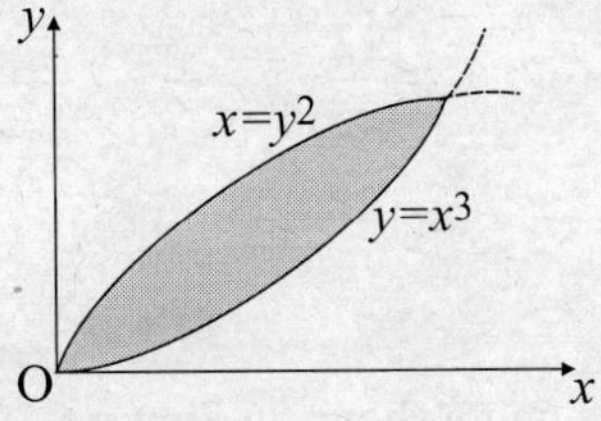

FIGURE 6.11(a) |

Solution. The equations of curves are

$$y = x^3 \tag{i}$$

$$x = y^2 \tag{ii}$$

Coordinates of intersection points are O(0,0) and C(1,1). Now consider a differential element of area $dA = dx\,dy$ shown in Figure 6.11(b). The x-coordinate of the centroid is given by

$$\bar{x} = \frac{\iint x\,dA}{\iint dA}$$

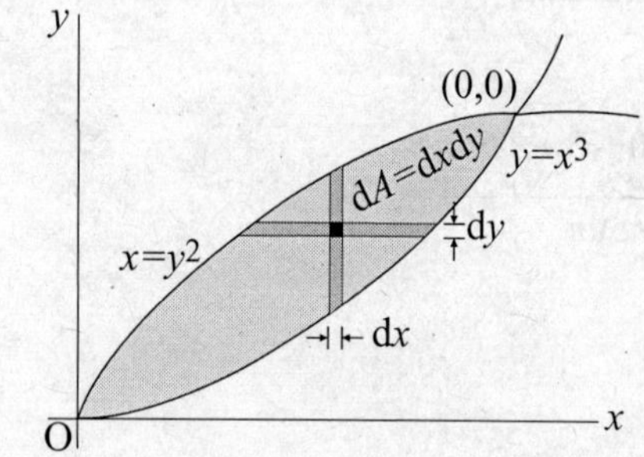

FIGURE 6.11(b) |

or
$$\bar{x} = \frac{\int_0^1 \int_{x^3}^{\sqrt{x}} x\,dx\,dy}{\int_0^1 \int_{x^3}^{\sqrt{x}} dx\,dy}$$

or
$$\bar{x} = \frac{\int_0^1 x(\sqrt{x} - x^3)dx}{\int_0^1 (\sqrt{x} - x^3)dx}$$

Integrating it gives,

$$\bar{x} = \frac{12}{25}$$

The y-coordinate of the centroid is given by

$$\bar{y} = \frac{\iint y\,dA}{\iint dA}$$

or
$$\bar{y} = \frac{\int_0^1 \int_{x^3}^{\sqrt{x}} y\,dx\,dy}{\int_0^1 \int_{x^3}^{\sqrt{3}} dx\,dy} = \frac{1}{2}\frac{\int_0^1 [(\sqrt{x})^2 - (x^3)^2]dx}{\int_0^1 (\sqrt{x} - x^3)dx}$$

Integrating it gives,

$$\bar{y} = \frac{3}{7}$$

Coordinates of the centroid of shaded area

$$\bar{x} = \frac{12}{25}, \bar{y} = \frac{3}{7}$$

PROBLEM 6.12

If the dimensions a and b of the plane figure shown in Figure 6.12(a) are fixed, find what the dimension c must be in order that the centroid of the shaded area will lie on FE.

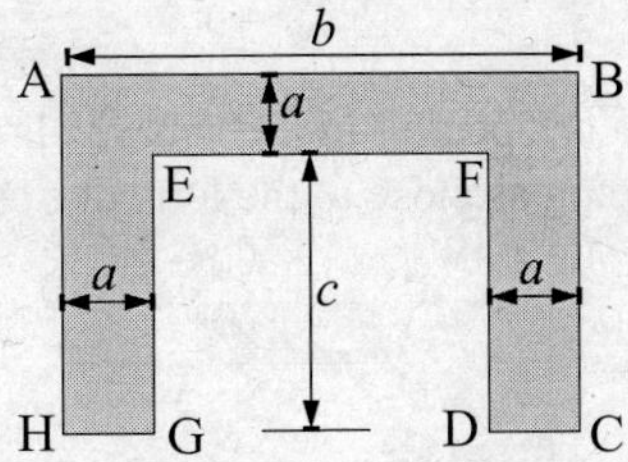

FIGURE 6.12(a) |

Solution. The given section can be divided into three parts as shown in Figure 6.12(b).

Part I

$$a_1 = a \times c$$

$$y_1 = \frac{c}{2}$$

Part II

$$a_2 = a \times b$$

$$y_2 = c + \frac{a}{2}$$

Part III

$$a_3 = a \times c$$

$$y_3 = \frac{c}{2}$$

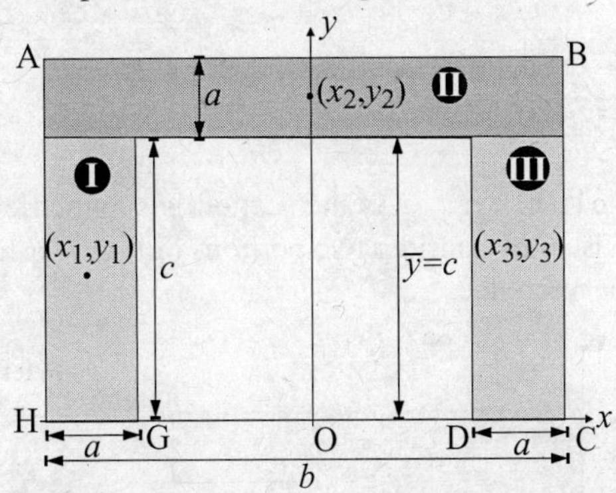

FIGURE 6.12(b) |

Due to symmetry of area about y-axis, the centroid lies on y-axis. The y-coordinate of centroid is given by

$$\overline{y} = \frac{\sum ya}{\sum a} = \frac{a_1y_1 + a_2y_2 + a_3y_3}{a_1 + a_2 + a_3}$$

$$\overline{y} = \frac{ac \times \left(\frac{c}{2}\right) + ab \times \left[c + \left(\frac{a}{2}\right)\right] + ac \times \left(\frac{c}{2}\right)}{ac + ab + ac}$$

$$= \frac{ac^2 + abc + \frac{a^2b}{2}}{ab + 2ac}$$

Given that

$$\overline{y} = c \Rightarrow c = \frac{ac^2 + abc + \frac{a^2b}{2}}{ab + 2ac}$$

or

$$abc + 2ac^2 = ac^2 + abc + \frac{a^2b}{2}$$

or

$$ac^2 = \frac{a^2b}{2}$$

$$c = \sqrt{\frac{ab}{2}}$$

PROBLEM 6.13

Find the dimension h of the rectangular opening in the square plate of side l which will result in the mass centre of the remaining plate being as close to the left edge as possible [Fig. 6.13(a)].

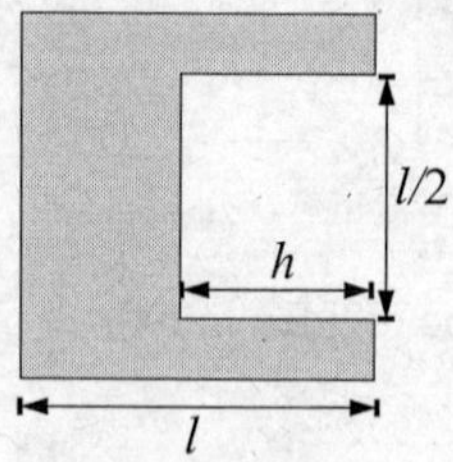

FIGURE 6.13(a) |

Solution. As the section is symmetrical about x-axis, therefore, its centroid will lie on this axis. Now consider two portions of Figure 6.13(b), viz. square ABCD and rectangle PQRS. Let O be the reference.

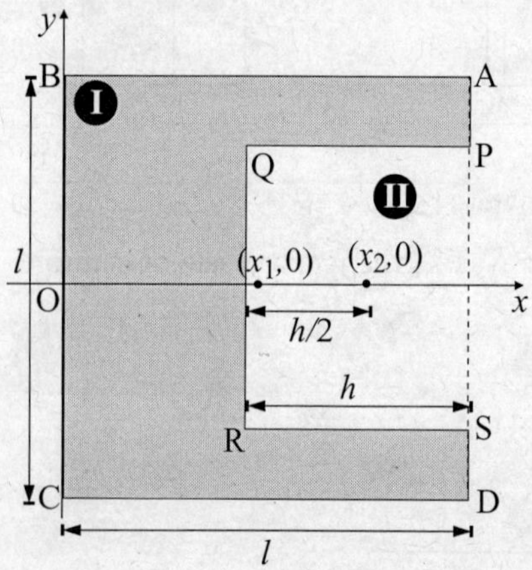

FIGURE 6.13(b) |

Part I Square ABCD

$$a_1 = l^2$$

$$x_1 = \frac{l}{2}$$

Part II Rectangle PQRS

$$a_2 = \frac{hl}{2}$$

$$x_2 = l - h + \frac{h}{2} = l - \frac{h}{2}$$

Distance of centroid from O is given by

$$\bar{x} = \frac{a_1 x_1 - a_2 x_2}{a_1 - a_2}$$

or
$$\bar{x} = \frac{l^2 \times \left(\frac{l}{2}\right) - \left(\frac{hl}{2}\right) \times \left[l - \left(\frac{h}{2}\right)\right]}{l^2 - \left(\frac{hl}{2}\right)} = \frac{2l^2 - 2hl + h^2}{2l - h}$$

or
$$\bar{x} = \frac{2l^2 - 2hl + h^2}{2l - h}$$

or
$$\frac{d}{dh}\bar{x} = 0 \Rightarrow \frac{(2l - h)(-2l + 2h) - (2l^3 - 2hl + l^2) \times -1}{(2l - h)^2} = 0$$

or
$$h^2 - 4hl + 2l^2 = 0$$

Solving the quadratic equation, we obtain

$$h = (2 \pm \sqrt{2})l = 0.58l, 3.41l$$

Possible solution is

$$h = 0.58l$$

PROBLEM 6.14

Find the moment of inertia about x-axis of an arc of mass m and radius r as shown in Figure 6.14(a).

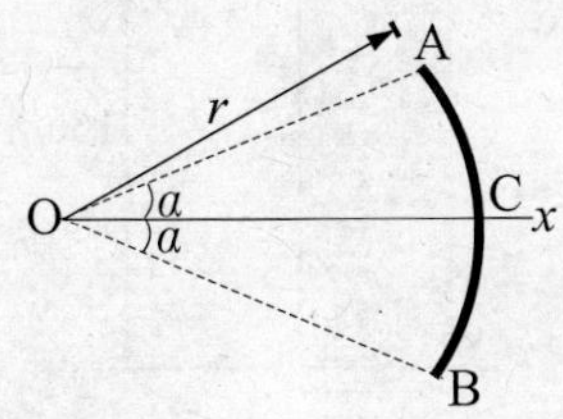

FIGURE 6.14(a)

Solution. Let O be the centre and C the middle point of the arc as shown in Figure 6.14(b). Consider an element $r\, d\theta$ at an angle θ, mass of the element is

$$\frac{m}{2ar} r\mathrm{d}\theta = \frac{m}{2a} \mathrm{d}\theta$$

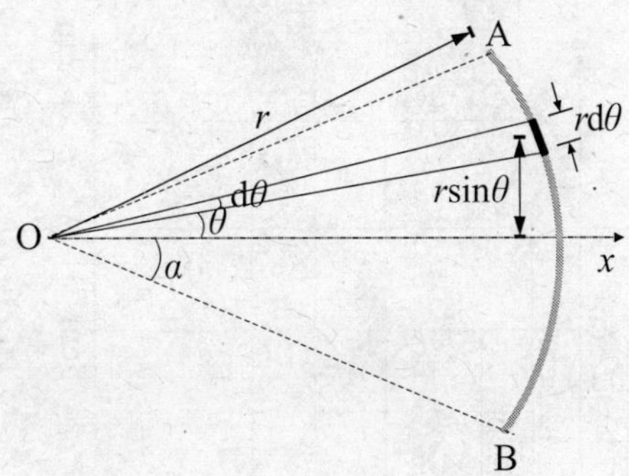

FIGURE 6.14(b)

Moment of inertia about OA,

$$\mathrm{d}I_x = \frac{M}{2a}\mathrm{d}\theta(r \sin \theta)^2$$

or $$\mathrm{d}I_x = \frac{m}{2a} r^2 \sin^2 \theta \mathrm{d}\theta$$

or $$I_x = \int_{-a}^{+a} \frac{m}{2a} r^2 \sin^2 \theta \mathrm{d}\theta$$

or $$I_x = \frac{m}{2a} r^2 \int_{-a}^{+a} \sin^2 \theta \mathrm{d}\theta$$

or $$I_x = \frac{m}{2a} r^2 \int_{-a}^{+a} \left(\frac{1 - \cos 2\theta}{2}\right) \mathrm{d}\theta$$

or $$I_x = \frac{m}{4a} r^2 \int_{-a}^{+a} \mathrm{d}\theta - \frac{m}{4a} r^2 \int_{-a}^{+a} \cos 2\theta \mathrm{d}\theta$$

$$I_x = mr^2 \left[\frac{1}{2} - \frac{\sin 2a}{4}\right]$$

PROBLEM 6.15

Find the moment of inertia of a T-section shown in Figure 6.15(a) about x-axis and y-axis through the centroid of the section.

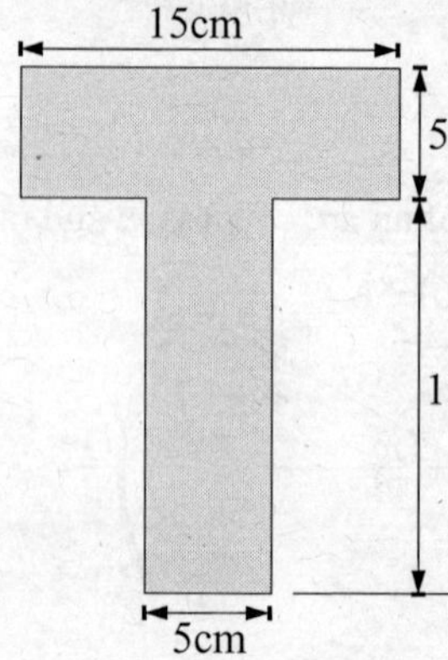

FIGURE 6.15(a)

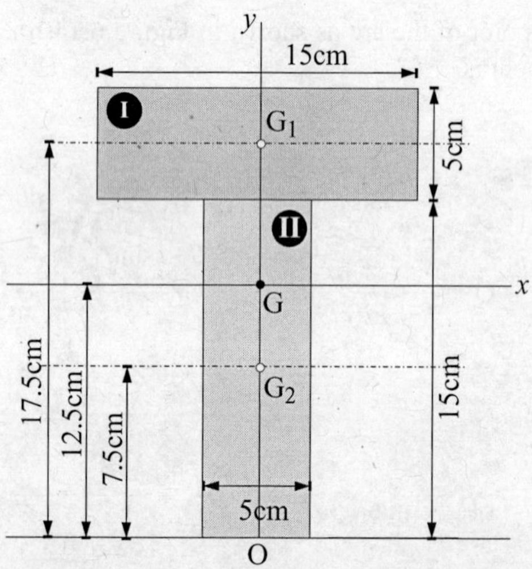

FIGURE 6.15(b)

Solution. The T-section is symmetrical about y-axis. T is divided into two rectangles, I and II, as shown in Figure 6.15(b).

Rectangle I

$$a_1 = 15 \times 5 = 75\text{cm}^2$$

$$y_1 = 17.5\text{cm},$$

$$I_{G_1} = \frac{15 \times 5^3}{12}\text{cm}^4$$

Rectangle II

$$a_2 = 5 \times 15 = 75\text{cm}^2$$

$$y_2 = 7.5\text{cm}$$

$$I_{G_2} = \frac{5 \times 15^3}{12}\text{cm}^4$$

The distance of centroid of the section from bottom of the flange,

$$\bar{y} = \frac{a_1 y_1 + a_2 y_2}{a_1 + a_2} = \frac{75 \times 17.5 + 75 \times 7.5}{75 + 75} = 12.5\text{cm}$$

$$\bar{y} = 12.5\text{cm}$$

Moment of inertia of the whole section about centroidal axis parallel to x-axis,

$$(I_G)_x = (I_{G_1} + a_1 h_1^2)_x + (I_{G_2} + a_2 h_2^2)_x$$

$$(I_G)_x = \left[\frac{15 \times 5^3}{12} + 75(17.5 - 12.5)^2\right] + \left[\frac{5 \times 15^3}{12} + 75(12.5 - 7.5)^2\right]$$

On simplifying, we gate

$$(I_G)_x = 5312.5\text{cm}^4$$

Moment of inertia of T-section about *yy*,

$$(I_G)_y = \frac{5 \times 15^3}{12} + \frac{15 \times 5^3}{12}$$

$$(I_G)_y = 1562.5\text{cm}^4$$

PROBLEM 6.16

Find the moment of inertia of a section, shown in Figure 6.16(a), about *x*-axis through the centroid of the section.

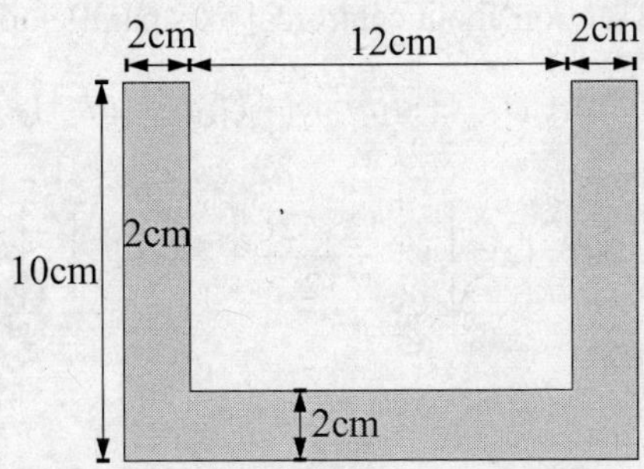

FIGURE 6.16(a) |

Solution. The section is symmetrical about *y*-axis, therefore its centroid will lie on this axis. Let us split up the area into three rectangles I, II and III, as shown in Figure 6.16(b).

Rectangle I

$$a_1 = 2 \times 10 = 20\text{cm}^2$$

$$y_1 = 5\text{cm}$$

$$(I_{G_1})_x = \frac{2 \times 10^3}{12}$$

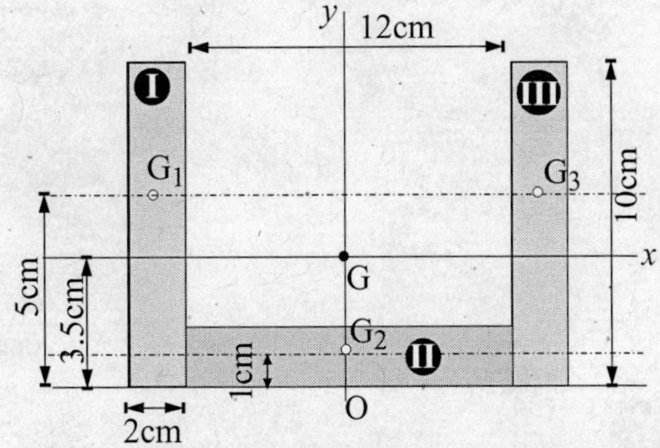

FIGURE 6.16(b) |

Rectangle II

$$a_2 = 12 \times 2 = 24\text{cm}^2$$

$$y_2 = 1\text{cm}$$

$$(I_{G_2})_x = \frac{12 \times 2^3}{12}$$

Rectangle III

$$a_3 = 2 \times 10 = 20\text{cm}^2$$

$$y_3 = 5\text{cm}$$

$$(I_{G_3})_x = \frac{2 \times 10^3}{12}$$

Distance of centroid from O is

$$\bar{y} = \frac{a_1 y_1 + a_2 y_2 + a_3 y_3}{a_1 + a_2 + a_3}$$

$$\bar{y} = \frac{20 \times 5 + 24 \times 1 + 20 \times 5}{20 + 24 + 20} = 3.5\text{cm}$$

$$\bar{y} = 3.5\text{cm}$$

$$h_1 = h_2 = 5 - 3.5 = 1.5\text{cm}, \qquad h_3 = 3.5 - 1 = 2.5\text{cm}$$

Moment of inertia of the whole section about centroidal axis parallel to x-axis is

$$(I_G)_x = (I_{G_1} + a_1 h_1^2)_x + (I_{G_2} + a_2 h_2^2)_x + (I_{G_3} + a_3 h_3^2)_x$$

or

$$(I_G)_x = \left[\frac{2 \times 10^3}{12} + 20 \times 1.5^2\right] + \left[\frac{12 \times 2^3}{12} + 24 \times 1^2\right] + \left[\frac{2 \times 10^3}{12} + 20 \times 1.5^2\right]$$

$$(I_G)_x = 581.33 \text{ cm}^4$$

PROBLEM 6.17

Find the moment of inertia of a hollow rectangular plane shown in Figure 6.17(a) about x-axis and y-axis through the centroid

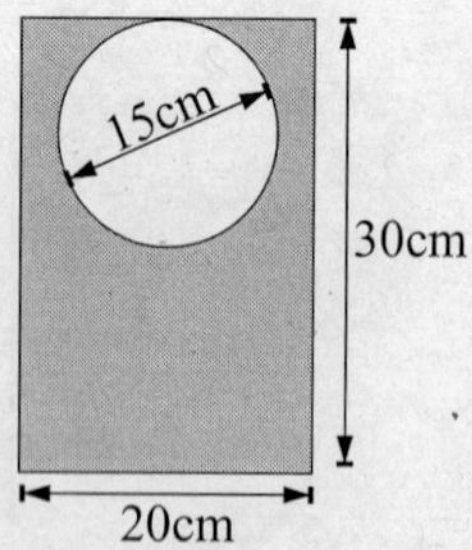

FIGURE 6.17(a) |

Solution. As the section is symmetrical about y-axis; therefore, centroid of the section will lie on this axis. Let bottom be the axis of reference [Fig. 6.17(b)].

Rectangle I

$$a_1 = 20 \times 30 = 600\text{cm}^2$$

$$y_1 = 15\text{cm}$$

$$(I_{G_1})_x = \frac{20 \times 30^3}{12}$$

Circular Hole II

$$a_2 = \frac{\pi}{4}(15)^2$$

$$y_2 = 22.5\text{cm}$$

$$(I_{G_2})_x = \frac{\pi}{64}(15)^4$$

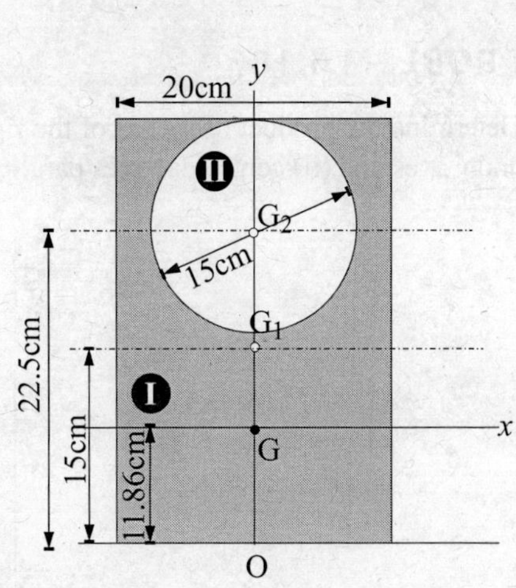

FIGURE 6.17(b) |

Distance of centroid from O is

$$\bar{y} = \frac{a_1 y_1 - a_2 y_2}{a_1 - a_2}$$

or
$$\bar{y} = \frac{60 \times 15 - \left(\frac{\pi}{4}\right) \times (15)^2 \times 22.5}{600 - \left(\frac{\pi}{4}\right) \times (15)^2}$$

or
$$\bar{y} = 11.86\text{cm}$$

and
$$h_1 = 15 - 11.86 = 3.13\text{cm}, \qquad h_2 = 10.64\text{cm}$$

Moment of inertia of the whole section about centroidal axis parallel to x-axis is

$$(I_G)_x = (I_{G_1} + a_1 h_1^2) - (I_{G_2} + a_2 h_2^2)$$

or
$$(I_G)_x = \left[\frac{20 \times 30^3}{12} + 600 \times 3.13^2\right] - \left[\frac{\pi}{64} \times (15)^4 + \frac{\pi}{4} \times (15)^2 \times (10.64)^2\right]$$

$$(I_G)_x = [45000 + 5878.14] - [2485.04 + 20005.78]$$

$$(I_G)_x = 28387.4\text{cm}^4$$

Moment of inertia of the whole section about centroidal axis parallel to y-axis,

$$(I_G)_y = \frac{30 \times 20^3}{12} - \frac{\pi(15)^4}{64}$$

$$(I_G)_y = 17514.9\text{cm}^4$$

PROBLEM 6.18

Determine the product of inertia of the right triangle shown in Figure 6.18(a) with respect to (i) the x and y axes and (ii) centroidal axes parallel to x and y axes.

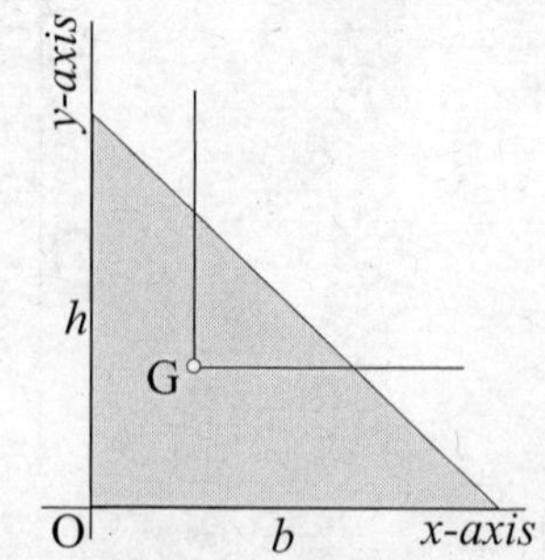

FIGURE 6.18(a) |

Solution. For calculation of the product of moment of inertia, a small element of area $dA = dx\,dy$ is chosen [Fig. 6.18(b)]. Thus,

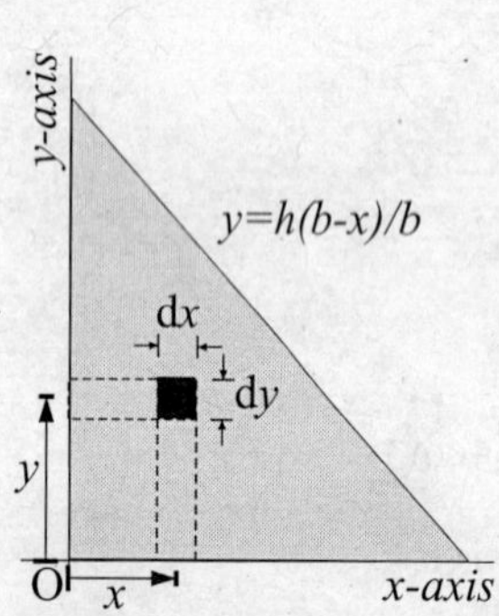

FIGURE 6.18(b) |

$$I_{xy} = \int xy\,dA = \int_0^b \int_0^{h(b-h)/b} xy\,dx\,dy$$

or

$$I_{xy} = \int_0^b x\,dx \left[\frac{y^2}{2}\right]_0^{h(b-h)/b}$$

$$I_{xy} = \frac{b^2h^2}{24}$$

The coordinates of the centroid of the triangle are $(b/3, h/3)$. Applying parallel axis theorem for product of inertia,

$$I_{xy} = \overline{I}_{x'y'} + \overline{xy}A$$

or

$$\frac{b^2h^2}{24} = \overline{I}_{x'y'} + \frac{b}{3} \times \frac{h}{3} \times \frac{1}{2}bh$$

$$\overline{I}_{x'y'} = -\frac{b^2h^2}{72}$$

PROBLEM 6.19

Find the product of inertia of the area of the quadrant of a circular area with respect to x and y axes.

Solution. Consider a small element of area $dA = r\,d\theta\,dr$ [Fig. 6.19(a)]. The product of inertia of the area of the quadrant of a circular area is.

$$I_{xy} = \int xy\,\mathrm{d}A$$

$$= \int_0^{\pi/2}\int_0^R (r\cos\theta)\times(r\sin\theta)r\,\mathrm{d}\theta\,\mathrm{d}r$$

$$= \int_0^{\pi/2}\int_0^R r^3\,\frac{\sin 2\theta}{2}\,\mathrm{d}r\,\mathrm{d}\theta$$

On simplifying,

$$I_{xy} = \frac{R^4}{8}$$

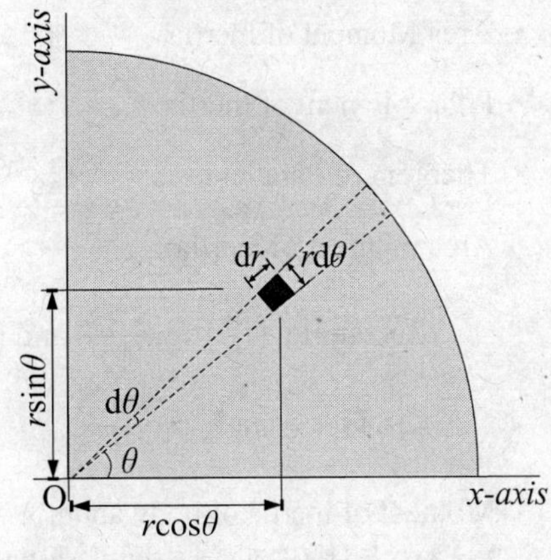

FIGURE 6.19(a)

CHAPTER HIGHLIGHTS

- Coordinates of centroid:

Mass $\bar{x} = \dfrac{\int x dm}{\mathrm{M}}, \bar{y} = \dfrac{\int y dm}{\mathrm{M}}\ \bar{z} = \dfrac{\int z dm}{\mathrm{M}}$, Length $\bar{x} = \dfrac{\int x dL}{L}, \bar{y} = \dfrac{\int y dL}{L}, \bar{z} = \dfrac{\int z dL}{L}$

Area $\bar{x} = \dfrac{\int x d\mathrm{A}}{\mathrm{A}}, \bar{y} = \dfrac{\int y d\mathrm{A}}{\mathrm{A}}, \bar{z} = \dfrac{\int z d\mathrm{A}}{\mathrm{A}}$, Volume $\bar{x} = \dfrac{\int x d\mathrm{V}}{\mathrm{V}}, \bar{y} = \dfrac{\int y d\mathrm{V}}{\mathrm{V}}, \bar{z} = \dfrac{\int z d\mathrm{V}}{\mathrm{V}}$

Composite bodies $\bar{x} = \dfrac{\sum_{i=1}^{i=n} x_i v_i}{\mathrm{V}}, \bar{y} = \dfrac{\sum_{i=1}^{i=n} y_i v_i}{\mathrm{V}}, \bar{z} = \dfrac{\sum_{i=1}^{i=n} z_i v_i}{\mathrm{V}}$

- Position of centroid:

Circular arc $\bar{x} = \dfrac{r\sin\alpha}{\alpha}$, Semi-circular arc $\bar{x} = \dfrac{2r}{\pi}$,

Area of triangle $\bar{y} = \dfrac{h}{3}$, Area of circular sector $\bar{x} = \dfrac{2}{3}R\dfrac{\sin\alpha}{\alpha}$,

Semi-circular area $\bar{x} = \dfrac{4r}{\pi}$, Right circular solid cone $\bar{x} = \dfrac{3}{4}h$,

Right circular hollow cone $\bar{x} = \dfrac{2}{3}h$, Solid hemisphere $\bar{x} = \dfrac{3}{8}r$,

Thin hollow hemisphere $\bar{x} = \dfrac{r}{2}$

- Pappus's theorem for volume: The volume of a body of revolution is equal to the generating area times the distance traveled by centroid of the area while the body is being generated.

$$V = \theta\bar{y}A,\ \mathit{for}\ 0 \le \theta \le 2\pi$$

- Pappus's theorem for surface areas: The area of a surface of revolution is equal to the length of the generating curve times the distance traveled by the centroid of the curve while the surface is being generated.

$$A = \bar{y}\theta L,\ \mathit{for}\ 0 \le \theta \le 2\pi$$

- Area Moment of Inertia $I_x = \int y^2 dA$ and $I_y = \int x^2 dA$,
- Polar Moment of Inertia $I_z = I_x + I_y$, Radius of gyration $k = \sqrt{\frac{I}{A}}$
- Theorem of Parallel axes: $I_{AB} = I_G + Ay_c^2$, Theorem of Perpendicular Axes: $I_z = I_x + I_y$
- Area moment of inertia:

  Rectangle $(I_G)_x = \frac{bd^3}{12}$ and $(I_G)_y = \frac{db^3}{12}$, Triangle $I_{base} = \frac{bh^3}{12}$ and $I_G = \frac{bh^3}{36}$

  Circular Area $I_x = \frac{\pi R^4}{4}$

- Moment of inertia of body about $x-axis$ is $(I_o)_x = \int(y^2 + z^2)dm$, Moment of inertia of body about $y-axis$ is $(I_o)_y = \int(x^2 + z^2)dm$, Moment of intertia of body about $z-axis$ is $(I_o)_z = \int(x^2 + y^2)dm$
- Moment of Inertia:

  Uniform rod $I_G = \frac{ml^2}{12}$, Circular ring $I_x = \frac{1}{2}mR^2, I_y = \frac{1}{2}mR^2$ and $I_z = mR^2$

  Disc $I_x = \frac{1}{4}mR^2, I_y = \frac{1}{4}mR^2$ and $I_z = \frac{1}{2}mR^2$

  Plate $(I_G)_x = \frac{md^2}{12}$, $(I_G)_y = \frac{mb^2}{12}$ and $(I_G)_z = m\left(\frac{b^2+d^2}{12}\right)$

  Solid sphere $(I_G)_x = \frac{2}{5}mR^2$, Cone $(I_G)_x = \frac{3}{10}mr^2$ $(I_o)_y = \frac{3}{5}m\left[\frac{r^2}{4} + h^2\right]$

- Product of Inertia $I_{xy} = \int xydA$
- Parallel axis theorem for product of inertia $I_{xy} = \overline{I}_{x'y'} + \overline{x}\,\overline{y}A$
- Transformation equations for moments and product of inertia

$$I_{x'} = \left(\frac{I_x + I_y}{2}\right) + \left(\frac{I_x - I_y}{2}\right)\cos 2\theta - I_{xy}\sin 2\theta$$

$$I_{y'} = \left(\frac{I_x + I_y}{2}\right) - \left(\frac{I_x - I_y}{2}\right)\cos 2\theta + I_{xy}\sin 2\theta, \quad I_{x'y'} = \left(\frac{I_x - I_y}{2}\right)\sin 2\theta + I_{xy}\cos 2\theta$$

- Moment of inertia about principal axes

$$I_{max} = \left(\frac{I_x + I_y}{2}\right) + \frac{1}{2}\sqrt{(I_x - I_y)^2 + 4I_{xy}} \text{ and } I_{min} = \left(\frac{I_x + I_y}{2}\right) - \frac{1}{2}\sqrt{(I_x - I_y)^2 + 4I_{xy}}$$

REVIEW QUESTIONS

Q1. Define the term centroid and centre of gravity.

Q2. Define moment of inertia of an area.

Q3. State and explain

(a) parallel axes theorem

(b) perpendicular axes theorem concerning moment of inertia for a plane area.

Q4. Define moment of inertia of a mass. How radius of gyration of a body is expressed in terms of its mass moment of inertia?

Q5. Prove that the moment of inertia of a circular section about a horizontal axis (in the plane of the circular section) and passing through the centre of gravity of the section is given by $\pi d^4/64$.

Q6. Determine the centroid of the surface of a hemisphere with respect to its base.

or

Determine the centroid of a thin hemispherical shell with respect to its base.

[Ans; $\frac{r}{2}$].

Q7. Determine the coordinates of the centroid of the shaded area in Figure 6.Q7. The area removed is a triangle.

[Ans; $\overline{x} = 0,\ \overline{x} = 23.4\text{mm}$].

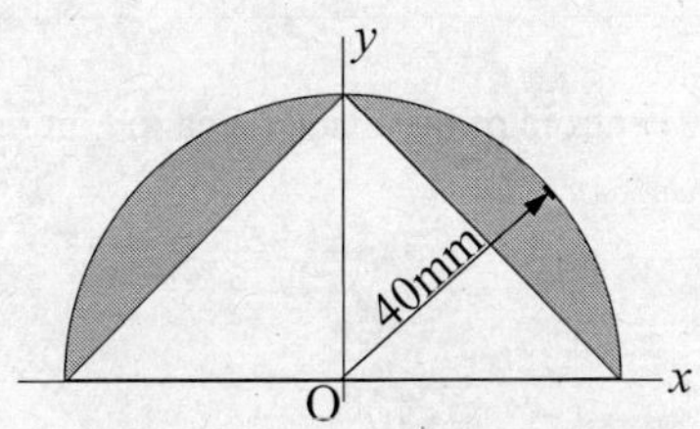

FIGURE 6.Q7 |

Q8. Determine the moment of inertia of a thin elliptical disk of mass m, shown in Figure 6.Q8.

$$\left[\text{Ans; } I_x = \frac{1}{4}mb^2, I_y = \frac{1}{4}ma^2, I_z = \frac{1}{4}m(a^2 + b^2)\right].$$

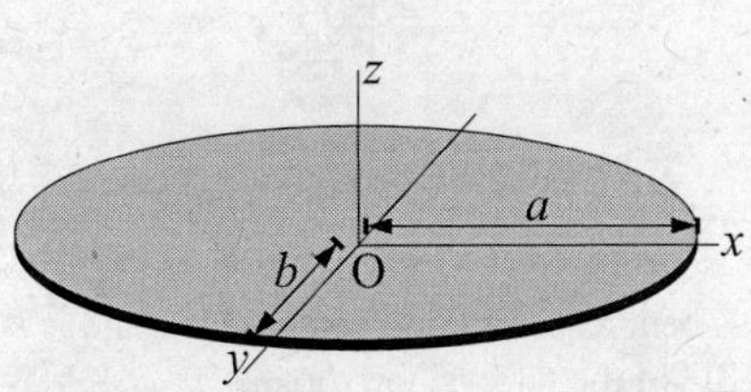

FIGURE 6.Q8 |

Q9. Determine the y-coordinate of the centroid of the shaded area shown in Figure 6.Q9.

$$\left[\text{Ans; } \overline{y} = \frac{14R}{9\pi}\right].$$

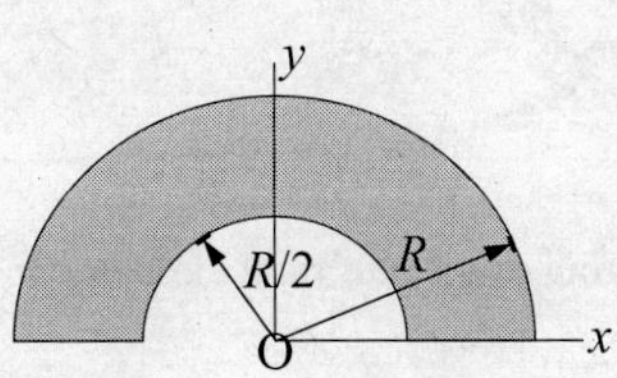

FIGURE 6.Q9 |

Q10. Determine the coordinates of the centroid of the shaded area in Figure 6.Q10.

[Ans; $\overline{x}$ = 19mm, $\overline{y}$ = 6mm].

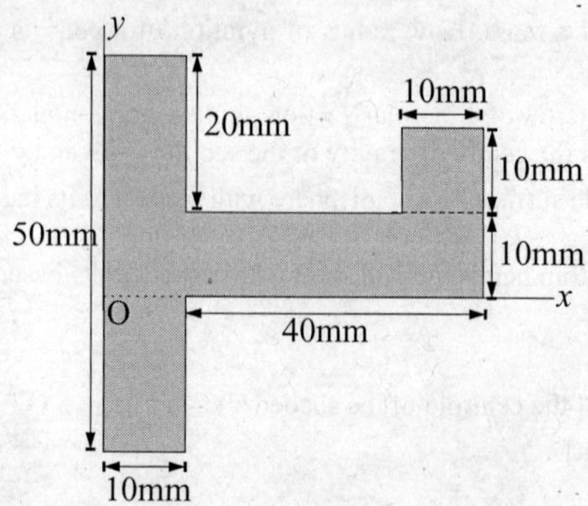

FIGURE 6.Q10 |

Q11. Determine the coordinates of the centroid of the shaded area in Figure 6.Q11 .

[Ans; $\overline{x}$ = 110.4mm, $\overline{y}$ = 0].

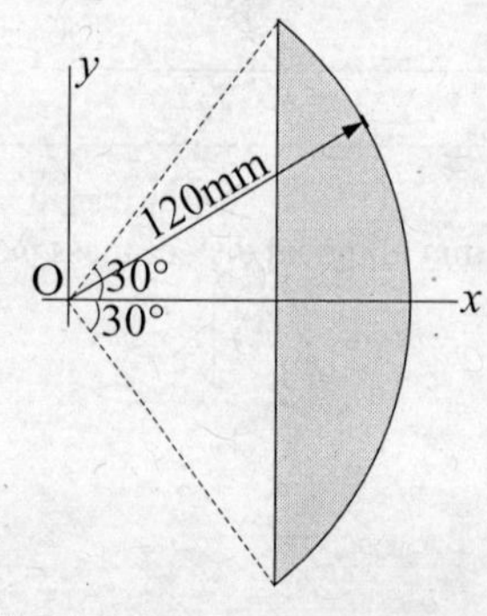

FIGURE 6.Q11 |

Q12. The homogeneous wire ABC is bent as shown in Figure 6.Q12 and is supported by a pin at B. Determine the length l for which portion BC of the wire is horizontal.

[Ans; $l = 2r$].

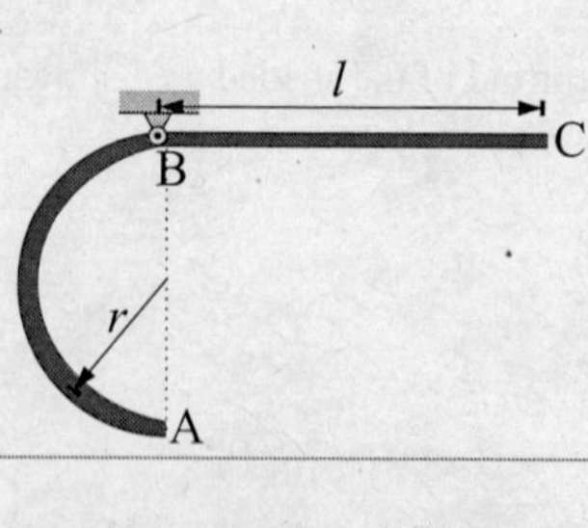

FIGURE 6.Q12 |

Q13. Determine the location of the centroid of the half right circular cone shown in Figure 6.Q13.

$$\left[\text{Ans; } \overline{x} = \left(\frac{3}{4}\right)h,\ \overline{y} = \left(\frac{r}{\pi}\right), \overline{z} = 0\right].$$

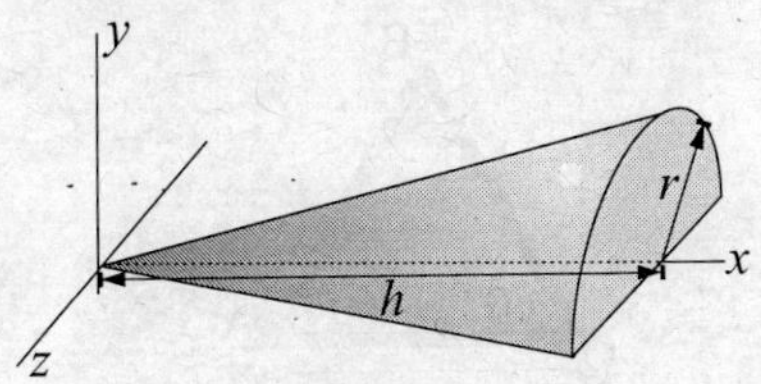

FIGURE 6.Q13 |

Q14. The homogeneous wire ABCD is bent as shown in Figure 6.Q14 and supported by a pin at C. Determine the length l for which the portion BCD of the wire is horizontal.

[Ans; 0.6 m].

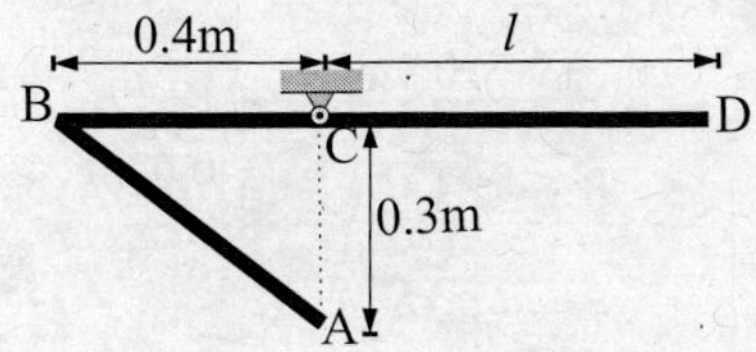

FIGURE 6.Q14 |

Q15. Determine the moment of inertia of a cylinder with respect to its y-axis (Fig. 6.Q15). The cylinder has a mass m.

$$\left[\text{Ans; } I = \left(\frac{1}{12}\right) m(3r^2 + 4h^2)\right].$$

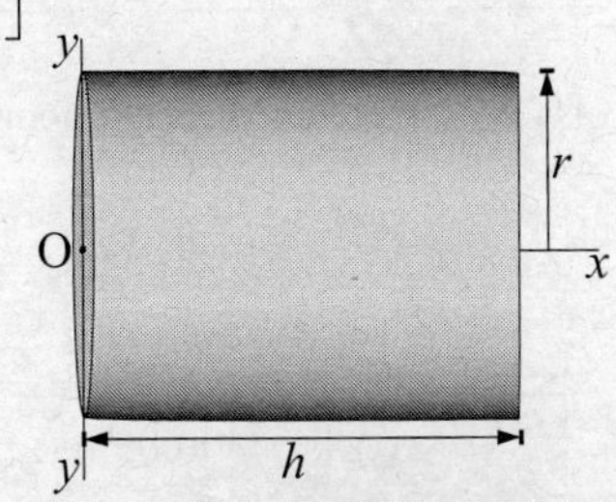

FIGURE 6.Q15 |

Q16. Determine the moment of inertia I_{xx} for a bent rod (Fig. 6.Q16). The rod has a mass of 2 kg/m.

[Ans; I_x = 0.341 kg m²].

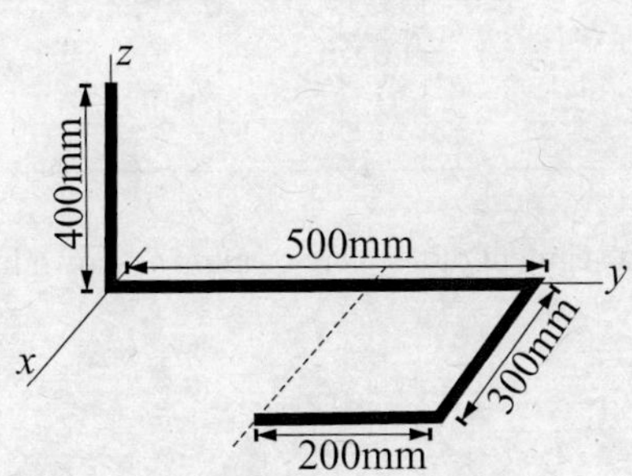

FIGURE 6.Q16 |

Q17. Determine the moment of inertia of the three-rod assembly shown in Figure 6.Q17 about an axis that is perpendicular to the page and passes through O. Mass of each rod is m.

$$\left[\text{Ans; } I = \left(\frac{3}{2}\right) ml^2\right].$$

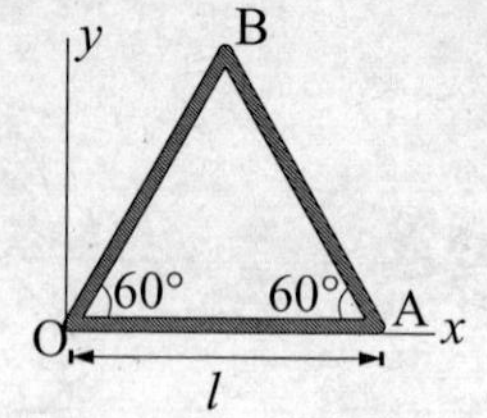

FIGURE 6.Q17 |

Q18. Find the area moment of inertia of a cross-section of beam about x-axis passing through a centroid (Figure 6.Q18).

[Ans; $I_x = 2744.3 \times 10^{-8}$ m^4].

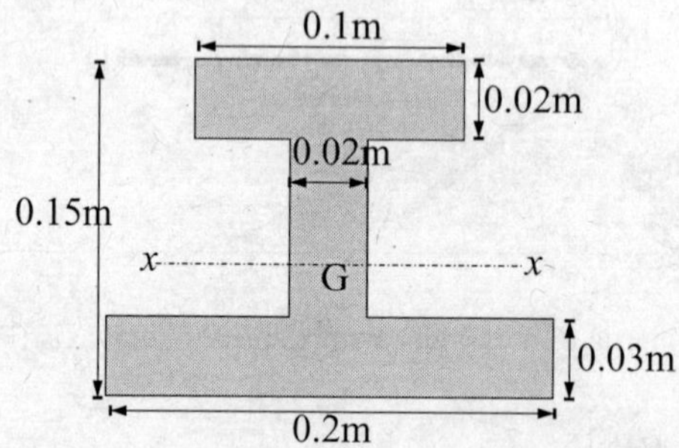

FIGURE 6.Q18 |

Q19. Find the area moment of inertia of a cross-section of beam about x-axis passing through a centroid (Fig. 6.Q19).

[Ans; $I_x = 5312.5 \times 10^{-8}$ m^4].

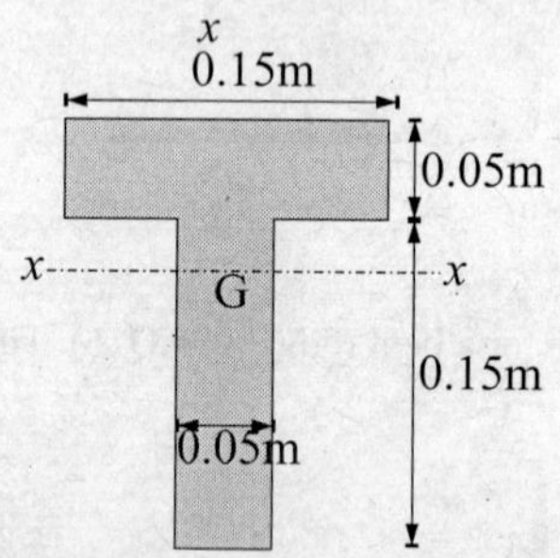

FIGURE 6.Q19 |

Q20. Find the moment of inertia of a segment of a sphere of mass m shown in Figure 6.Q20.

$$\left[\text{Ans; } I_x = \left(\frac{53}{200}\right) mr^2\right].$$

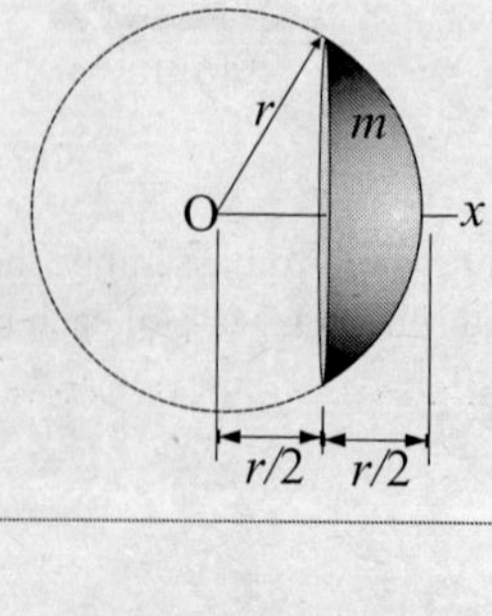

FIGURE 6.Q20 |

Q21. Find the moment of inertia of a solid cone, shown in Figure 6.Q21, about its z-axis.

$$\left[\text{Ans; } I_y = \frac{3}{5} m\left(\frac{r^2}{4} + h^2\right)\right].$$

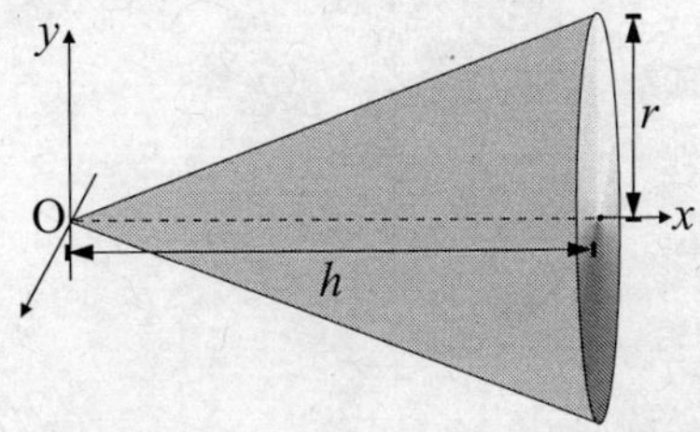

FIGURE 6.Q21 |

MULTIPLE CHOICE QUESTIONS

Q1. A thin wire of length L and uniform linear mass density ρ is bent into a circular loop with centre O as shown in Figure 6.MQ1. The moment of inertia of the loop about the x-axis is

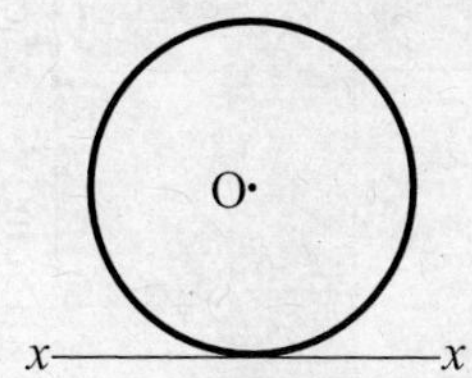

FIGURE 6.MQ1 |

(A) $\dfrac{\rho L^3}{8\pi^3}$ (B) $\dfrac{\rho L^3}{16\pi^3}$ (C) $\dfrac{5\rho L^3}{16\pi^2}$ (D) $\dfrac{3\rho L^3}{8\pi^2}$

[Ans: D]

Q2. A circular plate of diameter a is kept in contact with a square plate of edge a as shown in Figure 6.MQ2. The density of the material and the thickness are same everywhere. The centre of mass of the composite system will be

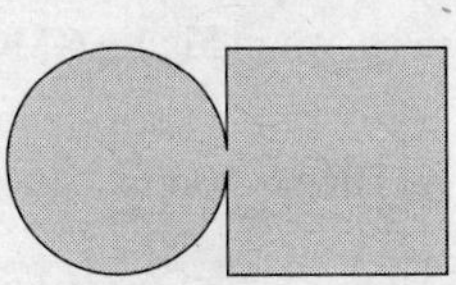

FIGURE 6.MQ2 |

(A) Inside the circulars plate (B) Inside the square plate
(C) At the point of contact (D) Outside the system

[Ans: B]

Q3. A square thin sheet of edge length L and uniform mass per unit are σ is used to form a hollow cylinder. The moment of inertia of this cylinder about the central axis is

(A) $\dfrac{2\sigma L^4}{\pi^2}$ (B) $\dfrac{\sigma L^4}{4\pi^2}$ (C) σL^2 (D) $\dfrac{\sigma L^4}{3\sqrt{2}\pi^2}$

[Ans: B]

Q4. A disc has mass $9M$. A hole of radius $R/3$ is cut from it as shown in Figure 6.MQ4. The moment of inertia of the remaining part about an axis passing through the centre O of the disc and perpendicular to the plane of the disc is

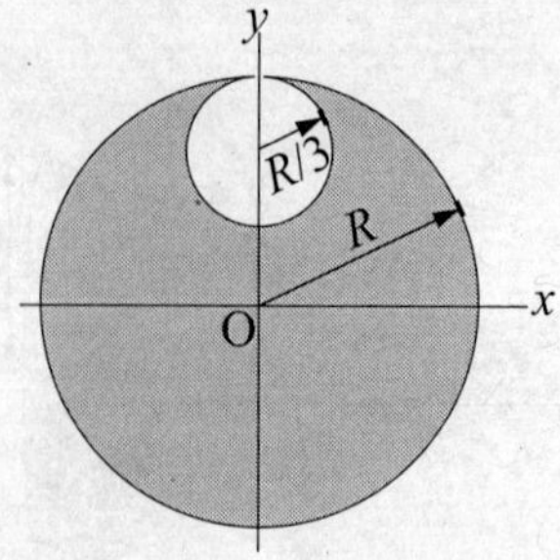

FIGURE 6.MQ4 |

(A) $8MR^2$ (B) $4MR^2$ (C) $\frac{40}{9}MR^2$ (D) None of these

[Ans: B]

Q5. The coordinate of centroid of shaded area shown in Figure 6.MQ5 is

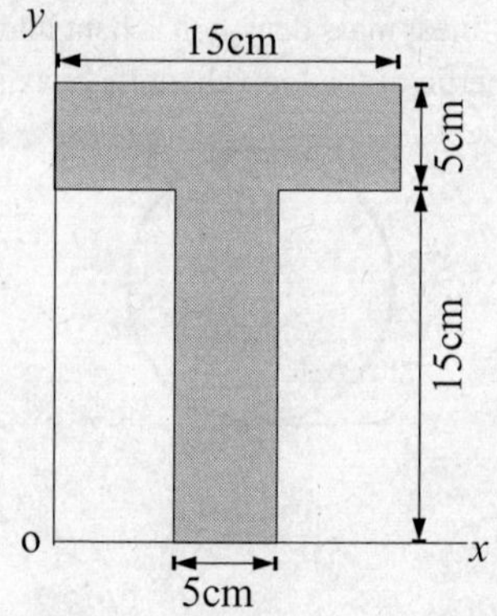

FIGURE 6.MQ5

(A) (7.5 cm, 12.9 cm) (B) (7.5 cm, 12.5 cm)
(C) (7.5 cm, 14 cm) (D) (7.5 cm, 8 cm)

[Ans: B]

Q6. The coordinate of centroid of shaded area shown in Figure 6.MQ6 is

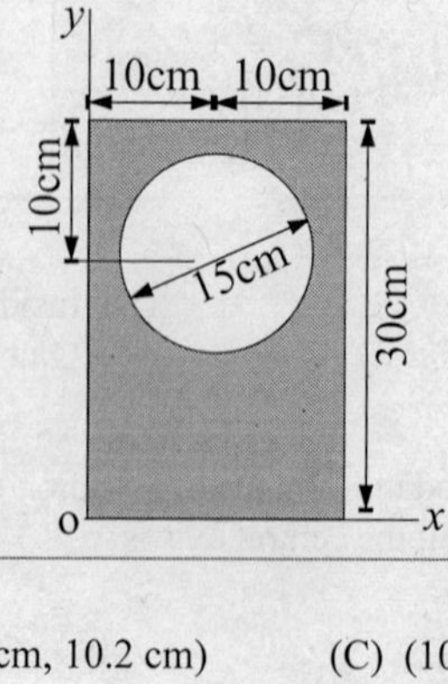

FIGURE 6.MQ6

(A) (10 cm, 12.9 cm) (B) (10 cm, 10.2 cm) (C) (10 cm, 9.5 cm) (D) (10 cm, 14.5 cm)

[Ans: A]

Q7. The coordinate of centroid of shaded area enclosed by a straight line and curve shown in Figure 6.MQ7 is

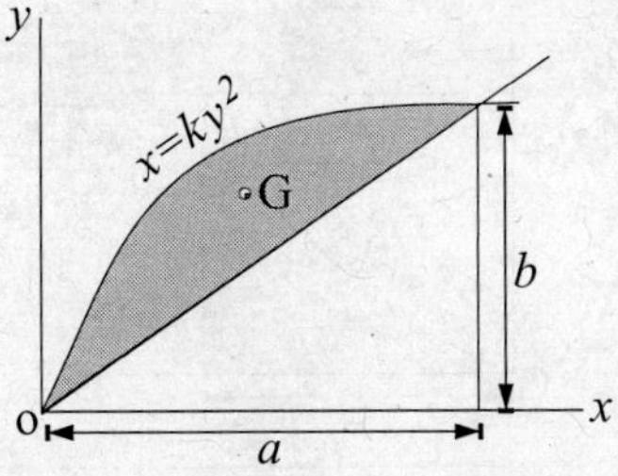

FIGURE 6.MQ7 |

(A) $\left(\frac{a}{5},\frac{b}{2}\right)$ (B) $\left(\frac{2a}{5},\frac{b}{2}\right)$ (C) $\left(\frac{2a}{5},\frac{3b}{2}\right)$ (D) $\left(\frac{6a}{5},\frac{3b}{2}\right)$

[Ans: B]

Q8. The *y*-coordinate of centroid of shaded area shown in Figure 6.MQ8 is

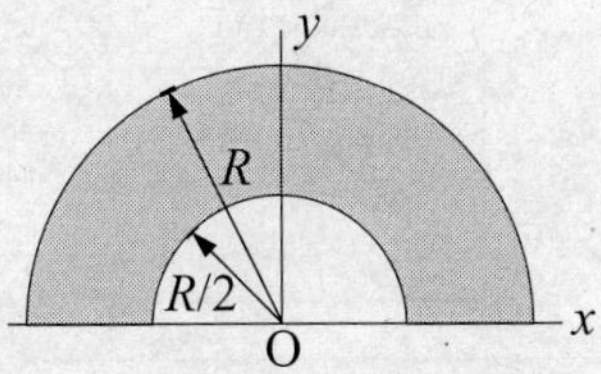

FIGURE 6.MQ8

(A) $\frac{14R}{9\pi}$ (B) $\frac{R}{9\pi}$ (C) $\frac{4R}{9\pi}$ (D) $\frac{5R}{9\pi}$

[Ans: A]

Q9. The moment of inertia of the shaded area in Figure 6.MQ9 about *x*-axis is

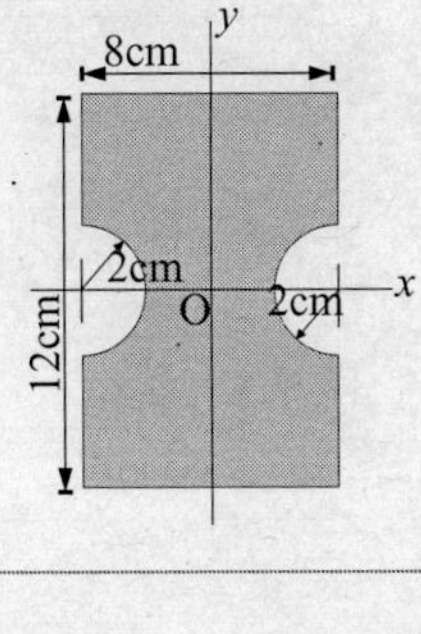

FIGURE 6.MQ9

(A) 739 cm^4 (B) 839.43 cm^4 (C) 129.43 cm^4 (D) 1139.43 cm^4

[Ans: D]

Q10. The moment of inertia of the area under the curve $y = kx^2$ shown in Figure 6.MQ10 about x-axis is

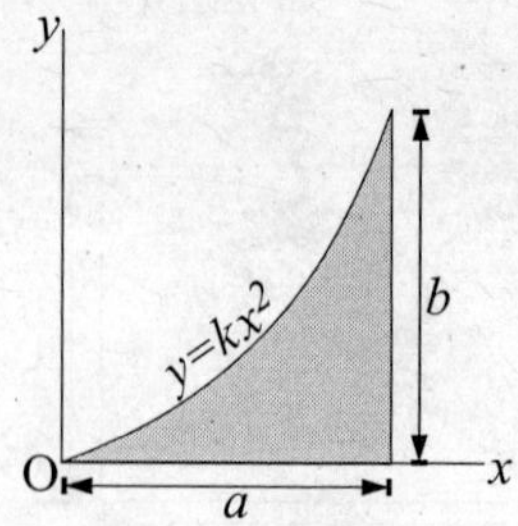

FIGURE 6.MQ10 |

(A) $I_x = \dfrac{ab^3}{21}$ (B) $I_x = \dfrac{ab^3}{12}$ (C) $I_x = \dfrac{ab^3}{9}$ (D) $I_x = \dfrac{ab^3}{4}$

[Ans: A]

Q11. The moment of inertia of the quarter circular area in Figure 6.MQ11 about y-axis is

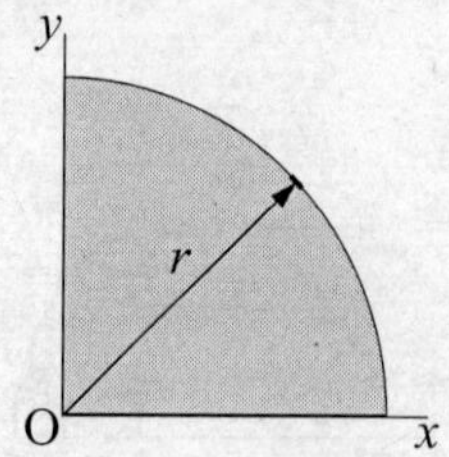

FIGURE 6.MQ11 |

(A) $I_y = \dfrac{\pi r^4}{4}$ (B) $I_y = \dfrac{\pi r^4}{16}$ (C) $I_y = \dfrac{\pi r^4}{2}$ (D) $I_y = \dfrac{\pi r^4}{32}$

[Ans: B]

Q12. The moment of inertia of the elliptical area in Figure 6.MQ12 about x-axis is

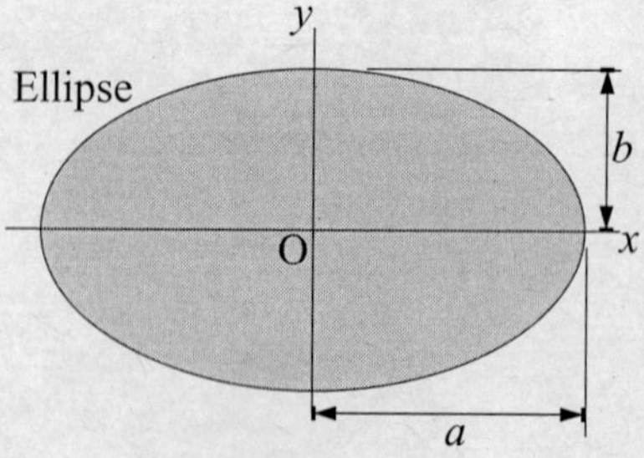

FIGURE 6.MQ12 |

(A) $I_x = \dfrac{\pi ab^3}{2}$ (B) $I_x = \dfrac{\pi ab^3}{3}$ (C) $I_x = \dfrac{\pi ab^3}{4}$ (D) $I_x = \dfrac{\pi ab^3}{12}$

[Ans: C]

Q13. The moment of inertia of the regular hexagonal area of side a (Figure 6.MQ13) about its central x-axis is

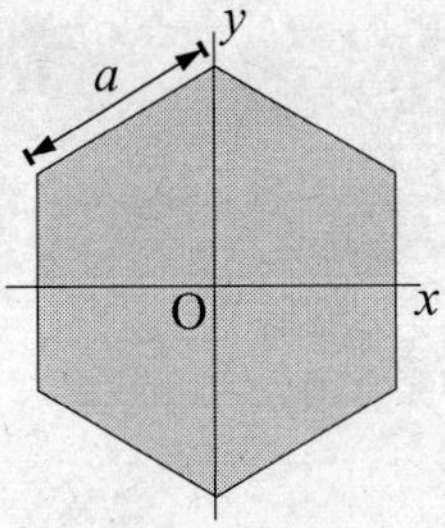

FIGURE 6.MQ13

(A) $I_x = \frac{4}{15}a^4$ (B) $I_x = \frac{5\sqrt{3}}{16}a^4$ (C) $I_x = \frac{5\sqrt{3}}{4}a^4$ (D) $I_x = \frac{15}{16}a^4$

[Ans: B]

Q14. A thin uniform rectangular plate of length 2 m and width 1 m is suspended from a corner as shown in Figure 6.MQ14. The angle θ will be

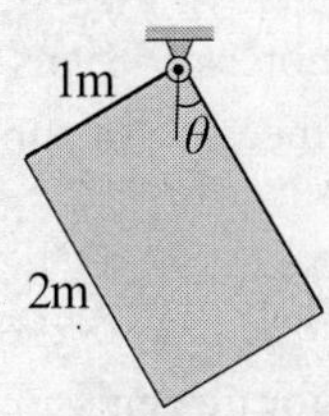

FIGURE 6.MQ14

(A) 45° (B) 26.5° (C) 40.7° (D) None of these

[Ans: B]

Q15. A uniform rod of length $3l$ and mass $3m$ is bent into an equilateral triangle shape as shown in Figure 6.MQ15. The moment of inertia of the rod about the axis passing through A and perpendicular to the plane will be

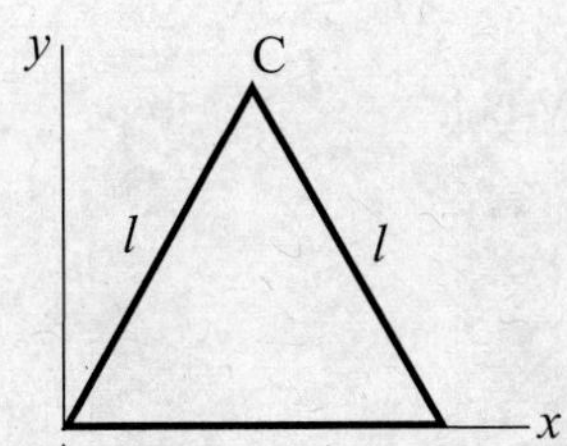

FIGURE 6.MQ15

(A) $(I_A)_z = \frac{5}{12}ma^2$ (B) $(I_A)_z = \frac{1}{2}ma^2$

(C) $(I_A)_z = \frac{3}{2}ma^2$ (D) $(I_A)_z = \frac{4}{3}ma^2$

[Ans: C]

7 Belt and Rope Drives

INTRODUCTION

Power is transmitted from one shaft to another shaft by means of belts, ropes, chains and gears. The salient features of belts and ropes are as follows:

- Belts and ropes are used in cases where the distance between the shafts is large. For small distances, gears are preferred.
- Belts and ropes are flexible type of connectors, i.e., they bend easily.
- Belts and ropes transmit power due to friction between them and the pulleys. If the power transmitted exceeds the force of friction, the belt or rope slips over the pulley.
- Belts and ropes are strained during motion as tensions are developed in them.
- Velocity ratio does not remain constant because of slip and creep.

TYPES OF BELTS

There are basically four types of belts used for transmission of power (Fig. 7.1):

1. flat belt,
2. V-belt,
3. circular belt or rope,
4. ribbed belt,
5. toothed or timing belt.

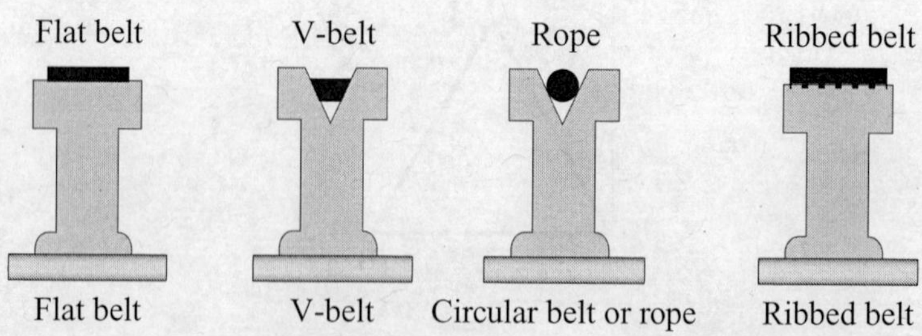

FIGURE 7.1 |

Of these, flat belts and V-belts are widely used. The selection of belt drive depends upon (i) power to be transmitted, (ii) speed of driver and driven shafts, (iii) reduction ratio, (iv) centre distance between shafts, (v) shaft relationship, (vi) positive drive requirements, (vii) service conditions and (viii) space available.

TYPES OF BELT DRIVES

OPEN BELT DRIVE

Open belt drive (Fig. 7.2) is used with shafts that are arranged in parallel and rotating in the same direction. When the shafts are far apart, tight side of the belt must be the lower one.

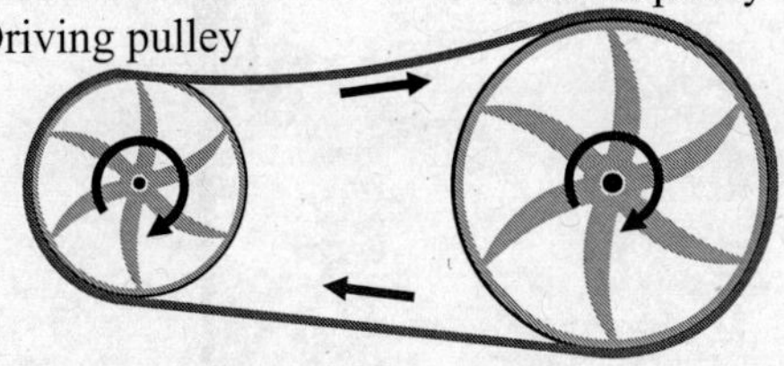

Open belt drive

FIGURE 7.2 |

CROSS OR TWIST BELT DRIVE

Cross or twist belt drive (Fig. 7.3) is used with shafts that are arranged in parallel and rotating in opposite directions. At the point where the belt crosses, it rubs against itself and wears. To avoid excessive wear, the shafts should have sufficient distance between them (a distance of more than $20b$, where b is the belt width) and operated at a speed of 15 m/s.

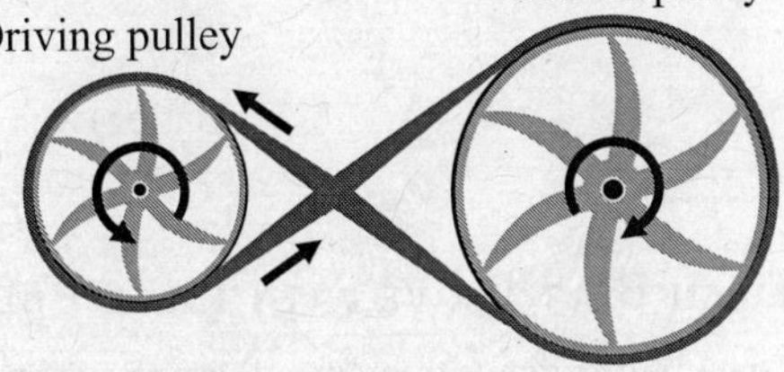

Cross belt drive

FIGURE 7.3 |

QUARTER TWIST BELT DRIVE

Quarter twist belt drive (Fig. 7.4) is used with shafts that are at right angles and rotating in one definite direction. In order to prevent the belt from leaving the pulleys, the latter should be at least $1.4b$ wide.

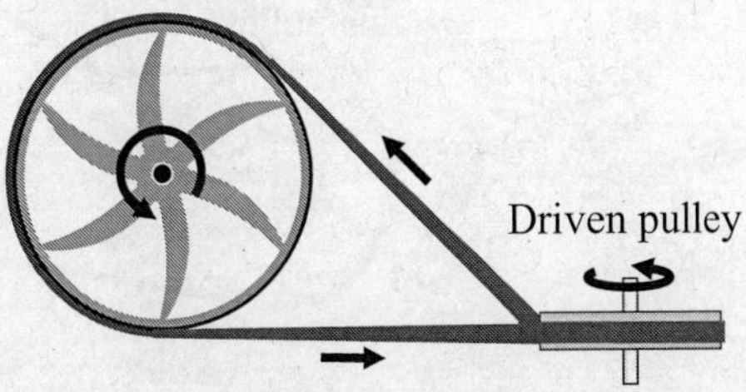

Quarter twist belt drive

FIGURE 7.4 |

OPEN BELT DRIVE WITH IDLER PULLEYS OR JOCKEY PULLEY DRIVE

Open belt drive with idler pulleys (Fig. 7.5) is used when an open belt drive cannot be employed because of the small arc of contact on the smaller pulley and when the required belt tension cannot be obtained by any other means.

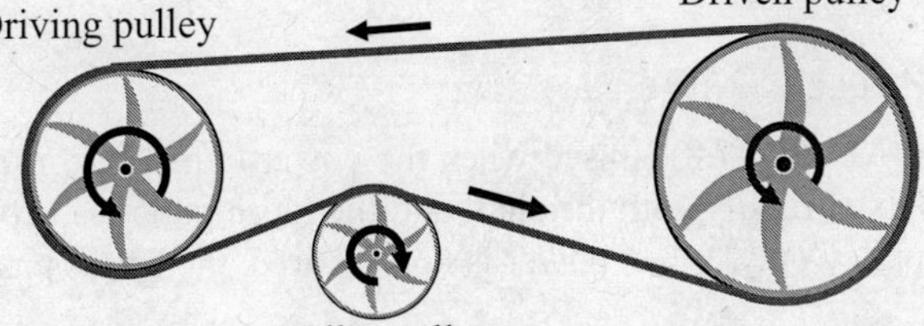

Open belt drive with an idler pulley

FIGURE 7.5 |

QUARTER TWIST BELT DRIVE WITH GUIDE PULLEY

Quarter twist belt drive with guide pulley (Fig. 7.6) is used with shafts that are at right angles when the pulleys cannot be arranged as in quarter twist belt drive or when a reversible action is required.

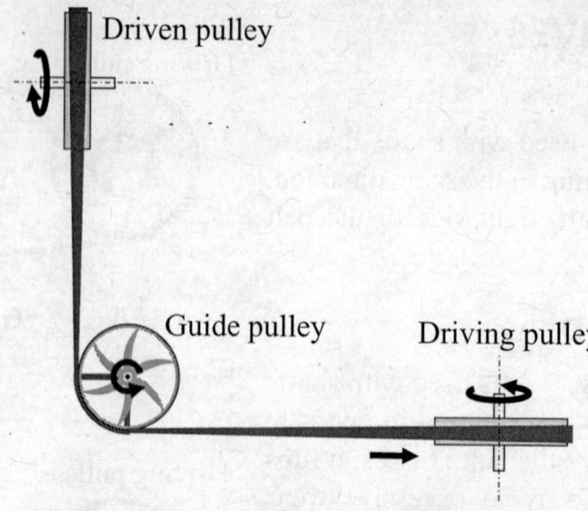

Quarter twist belt drive with guide pulley

FIGURE 7.6 |

OPEN BELT DRIVE WITH IDLER PULLEYS AND GUIDE PULLEYS

Open belt drive with many idler pulleys and driving pulleys (Fig. 7.7) is used to transmit power from one shaft to several shafts arranged in parallel.

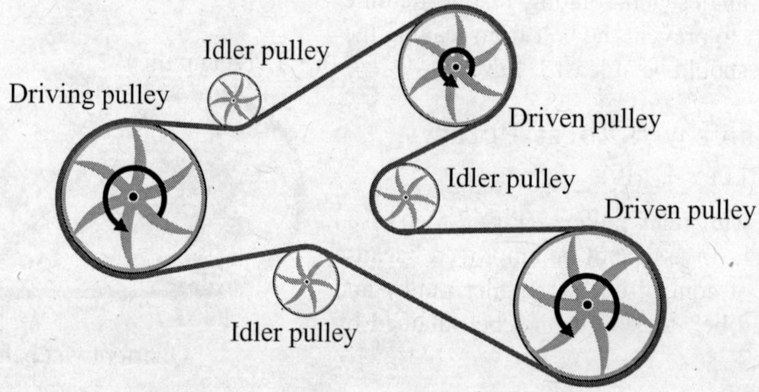

Open belt drive with many idler and driving pulleys

FIGURE 7.7 |

LOOSE AND FAST PULLEY DRIVE

Loose and fast pulley drive (Fig. 7.8) is used when the power transmitted is required to be intermittent. The fast pulley is fixed rigidly with the shaft and the loose pulley is kept loose. The power can be transmitted by fast pulley only. When power is not required, the belt is pushed to the loose pulley from the fast pulley.

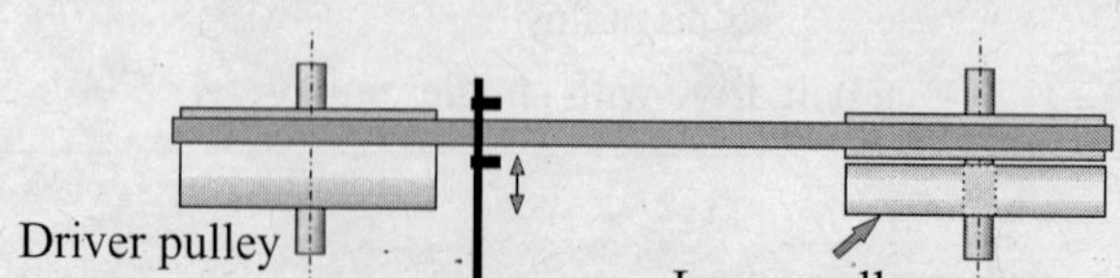

Loose and fast pulley drive

FIGURE 7.8 |

BELT MATERIALS

A belt material should have a high coefficient of friction, strength, flexibility and durability. Commonly used materials for belts are

- leather belts
- rubber belts
- batana belts
- fabric and canvas belts
- camel's hair belts.

VELOCITY RATIO

The ratio between the velocities of driver and driven is known as velocity ratio.

SIMPLE BELT DRIVE

Let N_1 be the rotational speed of the driving pulley, N_2 be the rotational speed of the driven pulley, d_1 be the diameter of the driving pulley, d_2 be the diameter of the driven pulley, and t be the thickness of the belt, as shown in Figure 7.9.

Neglecting any slip between the belt and the pulleys and also considering the belt to be inelastic, we can state that speed of belt on the driving pulley is equal to speed of belt on the driven pulley:

$$\left(d_1 + 2\frac{t}{2}\right)N_1 = \left(d_2 + 2\frac{t}{2}\right)N_2$$

or velocity ratio (VR) $= \dfrac{N_2}{N_1} = \dfrac{(d_1 + t)}{(d_2 + t)}$

$$\frac{N_2}{N_1} = \frac{d_1 + t}{d_2 + t}$$

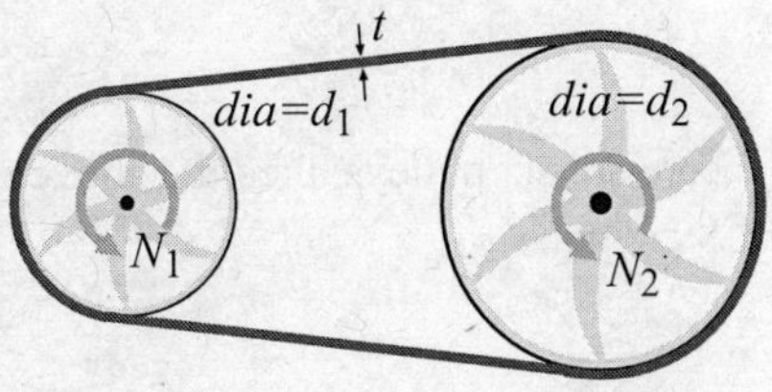

FIGURE 7.9 |

COMPOUND BELT DRIVE

The power is transmitted from one shaft to another through a number of pulleys, as shown in Figure 7.10. Such an arrangement is known as compound belt drive.

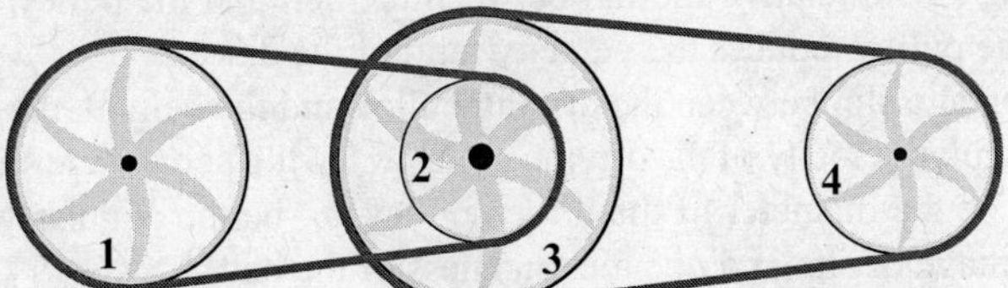

FIGURE 7.10 |

Figure 7.10 shows compound belt drive, in which pulley 1 drives pulley 2. However, pulleys 2 and 3 are keyed to the same shaft; therefore, pulley 1 also drives pulley 3. Pulley 3 drives pulley 4.

Let d_1 and N_1 be the diameter and speed of pulley 1 in rpm, respectively, d_2 and N_2 be the diameter and speed of pulley 2, respectively, d_3 and N_3 be the diameter and speed of pulley 3, and d_4 and N_4 be the diameter and speed of pulley 4, respectively.

We know that the velocity ratio of pulleys 1 and 2,

$$\frac{N_2}{N_1} = \frac{d_1}{d_2} \tag{i}$$

(thickness of belt is not considered).

Similarly, velocity ratio of pulleys 3 and 4.

$$\frac{N_4}{N_3} = \frac{d_3}{d_4} \tag{ii}$$

Multiplying Equations (i) and (ii), we have

$$\frac{N_2}{N_1} \times \frac{N_4}{N_3} = \frac{d_1 \times d_3}{d_2 \times d_4}$$

But $N_2 = N_3$, as pulleys 2 and 3 are keyed to the same shaft,

$$\frac{N_4}{N_1} = \frac{d_1 \times d_3}{d_2 \times d_4}$$

If there are six pulleys, then the above equation can be written as

$$\frac{N_6}{N_1} = \frac{d_1 \times d_3 \times d_5}{d_2 \times d_4 \times d_6}$$

General formula is given by

$$\frac{\text{speed of last follower}}{\text{speed of first driver}} = \frac{\text{product of diameter of drivers}}{\text{product of diameter of followers}},$$

$$\frac{N_{\text{last}}}{N_1} = \frac{d_1 \times d_3 \times d_5 \times \cdots}{d_2 \times d_4 \times d_6 \times \cdots}$$

EFFECT OF SLIP ON VELOCITY RATIO

The term slip is defined as the relative motion between the belt and the pulley. The presence of slip between the belt and the pulley reduces the velocity ratio of the drive.

Let s_1 be the percentage slip between the driving pulley and the belt, D_1 the diameter of the driving pulley, ω_1 be the angular velocity of the driving pulley, s_2 be the percentage slip between the driven pulley and the belt, d_2 be the diameter of the driven pulley, ω_2 be the angular velocity of the driven pulley, s be total percentage of slip and t be the thickness of the belt.

$$\text{Peripheral speed of driving pulley} = \omega_1\left(\frac{d_1 + t}{2}\right).$$

$$\text{Speed of belt on the driving pulley} = \omega_1\left(\frac{d_1 + t}{2}\right)\left(\frac{100 - s_1}{100}\right).$$

This is also the speed of the belt on the driven pulley,

$$\text{Peripheral speed of driving pulley} = \omega_1 \cdot \left(\frac{d_1 + t}{2}\right)\left(\frac{100 - s_1}{100}\right)\left(\frac{100 - s_2}{100}\right).$$

$$\text{Speed of belt on the driving pulley} = \omega_2 \left(\frac{d_2 + t}{2}\right).$$

From which,

$$\omega_2 \left(\frac{d_2 + t}{2}\right) = \omega_1 \left(\frac{d_1 + t}{2}\right)\left(\frac{100 - s_1}{100}\right)\left(\frac{100 - s_2}{100}\right)$$

or

$$\frac{\omega_2}{\omega_1} = \left(\frac{d_1 + t}{d_2 + t}\right)\left(\frac{100 - s_1}{100}\right)\left(\frac{100 - s_2}{100}\right)$$

$$\frac{N_2}{N_1} = \left(\frac{d_1 + t}{d_2 + t}\right)\left(1 - \frac{s_1}{100}\right)\left(1 - \frac{s_2}{100}\right)$$

If s is the total percentage slip of the driven and driver pulleys,

$$\frac{N_2}{N_1} = \left(\frac{d_1 + t}{d_2 + t}\right)\left(1 - \frac{s}{100}\right)$$

where

$$\left(1 - \frac{s}{100}\right) = \left(1 - \frac{s_1}{100}\right)\left(1 - \frac{s_2}{100}\right) \quad \text{or} \quad s = s_1 + s_2 - \frac{s_1 s_2}{100}$$

$$s = s_1 + s_2 - \frac{s_1 s_2}{100}$$

CROWNING OF PULLEYS

A slight misalignment of the shaft or a small defect in the belt induces the belt to slip off the pulleys. This can be avoided by the use of flanged flat pulleys or those with convex rims called crowned pulleys (Fig. 7.11). The former wear out edges of the belt due to rubbing and hence are unpopular.

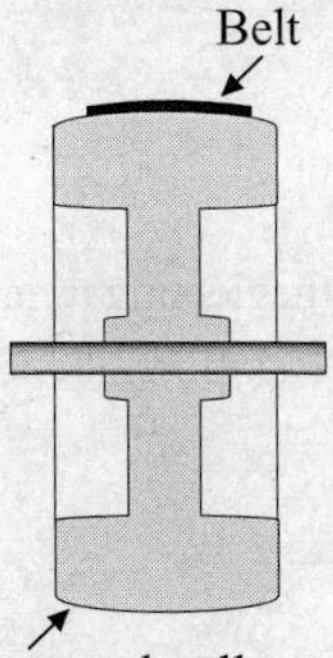

FIGURE 7.11

LENGTH OF BELT

LENGTH OF OPEN BELT DRIVE

Let r be the radius of the smaller pulley, R be the radius of the large pulley, C be the distance between O_1 and O_2 (Fig. 7.12).

Total length of the belt,

$$L = \text{arc AB} + \text{BC} + \text{arc CD} + \text{DA}$$

But arc AB = 2 arc AE, BC = AD and arc DC = 2 arc DF. Therefore,

$$L = 2(\text{arc AE} + \text{AD} + \text{arc DF})$$

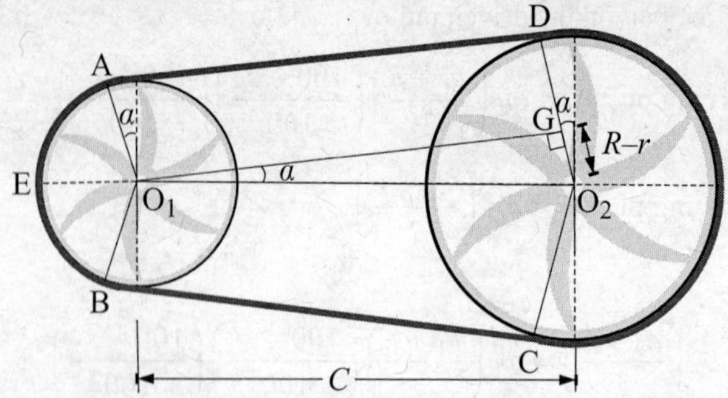

FIGURE 7.12 |

Substituting arc AE = $[(\pi/2) - \alpha]r$, arc DF = $[(\pi/2) + \alpha]R$ and AD = $C \cos \alpha$, we have

$$L = \pi(R + r) + 2\alpha(R - r) + 2C \cos \alpha$$

Through O_1 draw O_1G parallel to AD. From the geometry of the figure, we find O_1G will be perpendicular to O_2D. In triangle O_1O_2G $\sin \alpha = (R - r)/C$. Since the angle α is very small, $\sin \alpha \approx \alpha = (R - r)/C$:

$$AD = O_1G = \sqrt{(O_1O_2)^2 - (GO_2)^2}$$

$$= \sqrt{C^2 - (R - r)^2} = C\left[1 - \left(\frac{R - r}{C}\right)^2\right]^{-1/2}$$

Expanding by binomial theorem,

$$AD = C\left(1 - \frac{1}{2}\left(\frac{R - r}{C}\right)^2 + \cdots\right)$$

or

$$AD = C - \frac{1}{2}\frac{(R - r)^2}{C}$$

(neglecting terms of higher order).

Therefore, the length of the belt,

$$L = 2\left[\left(\frac{\pi}{2} - \alpha\right)r + C - \frac{1}{2}\frac{(R - r)^2}{C} + \left(\frac{\pi}{2} + \alpha\right)R\right]$$

or

$$L = \pi(R + r) + 2\alpha(R - r) + 2C - \frac{(R - r)^2}{C}$$

Substituting $\alpha = (R - r)/C$, we have

$$L = \pi(R + r) + 2\frac{(R - r)^2}{C} + 2C - \frac{(R - r)^2}{C}$$

$$L = \pi(R + r) + 2C + \frac{(R - r)^2}{C}$$

Length of Cross Belt Drive

Let r be radius of the smaller pulley, R be the radius of the large pulley, C be the distance between O_1 and O_2, as shown in Figure 7.13

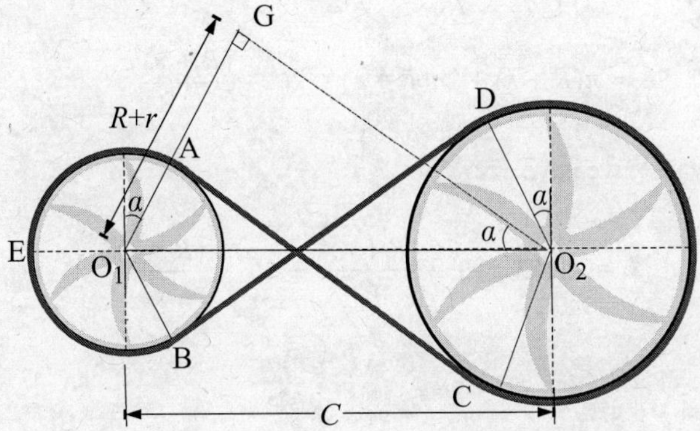

FIGURE 7.13 |

Total length of the belt

$$L = \text{arc AB} + \text{BD} + \text{arc DC} + \text{CA}$$

But arc AB = 2 arc AE, BD = AC and arc DC = 2 arc DF. Therefore,

$$L = 2(\text{arc AE} + \text{AD} + \text{arc DF})$$

Substituting arc AE $= [(\pi/2) + a]r$, arc DF $= [(\pi/2) + a]R$, and AD $= C\cos a$ we have

$$L = \pi(R + r) + 2a\,(R + r)2C\,\cos a$$

Through O_2 draw O_1G parallel to AC. From the geometry of the figure, we find O_1G will be perpendicular to O_2D. In triangle O_1O_2G,

$$\sin a = \frac{R + r}{C}$$

Since the angle α is very small,

$$\sin a \approx a = \frac{R - r}{C}$$

$$\text{AD} = O_2G = \sqrt{(O_1O_2)^2 - (GO_1)^2}$$

$$= \sqrt{C^2 - (R + r)^2} = C\left[1 - \left(\frac{R + r}{C}\right)^2\right]^{-1/2}$$

Expanding by binomial theorem,

$$\text{AD} = C\left[1 - \frac{1}{2}\left(\frac{R + r}{C}\right)^2 + \cdots\right]$$

or

$$\text{AD} = C - \frac{1}{2}\frac{(R + r)^2}{C}$$

(neglecting terms of higher order).
Therefore, the length of belt

$$L = 2\left[\left(\frac{\pi}{2} + a\right)r + C - \frac{1}{2}\frac{(R+r)^2}{C} + \left(\frac{\pi}{2} + a\right)R\right]$$

or

$$L = \pi(R+r) + 2a(R+r) + 2C - \frac{(R+r)^2}{C}$$

Substituting $a = (R-r)/C$, we have

$$L = \pi(R+r) + 2\frac{(R+r)^2}{C} + 2C - \frac{(R+r)^2}{C}$$

$$L = \pi(R+r) + 2C + \frac{(R+r)^2}{C}$$

CONE (STEPPED) PULLEYS

A stepped or cone pulley drive, as shown in Figure 7.14, is used for changing the speed of the driven shaft while the main or driving shaft runs at a constant speed. This is accomplished by shifting the belt from one part of the steps to the other.

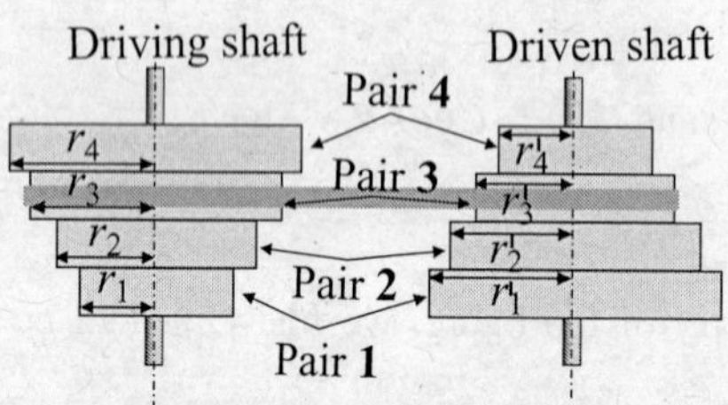

FIGURE 7.14 |

Let N be the speed of the driving shaft (constant), N'_n be the speed of the driven shaft when the belt is on the nth step, r_n be the radius of the nth step of the driving pulley, r'_n be the radius of the nth step of the driven pulley. The subscript n denotes 1, 2, 3, …, n.

The ratio of speeds of driving shaft to driven shaft is inversely proportional to the ratio of their pulley radii, i.e.,

$$\frac{N'_1}{N} = \frac{r_1}{r'_1}$$

Thus, to get speed N'_1 of the driven shaft from the first pair of steps pulleys, dimensions of r_1 and r'_1 can be chosen convenient to the design.

For the second pair of steps,

$$\frac{N'_2}{N} = \frac{r_2}{r'_2}$$

Similarly, for nth pair of steps,

$$\frac{N'_n}{N} = \frac{r_n}{r'_n}$$

However, it is always desired that the same belt is used on all the pairs of steps pulleys. To fulfil this condition, the length of the belt has to be same for all the pairs of steps, i.e.,

$$L_1 = L_2 = \cdots L_n$$

For Open Belt System

$$L = \pi(r_1 + r_1') + 2C + \frac{(r_1 - r_1')^2}{C} = \pi(r_2 + r_2') + 2C + \frac{(r_2 - r_2')^2}{C} = \cdots$$

$$\cdots = \pi(r_n + r_n') + 2C + \frac{(r_n - r_n')^2}{C}$$

For Cross-Belt System

$$L = \pi(r_1 + r_1') + 2C + \frac{(r_1 + r_1')^2}{C} = \pi(r_2 + r_2') + 2C + \frac{(r_2 + r_2')^2}{C} = \cdots$$

$$\cdots = \pi(r_n + r_n') + 2C + \frac{(r_n + r_n')^2}{C}$$

or
$$r_1 + r_1' = r_2 + r_2' = \cdots = r_n + r_n'$$

POWER TRANSMITTED

Power transmitted by a belt or a rope is given by a mathematical expression,

$$P = \frac{(T_1 - T_2)v}{1000},$$

where P is the power, in kW, T_1 and T_2 are the tensions on tight and slack sides of belt, respectively, in Newton, v is the linear velocity of belt, in m/s,

and
$$v = \frac{\pi dN}{60},$$

where d is the diameter of pulley, in m, and N is rpm of the pulley.

BELT TENSIONS RATIO

Flat Belt

Consider a flat belt passing over a driven pulley. Let T_1 and T_2 be the belt tensions on the tight and slack sides, respectively, when the belt is just about to slide.

Consider a small element AB of the belt. Figure 7.15 shows the free body diagram of the element of the belt. The tension increases from T at the angle α to $T + dT$ at the angle $a + da$. The friction force μ dR will act tangentially to the surface, resisting the slipping of the elementary belt on the drum (Fig. 7.16). We write the equation of equilibrium for the element AB of the belt.

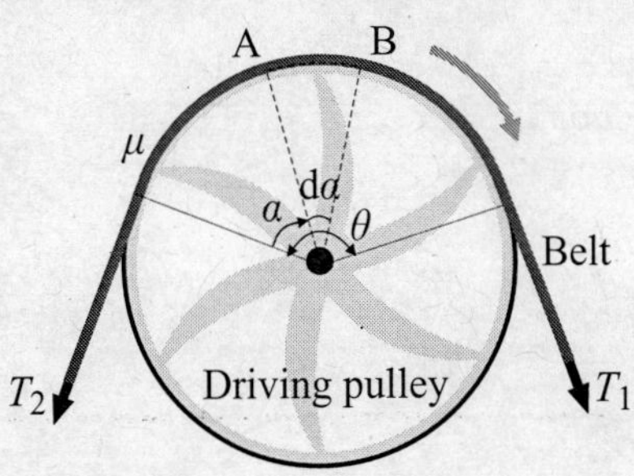

Figure 7.15 |

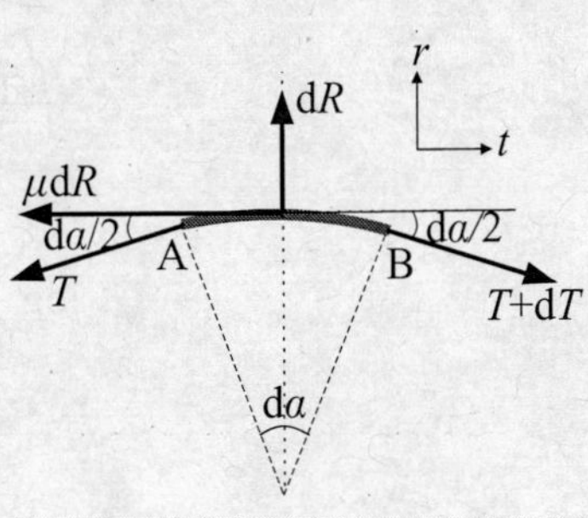

Figure 7.16 |

Resolving the forces in the tangential direction,

$$\sum F_t = 0 \Rightarrow \mu \mathrm{d}R + T\cos\frac{\mathrm{d}\alpha}{2} - (T + \mathrm{d}T)\cos\frac{\mathrm{d}\alpha}{2} = 0$$

For a small angle of dα, $\cos(\mathrm{d}\alpha/2) \approx 1$.
Therefore,

$$\mu \mathrm{d}R + T - (T + \mathrm{d}T) = 0$$

or

$$\mathrm{d}T = \mu \mathrm{d}R \quad \text{(i)}$$

Resolving the forces in the radial direction,

$$\sum F_r = 0 \Rightarrow \mathrm{d}R + T\sin\frac{\mathrm{d}\alpha}{2} - (T + \mathrm{d}T)\sin\frac{\mathrm{d}\alpha}{2} = 0$$

$$\left[\text{as } \mathrm{d}\alpha \text{ is small; } \sin\frac{\mathrm{d}\alpha}{2} \approx \frac{\mathrm{d}\alpha}{2}\right]$$

or

$$\mathrm{d}R - T\frac{\mathrm{d}\alpha}{2} - T\frac{\mathrm{d}\alpha}{2} - \mathrm{d}T\frac{\mathrm{d}\alpha}{2} = 0$$

or

$$\mathrm{d}R - T\mathrm{d}\alpha - \frac{\mathrm{d}T \times \mathrm{d}\alpha}{2} = 0$$

Neglecting the product of two small quantities, d$T \times$ dα, we have

$$\mathrm{d}R = T\mathrm{d}\alpha \quad \text{(ii)}$$

Substituting the value of dR from Equation (i), we have

$$\frac{\mathrm{d}T}{\mu} = T\mathrm{d}\alpha$$

or

$$\frac{\mathrm{d}T}{T} = \mu \mathrm{d}\alpha$$

Integrating between corresponding limits yields

$$\int_{T_2}^{T_1} \frac{\mathrm{d}T}{T} = \int_0^{\theta} \mu \mathrm{d}\alpha$$

or

$$\log\frac{T_1}{T_2} = \mu\theta$$

$$\frac{T_1}{T_2} = \mathrm{e}^{\mu\theta}$$

V-BELT OR ROPE

Consider a V-belt (or rope) passing over a driven pulley (Fig. 7.17). Let T_1 and T_2 be the belt tensions on the tight and slack sides, respectively, when the belt is just about to slide. Consider a small element AB of the belt. Figure 7.18 shows the free body diagram of the element of the belt. The tension increases from T at the angle α to $T + dT$ at the angle $\alpha + d\alpha$.

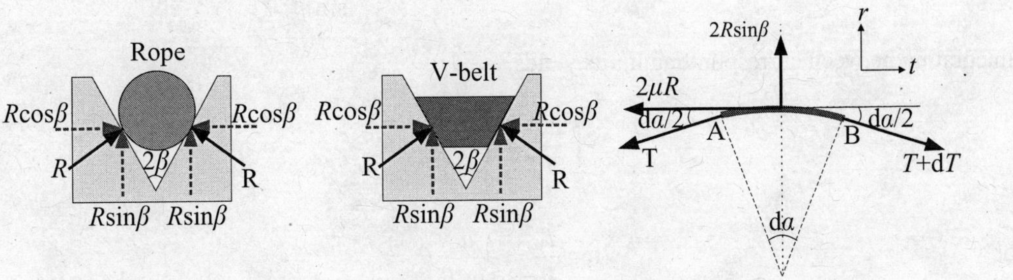

FIGURE 7.17 |

In case of V-belt or rope, there are two normal reactions as shown in the figure so that the radial reaction is equal to $2R \sin \beta$.

Thus, total frictional force = $2\mu R$.

We write the equation of equilibrium for the element AB of the belt.

Resolving the forces in the tangential direction,

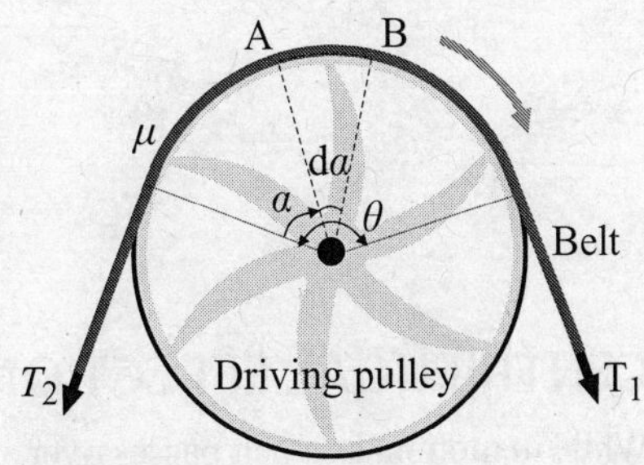

FIGURE 7.18 |

$$\sum F_t = 0 \Rightarrow 2\mu R + T \cos \frac{d\alpha}{2} - (T + dT) \cos \frac{d\alpha}{2} = 0$$

For small angle of $d\alpha$, $\cos(d\alpha/2) \approx 1$.

Therefore,

$$2\mu R + T - (T + dT) = 0$$

or

$$dT = 2\mu R \tag{i}$$

Resolving the forces in the radial direction,

$$\sum F_r = 0 \Rightarrow 2R \sin \beta + T \sin \frac{d\alpha}{2} - (T + dT) \sin \frac{d\alpha}{2} = 0$$

As $d\alpha$ is small,

$$\sin \frac{d\alpha}{2} \approx \frac{d\alpha}{2}$$

or

$$2R \sin \beta - T \frac{d\alpha}{2} - T \frac{d\alpha}{2} - dT \frac{d\alpha}{2} = 0$$

or

$$2R \sin \beta - T d\alpha - \frac{dT \times d\alpha}{2} = 0$$

Neglecting the product of two small quantities,

$$2R \sin \beta = T \, \mathrm{d}\alpha \qquad \text{(ii)}$$

Substituting value of dR from Equation (i), we have

$$\sin \beta \frac{\mathrm{d}T}{\mu} = T \, \mathrm{d}\alpha \quad \text{or} \quad \frac{\mathrm{d}T}{T} = \frac{\mu}{\sin \beta} \mathrm{d}\alpha$$

Integrating between corresponding limits yields

$$\int_{T_2}^{T_1} \frac{\mathrm{d}T}{T} = \int_0^{\theta} \frac{\mu}{\sin \beta} \mathrm{d}\alpha$$

or
$$\log \frac{T_1}{T_2} = \frac{\mu}{\sin \beta} \theta$$

or
$$\frac{T_1}{T_2} = \mathrm{e}^{(\mu/\sin \beta)\theta}$$

$$\boxed{\frac{T_1}{T_2} = \mathrm{e}^{(\mu/\sin \beta)\theta}}$$

CENTRIFUGAL EFFECT ON BELTS

While in motion, as a belt passes over a pulley, the centrifugal effect due to its own weight tends to lift the belt from the pulley. Due to symmetry, the centrifugal force produces equal tensions on the two sides of the belts, i.e., on the tight side as well as on the slack side.

Consider a small element AB of the belt. Figure 7.19 shows the free body diagram of the element of the belt.

Let ρ be the mass per unit length of the belt (kg/m), T_c be the centrifugal tension on tight and slack side of the element, r be the radius of the pulley, v be the velocity of the belt and $\delta\theta$ be the angle of lap of the element.

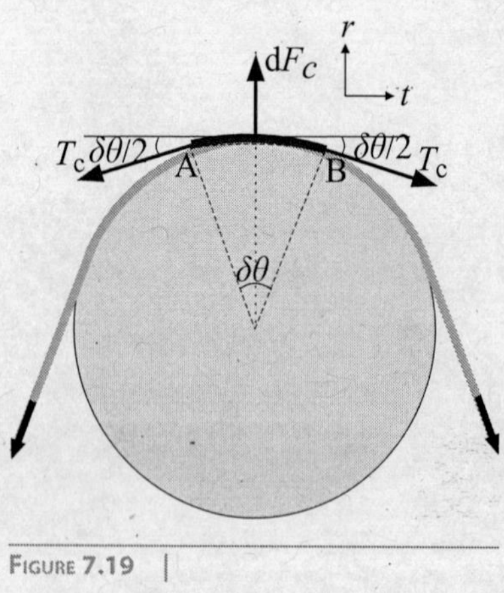

FIGURE 7.19 |

Resolving the forces in radial direction,

$$\sum F_r = 0 \Rightarrow \mathrm{d}F_c - T_c \sin \frac{\delta\theta}{2} - T_c \sin \frac{\delta\theta}{2} = 0$$

or
$$\mathrm{d}F_c = 2T_c \sin \frac{\delta\theta}{2}$$

Centrifugal force

$$\mathrm{d}F_c = \delta m \frac{v^2}{r} = 2T_c \sin \frac{\delta\theta}{2}$$

$$\rho r \delta\theta \frac{v^2}{r} = 2T_c \sin \frac{\delta\theta}{2}$$

As angle $\delta\theta$ is small,

$$\sin\frac{\delta\theta}{2} \approx \frac{\delta\theta}{2}$$

or

$$\rho r\delta\theta\frac{v^2}{r} = 2T_c \times \frac{\delta\theta}{2}$$

From which

$$T_c = \rho v^2 \quad \text{or} \quad T_c = \rho\omega^2 r^2,$$

where ω is the angular velocity.

$$\text{Centrifugal tension } T_c = \rho v^2$$

Thus, centrifugal tension is independent of the tight and slack side tensions and depends only on the velocity of the belt over pulley.

Total tension on the tight side,

$$T = \text{friction tension } (T_1) + \text{centrifugal tension } (T_c)$$

Total tension on tight side,

$$T = T_1 + T_c$$

Total tension on slack side,

$$T = T_2 + T_c$$

It can be shown that the power transmitted is reduced if centrifugal effect is considered for a given value of the total tight side tension T.

Centrifugal Tension Considered

Tension on tight side $T_1 = T - T_c$.

Let T_2 be the friction tension on the slack side. Then

$$\frac{T_1}{T_2} = e^{\mu\theta} = \text{constant} \quad \text{or} \quad T_2 = \frac{T_1}{e^{\mu\theta}}$$

Power

$$P = (T_1 - T_2)v \quad \text{or} \quad P = \left(T_1 - \frac{T_1}{e^{\mu\theta}}\right)v$$

$$P = T_1\left(1 - \frac{1}{e^{\mu\theta}}\right)v$$

Centrifugal Tension Neglected

Tension on tight side $T_1 = T$.

Let T_2' be the friction tension on the slack side. Then

$$\frac{T_1}{T_2'} = e^{\mu\theta} = \text{constant} \quad \text{or} \quad T_2' = \frac{T_1}{e^{\mu\theta}}$$

Power,

$$P = (T_1 - T_2')v \quad \text{or} \quad P = \left(T - \frac{T}{e^{\mu\theta}}\right)v$$

$$P = T\left(1 - \frac{1}{e^{\mu\theta}}\right)v$$

As T_1 is less than T, power transmitted is less when centrifugal force is taken into account.

MAXIMUM POWER TRANSMITTED BY A BELT

$$P = (T_1 - T_2)v$$

Substituting $T_2 = T_1/e^{\mu\theta}$, we have

$$P = \left(T_1 - \frac{T_1}{e^{\mu\theta}}\right)v = T_1\left(1 - \frac{1}{e^{\mu\theta}}\right)v = T_1 kv,$$

where $k = [1 - (1/e^{\mu\theta})] =$ constant,

$$P = T_1 kv = (T - T_c)kv,$$

where $T_1 = T - T_c$.
Substituting $T_c = \rho v^2$, we have

$$P = (T - \rho v^2)kv = Tkv - \rho kv^3$$

The maximum tension T in the belt should not exceed the permissible limit. Hence, treating it as a constant and differentiating the power with respect to v and equating the same equal to zero.

$$\frac{dP}{dv} = Tk - 3\rho kv^2 = 0 \quad \text{or} \quad T = 3\rho v^2 = 3T_c$$

$$T_c = \frac{T}{3} \Rightarrow v_{max} = \sqrt{\frac{T}{3\rho}}$$

Therefore, for maximum power transmission, centrifugal tension in the belt must be equal to one-third of the maximum allowable belt tension and the belt should be on the point of slipping.

Also $$T_1 = T - T_c = T - \frac{T}{3} = \frac{2}{3}T$$

$$T_c = \frac{T}{3}, \quad v_{max} = \sqrt{\frac{T}{3\rho^2}}, \quad T_1 = \frac{2}{3}T$$

INITIAL TENSION

When a belt is first fitted on pulleys, an initial tension T_0 is given to the belt when the system is stationary. When transmitting power, the tension on the tight side increases to T_1 and that on the slack side decreases to T_2. If it is assumed that the material of the belt is perfectly elastic, i.e., the strain in the belt is proportional to stress in it and the total length of the belt remains unchanged, the tension on the tight side will increase by the same amount as the tension on the slack side will decrease. If this change in the tension is δT, then

$$\text{tension on tight side, } T_1 = T_0 + \delta T,$$
$$\text{tension on slack side, } T_2 = T_0 - \delta T$$

Therefore,

$$\text{Initial tension, } T_0 = \frac{T_1 + T_2}{2},$$

If centrifugal tension is taken into consideration, then

$$\text{tension on tight side, } T_1 + T_c = T_0 + \delta T,$$
$$\text{tension on slack side, } T_2 + T_c = T_0 - \delta T$$

Therefore,

$$\text{initial tension, } T_0 = \frac{T_1 + T_2}{2} + T_c$$

or

$$T_1 + T_2 = 2(T_0 - T_c) \tag{i}$$

Tension ratio,

$$\frac{T_1}{T_2} = e^{\mu\theta}$$

$$T_1 = T_2 e^{\mu\theta} \tag{ii}$$

From Equations (i) and (ii), we have

$$T_1 = \frac{2e^{\mu\theta}(T_0 - T_c)}{1 + e^{\mu\theta}} \quad \text{and} \quad T_2 = \frac{2(T_0 - T_c)}{1 + e^{\mu\theta}}$$

When the belt drive is started, $v = 0$ and thus, $T_c = 0$,

$$T_1 = \frac{2e^{\mu\theta}T_0}{1 + e^{\mu\theta}}$$

Power transmitted, $P = (T - T_2)v$

$$P = \left(\frac{2e^{\mu\theta}(T_0 - T_c)}{1 + e^{\mu\theta}} - \frac{2(T_0 - T_c)}{1 + e^{\mu\theta}}\right)v$$

or

$$P = 2\left(\frac{e^{\mu\theta} - 1}{e^{\mu\theta} + 1}\right)(T_0 - T_c)v$$

$$P = 2\left(\frac{e^{\mu\theta} - 1}{e^{\mu\theta} + 1}\right)(T_0 - \rho v^2)v$$

To find the condition for maximum power transmitting, differentiating this expression with respect to v and equating the same to zero, i.e.,

$$\frac{d}{dv}P = 2\left(\frac{e^{\mu\theta} - 1}{e^{\mu\theta} + 1}\right)(T_0 - 3\rho v^2) = 0 \Rightarrow v = \sqrt{\frac{T_0}{3\rho}}$$

ADVANTAGE OF V-BELT OVER FLAT BELT

- Large powers can be transmitted by the convenient application of multiple V-belt system using sheaves with required number of grooves.
- Due to wedging action in the groove, V-belts transmit more power without slip as compared to flat belts.
- Due to wedging action in the groove, they require little adjustment to compensate for wear or stretch.
- They are more suitable for transmission of power between two shafts having a short centre distance. The distance may not be greater than the diameter of the larger sheave. However, the ideal centre distance of a V-belt drive is 1.5 to 1.25 times the diameter of the larger pulley.

CREEP

When the belt passes from the slack side to the tight side, it extends. On the other hand, when it passes from tight side to slack side, it contracts. Thus, the belt continuously extends and contracts. There is a relative motion between the belt and the pulley. This is known as creep. There is a loss of power because of it, and it reduces the velocity ratio.

The velocity ratio N_2/N_1 is given by

$$\frac{N_2}{N_1} = \frac{d_1}{d_2} \times \frac{E + \sqrt{\sigma_2}}{E + \sqrt{\sigma_1}},$$

where σ and σ_2 are the stresses in the belt on the tight and slack sides, respectively, E is the modulus of elasticity of belt material, and d_1 and d_2 are the diameters of driver and driven pulleys, respectively.

SOLVED PROBLEMS

PROBLEM 7.1

A prime mover running at 300 rpm drives an electric generator at 500 rpm by a belt drive. Diameter of the pulley on the output shaft of the prime mover is 600 mm. Assuming a slip of 3% on each pulley, determine the diameter of the generator pulley if the belt running over it is 6 mm thick.

Solution. Slip of belt $s = 3\%$, thickness of belt $t = 6$ mm, speed of prime mover $N_1 = 300$ rpm, speed of electric generator, $N_2 = 500$ rpm, diameter of the pulley on outer shaft of prime mover, $d_1 = 600$ mm.

Let d_2 be the diameter of generator pulley.

Velocity ratio,

$$\frac{N_2}{N_1} = \frac{d_1 + t}{d_2 + t}\left(1 - \frac{s}{100}\right)\left(1 - \frac{s}{100}\right)$$

or

$$\frac{500}{300} = \frac{600 + 6}{d_2 + 6}\left(1 - \frac{3}{100}\right)\left(1 - \frac{3}{100}\right)$$

or

$$d_2 + 6 = 342.11$$

$$d_2 = 336.11 \text{ mm}$$

or

$$d_2 = 336.11 \text{ mm}$$

PROBLEM 7.2

A flat belt is used to transmit a torque from pulley A to pulley B. The radius of each pulley is 75 mm and the coefficient of friction is 0.30. Determine the largest torque that can be transmitted if the allowable belt tension is 4 kN.

Solution.

Given $r_1 = r_2 = 75\text{mm},\ \mu = 0.3$ and $T_{\max} = 4\text{kN}$

Tension ratio,

$$\frac{T_1}{T_2} = e^{\mu\theta}$$

Substituting,

$$T_1 = T_{\max} = 4000\text{N},\ \theta = \pi, \quad \text{and} \quad \mu = 0.3$$

We have

$$\frac{4000}{T_2} = e^{0.3\pi} \quad \text{or} \quad T_2 = \frac{4000}{2.56}$$

$$T_2 = 1558.6\text{N}$$

Torque

$$\tau = (T_1 - T_2)r$$

$$\tau = (4000 - 1558.6)\frac{75}{1000} = 183.1\text{Nm}$$

$$\tau = 183.1\text{Nm}$$

PROBLEM 7.3

Determine the minimum force F needed so that pulley A can transmit a clockwise torque of 500 N m without slipping? The coefficient of friction between pulley A and the belt is 0.4 [Figs 7.3(a) and 7.3(b)].

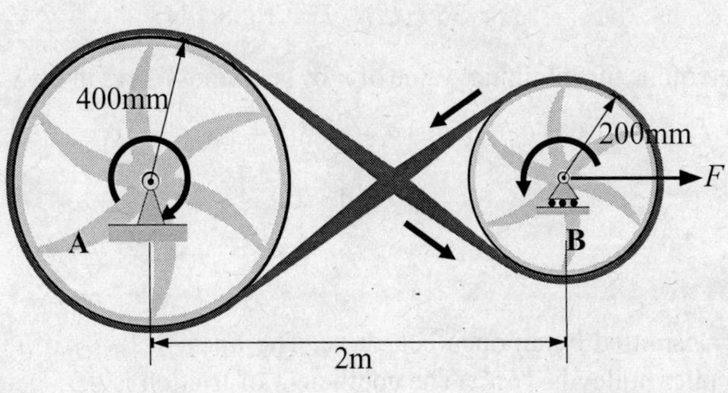

FIGURE 7.3(a)

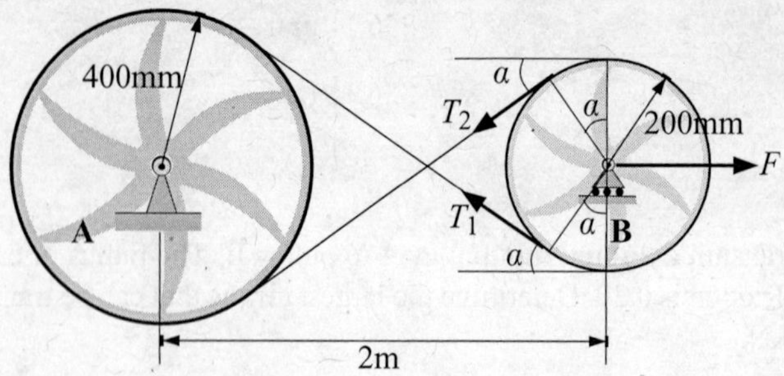

FIGURE 7.3(b) |

Solution.

Given: $\tau = 500$ N m, $\mu = 0.4$, $R = 0.4$ m, $r = 0.2$ m, and C = 2 m.

The angle of wrap,

$$\theta = \pi + 2a,$$

where

$$a = \sin^{-1}\left(\frac{R+r}{C}\right) = \sin^{-1}(0.3)$$

$$a = 17.45^\circ \quad \text{or} \quad 0.3\text{rad}$$

Therefore,

$$\theta = \pi + 2 \times 0.3 = 3.75\text{rad}$$

Torque,

$$\tau = (T_1 - T_2)r = 500$$

$$T_1 - T_2 = \frac{500}{0.4} = 1250\text{N} \tag{i}$$

Tension ratio,

$$\frac{T_1}{T_2} = e^{\mu\theta}$$

or

$$\frac{T_1}{T_2} = e^{0.4\times3.75} = 4.48 \tag{ii}$$

From Equations (i) and (ii), we get

$$T_2 = 359.02 \text{ N}, T_1 = 1608.4 \text{ N}$$

Finally, we can determine the minimum value of F by summing forces in the x-direction as follows:

$$\sum F_x = 0 \Rightarrow F_{\min} = T_1 \cos a + T_2 \cos a = (1608.4 + 359.02)\cos 17.4 = 1876.8\text{N}$$

$$F_{\min} = 1876.8\text{N}$$

PROBLEM 7.4

A power of 3 kW is transmitted by an open belt-drive. The linear velocity of the belt is 3 m/s. The angle of lap on the smaller pulley is 165°. The coefficient of friction is 0.3. Determine the effect on power transmission in the following cases:

(i) Initial tension in the belt is increased by 10%.

(ii) Angle of lap is increased by 8% by the use of an idler pulley, for the same speed and tension on the tight side.

(iii) Coefficient of friction is increased by 8% by suitable dressing to the friction surface of the belt.

Solution.

Given : $P = 3\times10^3\,\text{W},\ v = 3\text{m/s},\ \theta = 165^\circ = \dfrac{165\pi}{180} = 2.879\text{rad}$

and $\mu = 0.3$

Power transmitted:

$$P = (T_1 - T_2)v$$

$$3000 = (T_1 - T_2)3$$

or $$T_1 - T_2 = 1000\text{N} \qquad \text{(i)}$$

Tension ratio:

$$\frac{T_1}{T_2} = e^{\mu\theta} \Rightarrow \frac{T_1}{T_2} = e^{2.879\times0.3}$$

or $$T_1 = 2.37T_2 \qquad \text{(ii)}$$

Solving for T_1 and T_2 from Equations (i) and (ii),

$$T_1 = 1729.9\text{ N},\ T_2 = 729.9\text{ N}$$

Initial tension,

$$T_0 = \frac{T_1 + T_2}{2} = \frac{1729.9 + 729.9}{2} = 1229.9\text{N}$$

$$T_0 = 1229.9\text{N}$$

(i) When initial tension is increased by 10%:

$$T_0' = 1229.9\times1.10 = 1352.89\text{N}$$

or $$\frac{T_1 + T_2}{2} = 1352.89\text{N}$$

or $$T_1 + T_2 = 2705.78\text{N} \qquad \text{(iii)}$$

As μ and θ remain unchanged, $e^{\mu\theta}$ or $T_1 / T_2 = 2.37$ is same. Therefore,

$$2.37T_2 + T_2 = 2705.78\text{ N}$$

or $$T_2 = 802.9\text{ N}$$

Substituting $T_2 = 802.9$ N into Equation (iii), we have

$$T_1 = 1902.8\text{ N}$$

Power

$$P = (T_1 - T_2)v = (1902.8 - 802.9)3 = 3299.9\text{W}$$

Therefore, the increase in power is

$$\frac{(3299.9 - 3000)}{3000} \times 100 = 10\%$$

$$\boxed{\% \text{ change of power} = 10\%}$$

(ii) T_1 is the same as before, whereas θ increases by 8%.
New angle of lap,

$$\theta' = \theta \times 1.08 = 2.879 \times 1.08 = 3.109 \text{ rad}$$

Tension ratio,

$$\frac{T_1}{T_2} = e^{\mu\theta} \Rightarrow \frac{1729.9}{T_2} = e^{0.3 \times 3.109} = 2.541$$

$$T_2 = 680.7 \text{ N}$$

Power

$$P = (T_1 - T_2)v = (1729.9 - 680.7)3$$

$$P = 3147.6\text{W}$$

Increase in power is

$$\frac{(3147.6 - 3000)}{3000} \times 100 = 4.92\%,$$

$$\boxed{\% \text{ change power} = 4.92\%}$$

(iii) Coefficient of friction is increased by 8%.
New coefficient of friction $\mu' = 0.3 \times 1.08 = 0.324$
Tension ratio,

$$\frac{T_1}{T_2} = e^{0.32 \times 2.879} = 2.54$$

$$T_1 = 2.54T_2$$

and

$$T_1 + T_2 = 2459.8$$

From which,

$$T_2 = 694.8\text{N}, \ T_1 = 1764.9\text{N}$$

Power,

$$P = (T_1 - T_2)v = (1764.9 - 694.8) \times 3$$

$$P = 3210.4 \text{ W}$$

Therefore, the increase in power is

$$\frac{(3210.4 - 3000)}{3000} \times 100 = 7\%,$$

$$\boxed{\% \text{ change power} = 7\%}$$

PROBLEM 7.5

A flat belt connects a 1.20 m diameter pulley on a shaft running at 25 rad/s with another pulley running at 50 rad/s, the angle of lap on the latter pulley being 150°. The maximum permissible load on the belt is 1200 N and the coefficient of friction between the belt and the pulley is 0.25. If the initial tension in the belt may have any value between 800 N and 960 N, what is the maximum power that the belt can transmit?

Solution.

Given:

$$d_1 = 1.2\text{m}, \omega_1 = 25\text{rad/s}, \omega_2 = 50\text{rad/s},$$

$$\theta = 150^\circ = 2.61\text{rad (smaller pulley)},$$

$$T_{max} = 1200\text{N}, \mu = 0.25, T_0 = 800\text{N to } 960\text{N}$$

Velocity of belt,

$$v = \omega_1 r_1 = 25 \times \frac{1.2}{2} = 15\text{m/s}$$

Initial tension

$$T_0 = \frac{T_1 + T_2}{2}$$

(i) When initial tension is $T_0 = 800$ N:

$$800 = \frac{T_1 + T_2}{2} \quad \text{or} \quad T_1 + T_2 = 1600 \qquad \text{(i)}$$

Tension ratio,

$$\frac{T_1}{T_2} = e^{0.25\times 2.61} = 1.924$$

$$T_1 = 1.924T_2 \qquad \text{(ii)}$$

Solving Equations (i) and (ii) for T_1 and T_2,

$$T_1 = 1052.82\text{N} \quad \text{and} \quad T_2 = 547.20\text{N}$$

Power transmitted,

$$P = (T_1 - T_2)v$$

$$P = (1052.80 - 547.20)15 = 7584 \text{ W}$$

(ii) When initial tension is $T_0 = 960$ N,

$$960 = \frac{T_1 + T_2}{2} \quad \text{or} \quad T_1 + T_2 = 960 \times 2 = 1920\text{N} \qquad \text{(iii)}$$

Tension ratio,

$$\frac{T_1}{T_2} = e^{0.25\times 2.61} = 1.924$$

$$T_1 = 1.924T_2 \qquad \text{(iv)}$$

Solving Equations (iii) and (iv) for T_1 and T_2,

$$T_2 = 656.63\text{N}$$

$T_1 = 1263.37$ N, which is greater than 1200 N (not permissible).
So let us take $T_1 = 1200\text{N}$. Then

$$T_2 = \frac{T_1}{1.924} = \frac{1200}{1.924} = 623.7\text{N}$$

So the initial tension,

$$T_0 = \frac{T_1 + T_2}{2} = \frac{1200 + 623.7}{2} = 911.85\text{N},$$

which is within the limit.
Power transmitted, $P = (T_1 - T_2)v$,

$$P = (1200 - 623.7)15 = 8644.5 \text{ W}$$

$$P = 8644.5 \text{ W}$$

This is the maximum power transmitted for which the initial tension is $T_0 = 911.85$ N.

PROBLEM 7.6

A V-belt of 6.0 cm² cross-section has a groove angle of 40° and an angle of lap of 165°. The mass per unit length of belt is 1.2 kg/s. The coefficient of friction between belt and pulley is 0.1. The maximum allowable stress in the belt is 850 N/cm². Determine the power that can be transmitted at a belt speed of 30 m/s.

Solution.

Given: $2\beta = 40^\circ \Rightarrow \beta = 20^\circ, \theta = 165^\circ = 2.879\text{rad},$

$$\rho = 1.2\text{kg/m}, \mu = 0.1, \sigma_{\max} = 850\text{N/cm}^2 = 850 \times 10^4 \text{N/m}^2,$$

$$v = 30\text{m/s}, a = 6.0\text{cm}^2 = 6 \times 10^{-4}\text{m}^2$$

Maximum tension,

$$T = \sigma a = 850 \times 10^4 \times 6 \times 10^{-4} = 5100\text{N}$$

Tension ratio

$$\frac{T_1}{T_2} = e^{[(\mu/\sin\beta)\theta]} = e^{[(0.1/\sin 20)\times(165\pi/180)]}$$

or

$$\frac{T_1}{T_2} = 2.32$$

Centrifugal tension

$$T_c = \rho v^2 = 1.2 \times (30)^2 = 1080\text{N}$$

Therefore, tension on tight side

$$T_1 = T - T_c = 5100 - 1080 = 4020\text{N}$$

Tension on slack side

$$T_2 = \frac{T_1}{2.32} = \frac{4020}{2.32} = 1732.7\text{N}$$

Power transmitted

$$P = (T_1 - T_2)v = (4020 - 1732.7) \times 30 = 68.6\text{kW}$$

$$P = 68.6 \text{ W}$$

PROBLEM 7.7

The initial tension in a belt drive is found to be 500 N and the ratio of friction tension 1.5. The mass of the belt per unit length is 0.8 kg/m. Determine (a) the velocity of the belt for the maximum power transmission, (b) tension on the tight side of the belt when it is started, and (c) tension on the tight side of the belt when running at the maximum speed.

Solution.

Given: $T_0 = 500\text{N}, \dfrac{T_1}{T_2} = 1.5 = e^{\mu\theta}$ and $\rho = 0.8\text{kg/m}$

From equation for maximum power transmission with consideration of initial tension,

(a)
$$v = \sqrt{\frac{T_0}{3\rho}} = \sqrt{\frac{500}{3 \times 0.8}} = 14.4\text{m/s}$$

$$v = 14.4 \text{ m/s}$$

(b) Tension on the tight side of the belt when it is started is

$$T_1 = \frac{2e^{\mu\theta}T_0}{e^{\mu\theta} + 1}$$

or
$$T = \frac{2 \times 1.5 \times 500}{1.5 + 1} = 600\text{N}$$

$$T_1 = 600\text{N}$$

(c) Tension on the tight side of the belt when running at maximum speed, centrifugal tension

$$T_c = \rho v^2 = 0.8 \times (14.4)^2$$

$$T_c = 165.89\text{N}$$

Initial tension

$$T_0 = \frac{T_1 + T_2}{2} + T_c$$

or
$$T_0 = \frac{T_1\left[1 + \left(\frac{T_2}{T_1}\right)\right]}{2} + T_c$$

or
$$500 = \frac{T_1\left[1 + \left(\frac{1}{1.5}\right)\right]}{2} + 165.89$$

$$T_1 = 400.9\text{N}$$

It can also be found by applying the relation

$$T_1 = \frac{2e^{\mu\theta}(T_0 - T_c)}{1 + e^{\mu\theta}} = \frac{2 \times 1.5(500 - 165.89)}{1 + 1.5} = 400.9\text{N}$$

PROBLEM 7.8

A rope-drive is used for transmitting 250 kW at 300 rpm of a pulley of diameter 1.25 m. The angle of lap is 180° and groove angle is 45°. Permission tension in each rope is 200 N, mass per unit length of the rope is 1.27 kg/m and coefficient of friction between the rope and the pulley is 0.3. Calculate the number of ropes.

Solution.

Given:

$$P = 250\text{ kW}, N = 300\text{ rpm}, d = 1.25\text{ m}, \theta = 180^\circ = \pi, 2\beta = 45^\circ$$

$$T_{\max} = 2000\text{N}, r = 1.27\text{kg/m} \quad \text{and} \quad \mu = 0.3$$

Velocity of rope

$$v = \frac{\pi dN}{60} = \frac{\pi \times 1.25 \times 300}{60} = 19.63\text{m/s}$$

Centrifugal tension

$$T_c = \rho v^2 = 1.27 \times (19.63)^2 = 489.37\text{N}$$

Tension on tight side

$$T_1 = T_{\max} - T_c = 2000 - 489.37 = 1510.63\text{N}$$

Since

$$\frac{T_1}{T_2} = e^{(\mu/\sin\beta)\theta} = e^{(0.3/\sin 22.5)\times\pi} = 11.73$$

$$T_2 = \frac{T_1}{11.73} = \frac{1510.63}{11.73} = 128.69\text{N}$$

Power transmitted per rope,

$$P = (T_1 - T_2)v = \frac{(1510.63 - 128.69) \times 19.63}{1000}\text{kW}$$

$$\text{P} = 27.127\text{ kW}$$

Therefore, number of ropes

$$n = \frac{250}{27.127} = 9.21$$

Number of ropes $n = 10 \approx 9.21$

PROBLEM 7.9

Find the length of an open belt driving two parallel shafts 3 m apart, connected by two pulleys of 30 cm and 60 cm diameters. What is the additional length of belt required if a crossed type drive is used?

Solution.

Given: $C = 3\text{ m}, R = 0.3\text{ m}, \quad \text{and} \quad r = 0.15\text{ m}$

For Open Belt

$$a = \sin^{-1}\left(\frac{R-r}{C}\right) = \sin^{-1}\left(\frac{0.3-0.15}{3}\right)$$

$$= \sin^{-1}(0.05) = 2.86^\circ = 0.05\text{rad}$$

As the angle α is very small, the approximate relation can be used.

Length of belt,

$$L_{\text{open}} = \pi(R+r) + \frac{(R-r)^2}{c} + 2C$$

or

$$L_{\text{open}} = \pi(0.3+0.15) + \frac{(0.3-0.15)^2}{3} + 2\times 3$$

$$L_{\text{open}} = 7.42\text{m}$$

For Cross Belt

$$a = \sin^{-1}\left(\frac{R+r}{c}\right) = \sin^{-1}\left(\frac{0.3+0.15}{3}\right)$$

$$= \sin^{-1}(0.15) = 8.62^\circ = 0.15\text{rad}$$

Length of belt,

$$L_{\text{cross}} = (\pi + 2a)(R+r)2\cos a$$

or

$$L_{\text{cross}} = (\pi + 2\times 0.15)(0.3+0.15) + 2\times 3\cos 8.62$$

$$L_{\text{cross}} = 7.48\text{m}$$

Required additional length of belt,

$$\Delta L = 7.48 - 7.42 = 0.06\text{m}$$

$$\Delta L = 6\text{cm}$$

PROBLEM 7.10

Design a set of stepped pulleys to drive a machine from a counter-shaft that runs at 220 rpm. The distance between centres of the two sets of pulleys is 2 m, the radius of the smallest step on the counter-shaft is 80 mm. The machine is to run at 80, 100 and 130 rpm and should be able to rotate in either direction.

Solution. As the driven shaft is to rotate in either direction, both the cases of a crossed-belt and an open belt are to be considered.

For Crossed-belt System

The smallest step on the counter-shaft will correspond to the biggest step on the machine shaft (or the minimum speed of the machine shaft).

$$N_1 = N_2 = N_3 = 220\text{rpm}, r_1 = 80\text{mm}$$

$$N_1', N_2', N_3' = 80, 100, 130\text{rpm, respectively}$$

(a) For first step:

$$\frac{r_1'}{r_1} = \frac{N_1}{N_1'} \quad \text{or} \quad \frac{r_1'}{80} = \frac{220}{80}$$

$$r_1' = 220\text{mm}$$

(b) For second step:

$$\frac{r_2'}{r_2} = \frac{N_2}{N_2'} = \frac{220}{100} \quad \text{or} \quad r_2' = 2.2r_2$$

also, $\quad r_2' + r_2 = r_1' + r_1$

or $\quad 2.2r_2 + r_2 = 220 + 80 \quad \text{or} \quad 3.2r_2 = 300$

or $\quad r_2 = 93.75\text{mm}, r_2' = 93.7 \times 2.2 = 206.3\text{mm}$

$$r_2 = 93.75\text{mm}$$
$$r_2' = 206.3\text{mm}$$

(c) For third step:

$$\frac{r_3'}{r_3} = \frac{220}{130} \quad \text{or} \quad r_3' = 1.69r_3$$

also $\quad r_3' + r_3 = r_1' + r_1$

or $\quad 1.69r_3 + r_3 = 220 + 80 = 300$

or $\quad r_3 = 111.5\text{mm}, r_3' = 111.5 \times 1.69 = 188.5\text{mm}$

$$r_3 = 111.5\text{mm}$$
$$r_3' = 188.5\text{mm}$$

For Open Belt System

(a) For first step:

$$\frac{r_1'}{r_1} = \frac{N_1}{N_1'} \quad \text{or} \quad \frac{r_1'}{80} = \frac{220}{80}$$

$$r_1' = 220\text{mm}$$

(b) For second step:

$$\pi(r_2' + r_2) + \frac{(r_2' - r_2)^2}{c} = \pi(r_1' + r_1) + \frac{(r_1' - r_1)^2}{c}$$

or $\quad \pi(2.2r_2 + r_2) + \dfrac{(2.2r_2 - r_2)^2}{2} = \pi(0.22 + 0.08) + \dfrac{(0.22 + 0.08)^2}{2}$

or
$$10.05r_2 + \frac{1.44}{2}r_2^2 = 0.9523$$

or
$$r_2^2 + 13.958r_2 = 1.323$$

From which,
$$r_2 = 94\text{mm}$$

or
$$r_2' = 2.2 \times 94 = 206.8\text{mm}$$
$$r_2' = 206.8\text{mm}$$

(c) For third step:
$$\pi(1.69r_3 + r_3) + \frac{(1.69r_3 - r_3)^2}{2} = 0.9523$$

or
$$8.451r_3 + \frac{0.476}{2}r_3^2 = 0.9523 \quad \text{or} \quad r_3^2 + 35.508r_3 = 4.001$$

From which
$$r_3 = 112\text{mm}$$

or
$$r_3' = 112 \times 1.69$$
$$r_3' = 189.3\text{mm}$$

PROBLEM 7.11

The driving pulley of an open belt drive is of 720 mm diameter and rotates at 300 rpm while transmitting power to a driven pulley having 225 mm diameter. The modulus of elasticity of the belt material is 110 N/mm^2. Determine the speed lost by the driven pulley due to creep if the stresses in the tight and slack sides of the belt are found to be 0.8 N/mm^2 and 0.32 N/mm^2, respectively.

Solution. If creep is neglected,
$$\frac{N_2}{N_1} = \frac{D_1}{D_2}$$

or
$$N_2 = N_1 \times \frac{D_1}{D_2} = 300 \times \frac{720}{225} = 960\text{rpm}$$

When creep is considered,
$$\frac{N_2}{N_1} = \frac{D_1}{D_2}\left[\frac{E + \sqrt{\sigma_2}}{E + \sqrt{\sigma_1}}\right]$$

or
$$N_2 = 300 \times \frac{720}{225}\left[\frac{110 + \sqrt{0.32}}{110 + \sqrt{0.8}}\right] = 957\text{rpm}$$

$$\text{speed lost} = 960 - 957 = 3\text{rpm}$$
$$\text{speed lost} = 3\text{rpm}$$

CHAPTER HIGHLIGHTS

- Belts and Ropes are used to transmission of power from one shaft to another shaft.
- Types of belts: (1) Flat belt, (2) V-belt, (3) Circular belt or rope, (4) Ribbed belt and (5) Toothed belt.
- Types of belt drives: (1) Open belt drive, (2) Cross or Twist belt drive, (3) Quarter twist belt drive, (4) Open belt drive with idler pulleys or jockey pulley drive, (5) Quarter twist belt drive with guide pulley, (6) Open belt drive with idler pulleys and guide pulleys and (7) Loose and fast pulley drive
- Velocity ratio:

  Simple belt drive

  $$\frac{N_2}{N_1} = \frac{d_1}{d_2}, \qquad \frac{N_2}{N_1} = \frac{(d_1 + t)}{(d_2 + t)},$$

  $$\frac{N_2}{N_1} = \left(\frac{d_1 + t}{d_2 + t}\right)\left(1 - \frac{s_1}{100}\right)\left(1 - \frac{s_2}{100}\right) \text{ and } \frac{N_2}{N_1} = \frac{d_1}{d_2} \times \frac{E + \sqrt{\sigma_2}}{E + \sqrt{\sigma_1}}$$

  Compound belt drive $\dfrac{N_{last}}{N_1} = \dfrac{d_1 \times d_3 \times d_5 \times - - -}{d_2 \times d_4 \times d_6 \times - - -}$
- Length of an open belt $L = \pi(R + r) + 2C + \dfrac{(R - r)^2}{C}$
- Length of cross belt $L = \pi(R + r) + 2C + \dfrac{(R + r)^2}{C}$
- Power transmitted by a belt or rope is $P = \dfrac{(T_1 - T_2)\,v}{1000}$
- Tension ratio of flat belt: $\dfrac{T_1}{T_2} = e^{\mu\theta}$, Tension ratio of V-Belt (or Rope): $\dfrac{T_1}{T_2} = e^{\frac{\mu}{\sin\beta}\theta}$
- Centrifugal tension of belt $T_c = \rho v^2$
- Power transmitted by a belt is (consider centrifugal tension) $P = T_1\left(1 - \dfrac{1}{e^{\mu\theta}}\right)v$
- Condition for maximum power transmitted by a belt $T_c = \dfrac{T}{3}$ or $v_{max} = \sqrt{\dfrac{T}{3\rho}}$.
- Initial tension $T_0 = \dfrac{T_1 + T_2}{2}$, $T_0 = \dfrac{T_1 + T_2}{2} + T_c$ (consider centrifugal tension)

REVIEW QUESTIONS

Q1. What are the various types of belts?

Q2. Explain the various types of belt drives.

Q3. What materials are used for making belts?

Q4. Derive an expression for velocity ratio of the belt drive in the following cases:

(i) Without considering the thickness of belt,

(ii) considering the thickness of belt,

(iii) considering slip of belt on each pulley,

(iv) considering creep in belt.

Q5. Why is the face of the pulley crowned?

Q6. Derive an expression for the length of belt for an open belt drive.

Q7. Derive an expression for the length of belt for a cross belt drive.

Q8. Derive an expression for the limiting ratio of tensions in a flat belt over pulley.

Q9. Derive an expression for the limiting ratio of tensions in a V-belt or rope drive.

Q10. Write short notes on:

(a) cone pulley drive,
(b) fast and loose pulley drive,
(c) crowning of flat belt pulley,
(d) idler pulley.

Q11. Derive the expression for optimum speed of a flat belt for the transmission of maximum power, considering the effect of centrifugal tension.

Q12. Explain initial tension in belt drive system.

Q13. Give the advantages of V-belt drives over flat belt drives.

Q14. A pulley is driven by a flat belt, running at 600 m/min. The coefficient of friction between the pulley and the belt is 0.3. The centre distance between the shafts is 1.425 m and the diameters of pulleys are 1000 and 500 mm. Maximum tension in the belt is 700 N. Calculate the power transmitted.

[Ans; 3.97 kW].

Q15. In a rope drive, 100 kW is transmitted by means of ropes of mass 1.2 kg/m. The groove angle is $50°$ and the angle of lap is $170°$, $\mu = 0.3$. If the initial tension in each rope is 900 N, find the number of ropes required.

[Ans; 8].

Q16. An open belt drive connects two pulleys of diameter 60 and 30 cm by a flat belt. The distance between centres of two pulleys is 2.10 m. The coefficients of friction are 0.28 and 0.22 between both pulleys and the belt. The maximum allowable tension in the belt is 10 kN. Determine the largest couple that can be transmitted.

[Ans; 723 N m].

MULTIPLE CHOICE QUESTIONS

Q1. If the percentage slip is same on both the driving and driven pulleys, then the speed ratio will

(A) increase (B) decrease (C) remain same (D) unpredictable.

[Ans: A]

Q2. With the same set of pulleys, belt and centre distance, the maximum power transmitted by

(A) a cross belt is more than an open belt
(B) a cross belt is less than an open belt
(C) cross and open belts is same
(D) cross and open belts depends upon pulley diameter

[Ans: A]

Q3. The centrifugal tension in belts

(A) reduces power transmission
(B) increases power transmission
(C) does not affect power transmission
(D) increases or decreases power transmission depending on speed

[Ans: A]

Q4. The crowning of pulleys is done to

(A) make the belt run in the centre of the pulley face width
(B) strengthen the pulley
(C) give better shape to the pulley
(D) decrease slip

[Ans: A]

Q5. If the initial tension in the belt is increased, then the maximum power transmitted by the belt

(A) reduces (B) increases (C) remains same (D) depends on speed

[Ans: B]

Q6. Considering centrifugal tension in a belt, maximum linear velocity of the belt is proportional to

(A) cubic root of maximum tension (B) square root of maximum tension
(C) maximum tension (D) reciprocal of maximum tension

[Ans: B]

Q7. The ratio of tensions in the tight and slack sides of a belt drive is

(A) $\mu\theta$ (B) $e^{\mu\theta}$ (C) $\dfrac{1}{\mu\theta}$ (D) $e^{(1/\mu\theta)}$

[Ans: B]

Q8. The difference between tensions on the tight and slack sides of a belt drive is 3000 N. If the belt speed is 15 m/s, the transmitted power in kW is

(A) 45 (B) 22.5 (C) 90 (D) 100

[Ans: A]

Q9. The percentage improvement in power capacity of a flat belt drive, when the wrap angle at the driving pulley is increased from 150° to 210° by an idler arrangement, for a friction coefficient of 0.3 is

(A) 25.21 (B) 392 (C) 40.17 (D) 67.85

[Ans: D]

Q10. Given that T_1 and T_2 are the tensions on the tight and slack sides of the belt, respectively, the initial tension of the belt taking in to account the centrifugal tension T_c is equal to

(A) $\dfrac{T_1 + T_2 + T_c}{3}$ (B) $\dfrac{T_1 + T_2 + 2T_c}{2}$

(C) $\dfrac{T_1 + T_2 + 3T_c}{3}$ (D) $\dfrac{T_1 - T_2 + 3T_c}{3}$

[Ans: B]

Q11. Length of the belt in a cross belt drive is

(A) $2C + \dfrac{\pi}{2}(D + d) + \dfrac{(D + d)^2}{4C}$ (B) $2C + \dfrac{\pi}{2}(D - d) + \dfrac{(D + d)^2}{4C}$

(C) $2C + \dfrac{\pi}{2}(D + d) + \dfrac{(D - d)^2}{4C}$ (D) $2C + \dfrac{\pi}{2}(D - d) + \dfrac{(D - d)^2}{4C}$

[Ans: A]

Q12. By increasing the centre distance, the ratio of tensions in an open belt drive is

(A) increased (B) decreased
(C) not affected (D) none of the above

[Ans: A]

Q13. Ratio of tension T_1 / T_2 in a V belt drive is given by

(A) $e^{\mu\theta}$ (B) $e^{\mu\theta \operatorname{cosec}\beta}$ (C) $e^{\mu\theta\cos\beta}$ (D) $e^{\mu\theta\cos\beta/2}$

where θ is the angle of lap and 2β is the groove angle.

[Ans: B]

Q14. In an open belt drive the lower side is used on the

(A) slack side (B) tight side (C) none of the above

[Ans: B]

Q15. The power transmitted in a belt drive depends on the

(A) revolutions per minute (B) T_1 and T_2
(C) arc of contact (D) all of the above

[Ans: D]

8 Lifting Machines

INTRODUCTION

Lifting machines are those machines that are used for lifting loads. The force (or effort) is applied at one point of the machine and weight (or load) is lifted at the other point of the machine. Pulleys used to lift water from a well and screw jacks used to lift buses are some of the common examples of lifting machines.

DEFINITIONS

The terms commonly used while dealing with lifting machines are defined below.

Mechanical Advantage

It is the ratio of load (or weight) lifted to the effort applied. Thus, mathematically, mechanical advantage is

$$\text{MA} = \frac{\text{Weight lifted}}{\text{Effort applied}} = \frac{W}{P}$$

$$\boxed{\text{MA} = \frac{W}{P}}$$

Velocity Ratio

It is the ratio of the distance moved by the effort to the distance moved by the load. If y is the distance moved by effort P and x is the distance moved by load W, then velocity ratio,

$$\text{VR} = \frac{\text{Distance moved by effort}}{\text{Distance moved by load}} = \frac{y}{x}$$

$$\boxed{\text{VR} = \frac{y}{x}}$$

Input of Machine

Input of machine is defined as the work done on the machine. In a lifting machine, it is the product of the effort and distance moved by the effort. Mathematically,

$$\text{Input} = \text{Effort} \times \text{distance moved by the effort} = P \times y$$

$$\boxed{\text{Input} = P \times y}$$

Output of Machine

Output of machine is defined as the actual work done by the machine. In a lifting machine, it is measured by the product of the load lifted and the distance through which load is lifted. Mathematically,

$$\text{Output} = \text{Load lifted} \times \text{distance through which load is lifted} = W \times x$$

$$\boxed{\text{Output} = W \times x}$$

Efficiency of Machine

Efficiency of machine is defined as the ratio of output to input of machine. Mathematically,

$$\eta = \frac{\text{Output}}{\text{Input}} = \frac{W \times x}{P \times y}$$

or $$\eta = \frac{W/P}{y/x}$$

or $$\eta = \frac{\text{Mechanical advantage}}{\text{Velocity ratio}}$$

where $\frac{W}{P}$ is the mechanical advantage and $\frac{y}{x}$ is the velocity ratio.

$$\boxed{\eta = \frac{\text{MA}}{\text{VR}}}$$

Ideal Machine

If the efficiency of a machine is 1 (100%), the machine is called ideal machine. Thus, for an ideal machine,

$$\eta = 1 = \frac{\text{Output}}{\text{Input}}$$

$$\boxed{\text{Output} = \text{Input}}$$

or $$Wx = Py \Rightarrow \frac{W}{P} = \frac{y}{x}$$

$$\boxed{\text{Mechanical advantage (MA)} = \text{Velocity ratio (VR)}}$$

IDEAL EFFORT AND IDEAL LOAD

Ideal effort is the effort required to lift the weight by the ideal machine. Ideal load is the load that can be lifted using the given effort by the ideal machine.

For the ideal machine,

$$\eta = 1 \Rightarrow \text{Mechanical advantage (MA)} = \text{Velocity ratio (VR)}$$

If P_i is the ideal effort, then

$$\text{MA} = \frac{W}{P_i} = \text{VR}$$

Hence, ideal effort is

$$P_i = \frac{W}{\text{VR}}$$

If W_i is the ideal load, then

$$\text{MA} = \frac{W_i}{P} = \text{VR}$$

Hence, ideal load is

$$W_i = P \times \text{VR}$$

Friction of the machine in terms of effort is

$$F_{\text{p}} = P - P_i = P - \frac{W}{\text{VR}}$$

$$F_{\text{p}} = P - \frac{W}{\text{VR}}$$

Friction of the machine in terms of load is

$$F_{\text{W}} = W_i - W = P \times \text{VR} - W$$

$$F_{\text{W}} = P \times \text{VR} - W$$

REVERSIBILITY OF MACHINE

If the effort is removed, the load may or may not fall down to the initial position. If the load falls down, the machine is said to be reversible, and if the load does not fall down, the machine is said to be *irreversible* or *self-locking* machine.

Let P be the effort applied, W be the load lifted, y be the distance moved by the effort, x be the distance moved by the load.
Then,

$$\text{Input of machine} = Py,$$

$$\text{Output of machine} = Wx$$

We know that

$$\text{Input} - \text{Output} = \text{Work lost in friction}$$

or

$$\text{Work lost in friction} = Py - Wx$$

When the effort P is removed, the load W will not fall down if the work done by the force of friction is more than the output of the machine. And if the load W does not fall down, the machine is said to be *irreversible*. Hence,

Condition for irreversible or self-locking of machine is

Work done by friction force > output of machine

$$Py - Wx > Wx$$

or

$$Py > 2Wx$$

or

$$\frac{W/P}{y/x} < \frac{1}{2}$$

or
$$\frac{\text{Mechanical advantage}}{\text{Velocity ratio}} < \frac{1}{2}$$

$$\text{Efficiency, } \eta < \frac{1}{2} \text{ (50\%)}$$

$$\eta < 50\%$$

Hence, the condition for a machine to be *irreversible* or *self-locking* is that its efficiency should be less than 50%.

LAW OF MACHINE

The relationship between the load lifted (W) and the effort required (P) is called the law of machine.

The relation between effort P and load W is found by conducting experiments in which efforts required for different loads are determined and then a graph is plotted between loads and efforts. Figure 8.1 shows the graph between effort and load. For an ideal machine, the straight line will pass through the origins. But for an actual machine, the straight line will intercept the y-axis at A as shown in Figure 8.1. Mathematically, the law of a machine is given by the relation

$$P = mW + c$$

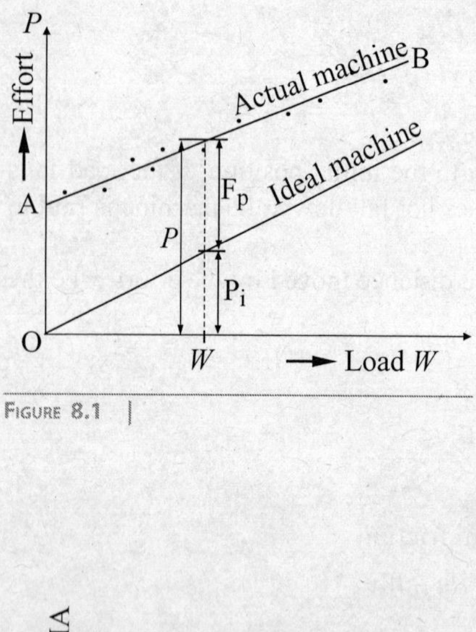

FIGURE 8.1 |

where P is the effort, W is the load (or weight), c is a constant, which represents the effort required to overcome friction, and m is the slope of the line and is equal to a constant which is known as coefficient of friction.

VARIATION OF MECHANICAL ADVANTAGE

Mechanical advantage MA $= \dfrac{W}{P}$. Substituting $P = m\,W + c$, we have

$$\text{MA} = \frac{W}{mW + c} = \frac{1}{m + \left(\dfrac{c}{W}\right)}$$

$$\text{MA} = \frac{1}{m + \left(\dfrac{c}{W}\right)}$$

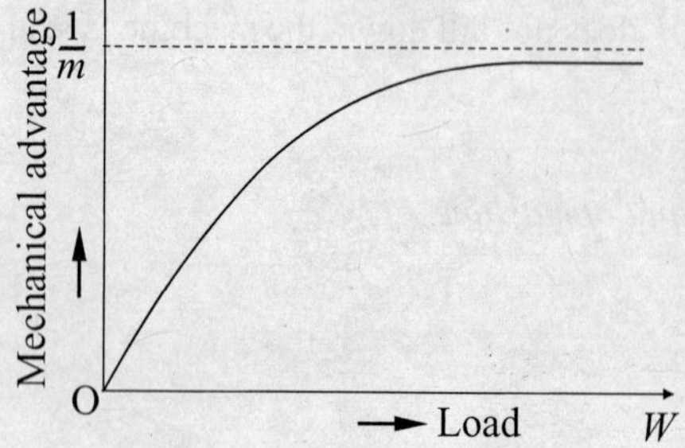

FIGURE 8.2 |

As the load increases, c/W, which is in denominator, decreases and hence mechanical advantage increases. If the load W is so large, that the value c/W becomes negligible, then mechanical advantage will be maximum (Fig. 8.2).

$$W \rightarrow \text{large}, \frac{c}{W} \rightarrow \text{very small}$$

$$(\text{MA})_{\text{max}} = \frac{1}{m}$$

Variation of Efficiency

Efficiency of machine

$$\eta = \frac{\text{MA}}{\text{VR}} = \frac{\frac{W}{P}}{\text{VR}}$$

$$\eta = \frac{W}{(\text{VR}) \times P}$$

Substituting $P = mW + c$, we have

$$\eta = \frac{W}{\text{VR} \times (mW + c)}$$

or

$$\eta = \frac{1}{\text{VR} \times \left[m + \left(\frac{c}{W} \right) \right]}$$

since the velocity ratio (VR) is constant for a machine. If the load W is so large that the value of c/W becomes negligible, then the efficiency will be maximum (Fig. 8.3):

$$\eta_{\max} = \frac{1}{\text{VR} \times m}$$

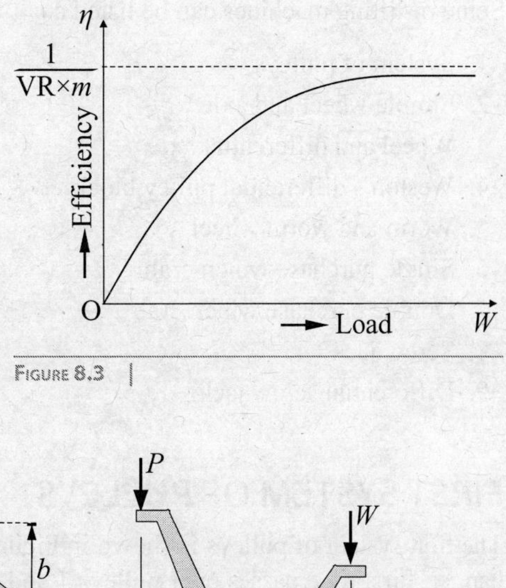

FIGURE 8.3 |

LEVER

A lever is a device that changes the effect of a force. It consists of a rigid bar, straight or curved, that is hinged at one point. The hinged point is called the fulcrum. Levers are used in many engineering applications. Figure 8.4 shows different types of levers.

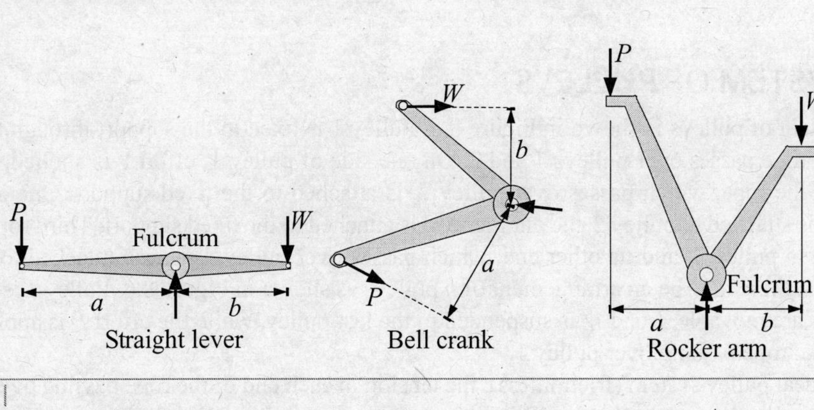

FIGURE 8.4 |

Equilibrium of moment about fulcrum

$$W \times b - P \times a = 0 \Rightarrow \frac{W}{P} = \frac{a}{b}$$

Mechanical advantage

$$\text{MA} = \frac{W}{P}$$

Velocity ratio

$$\text{VR} = \frac{a}{b}$$

PULLEY

A pulley is a device for changing the direction of a force or magnifying the effect of a force. It is a wheel that is supported by an axle or shaft. The outer surface of the wheel's rim may be grooved or flat, depending on its use. A rope (belt, flexible cable or chain) is turned around the wheel. When a pulley used as shown in Figure 8.5, it is referred to as a simple pulley.

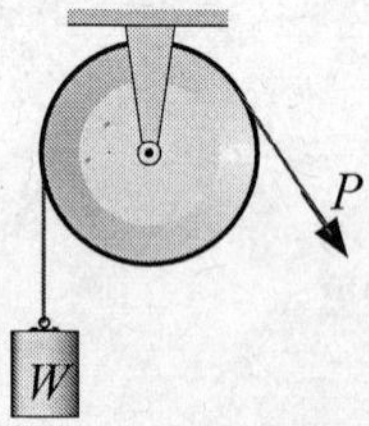

Simple pulley

FIGURE 8.5

TYPES OF LIFTING MACHINES

Some of lifting machines can be listed as

1. System of pulleys
2. Simple wheel and axle
3. Wheel and differential axle
4. Weston's differential pulley block
5. Worm and worm wheel
6. Single purchase winch crab
7. Double purchase winch crab
8. Screw jack
9. Differential screw jack.

FIRST SYSTEM OF PULLEYS

The first system of pulleys is shown in Figure 8.6. Pulley 1 is fixed to the support through the pulley hanger. First rope passes over pulleys 1 and 2. On one side of pulley 1, effort P is applied, while the other end of the rope, which passes over pulley 2, is attached to the fixed support. One end of the second rope is attached to pulley 2 and another end is attached to the fixed support. Third rope has one end attached to pulley 3, and its other end, which passes over pulley 4, is also attached to the fixed support. Thus, there may be an arrangement of n pulleys as shown in Figure 8.6. Pulley 1 is fixed and other pulleys are movable. Load W is suspended to the last pulley, while the effort P is applied to the free end of the rope passing over pulley 1.

For an ideal pulley system (frictionless), the tension in each end of the rope passing over pulley 4 is $W/2$. The load at the centre of pulley 3 is $W/2$, and the tension in each end of the rope passing over pulley 3 is $W/4$. Similarly, tension in each end of the rope passing over pulley 2 is $W/8$ and that in the rope passing over pulley 1 is also $W/8$.

Hence, effort P is equal to $W/8$.

Mechanical advantage

$$\text{MA} = \frac{W}{P} = \frac{W}{\left(\frac{W}{8}\right)} = 8$$

When there are three movable pulleys, MA=2^3.

Since, the pulley system is ideal, $\eta = 100\%$, and

Velocity ratio (VR) = Mechanical advantage (MA) = 2^3.

If there are n movable pulleys, and the pulley system is ideal,

$$\text{VR} = \text{MA} = 2^3$$

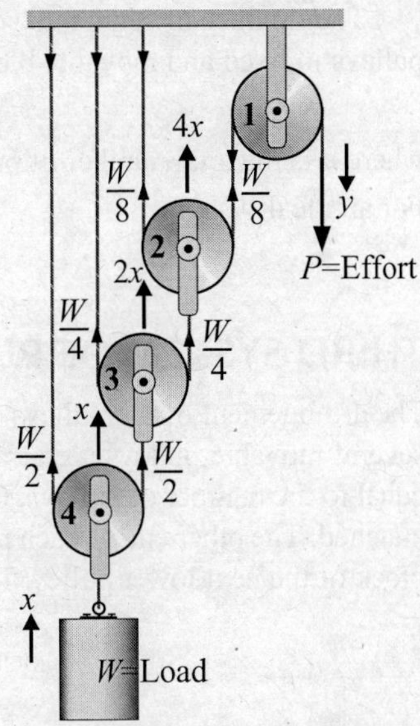

First system of pulleys

FIGURE 8.6 |

SECOND SYSTEM OF PULLEYS

Figure 8.7 shows the second system of pulleys, which consists of two blocks each containing a number of pulleys. The lower block is movable and the upper block is fixed to the support. Either both the blocks carry equal number of pulleys or the upper block may have one pulley more than the lower one.

The same rope passes around all the pulleys. One end of the rope is fixed to the lower block (or upper block, depends on the number of pulleys), and the other end of the rope is free and effort is applied to this free end. Let an effort P be applied to lift a load W.

Neglecting weight of pulleys and friction, the tension in the rope is $T = P$. Take the section along xx and consider the equilibrium of the bottom pulley block. Therefore,

$$6T - W = 0$$

Tension of rope $T = P$, that is

$$6P = W$$

Mechanical advantage

$$\text{MA} = \frac{W}{P} = 6$$

Since the system is ideal (i.e., without friction, $\eta = 100\%$), hence

$$\text{Velocity ratio} = \text{Mechanical advantage}$$

$$\text{VR} = 6$$

In general, in the second system of pulleys

$$\text{MA} = \frac{W}{P}$$

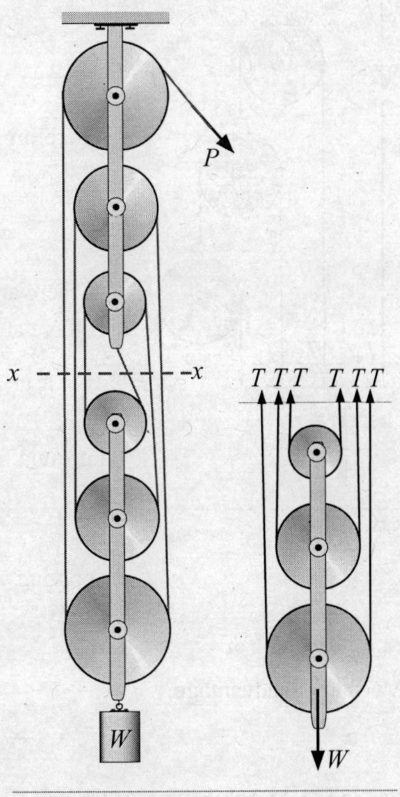

FIGURE 8.7 |

The number of segments supporting the movable block or load is equal to the total number of pulleys in fixed and movable blocks:

$$\text{MA} = n,$$

where n is the total number of pulleys.

For an ideal system,

$$\text{MA} = \text{VR} = n$$

THIRD SYSTEM OF PULLEYS

The arrangement of the pulleys in the third system of pulleys is shown in Figure 8.8. In this system, several movable pulleys are arranged, keeping the top-most pulley as fixed. The number of ropes is equal to the number of pulleys. One end of each rope is attached to a common block, to which load is attached. The other end of each rope, which passes round the periphery of the pulley, is fastened to the block of the next lower pulley. The effort is applied to the free end of the lower most pulley.

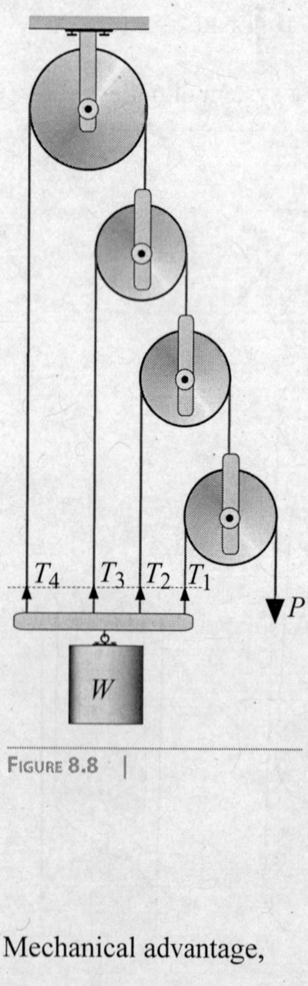

FIGURE 8.8 |

Neglecting the weight of pulleys and friction, the equilibrium of the system gives

$$T_1 + T_2 + T_3 + T_4 = W,$$

where

$$T_1 = P$$
$$T_2 = 2T_1 = 2P$$
$$T_3 = 2T_2 = 2 \times 2P = 2^2 P$$
$$T_4 = 2T_3 = 2 \times 2 \times 2P = 2^3 P$$

Therefore,

$$P + 2P + 2^2 P + 2^3 P = W$$

$$\frac{W}{P} = 1 + 2 + 2^2 + 2^3$$

as there are four pulleys in this case.

If there are n pulleys, then

$$\frac{W}{P} = 1 + 2 + 2^2 + 2^3 + \cdots + 2^{n-1}$$

Formula for the sum of geometric series is

$$a + ar + ar^2 + \cdots + ar^{n-1} = \frac{a(r^n - 1)}{r - 1},$$

applying which, we get

$$\frac{W}{P} = \frac{2^n - 1}{2 - 1} = 2^n - 1$$

Mechanical advantage,

$$\text{MA} = \frac{W}{P} = 2^n - 1$$

For an ideal machine ($\eta = 100\%$), mechanical advantage is equal to velocity ratio:

$$\text{VR} = 2^n - 1$$

SIMPLE WHEEL AND AXLE

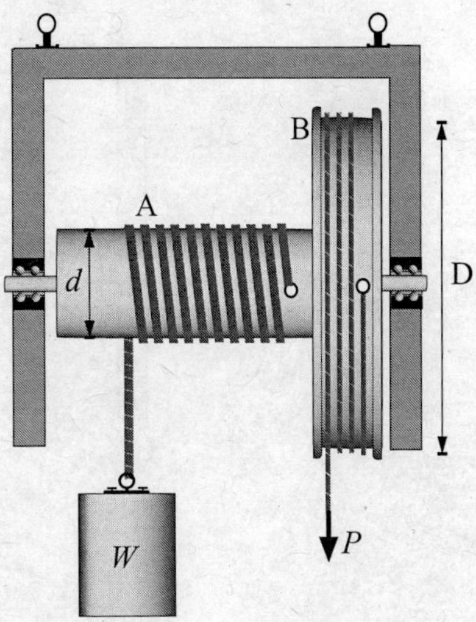

FIGURE 8.9

Figure 8.9 shows a simple wheel and axle that consists of an axle A and wheel B, co-axially fitted with key to the shaft. The whole assembly is mounted on ball bearing so that wheel and axle can be rotated.

A rope is wound round axle A, which carries the load W to be lifted. A second rope is wound round wheel B, to which effort P is applied. The two ropes are wound in the opposite directions. Hence, a downward motion of P will raise the load W.

Let P be the effort applied, W be the load lifted, D be the diameter of the wheel, and d be the diameter of the axle. Since the wheel and axle are keyed to the same shaft, when the wheel makes one revolution, the axle will also make one revolution.

Then, the distance moved by the effort is πD and the distance moved by the load is πd. Therefore, velocity ratio

$$\text{VR} = \frac{\text{Distance moved by effort}}{\text{Distance moved by load}} = \frac{\pi D}{\pi d}$$

$$\text{VR} = \frac{D}{d}$$

Mechanical advantage

$$\text{MA} = \frac{\text{Load lifted}}{\text{Effort applied}} = \frac{W}{P}$$

$$\text{MA} = \frac{W}{P}$$

Efficiency of machine

$$\eta = \frac{\text{MA}}{\text{VR}} = \frac{W/P}{D/d}$$

$$\eta = \frac{W}{P} \times \frac{d}{D}$$

WHEEL AND DIFFERENTIAL AXLE

Figure 8.10 shows a wheel and differential axle. It consists of a differential axle of diameter d_1 and d_2 ($d_1 > d_2$) and a wheel of diameter D fixed co-axially. One rope is wound round wheel A to which effort P is applied. The second rope is wound round axles B and C. This rope goes around a pulley to which the load W is attached. This rope is wound on axles B and C in such a way that as the shaft rotates, the rope unwinds on axle C and winds at the same time on axle B, thus lifting the weight W.

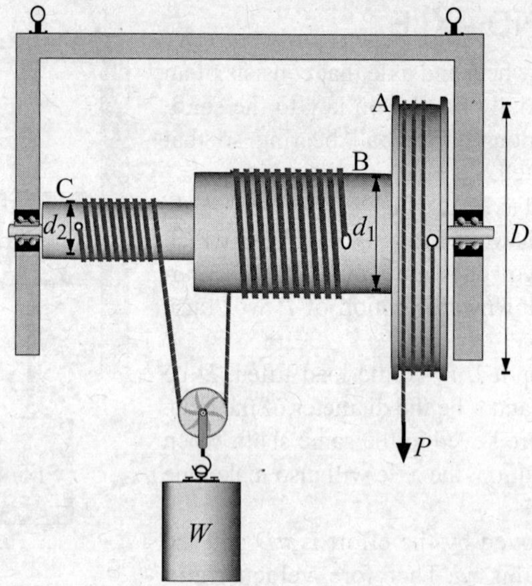

FIGURE 8.10 |

Let P be effort applied, W be the load lifted, D be the diameter of the wheel, d_1 be the diameter of axle B, d_2 be the diameter of axle C. Suppose the whole system makes one complete revolution due to the applied effort. Then the distance moved by effort is πD, the length of rope that winds on axle B is πd_1 and the length of rope that unwinds on axle C is πd_2

As

$$d_1 > d_2, \; \pi d_1 > \pi d_2$$

Therefore, the net length of rope which will wind is $\pi d_1 - \pi d_2$

Distance moved by load is

$$\frac{\pi d_1 - \pi d_2}{2}$$

Velocity ratio

$$\text{VR} = \frac{\text{Distance moved by effort}}{\text{Distance moved by load}} = \frac{\pi D}{\frac{1}{2}(\pi d_1 - \pi d_2)}$$

$$\text{VR} = \frac{2D}{d_1 - d_2}$$

Mechanical advantage

$$\text{MA} = \frac{\text{Weight lifted}}{\text{Effort applied}} = \frac{W}{P}$$

Efficiency

$$\eta = \frac{\text{MA}}{\text{VR}} = \frac{W}{P} \times \left(\frac{d_1 - d_2}{2D}\right)$$

$$\eta = \frac{W}{P} \times \left(\frac{d_1 - d_2}{2D}\right)$$

WESTON'S DIFFERENTIAL PULLEY BLOCK

Figure 8.11 shows a Weston's differential pulley block. This is a special type of simple pulley system. It consists of two pulley blocks: one at the top attached to the support and the other at the bottom hanging in the chain loop. The upper block consists of two pulleys of different diameters, but which are fixed co-axially. These pulleys turn together as one pulley i.e., both of them behave as one pulley with two grooves. The bottom block is a simple pulley to which the load W is attached.

An endless chain is wound around the pulley system as shown in Figure 8.11.All the pulleys are made with teeth so as to accommodate the links of the chain.

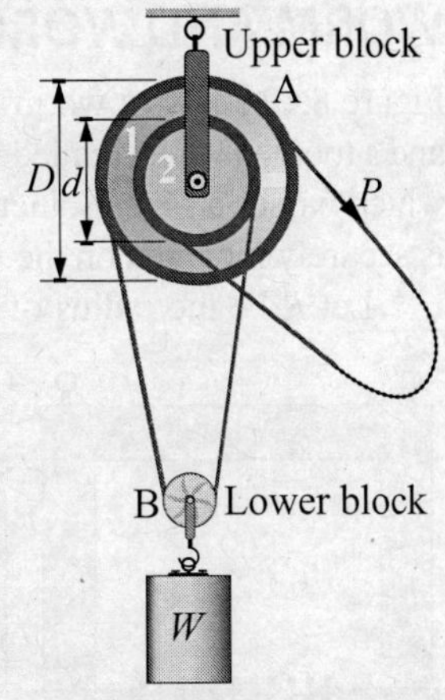

FIGURE 8.11

The effort P is applied to the chain that passes over the bigger pulley of block A. As a result, the chain unwinds from pulley 1 and winds on pulley 2. This results in lengthening of the chain on the effort side and shortening of the chain on the load side, thus raising the load.

Let P be the effort applied, D be the diameter of pulley 1, d be the diameter of pulley 2, W be the load lifted, and P the effort applied. For one revolution of the upper block, displacement of the effort is πD. This is also equal to the length of the chain that is pulled over the larger pulley. Since the smaller pulley also turns with the larger one, the length of the chain released by the smaller pulley is πd.

Net shorting of the chain is $\pi D - \pi d$ per revolution of upper block.

This shorting of the chain is divided equally between two segments of the chain supporting the load. Therefore,

The distance through which the load will move up is $= \dfrac{\pi}{2}(D - d)$

Velocity ratio

$$\text{VR} = \frac{\text{Distance moved by effort}}{\text{Distance moved by load}} = \frac{\pi D}{\frac{\pi}{2}(D-d)}$$

or

$$\text{VR} = \frac{2D}{D-d}$$

Efficiency

$$\eta = \frac{\text{MA}}{\text{VR}} = \frac{W}{P} \times \left(\frac{D-d}{2D}\right)$$

$$\eta = \frac{W}{P} \times \left(\frac{D-d}{2D}\right)$$

Sometimes, the size of upper pulleys is expressed in terms of number of teeth.

Let T_1 = number of teeth of the larger pulley, and

T_2 = be the number of teeth of the smaller pulley. [*No. of teeth* "T" $\propto$ *Diameter of the pulley* "d"]

$$\text{VR} = \frac{2T_1}{T_1 - T_2}$$

$$\eta = \frac{W}{P} \times \left(\frac{T_1 - T_{21}}{2T}\right)$$

WORM AND WORM WHEEL

Figure 8.12 shows a worm and worm wheel. It consists of a square threaded screw (known as worm) and a toothed wheel (called worm wheel) geared with each other. Wheel B is attached to the worm, over which passes a rope. Sometimes a handle is also fixed to the worm instead of the wheel. A load drum is securely mounted on the worm wheel. The load is attached to the rope wound round the load drum.

Let R be the radius of the effort wheel, r be the radius of load drum, T be the number of teeth on the worm wheel, W be the load lifted, and P be the effort applied.

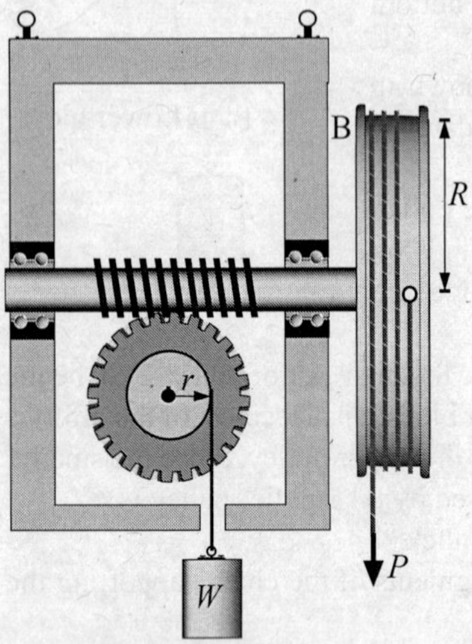

FIGURE 8.12 |

In one revolution of the effort wheel, the distance moved by the effort is $2\pi R$.

If the worm is n-threaded, then for one revolution of the effort wheel, the worm will also complete one revolution. The worm will push the worm wheel through n-teeth.

The load drum will move through $\frac{n}{T}$ revolution.

Therefore,

Distance moved by the load is $= 2\pi r\left(\frac{n}{T}\right)$

Velocity ratio,

$$\text{VR} = \frac{\text{Distance moved by effort}}{\text{Distance moved by load}} = \frac{2\pi R}{2\pi r\frac{n}{T}}$$

$$\text{VR} = \frac{RT}{nr}$$

Mechanical advantage

$$\text{MA} = \frac{\text{Weight lifted}}{\text{Effort applied}} = \frac{W}{P}$$

Efficiency

$$\eta = \frac{\text{MA}}{\text{VR}} = \frac{W}{P} \times \frac{nr}{RT}$$

$$\eta = \frac{W}{P} \times \frac{nr}{RT}$$

SINGLE PURCHASE WINCH CRAB

Figure 8.13 shows the single purchase winch crab. It consists of an effort axle and a load axle. On effort axle, a small gear known as pinion is mounted. On load axle, a large gear is mounted in such a way that the spur wheel meshes with the pinion. One end of rope is fixed to the load drum and wound a few turn around it. The other end of the rope carries the load W.

Let T_1 be the number of teeth on the pinion, T_2 be the number of teeth on the spur wheel, L be the length of lever arm, D be the diameter of load drum, P be the effort applied, and W be the load lifted.

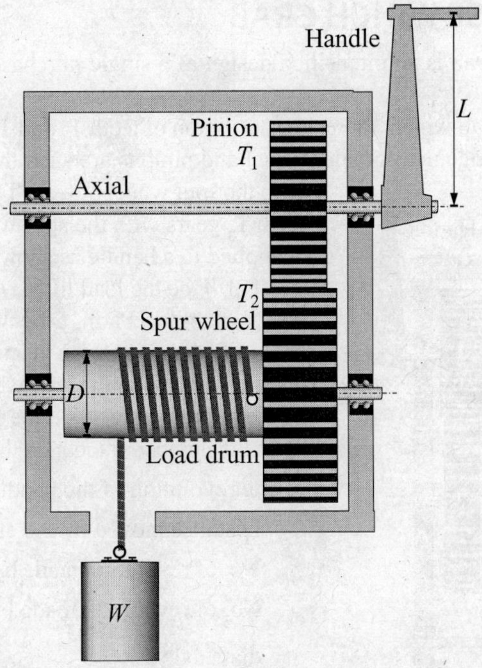

FIGURE 8.13 |

Consider one revolution of the handle.

Distance moved by the effort = $2\pi L$.

Number of revolution made by the pinion = 1

Number of revolution made by the spur wheel $= \dfrac{T_1}{T_2}$

$\therefore$ Distance moved by load $= \pi D \times \dfrac{T_1}{T_2}$

Velocity ratio

$$\text{VR} = \frac{\text{Distance moved by effort}}{\text{Distance moved by load}} = \frac{2\pi L}{\pi D \times \left(\dfrac{T_1}{T_2}\right)}$$

$$\boxed{\text{VR} = \frac{2LT_2}{DT_1}}$$

Mechanical advantage

$$\text{MA} = \frac{\text{Weight lifted}}{\text{Effort applied}} = \frac{W}{P}$$

Efficiency of machine

$$\eta = \frac{\text{MA}}{\text{VR}} = \frac{W}{P} \times \frac{DT_1}{2LT_2}$$

$$\boxed{\eta = \frac{W}{P} \times \frac{DT_1}{2LT_2}}$$

DOUBLE PURCHASE WINCH CRAB

A double purchase winch crab is an intensified design of a single purchase winch crab and obtains a higher value of velocity ratio.

In a double purchase crab winch, there are two pinion of teeth T_1 and T_3 as well as two spur wheel of teeth T_2 and T_4. The arrangement of spur wheel and pinions are such that the pinion with T_1 gears with the spur wheel of teeth T_2. Similarly, the pinion with teeth T_3 gears with the spur wheel of teeth T_4. The effort is applied to a handle as shown in Figure 8.14.

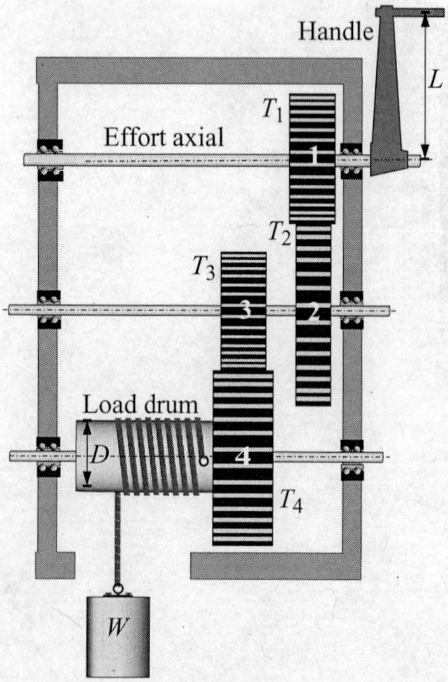

FIGURE 8.14 |

Let W be the load lifted, P be the effort applied, L be the length of lever arm, D be the diameter of load drum, T_1 be the number of teeth on the pinion of effort axle, T_2 be the number of teeth on the spur wheel of intermediate axle, T_3 be the number of teeth on the pinion of intermediate axle, T_4 be the number of teeth on the spur wheel of load axle.

In one revolution of the handle,

Distance moved by the effort = $2\pi L$.

No. of revolution made by pinion 1 of effort axle = 1

No. of revolution made by spur wheel 2 of intermediate axle $= \dfrac{T_1}{T_2}$.

No. of revolution made by pinion 3 of intermediate axle is $\dfrac{T_1}{T_2}$.

No. of revolution made by spur wheel 4 of load axle

$$= \frac{T_1}{T_2} \times \frac{T_3}{T_4}$$

Therefore

$$\text{Distance moved by the load} = \pi D \times \frac{T_1}{T_2} \times \frac{T_3}{T_4}$$

Velocity ratio

$$\text{VR} = \frac{\text{Distance moved by effort}}{\text{Distance moved by load}} = \frac{2\pi L}{\pi D \times \left(\dfrac{T_1}{T_2} \times \dfrac{T_3}{T_4}\right)}$$

$$\text{VR} = \frac{2L}{D}\left(\frac{T_2}{T_1} \times \frac{T_4}{T_3}\right)$$

$$\text{Mechanical advantage MA} = \frac{\text{Weight lifted}}{\text{Effort applied}} = \frac{W}{P}$$

$$\text{Efficiency of machine } \eta = \frac{\text{MA}}{\text{VR}} = \frac{W}{P} \times \left(\frac{D}{2L} \times \frac{T_1}{T_2}\frac{T_3}{T_4}\right)$$

$$\eta = \frac{W}{P} \times \left(\frac{D}{2L} \times \frac{T_1}{T_2}\frac{T_3}{T_4}\right)$$

SCREW JACK

A screw jack is used for lifting heavy weights or loads with the help of a little effort. Figure 8.15 shows a simple screw jack. It consists of a nut, a screw with square threads, and a handle fitted to the head of the screw. The load to be lifted is placed directly on the head of the screw or on a specially designed platform, which is attached to the screw. An effort P is applied at the end of the handle that is attached to the screw head. When the handle is rotated through one complete turn, the screw is also rotated through one turn and hence the screw will have vertical displacement by height p (equal to pitch) of screw.

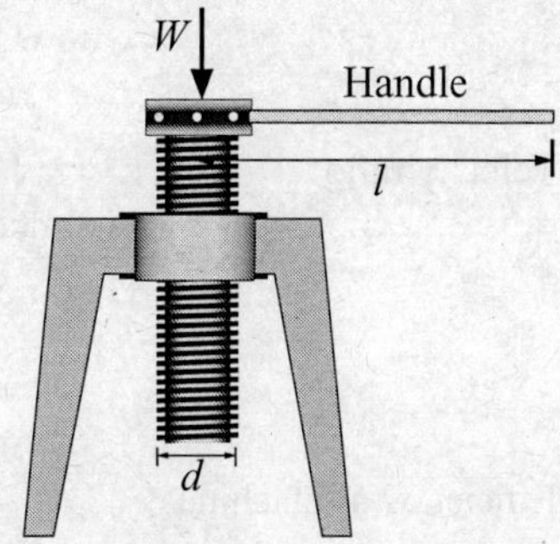

FIGURE 8.15 |

Let W be the load lifted, P be the effort applied at the end of the lever, l be the length of the handle, p be the pitch of the screw, d be the mean diameter of the screw, α be the Helix angle of screw, μ be the coefficient of friction between nut and screw (tan ϕ).

The development of one complete turn of the screw is shown in Figure 8.16. The developed surface is similar to the inclined plane. The distance AB will be equal to the circumference and the distance BC will be equal to the pitch of the screw.

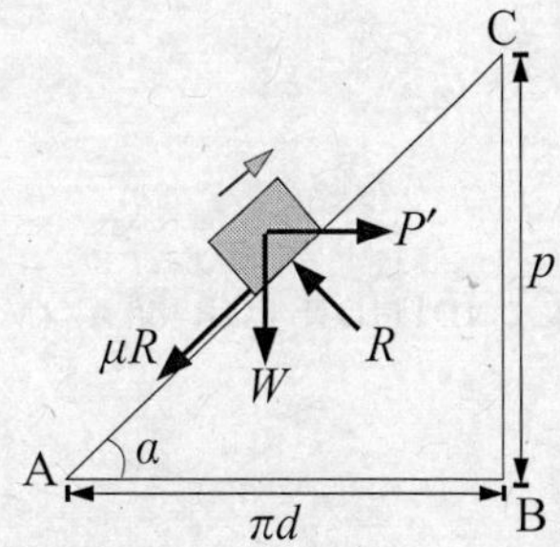

FIGURE 8.16 |

P′ is the effort applied horizontally at the mean radius of the screw, r be the mean radius of screw (d/2).

Resolving the forces along and normal to the inclined plane,

$$W \sin \alpha + \mu R = P' \cos \alpha \qquad \text{(i)}$$

or

$$W \cos \alpha + P' \sin \alpha = R \qquad \text{(ii)}$$

Substituting the value of R, in Equation (i), we get

$$W \sin \alpha + \mu(W \cos \alpha + P' \sin \alpha) = P' \cos \alpha$$

$$W(\sin \alpha + \mu \cos \alpha) = P'(\cos \alpha - \mu \sin \alpha)$$

Substituting

$$\mu = \tan \phi = \frac{\sin \phi}{\cos \phi}$$

$$W\left(\sin \alpha + \frac{\sin \phi}{\cos \phi} \cos \alpha\right) = P'\left(\cos \alpha - \frac{\sin \phi}{\cos \phi} \sin \alpha\right)$$

On simplifying, we get

$$P' = W \frac{\sin \alpha \cos \phi + \sin \phi \cos \alpha}{\cos \alpha \cos \alpha - \sin \phi \sin \alpha}$$

$$P' = W \tan(\alpha + \phi)$$

Moment of P' about the axis of the screw is equal to the moment of force P about axis of screw:

$$P' \times \frac{d}{2} = P \times l \quad \text{or} \quad P = \frac{d}{2l} \times P' = \frac{Wd}{2l} \tan(\alpha + \phi)$$

Mechanical advantage

$$\text{MA} = \frac{\text{Weight lifted}}{\text{Effort applied}} = \frac{W}{P} = \frac{W}{\frac{Wd}{2l} \tan(\alpha + \phi)}$$

$$\text{MA} = \frac{2l}{d\tan(\alpha+\phi)} = \frac{2l}{d}\cot(\alpha+\phi)$$

Velocity ratio

$$\text{VR} = \frac{\text{Distance moved by effort}}{\text{Distance moved by load}} = \frac{2\pi l}{P} = \frac{2l}{d}\cot\alpha \left(\tan\alpha = \frac{p}{\pi d}\right)$$

$$\text{VR} = \frac{2l}{d}\cot\alpha$$

Efficiency of machine

$$\eta = \frac{\text{MA}}{\text{VR}} = \frac{\frac{2l}{d}\cot(\alpha+\phi)}{\frac{2l}{d}\cot\alpha}$$

$$\eta = \frac{\tan\alpha}{\tan(\alpha+\phi)}$$

Condition for Maximum Efficiency

$$\frac{\mathrm{d}}{\mathrm{d}\alpha}\eta = 0 \Rightarrow \frac{\mathrm{d}}{\mathrm{d}\alpha}\left(\frac{\tan\alpha}{\tan(\alpha+\phi)}\right) = 0$$

$$\frac{\tan(\alpha+\phi)\sec^2\alpha . - \tan\alpha\sec^2(\alpha+\phi)}{\tan^2(\alpha+\phi)} = 0$$

On simplifying, we get

$$\sin(2\alpha+2\phi) = \sin 2\alpha$$

Solution of above equation

$$2\alpha+2\phi = \pi - 2\alpha$$

$$\alpha = \frac{\pi}{4} - \frac{\phi}{2}$$

Therefore, maximum efficiency of screw jack

$$\eta = \frac{\tan\left(\frac{\pi}{4}-\frac{\phi}{2}\right)}{\tan\left(\frac{\pi}{4}-\frac{\phi}{2}+\phi\right)} = \frac{\tan\left(\frac{\pi}{4}-\frac{\phi}{2}\right)}{\tan\left(\frac{\pi}{4}+\frac{\phi}{2}\right)}$$

On simplifying, we get

$$\eta_{\max} = \frac{1-\sin\phi}{1+\sin\phi}$$

DIFFERENTIAL SCREW JACK

Figure 8.17 shows a differential screw jack. It is an improvement over simple screw jack. It consists of two screws S_A and S_B. Both S_A and S_B have threads in the same direction (right handed). S_A is threaded on the outside only, whereas S_B is threaded both on inside and outside. The external threads of S_B mesh with the thread of nut. The internal threads of S_B gear with the external thread of S_A. The screw S_A does not rotate but free to move in vertical direction. It carries the load when the effort is applied at the end of the lever. S_B rises up and S_A goes down making the lift of the load as the algebraic sum of the motion of the screws S_A and S_B.

Let L be the length of lever arm, W be the load lifted, P be the effort applied, p_A be the pitch of the screw S_A, and p_B be the pitch of the screw S_B ($p_B > p_A$).

Consider one revolution of the lever arm. Distance moved by the effort is $2\pi L$. Upward distance moved by S_B is p_B and downward distance moved by S_A is p_A.

Velocity ratio

$$\text{VR} = \frac{\text{Distance moved by effort}}{\text{Distance moved by load}} = \frac{2\pi L}{p_B - p_A}$$

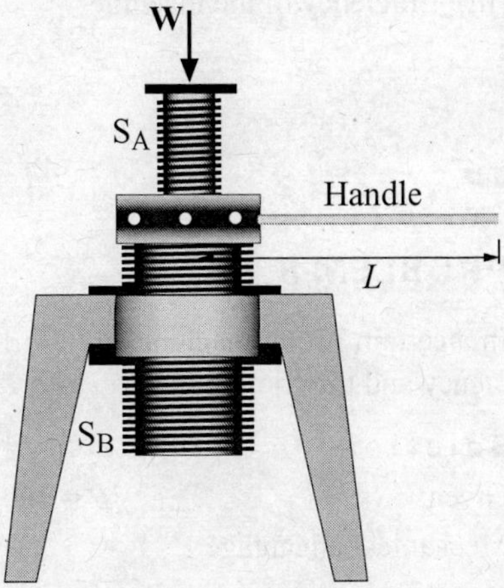

FIGURE 8.17 |

$$\text{VR} = \frac{2\pi L}{p_B - p_A}$$

Mechanical advantage

$$\text{MA} = \frac{\text{Weight lifted}}{\text{Effort applied}} = \frac{W}{P}$$

Efficiency of machine

$$\eta = \frac{\text{MA}}{\text{VR}} = \frac{W}{P}\left(\frac{p_B - p_A}{2\pi L}\right)$$

$$\eta = \frac{W}{P}\left(\frac{p_B - p_A}{2\pi L}\right)$$

SOLVED PROBLEMS

PROBLEM 8.1

A weight of 450 N is to be raised by means of wheel and axle. The axle is 100 mm diameter and the wheel is 500 mm diameter. If a force of 120 N has to be applied to the wheel, find (i) mechanical advantage, (ii) velocity ratio and (iii) efficiency of the machine.

Solution. Given: weight, $W = 450$ N, effort $P = 120$ N, diameter of axial, $d = 100$ mm, diameter of wheel, $D = 500$ mm.

(i) Mechanical advantage:

$$\text{MA} = \frac{W}{P} = \frac{450}{120} = 3.75$$

$$\text{MA} = 3.75$$

(ii) Velocity ratio:

$$\text{VR} = \frac{D}{d} = \frac{500}{100} = 5$$

$$\text{VR} = 5$$

(iii) Efficiency of the machine

$$\eta = \frac{\text{MA}}{\text{VR}} = \frac{3.75}{5} = 0.75 = 75\%$$

$$\eta = 75\%$$

PROBLEM 8.2

In a certain lifting machine, an effort of 100 N is just able to lift a load of 850 N. Calculate the efficiency and the friction both on effort and load sides, if the velocity ratio of the machine is 10.

Solution.

Given: $P = 100$ N, $W = 850$ N, VR = 10

Mechanical advantage

$$\text{MA} = \frac{W}{P} = \frac{850}{100} = 8.5$$

Efficiency

$$\eta = \frac{\text{MA}}{\text{VR}} \Rightarrow \eta = \frac{8.5}{10}$$

$$\eta = 85\%$$

Friction of the machine in terms of effort is

$$F_\text{p} = P - \frac{W}{VR}$$

or

$$F_\text{p} = 100 - \frac{850}{10} = 15$$

$$F_\text{p} = 15\text{N} \qquad \text{(i)}$$

Friction of the machine in terms of load is

$$F_\text{w} = W_i - W = P \times VR - W$$

or

$$F_\text{w} = 100 \times 10 - 850$$

$$F_\text{w} = 150\text{N} \qquad \text{(ii)}$$

It may be noted from Equations (i) and (ii) that an effort of 15 N is required to overcome the friction, or in other words, this can lift an additional load of 150 N.

PROBLEM 8.3

There are four pulleys in a third system of pulleys. An effort of 150 N is required to lift an unknown weight. If the efficiency of this machine is 75%, find the weight lifted.

Solution.

Given: number of pulleys, $n = 4$, effort, P = 150 N, efficiency, $\eta = 0.75$

Let W be the weight lifted. Velocity ratio of the third system of pulley is given by

$$\text{VR} = 2^n - 1$$

or

$$\text{VR} = 2^4 - 1 = 15$$

Mechanical advantage

$$\text{MA} = \frac{W}{P}$$

Efficiency of machine

$$\eta = \frac{\text{MA}}{\text{VR}} = \frac{\frac{W}{P}}{\text{VR}} \Rightarrow W = \eta \times \text{VR} \times P$$

or

$$W = 0.75 \times 15 \times 150 = 1687.5$$

$$W = 1687.5\text{N}$$

PROBLEM 8.4

With a wheel and differential axle, an effort of 10 N raised a load of 50 N. If the efficiency of the machine at this load is 75%, find the velocity ratio of the machine.

If the diameter of the effort wheel is 400 mm, determine the difference between the diameters of the axles. If the sum of the diameters of the axles is 300 mm, determine the diameter of each axle.

Solution.

Given: $P = 10$ N, $W = 50$ N, $\eta = 0.75$, $D = 400$ mm, and $d_1 + d_2 = 300\text{mm}$

Mechanical advantage

$$\text{MA} = \frac{W}{P} = \frac{50}{10} = 5$$

Efficiency

$$\eta = \frac{\text{MA}}{\text{VR}} \Rightarrow 0.75 = \frac{5}{\text{VR}}$$

Velocity ratio

$$\text{VR} = \frac{5}{0.75} = 6.66$$

$$\text{VR} = 6.66$$

Velocity ratio of wheel and differential axle is given by

$$\text{VR} = \frac{2D}{d_1 - d_2}$$

or

$$6.66 = \frac{2 \times 400}{d_1 - d_2}$$

or

$$d_1 - d_2 = 120.01 \qquad \text{(i)}$$

and given

$$d_1 + d_2 = 300\text{mm} \qquad \text{(ii)}$$

Solving from Equations (i) and (ii), we get

$$d_1 = 210\text{mm} \quad \text{and} \quad d_2 = 90\text{mm}$$

PROBLEM 8.5

A weight of 25 kN is to be lifted by a wheel and differential axle. It consists of differential axle 200 mm and 240 mm diameter and the wheel diameter is 1000 mm. Find the effort required, if the efficiency of the machine is 60%.

Solution.

Given: W = 25 kN, $d_1 = 240$ mm, $d_2 = 200$ mm, $D = 1000$ mm and $\eta = 0.6$

Velocity ratio of wheel and differential axle is given by

$$\text{VR} = \frac{2D}{d_1 - d_2} = \frac{2 \times 1000}{240 - 200} = 50$$

$$\text{VR} = 50$$

$$\text{Efficiency } \eta = \frac{\text{Mechanical advantage (MA)}}{\text{Velocity ratio (VR)}}$$

or
$$0.6 = \frac{\text{MA}}{50}$$

or
$$\text{MA} = 30 = \frac{W}{P}$$

or
$$\frac{25}{P} = 30$$

or
$$P = \frac{25}{30} = 8.33\text{kN}$$

$$P = 8333\text{N}$$

PROBLEM 8.6

The efficiency of a Weston's differential pulley block is 50%. The upper block has two pulleys of diameter 25 cm and 20 cm. Find the load lifted by this machine if effort applied is 20 N.

Solution. Given: effort applied, $P = 20$ N, diameter of pulley 1, $D = 25$ cm, diameter of pulley 2, $d = 20$ cm, and efficiency of the machine, $\eta = 50\%$.

Let W be load lifted.

Velocity ratio is given by

$$\text{VR} = \frac{\text{Distance moved by effort}}{\text{Distance moved by load}} = \frac{\pi D}{\frac{\pi}{2}(D - d)}$$

or
$$\text{VR} = \frac{2D}{D - d},$$

and efficiency

$$\eta = \frac{\text{MA}}{\text{VR}} = \frac{W}{P} \times \left(\frac{D-d}{2D}\right)$$

or

$$0.5 = \frac{W}{20} \times \left(\frac{25-20}{2 \times 25}\right) \Rightarrow W = \frac{0.5 \times 20 \times 2 \times 25}{25-20}$$

$$\boxed{W = 100\text{N}}$$

PROBLEM 8.7

The number of teeth on the worm wheel of a single worm and worm wheel is 60. Calculate the velocity ratio if the diameter of effort wheel is 25 cm and that of load drum is 12.5 cm. The effort required to lift a load of 600 N by this machine is 20 N. Find the efficiency of the machine.

Solution. Given: radius of the effort wheel, $R = 25/2 = 12.5\text{cm}$, radius of load drum, $R = 12.5/2 = 6.25\text{cm}$, number of teeth on the worm wheel, $T = 60$, load lifted, $W = 600$ N, effort applied, $P = 20\text{N}$.

Velocity ratio for a single-threaded worm ($n = 1$) and worm wheel is given by

$$\text{VR} = \frac{\text{Distance moved by effort}}{\text{Distance moved by load}} = \frac{2\pi R}{2\pi r \dfrac{n}{T}}$$

or

$$\text{VR} = \frac{RT}{nr} = \frac{12.5 \times 60}{1 \times 6.25} = 120$$

$$\boxed{\text{VR} = 120}$$

Efficiency is given by

$$\eta = \frac{\text{MA}}{\text{VR}} = \frac{\dfrac{W}{P}}{\text{VR}}$$

or

$$\eta = \frac{600/20}{120} = 0.25$$

$$\boxed{\eta = 25\%}$$

PROBLEM 8.8

A single purchase winch crab has the following particulars: number of teeth on the pinion is 16, number of teeth on the spur wheel is 96, length of lever arm is 70 cm and diameter of load drum is 20 cm.

It is observed that an effort of 60 N lifts a load of 1800 N and an effort of 120 N lifts a load of 3960 N.

(i) Establish the law of the machine.

(ii) Find efficiency in the two cases.

(iii) Determine the loss of load and the loss of effort in friction when lifting the load of 3960 N.

Solution.

Given $T_2 = 96$, $T_1 = 16$, $L = 70$ cm and $D = 20$ cm.

(i) Law of the machine

Here, when $P_1 = 60$ N, $W_1 = 1800$ N and when $P_2 = 120$ N, $W_2 = 3960$ N, substituting these values of P and W in the law of the machine, $P = mW + c$ we have

$$60 = m \times 1800 + c \qquad \text{(i)}$$

$$20 = m \times 3960 + c \qquad \text{(ii)}$$

From Equations (i) and (ii), we get

$$c = 10 \quad \text{and} \quad m = \frac{1}{36}$$

Substituting the value of c and m in the law of machine, we have

$$P = \frac{1}{36}W + 10$$

(ii) Efficiencies of machine in the two cases: velocity ratio of single purchase winch crab is given by

$$\text{VR} = \frac{\text{Distance moved by effort}}{\text{Distance moved by load}} = \frac{2\pi L}{\pi D \times \dfrac{T_1}{T_2}}$$

$$\text{VR} = \frac{2LT_2}{DT_1} = \frac{2 \times 70 \times 96}{20 \times 16} = 42$$

$$\text{VR} = 42$$

Mechanical advantage in the first case

$$\text{MA}_1 = \frac{W_1}{P_1} = \frac{1800}{60} = 30$$

Therefore, efficiency in first case

$$\eta_1 = \frac{\text{MA}_1}{\text{VR}} = \frac{30}{42} = 0.714$$

$$\eta_1 = 71.4\%$$

Mechanical advantage in the second case

$$\text{MA}_2 = \frac{W_2}{P_2} = \frac{3960}{120} = 33$$

Therefore, efficiency in the second case

$$\eta_2 = \frac{\text{MA}_2}{\text{VR}} = \frac{33}{42} = 0.786$$

$$\eta_2 = 78.6\%$$

(iii) The loss of load, and the loss of effort in friction: here $P = 120$ N, $W = 3960$ N.

Friction of the machine in terms of load is

$$F_{\mathrm{w}} = W_i - W = P \times VR - W$$

or

$$F_{\mathrm{w}} = 120 \times 42 - 3690 = 1080$$

$$F_{\mathrm{w}} = 1080\mathrm{N}$$

Friction of the machine in terms of effort is

$$F_{\mathrm{p}} = P - P_i = P - \frac{W}{\mathrm{VR}}$$

or

$$F_{\mathrm{p}} = 120 - \frac{3960}{42} = 25.7$$

$$F_{\mathrm{p}} = 25.7\mathrm{N}$$

PROBLEM 8.9

In a double purchase winch crab, the pinions have 15 and 20 teeth, while the spur wheels have 45 and 40 teeth. The effort handle is 40 cm while the effective diameter of the drum is 15 cm. If the efficiency is 40% and load lifted is 250 N, then what effort will be applied at the end of the handle.

Solution. Given: load lifted $W = 250$ N, length of lever arm $L = 40$ cm, diameter of load drum, $D = 15$ cm, efficiency $\eta = 40\%$, number of teeth on the pinion of effort axle $T_1 = 15$, number of teeth on the pinion of intermediate axle $T_3 = 20$, number of teeth on the spur wheel of intermediate axle $T_2 = 45$, number of teeth on the spur wheel of load axle $T_4 = 40$.

Let P be the effort applied. Velocity ratio is given by

$$\mathrm{VR} = \frac{\text{Distance moved by effort}}{\text{Distance moved by load}}$$

$$= \frac{2\pi L}{\pi D \times \left(\frac{T_1}{T_2}\right) \times \left(\frac{T_3}{T_4}\right)}$$

or

$$\mathrm{VR} = \frac{2\pi \times 40}{\pi \times 15 \times [(15/45) \times (20/40)]} = 32$$

$$\mathrm{VR} = 32$$

Efficiency

$$\eta = \frac{\mathrm{MA}}{\mathrm{VR}} = \frac{\frac{W}{P}}{\mathrm{VR}}$$

or

$$P = \frac{W}{\eta \times \mathrm{VR}} = \frac{250}{0.4 \times 32}$$

$$P = 19.53\mathrm{N}$$

PROBLEM 8.10

A load of 10 kN is raised by means of a screw jack, having a square threaded screw of 12 mm pitch and of mean diameter 50 mm. If an effort of 100 N is applied at the end of a lever to raise the load, what should be the length of the lever used? Coefficient of friction is 0.15. State whether the screw is self-locking.

Solution.

Given: $W = 10$ kN, $p = 12$ mm, $d = 50$ mm, $P = 100$ N, $\mu = 0.15$.

Let l be the length of lever. Friction angle

$$\phi = \tan^{-1}\mu = \tan^{-1} 0.15$$

$$\phi = 8.53^\circ$$

Helix angle of screw

$$a = \tan^{-1}\frac{p}{\pi d} = \tan^{-1}\frac{12}{\pi \times 50}$$

$$a = 4.36^\circ$$

Therefore, effort required at the circumference of the screw to raise the load

$$P' = W\tan(a + \phi)$$

or $$P' = 10\tan(8.53 + 4.36)$$

or $$P' = 2.29\text{kN}$$

and torque required to overcome friction,

$$T' = P' \times \frac{d}{2}$$

or $$T' = 2.29 \times 1000 \times \frac{50}{2 \times 1000}$$

or $$T' = 57.25\text{Nm}$$

We know that torque applied at the end of the lever is

$$P \times l = P' \times \frac{d}{2} = T'$$

or $$l = \frac{T'}{P}$$

or $$l = \frac{57.25}{100} = 0.5725\text{m}$$

$$l = 57.25\text{cm}$$

Efficiency of the screw jack,

$$\eta = \frac{\tan \alpha}{\tan(\alpha + \phi)}$$

or

$$\eta = \frac{\tan 4.36}{\tan(4.36 + 8.53)} = 0.333$$

$$\eta = 33.3\%$$

Since the efficiency of the screw jack is less than 50%, the screw is a self-locking screw.

PROBLEM 8.11

A differential screw jack has the following specifications: pitch of larger screw is 12 mm, pitch of smaller screw is 8 mm and length of lever from centre of screw is 400 mm.

Find the velocity ratio. If an effort of 75 N applied at the end of the lever lifts a load of 10 kN, determine the efficiency of the differential screw jack.

Solution. Given: pitch of larger screw $p_B = 12$ mm, pitch of smaller screw $p_A = 8$ mm, length of lever $L = 400$ mm, effort applied, $P = 75$ N and load lifted $W = 10$ kN $= 10000$ N.

Velocity ratio of differential screw jack is given by

$$\text{VR} = \frac{2\pi L}{p_B - p_A}$$

or

$$\text{VR} = \frac{2\pi \times 400}{12 - 8} = 628$$

$$\text{VR} = 628$$

Mechanical advantage of differential screw jack,

$$\text{MA} = \frac{W}{P} = \frac{10000}{75} = 133.3$$

Efficiency,

$$\eta = \frac{\text{MA}}{\text{VR}} = \frac{W}{P}\left(\frac{p_B - p_A}{2\pi L}\right)$$

or

$$\eta = \frac{10000}{75}\left(\frac{12 - 8}{2\pi \times 400}\right) = 0.212$$

$$\eta = 21.2\%$$

CHAPTER HIGHLIGHTS

- Mechanical advantage (MA) is the ratio of load (*or weight*) lifted to the effort applied.

$$\text{MA} = \frac{\text{Weight lifted}}{\text{Effort applied}} = \frac{W}{P}$$

- Velocity ratio (VR) is the ratio of the distance moved by the effort to the distance moved by the load.

$$\text{VR} = \frac{\text{Distance moved by effort}}{\text{Distance moved by load}} = \frac{y}{x}$$

- Input of machine is defined as the work done on the machine.
- Output of machine is defined as the actual work done by the machine.
- Efficiency of machine is the ratio of output to input of machine.
- If the efficiency of machine is 1 (100%), the machine is called as *Ideal Machine*.
- Ideal effort "P_i" is the effort required to lift the weight by ideal machine and ideal load "W_i" is the load that can be lifted using the given effort by the *ideal machine*.

$$\text{Ideal effort } P_i = \frac{W}{\text{VR}} \quad \text{and} \quad \text{Ideal load } W_i = P \times \text{VR}$$

- Friction of the machine in terms of effort is $F_\text{p} = P - \dfrac{W}{\text{VR}}$
- Friction of the machine in terms of load is $F_\text{W} = P \times \text{VR} - W$
- If the effort is removed, the load may or may not fall down to the initial position. If the load falls down, the machine is said to be *reversible*, and if the load does not fall down, the machine is said to be *irreversible* or *self-locking* machine.
- The relationship between the load lifted (W) and the effort required (P) is called the *law of machine*. $P = m\,\text{W} + c$
- $\text{MA} = \dfrac{1}{m + \left(\dfrac{c}{W}\right)}$, $(\text{MA})_\text{max} = \dfrac{1}{m}$, $\eta = \dfrac{1}{\text{VR} \times \left[m + \left(\dfrac{c}{W}\right)\right]}$ and $\eta_\text{max} = \dfrac{1}{\text{VR} \times m}$
- Lifting Machines:

1. First System of Pulleys VR = 2^n,
2. Second System of Pulleys VR = n
3. Third System of Pulleys VR = $2^n - 1$
4. Simple Wheel and Axle $\text{VR} = \dfrac{D}{d}$ and $\eta = \dfrac{W}{P} \times \dfrac{d}{D}$
5. Wheel and Differential Axle $\text{VR} = \dfrac{2D}{d_1 - d_2}$ and $\eta = \dfrac{W}{P} \times \left(\dfrac{d_1 - d_2}{2D}\right)$
6. Weston's Differential Pulley Block $\text{VR} = \dfrac{2D}{D - d}$ and $\eta = \dfrac{W}{P} \times \left(\dfrac{D - d}{2D}\right)$
7. Worm and Worm Wheel $\text{VR} = \dfrac{RT}{nr}$ and $\eta = \dfrac{W}{P} \times \dfrac{nr}{RT}$

8. Single Purchase Winch Crab $VR = \frac{2LT_2}{DT_1}$ and $\eta = \frac{W}{P} \times \frac{DT_1}{2LT_2}$

9. Double Purchase Winch Crab $VR = \frac{2L}{D}\left(\frac{T_2}{T_1} \times \frac{T_4}{T_3}\right)$ and $\eta = \frac{W}{P} \times \left(\frac{D}{2L} \times \frac{T_1T_3}{T_2T_4}\right)$

10. Screw Jack $VR = \frac{2l}{d}\cot a$, $\eta = \frac{\tan a}{\tan(a + \phi)}$ and $\eta_{max} = \frac{1 - \sin\phi}{1 + \sin\phi}$

11. Differential Screw Jack: $VR = \frac{2\pi L}{p_B - p_A}$ and $\eta = \frac{W}{P}\left(\frac{p_B - p_A}{2\pi L}\right)$

REVIEW QUESTIONS

Q1. Define velocity ratio, mechanical advantage and efficiency of machine.

Q2. Define the efficiency of machine. Derive an expression for efficiency of a machine in terms of mechanical advantage and velocity ratio.

Q3. What is an ideal load in terms of velocity ratio?

Q4. Prove that the friction of a machine is given by $F_p = P - (W/VR)$.

Q5. Explain the difference between a reversible machine and a self-locking machine.

Q6. Prove that the condition for a self-locking machine is that its efficiency is less than 50%.

Q7. What is law of machine?

Q8. How does the efficiency of a machine depend on load? Obtain an expression for the maximum efficiency of a machine.

Q9. How does the mechanical advantage of a machine depend on load? Obtain an expression for the maximum mechanical advantage of a machine.

Q10. Describe the first, second and third systems of pulleys.

Q11. Obtain the expression for velocity ratio for the following system

(i) simple pulley

(ii) first system of pulleys

(iii) second system of pulleys

(iv) third system of pulleys.

Q12. Describe the working of simple wheel and axle. Obtain an expression for velocity ratio and efficiency.

Q13. Describe the working of wheel and differential axle. Obtain an expression for velocity ratio and efficiency.

Q14. Describe the working of Weston's differential pulley block. Obtain an expression for velocity ratio and efficiency.

Q15. Describe the working of worm and worm wheel. Obtain an expression for velocity ratio and efficiency.

Q16. Describe the working of single purchase winch crab. Obtain an expression for velocity ratio and efficiency.

Q17. Describe the working of double purchase winch crab. Obtain an expression for velocity ratio and efficiency.

Q18. Describe the working of screw jack. Obtain an expression for velocity ratio and efficiency of machine.

Q19. Describe the working of differential screw jack. Obtain an expression for velocity ratio and efficiency of machine.

Q20. Prove that the efficiency of a screw jack for raising a load is given by $\eta = (\tan \alpha)/\tan(\alpha + \phi)$.

Q21. Derive an expression for maximum efficiency of screw jack.

Q22. Prove that the effort applied at the end of the handle of a screw jack to lift a load W is given by $P = \left(\frac{Wd}{2l}\right)\tan(\alpha + \phi)$, where α is the angle of screw, ϕ is the friction angle, d is the mean diameter of the screw, l is the length of handle.

Q23. The following observations were made in an experiment on a lifting machine:

Effort P (N)	26	47	76	95	105	127
Load W (N)	500	1000	1500	2000	2500	3000

Draw the load versus effort graph and determine the law of machine. If the velocity ratio is 30, determine the efficiency while lifting a load of 1800 N, what is the maximum efficiency of the machine.

[Ans; $P = 0.04W + 6$; $\eta = 51.28\%$; $\eta_{max} = 83.33\%$].

Q24. The law of a machine is $P = \left(\frac{W}{25}\right) + 20$ N. If the velocity ratio of the machine is 50, determine its maximum efficiency and maximum mechanical advantage.

[Ans; 50%, 25].

Q25. A Weston's differential pulley block of diameters 400 mm and 80 mm is used to lift a load of 40,000 N. Find the effort required if efficiency of the system is 60%.

[Ans; 26667 N].

MULTIPLE CHOICE QUESTIONS

Q1. In actual machines

(A) Mechanical advantage is greater than velocity ratio

(B) Mechanical advantage is equal to velocity ratio

(C) Mechanical advantage is less than velocity ratio

(D) None of the above

[Ans: C]

Q2. The velocity ratio in case of an inclined plane inclined at angle θ to horizontal and weight being pulled up the inclined plane by effort, is

(A) $\sin \theta$ (B) $\frac{1}{\sin \theta}$ (C) $\frac{1}{\cos \theta}$ (D) $\tan \theta$

[Ans: B]

Q3. In a simple screw jack, the pitch of the screw is 9 mm and the length of the handle operating the screw is 45 cm, then the velocity ratio of system will be

(A) 5 (B) 157 (C) 314.14 (D) 15. 73

[Ans: C]

Q4. For a machine to be self-locking, its efficiency should be

(A) 100% (B) less than 50% (C) greater than 50% (D) between 50 % and 100%

[Ans: B]

Q5. The efficiency of a screw jack with Helix angle α and friction angle ϕ is equal to

(A) $\frac{\tan\alpha}{\tan(\alpha-\phi)}$ (B) $\frac{\tan(\alpha+\phi)}{\tan\alpha}$ (C) $\frac{\tan(\alpha-\phi)}{\tan\alpha}$ (D) $\frac{\tan\alpha}{\tan(\alpha+\phi)}$

[Ans: D]

Q6. The efficiency of a screw jack is maximum when its helix angle α and friction angle ϕ is equal to

(A) $\frac{\pi}{4}-\frac{\phi}{2}$ (B) $\frac{\pi}{4}+\frac{\phi}{2}$ (C) $\frac{\pi}{4}-\phi$ (D) $\frac{\pi}{4}-\frac{\phi}{4}$

[Ans: A]

Q7. The maximum efficiency of a screw jack for angle of friction ϕ is

(A) $\frac{1+\sin\phi}{1-\sin\phi}$ (B) $\frac{1-\sin\phi}{1+\sin\phi}$ (C) $\frac{1-\sin\phi}{\sin\phi}$ (D) $\frac{\sin\phi}{1+\sin\phi}$

[Ans: B]

Q8. A Weston's pulley block consists of

(A) One big diameter wheel and one axle
(B) One big diameter wheel and two axles of different diameters
(C) Two big wheels and two axles of different diameters
(D) Two big wheels of different diameters and one axle

[Ans: D]

Q9. The maximum efficiency of a screw jack with square threads and friction angle of 30° can be

(A) 100% (B) 50% (C) 33.33% (D) 11.11%

[Ans: C]

Q10. A machine is said to be irreversible if its efficiency is

(A) 100% (B) less than 50%
(C) greater than 50% (D) between 50% and 100%

[Ans: B]

Q11. In a simple screw jack, with l as the length of the handle and p as pitch of the screw, its velocity ratio is

(A) $\frac{2\pi l}{p}$ (B) $\frac{\pi l}{2p}$ (C) $\frac{2\pi p}{l}$ (D) $\frac{\pi p}{l}$

[Ans: A]

Q12. If a Weston's differential pulley block has larger and smaller diameter of 100 mm and 80 mm, respectively, its velocity ratio is

(A) 5 (B) 10 (C) 20 (D) 40

[Ans: B]

Q13. The velocity ratio of a first system of pulleys with four pulleys is

(A) 4 (B) 8 (C) 16 (D) 15

[Ans: C]

Q14. If the law of machine is $P = 0.05W + 50$, then the maximum value of mechanical advantage is

(A) 5 (B) 10 (C) 20 (D) 40

[Ans: C]

Q15. If the law of machine is $P = 0.01W + 50$ and velocity ratio is 8, then the maximum efficiency of machine is

(A) 0 (B) 5.6 (C) 32.8 (D) 12.5

[Ans: D]

9 Kinematics of Particles and Rigid Bodies

INTRODUCTION

Dynamics is part of mechanics that deals with the analysis of bodies in motion. Dynamics is divided into two parts, (1) kinematics and (2) kinetics. Kinematics is the study of the geometry of motion. It is used to relate displacement, velocity, acceleration and time, without reference to the cause of the motion. Kinetics is the study of the relation existing between the forces acting on a body, the mass of the body and the motion of the body. It is used to predict the motion caused by given forces or to determine the forces required to produce a given motion.

IMPORTANT TERMS

Rest: When a body occupies a fixed position for any length of time, it is said to be at rest.

Motion: A body is said to be in motion when it changes its position with respect to some other fixed object.

Displacement: The change in position of a body in a definite direction is known as displacement, and it is measured by the straight distance between the initial and the final positions of the moving body. The displacement represents both magnitude and direction followed by the body. Hence, it is a vector quantity.

Speed: The speed of a moving body is the rate of change of its position. It is calculated as

$$\text{speed} = \frac{\text{distance travelled}}{\text{time taken}}$$

When the distances travelled in equal interval of time are equal, then we say that the body is moving with a uniform speed.

Velocity: The velocity of a moving body is the rate of change of its displacement in a specific direction:

$$\text{velocity} = \frac{\text{change in displacement}}{\text{time taken}}$$

Acceleration: The rate of change of velocity of a body is defined as its acceleration:

$$\text{acceleration} = \frac{\text{change in velocity}}{\text{time taken}}$$

TYPES OF MOTION

Rectilinear motion: The motion of a body is called rectilinear when it moves in a straight line.

Curvilinear motion: The motion of a body is called curvilinear when it moves along a curved line.

Circular motion or rotary motion: The motion of a body is called circular when it moves along a circular path. Circular motion is a special case of curvilinear motion.

RECTILINEAR MOTION: VELOCITY AND ACCELERATION

Consider the motion of a particle moving along a straight line O-x, on which O is a fixed point (Fig. 9.1). Let P and Q be the positions of the particle at time t and $t + \Delta t$, respectively, such that OP = s and $OQ = s + \Delta s$. Then the displacement moved in time Δt is PQ = Δs. Therefore, the average rate of displacement of particle $\Delta s/\Delta t$, or average velocity

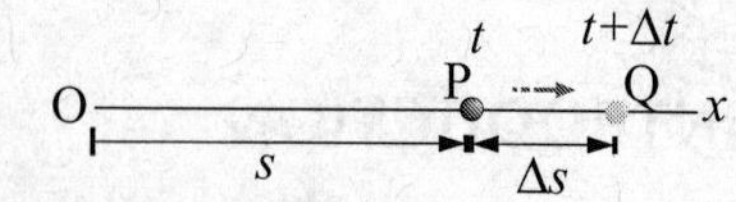

FIGURE 9.1

$$v_{avg} = \frac{\Delta s}{\Delta t}$$

If this ratio has the same value for all intervals of time, then the velocity of the particle is said to be uniform; otherwise, variable.

As Δt becomes smaller and approaches zero in the limit, the average velocity approaches the instantaneous velocity of the particle, which is

$$v = \lim_{\Delta t \to 0} \frac{\Delta s}{\Delta t} = \frac{ds}{dt}$$

$$v = \frac{ds}{dt}$$

The average acceleration of the particle during the interval Δt is the change in its velocity divided by the time interval, or $a_{avg} = \Delta v/\Delta t$. As Δt becomes smaller and approaches zero in the limit, the average acceleration approaches the instantaneous acceleration of the particle, which is

$$a = \lim_{\Delta t \to 0} \frac{\Delta v}{\Delta t} = \frac{dv}{dt}$$

$$a = \frac{dv}{dt} \quad \text{or} \quad a = \frac{d^2 s}{dt^2} \quad \text{or} \quad a = \frac{dv}{ds}\frac{ds}{dt} = v\frac{dv}{ds}$$

EQUATIONS OF MOTION: RECTILINEAR

When the acceleration is constant, each of the three equations $a = dv/dt$, $v = ds/dt$ and $a = v(dv/ds)$ may be integrated to obtain formulas that relate a, v, s, and t

VELOCITY AS A FUNCTION OF TIME

Integrate $a = dv/dt$ or $dv = a dt$ assuming that initially $v = u$ at $t = 0$:

$$\int_u^v dv = \int_0^t a dt$$

or

$$v - u = at$$

$$v = u + at$$

Position as a Function of Time

Integrate $v = \mathrm{d}s/\mathrm{d}t$ or $\mathrm{d}s = v\mathrm{d}t$, assuming that initially $s = 0$ at $t = 0$.

$$\int_0^s \mathrm{d}s = \int_0^t v\mathrm{d}t$$

Substituting $v = u + at$, we have

$$\int_0^s \mathrm{d}s = \int_0^t (u + at)\mathrm{d}t$$

$$s = ut + \frac{1}{2}at^2$$

Position as a Function of Time

Integrate $a = v(\mathrm{d}v/\mathrm{d}s)$, or $v\mathrm{d}v = a\mathrm{d}s$, assuming that initially $v = u$ at $s = 0$.

$$\int_u^v v\mathrm{d}v = \int_0^s a\mathrm{d}s$$

or

$$\frac{v^2}{2} - \frac{u^2}{2} = as$$

$$v^2 = u^2 + 2as$$

CURVILINEAR MOTION: VELOCITY AND ACCELERATION

Consider a continuous motion of a particle along a curve path as represented in Figure 9.2. At time t, the particle is at position P, which is located by the position vector $\mathbf{r} = x\hat{i} + y\hat{j} + z\hat{k}$ measured from the fixed origin O. At time $t + \Delta t$, the particle is at Q, located by the position vector $\mathbf{r} + \Delta\mathbf{r}$. The displacement of the particle during time Δt is the vector $\Delta\mathbf{r}$, which represents the vector change of position. The average velocity of particle over the time interval Δt is defined as

$$\mathbf{v}_{\text{avg}} = \frac{\Delta\mathbf{r}}{\Delta t}$$

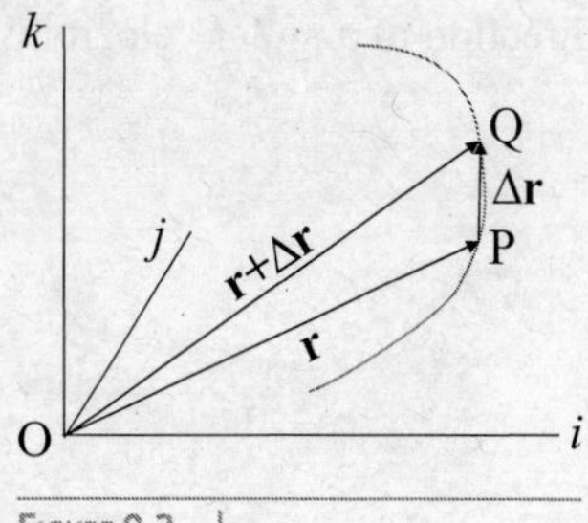

Figure 9.2 |

The instantaneous velocity $\mathbf{v}$ of the particle is defined as the limiting value of the average velocity as the time interval approaches zero. Thus,

$$\mathbf{v} = \lim_{\Delta t \to 0} \frac{\Delta\mathbf{r}}{\Delta t} = \frac{\mathrm{d}\mathbf{r}}{\mathrm{d}t}$$

$$\mathbf{v} = \frac{\mathrm{d}\mathbf{r}}{\mathrm{d}t}$$

The average acceleration of the particle between P and Q is defined as $\mathbf{a}_{\text{avg}} = \Delta\mathbf{v}/\Delta t$.

The instantaneous acceleration **a** of the particle is defined as the limiting value of the average acceleration as the time interval approaches zero. Thus,

$$\mathbf{a} = \lim_{\Delta t \to 0} \frac{\Delta \mathbf{v}}{\Delta t} = \frac{\mathrm{d}\mathbf{v}}{\mathrm{d}t}$$

$$\mathbf{a} = \frac{\mathrm{d}\mathbf{v}}{\mathrm{d}t}$$

RECTANGULAR COMPONENTS OF VELOCITY AND ACCELERATION

When the position of a particle is defined at any instant by its rectangular coordinate x, y and z, it is convenient to resolve the velocity $\vec{\mathbf{v}}$ and the acceleration $\vec{a}$ of the particle into rectangular components. We write $\mathbf{r} = x\hat{i} + y\hat{j} + z\hat{k}$, where x, y, and z are function of time t.

Differentiating with respect to time,

$$\frac{\mathrm{d}\mathbf{r}}{\mathrm{d}t} = \frac{\mathrm{d}x}{\mathrm{d}t}\hat{i} + \frac{\mathrm{d}y}{\mathrm{d}t}\hat{j} + \frac{\mathrm{d}z}{\mathrm{d}t}\hat{k}$$

$$\mathbf{v} = v_x\hat{i} + v_y\hat{j} + v_z\hat{k}$$

Again differentiating with respect to time.

$$\frac{\mathrm{d}\mathbf{v}}{\mathrm{d}t} = \frac{\mathrm{d}v_x}{\mathrm{d}t}\hat{i} + \frac{\mathrm{d}v_y}{\mathrm{d}t}\hat{j} + \frac{\mathrm{d}v_z}{\mathrm{d}t}\hat{k}$$

$$\mathbf{a} = a_x\hat{i} + a_y\hat{j} + a_z\hat{k}$$

ANGULAR VELOCITY AND ANGULAR ACCELERATION

Angular Velocity

The angular velocity ω is the rate at which the angular position θ changes with respect to time t. The direction of angular velocity is determined by the right-hand rule:

$$\omega = \dot{\theta} = \frac{\mathrm{d}\theta}{\mathrm{d}t}$$

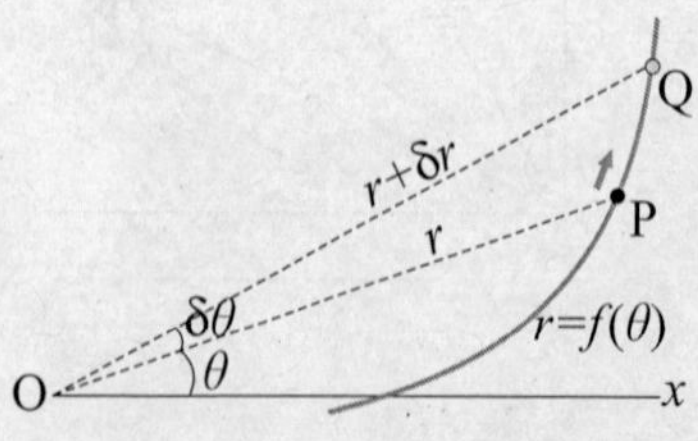

FIGURE 9.3 |

Angular Acceleration

The angular acceleration a is the rate at which the angular velocity ω changes with respect to time t (Fig. 9.3):

$$a = \ddot{\theta} = \frac{\mathrm{d}\omega}{\mathrm{d}t}$$

Other form of angular acceleration

$$a = \frac{\mathrm{d}\omega}{\mathrm{d}t} = \frac{\mathrm{d}}{\mathrm{d}t}\left(\frac{\mathrm{d}\theta}{\mathrm{d}t}\right) = \frac{\mathrm{d}^2\theta}{\mathrm{d}t^2},\ a = \frac{\mathrm{d}\omega}{\mathrm{d}\theta}\frac{\mathrm{d}\theta}{\mathrm{d}t} = \omega\frac{\mathrm{d}\omega}{\mathrm{d}\theta}$$

$$a = \frac{\mathrm{d}^2\theta}{\mathrm{d}t^2}, \quad a = \omega\frac{\mathrm{d}\omega}{\mathrm{d}\theta}$$

EQUATIONS OF MOTION: CIRCULAR

When the angular acceleration is constant, each of the three equations $a = \mathrm{d}\omega/\mathrm{d}t$, $\omega = \mathrm{d}\theta/\mathrm{d}t$ and $a = \omega(\mathrm{d}\omega/\mathrm{d}\theta)$ may be integrated to obtain formulas that related a, ω, θ and t.

Angular Velocity as a Function of Time

Integrate $a = \mathrm{d}\omega/\mathrm{d}t$ or $\mathrm{d}\omega = a\mathrm{d}ot$, assuming that initially $\omega = \omega_0$ when $t = 0$,

$$\int_{\omega_0}^{\omega} \mathrm{d}\omega = \int_0^t a\mathrm{d}t \quad \text{or} \quad \omega - \omega_0 = at$$

$$\omega = \omega_0 + at$$

Angular Displacement as a Function of Time

Integrate $\omega = \mathrm{d}\theta/\mathrm{d}t$ or $\mathrm{d}\theta = \omega\mathrm{d}ot$, assuming that initially $\theta = 0$ when $t = 0$

$$\int_0^{\theta} \mathrm{d}\theta = \int_0^t \omega\mathrm{d}t$$

Substituting $\omega = \omega_0 + at$, we have

$$\int_0^{\theta} \mathrm{d}\theta = \int_0^t (\omega_0 + at)\,\mathrm{d}t$$

$$\theta = \omega_0 t + \frac{1}{2}at^2$$

Angular Velocity as a Function of Angular Displacement

Integrate $a = \omega(\mathrm{d}\omega/\mathrm{d}\theta)$ or $\omega\mathrm{d}\omega = a\mathrm{d}\theta$, assuming that initially $\omega = \omega_0$ at $\theta = 0$

$$\int_{\omega_0}^{\omega} \omega\mathrm{d}\omega = \int_0^{\theta} a\mathrm{d}\theta$$

or

$$\frac{\omega^2}{2} - \frac{{\omega_0}^2}{2} = a\theta$$

$$\omega^2 = \omega_0^2 + 2a\theta$$

RADIAL AND TRANSVERSE VELOCITIES

Consider a particle moving in a plane along path $r = f(\theta)$. The position of the particle at any instant of time t is P of coordinate (r, θ). At time $t + \delta t$, the particle has moved to Q and its coordinate becomes $(r + \delta r, \theta + \delta\theta)$. The change in radial position or radial displacement is PN and transverse displacement QN during the interval δt (Fig. 9.4). Thus,

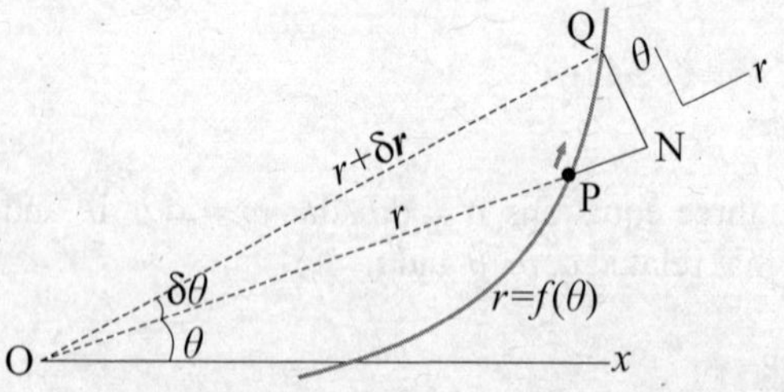

FIGURE 9.4

radial velocity at P

$$\text{or} \quad v_r = \lim_{\delta t \to 0} \frac{\text{displacement along OP in time } \delta t}{\delta t}$$

$$\text{or} \quad v_r = \lim_{\delta t \to 0} \frac{\text{PN}}{\delta t}$$

$$\text{or} \quad v_r = \lim_{\delta t \to 0} \frac{(r + \delta r)\cos\delta\theta - r}{\delta t}$$

Since $\delta\theta$ is small, coos $\delta\theta = 1$,

or
$$v_r = \lim_{\delta t \to 0} \frac{\delta r}{\delta t} = \frac{dr}{dt}$$

radial velocity
$$v_r = \frac{dr}{dt} = \dot{r}$$

Transverse velocity at P

or
$$v_\theta = \lim_{\delta t \to 0} \frac{\text{displacement perpendicular to OP in time } \delta t}{\delta t}$$

$$= \lim_{\delta t \to 0} \frac{\text{QN}}{\delta t}$$

or
$$v_\theta = \lim_{\delta t \to 0} \frac{(r + \delta r)\sin\delta\theta}{\delta t}$$

Since $\delta\theta$ is small, $\sin\delta\theta \approx \delta\theta$ and neglecting the term $\delta r\delta\theta$,

or
$$v_\theta = \lim_{\delta t \to 0} r\frac{\delta\theta}{\delta t} = r\frac{d\theta}{dt}$$

$$\text{transverse velocity, } v_\theta = r\frac{d\theta}{dt} = r\dot{\theta}$$

Resultant velocity of particle at P(r, θ)

$$v = \sqrt{v_r^2 + v_\theta^2}$$

$$v = \sqrt{\left(\frac{dr}{dt}\right)^2 + \left(r\frac{d\theta}{dt}\right)^2}$$

RADIAL AND TRANSVERSE ACCELERATIONS

Consider a particle moving along path $r = f(\theta)$. The position, radial velocity and transverse velocity of particle at any instant of time t are $P(r, \theta)$, V_R and v_θ, respectively (Fig. 9.5). At the time $t + \delta t$, the particle has moved to $Q(r + \delta r, \theta + \delta\theta)$ and its radial and transverse velocities become $v_r + \delta v_r$ and $v_\theta + \delta v_\theta$, respectively. Thus radial acceleration at P

$$a_r = \lim_{\delta t \to 0} \frac{\text{change in velocity along radial direction (OP) in time } \delta t}{\delta t}$$

$$a_r = \lim_{\delta t \to 0} \frac{[(v_r + \delta v_r)\cos\delta\theta - (v_\theta + \delta v_\theta)\sin\delta\theta] - v_r}{\delta t}$$

Since $\delta\theta$ is small, $\sin\delta\theta \approx \delta\theta$ and $\cos\delta\theta = 1$.

$$a_r = \lim_{\delta t \to 0} \frac{\delta v_r - v_\theta\delta\theta - \delta v_\theta\delta\theta}{\delta t} = \frac{dv_r}{dt} - v_\theta \frac{d\theta}{dt}$$

(neglect $\delta\theta\delta r$).

Substituting $v_r = dr/dt$ and $v_\theta = r(d\theta/dt)$, we have

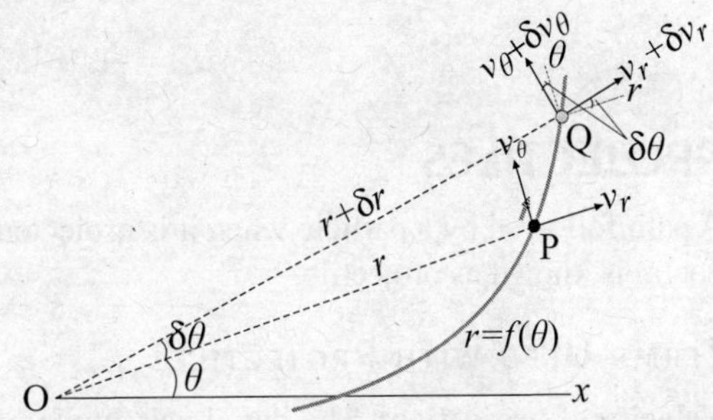

FIGURE 9.5

$$a_r = \frac{d^2r}{dt^2} - r\left(\frac{d\theta}{dt}\right)^2 \quad \text{or} \quad a_r = \ddot{r} - r\dot{\theta}^2$$

Transverse acceleration at P

$$a_\theta = \lim_{\delta t \to 0} \frac{\text{change in velocity perpendicular to radial direction in time } \delta t}{\delta t}$$

$$a_\theta = \lim_{\delta t \to 0} \frac{\left[(v_r + \delta v_r)\sin\delta\theta - (v_\theta + \delta v_\theta)\cos\delta\theta\right] - v_\theta}{\delta t}$$

Since $\delta\theta$ is small, $\sin\delta\theta \approx \delta\theta$ and $\cos\delta\theta = 1$,

$$a_\theta = \lim_{\delta t \to 0} \frac{(v_r + \delta v_r)\delta\theta - (v_\theta + \delta v_\theta) - v_\theta}{\delta t}$$

Neglecting the term $\delta\theta\delta v_r$

$$a_\theta = v_r \frac{d\theta}{dt} + \frac{dv_\theta}{dt}$$

Substituting $v_r = dr/dt$ and $v_\theta = r(d\theta/dt)$, we have

$$a_\theta = r\frac{d^2\theta}{dt^2} + 2\left(\frac{dr}{dt}\right)\left(\frac{d\theta}{dt}\right) = \frac{1}{r}\frac{d}{dt}\left(r^2\frac{d\theta}{dt}\right)$$

$$a_\theta = \frac{1}{r}\frac{d}{dt}\left(r^2\frac{d\theta}{dt}\right)$$

Resultant acceleration,

$$a = \sqrt{a_r^2 + a_\theta^2}$$

Special case: When the path of particle is circular, radius of circle, r, is constant.
Radial acceleration,

$$a_r = 0 - r\left(\frac{d\theta}{dt}\right)^2 = -r\omega^2, \quad \text{where} \quad \omega = \frac{d\theta}{dt}$$

Transverse acceleration,

$$a_\theta = r\left(\frac{d^2\theta}{dt^2}\right) = r\alpha \quad \text{where} \quad \alpha = \frac{d^2\theta}{dt^2}$$

$$a_r = -\omega^2 r \quad \text{and} \quad a_\theta = r\alpha$$

PROJECTILES

A path followed by a particle when it is projected into the space making a certain angle with the horizontal is known as projectile.

TERMS USED WITH PROJECTILES

Velocity of projection: The initial velocity by which the particle is thrown in the space is known as velocity of projection.

Angle of projection: The angle with the horizontal made by the particle thrown into the space is known as angle of projection.

Trajectory: The path traced by a projectile in space is known as trajectory.

Range: The distance along the plane between the point of projection and the point, where the particle strikes the plane at the end of the flight is known as range.

Time of flight: This is the duration for which the particle is in motion or this is the interval of time since the instant of projection to the instant the particle meets the ground.

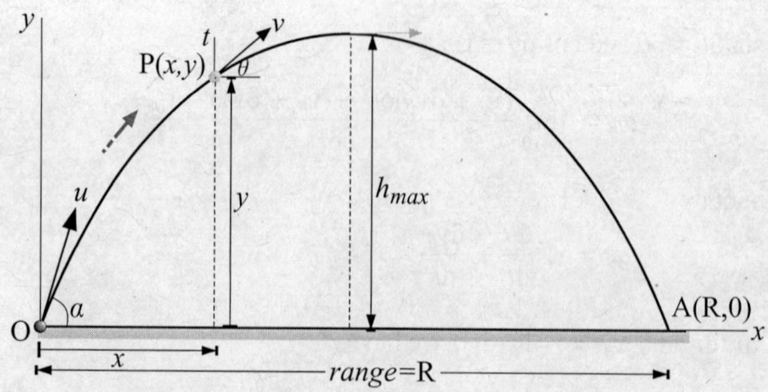

FIGURE 9.6 |

Consider a particle or a body projected upwards from a point O, with an initial velocity u at an angle a with the horizontal, as shown in Figure 9.6. After a time t, the particle reached the point P(x, y)

with a velocity v and making an angle q with the horizontal. Applying the equation of motion between O and P in horizontal and vertical directions, we get equations as shown in Table 9.1.

TABLE 9.1 | Equations of motion

	Horizontal motion ($a_x = 0$, $u_x = u \cos \alpha$, $v_x = v \cos \theta$)		**Vertical motion ($a_y = -g$, $u_y = u \sin \alpha$, $v_y = v \sin \theta$)**	
$v = u + at \Rightarrow$	$v\cos \theta = u\cos \alpha$	(i)	$v\sin \theta = u\sin \alpha - gt$	(ii)
$s = ut + \frac{1}{2} at^2 \Rightarrow$	$x = u\cos \alpha t$	(iii)	$y = u\sin \alpha t - \frac{1}{2} gt^2$	(iv)
$v^2 = u^2 + 2as \Rightarrow$	$v^2 \cos^2 \theta = u^2 \cos^2 \alpha$ [same as Equation (i)]		$v^2 \sin^2 \theta = u^2 \sin^2 \alpha - 2gy$	(v)

EQUATION OF TRAJECTORY

Eliminating t from equations (iii) and (iv), we obtain

$$y = u \sin \alpha \left(\frac{x}{u \cos \alpha} \right) - \frac{1}{2} g \left(\frac{x}{u \cos \alpha} \right)^2$$

$$y = x \tan \alpha - \frac{1}{2} \frac{gx^2}{u^2 \cos^2 \alpha}$$

This is the equation of parabola.

VELOCITY AND DIRECTION OF MOTION

Squaring equation (i) and then adding equation (ii), we have

$$v^2 \cos^2 \theta + v^2 \sin^2 \theta = (u \sin \alpha)^2 + (u \sin \alpha - gt)^2$$

From which,

$$v = \sqrt{u^2 - 2ugt \sin \alpha + g^2 t^2}$$

Dividing equation (ii) by equation (i), we have

$$\tan \theta = \frac{u \sin \alpha - gt}{u \cos \alpha}$$

$$\theta = \tan^{-1} \left(\frac{u \sin \alpha - gt}{u \cos \alpha} \right)$$

Squaring equation (i) and then adding equation (v), we have

$$v^2 \sin^2 \theta + v^2 \cos^2 \theta = u^2 \cos^2 \alpha + u^2 \sin^2 \alpha - 2gy$$

From which,

$$v = \sqrt{u^2 - 2gy}$$

Fist squaring equation (i) then dividing from equation (v), we have

$$\tan^2 \theta = \frac{u^2 \sin^2 a - 2gy}{u^2 \cos^2 a}$$

$$\theta = \tan^{-1}\left(\frac{\sqrt{u^2 \sin^2 a - 2gy}}{u \cos a}\right)$$

Maximum Height Attained

Vertical component of velocity becomes zero ($v_y = v\sin \theta = 0$) at maximum height.
From Equation (v),

$$0 = u^2 \sin^2 a - 2gh_{\max} \qquad (y = h_{\max})$$

$$h_{\max} = \frac{u^2 \sin^2 a}{2g}$$

Time of Flight

When the projectile reaches at A, $y = 0$ and $t = T$.
From Equation (v),

$$0 = u \sin aT - \frac{1}{2} gT^2$$

$$T = \frac{2u \sin a}{g}$$

Range

The horizontal distance covered by the projectile till the end of its flight is called its range R. As the horizontal components of velocity of projectile u_x remains constant, the range is given by

$$R = u_x \times T \Rightarrow R = u \cos a \times \frac{2u \sin a}{g}$$

or

$$R = \frac{2u^2 \sin a \cos a}{g}$$

$$R = \frac{u^2 \sin 2a}{g}$$

Maximum value of range is

$$R_{\max} = \frac{u^2}{g} \qquad \left(\sin 2a = 1 \Rightarrow a = \frac{\pi}{4}\right)$$

INCLINED PLANE

Range of Inclined Plane The range of an inclined plane is given by the distance OA (Fig. 9.7). The coordinate of point P is $(R \cos \beta, R \sin \beta)$. Substituting the coordinate of point P in the path of projectile

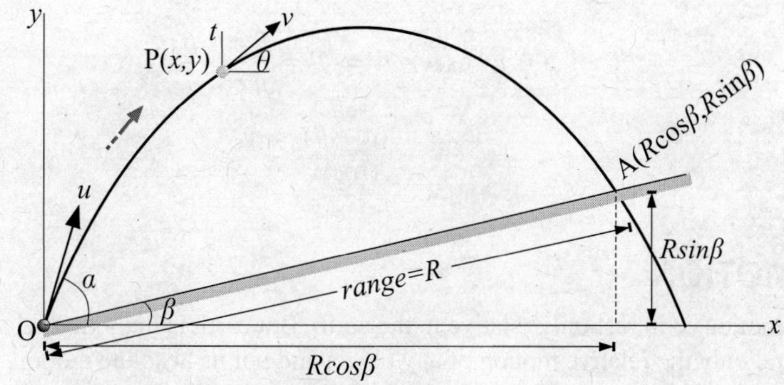

FIGURE 9.7 |

$$y = x \tan \alpha - \frac{1}{2} \frac{gx^2}{u^2 \cos^2 \alpha}$$

or
$$R \sin \beta = R \cos \beta \tan \alpha - \frac{1}{2} \frac{g(R \cos \beta)^2}{u^2 \cos^2 \alpha}$$

or
$$\sin \beta = \cos \beta \tan \alpha - \frac{1}{2} \frac{gR \cos^2 \beta}{u^2 \cos^2 \alpha}$$

$$R = \frac{2u^2 \cos \alpha}{g} \left(\frac{\sin \alpha \cos \beta - \cos \alpha \sin \beta}{\cos^2 \beta} \right)$$

$$= \frac{2u^2 \cos \alpha \sin(\alpha - \beta)}{g \cos^2 \beta}$$

[From trigonometry, $2 \sin(\alpha - \beta) \cos \beta = \sin(2\alpha - \beta) - \cos \beta$]

$$R = \frac{u^2}{g} \frac{[\sin(2\alpha - \beta) - \cos \beta]}{\cos^2 \beta}$$

Maximum Range of Inclined Plane R will be maximum when $\sin(2\alpha - \beta) = 1 \Rightarrow \sin(2\alpha - \beta) = \sin(\pi/2)$:

$$R_{\max} = \frac{u^2}{g(1 + \cos \beta)} \quad \text{at} \quad \alpha = \frac{\pi}{4} + \frac{\beta}{2}$$

TIME OF FLIGHT

It is the time taken by the projectile to travel from O to point A ($R\cos\beta$, $R\cos\beta$).
From Equation (iii),

$$x = u\cos at$$

$$R\cos\beta = u\cos a \times T \Rightarrow T = \frac{R\cos\beta}{u\cos a}$$

$$T = \frac{2u\sin(a-\beta)}{g\cos\beta}$$

RELATIVE MOTION

Everything is in motion or in rest only relative to the earth. But earth is in motion relative to the sun. Hence, we can find only the relative motion of any object and not its absolute motion.

When the velocity of a moving body is obtained relative to another moving body, then that velocity is known as its relative velocity.

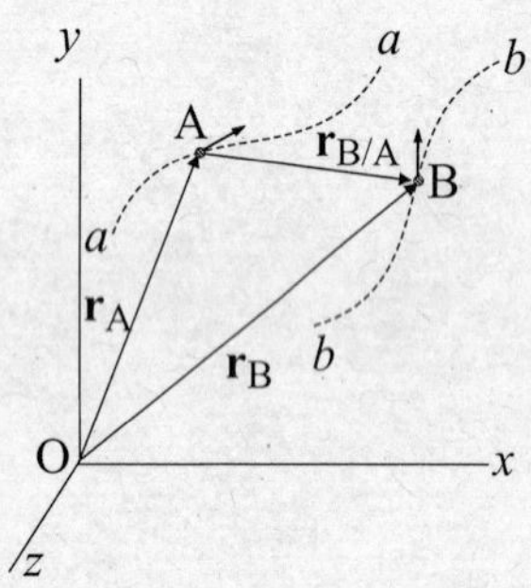

FIGURE 9.8 |

Consider two particles A and B, which move along the arbitrary path *aa* and *bb* in a reference fixed frame in Figure 9.8. Let the position vector of particles A and B at any instant be $\mathbf{r}_A$ and $\mathbf{r}_B$, respectively. The relative position of particle B with respect to particle A is designated by $\mathbf{r}_{B/A}$. By using vector diagram

$$\mathbf{r}_B = \mathbf{r}_A + \mathbf{r}_{B/A}$$

$$\mathbf{r}_{B/A} = \mathbf{r}_B - \mathbf{r}_A$$

Taking time derivative,

$$\frac{d}{dt}\mathbf{r}_{B/A} = \frac{d}{dt}\mathbf{r}_B - \frac{d}{dt}\mathbf{r}_A$$

$$\mathbf{v}_{B/A} = \mathbf{v}_B - \mathbf{v}_A$$

Here $\mathbf{v}_{B/A} = \frac{d}{dt}\mathbf{r}_{B/A}$ is the relative velocity of B with respect to A, $\frac{d}{dt}\mathbf{r}_A = \mathbf{v}_A$ is the velocity of A, and $\frac{d}{dt}\mathbf{r}_B = \mathbf{v}_B$ is the velocity of B.

Again taking time derivative, we obtain

$$\mathbf{a}_{B/A} = \mathbf{a}_B - \mathbf{a}_A$$

Relative acceleration of particle B with respect to particle A is $\mathbf{a}_{B/A} = \mathbf{a}_B - \mathbf{a}_A$.

SOLVED PROBLEMS

PROBLEM 9.1

A bus starts from rest at $x = 0$ and is subjected to an acceleration shown by the a–x graph [Fig. 9.1(a)]. Draw the v–x graph and determine the time needed to travel 60 m.

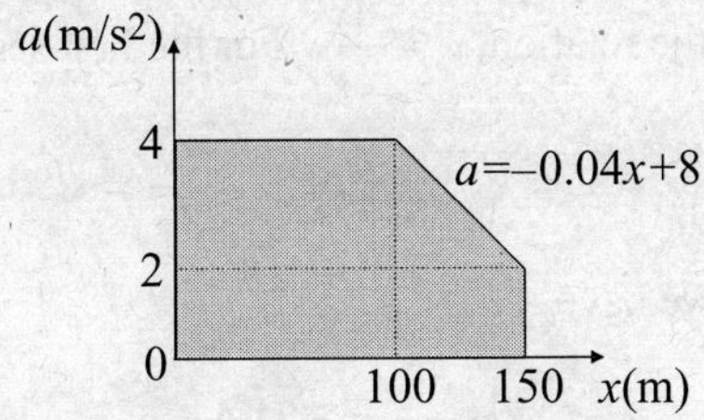

FIGURE 9.1(a) |

Solution. Since the equations of the a–x graph is given, the v–x graph [Fig. 9.1(b)] can be determined using the equation

$$a = v\,\frac{dv}{dx}$$

We write

$$\int v\,dv = \int a\,dx$$

For

$$0 \le x \le 100,$$

$$\int_0^v v\,dv = \int_0^x a\,dx$$

or

$$\int_0^v v\,dv = \int_0^x 4\,dx \Rightarrow \frac{v^2}{2} = 4x$$

$$v = \sqrt{8x}$$

Velocity at $x = 100\ m$; $v = \sqrt{8 \times 100}$

$$v_{x=100} = 28.28\text{m/s}$$

For $100 \le x \le 150$,

$$\int_{28.28}^{v} v\,dv = \int_{100}^{x} (-0.04x + 8)\,dx$$

or

$$\frac{v^2 - 28.28^2}{2} = -0.04\left(\frac{x^2 - 100^2}{2}\right) + 8(x - 100)$$

or

$$v^2 - 28.28^2 = -0.04(x^2 - 100^2) + 2 \times 8(x - 100)$$

$$v = \sqrt{-0.04x^2 + 16x - 400}$$

Velocity at $x = 150$m

$$v = \sqrt{-0.04 \times 150^2 + 16 \times 150 - 400}$$

$$v = 33.16\text{m/s}$$

The time can be obtained using the relation, $v = \dfrac{dx}{dt}$. For the first segment of motion

$$\frac{dx}{dt} = v = \sqrt{8x} \quad \text{or} \quad \frac{dx}{\sqrt{x}} = \sqrt{8}dt$$

Integration from $x = 0$ to 60 m, we have

$$\int_0^{60} \frac{dx}{\sqrt{x}} = \sqrt{8}\int_0^t dt$$

or

$$\left[2\sqrt{x}\right]_0^{60} = \sqrt{8}t \Rightarrow t = \frac{2\sqrt{60}}{\sqrt{8}}$$

$$t = 5.47\text{s}$$

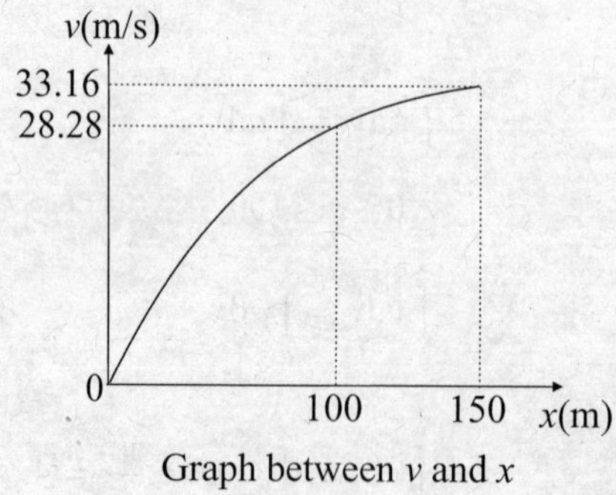

FIGURE 9.1(b) |

PROBLEM 9.2

A small projectile is fired into a viscous fluid with an initial velocity $u = 360$ m/s and travels 100 mm before coming to rest [Fig. 9.2(a)]. Assuming that the velocity of the projectile was defined by the relation $v = u - \lambda x$, where v is in m/s and x in metres, determine (a) the initial acceleration of the projectile and (b) the time required for the projectile to penetrate 50 mm into the fluid.

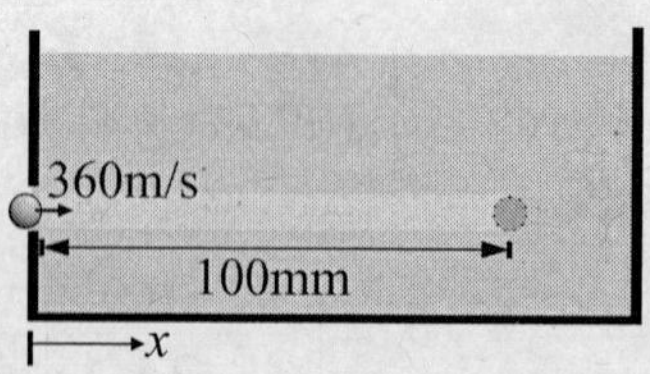

FIGURE 9.2(a) |

Solution. Given that $u = 360$ m at $x = 0$ and $v = 0$ at $x = 0.1$ m, thus

$$v = u - \lambda x \Rightarrow 0 = 360 - \lambda \times 0.1$$

or

$$\lambda = 3600$$

Hence, $v = 360 - 3600x$.

(a) Acceleration of projectile

$$a = \frac{dv}{dt}$$

or $$a = \frac{d}{dt}(360 - 3600x) = -3600\frac{dx}{dt}$$

or $$a = -3600v = -3600(360 - 3600x)$$

Initial acceleration $a_{x=0} = -3600(360 - 3600 \times 0) = -3600 \times 360$

$$a_{x=0} = 1.296 \times 10^6 \text{ m/s}^2$$

(b) Velocity of projectile $$v = \frac{dx}{dt} = 360 - 3600x$$

or $$\int_0^{0.05} \frac{dx}{360 - 3600x} = \int_0^t dt$$

or $$t = -\frac{1}{3600}\log_e\left(\frac{180}{360}\right)$$

$$t = 1.92 \times 10^{-4} \text{ s}$$

PROBLEM 9.3

A cam-follower arrangement is shown in Figure 9.3(a). The cam moves horizontally to the left with a constant speed v_0, while the follower is constrained to move vertically. Counting time t from the instant when the follower is in its lowest position, find the acceleration of the follower as a function of time. Assume cam profile is parabola.

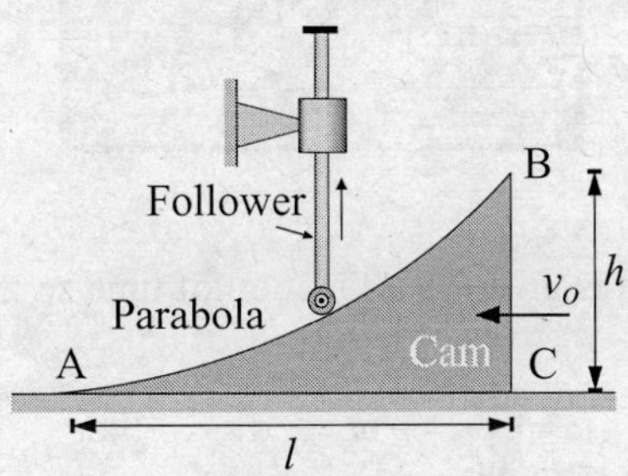

FIGURE 9.3(a) |

Solution. Taking the lowest position A of the cam as shown in Figure 9.3(b), The equation of cam surface will be

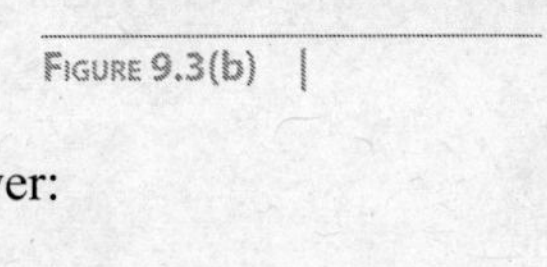

FIGURE 9.3(b) |

$$y = ax^2, \quad \text{where} \quad a = \frac{y}{x^2} = \frac{h}{l^2}$$

$$y = \frac{h}{l^2}x^2$$

Substituting $x = v_0 t$, we obtain displacement–time equation of the follower:

$$y = \frac{h}{l^2}v_0^{\,2}t^2$$

Differentiating successively with respect to time, we obtain

$$\frac{dy}{dt} = \frac{d}{dt}\left(\frac{h}{l^2}v_0^2t^2\right) = \frac{2hv_0^2t}{l^2}$$

Velocity of follower

$$v_{\text{follower}} = \frac{2hv_0^2t}{l^2}$$

and

$$\frac{d^2y}{dt^2} = \frac{d}{dt}\left(\frac{dy}{dt}\right) = \frac{d}{dt}\left(\frac{2hv_0^2t}{l^2}\right) = \frac{2hv_0^2}{l^2}$$

Acceleration of follower

$$a_{\text{follower}} = \frac{d^2y}{dt^2} = \frac{2hv_0^2}{l^2}$$

$$a_{\text{follower}} = \frac{2hv_0^2}{l^2}$$

PROBLEM 9.4

A sphere is fired horizontally into a viscous liquid with an initial velocity of 27 m/s [Fig. 9.4(a)]. If it experiences a deceleration $a = -6t$ m/s$_2$, where t is in seconds, determine the distance travelled before it stops.

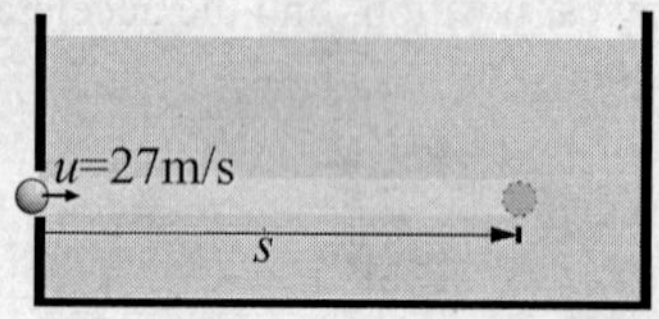

FIGURE 9.4(a)

Solution. The acceleration is given as a function of time so that the velocity can be obtained as follows:

$$a = \frac{dv}{dt} = -6t \Rightarrow dv = -6t\,dt$$

Given: $v = 27$ m/s at $t = 0$,

$$\int_{27}^{v} dv = -\int_{0}^{t} 6t\,dt \Rightarrow v - 27 = -3t^2$$

or

$$v = 27 - 3t^2$$

Velocity of sphere becomes zero at $t = \sqrt{(27-0)/3} = 3$s

Knowing the velocity as a function of time, we can write

$$v = \frac{ds}{dt} = 27 - 3t^2$$

or

$$ds = (27 - 3t^2)dt$$

or

$$\int_{0}^{s} ds = \int_{0}^{3} (27 - 3t^2)dt$$

or $$s = [27t - t^3]_0^3 = 27(3 - 0) - (3^3 - 0^3)$$

Distance travelled, s = 54 m

PROBLEM 9.5

A particle is projected vertically upwards from the earth surface with an initial velocity u. The variation of the acceleration of gravity with respect to altitude h must be taken into account. Neglecting the air resistance, determine (a) the minimum initial velocity of particle so that it does not fall back to the earth; (b) time to reach a height H.

Solution. Let O be the centre of the earth and A is the point of projection on its surface [Fig. 9.5(a)]. Let P be the position of the particle at time t, such that OP = r. We know that outside the surface of the earth, the acceleration is given by

$$a = -\frac{R^2}{(R+h)^2}g = -\frac{R^2}{r^2}g$$

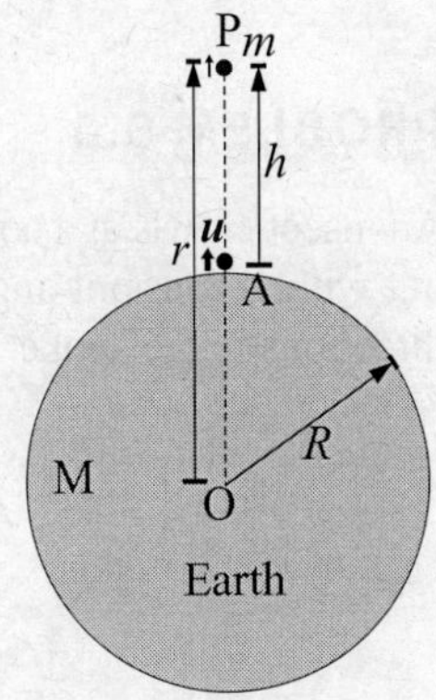

FIGURE 9.5(a)

Since the acceleration is related to position, we can write

$$a = v\frac{dv}{dr} = -\frac{R^2}{r^2}g \Rightarrow v dv = -\frac{R^2}{r^2}g dr$$

Let u be the initial velocity at $r = R$;

$$\int_u^v v dv = -gR^2 \int_R^{R+h} \frac{1}{r^2} dr$$

or $$\frac{v^2 - u^2}{2} = gR^2 \left[\frac{1}{r}\right]_R^{R+h} = -gR^2\left(\frac{h}{R(R+h)}\right)$$

or $$v^2 = u^2 - 2gR\frac{h}{(R+h)}$$

(a) Given that $v = 0$ at $r = \infty$ (or $h = \infty$):

$$0^2 = u^2 - 2gR\frac{1}{[(R/\infty)+1]} \Rightarrow u^2 = 2gR$$

Minimum initial velocity $u = \sqrt{2gR}$

(b) Velocity of particle

$$v = \sqrt{u^2 - 2gR\frac{h}{(R+h)}} = \sqrt{2gR - 2gR\frac{h}{(R+h)}}$$

or $$v = \sqrt{\frac{2gR^2}{(R+h)}}$$

or $$v = \sqrt{\frac{2gR^2}{r}}$$

Velocity of particle

$$v = \frac{dr}{dt} = \sqrt{\frac{2gR^2}{r}}$$

or
$$\sqrt{r}dr = \sqrt{2gR^2}dt$$

or
$$\int_R^{R+H} \sqrt{r}dr = \sqrt{2gR^2}\int_0^T dt$$

or
$$\frac{2}{3}\left(r^{3/2}\right)_R^{R+H} = \sqrt{2gR^2T}$$

or
$$\frac{2}{3}\left[(R+H)^{3/2} - R^{3/2}\right] = \sqrt{2gR^2}T$$

$$T = \frac{2[(R+H)^{3/2} - R^{3/2}]}{3\sqrt{2gR^2}}$$

PROBLEM 9.6

An insect crawls at a constant speed u along the spoke of a bicycle wheel [Fig. 9.6(a)], which is rotating with a constant angular velocity ω. Find the acceleration of insect in radial and perpendicular directions to the spoke.

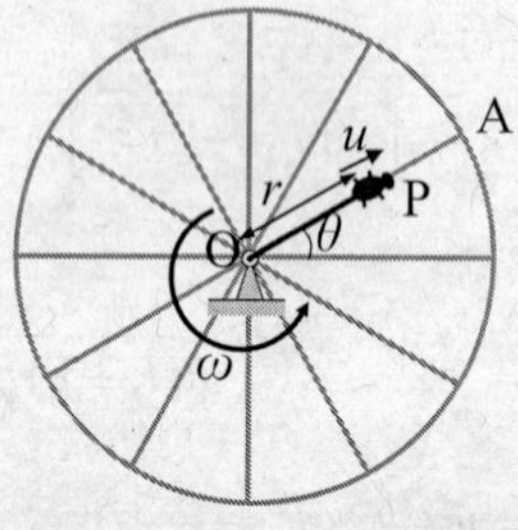

FIGURE 9.6(a) |

Solution. Let the initial position of the insect be at the centre O and the spoke be OA. Let P be the position ($r = ut$) of the insect after time t. Therefore, angular velocity

$$\frac{d\theta}{dt} = \omega$$

Acceleration along the spoke direction is given by

$$a_r = \frac{d^2r}{dt^2} - r\left(\frac{d\theta}{dt}\right)^2$$

or
$$a_r = \frac{d^2}{dt^2}(ut) - r\omega^2 = -r\omega^2$$

$$a_r = -r\omega^2$$

Acceleration perpendicular to the spoke is given by

$$a_t = \frac{1}{r}\frac{d}{dt}\left(r^2\frac{d\theta}{dt}\right)$$

or
$$a_t = \frac{1}{ut}\frac{d}{dt}u^2t^2\omega = 2u\omega$$

$$a_t = 2u\omega$$

PROBLEM 9.7

A projectile is fired with an initial velocity of $u = 400$ m/s at target B located 1500 m above the gun and at a horizontal distance of 5000 m, as shown in Figure 9.7(a). Determine the two angles of elevation θ that will permit the projectile to hit the mountain target B.

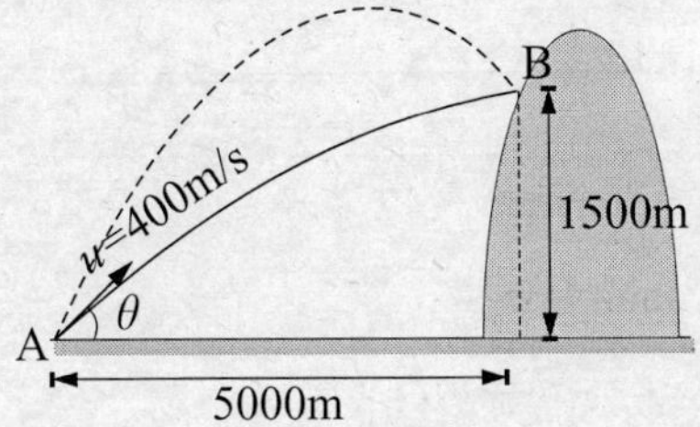

FIGURE 9.7(a) |

Solution. Let q be the angle of projection. Applying the equation of motion of projectile along x-direction,

$$5000 = 400 \cos \theta t$$

or
$$t = \frac{5000}{400 \cos \theta} \qquad \text{(i)}$$

Applying the equation of motion of projectile along y-direction,

$$1500 = 400 \sin \theta t - \frac{1}{2} g t^2 \qquad \text{(ii)}$$

Substituting the value of t from Equation (i) into Equation (ii), we have

$$1500 = 400 \sin \theta \left(\frac{5000}{400 \cos g\theta} \right) - \frac{1}{2} g \left(\frac{5000}{400 \cos \theta} \right)^2$$

or
$$1500 = 5000 \tan \theta - 766.4 \sec^2 \theta$$

or
$$1500 = 5000 \tan \theta - 766.4(1 + \tan^2 \theta)$$

or
$$\tan^2 \theta - 6.52 \tan \theta + 2.95 = 0$$

Solving this quadratic equation for $\tan \theta$, we have

$$\tan \theta = 0.49, 6.01 \Rightarrow \theta = 26.1^\circ, 80.6^\circ$$

$$\theta_1 = 26.1^\circ, \theta_2 = 80.6^\circ$$

PROBLEM 9.8

A ball is thrown horizontally from a tower at A and hits the ground 3.5 s later at B [Fig. 9.8(a)]. The line of sight from A to B makes an angle of 50° with the horizontal. Find the initial velocity u of the ball.

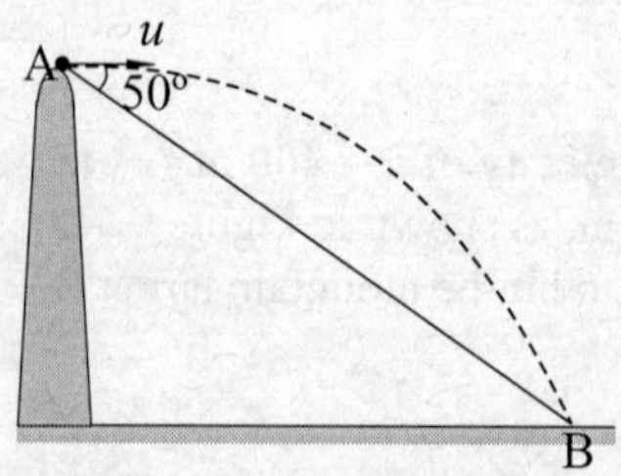

FIGURE 9.8(a) |

Solution. x-coordinate of point B is

$$d = u_x t$$

$$d = u \times 3.5 \tag{i}$$

y-coordinate of point B

$$-h = u_y t - \frac{1}{2} g t^2$$

or

$$-h = 0 \times t - \frac{1}{2} 9.81(3.5)^2$$

$$h = 60.08\text{m} \tag{ii}$$

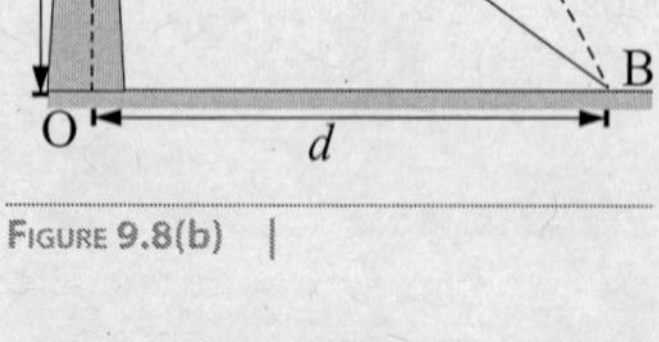

FIGURE 9.8(b) |

In triangle AOB [Fig. 9.8(b)],

$$\tan 50 = \frac{h}{d} = \frac{60.08}{d}$$

$$d = 50.4\text{m}$$

From Equation (i), we have

$$50.4 = u \times 3.5$$

$$u = 14.40\text{m/s}$$

PROBLEM 9.9

A projectile is projected from a point on the level of the ground and its height is h when it is at a horizontal distance l and $2l$ from its point of projection. Prove that the velocity of projection u is given by

$$u^2 = \frac{g}{4}\left(\frac{4l^2}{h} + 9h\right)$$

Solution. Let θ be the angle of projection and O be the point of projection [Fig. 9.9(a)]. The coordinates of the points P and Q whose height is h above x-axis are (l, h) and $(2l, h)$, respectively.

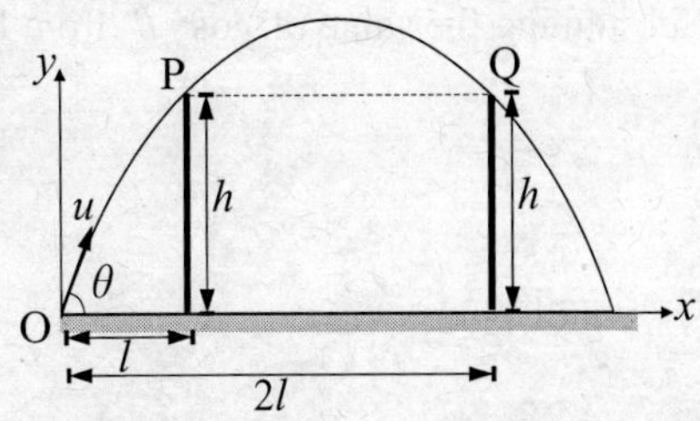

FIGURE 9.9(a)

The equation of the path of the particle is

$$y = x \tan\theta - \frac{gx^2}{2u^2 \cos^2\theta}$$

Points $P(l, h)$ and $Q(2l, h)$ are on the path of projectile,

$$h = l \tan\theta - \frac{gl^2}{2u^2 \cos^2\theta} \tag{i}$$

$$h = 2l \tan\theta - \frac{4gl^2}{2u^2 \cos^2\theta} \tag{ii}$$

Subtracting Equation (i) from equation (ii), we get

$$l \tan\theta = \frac{3gl^2}{2u^2 \cos^2\theta} \tag{iii}$$

From Equation (i),

$$h = l \tan\theta - \frac{gl^2}{2u^2 \cos^2\theta} = \frac{3gl^2}{2u^2 \cos^2\theta} - \frac{gl^2}{2u^2 \cos^2\theta}$$

or $$h = \frac{gl^2}{u^2 \cos^2\theta}$$

or $$\cos^2\theta = \frac{gl^2}{hu^2} \tag{iv}$$

Substituting $\cos^2\theta = gl^2/hu^2$ into equation (i), we have

$$h = l \tan\theta - \frac{gl^2}{2u^2\left(\dfrac{gl^2}{hu^2}\right)} = l \tan\theta - \frac{h}{2}$$

or $$\tan\theta = \frac{3h}{2l} \Rightarrow \tan^2\theta = \frac{9h^2}{4l^2}$$

or $$\sec^2\theta - 1 = \frac{9h^2}{4l^2}$$

or $$\frac{1}{\cos^2\theta} = \frac{9h^2}{4l^2} + 1$$

Substituting the value of $\cos^2 \theta$ from Equation (iv), we obtain

$$\frac{1}{\dfrac{gl^2}{hu^2}} = \frac{9h^2}{4l^2} + 1$$

On simplifying

$$u^2 = \frac{g}{4}\left(9h + \frac{4l^2}{h}\right)$$

PROBLEM 9.10

A particle is projected at an angle of elevation a and after t seconds, it appears to have an of elevation β as seen from the point of projection. Find the initial velocity of particle.

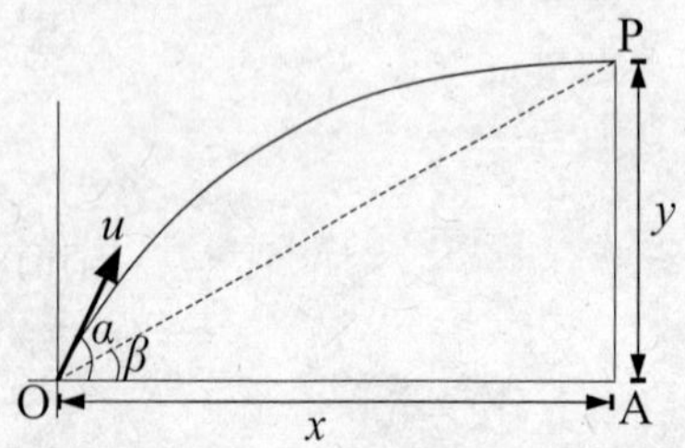

FIGURE 9.10(a) |

Solution. Let u be the initial velocity of projection and coordinate (x, y) be the position after t seconds. The equations of motion of projectile are

$$y = u \sin at - \frac{1}{2} gt^2 \quad \text{(i)}$$

$$x = u \cos at \quad \text{(ii)}$$

in triangle OAP [Fig. 9.10(a)].

$$\tan \beta = \frac{y}{x}$$

or

$$\tan \beta = \frac{u \sin at - \frac{1}{2} gt^2}{u \cos at}$$

or

$$\tan \beta = \tan a - \frac{1}{2}\frac{g}{u \cos a} t$$

or

$$u = \frac{gt}{2 \cos a(\tan a - \tan \beta)} = \frac{gt \cos a \cos \beta}{2 \cos a \sin(a - \beta)}$$

$$u = \frac{gt \cos \beta}{2 \sin(a - \beta)}$$

PROBLEM 9.11

A girl throws a ball in the air with a speed u at an angle θ_1. If another girl at the same time and same place throws another ball at the same speed u at an angle θ_2 as shown in Figure 9.11(a), determine the time between the throws so the balls collide in mid air at P.

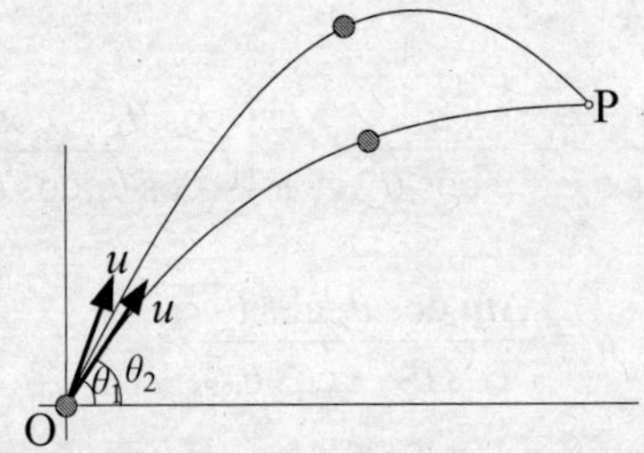

FIGURE 9.11(a) |

Solution. Let (a, h) be the coordinate of point P [Fig. 9.11(b)]. The equation of path of projectile I is

$$y = x\tan\theta_1 - \frac{gx^2}{2u^2\cos^2\theta_1}$$

Substituting coordinate point $P(a, h)$ in the above equation,

$$h = a\tan\theta_1 - \frac{ga^2}{2u^2\cos^2\theta_1} \quad \text{(i)}$$

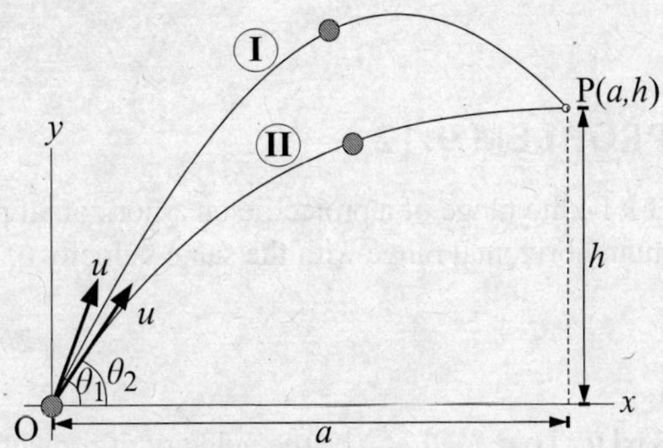

FIGURE 9.11(b) |

Similarly, for the projectile II,

$$h = a\tan\theta_2 - \frac{ga^2}{2u^2\cos^2\theta_2} \quad \text{(ii)}$$

From Equations (i) and (ii), we get

$$h = a\tan\theta_1 - \frac{ga^2}{2u^2\cos^2\theta_1} = a\tan\theta_2 - \frac{ga^2}{2u^2\cos^2\theta_2}$$

or

$$a(\tan\theta_1 - \tan\theta_2) = \frac{ga^2}{2u_2}\left(\frac{\cos^2\theta_2 - \cos^2\theta_1}{\cos^2\theta_1\cos^2\theta_2}\right)$$

or

$$\sin(\theta_1 - \theta_2) = \frac{ga}{2u^2}\left(\frac{\cos^2\theta_2 - \cos^2\theta_1}{\cos^2\theta_1\cos^2\theta_2}\right) \quad \text{(iii)}$$

Let t and $t + \Delta t$ be the time to reach the ball at P by paths II and I, respectively.
The equations of motion are: for path I

$$a = u\cos\theta_1(t + \Delta t) \quad \text{(iv)}$$

and for path II

$$a = u\cos\theta_2 t \quad \text{(v)}$$

From Equations (iv) and (v), we have

$$\Delta t = \frac{a}{u \cos \theta_1} - \frac{a}{u \cos \theta_2} = \frac{a}{u}\left(\frac{\cos \theta_2 - \cos \theta_1}{\cos \theta_1 \cos \theta_2}\right)$$

or
$$a = \frac{\Delta t u \cos \theta_1 \cos \theta_2}{\cos \theta_2 - \cos \theta_1} \quad \text{(vi)}$$

Substituting the value of a in Equation (iii), we have

$$\sin(\theta_1 - \theta_2) = \frac{g}{2u^2} \times \frac{\Delta t u \cos \theta_1 \cos \theta_2}{\cos \theta_2 - \cos \theta_1} \times \left(\frac{\cos^2 \theta_2 - \cos^2 \theta_1}{\cos \theta_1 \cos \theta_2}\right)$$

$$\Delta t = \frac{2u}{g}\left(\frac{\sin(\theta_1 - \theta_2)}{\cos \theta_2 + \cos \theta_1}\right)$$

PROBLEM 9.12

If R be the range of a projectile on a horizontal plane and h its maximum height, show that the maximum horizontal range with the same velocity of projection is

$$\left(2h + \frac{R^2}{8h}\right)$$

Solution. Let u be the velocity of projection and θ be the angle of projection. Range

$$R = \frac{u^2}{g} \sin 2\theta \quad \text{(i)}$$

and height

$$h = \frac{u^2 \sin^2 \theta}{2g} \quad \text{(ii)}$$

From Equations (i) and (ii), we obtain

$$\frac{h}{R} = \frac{u^2 \sin^2 \theta}{2g} \times \frac{g}{u^2 \sin 2\theta} = \frac{1}{4} \tan \theta$$

or
$$\tan \theta = \frac{4h}{R}$$

or
$$\sin \theta = \frac{4h}{\sqrt{R^2 + 16h^2}} \quad \text{(iii)}$$

Maximum range

$$R_{max} = \frac{u^2}{g} \quad \text{(iv)}$$

From Equations (ii) and (iii), we obtain

$$R_{max} = \frac{u^2}{g} = \frac{2h}{\sin^2 \theta}$$

Substituting the value of sin θ from Equation (iii), we write

$$R_{\max} = \frac{2h}{\left(\dfrac{4h}{\sqrt{R^2 + 16h^2}}\right)^2}$$

or

$$R_{\max} = \frac{R^2 + 16h^2}{8h} = 2h + \frac{R^2}{8h}$$

$$\boxed{R_{\max} = 2h + \frac{R^2}{8h}}$$

PROBLEM 9.13

A paper of thickness h is being wound onto the reel, which is turned at a constant angular velocity ω [Fig. 9.13(a)]. Determine the acceleration of the paper as it enters the rollers.

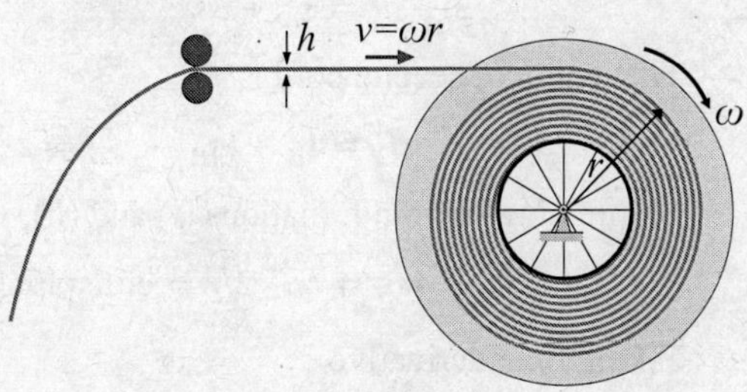

FIGURE 9.13(a) |

Solution. Let r be the radius of the paper reel at time t and r' be the radius of the paper at time $t + \mathrm{d}t$ and after θ rotation of reel. Thus,

$$r' = r + \frac{h}{2\pi}\theta$$

Velocity of the paper

$$v = \omega r'$$

Taking time derivative

$$\frac{\mathrm{d}v}{\mathrm{d}t} = \omega \frac{\mathrm{d}r'}{\mathrm{d}t},$$

$$\frac{\mathrm{d}v}{\mathrm{d}t} = \omega \frac{\mathrm{d}}{\mathrm{d}t}\left(r + \frac{h}{2\pi}\theta\right) = \frac{\omega h}{2\pi}\frac{\mathrm{d}\theta}{\mathrm{d}t}$$

Acceleration

$$a = \frac{\mathrm{d}v}{\mathrm{d}t} = \frac{\omega h}{2\pi} \times \omega = \frac{\omega^2 h}{2\pi}$$

or

$$a = \frac{(v/r)^2 h}{2\pi} = \frac{v^2 h}{2\pi r^2}$$

$$\boxed{a = \frac{v^2 h}{2\pi r^2}}$$

PROBLEM 9.14

If block A of the pulley system shown in Figure 9.14(a) is moving downward at v_A, determine the velocity of block B.

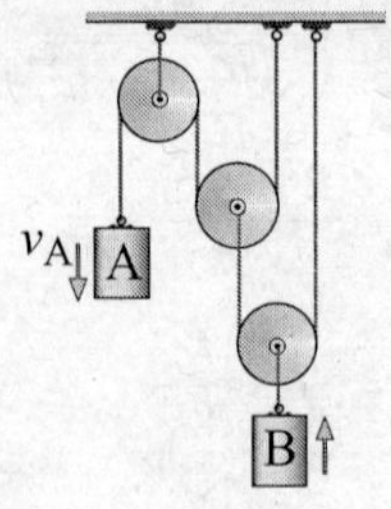

FIGURE 9.14(a)

Solution. The position of blocks A and B are defined by *s*A and *s*B [Fig. 9.14(b)]. Since the system has two cords, the length of first cord is

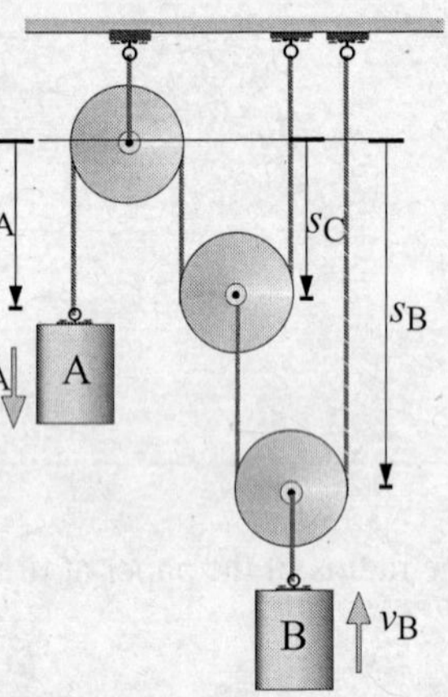

FIGURE 9.14(b)

$$l_1 = s_A + 2s_C \quad \text{(i)}$$

The length of second cord is

$$l_2 = s_B + (s_B - s_C) = 2s_B - s_C \quad \text{(ii)}$$

Eliminating s_C from Equations (i) and (ii), we have

$$s_A + 4s_B = l_1 + 2l_2 = \text{constant}$$

Taking time derivative

$$\frac{ds_A}{dt} + 4\frac{ds_B}{dt} = 0$$

or

$$v_A + 4v_B = 0$$

$$v_B = \frac{v_A}{4} \text{ (upward)}$$

PROBLEM 9.15

The link shown in Figure 9.15(a) is guided by two blocks at A and B, which move in the fixed slots. If the velocity of A is 5 m/s to right, determine the velocity of B and angular velocity of the link at the instant $\theta = 30°$.

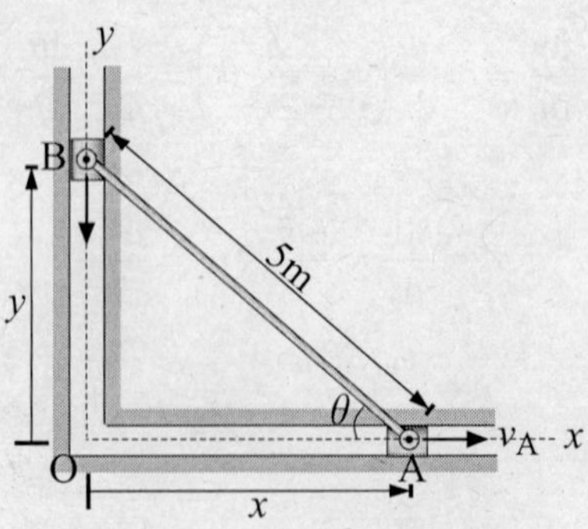

FIGURE 9.15(a)

Solution. With the x–y coordinate chosen as shown in the figure, the given velocity of motion of A is

$$v_A = \frac{dx}{dt} = 5\text{m/s}$$

$$x^2 + y^2 = 5^2$$

Differentiating with respect to time

$$x\frac{dx}{dt} + y\frac{dy}{dt} = 0$$

or
$$xv_A + yv_B = 0$$

From the geometry,

$$x = 5\cos\theta \quad \text{and} \quad y = 5\sin\theta$$

or
$$v_B = -\frac{x}{y}v_A = -\frac{5\cos\theta}{5\sin\theta}\times 5$$

or
$$v_B = -5\cos\theta = -5\cot 30$$

$$v_B = 8.66\text{m/s (downward)}$$

Now
$$x = 5\cos\theta$$

Differentiating with respect to time

$$\frac{dx}{dt} = -5\sin\theta\frac{d\theta}{dt}$$

or
$$\frac{d\theta}{dt} = \frac{\frac{dx}{dt}}{-5\sin\theta} = \frac{v_A}{-5\sin\theta}$$

or
$$\frac{d\theta}{dt} = \frac{5}{-5\sin 30} = -2$$

$$\omega = 2\text{rad/s (counter-clockwise)}$$

PROBLEM 9.16

The motion of a slider C is defined by the relations $r = 3t - t^2$ and $\theta = 2t$, where r is expressed in meters, t in seconds and θ in radians [Fig. 9.16(a)]. Determine the radial and transverse acceleration of the slider when $t = 2$ s.

FIGURE 9.16(a) |

Solution. Radial acceleration of slider is given by

$$a_r = \frac{d^2 r}{dt^2} - r\left(\frac{d\theta}{dt}\right)^2$$

or
$$a_r = \frac{d^2}{dt^2}(3t - t^2) - (3t - t^2)\left(\frac{d}{dt}2t\right)^2$$

or
$$a_r = -2 - 4(3t - t^2)$$

or
$$a_r = 4t^2 - 12t - 2$$

$$(a_r)_{t=2s} = -10\text{m/s}^2$$

Transverse acceleration of slider is given by

$$a_t = \frac{1}{r}\frac{d}{dt}\left(r^2 \frac{d\theta}{dt}\right)$$

or
$$a_t = \frac{1}{(3t - t^2)}\frac{d}{dt}\left((3t - t^2)^2 \frac{d}{dt}2t^2\right)$$

or
$$a_t = \frac{1}{(3t - t^2)} \times 2(3t - t^2)^2 \times (3t - t^2) = 4(3 - 2t)$$

$$a_t = -4\text{m/s}^2$$

PROBLEM 9.17

In the pulley configuration shown in Figure 9.17(a), block A has a downward velocity of 1.2 m/s. Determine the velocity of B as a function of *y*. Neglect the size of pulleys.

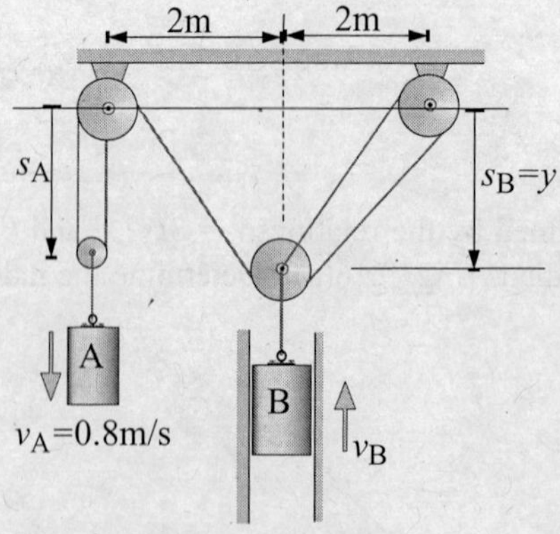

FIGURE 9.17(a) |

Solution. The position of block A is defined by s_A, and the position of block is specified by s_B from the fixed horizontal datum as shown in the figure. Since the system consists of one cable, the length of cable can be expressed as

$$l = 2s_A + \sqrt{2^2 + s_B^2} + 2\sqrt{2^2 + s_B^2}$$

$$l = 2s_A + 3\sqrt{2^2 + s_B^2}$$

Differentiating with time gives

$$0 = 2\frac{ds_A}{dt} + \frac{3}{2}\frac{2s_B}{\sqrt{2^2 + s_B^2}}\frac{ds_B}{dt}$$

or

$$0 = 2v_A + \frac{3s_B}{\sqrt{2^2 + s_B^2}} v_B$$

or

$$v_B = 2v_A \frac{\sqrt{2^2 + s_B^2}}{3s_B} = \frac{2}{3} v_A \sqrt{1 + \left(\frac{2}{s_B}\right)^2}$$

or

$$v_B = \frac{2}{3} \times 1.2\sqrt{1 + \left(\frac{2}{y}\right)^2} = 0.8\sqrt{1 + \left(\frac{2}{y}\right)^2}$$

$$v_B = 0.8\sqrt{1 + \left(\frac{2}{y}\right)^2}$$

PROBLEM 9.18

The collar A moves down with a constant velocity v_A as shown in Figure 9.18(a). Determine the velocity of block B.

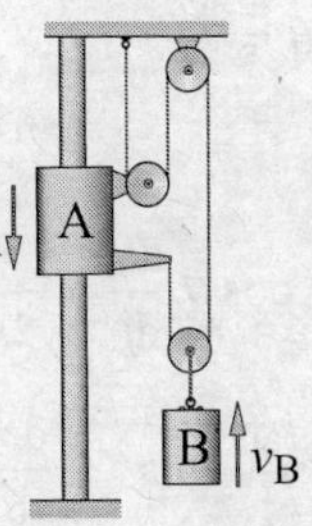

FIGURE 9.18(a) |

Solution. The position of collar A is defined by sA and the position of block B is specified by sB. Both coordinates are measured from a horizontal datum passing through the fixed pulley as shown in Figure 9.18(b). The length of the cable

$$l = 2s_A + (2s_B - S_A - h)$$

$$l = s_A + 2s_B - h$$

Taking time derivative

$$0 = \frac{ds_A}{dt} + 2\frac{ds_B}{dt}$$

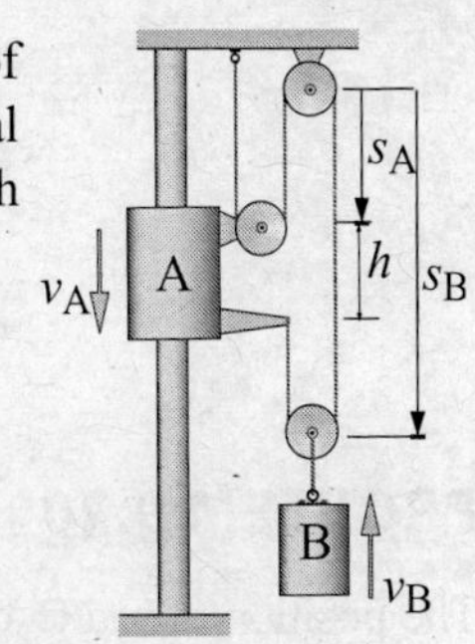

FIGURE 9.18(b) |

or
$$0 = v_A + 2v_B$$

$$v_A = -\frac{v_A}{2}$$

PROBLEM 9.19

A semicircular wire of radius r is supported in its own vertical plane by a hinge at O and a smooth peg P as shown in Figure 9.19(a). If peg P starts from O and move with a constant speed v along the horizontal axis through O, find the angular velocity of the wire.

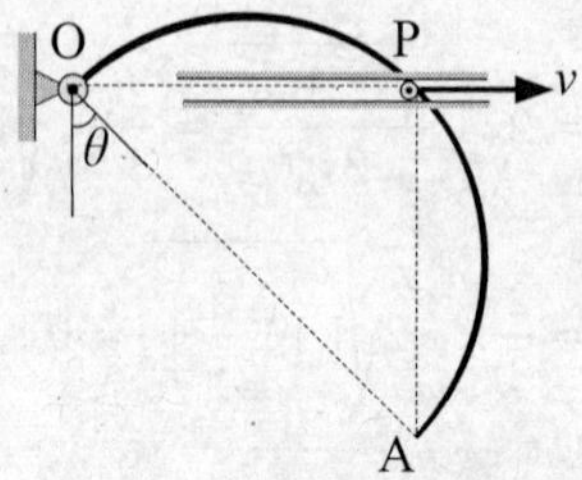

FIGURE 9.19(a) |

Solution. Horizontal displacement of the peg P is OP $= x = vt$.

OA is the diameter of semicircular, and from property of circle, OPA = 90°. Therefore, PA is always vertical:

$$x = vt = 2r \sin\theta$$

or
$$\sin\theta = \frac{vt}{2r}$$

Differentiating with respect to t gives

$$\cos\theta \frac{d\theta}{dt} = \frac{v}{2r}$$

or
$$\sqrt{1-\sin^2\theta}\,\frac{d\theta}{dt} = \frac{y}{2r}$$

or
$$\frac{d\theta}{dt} = \frac{v}{2r} \times \frac{1}{\sqrt{1-\sin^2\theta}} = \frac{v}{2r} \times \frac{1}{\sqrt{1-\left(\frac{vt}{2r}\right)^2}}$$

$$\theta = \frac{v}{\sqrt{4r^2 - v^2t^2}}$$

PROBLEM 9.20

The position of rod OA is controlled by yoke BC [Fig. 9.20(a)]. Knowing that the yoke moves horizontally with a constant velocity v, derive an expression for the angular velocity of rod OA.

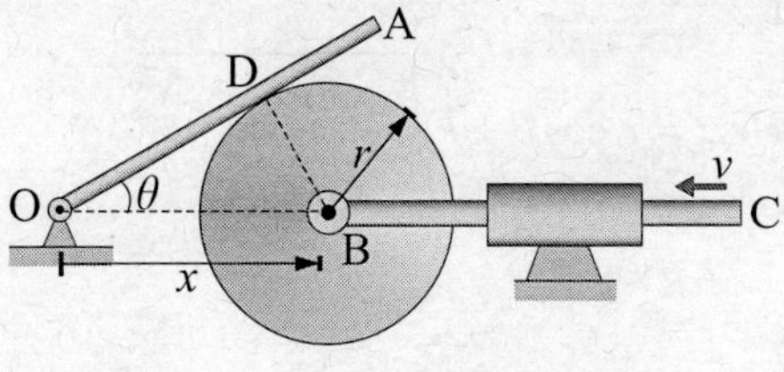

FIGURE 9.20(a)

Solution. Coordinates x and θ are chosen for analysis in order to express the rotational motion of rod OA. These coordinates are measured from the fixed point O and may be related to each other by using trigonometry. Since BD = r, then

$$x = r\operatorname{cosec}\theta$$

Differentiating with respect to time, we have

$$\frac{dx}{dt} = \frac{d}{dt} r\operatorname{cosec}\theta = -r\cot\theta\operatorname{cosec}\theta\frac{d\theta}{dt}$$

or

$$-v = -r\cot\theta\operatorname{cosec}\theta\frac{d\theta}{dt}$$

or

$$\frac{d\theta}{dt} = \frac{v}{r}\tan\theta\sin\theta$$

$$\omega = \frac{v}{r}\sin^2\theta\sec\theta$$

PROBLEM 9.21

For the slider-crank mechanism shown in Figure 9.21(a), determine (a) the velocity of the slider; and (b) the angular velocity of the connecting rod.

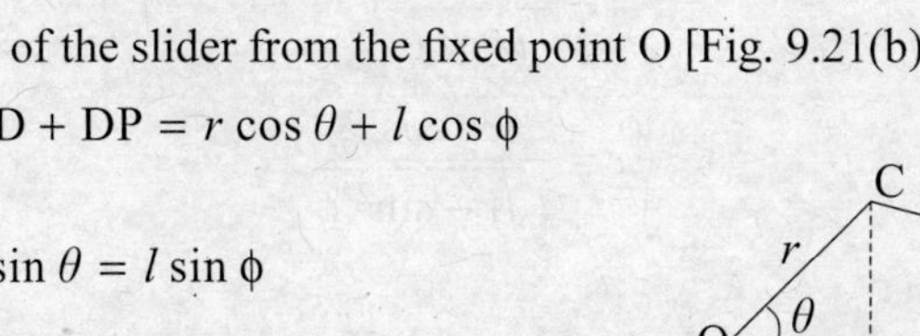

FIGURE 9.21(a)

Solution. Position of the slider from the fixed point O [Fig. 9.21(b)] is

$$x = \text{OD} + \text{DP} = r\cos\theta + l\cos\phi$$

Now

$$\text{CD} = r\sin\theta = l\sin\phi$$

or

$$\cos\phi = \sqrt{1-\left(\frac{r}{l}\sin\theta\right)^2}$$

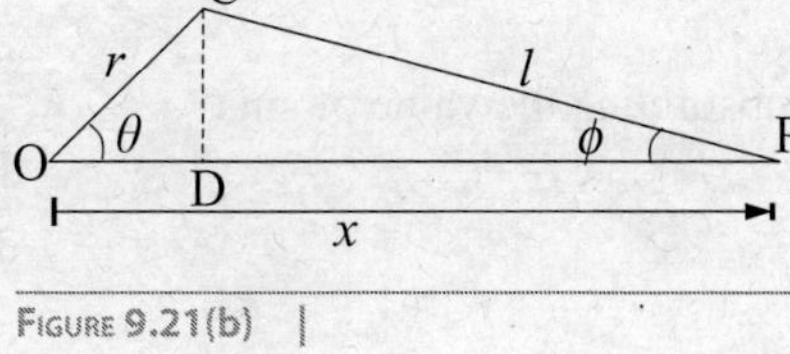

FIGURE 9.21(b)

Therefore, position of the slider

$$x = r\cos\theta + l\sqrt{1-\left(\frac{r}{l}\sin\theta\right)^2}$$

Velocity of the slider

$$v = \frac{dx}{dt}$$

or $$v = \frac{d}{dt}\left(r\cos\theta + 1\sqrt{1-\left(\frac{r}{l}\sin\theta\right)^2}\right)$$

or $$= \frac{d}{d\theta}\left(r\cos\theta + 1\sqrt{1-\left(\frac{r}{l}\sin\theta\right)^2}\right)\times\frac{d\theta}{dt}$$

or $$v = \left(-r\sin\theta - \frac{r^2\sin\theta\cos\theta}{l\sqrt{1-\left(\frac{r}{l}\sin\theta\right)^2}}\right)\times\frac{d\theta}{dt}$$

$$v = \omega r\left[\sin\theta + \frac{r\sin 2\theta}{2l\sqrt{1-\left(\frac{r}{l}\sin\theta\right)^2}}\right]$$

where $$\omega = -\frac{d\theta}{dt}$$

(*b*) $$CD = r\sin\theta = l\sin\phi$$

$$\sin\phi = \frac{r}{l}\sin\theta$$

Taking time derivative,

$$\cos\phi\frac{d\phi}{dt} = \frac{r}{l}\cos\theta\frac{d\theta}{dt}$$

Angular velocity of the connecting rod

$$\frac{d\phi}{dt} = \frac{\omega r\cos\theta}{l\cos\phi}$$

or $$\frac{d\phi}{dt} = \frac{\omega r\cos\theta}{l\sqrt{1-\sin^2\phi}}$$

Substituting the value of $\sin\phi$

$$\frac{d\phi}{dt} = \frac{\omega r\cos\theta}{1\sqrt{1-\left(\frac{r}{l}\sin\theta\right)^2}}$$

$$\omega_{\text{connecting rod}} = \frac{\omega \cos \theta}{\sqrt{\left(\frac{l}{r}\right)^2 - \sin^2 \theta}}$$

PROBLEM 9.22

The piston rod of the hydraulic cylinder gives a velocity v to the point B as shown in Figure 9.22(a). Determine the magnitude v_c of the velocity of end C of the link ABC in terms of θ.

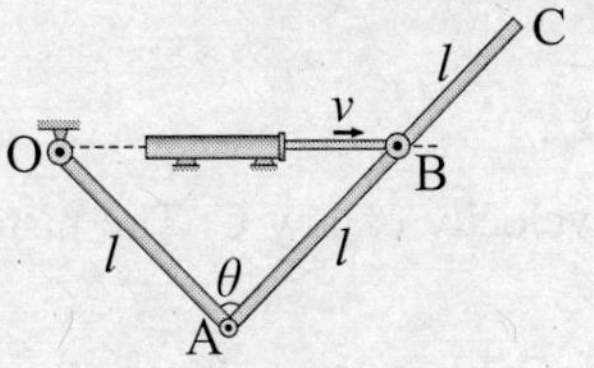

FIGURE 9.22(a) |

Solution. The coordinate system of C [Fig. 9.22(b)] may be expressed in terms of angle θ;

$$x = 3l \sin \frac{\theta}{2} \qquad \text{(i)}$$

$$y = l \cos \frac{\theta}{2} \qquad \text{(ii)}$$

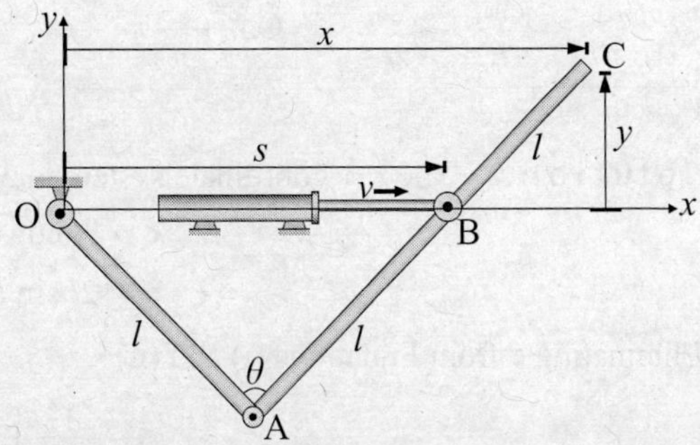

FIGURE 9.22(b) |

Differentiating Equations (i) and (ii) with respect to time, we write

$$\frac{dx}{dt} = \frac{3}{2} l \cos \frac{\theta}{2} \frac{d\theta}{dt}$$

$$\frac{dy}{dt} = -\frac{1}{2} l \sin \frac{\theta}{2} \frac{d\theta}{dt}$$

Velocity of end C

$$v_c = \sqrt{\left(\frac{dx}{dt}\right)^2 + \left(\frac{dy}{dt}\right)^2}$$

$$v_c = \sqrt{\left(\frac{3}{2} l \cos \frac{\theta}{2} \frac{d\theta}{dt}\right)^2 + \left(-\frac{1}{2} l \sin \frac{\theta}{2} \frac{d\theta}{dt}\right)^2}$$

$$v_c = \sqrt{9 \cos^2 \frac{\theta}{2} + \sin^2 \frac{\theta}{2}} \left(\frac{1}{2} \frac{d\theta}{dt}\right) \qquad \text{(iii)}$$

Velocity of point B

$$v = \frac{ds}{dt} = \frac{d}{dt}\left(2l \sin \frac{\theta}{2}\right) = l \cos \frac{\theta}{2} \frac{d\theta}{dt}$$

or

$$\frac{d\theta}{dt} = \frac{v}{l \cos\left(\frac{\theta}{2}\right)} \qquad \text{(iv)}$$

Substituting the value of $\frac{d\theta}{dt}$ in Equation (iii), we have

$$v_c = \sqrt{9\cos^2\frac{\theta}{2} + \sin^2\frac{\theta}{2}}\left(\frac{l}{2} \times \frac{v}{l\cos\left(\frac{\theta}{2}\right)}\right)$$

$$v_c = \frac{v}{2}\sqrt{9 + \tan^2\frac{\theta}{2}}$$

PROBLEM 9.23

Derive the equation for horizontal velocity of jaw C. The piston rod of the hydraulic cylinder is extending at the rate v [Fig. 9.23(a)].

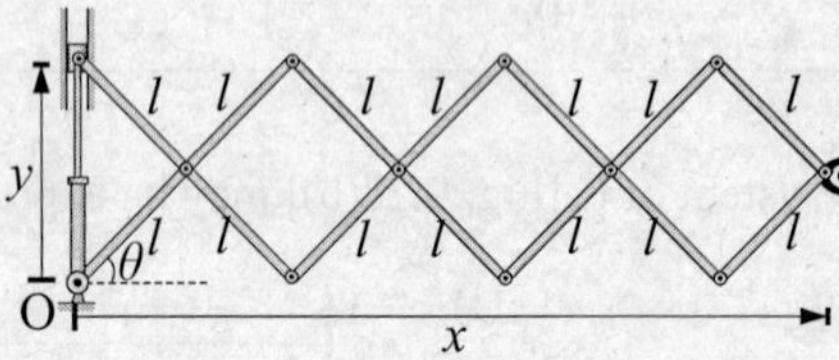

FIGURE 9.23(a) |

Solution. The x–y coordinate system may be expressed in terms of angle θ;

$$x = 7l\cos\theta \tag{i}$$

$$y = 2l\sin\theta \tag{ii}$$

Eliminating θ from Equations (i) and (ii),

$$\frac{x^2}{49} + \frac{y^2}{4} = l^2$$

Differentiating with respect to time

$$\frac{2x}{49}\frac{dx}{dt} + \frac{2y}{4}\frac{dy}{dt} = 0$$

or

$$\frac{x}{49}v_c + \frac{y}{4}v = 0$$

Velocity of the jaw

$$v_c = \frac{-49}{4}\frac{y}{x}v$$

or

$$v_c = \frac{-49}{4}\frac{2l\sin\theta}{7l\cos\theta}v = -\frac{7}{2}\tan\theta$$

$$v_c = -\frac{7}{2}\tan\theta v$$

PROBLEM 9.24

A rocket is fired vertically with a constant velocity from the station. Its flight is tracked by radar from point O [Fig. 9.24(a)]. Determine the angular velocity and angular acceleration of the radar.

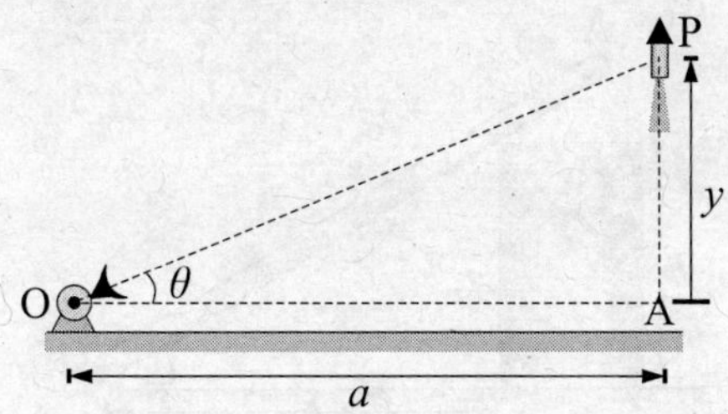

FIGURE 9.24(a) |

Solution. Vertical position of the rocket from the surface is $y = vt$.
From the geometry

$$\tan\theta = \frac{y}{a} \Rightarrow y = a\tan\theta$$

Taking time derivative gives

$$\frac{dy}{dt} = a\sec^2\theta\,\frac{d\theta}{dt}$$

or

$$\frac{d\theta}{dt} = \frac{1}{a\sec^2\theta}\frac{dy}{dt}$$

Angular velocity of radar

$$\omega = \frac{d\theta}{dt} = \frac{1}{a\sec^2\theta}\frac{dy}{dt} = \frac{1}{a(1+\tan^2\theta)}\frac{dy}{dt}$$

or

$$\omega = \frac{1}{a\left[1+\left(\frac{y}{a}\right)^2\right]}v$$

$$\omega = \frac{av}{a^2+y^2}$$

Angular acceleration of radar

$$\alpha = \frac{d\omega}{dt} = \frac{d}{dy}\left(\frac{av}{a^2+y^2}\right)\frac{dy}{dt}$$

or

$$\alpha = \frac{-ay}{(a^2+y^2)^2}\times 2y\times\frac{dy}{dt} = \frac{-2ayv^2}{(a^2+y^2)^2}$$

$$\alpha = \frac{-2ayv^2}{(a^2+y^2)^2}$$

PROBLEM 9.25

A box of height h moves to right with a constant velocity v [Fig. 9.25(a)]. Determine the angular velocity and angular acceleration of lever OA as a function of θ.

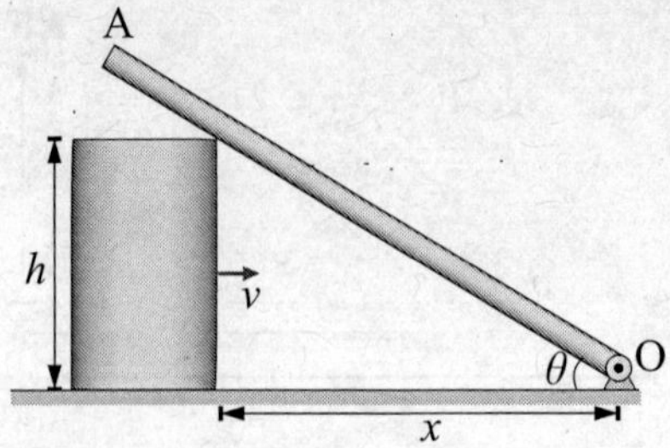

FIGURE 9.25(a) |

Solution. Let x be the distance of block from O and q be the angle of the rod at any instant during the motion. Since

$$\tan\theta = \frac{h}{x}$$

or

$$x = h\cot\theta$$

Differentiating with respect to time, we have

$$\frac{dx}{dt} = \frac{d}{dt}h\cot\theta = -h\text{cosec}^2\theta\frac{d\theta}{dt}$$

$$\frac{d\theta}{dt} = \frac{v}{-h\text{cosec}^2\theta} = -\frac{v}{h}\sin^2\theta$$

Angular velocity of the rod

$$\omega = \frac{v}{h}\sin 2\theta \qquad \text{or} \qquad \omega = \frac{vh}{v^2 + h^2}$$

Again differentiating with respect to time, we have

$$a = \frac{d\omega}{dt} = \frac{v}{h}\frac{d}{dt}\sin^2\theta$$

or

$$a = \frac{v}{h}2\sin\theta\cos\theta\frac{d\theta}{dt} = \frac{\omega v}{h}\sin 2\theta$$

or

$$a = \frac{v}{h}\left(\frac{v}{h}\sin^2\theta\right)\sin 2\theta$$

Angular acceleration of the rod

$$a = \left(\frac{v}{h}\right)^2\sin^2\theta\sin 2\theta$$

PROBLEM 9.26

Derive an expression for the velocity of slider A in its horizontal guide in terms of θ [Fig. 9.26(a)]. The piston rod of the hydraulic cylinder is extending at a rate v.

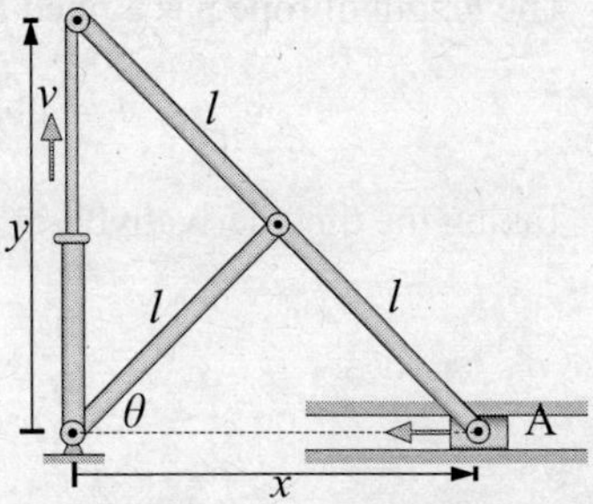

FIGURE 9.26(a) |

Solution. With the x–y coordinates chosen as shown in the figure, velocity of the given motion of C is $v_c = \dot{y} = v$. From the diagram,

$$x = 2l \cos \theta \tag{i}$$

and

$$y = 2l \sin \theta \tag{ii}$$

Eliminating θ from Equations (i) and (ii), we have

$$x^2 + y^2 = 4l^2$$

Differentiating with respect to time

$$x \frac{dx}{dt} + y \frac{dy}{dt} = 0$$

Velocity of slider A

$$v_A = \frac{dx}{dt} = -\frac{y}{x}\frac{dy}{dt}$$

or

$$v_A = -\frac{2l \sin \theta}{2l \cos \theta} v = -\tan \theta v$$

$$v_A = -\tan \theta v$$

PROBLEM 9.27

If end B of a rod is moving with a constant velocity v, determine the velocity of block A in terms of v, x and h.

Solution. As shown in Figure 9.27(a), the position of block A is defined by x and y. Velocity of block A in x-direction

$$(v_A)_x = v_B = v = \frac{dx}{dt} \rightarrow$$

Velocity of block A in y-direction

$$(v_A)_y = \frac{dy}{dt} \uparrow$$

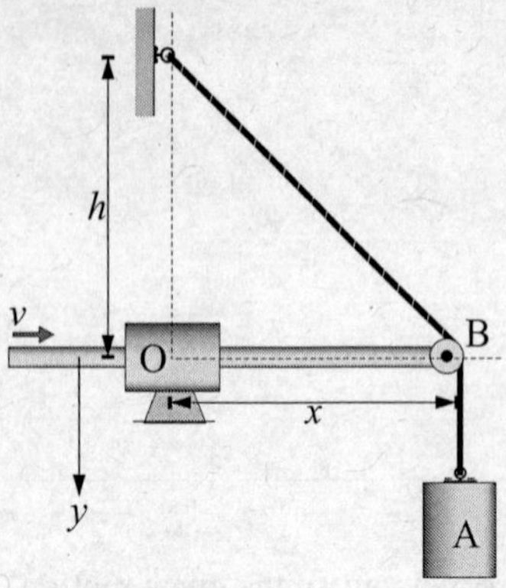

FIGURE 9.27(a) |

The length of rope has a fixed length

$$l = y + \sqrt{x^2 + h^2}$$

Taking the time derivative,

$$0 = \frac{dy}{dt} + \frac{x}{\sqrt{x^2 + h^2}} \frac{dx}{dt}$$

or

$$\frac{dy}{dt} = -\frac{x}{\sqrt{x^2 + h^2}} v_B$$

The resultant velocity of block A is

$$v = \sqrt{\left(\frac{dx}{dt}\right)^2 + \left(\frac{dy}{dt}\right)^2}$$

or

$$v = \sqrt{{v_B}^2 + \left(-\frac{x}{\sqrt{x^2 + h^2}} v_B\right)^2}$$

$$v = v_B \sqrt{\frac{2x^2 + h^2}{x^2 + h^2}}$$

CHAPTER HIGHLIGHTS

- Displacement is the change in position of a body in a definite direction.
- The velocity of a body is the rate of change of its displacment.
- Acceleration of body is the rate of change of its velocity.
- angular velocity ω is the rate at which the angular position θ changes with respect to time t.
  $$\omega = \frac{d\theta}{dt}$$
- Angular acceleration a is the rate at which the angular velocity ω changes with respect to time t.
  $$a = \frac{d\omega}{dt} = \frac{d^2\theta}{dt^2} = \omega \frac{d\omega}{d\theta}$$
- Velocity and Acceleration

  Rectilinear Motion: $v = \frac{ds}{dt}$, $a = \frac{dv}{dt} = \frac{d^2 s}{dt^2} = v\frac{dv}{ds}$

  Curvilinear Motion: $\mathbf{v} = \frac{d\mathbf{r}}{dt}$, $\mathbf{a} = \frac{d\mathbf{v}}{dt}$

  In polar coordinates: $v_r = \frac{dr}{dt}$, $a_r = \frac{d^2 r}{dt^2} - r\left(\frac{d\theta}{dt}\right)^2$

  $v_\theta = r\frac{d\theta}{dt}$, $a_\theta = \frac{1}{r}\frac{d}{dt}\left(r^2 \frac{d\theta}{dt}\right)$

- Equations of Motion

Rectilinear Motion: $v = u + at,\quad s = ut + \frac{1}{2}at^2,\quad v^2 = u^2 + 2as$

Circular Motion: $\omega = \omega_0 + at,\quad \theta = \omega_0 t + \frac{1}{2}at^2,\quad \omega^2 = \omega_0{}^2 + 2a\theta$

- Projectile Motion

A path followed by a particle when it is projected into the space making certain angle with the horizontal is known as projectile.

Equation of Trajectory $y = x\tan\alpha - \frac{1}{2}\frac{gx^2}{u^2\cos^2\alpha}$

Velocity and Direction of projectile:

$$v = \sqrt{u^2 - 2ugt\sin\alpha + g^2t^2} \quad \text{and} \quad \theta = \tan^{-1}\left(\frac{u\sin\alpha - gt}{u\cos\alpha}\right)$$

$$v = \sqrt{u^2 - 2gy} \quad \text{and} \quad \theta = \tan^{-1}\left(\frac{u^2\sin^2\alpha - 2gy}{u\cos\alpha}\right)$$

Maximum Height $h_{max} = \frac{u^2\sin^2\alpha}{2g}$ Time of Flight $T = \frac{2u\sin\alpha}{g}$

Range $R = \frac{u^2\sin 2\alpha}{g}$ Maximum range $R_{max} = \frac{u^2}{g}$

Range on inclined plane $R = \frac{u^2}{g}\frac{[\sin(2\alpha - \beta) - \cos\beta]}{\cos^2\beta}$

Maximum range on inclined plane $R_{max} = \frac{u^2}{g(1+\cos\beta)}$ at $\alpha = \frac{\pi}{4} + \frac{\beta}{2}$

Time of flight on inclined plane $T = \frac{2u\sin(\alpha - \beta)}{g\cos\beta}$

- Relative position of particle B with respect to particle A is $\mathbf{r}_{B/A} = \mathbf{r}_B - \mathbf{r}_A$
- Relative velocity of particle B with respect to particle A is $\mathbf{v}_{B/A} = \mathbf{v}_B - \mathbf{v}_A$
- Relative acceleration of particle B with respect to particle A is $\mathbf{a}_{B/A} = \mathbf{a}_B - \mathbf{a}_A$

REVIEW QUESTIONS

Q1. Obtain an equation for the trajectory of a projectile, and show that it is a parabola.

Q2. How would you find out (i) time of flight and (ii) range of a projectile, when projected upwards on an inclined plane?

Q3. Derive a relation for the velocity and direction of motion of a projectile (i) after a given interval of time t from the instant of projection and (ii) at a given height h above the point of projection.

Q4. Derive a relation for the velocity of a piston in a crank and connecting rod mechanism.

Q5. Derive the expression of radial and transverse velocity of a particle moving along a path $r = f(\theta)$.

Q6. Derive the expression of radial and transverse acceleration of a particle moving along a path $r = f(\theta)$.

Q7. A cam-follower system is shown in Figure 9.Q7. Determine the velocity v of the rod for any angle θ of the cam if the cam rotates with a constant angular velocity ω.

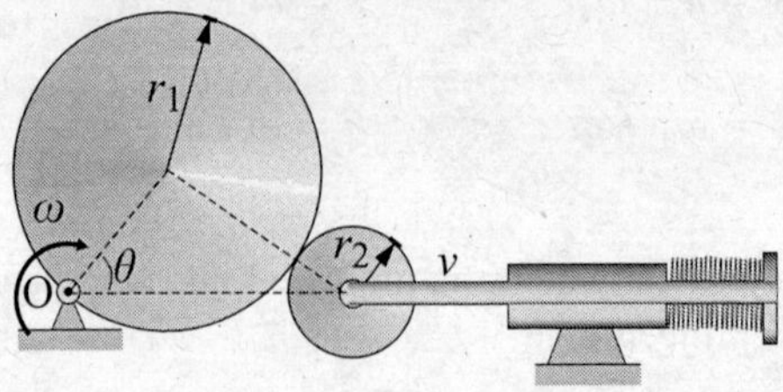

FIGURE 9.Q7

$$\left[\text{Ans; } v = \omega r_1 \sin\theta \left[1 + \frac{r_1 \cos\theta}{\sqrt{(r_1\cos\theta)^2 + 2r_1r_2 + r_2^2}}\right]\right]$$

Q8. An army vehicle travelling at 45 km/hr has wheels of 1 m diameter. Find the velocity of the top of the wheel relative to a person (i) sitting in the carriage and (ii) standing on the ground.

[Ans; (i) 12.5 m/s, (ii) 25 m/s].

Q9. A ladder of length l makes an angle θ with a vertical wall, as shown in Figure 9.Q9. The foot of the ladder moves to the right with a constant speed v_A. Determine angular velocity and angular acceleration of the ladder in terms of v_A, l and θ.

$$\left[\text{Ans; } \omega = \frac{v_A}{l}\sec\theta,\ a = \frac{v_A^2 \tan\theta}{l^2\cos^2\theta}\right].$$

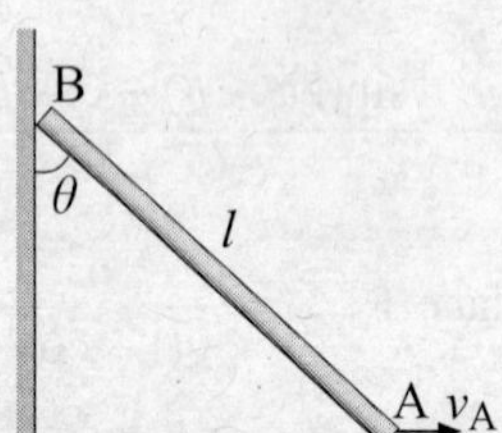

FIGURE 9.Q9

Q10. A train, starting from rest, is uniformly accelerated. The acceleration at any instant is $10/(1 + v)\text{m/s}^2$, where v is in m/s. Find the distance in which the train will attain a velocity of 10 m/s.

[Ans; $s = 38.33$ m].

Q11. A car starts from rest at $x = 0$ and travels along a straight road with the speed shown by the v–t graph shown in Figure 9.Q11. Determine the car's acceleration and position when $t = 8$ s and $t = 12$ s.

[Ans; $a_{t=8} = 0$, $x_{t=8} = 30$ m, $a_{t=12} = -1$ m/s^2, $x_{t=12} = 48$ m].

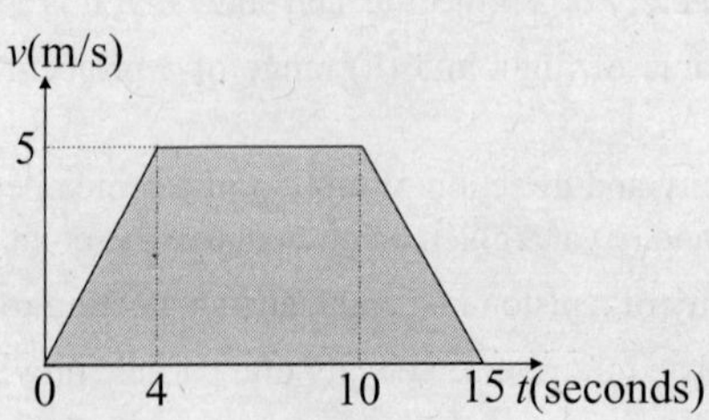

FIGURE 9.Q11

Q12. A car starts from rest at $s = 0$ and is subjected to the acceleration shown in a–x graph shown in Figure 9.Q12. Determine the speed of the car when it has travelled a distance 200 m. Also, how much time is required for it to travel the distance 200 m.
[Ans; $v = 155$ m/s, $t = 2.394$ s].

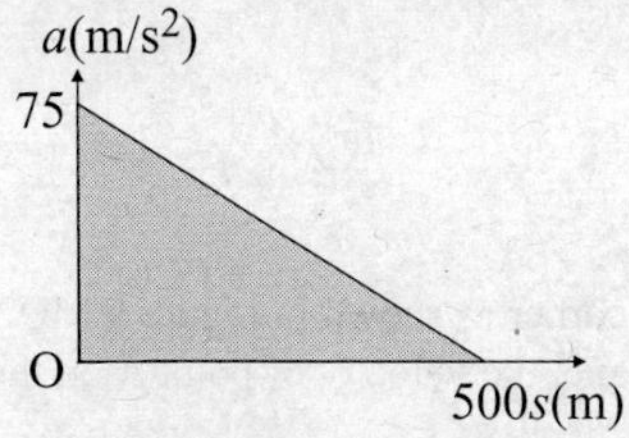

FIGURE 9.Q12 |

Q13. A particle, starting from rest, moves in a straight line, whose acceleration is given by the equation $a = 100 - 0.006s^2$ where a is in m/s² and s in m. Determine (i) the velocity of the particle when it has travelled 50 m and (ii) the distance travelled by the particle when it comes to rest.
[Ans; (i) $v = 22.36$ m/s, (ii) 70.7 m].

Q14. An inclined plane moves to left with a constant velocity v as shown in Figure 9.Q14. Determine the angular velocity and angular acceleration of the rod of length l. The rod hinges at one end O, and the other end A slides on the inclined plane.

$$\left[\text{Ans; } \omega = \frac{v \sin\phi}{l\cos(\phi - \theta)},\ \alpha = \frac{v^2 \sin^2\phi \sin(\phi - \theta)}{l^2 \cos^3(\phi - \theta)}\right].$$

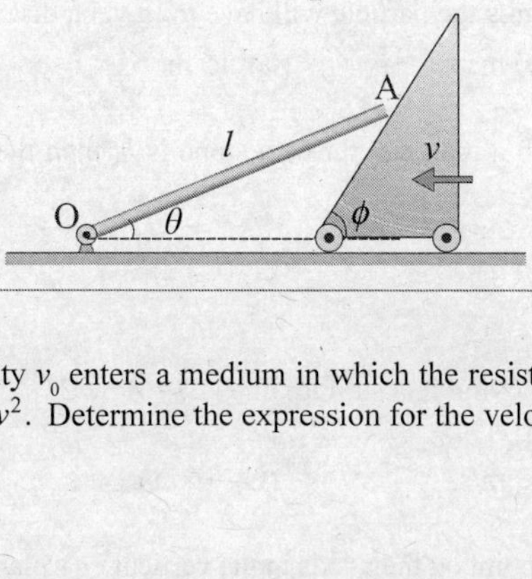

FIGURE 9.Q14 |

Q15. A body moving with a velocity v_0 enters a medium in which the resistance is proportional to the velocity squared. This means $a = -\lambda v^2$. Determine the expression for the velocity in terms of time t.

$$\left[\text{Ans; } v = \frac{1}{[kt + (1/v_0)]}\right].$$

Q16. A slotted lever is rotating about O at a constant angular velocity $\omega = 3\text{rad/s}$ as shown in Figure 9.Q16. Determine the radial and transverse components of the velocity and acceleration of pin P at the instant $\theta = 2\pi$. The path is defined by the spiral groove $r = [5 + (\theta/\pi)]\text{cm/s}$ where θ is in radians.

[Ans; $v_r = 0.955$ cm/s, $v_\theta = 21$ cm/s, $a_r = -63$ cm/s², $a_\theta = 5.73$ cm/s²].

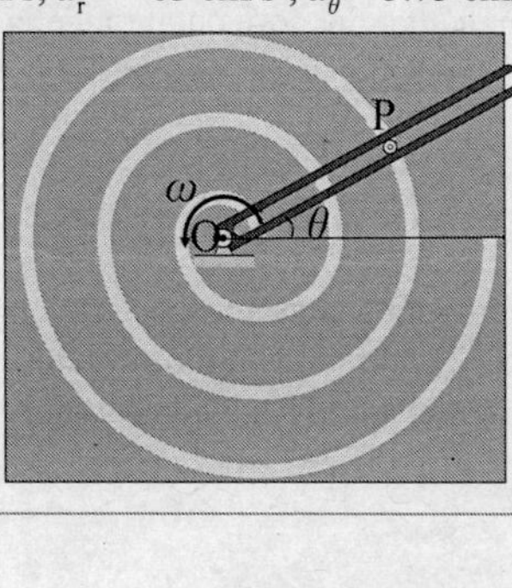

FIGURE 9.Q16 |

MULTIPLE CHOICE QUESTIONS

Q1. A particle of mass m is projected with a velocity v making an angle of 45° with the horizontal. The magnitude of the angular momentum of the projectile about the point of projection when the particle is at its maximum height h is

(A) zero (B) $\frac{mv^3}{g\sqrt{32}}$ (C) $\frac{mv^3}{g\sqrt{2}}$ (D) $m\sqrt{2gh^3}$

[Ans: B]

Q2. A rod of length 1 m is sliding in a corner as shown in Figure 9.MQ2. At an instant when the rod makes an angle of 60° with the horizontal plane, the velocity of point A on the rod is 1 m/s. The angular velocity of the rod at this instant is

(A) 2 rad/s (B) 1.5 rad/s (C) 0.5 rad/s (D) 0.75 rad/s

[Ans: A]

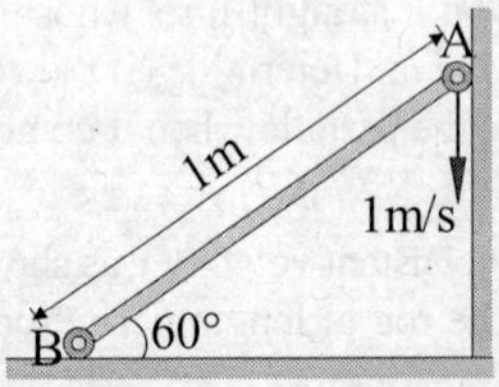

FIGURE 9.MQ2 |

Q3. The motion of a particle is given by the equation $S = 2t^3 + 3t$ (S is in m and t is time in s). Starting from $t = 0$, to attain a velocity of 9 m/s the particle will have to travel a distance of

(A) 5 m (B) 10 m (C) 15 m (D) 20 m

[Ans: A]

Q4. If the greatest height to which a man can throw a stone is h, then the greatest distance to which he can throw it will be

(A) $\frac{h}{2}$ (B) h (C) $2h$ (D) $4h$

[Ans: C]

Q5. The displacement of a point moving in a straight line is $S = 8t^2 + 3t - 5$ (S is in m and t in s). The velocity, when the displacement is zero, is

(A) 3 m/s (B) 13 m/s (C) 16 m/s (D) 12 m/s

[Ans: B]

Q6. A particle is projected from a point on the x-axis in the vertical $x-y$ plane. If the trajectory above the x-axis is given by $x^2 + 4y - 8 = 0$, then the velocity of projection is

(A) $\sqrt{3g}$ (B) $\sqrt{4g}$ (C) $\sqrt{5g}$ (D) $\sqrt{6g}$

[Ans: D]

Q7. If a body is projected at such an angle that the horizontal range is four times the greatest height, then the angle of projection is

(A) 45° (B) $\tan^{-1}(2)$ (C) $\cot^{-1}(2)$ (D) $\tan^{-1}\left(\frac{1}{2}\right)$

[Ans: A]

Q8. A train starts from rest from a station with a constant acceleration for 2 min and attains a constant speed. If it then runs for 11 min at this speed and retards uniformly during the next 3 min and stops at the next station, which is 9 km off. The maximum speed (in km/hr) attained by the train is

(A) 30 (B) 35 (C) 40 (D) 45

[Ans: C]

Q9. A particle moves along a space curve such that its position vector **r** (t) at time t is given by **r** $(t) = (2\cos)i + (2\sin)j + 3t^2k$ then the particle has

(A) constant speed

(B) constant acceleration

(C) speed which continuously decreases with t

(D) acceleration which continuously increases with t

[Ans: B]

Q10. The range of a projectiles fired at an angle of 15° is 50 m. If it fired with the same speed at an angle of 45°, its range will be

(A) 25 m (B) 37 m (C) 50 m (D) 100 m

[Ans: D]

Q11. Two balls are projected simultaneously with the same velocity from the top of a tower, one vertically upwards and the other vertically downwards. If they reach the ground in times t_1 and t_2, then the height of the tower is

(A) $\frac{1}{2}gt_1t_2$ (B) $\frac{1}{2}g(t_1^2 + t_2^2)$ (C) $\frac{1}{2}g(t_2^2 - t_1^2)$ (D) $\frac{1}{2}g(t_1 + t_2)^2$

[Ans: A]

Q12. α and β $(\alpha \neq \beta)$ be the two angles of projection of a projectile to reach a point at a distance R on the horizontal through the point of projection, u being the speed of projection and we have $u^2 > gR$. If the greatest heights attained by the projectile in the two trajectories be h_1 and h_2, respectively, then

(A) $h_1 + h_2 = \frac{u^2}{2g}$ (B) $h_1 + h_2 = \frac{u^2}{4g}$

(C) $h_1 + h_2 = \frac{u^2}{g}(\sin^2 \alpha + \sin^2 \beta)$ (D) $h_1 + h_2 = \frac{u^2}{4g}(\cos^2 \alpha + \cos^2 \beta)$

[Ans: A]

Q13. A shell is fired from a cannon. At the instant the shell is just about to leave the barrel, its velocity relative to the barrel is 3 m/s, while the barrel is swinging upwards with a constant angular velocity of 2 rad/s (Fig. 9.MQ13). The magnitude of the absolute velocity of the shell is

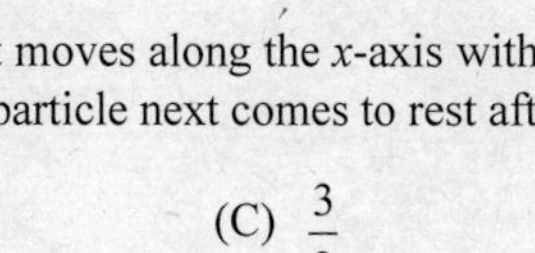

Figure 9.MQ13 |

(A) 3 m/s (B) 4 m/s (C) 5 m/s (D) 7 m/s

[Ans: C]

Q14. A particle is at rest at the origin. It moves along the x-axis with an acceleration $x - x^2$, where x is the distance of the particle at time t. The particle next comes to rest after it has covered a distance

(A) 1 (B) $\frac{1}{2}$ (C) $\frac{3}{2}$ (D) 2

[Ans: C]

Q15. A person travelling on a straight line moves with a uniform velocity v^1 for a distance x and with a uniform velocity v^2 for the next equal distance. The average velocity v is given by

(A) $v = \frac{v_1 + v_2}{2}$ (B) $v = \sqrt{v_1v_2}$ (C) $\frac{2}{v} = \frac{1}{v_1} + \frac{1}{v_2}$ (D) $\frac{1}{v} = \frac{1}{v_1} + \frac{1}{v_2}$

[Ans: C]

10

Kinetics of Particles and Rigid Bodies

INTRODUCTION

The mechanics of rigid bodies may be divided into statics and dynamics. Dynamics is further divided into kinematics and kinetics. Kinematics is concerned with rates of change of geometrical quantities in a moving system; it does not involve the concept of force. Kinetics treats the causes of motion and the nature of motion that results from specified forces. The relationships among position, velocity and acceleration of a moving body, such as a thrown ball, are defined by kinematics. The relationship between the forces that act on the ball and the motion of the ball involves kinetics.

NEWTON'S LAW OF GRAVITATIONAL ATTRACTION

Any two bodies whose dimensions are negligible compared to the distance separating them attract each other with equal and oppositely directed forces along the line joining them. The magnitude of the forces is directly proportional to the product of their masses and inversely proposal to the square of the distance separating them:

$$F = \frac{Gm_1m_2}{r^2},$$

where F is the mutual force of attraction between two particles, G is the universal constant called the constant of gravitation, m_1 and m_2 are the masses of the particles and r is the distance between the centres of the particles.

WEIGHT

The gravitational force of attraction that the earth exerts on an object is called the weight of the object. Since weight is a force, it is a vector quantity; it is directed towards the centre of the earth (Fig. 10.1), i.e.,

$$F = W = \frac{GMm}{r^2}$$

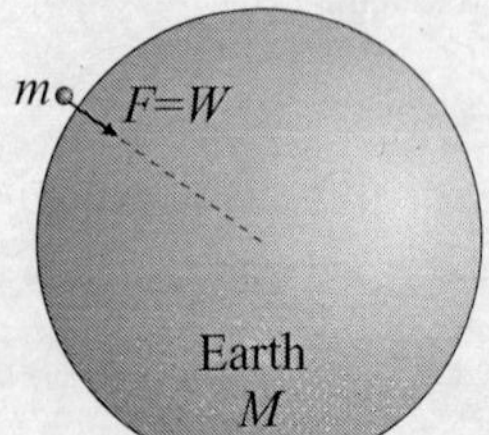

FIGURE 10.1

Weight of object

$$W = mg,$$

where $g = GM/r^2$ is magnitude of the acceleration of gravity, $m_1 = M$ is the mass of earth, $m_2 = m$ is the mass of the particle and $g = GM/R^2 = 9.81\text{m/s}^2$ on the surface of earth.

NEWTON'S LAWS OF MOTION

First law: A particle originally at rest, or moving in a straight line with a constant velocity, will remain in this state, provided the particle is not subjected to an unbalanced force.

Second law: If the resultant force acting on a particle is not zero, the particle will have an acceleration proportional to the magnitude of the resultant force and in the direction of this resultant force.

Third law: The mutual forces of action and reaction between two particles are equal, opposite and collinear.

NEWTON'S SECOND LAW

Consider a particle is subjected a force $\mathbf{F}_1$ of constant magnitude F_1 and constant direction. Due to the force, the particle will be observed to move in a straight line and in the direction of force. By determining the position of the particle at various instants, we find that its acceleration has a constant magnitude a_1. If the experiment is repeated with forces $\mathbf{F}_2$, $\mathbf{F}_3$, $\mathbf{F}_4$, etc. (Fig. 10.2), we find each time the particle moves in the direction of the force acting on it and that the magnitudes a_1, a_2, a_3 of the accelerations are proportional to the magnitudes F_1, F_2, F_3, etc. of the corresponding forces,

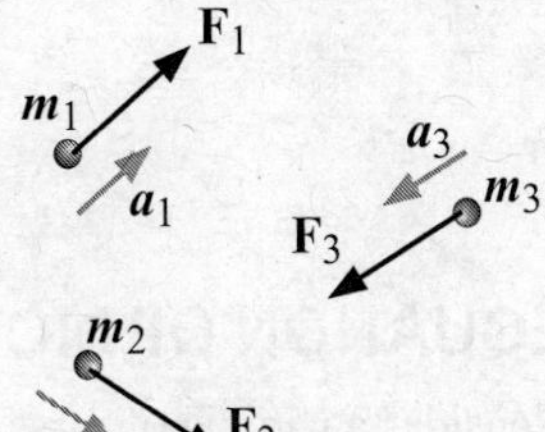

FIGURE 10.2

$$\frac{F_1}{a_1} = \frac{F_2}{a_2} = \frac{F_3}{a_3} = \cdots = C,$$

where C is a constant.

We conclude that the constant C is a measure of some invariable property of the particle. This property is the inertia of the particle, which is its resistance to rate of change of velocity. It is called the mass of particle and is denoted by m.

Newton's second law of motion may be written in the mathematical form as

$$\mathbf{F} = m\mathbf{a}$$

This equation is referred to as the equation of motion.

When a particle is subjected simultaneously to several forces, as shown in Figure 10.3, this equation should be replaced by

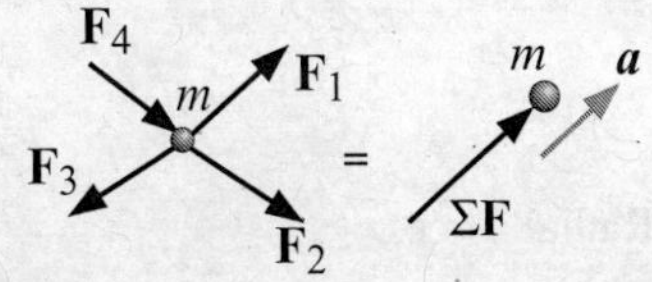

FIGURE 10.3

$$\Sigma\mathbf{F} = m\mathbf{a},$$

where ΣF represents the sum, or the resultant, of all the forces acting on the particle.

Consider a system of particles (rigid body) of total mass m acted upon by several forces (Fig. 10.4). Newton's second law may be expressed by writing the equation

$$\Sigma\mathbf{F} = \Sigma m\mathbf{a}_G$$

Note: Whenever the equation of motion is applied, it is required that measurements of the acceleration be made from Newtonian or inertial frame of reference. Such a coordinate system does not rotate. Either it is fixed or it translates in a given direction with a constant velocity (zero acceleration)

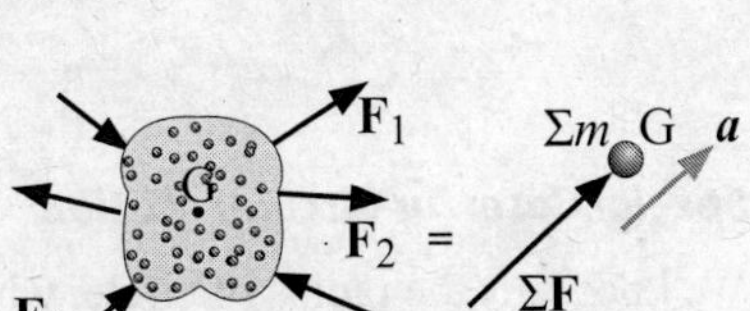

FIGURE 10.4

EQUATION OF MOTION: RECTANGULAR COORDINATES

Consider a particle of mass m acted upon by several forces. Newton's second law may be expressed by writing the equation

$$\Sigma \mathbf{F} = m\mathbf{a}$$

If the forces and the acceleration are resolved into components parallel to rectangular x, y, and z coordinate axes, it is expressed by three scalar equations:

$$\Sigma(F_x i + F_y j + F_z k) = m(a_x i + a_y j + a_z k)$$

From which it follows that

$$\Sigma F_x = ma_x \qquad \Sigma F_y = ma_y \qquad \Sigma F_z = ma_z$$

or

$$\Sigma F_x = m\ddot{x} \qquad \Sigma F_y = m\ddot{y} \qquad \Sigma F_z = m\ddot{z}$$

EQUATION OF MOTION: RADIAL AND TRANSVERSE COMPONENTS

Consider a particle of mass m, of polar coordinates(r,θ), which moves in a plane under the action of several forces (Fig. 10.5). Resolving the forces and the acceleration of the particle into radial and transverse components, we can express Newton's second law by writing the equations

$$\Sigma F_r = ma_r$$

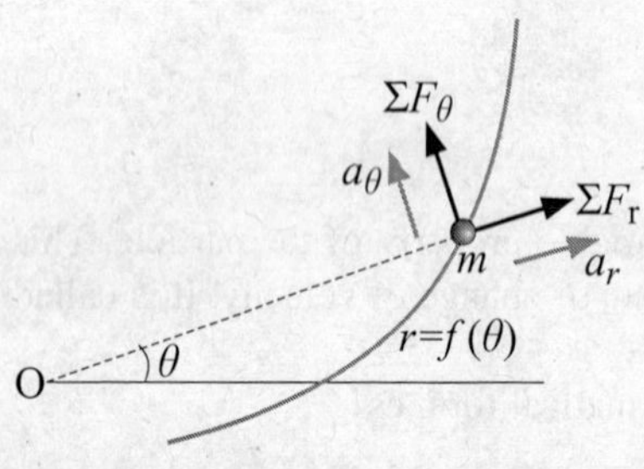

FIGURE 10.5 |

and

$$\Sigma F_\theta = ma_\theta$$

We know,

Radial acceleration

$$a_r = \frac{d^2r}{dt^2} - r\left(\frac{d\theta}{dt}\right)^2$$

and transverse acceleration

$$a_\theta = r\frac{d^2\theta}{dt^2} + 2\frac{dr}{dt}\frac{d\theta}{dt}$$

Radial force,

$$\Sigma F_r = m\left[\frac{d^2r}{dt^2} - r\left(\frac{d\theta}{dt}\right)^2\right]$$

Transverse force,

$$\Sigma F_\theta = m\left[r\frac{d^2\theta}{dt^2} + 2\frac{dr}{dt}\frac{d\theta}{dt}\right]$$

Special Case: for Circular Motion

We know r is the radius of circle, which is a constant. Therefore,

$$\frac{dr}{dt} = 0, \qquad \text{and} \qquad \frac{d^2r}{dt^2} = 0$$

Radial force,

$$\Sigma F_r = -mr\left(\frac{d\theta}{dt}\right)^2$$

The negative sign indicates that the direction of radial force is inward.

Centrifugal force,

$$F_c = m\omega^2 r$$

where $\frac{d\theta}{dt} = \omega$ is the angular velocity.

Transverse force,

$$\Sigma F_\theta = mr\frac{d^2\theta}{dt^2}$$

EQUATION OF MOTION IN PLANE FOR A RIGID BODY

Consider a rigid body (the system of particles), acted upon by several external forces $\mathbf{F}_1$, $\mathbf{F}_2$, $\mathbf{F}_3$, ..., etc., as shown in Figure 10.6. The equation of motion, written for the rigid body (or system of particles), becomes

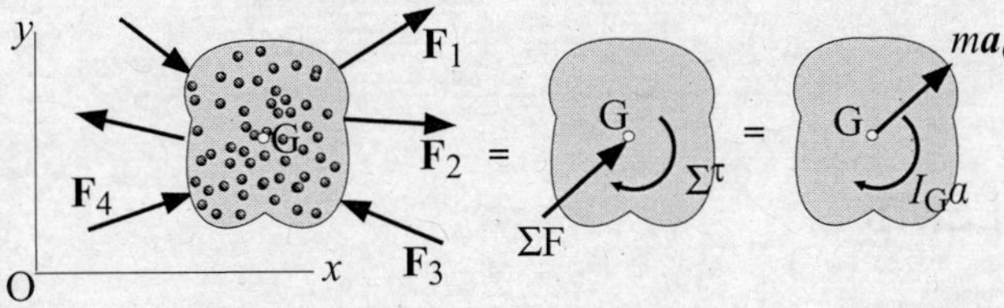

FIGURE 10.6 |

$$\sum \mathbf{F} = \sum m\mathbf{a}_G$$

or

$$\sum \mathbf{F} = m\mathbf{a}_G$$

and

$$\sum \tau = I_G \alpha,$$

where $\mathbf{a}_G$ is the acceleration of the mass centre G, $\Sigma m = m$ is the total mass of the body (or system of particle), ΣF is the resultant force, $\Sigma\tau$ is the resultant moment, I_G is the moment of inertia and α is the angular acceleration.

STEPS TO SOLVE PROBLEMS

When a body undergoes plane motion, it translates and rotates at the same time.

1. Sketch a free body diagram of the body, including all forces and couples that act on the body.
2. Select coordinate axes on which to sum forces and moments. Be sure to indicate the positive direction of motion for the body.
3. Apply the equations of motion, $\Sigma F_x = m\ddot{x}$, $\Sigma F_y = m\ddot{y}$, $\Sigma F_z = m\ddot{z}$ and $\Sigma\tau = I_G\alpha$ for the body to the free body diagram.
4. Solve the equations derived in step 3 for the unknowns.

D'ALEMBERT'S PRINCIPLE

In 1743, the French mathematician D'Alembert proposed a method of analysis that would transform a dynamics problem into an equivalent problem in statics. The essence of his idea, which is known as D'Alembert's principle, is to consider *ma* as the force acting on a particle, called the inertial force. Newton's second law $\Sigma\mathbf{F} - m\mathbf{a} = 0$ can then be treated as an equilibrium equation (the term dynamic equilibrium is often used) that involves the applied forces together with the inertial force.

For example, suppose that a particle P is subjected to three forces $\mathbf{F}_1$, $\mathbf{F}_2$ and $\mathbf{F}_3$ (Fig. 10.7). By D'Alembert's principle, the polygon of the four forces, $\mathbf{F}_1$, $\mathbf{F}_2$, $\mathbf{F}_3$ and inertia force *ma*, is closed, a shown in the figure, because, when the inertial force *ma* is taken into account, equilibrium conditions always exist. In other words, we may consider the inertial force *ma* as the equilibrant of the real forces $\mathbf{F}_1$, $\mathbf{F}_2$ and $\mathbf{F}_3$. In essence, the inertial force *ma* puts the particle in equilibrium in an accelerating frame. This case contains all the implications of Newton's second law. An inertial force is represented by a dashed vector.

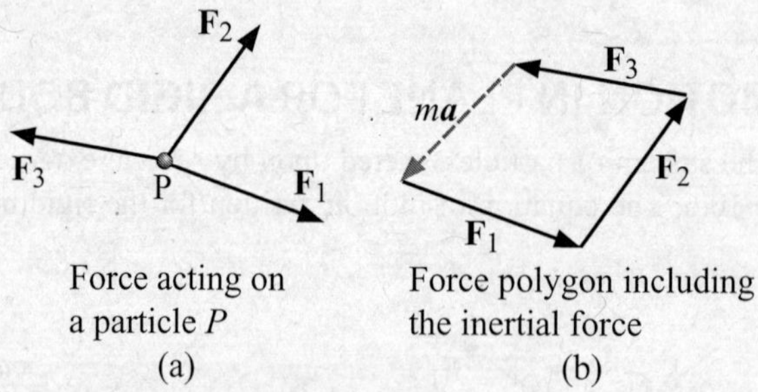

FIGURE 10.7 |

SOLVED PROBLEMS

PROBLEM 10.1

A block shown in Figure 10.1(a) is observed to have $v_1 = 5$ m/s as it passes point A and a velocity $v_2 = 2$ m/s as it passes point B on the incline. Calculate the coefficient of kinetic friction μ between the block and the incline if $L = 10$ m and $\theta = 15°$.

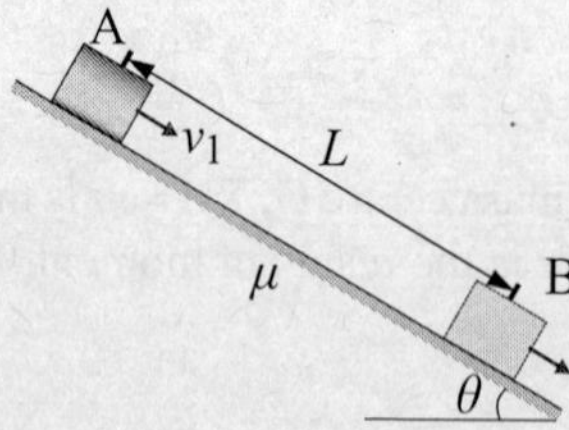

FIGURE 10.1(a) |

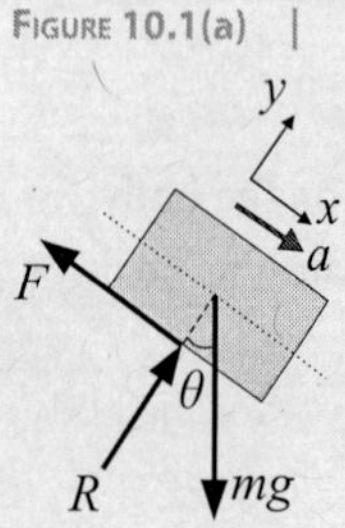

FIGURE 10.1(b) |

Solution. As shown in Figure 10.1(b), the friction force is directed opposite to sliding motion and has a magnitude $F = \mu R$. By applying the equation of motion in the *x* direction, we get

$$\Sigma F_x = ma_x \Rightarrow mg\sin\theta - F = ma$$

or $$mg\sin\theta - \mu R = mv\frac{dv}{dx}, \tag{i}$$

where equilibrium in the *y*-direction requires $R = mg\cos\theta$.

$$mg \sin \theta - \mu mg \cos \theta = mv \frac{dv}{dx}$$

or
$$vdv = g(\sin \theta - \mu \cos \theta)dx$$

Integrating from $v_1 = 5$ m/s, where $x = 0$ to $v_2 = 2$ m/s, where $x = 10$ m, we have

$$\int_5^2 vdv = \int_0^{10} g(\sin \theta - \mu \cos \theta)dx$$

or
$$\left[\frac{v^2}{2}\right]_5^2 = g(\sin \theta - \mu \cos \theta)[x]_0^{10}$$

or
$$\frac{4-25}{2} = 9.81[\sin 15 - \mu \cos 15] \times 10$$

or
$$-10.5 = 98.1(0.2588 - 0.9655\mu)$$

From which
$$\mu = 0.379$$

PROBLEM 10.2

A ball of mass m is fired vertically downward into a viscous fluid with an initial velocity v_0 as shown in Figure 10.2(a). If the fluid develops a drag resistance on the ball which is proportional to square of its velocity ($F_D = \lambda v^2$), determine the velocity v of the ball as a function of y.

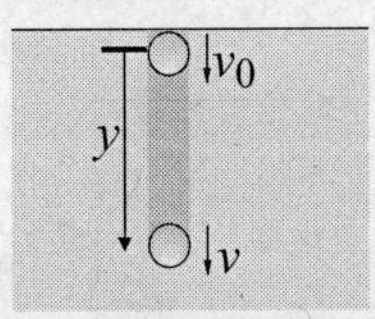

FIGURE 10.2(a) |

Solution. The free body diagram [Fig. 10.2(b)] of the ball shows the gravitational force mg acting down and the retarding force F_D acting up. Assuming that y is positive in downward direction, the equation of motion is

$$\Sigma F_y = ma_y \Rightarrow mg - F_D = mv \frac{dv}{dy}$$

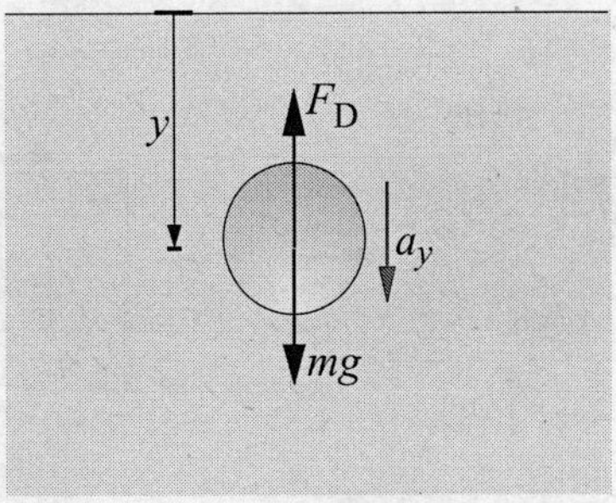

FIGURE 10.2(b) |

or
$$mv \frac{dv}{dy} = mg - \lambda v^2$$

or
$$v \frac{dv}{g - \left(\frac{\lambda}{m}\right) v^2} = dy$$

Integrating,

$$\int_{v_0}^{v} \frac{vdv}{g - \left(\frac{\lambda}{m}\right) v^2} = \int_0^y dy$$

or $$\frac{-m}{2\lambda}\left[\log\left(g-\frac{\lambda}{m}v^2\right)\right]_{v_0}^{v} = y$$

or $$\log\frac{g-\left(\frac{\lambda}{m}\right)v^2}{g-\left(\frac{\lambda}{m}\right)v_0^2} = -\frac{2\lambda}{m}y$$

or $$\frac{g-\left(\frac{\lambda}{m}\right)v^2}{g-\left(\frac{\lambda}{m}\right)v_0^2} = e^{-(2\lambda/m)y}$$

Simplifying

$$v^2 = \frac{mg}{\lambda} - \left(\frac{mg}{\lambda} - v_0^2\right)e^{(-2\lambda/m)y}$$

From which

$$v = \sqrt{\frac{mg}{\lambda}\left(1 - e^{(-2\lambda/m)y}\right) + v_0^2 e^{(-2\lambda/m)y}}$$

PROBLEM 10.3

Determine the largest horizontal force P that can be applied to B so that A will not move relative to B [Fig. 10.3(a)]. All surfaces are smooth.

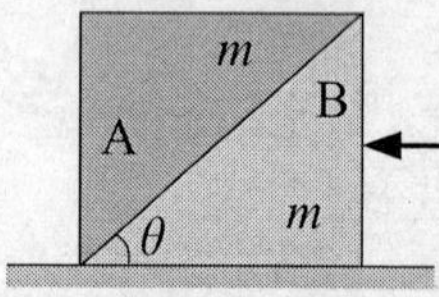

FIGURE 10.3(a) |

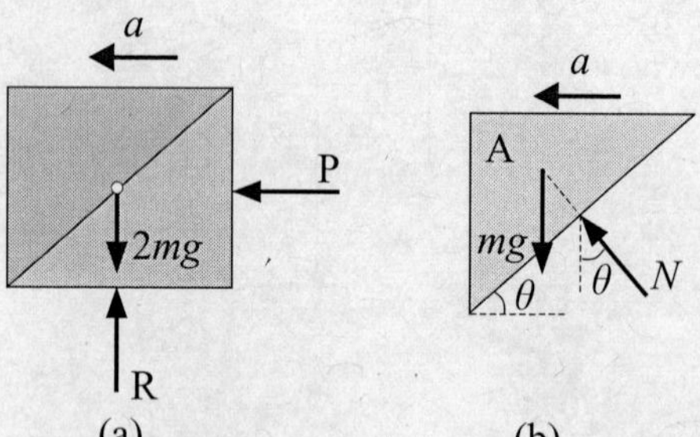

FIGURE 10.3(b) |

Solution. Let a be acceleration of the blocks, and N be the reaction between blocks A and B. Considering the motion of combined blocks [Fig. 10.3(b)]. Applying equation of motion

$$\Sigma F_x = (\Sigma m)a \Rightarrow P = 2ma$$

$$a = \frac{P}{2m} \tag{i}$$

Now considering block A in part b of Figure 10.3(b),

$$\Sigma F_x = ma \Rightarrow N\sin\theta = ma \tag{ii}$$

$$\Sigma F_y = 0 \Rightarrow N\cos\theta = mg \tag{iii}$$

Eliminating N from Equations (ii) and (iii), we have

$$\tan\theta = \frac{ma}{mg}$$

Substituting the value of a from Equation (i), we have

$$\tan\theta = \frac{m}{mg}\left(\frac{P}{2m}\right)$$

From which

$$P = 2mg\tan\theta$$

PROBLEM 10.4

Two bodies of weights 50 kN and 20 kN [Fig. 10.4(a)] are hung to the ends of an inextensible long string passing over an ideal pulley. Determine the acceleration of the body of weight 50 kN and tension in the string.

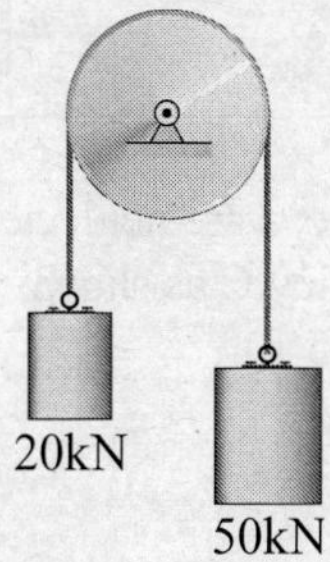

FIGURE 10.4(a) |

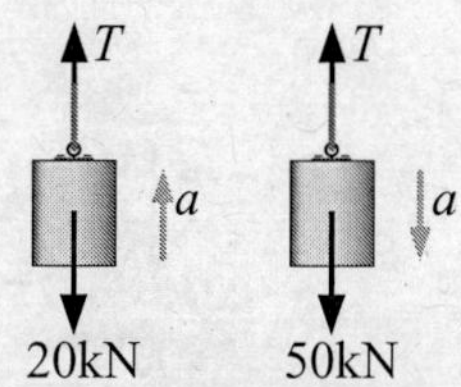

FIGURE 10.4(b) |

Solution. Let T be the tension of the string, and a be the acceleration of the block. First we draw the free body diagrams of 50 kN and 20 kN bodies as in Figure 10.4(b). Applying Newton's second law to both bodies, we get

Motion of 50 kN body $\sum F_y = 50\times10^3 - T = \dfrac{50\times10^3}{g}a$

or $$50000 - T = 5096.84a \quad \text{(i)}$$

Motion of 20 kN body

$$\sum F_y = T - 20\times10^3 = \frac{20\times10^3}{g}a$$

or $$T - 20000 = 2038.73a \quad \text{(ii)}$$

Now adding Equations (i) and (ii), we get

$$30000 = 7135.5a$$

$$a = 4.2\text{m/s}^2$$

Substituting the value of $a = 4.2$ m/s^2 in Equation (ii), we get

$$T = 28571.4\text{N}$$

$$T = 28.57\text{N}$$

PROBLEM 10.5

Determine the acceleration of blocks A and B and the tension of a rod connected to block B, as shown in Figure 10.5(a). Neglect the mass of pulleys and cords. Assume the surface at B is smooth.

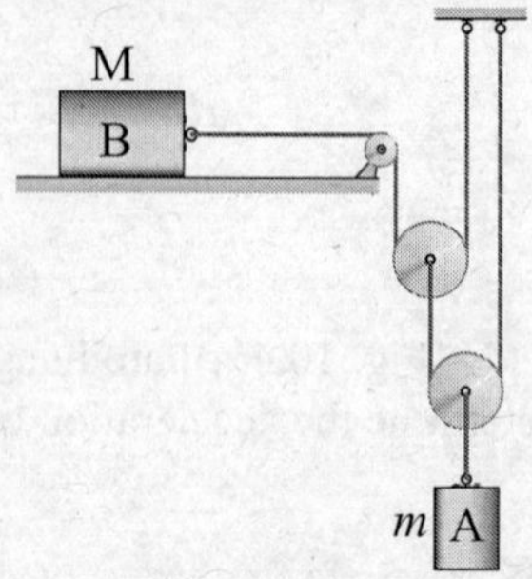

FIGURE 10.5(a) |

Solution. Since the mass of the pulleys are neglected, the effect of their inertia is zero and we can apply the equilibrium condition for pulley C as shown in Figure 10.5(b). The free body diagrams for blocks A and B are shown in Figure 10.5(b).

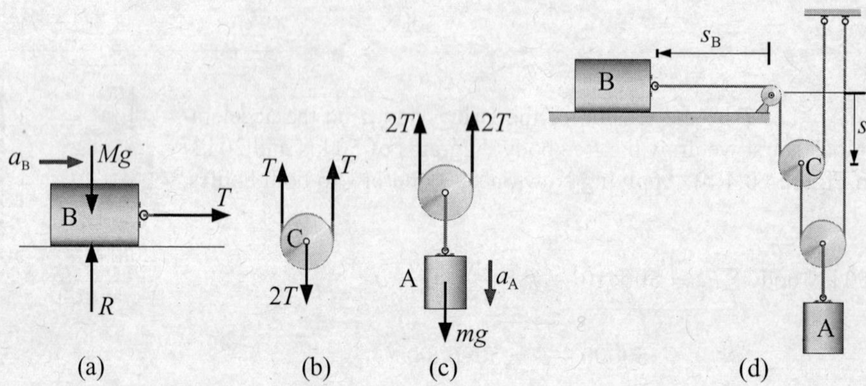

FIGURE 10.5(b) |

Considering the motion of block A [Fig. 10.5(b)],

$$F_y = ma_A \Rightarrow mg - 2T - 2T = ma_A \quad \text{(i)}$$

Considering the motion of block B [Fig. 10.5(b)],

$$\Sigma F_x = Ma_B \Rightarrow T - 0 = Ma_B \quad \text{(ii)}$$

Eliminating T from Equations (i) and (ii), we have

$$mg - 4Ma_B = Ma_A \quad \text{(iii)}$$

The position of blocks A, B, and pulley C are defined by s_A, s_B and s_C from fixed datum as shown in Figure 10.5(b). Since the system consists of two cables, the lengths of cables are

$$l_1 = 2s_A - s_C + h$$

and $$l_2 = 2s_C + s_B + h$$

$$4s_A + s_B + 3h = 2l_1 + l_2 = \text{constant.}$$

Differentiating twice with respect to time gives

$$4a_A + a_B = 0$$

$$a_B = 4a_A \text{ (by magnitude)}$$

Substituting the value of a_B into Equation (iii), we have

$$mg - 16Ma_A = ma_A$$

$$a_A = \frac{mg}{m + 16M} \quad \text{and} \quad a_B = \frac{4mg}{m + 16M}$$

From Equation (i), we have

$$T = \frac{4Mmg}{m + 16M}$$

PROBLEM 10.6

A fighter plane [Fig. 10.6(a)] is moving on the ground at a speed of 360 km/hr when the pilot deploys the braking parachute. How far does the plane move to get down to a speed of 240 km/hr? The plane has a mass of 8000 kg. The drag is $25v^2$, with v in m/s.

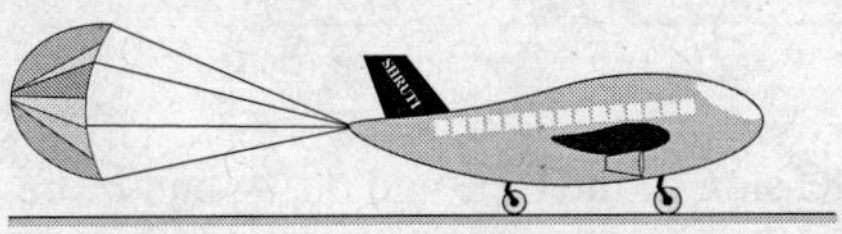

FIGURE 10.6(a)

Solution. The free body diagram for the plane when it is at a velocity v is shown in Figure 10.6(b). The drag force $F_D = 25v^2$ and acts opposite to the direction of motion. Applying the equation of motion,

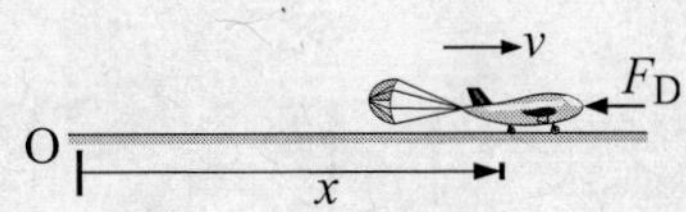

FIGURE 10.6(b)

$$\Sigma F_x = ma_x \Rightarrow -F_D = ma$$

or $$-25v^2 = mv\frac{dv}{dx}$$

or $$m\frac{dv}{v} = -25dx$$

Integrating from,

$$v_1 = \frac{360 \times 1000}{3600} = 100\text{m/s to } v_2 = \frac{240 \times 1000}{3600} = 66.66\text{m/s}$$

$$8000\int_{100}^{66.66} \frac{dv}{v} = -25\int_0^d dx$$

or
$$8000 \log \frac{66.66}{100} = -25d$$

or
$$d = \frac{8000 \log 0.666}{-25} = 129.75$$

$$d = 129.75\text{m}$$

PROBLEM 10.7

Blocks A, B and C shown in Figure 10.7(a) have a mass of 5 kg, 3 kg and 2 kg, respectively. Determine the tension of rope and acceleration of each block. Neglect the mass and size of pulleys

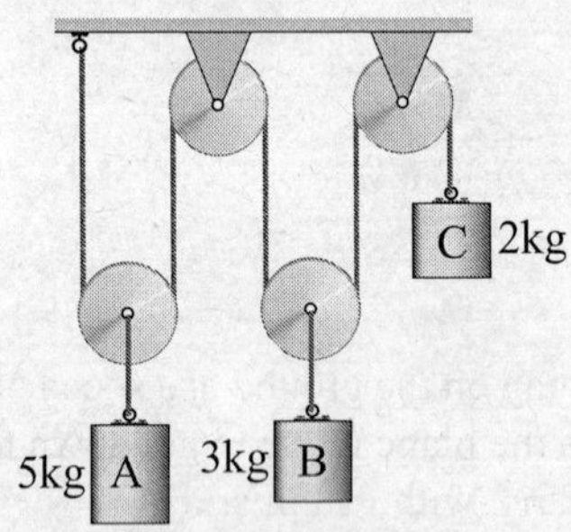

FIGURE 10.7(a) |

Solution. Let T be the tension of the rope and a_A, a_B and a_C be the downward acceleration of blocks A, B and C, respectively. Considering the motion of block A [Fig. 10.7(b)]

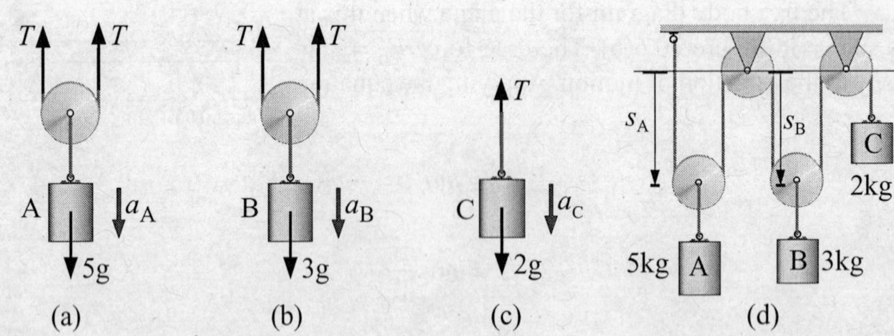

FIGURE 10.7(b) |

$$\Sigma F_y = ma_y \Rightarrow 5g - 2T = 5a_A$$

$$a_A = \frac{5g - 2T}{5} \tag{i}$$

Considering the motion of block B [Fig. 10.7(b)]

$$\Sigma F_y = ma_y \Rightarrow 3g - 2T = 3a_B$$

$$a_B = \frac{3g - 2T}{3} \tag{ii}$$

Considering the motion of block C [Fig. 10.7(b)],

$$\Sigma F_y = ma_y \Rightarrow 2g - T = 2a_C$$

$$a_C = \frac{2g - T}{2} \tag{iii}$$

At any instant, the total length l of all the vertical segments of the cable may be expressed in terms of the position coordinates s_A, s_B and s_C:

$$2s_A + 2s_B + s_C = l$$

Differentiating this expression twice with respect to time yields

$$2a_A + 2a_B + a_C = 0$$

Substituting the value of a_A, a_B and a_C from Equations (i), (ii), and (iii), we have

$$2\left(\frac{5g - 2T}{5}\right) + 2\left(\frac{3g - 2T}{3}\right) + \left(\frac{2g - T}{2}\right) = 0$$

or

$$150g - 79T = 0$$

$$\boxed{T = 18.62\text{N}}$$

From Equation (i)

$$a_A = \frac{5g - 2\text{T}}{5} = \frac{5 \times 9.81 - 2 \times 18.62}{5} = 2.35$$

$$\boxed{a_A = 2.35\text{m/s}^2 \text{ (downward)}}$$

From Equation (ii)

$$a_B = \frac{3g - 2T}{3} = \frac{3 \times 9.81 - 2 \times 18.62}{3} = -2.6$$

$$\boxed{a_B = -2.6\text{m/s}^2 \text{ (upward)}}$$

From Equation (iii)

$$a_C = \frac{2g - T}{2} = \frac{2 \times 9.81 - 18.62}{2} = 0.496$$

$$\boxed{a_C = 0.496\text{m/s}^2 \text{ (downward)}}$$

PROBLEM 10.8

A flexible chain of length L and mass m on a table with length c is overhanging as shown in Figure 10.8(a). The chain, initially at rest, is released from the configuration shown. The coefficient of friction between table and chain is μ. Find the velocity of the chain as a function of y. Assume that $c > \mu(L - c)$ and chain maintains contact with the table surface.

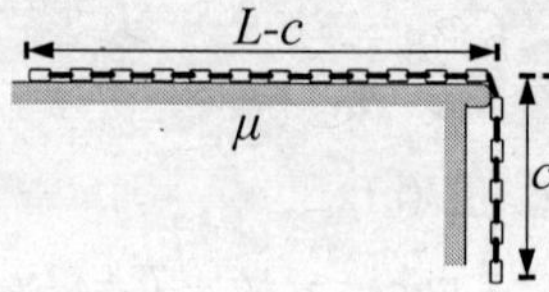

FIGURE 10.8(a) |

Solution. The free body diagrams of chain in two segments (i.e., vertical and horizontal), when it is located at the arbitrary position y is shown in Figure 10.8(b)

Considering the motion of part I (vertical) of chain [Fig. 10.8(c)],

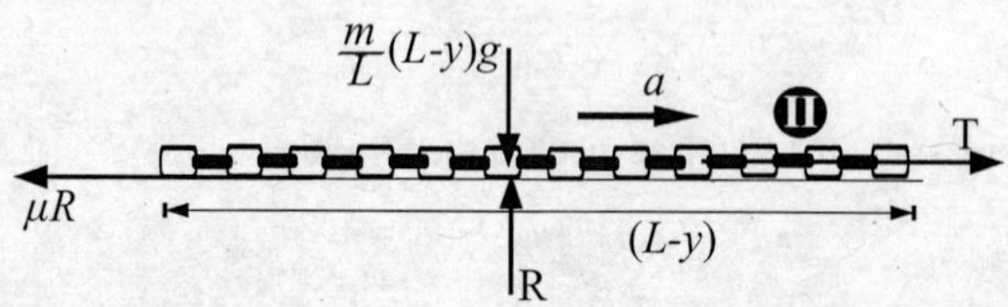

FIGURE 10.8(b) |

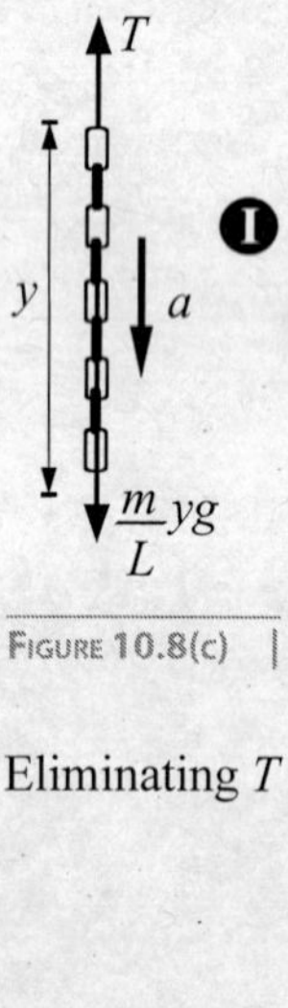

FIGURE 10.8(c) |

$$\Sigma F_y = ma_y \Rightarrow m_{\text{I}} g - T = m_{\text{I}} a$$

or $$\left(\frac{m}{L} y\right) g - T = \left(\frac{m}{L} y\right) a \qquad \text{(i)}$$

Considering the motion of part II (horizontal) of chain

$$\Sigma F_x = ma_x \Rightarrow T - \mu R = m_{\text{II}} a$$

or $$T - \mu m_2 g = m_{\text{II}} a$$

or $$T - \mu\left[\frac{m}{L}(L - y)\right] g = \left[\frac{m}{L}(L - y)\right] a \qquad \text{(ii)}$$

Eliminating T from Equations (i) and (ii), we have

$$\left(\frac{m}{L} y\right) g - \mu\left[\frac{m}{L}(L - y)\right] g = ma$$

or $$\frac{m}{L} g(y - \mu L + \mu y) = mv\frac{\mathrm{d}v}{\mathrm{d}y}$$

Integrating,

$$\int_0^v v\,\mathrm{d}v = \int_c^y \frac{g}{L}(y - \mu L + \mu y)\mathrm{d}y$$

or $$v^2 = \frac{2g}{L}(y - c)\left[\frac{1+\mu}{2}(y + c) - \mu L\right]$$

From which

$$v = \sqrt{\frac{2g}{L}(y - c)\left[\frac{1+\mu}{2}(y + c) - \mu L\right]}$$

PROBLEM 10.9

Two blocks A and B of masses 5 kg and 10 kg are connected by a string as shown in Figure 10.9(a). The coefficient of friction between blocks and the turntable which rotates about a vertical axis with a constant speed. Determine the angular speed at which the blocks begin to slide. Also find the tension in the string.

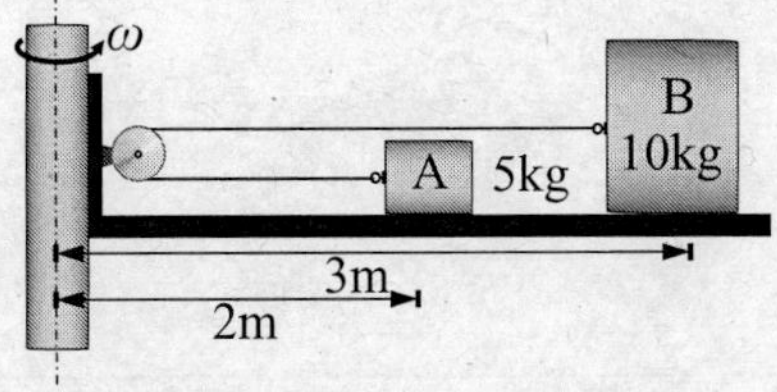

FIGURE 10.9(a)

Solution. Let T be the tension of string. Figures 10.9(b) and 10.9(c) show the free body diagrams of block A and B when blocks are just about to slide.

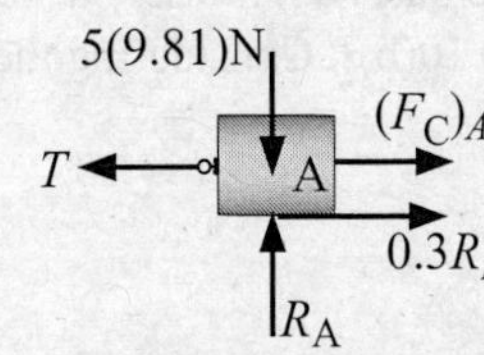

FIGURE 10.9(b)

Equilibrium equations of block A are

$$(\Sigma F_x)_A = 0 \Rightarrow (F_c)_A + 0.3R_A - T = 0$$

$$(\Sigma F_y)_A = 0 \Rightarrow R_A = 5(9.81)$$

Centrifugal force

$$(F_c)_A = 5 \times \omega^2 \times 2 = 10\omega^2$$

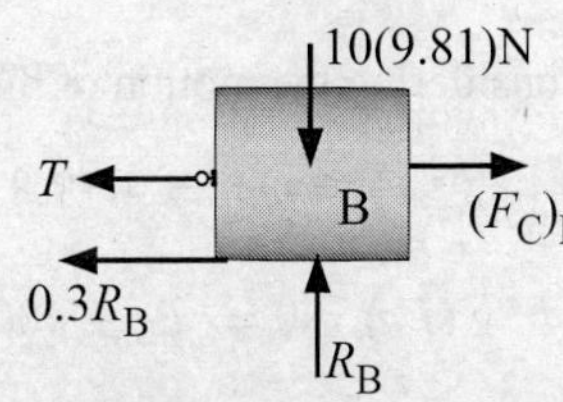

FIGURE 10.9(c)

From which

$$10\omega^2 + 0.3 \times 5(9.81) - T = 0 \quad \text{(i)}$$

Equilibrium equations of block B are

$$(\Sigma F_x)_B = 0 \Rightarrow (F_c)_B - T - 0.3R_B = 0$$

$$(\Sigma F_y)_B = 0 \Rightarrow R_B = 10(9.81)$$

Centrifugal force

$$(F_c)_B = 10\omega^2 \times 3 = 30\omega^2$$

From which

$$30\omega^2 - T - 0.3 \times 10(9.81) = 0 \quad \text{(ii)}$$

Eliminating T from Equations (i) and (ii), we have

$$30\omega^2 - 10\omega^2 = 0.3 \times 10(9.81) + 0.3 \times 5(9.81)$$

or

$$\omega^2 = 2.207$$

$$\omega = 1.485 \text{rad/s}$$

PROBLEM 10.10

Two blocks A and B slide down the inclined plan OC that makes with horizontal an angle $\theta = 30°$ as shown in Figure 10.10(a). If the mass of the blocks $m_A = 20$ kg and $m_B = 40$ kg and the coefficients of friction between them and the inclined plane $\mu_A = 0.10$ and $\mu_B = 0.20$, find the reaction force between the blocks during the motion.

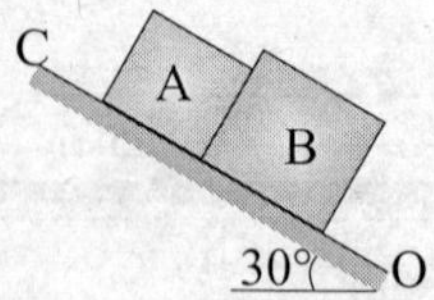

FIGURE 10.10(a) |

Solution. Let a be the common acceleration and T be the reaction between A and B [Fig. 10.10(b)]. Considering the motion of block A.

$$\Sigma F_x = ma \Rightarrow 20(9.81) \sin 30 - T - 0.1 R_A = 20a \quad \text{(i)}$$

$$\Sigma(F_y) = 0 \Rightarrow R_A - 20(9.81) \cos 30 = 0 \quad \text{(ii)}$$

Eliminating R_A from Equations (i) and (ii)

$$a = \frac{98.1 - T - 16.99}{20} \quad \text{(iii)}$$

Considering the motion of block B

$$\Sigma F_x = ma \Rightarrow 40(9.81) \sin 30 + T - 0.2 R_B = 40a \quad \text{(iv)}$$

$$\Sigma(F_y) = 0 \Rightarrow R_B - 40(9.81) \cos 30 = 0 \quad \text{(v)}$$

Eliminating R_B from Equations (iv) and (v), we have

$$a = \frac{98.1 - T - 16.99}{20} \quad \text{(vi)}$$

From Equations (iii) and (iv)

$$\frac{98.1 - T - 16.99}{20} = \frac{196.2 + T - 67.96}{40}$$

or $$196.2 - 2T - 33.98 = 196.2 + T - 67.96$$

or $$3T = 33.98$$

$$T = 11.32\text{N}$$

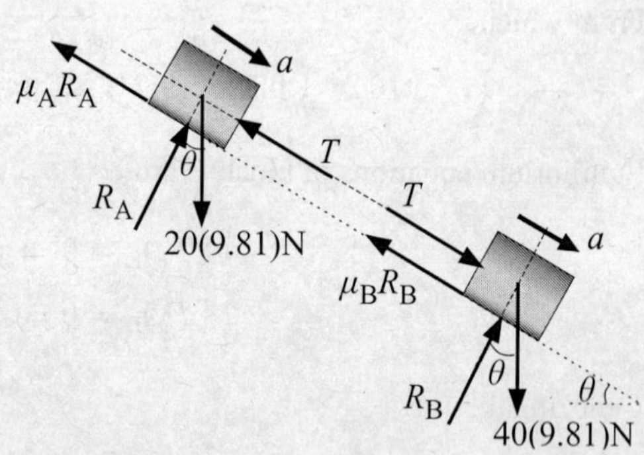

FIGURE 10.10(b) |

PROBLEM 10.11

A uniform rod AB of mass 500 kg rests on block C of mass 50 kg. A force F of 10 kN is applied suddenly to block C at the position shown in Figure 10.11(a). If the coefficient of friction is 0.4 for all contact surfaces, what is the speed of block C when it has moved 3 m from the end of the rod?

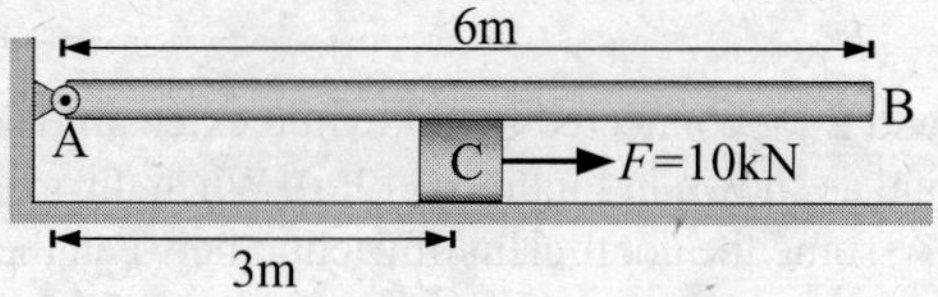

FIGURE 10.11(a) |

Solution. Figure 10.11(b) shows the free body diagram of rod AB when block C is at arbitrary position x from end A. Taking moment about A,

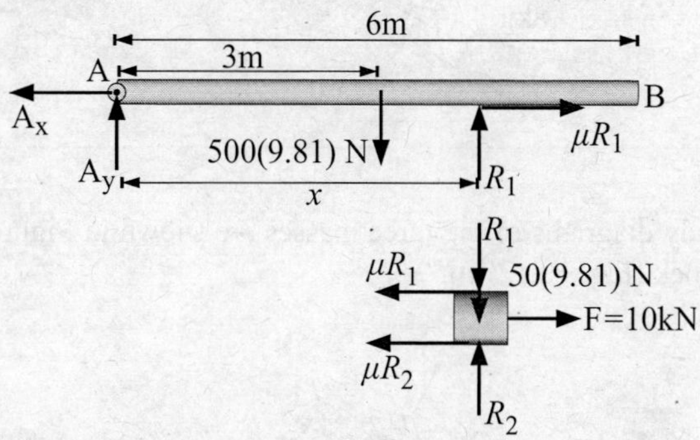

FIGURE 10.11(b) |

$$\Sigma M_A = 0 \Rightarrow 500(9.81) \times 3 - R_1 x = 0$$

or $$R_1 = \frac{14715}{x} \tag{i}$$

Considering the motion of block C,

$$\Sigma F_x = ma_x \Rightarrow F - \mu R_1 - \mu R_2 = mv\frac{dv}{dx}$$

Substituting reaction $R_1 = 14715/x$ and $R_2 = R_1 +$ weight of block $C = (14715/x) + 490.5$, we have

$$10 \times 1000 - 0.4 \times \frac{14715}{x} - 0.4\left[\frac{14715}{x} + 490.5\right] = 50v\frac{dv}{dx}$$

or $$9803.8 - \frac{11772}{x} = 50v\frac{dv}{dx}$$

or $$v dv = \left(196.07 - \frac{235.44}{x}\right)dx$$

Integrating,

$$\frac{v^2}{2} = 196.07(6-3) - 235.44\log_e\left(\frac{6}{3}\right)$$

On simplifying

$$v = 29.15\text{m/s}$$

PROBLEM 10.12

Three masses of 8 kg, 4 kg and 2 kg connected by a flexible extensible cord rest on a smooth plane inclined 45° with the horizontal as shown in Figure 10.12(a). When the masses are released, what will be the tensions in the cords? Assume the coefficient of friction between the plane and the 8 kg mass is 0.2, between plane and the 4 kg mass is 0.4 and between plane and the 2 kg mass is 0.6.

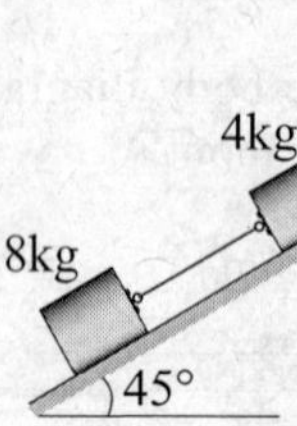

FIGURE 10.12(a) |

Solution. Free body diagrams of the three masses are shown in Figure 10.12(b) Considering the motion of the 8 kg block [Fig. 10.12(b)],

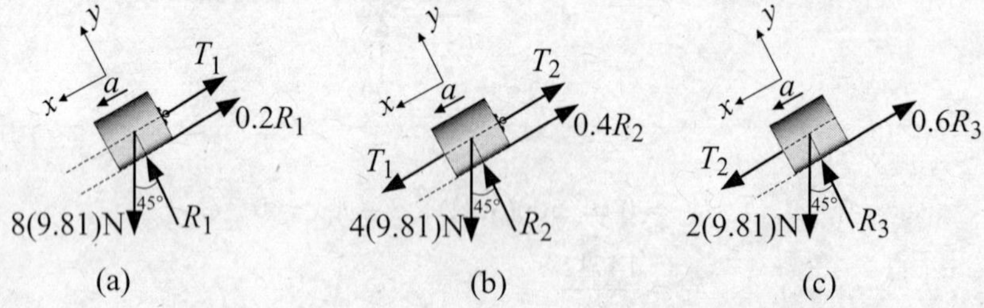

FIGURE 10.12(b) |

$$\Sigma F_x = ma \Rightarrow 8(9.81)\sin\ 45 - T_1 - 0.2R_1 = 8a \tag{i}$$

$$\Sigma F_y = 0 \Rightarrow R_1 - 8(9.81)\cos\ 45 = 0 \tag{ii}$$

From Equations (i) and (ii), we have,

$$8(9.81)\sin\ 45 - T_1 - 0.2 \times 8(9.81)\cos 45 = 8a \tag{iii}$$

Considering the motion of the 4 kg block [Fig. 10.12(b)],

$$\Sigma F_x = ma \Rightarrow T_1 + 4(9.81)\sin 45 - T_2 - 0.4R_2 = 4a \tag{iv}$$

$$(\Sigma F_y) = 0 \Rightarrow R_2 - 4(9.81)\cos 45 = 0 \tag{v}$$

From Equations (iv) and (v), we have

$$T_1 + 4(9.81)\sin\ 45 - T_2 - 0.4 \times 4(9.81)\cos 45 = 4a \tag{vi}$$

Considering the motion of the 2 kg block [Fig. 10.12(b)],

$$\Sigma F_x = ma \Rightarrow T_2 + 2(9.81)\sin 45 - 0.6R_3 = 2a \tag{vii}$$

$$(\Sigma F_y) = 0 \Rightarrow R_3 - 2(9.81)\cos 45 = 0 \tag{viii}$$

From Equations (vii) and (viii), we have

$$T_2 + 2(9.81)\sin 45 - 0.6 \times 2(9.81)\cos 45 = 2a \tag{ix}$$

Summing Equations (iii), (vi) and (ix), we obtain

$$[8(9.81) + 4(9.81) + 2(9.81)]\sin 45 - (0.2 \times 8(9.81) + 0.4 \times 4(9.81) + 0.6 \times 2(9.81)]\cos 45 = (8 + 4 + 2)a$$

or

$$97 - 30 = 14a$$

$$a = 4.74\text{m/s}^2$$

Substituting the value of a in Equation (iii), we have

$$T_1 = 8(9.81)\sin 45 - 0.2 \times 8 \times (9.81)\cos 45 - 8 \times 4.74$$
$$= 55.49 - 11.09 - 37.92 = 6.48$$

$$T_1 = 6.48\text{N}$$

Substituting the value of $a = 4.74$ m/s$_2$ and $T_1 = 6.48$ N in Equation (vi), we have

$$T_2 = 3.96N$$

PROBLEM 10.13

A simple device for measuring reasonably uniform accelerations is the simple pendulum [Fig. 10.13(a)]. Calibrate θ of the pendulum for vehicle acceleration of $g/4$, $g/2$ and $3g/4$.

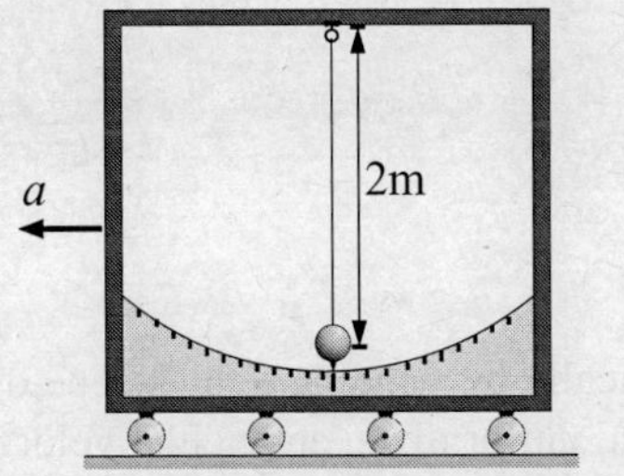

FIGURE 10.13(a) |

Solution. Applying the equation of motion along t–t' (tangential) direction [Fig. 10.13(b)],

$$\Sigma F_t = ma_t \Rightarrow ma\cos\theta - mg\sin\theta = ma_t$$

$$ma\cos\theta - mg\sin\theta = mv\frac{dv}{ds}$$

Substituting $ds = rd\theta$

$$v\frac{dv}{rd\theta} = a\cos\theta - g\sin\theta$$

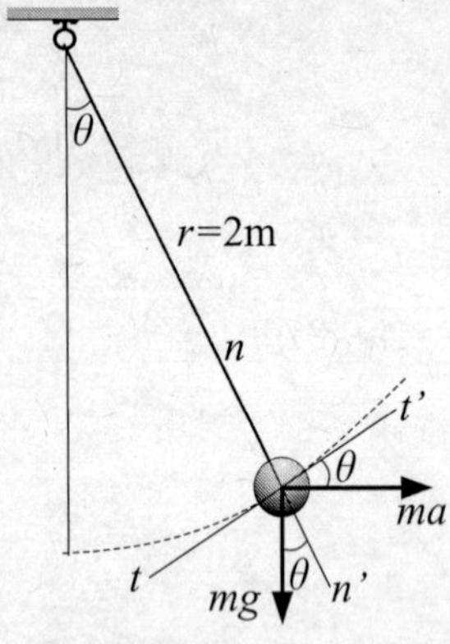

FIGURE 10.13(b) |

or $$v\,dv = (a\cos\theta - g\sin\theta)r\,d\theta$$

or $$\int_0^v v\,dv = \int_0^\theta (a\cos\theta - g\sin\theta)r\,d\theta$$

or $$\frac{v^2}{2} = r(a\sin\theta + g\cos\theta - g)$$

Velocity becomes zero at $\theta = \theta_{max}$, therefore

$$a\sin\theta_{max} + g\cos\theta_{max} - g = 0$$

or $$a\sin\theta_{max} = g(1 - \cos\theta_{max})$$

Squaring the equation, we get

$$a^2\sin^2\theta_{max} = g^2(1 - \cos\theta_{max})^2$$

or $$a^2(1 - \cos\theta_{max})(1 + \cos\theta_{max}) = g^2(1 - \cos\theta_{max})^2$$

From which, we obtain

$$\cos\theta_{max} = \frac{g^2 - a^2}{g^2 + a^2} \Rightarrow \theta_{max} = \cos^{-1}\frac{(g/a)^2 - 1}{(g/a)^2 + 1}$$

$a = \frac{g}{4}$	$\theta_{max} = \cos^{-1}\left(\frac{15}{17}\right)$	$\theta_{max} = 28.07°$
$a = \frac{g}{2}$	$\theta_{max} = \cos^{-1}\left(\frac{3}{5}\right)$	$\theta_{max} = 53.13°$
$a = \frac{3g}{4}$	$\theta_{max} = \cos^{-1}\left(\frac{0.77}{2.77}\right)$	$\theta_{max} = 73.73°$

PROBLEM 10.14

A weight of mass m is raised vertically by winding a thick rope of diameter d on a reel as shown in Figure 10.14(a). If the reel is turned uniformly at an angular velocity ω, what will be the tension T in the rope? Neglect inertia of the rope and slight lateral motion of the suspended weight.

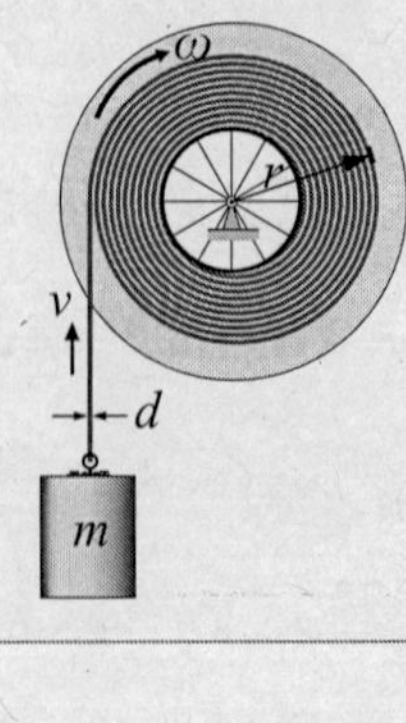

FIGURE 10.14(a) |

Solution. Let r be the radius of the paper reel after θ rotation of the reel, thus

$$r = \frac{d}{2\pi}\theta$$

Velocity of weight

$$v = \omega r$$

$$v = \omega \frac{d}{2\pi}\theta$$

Taking time derivative

$$\frac{dv}{dt} = \omega \frac{d}{2\pi}\frac{d\theta}{dt}$$

$$a = \frac{dv}{dt} = \omega \frac{d}{2\pi}\omega$$

$$a = \frac{\omega^2 d}{2\pi}$$

From the free body diagram of weight [Fig. 10.14(b)]

$$T - mg = ma \Rightarrow T = mg + ma$$

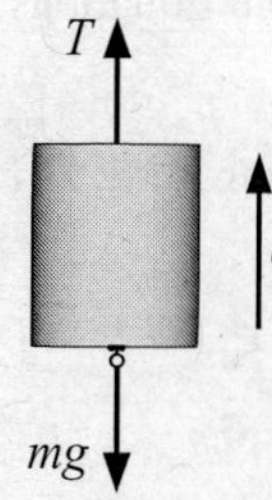

FIGURE 10.14(b) |

Substituting the value of a, we obtain

$$T = mg + m\frac{\omega^2 d}{2\pi}$$

$$\boxed{T = m\left(g + \frac{\omega^2 d}{2\pi}\right)}$$

PROBLEM 10.15

A thin hoop and a solid cylinder of equal mass m and radii r are connected by a bar AB and roll down without slipping on an inclined plane as shown in Figure 10.15(a). Find the acceleration of the system down the plane and also the force T in bar AB. Neglect the mass of the bar.

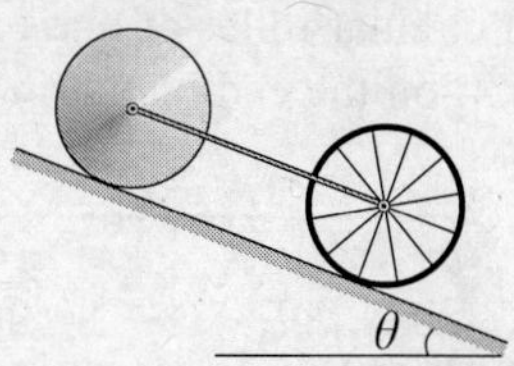

FIGURE 10.15(a) |

Solution. Let a be the common linear acceleration of the system and T be the force (compressive) in bar AB [Fig. 10.15(b)]. Considering the motion of the cylinder

$$\Sigma F_x = ma_x \Rightarrow mg\sin\theta + T - f_1 = ma \quad \text{(i)}$$

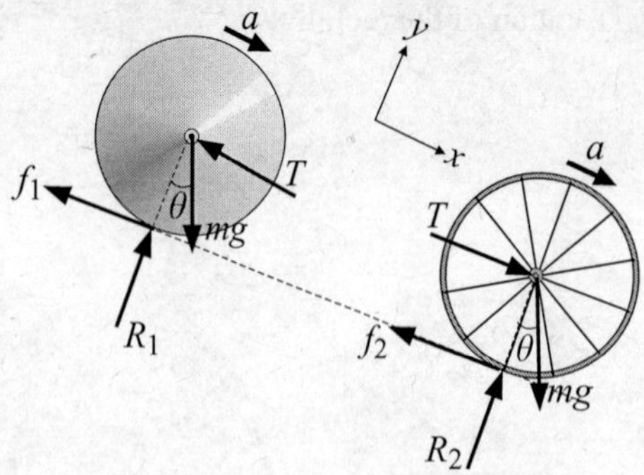

FIGURE 10.15(b) |

$$\Sigma\tau = I\alpha \Rightarrow f_1 \times r = I_1\alpha = I_1 \frac{a}{r}$$

or $$f_1 = I_1 \frac{a}{r^2} = \frac{mr^2}{2}\frac{a}{r^2} = \frac{ma}{2} \quad \text{(ii)}$$

From Equations (i) and (ii), we have

$$mg \sin\theta + T = \frac{3}{2} ma \quad \text{(iii)}$$

Considering the motion of the hoop

$$\Sigma F_x = ma_x \Rightarrow mg \sin\theta - T - f_2 = ma \quad \text{(iv)}$$

$$\Sigma\tau = I\alpha \Rightarrow f_2 \times r = I_2\alpha = I_2 \frac{a}{r}$$

or $$f_2 = I_2 \frac{a}{r^2} = mr^2 \frac{a}{r^2} = ma \quad \text{(v)}$$

From Equations (iv) and (v)

$$mg \sin\theta - T = 2ma \quad \text{(vi)}$$

Eliminating T from Equations (iii) and (vi), we have

$$2mg \sin\theta = \frac{7}{2} ma$$

or $$a = \frac{4}{7} g \sin\theta$$

Substituting the value of a into Equation (iii), we have

$$mg \sin\theta + T = 2m \times \frac{4}{7} g \sin\theta$$

$$T = \frac{1}{7} mg \sin\theta$$

PROBLEM 10.16

Find the acceleration of the solid right circular roller of mass m, which is pulled along a horizontal plane by means of the weight of mass M on the end of a string wound around the circumference of the roller [Fig. 10.16(a)].

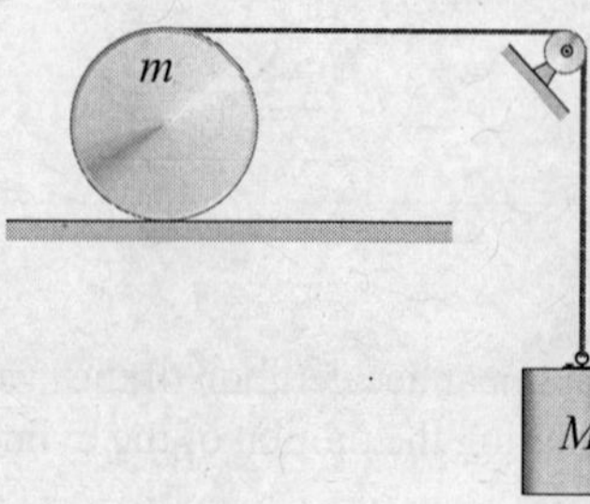

FIGURE 10.16(a) |

Solution. Let a be the linear acceleration of centre of roller and T be the tension of rope [Fig. 10.16(b)]. Considering motion of weight M

$$\Sigma F_y = Ma_y \Rightarrow Mg - T = Ma_y = M(2a) \quad \text{(i)}$$

Considering motion of roller

$$\Sigma F_x = ma_x \Rightarrow T - f = ma \quad \text{(ii)}$$

$$\Sigma \tau = I\alpha \Rightarrow Tr + fr = I\alpha$$

or

$$T + f = \frac{Ia}{r^2} \quad \text{(iii)}$$

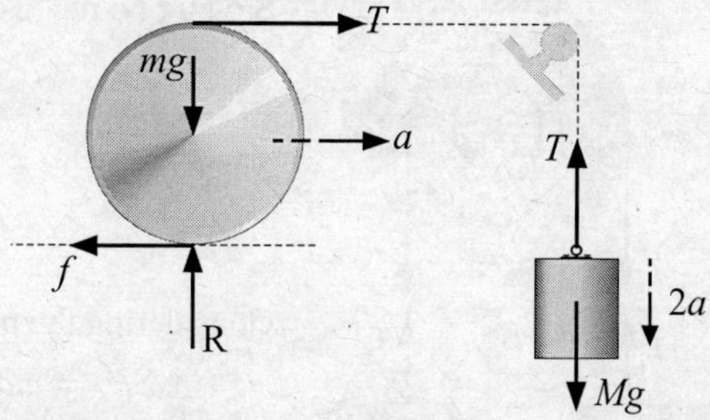

FIGURE 10.16(b)

From Equations (ii) and (iii)

$$2T = ma + \frac{Ia}{r^2}$$

or

$$T = \frac{a}{2}\left(m + \frac{I}{r^2}\right)$$

Substituting the value of T into Equation (i), we have

$$Mg - \frac{a}{2}\left(m + \frac{I}{r^2}\right) = 2Ma$$

or

$$a\left[2M + \frac{m}{2} + \frac{I}{2r^2}\right] = Mg$$

or

$$a = \frac{Mg}{\left[2M + \left(\frac{m}{2}\right) + \left(\frac{I}{2r^2}\right)\right]}$$

Since the moment of inertia of cylinder $I = (1/2)mr^2$,

$$a = \frac{4Mg}{8M + 3m}$$

PROBLEM 10.17

The two-step pulley shown in Figure 10.17(a) has mass M and radius of gyration k. Derive a formula for the downward acceleration of the falling weight A.

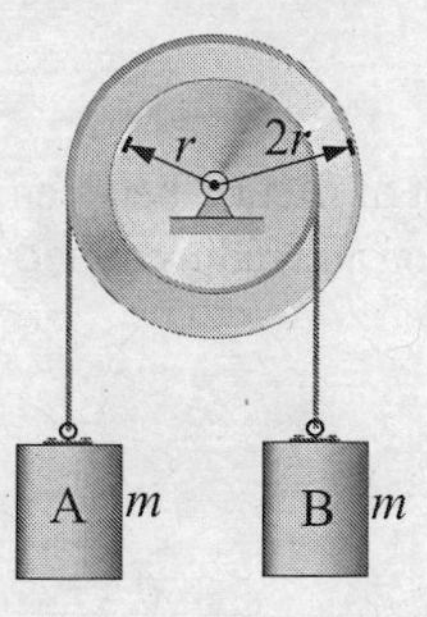

FIGURE 10.17(a)

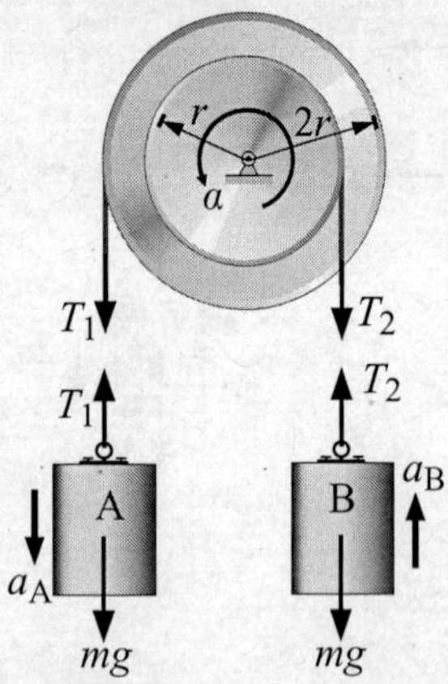

FIGURE 10.17(b) |

Solution. Considering the motion of weight A [Fig. 10.17(b)],

$$\Sigma F_y = ma_y \Rightarrow mg - T_1 = ma_A$$

Since $$a_A = 2r\alpha \Rightarrow \alpha = \frac{a_A}{2r} = \frac{a_B}{r},$$

$$mg - T_1 = m2r\alpha \tag{i}$$

Considering the motion of weight B,

$$\Sigma F_y = ma_y \Rightarrow T_2 - mg = ma_B$$

or $$T_2 - mg = mr\alpha \tag{ii}$$

Considering the motion of pulley

$$\Sigma\tau = I\alpha \Rightarrow T_1 2r - T_2 r = I\alpha$$

or $$2T_1 - T_2 = \frac{I\alpha}{r} \tag{iii}$$

From Equations (i) and (ii)

$$2mg - (2T_1 - T_2) - mg = 4mr\alpha + mr\alpha$$

or $$mg - \frac{I\alpha}{r} = 5mr\alpha$$

or $$\alpha = \frac{mg}{\left(\frac{I}{r}\right) + 5mr}$$

or $$\alpha = \frac{mg}{\left(\frac{Mk^2}{r}\right) + 5mr}$$

Acceleration of block A

$$a_A = 2r\alpha = \frac{2rmg}{\left(\frac{Mk^2}{r}\right) + 5mr}$$

$$a_A = \frac{2mg}{5m + M\left(\frac{k^2}{r^2}\right)}$$

PROBLEM 10.18

Find the minimum value of the coefficient of friction between a body and a plane, so that the body may roll without slipping. The radius of gyration and radius of body are k and r, respectively [Fig. 10.18(a)].

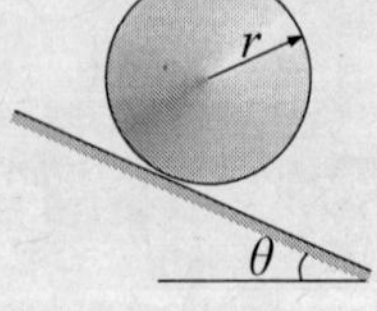

FIGURE 10.18(a) |

Solution. The free body diagram [Fig. 10.18(b)] shows the weight mg, the normal force R and friction force F_f. Applying the equations of motion,

$$\Sigma F_x = ma_x \Rightarrow mg\sin\theta - F_f = ma \quad \text{(i)}$$

$$\Sigma\tau = I\alpha \Rightarrow F_f \times r = I\alpha = I\frac{a}{r}$$

or
$$F_f = \frac{Ia}{r^2} \quad \text{(ii)}$$

$$\Sigma F_y = 0 \Rightarrow R = mg\cos\theta \quad \text{(iii)}$$

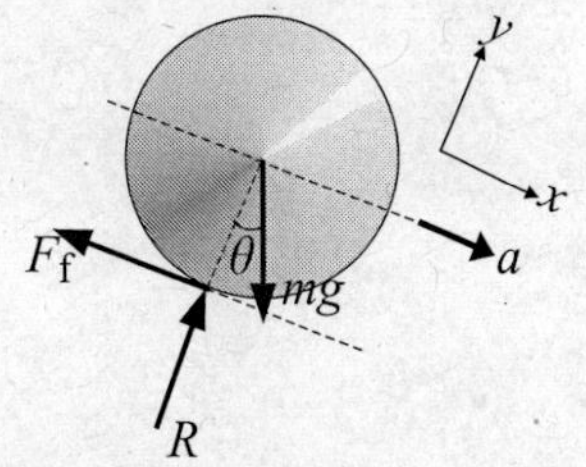

FIGURE 10.18(b)

Eliminating F_f from Equations (i) and (ii), we have

$$mg\sin\theta - \frac{Ia}{r^2} = ma$$

or
$$a\left(m + \frac{I}{r^2}\right) = mg\sin\theta$$

Acceleration of body

$$a = \frac{mg\sin\theta}{\left[m + \left(\frac{I}{r^2}\right)\right]}$$

Subst ituting the value of a into Equation (ii), we have

$$F_f = \frac{I}{r^2} \times \frac{mg\sin\theta}{\left[m + \left(\frac{I}{r^2}\right)\right]}$$

Friction force

$$F_f = \frac{mg\sin\theta}{1 + \left(\frac{mr^2}{I}\right)}$$

But the maximum possible friction force is

$$F_f \le \mu R \Rightarrow \frac{mg\sin\theta}{1 + \left(\frac{mr^2}{I}\right)} \le \mu mg\cos\theta$$

or

$$\mu \ge \frac{\tan\theta}{1 + \left(\frac{mr^2}{I}\right)}$$

or

$$\mu \ge \frac{\tan\theta}{1 + \left(\frac{r^2}{k^2}\right)}$$

PROBLEM 10.19

In Figure 10.19(a), a flywheel of mass M and radius r is mounted on a horizontal shaft and set in rotation by a falling block having mass m. If the system is released from rest, find the velocity of the block after it has fallen through a distance of h.

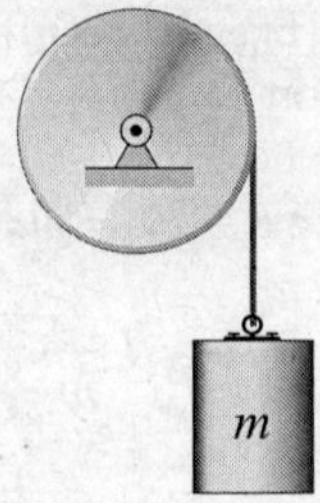

FIGURE 10.19(a) |

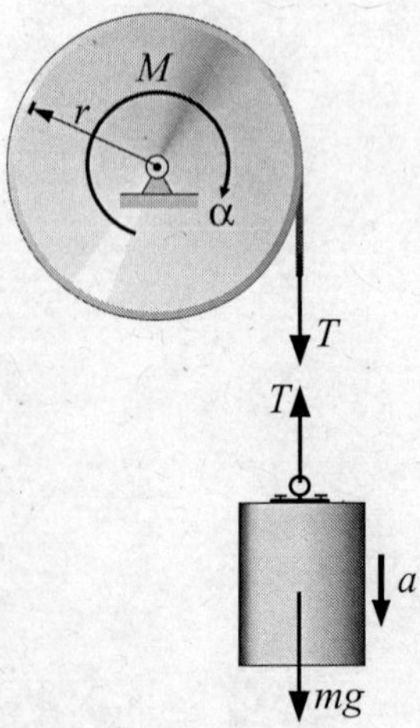

FIGURE 10.19(b) |

Solution. Here we will consider the flywheel and block separately as shown in Figure 10.19(b). Assuming the block accelerates downward at a, it creates a clockwise angular acceleration α of the flywheel. Considering the motion of block

$$\Sigma F_y = ma_y \Rightarrow mg - T = ma \tag{i}$$

Considering the motion of flywheel

$$\Sigma\tau = I\alpha \Rightarrow Tr = I\alpha = \frac{Ia}{r}$$

$$T = \frac{Ia}{r^2} \tag{ii}$$

From Equations (i) and (ii), we have

$$mg - \frac{Ia}{r^2} = ma$$

or

$$a = \frac{mg}{m + \left(\dfrac{I}{r^2}\right)} \tag{iii}$$

Velocity of block

$$v = \sqrt{u^2 + 2as}$$

Given initial velocity $u = 0$, $s = h$

$$v = \sqrt{2ah}$$

Substituting the value of a from Equation (iii), we obtain

$$v = \sqrt{\frac{2mgh}{m + \left(\dfrac{I}{r^2}\right)}} = \sqrt{\frac{2gh}{1 + \left(\dfrac{k^2}{r^2}\right)}}$$

$$\boxed{v = \sqrt{\frac{2gh}{1 + \left(\dfrac{k^2}{r^2}\right)}}}$$

PROBLEM 10.20

Referring to Figure 10.20(a), find the velocity a rotor of mass m, radius r and radius of gyration k will acquire after falling from rest through a vertical distance h.

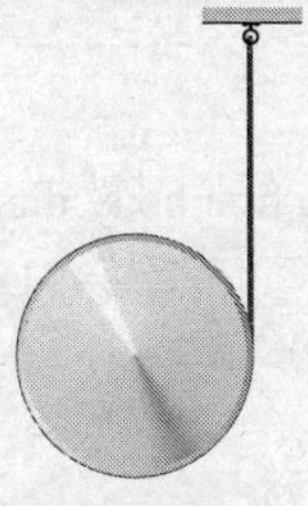

FIGURE 10.20(a) |

Solution. The free body diagram [Fig. 10.20(b)] shows the weight mg and tension T. Applying the equations of motion in y-direction.

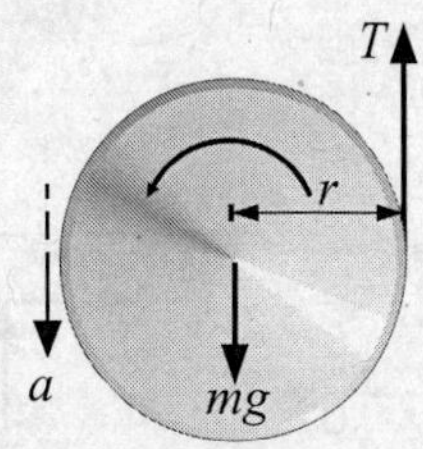

FIGURE 10.20(b) |

$$\Sigma F_y = ma_y \Rightarrow mg - T = ma \qquad \text{(i)}$$

$$\Sigma\tau = I\alpha \Rightarrow Tr = I\alpha = I\frac{a}{r}$$

or

$$T = \frac{Ia}{r^2} \qquad \text{(ii)}$$

From Equations (i) and (ii), we have

$$mg - \frac{Ia}{r^2} = ma$$

Acceleration of rotor

$$a = \frac{mg}{m + \left(\frac{I}{r^2}\right)}$$

Velocity of rotor

$$v = \sqrt{u^2 + 2as}$$

Given; Initial velocity $u = 0$, $s = h$,

$$v = \sqrt{2ah}$$

Substituting the value of a from Equation (iii), we obtain

$$v = \sqrt{\frac{2mgh}{m + \left(\frac{I}{r^2}\right)}} = \sqrt{\frac{2gh}{1 + \left(\frac{k^2}{r^2}\right)}}$$

$$v = \sqrt{\frac{2gh}{1 + \left(\frac{k^2}{r^2}\right)}}$$

PROBLEM 10.21

A flat plate of mass M rests on two rollers, each of mass m and radius r, and is pulled along a horizontal plane by a force P as shown in Figure 10.21(a). Assuming that there is no slipping, and treating the rollers as solid right circular cylinders, find the acceleration of the plate.

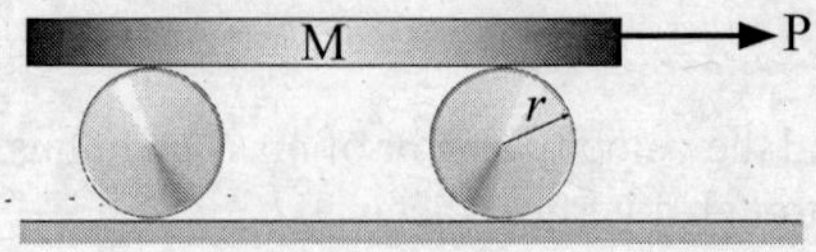

FIGURE 10.21(a) |

Solution. Figure 10.21(b) shows the free body diagrams of the plate and the roller. Let a be the acceleration of the plate.

Considering the motion of plate

$$\Sigma F_x = ma_x \Rightarrow \mathrm{P} - 2\mathrm{F} = \mathrm{M}a \tag{i}$$

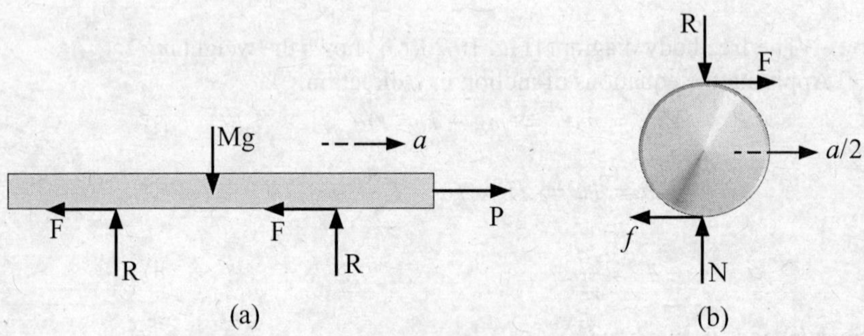

FIGURE 10.21(b) |

Considering the motion of roller

$$\Sigma F_x = ma_x \Rightarrow F - f = m\frac{a}{2} \tag{ii}$$

$$\Sigma\tau = I\alpha \Rightarrow F \times r + f \times r = I\alpha = I\frac{a}{2}$$

$$F + f = I\frac{a}{2r^2} = \frac{1}{2}mr^2 \times \frac{a}{2r^2} = \frac{1}{4}ma \tag{iii}$$

Summing Equations (i), (ii) and (iii), we get

$$P = Ma + \frac{m}{2}a + \frac{m}{4}a$$

$$a = \frac{4P}{(4M + 3m)}$$

PROBLEM 10.22

A cylinder of mass m and radius of gyration k has a string wrapped around a groove of radius r, as shown in Figure 10.22(a). Determine the acceleration of the centre of mass if pure rolling is assumed.

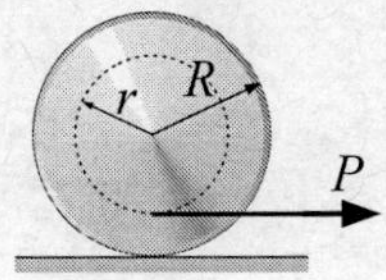

FIGURE 10.22(a)

Solution. A free body diagram of the cylinder is shown in Figure 10.22(b). Friction force F_f is assumed acting to the left. The equations of motion are

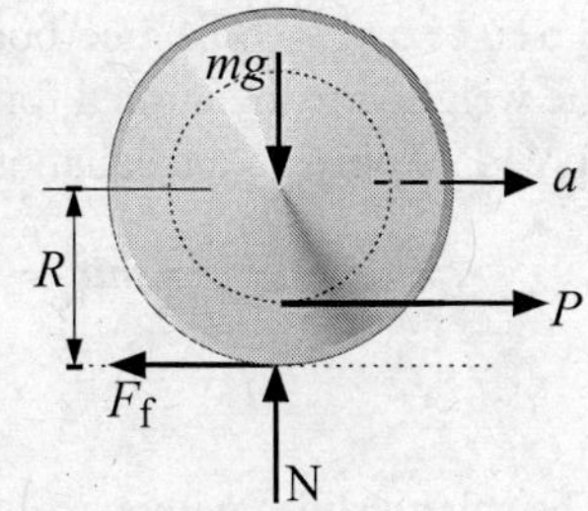

FIGURE 10.22(b)

$$\Sigma F_x = ma_x \Rightarrow P - F_f = ma_x \qquad \text{(i)}$$

$$\Sigma F_y = 0 \Rightarrow N - mg = 0 \qquad \text{(ii)}$$

$$\Sigma \tau = I\alpha \Rightarrow -Pr + F_f R = I\alpha \qquad \text{(iii)}$$

Eliminating F_f from Equations (i) and (iii), we have

$$P - \left(\frac{I\alpha + Pr}{R}\right) = ma_x$$

or

$$P\frac{(R-r)}{R} - \frac{I\alpha}{R} = ma_x$$

Since $\alpha = a_x/R$, the above equation may be written

$$P\frac{(R-r)}{R} - \frac{I}{R} \times \frac{a_x}{R} = ma_x$$

or

$$a_x\left(m + \frac{I}{R^2}\right) = P\frac{(R-r)}{R}$$

or

$$a_x = \frac{PR(R-r)}{(I + mR^2)} = \frac{PR(R-r)}{(mk^2 + mR^2)}$$

$$a_x = \frac{PR(R-r)}{m(k^2 + R^2)}$$

PROBLEM 10.23

A homogeneous sphere of mass m and radius r slides along the floor under the action of a horizontal force P applied to a string, as shown in Figure 10.23(a). Find the height h during this motion if the coefficient of friction between the sphere and the floor is μ.

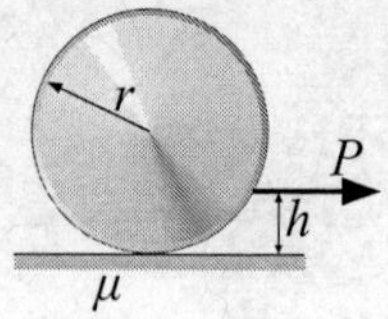

FIGURE 10.23(a) |

Solution. The free body diagram [Fig. 10.23(b)] shows the weight mg, the normal force R, and the friction force μR and force P. Applying the equation of motion

$$\Sigma F_x = ma_x \Rightarrow P - \mu R = ma \quad \text{(i)}$$

$$\Sigma F_y = 0 \Rightarrow R - mg = 0 \quad \text{(ii)}$$

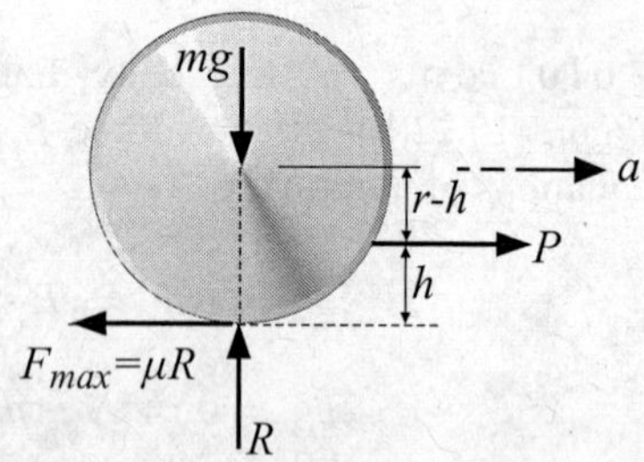

FIGURE 10.23(b) |

The sphere slides without rolling, if $\Sigma\tau = 0$;

$$\sum \tau = 0 \Rightarrow +\mu R \times r - P(r - h) = 0$$

or

$$h = r - \frac{\mu R r}{P} \quad \text{(iii)}$$

From Equations (ii) and (iii), we obtain

$$h = r - \frac{\mu m g r}{P}$$

$$h = r\left[1 - \frac{\mu m g}{P}\right]$$

PROBLEM 10.24

A homogeneous disk of mass m and radium r is at rest on a horizontal plane when a couple M is applied as shown in Figure 10.24(a). Determine the coefficient of friction between the disk and the plane so that rolling will occur.

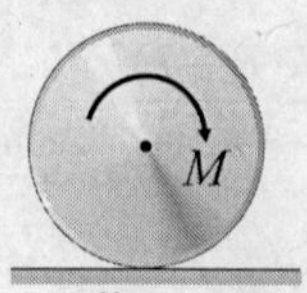

FIGURE 10.24(a) |

Solution. Applying the equations of motion [Fig. 10.24(b)],

$$\sum F_x = ma \Rightarrow \mu m g = ma$$

or

$$a = \mu g \quad \text{(i)}$$

$$\sum \tau = I\alpha \Rightarrow M - f_{\max} r = I\alpha$$

or
$$M - \mu mgr = \frac{Ia}{r} \quad \text{(ii)}$$

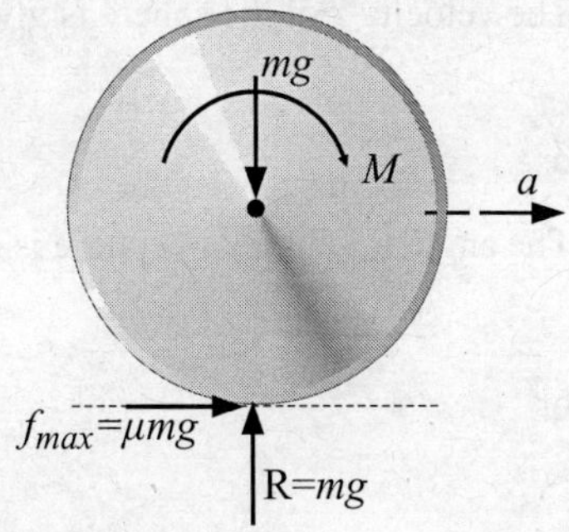

FIGURE 10.24(b) |

Substituting the value of a, we have

$$M - \mu mgr = \frac{I}{r}\mu g$$

or
$$M - \mu mgr = \frac{mr^2}{2r}\mu g$$

or
$$M = \frac{3}{2}\mu mgr$$

$$\mu = \frac{2M}{3mgr}$$

PROBLEM 10.25

A solid homogenous sphere of mass m and radius r, initially spinning about its geometric axis with an angular velocity ω, is suddenly set down on a horizontal plane without any forward velocity, as shown in Figure 10.25(a). What velocity along the plane will the sphere finally acquire, if the coefficient of friction at the point of contact is μ?

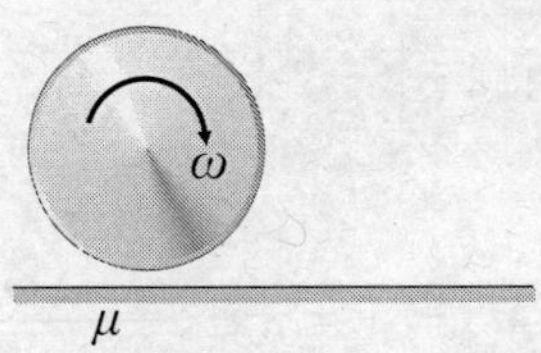

FIGURE 10.25(a) |

Solution. A free body diagram in Figure 10.25(b) shows the friction force μR, the normal force R, and weight mg acting on the sphere. The equations of motion are

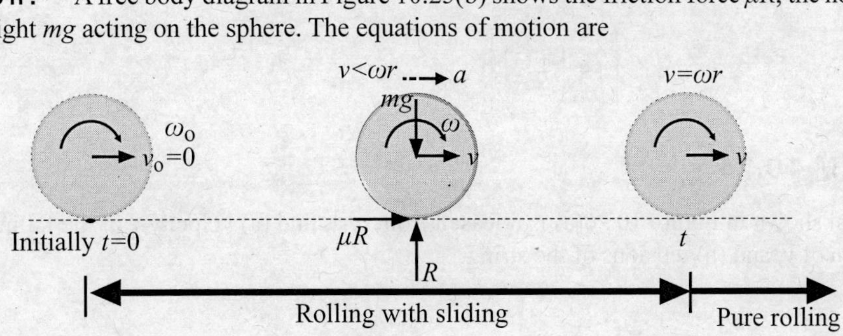

FIGURE 10.25(b) |

$$\sum F_x = ma_x \Rightarrow \mu R = ma \quad \text{(i)}$$

$$\sum F_y = 0 \Rightarrow R = mg \quad \text{(ii)}$$

$$\sum \tau = I\alpha \Rightarrow \mu Rr = I\alpha \quad \text{(iii)}$$

From Equations (i), (ii) and (iii), we obtain

$$a = \mu g \text{ and } \alpha = \frac{\mu mgr}{I}$$

The velocity v of the sphere is given by

$$v = v_o + at$$

or

$$v = 0 + \mu g t$$

The angular velocity of sphere is given by

$$\omega = \omega_o - at$$

or

$$\omega = \omega_o - \frac{\mu mgr}{I} t$$

Sliding stops and pure rolling starts at t = t', $v = \omega r \Rightarrow \mu g t' = \left(\omega_o - \frac{\mu mgr}{I} t'\right) r$

$$\left(\mu g + \frac{\mu mgr^2}{I}\right) t' = \omega_o r$$

or

$$t' = \frac{\omega_o r}{\left[\mu g + \left(\frac{\mu mgr^2}{I}\right)\right]} = \frac{\omega_o r}{\mu g\left[1 + \left(\frac{mr^2}{I}\right)\right]}$$

$$t' = \frac{\omega_o r}{\mu g\left[1 + \left(\frac{r^2}{k^2}\right)\right]}$$

Final velocity

$$v = 0 + \mu g t' \Rightarrow v = \mu g \times \frac{\omega_o r}{\mu g\left[1 + \left(\frac{r^2}{k^2}\right)\right]}$$

$$v = \frac{\omega_o r}{\left[1 + \left(\frac{r^2}{k^2}\right)\right]}$$

PROBLEM 10.26

If the system shown in Figure 10.26(a) is released from rest, find (a) velocity v of the falling block A as a function of y, and (b) tensions of the string.

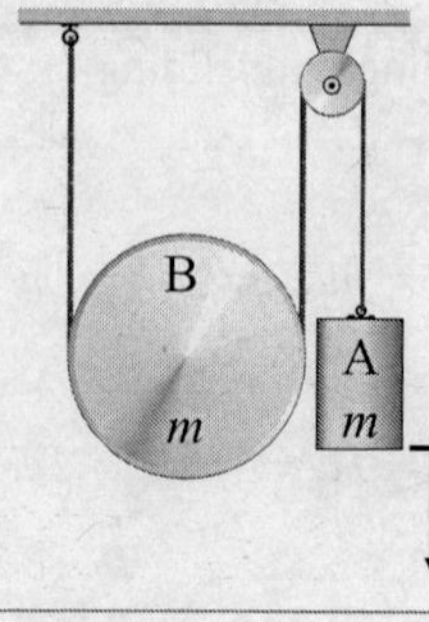

FIGURE 10.26(a) |

Solution. Consider the motion of block A [Fig. 10.26(b)]:

$$\sum F_y = ma_y \Rightarrow mg - T_1 = ma \quad \text{(i)}$$

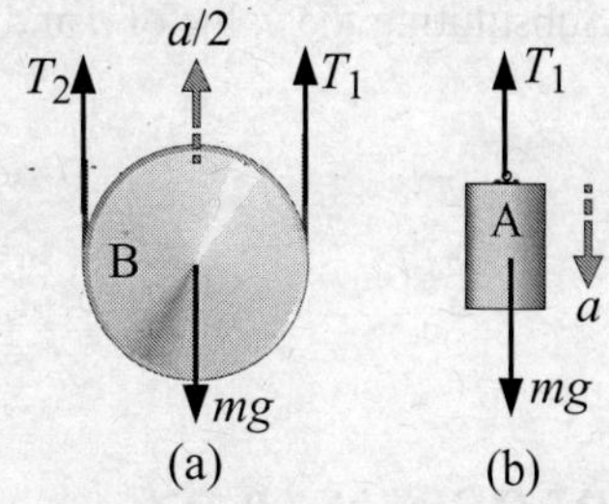

FIGURE 10.26(b) |

Consider the motion of disk B [Fig. 10.26(b)]:

$$\sum F_y = ma_y \Rightarrow T_1 + T_2 - mg = \frac{ma}{2} \quad \text{(ii)}$$

$$\sum \tau = I\alpha \Rightarrow (T_1 - T_2)r = I\alpha = \frac{Ia}{2r}$$

or

$$T_1 - T_2 = \frac{Ia}{2r^2} \quad \text{(iii)}$$

From Equations (ii) and (iii), we have

$$2T_1 = mg + \frac{ma}{2} + \frac{Ia}{2r^2}$$

or

$$T_1 = \frac{mg}{2} + \frac{ma}{4} + \frac{Ia}{4r^2} \quad \text{(iv)}$$

From Equations (i) and (iv), we get

$$mg = ma + \frac{mg}{2} + \frac{ma}{4} + \frac{Ia}{4r^2}$$

or

$$\left(m + \frac{m}{4} + \frac{I}{4r^2}\right)a = \frac{mg}{2}$$

or

$$\left(m + \frac{m}{4} + \frac{mr^2}{2} \times \frac{1}{4r^2}\right)a = \frac{mg}{2}$$

or

$$\frac{11}{8}ma = \frac{mg}{2}$$

$$a = \frac{4}{11}g$$

Velocity of block

$$v = \sqrt{u^2 + 2ay}$$

$$\boxed{v = \sqrt{\frac{8}{11}gy}}$$

Substituting the value of a into Equation (i), we have

$$T_1 = mg - ma = mg - \frac{4mg}{11} = \frac{7}{11}mg$$

$$\boxed{T_1 = \frac{7}{11}mg}$$

Substituting the value of a and T_1 into Equation (ii), we have

$$T_2 = \frac{ma}{2} + mg - T_1 = \frac{m}{2} \times \frac{7}{11} g + mg - \frac{7}{11} mg$$

$$T_2 = \frac{15}{22} mg$$

PROBLEM 10.27

Two flywheels having mass moments of inertia I_1 and I_2 are mounted on a bearing and geared together as shown in Figure 10.27(a). A cord wound around the circumference of one of the flywheels carries a weight of mass M at its free end. Neglecting friction, find the acceleration of the weight if it is allowed to fall vertically downward.

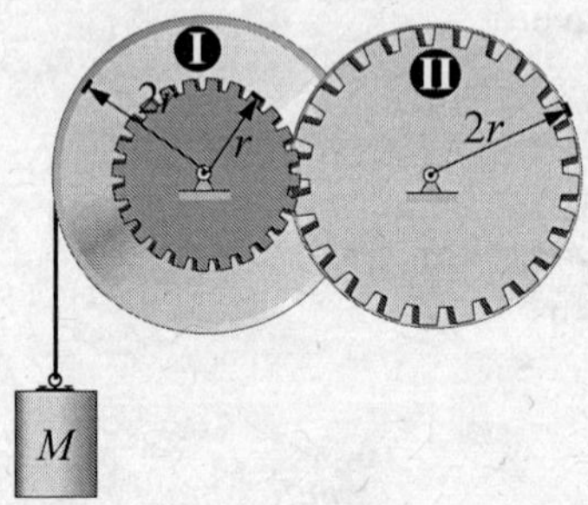

FIGURE 10.27(a) |

Solution. Let a be the acceleration of weight M.

Considering the motion of weight M [Fig. 10.27(b)]:

$$\sum F_y = Ma \Rightarrow Mg - T = Ma \quad \text{(i)}$$

Considering the motion of flywheel I [Fig. 10.27(b)]:

$$\sum \tau = I\alpha \Rightarrow T \times 2r - F \times r = I_1 \alpha_1$$

$$2T - F = \frac{I_1 \alpha_1}{r} \quad \text{(ii)}$$

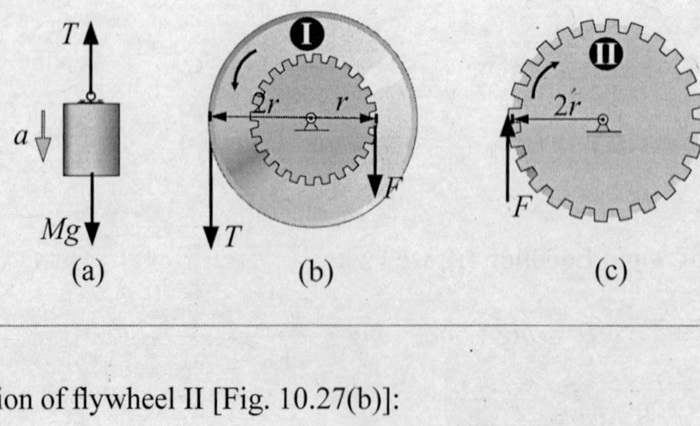

FIGURE 10.27(b) |

Considering the motion of flywheel II [Fig. 10.27(b)]:

$$\sum \tau = I\alpha \Rightarrow F \times 2r = I_2 \alpha_2$$

or $$F = \frac{I_2 \alpha_2}{2r} \qquad \text{(iii)}$$

Eliminating T and F from Equations (i), (ii) and (iii), we have

$$Mg - \frac{1}{2}\left[\frac{I_1 \alpha_1}{r} + \frac{I_2 \alpha_2}{2r}\right] = Ma$$

Since $a = 2r\alpha_1$ and $r\alpha_1 = 2r\alpha_2$, $\alpha_2 = (\alpha_1)/2$.

$$Mg - \frac{1}{2}\left[\frac{I_1 a}{2r^2} + \frac{I_2 a}{8r^2}\right] = Ma$$

or $$Mg = \left[M + \frac{I_1}{4r^2} + \frac{I_2}{16r^2}\right]a$$

From which

$$a = \frac{Mg}{\left[M + \left(\frac{I_1}{4r^2}\right) + \left(\frac{I_2}{16r^2}\right)\right]}$$

PROBLEM 10.28

A collar of mass m is free to slide up along a vertical shaft under the action of force F of constant magnitude but variable direction [Fig. 10.28(a)]. If $\theta = ct$, where c is a constant and if the collar starts from rest with $\theta = 0$, determine the magnitude F of the force which will result in the collar coming to rest as θ reaches $\pi/2$. The coefficient of friction between the collar and the shaft is μ.

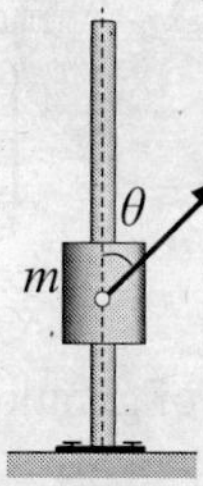

FIGURE 10.28(a)

Solution. Figure 10.28(b) shows free body diagram of the collar with friction force μR, weight mg, reaction R and force F for motion up. Equilibrium of collar in x-direction gives

$$\Sigma F_x = 0 \Rightarrow R = F \sin\theta$$

Equation of motion in y-direction

$$\Sigma F_y = ma_y \Rightarrow F\cos\theta - mg - \mu R = m\frac{dv}{dt}$$

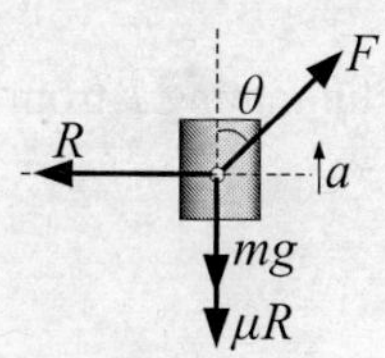

FIGURE 10.28(b)

Substituting $R = F\sin\theta$ and $\theta = ct$ and integrating

$$\int_0^v m\mathrm{d}v = \int_0^t (F\cos ct - mg - \mu F\sin ct)\mathrm{d}t$$

or
$$mv = \frac{F}{c}\sin ct - mgt + \frac{\mu F}{c}\cos ct - \frac{\mu F}{c}$$

for $\theta = \pi/2$, the time $t = \pi/2c$ and $v = 0$, so that

$$0 = \frac{F}{c}\sin\frac{\pi}{2} - mg\frac{\pi}{2c} + \frac{\mu F}{c}\cos ct - \frac{\mu F}{c}$$

or
$$\frac{F}{c} - mg\frac{\pi}{2c} - \frac{\mu F}{c} = 0$$

or
$$F(1-\mu) = \frac{mg\pi}{2}$$

$$F = \frac{mg\pi}{2(1-\mu)}$$

PROBLEM 10.29

A roll of paper is pin-supported at both ends by two brackets as shown in Figure 10.29(a). If the roll rests against a wall for which the coefficient of kinetic friction is $\mu = 0.20$ and a vertical 60 N force is applied to the end of the paper, determine the angular acceleration of the roll as the paper unrolls.

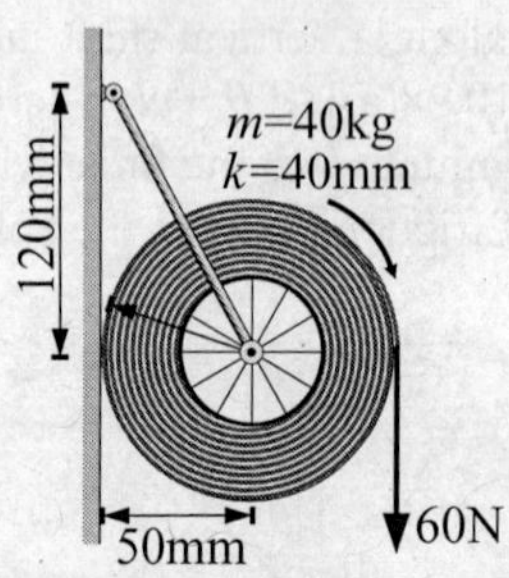

FIGURE 10.29(a) |

The paper roll has a mass 40 kg and rad ius of gyration 40 mm.

Solution. Figure 10.29(b) shows free body diagram of roll the friction force $0.2R$, weight 40(9.81) N, reaction R, applied force 60 N and T. The equilibrium equations are

$$\Sigma F_y = 0 \Rightarrow T\sin\theta - 0.2R - 40(9.81) - 60 = 0 \quad \text{(i)}$$

$$\Sigma F_x = 0 \Rightarrow R - T\cos\theta = 0 \quad \text{(ii)}$$

Eliminating T from Equations (i) and (ii), we have

$$R\tan\theta - 0.2R - 40(9.81) - 60 = 0$$

or
$$R\left(\frac{120}{50} - 0.2\right) = 452.4$$

or

$$R = 205.6N$$

Now applying the equation of motion,

$$\Sigma\tau = I\alpha \Rightarrow 60 \times \frac{50}{1000} - 0.2R \times \frac{50}{1000} = 40 \times \left(\frac{40}{1000}\right)^2 \alpha$$

or

$$3 - 2.056 = 0.064\alpha$$

$$\alpha = 14.75\text{rad/s}^2$$

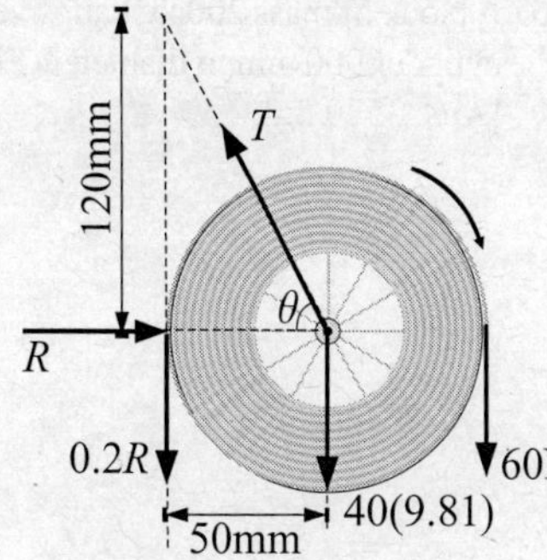

FIGURE 10.29(b) |

CHAPTER HIGHLIGHTS

- Weight is a gravitational force of attraction that the earth exerts on an object.

$$\text{Weight of object } W = F = \frac{GMm}{r^2} = mg \downarrow$$

- Newton's Laws of Motion

First Law: A particle originally at rest, or moving in a straight line with a constant velocity, will remain in this state provided the particle is not subjected to an unbalanced force.
Second Law: If the resultant force acting on a particle is not zero, the particle will have an acceleration proportional to the magnitude of the resultant and in the direction of this resultant force.
Third Law: The mutual forces of action and reaction between two particles are equal, opposite, and collinear.

- Newton's second law of motion (or *equation of motion*)in the mathematical form is

$\mathbf{F} = m\mathbf{a}$
$\Sigma\mathbf{F} = m\mathbf{a}$; for a particle subjected to several forces
$\Sigma\mathbf{F} = \Sigma m\mathbf{a}_G$; for a system of particles(or rigid body) subjected to several forces

- Equation of Motion in radial coordinates
- Radial force $\sum F_r = m\left[\frac{d^2r}{dt^2} - r\left(\frac{d\theta}{dt}\right)^2\right]$, Transverse force $\sum F_\theta = m\left[r\frac{d^2\theta}{dt^2} + 2\frac{dr}{dt}\frac{d\theta}{dt}\right]$
- D'Alembert Principle: $\sum \mathbf{F} - m\mathbf{a} = 0$

REVIEW QUESTIONS

Q1. State and explain D'Alembert's principle.

Q2. Define and explain Newton's laws of motion for linear motion and rotational motion.

Q3. Two blocks of masses m_1 and m_2 are connected by a light and inextensible string that passes over a smooth pulley. If $m_1 > m_2$, prove that the acceleration a of the system and tension T in the string are given by

$$a = g\frac{m_1 - m_2}{m_1 + m_2} \quad \text{and} \quad T = \frac{2m_1m_2}{m_1 + m_2}g$$

Q4. Two masses weighing 40 kg and 35 kg, respectively, are attached by a cord that passes over a frictionless pulley. If the masses start from rest, find the distance covered by either mass in 6 s.
[Ans; 11.8 m].

Q5. A block of mass 200 kg hangs vertically downward from the end of a mass-less rope that is wrapped around a cylinder 900 mm in diameter (Fig. 10.Q4). The weight descends 8 m in 4 s. What is the mass of the cylinder? [Ans; 3520 kg].

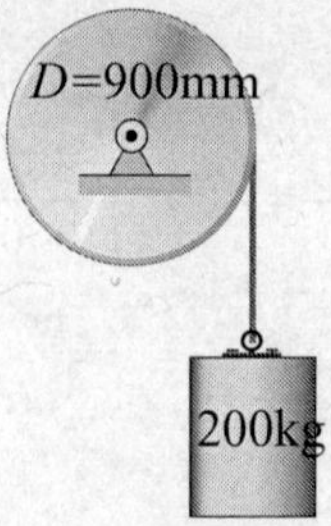

FIGURE 10.Q5 |

Q6. A uniform bar AB has a mass of 5 kg and is fastened by a frictionless pin at A and rests against the smooth vertical part of a cart (Fig. 10.Q5). The cart has a mass of 50 kg and is pulled to the right by a horizontal force F. At what value of F will the bar exert no force on the cart at B?

[Ans; $F = 642$ N].

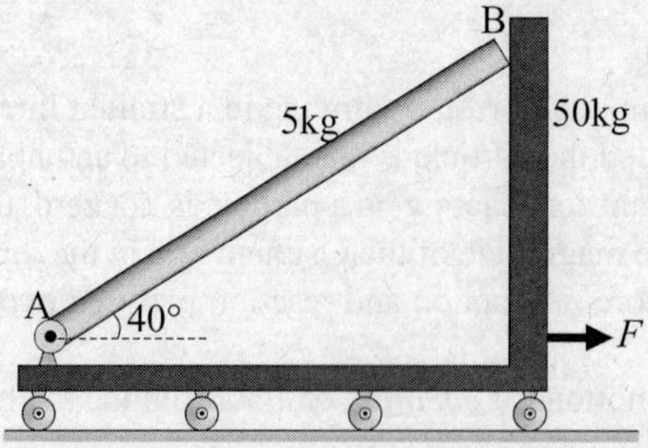

FIGURE 10.Q6 |

Q7. A homogeneous cylinder of mass 3 kg and radius 60 mm is rotating at 8 rad/s about its horizontal axis (Fig. 10.Q6), when dropped on a horizontal plane for which the coefficient of friction is 0.25. How far does the centre travel before rolling begins and skidding stops?

[Ans; $s = 5.2$ mm].

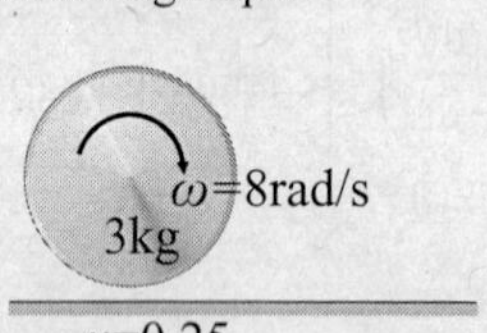

FIGURE 10.Q7 |

Q8. A cylinder of mass m and radius of gyration k has a rope wrapped in a groove of radius r as shown in Figure 10.Q7. Determine the acceleration of the cylinder if pure rolling is assumed.

$$\left[\text{Ans; } a = \frac{PR(R-r)}{m(R^2+k^2)}\right].$$

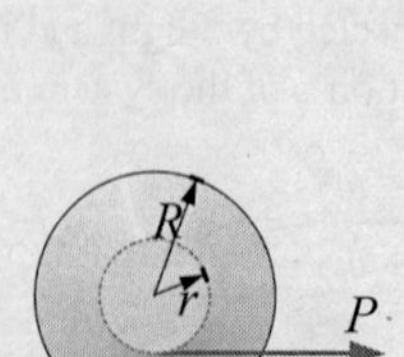

FIGURE 10.Q8 |

Q9. A solid circular rotor of radius *r* and mass *m* which can rotate about its geometric axis is braked by the device shown in Figure 10.Q8. If the coefficient of friction between rotor and brake shoe is μ, and the rotor has an initial angular velocity ω before the brake is applied, how many revolutions will it make before coming to rest? Neglect friction in the bearing.

$$\left[\text{Ans; } n = \frac{\omega^2 arm}{8\pi\mu lmP}\right].$$

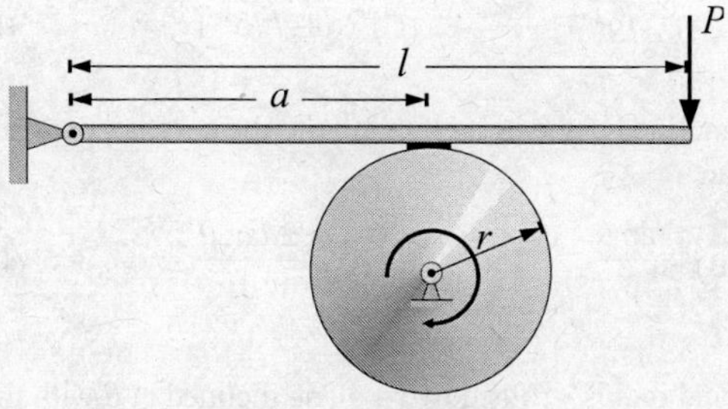

FIGURE 10.Q9 |

Q10. Find the acceleration of each weight and tension of rope in the system shown in Figure 10.Q9.

$$\left[\text{Ans; } T = 75\text{N}, a_A = \frac{g}{2}\uparrow, a_B = \frac{g}{2}\downarrow, a_C = 0\right].$$

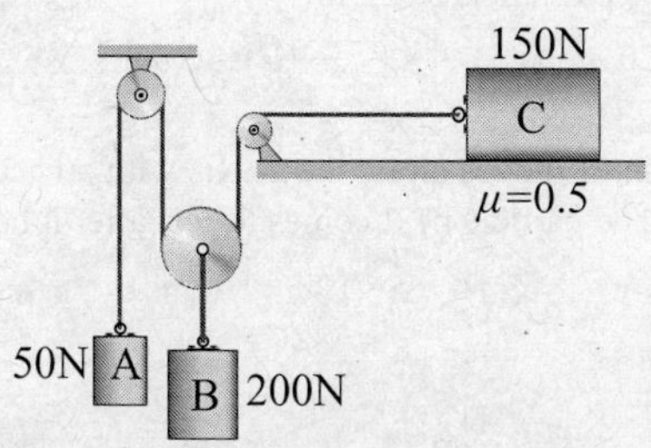

FIGURE 10.Q10 |

Q11. A sphere rolls down a plane inclined at 30° to the horizontal. Find the minimum value of the coefficient of friction between the sphere and the plane, so that the sphere may roll without slipping.

[Ans; $m = 0.165$].

MULTIPLE CHOICE QUESTIONS

Q1. A stone tied to a string of length *l* is whirled in a vertical circle with the other end of the string at the centre. At a certain instant of time, the stone is at its lowest position and has a speed *u*. The magnitude of the velocity as it reaches a position where the string is horizontal is

(A) $\sqrt{u^2 - 2gl}$ (B) $\sqrt{2gl}$ (C) $\sqrt{u^2 + 2gl}$ (D) $\sqrt{2(u^2 - 2gl}$

[Ans: A]

Q2. A cord is wrapped around a cylinder of radius *r* and mass *m* as shown in Figure 10.MQ2. If the cylinder is released for the rest, the velocity of the cylinder after it has moved through a distance *h* will be

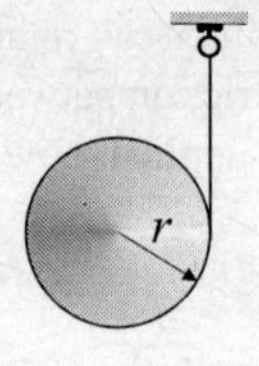

FIGURE 10.MQ2 |

(A) $\sqrt{2gh}$ (B) $\sqrt{gh}$ (C) $\sqrt{4gh/3}$ (D) $\sqrt{gh/3}$

[Ans: C]

Q3. A cylinder will slip on an inclined plane of inclination θ if the coefficient of static friction between the plane and the cylinder is less than

(A) $\frac{\sin\theta}{3}$ (B) $\frac{\tan\theta}{3}$ (C) $\frac{2\tan\theta}{3}$ (D) $\frac{3\sin\theta}{2}$

[Ans: B]

Q4. A solid sphere of mass m and radius r rolls down a plane inclined at θ with the horizontal. The acceleration of the sphere will be

(A) $\frac{2}{3}g\sin\theta$ (B) $\frac{5}{7}g\sin\theta$ (C) $\frac{3}{7}g\sin\theta$ (D) None of these

[Ans: B]

Q5. A particle of mass 2 units moving along the x-axis is attracted towards the origin by a force whose magnitude is $8x$, when the particle is at a distance x from the origin. If the particle is at rest at $x = 20$, then the maximum speed attained by the particle is equal to

(A) 10 units (B) 20 units (C) 30 units (D) 40 units

[Ans: D]

Q6. A particle is at rest at the origin. It moves along the x-axis with an acceleration $x - x^2$, where x is the distance of the particle at time t. The particle next comes to rest after it has covered a distance

(A) 1 (B) $\frac{1}{2}$ (C) $\frac{3}{2}$ (D) 2

[Ans: C]

Q7. A particle with mass m is tied to one end of light inextensible string of length l is displaced from its vertical position of equilibrium with a velocity u then

(A) $\frac{T}{m} = \frac{u^2 + lg - 3gh}{l}$

(B) the particle will oscillate if u^2 is greater than $5lg$

(C) the particle will leave the circular path $5lg > u^2 > lg$

(D) the particle will make revolutions if $u^2 = 2lg$

(Given: T = tension at a height h as order above the lowest point)

[Ans: A]

Q8. Two blocks of mass M and $3M$ are connected by a light cord which passes over a light frictionless pulley, the blocks are released from rest and are at the same height at $t = 0$. The magnitude of acceleration of the centre of mass of the two-block system is

(A) $\frac{g}{2}$ (B) $\frac{g}{4}$ (C) $\frac{3g}{4}$ (D) g

[Ans: A]

Q9. Two masses of 5 kg are 9 kg are fastened to ends of a cord passing over a frictionless pulley. The acceleration of the resulting motion is

(A) 2.8 m/s² (B) 2.5 m/s² (C) 5.6 m/s² (D) 5.0 m/s²

[Ans: A]

Q10. A particle of mass m moves in a straight line under an attractive force $m\mu x$ towards a fixed point O on the line, x being the distance of the particle from O. If $x = a$ at time $t = 0$, then the velocity of the particle at a distance x is given by

(A) $\mu(a - x)$ (B) $\mu(x - a)$ (C) $\sqrt{\mu(a^2 - x^2)}$ (D) $-\sqrt{\mu(a^2 - x^2)}$

[Ans: C]

Q11. Consider the following statements: if a particle moves under the action of a force $f(r)$ towards a fixed point, r being the distance of the particle from the fixed point, then the

1. Linear momentum of the particle will be a constant.
2. Angular momentum of the particle about the centre of force will be constant.
3. Orbital path will necessarily be a plane curve.
4. Path will always be closed.

Which of the above statements are correct?

(A) 1 and 2 (B) 2 and 3 (C) 3 and 4 (D) 2 and 4

[Ans: B]

Q12. A particle of unit mass is travelling along the x-axis such that at $t = 0$, it is located at $x = 0$ and has velocity v_0. If the particle is acted upon by a force, which opposes the motion and has a magnitude proportional to the square of the instantaneous speed, then the speed at time t is proportional to

(A) $1 + kt$ (B) $(1 + t)^2$ (C) $\frac{1}{t}$ (D) $\frac{1}{1 + kt}$

[Ans: D]

Q13. A block of mass m is placed on a smooth wedge of inclination a. The whole system is accelerated horizontally so that the block does not slip on the wedge [Fig. 10.MQ13]. The force exerted by the wedge on the block has a magnitude

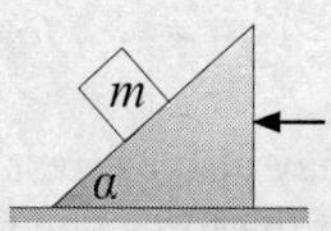

FIGURE 10.MQ13 |

(A) mg (B) $mg \sec a$ (C) $mg \cos a$ (D) $mg \tan a$

[Ans: C]

Q14. Starting from rest, a body slides down a 45° inclined plane in twice the time it takes to slide down the same distance in the absence of friction. The coefficient of friction between the body and the inclined plane is

(A) 0.33 (B) 0.25 (C) 0.75 (D) 0.80

[Ans: C]

Q15. A thin circular ring of mass per unit length ρ and radius r is rotating at an angular speed ω as shown in Figure 10.MQ15. The tension in the ring is

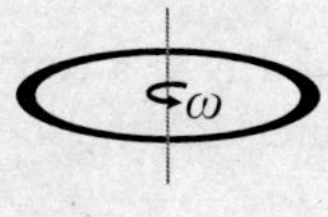

FIGURE 10.MQ15 |

(A) $\rho\omega^2 r^2$ (B) $\frac{1}{2}\rho\omega^2 r^2$ (C) $\frac{1}{3}\rho\omega^2 r^2$ (D) None of these

[Ans: A]

11 Work, Energy and Power

WORK OF A FORCE

The work done by a force on a moving body is defined as the product of the force and the distance moved in the direction of the force.

Consider a particle of mass m which moves from a point A to a neighbouring point A′, as shown in Figure 11.1. If **r** be the position vector corresponding to point A, the small vector joining A to A′ may be denoted by the differential **dr**; the vector **dr** is called the displacement of particle. Now let us assume that a force F is acting on the particle. The work of the force F corresponding to the displacement **dr** is defined as

$$dU = \mathbf{F} \cdot \mathbf{dr}$$

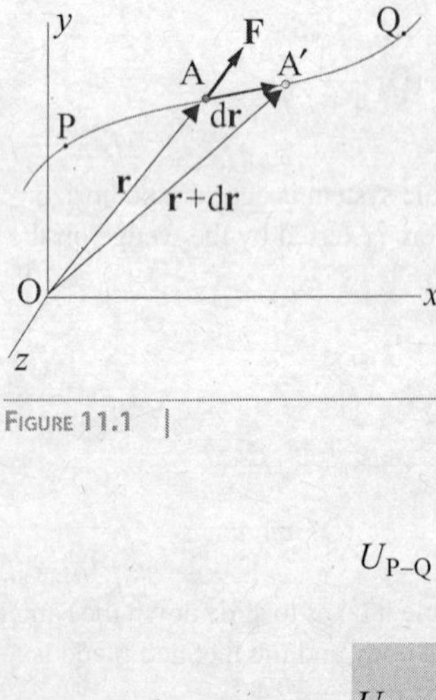

FIGURE 11.1

The work of force F during a finite displacement of the particle from P to Q is obtained by integrating the equation along the path described by the particle,

$$U_{P-Q} = \int_P^Q \mathbf{F} \cdot \mathbf{dr}$$

We may also express the work in terms of the rectangular components of the force and the displacement:

$$U_{P-Q} = \int_P^Q (F_x i + F_y j + F_z k) \cdot (\mathrm{d}xi + \mathrm{d}yj + \mathrm{d}xk)$$

$$U_{P-Q} = \int_P^Q (F_x \mathrm{d}x + F_y \mathrm{d}y + F_z \mathrm{d}z)$$

WORK OF A CONSTANT FORCE IN RECTILINEAR MOTION

If the force F has a constant magnitude and acts at a constant angle θ from its straight line path as shown in Figure 11.2, the formula yields

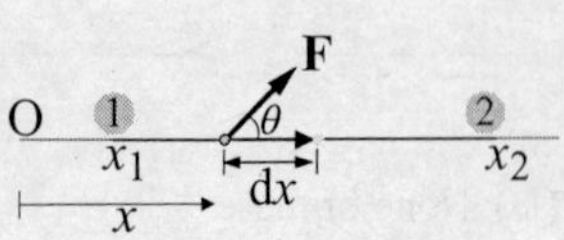

FIGURE 11.2

$$U_{1-2} = \int_1^2 \mathbf{F} \cdot \mathbf{dr} = F \cos\ \theta \int_{x_1}^{x_2} \mathrm{d}x$$

$$U_{1-2} = F \cos\theta(x_2 - x_1)$$

WORK OF THE WEIGHT

Consider a particle of weight W that moves up along the path as shown in Figure 11.3 from 1 to 2.

Position vector of positions 1 and 2 are $\mathbf{r}_1 = x_1 i + y_1 j + z_1 k$ and $\mathbf{r}_2 = x_2 i + y_2 j + z_2 k$, respectively. Since weight $\boldsymbol{W} = -Wj$, we write

$$U_{1-2} = \int \mathbf{F}\cdot \mathbf{dr} = \int_{r_1}^{r_2} (-Wj).(\mathrm{d}xi + \mathrm{d}yj + \mathrm{d}zk)$$

$$U_{1-2} = -\int_{y_1}^{y_2} W\mathrm{d}y = -W(y_2 - y_1)$$

$$U_{1-2} = -W\Delta y$$

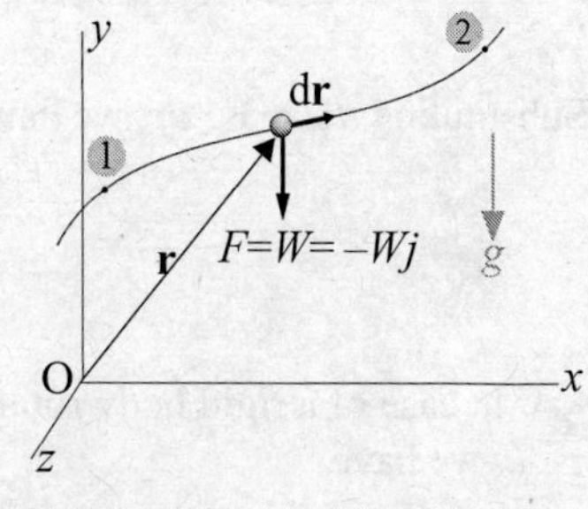

FIGURE 11.3 |

where Δy is the vertical displacement from ① to ②.

WORK OF A SPRING FORCE

A body of mass m is attached to the spring as shown in Figure 11.4. The position O is the equilibrium. The magnitude of spring force $F_s = kx$, where k is the spring stiffness. If the spring is elongated (or compressed) from x_1 to x_2, the work done on the spring by force F_s is

$$U_{1-2} = \int_{x_1}^{x_2} F_s \mathrm{d}x = \int_{x_1}^{x_2} kx\mathrm{d}x = \frac{1}{2}k(x_2^2 - x_1^2)$$

$$U_{1-2} = \frac{1}{2}k(x_2^2 - x_1^2)$$

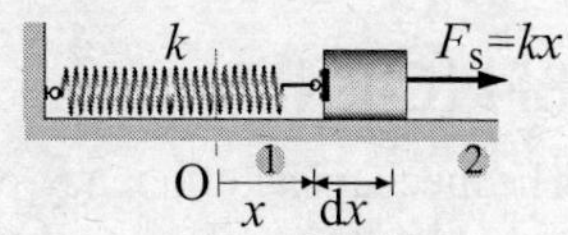

FIGURE 11.4 |

WORK OF A COUPLE

Consider a body, as shown in Figure 11.5, which is subjected to a couple having a magnitude of $M = Fr$. Any small displacement of the rigid body bringing P′ and Q, respectively, into P′ and Q″ may be divided into two parts, one in which point P and Q undergo equal displacement dr_1, the other in which P′ remains fixed while Q′ move into Q″ through displacement dr_2 of magnitude $r\mathrm{d}\theta$. In the first part of the motion, the net work $F\mathrm{d}r - F\mathrm{d}r = 0$.

In the second part of the motion, only force F works and its work is

$$\mathrm{d}U = Fr\mathrm{d}\theta \quad \text{or} \quad \mathrm{d}U = M\mathrm{d}\theta,$$

where couple $M = Fr$.

$$\mathrm{d}U = M\mathrm{d}\theta$$

When the body rotates in the plane through a finite angle θ, from θ_1 to θ_2, the work of a couple is

$$U = \int_{\theta_1}^{\theta_2} M\mathrm{d}\theta$$

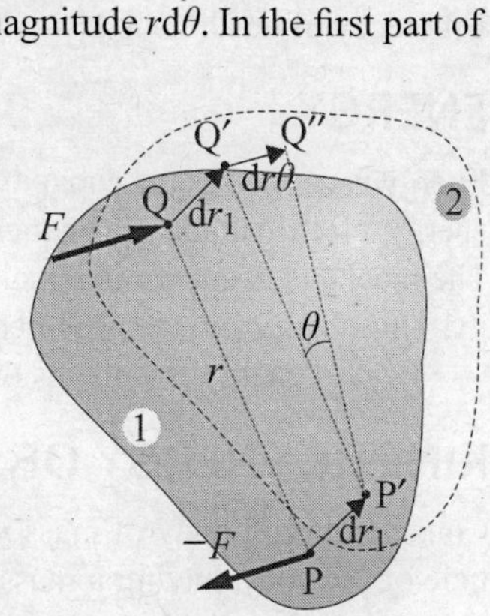

FIGURE 11.5 |

POWER

Power is the time rate at which work is performed. Hence, the power generated by a machine or an engine that performs an amount of work dU within the time interval dt is

$$P = \frac{dU}{dt}$$

Substituting $dU = \mathbf{F} \cdot d\mathbf{r}$, we have

$$P = \frac{\mathbf{F} \cdot d\mathbf{r}}{dt} = \mathbf{F} \cdot \frac{d\mathbf{r}}{dt}$$

$$P = \mathbf{F} \cdot \mathbf{v}$$

In case of a rigid body rotating with an angular velocity ω and acted upon by a couple τ, $dU = \tau \cdot d\theta$, we have

$$P = \frac{\tau \cdot d\theta}{dt} = \tau \cdot \frac{d\theta}{dt}$$

$$P = \tau \cdot \omega$$

The unit of power in SI system is the Watt (W):

$$1\text{W} = 1\text{J/s} = 1\text{N.m/s}$$

EFFICIENCY

The mechanical efficiency of a machine is defined as the ratio of the output of the useful power produced by the machine to the input power supplied to the machine (Fig. 11.6). Hence,

$$\eta = \frac{\text{power output}}{\text{power input}}$$

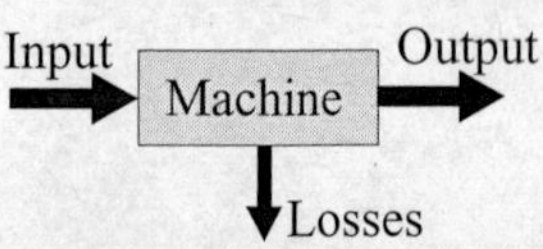

FIGURE 11.6 |

It may also be expressed in terms of the ratio of output to input work, i.e.,

$$\eta = \frac{\text{output work}}{\text{input work}}$$

Efficiency of machine is always less than 1.

ENERGY

Energy is defined as the capacity to do work. It exists in many forms like mechanical energy, chemical energy, electrical energy, heat energy, light energy, etc. Here we are interested in mechanical energy. The mechanical energy is equal to the work done on a body in altering either its motion or its velocity. This energy can be classified into three types: kinetic energy, potential energy and strain energy.

Unit of energy is same as that of work, i.e., Joules.

KINETIC ENERGY OF RIGID BODY

Consider a rigid body of mass m moving in $x-y$ plane as shown in Figure 11.7. An arbitrary ith particle of the body, having a mass dm, is located at r from O. If at the instant shown, the particle has a velocity $\mathbf{v}_i$, then the kinetic energy of particle is $T_i = (1/2)\, dmv_i^2$. The Kinetic energy of the whole body is represented by the equation

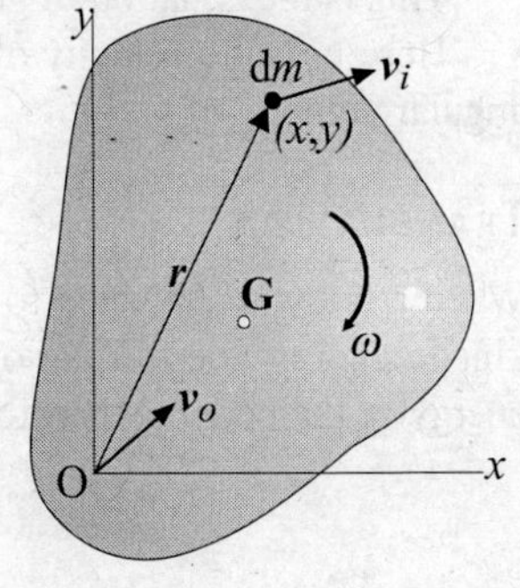

FIGURE 11.7 |

$$T = \int \frac{1}{2} \mathrm{d}mv_i^2$$

Let v_o be the velocity of point O and ω be the angular velocity of body about z-axis, then we have

$$\mathbf{v}_i = \mathbf{v}_o + \mathbf{v}_{i/p}$$

Substituting $\mathbf{v}_o$ and $\mathbf{v}_{i/p}$, where the velocity of point O

$$\mathbf{v}_o = (v_o)_x i + (v_o)_y j$$

and the relative velocity

$$\mathbf{v}_{i/p} = \omega k \times \mathbf{r} = \omega k \times (xi + yj) = \omega xj - \omega yi$$

we obtain

$$\mathbf{v}_i = \{(v_o)_x - \omega y\} i + \{(v_o)_y + \omega x\} j$$

On squaring of the magnitude of $\mathbf{v}_i$

$$v_i^2 = \mathbf{v}_i \cdot \mathbf{v}_i = \left[\left\{(v_o)_x - \omega y\right\} i + \left\{(v_o)_y + \omega x\right\} j\right]$$

$$\left[\left\{(v_o)_x - \omega y\right\} i + \left\{(v_o)_y + \omega x\right\} j\right]$$

or $$v_i^2 = \left\{(v_o)_x - \omega y\right\}^2 + \left\{(v_o)_y + \omega x\right\}^2$$

or $$v_i^2 = (v_o)_x^{\,2} + (v_o)_y^{\,2} + \omega^2 (x^2 + y^2) - 2(v_o)_x \omega y + 2(v_o)_y \omega x$$

or $$v_i^2 = v_o^{\,2} + \omega^2 r^2 - 2(v_o)_x \omega y + 2(v_o)_x \omega x$$

Thus, kinetic energy

$$T = \int \frac{1}{2} \mathrm{d}mv_i^2 = \frac{1}{2}\int \left\{v_o^{\,2} + \omega^2 r^2 - 2(v_o)_x \omega y + 2(v_o)_y \omega x\right\} \mathrm{d}m$$

or $$T = \frac{1}{2} v_o^{\,2} \int \mathrm{d}m + \frac{1}{2}\omega^2 \int r^2 \mathrm{d}m - (v_o)_x \omega \int y\mathrm{d}m + (v_o)_y \omega \int x\mathrm{d}m$$

$$T = \frac{1}{2} mv_o^{\,2} + \frac{1}{2} I_o \omega^2 - (v_o)_x \omega \bar{y} m + (v_o)_y \omega \bar{x} m$$

where

$$m = \int \mathrm{d}m, I_o = \int r^2 \mathrm{d}m, m\bar{y} = \int y\mathrm{d}m \quad \text{and} \quad m\bar{x} = \int x\mathrm{d}m$$

This equation reduces to a simpler form if point O coincides with the mass centre G for the body, in which case $x = y = 0$, and therefore,

$$T = \frac{1}{2} mv_G^{\,2} + \frac{1}{2} I_G \omega^2$$

This is the equation of general plane motion.

Here v_G is the velocity of centre of mass, I_G moment of inertia about centre of mass, and ω is the angular velocity of body.

TRANSLATION

When a body is subjected to either rectilinear or curvilinear translation, the kinetic energy due to rotation is zero, since $\omega = 0$ (Fig. 11.8). The kinetic energy of the body is, therefore,

$$T = \frac{1}{2} m v_G^2$$

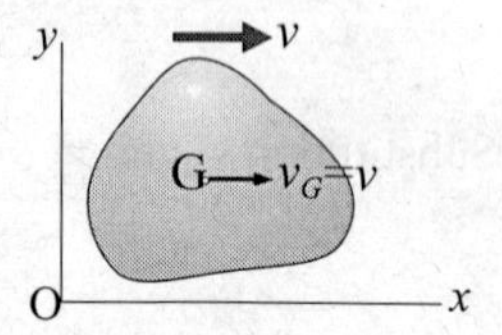

FIGURE 11.8 |

ROTATION ABOUT FIXED AXIS

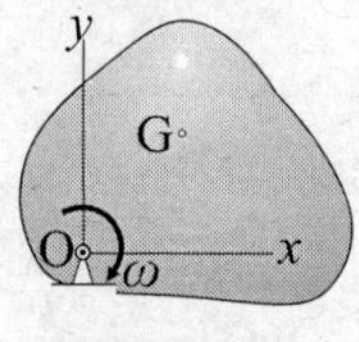

FIGURE 11.9 |

When a body is rotated about a fixed axis O (Fig. 11.9), the kinetic energy of the body from the equation is

$$T = \frac{1}{2} m(0)^2 + \frac{1}{2} I_o \omega^2 - 0 \times \omega \bar{y} m + 0 \times \omega \bar{x} m$$

$$T = \frac{1}{2} I_o \omega^2$$

PRINCIPLE OF WORK AND ENERGY

The work done by a system of forces acting on a body during a displacement is equal to the change in kinetic energy of the body during the same displacement.

Consider a particle of mass m acted upon by external forces of resultant F and moving along a path as shown in Figure 11.10. From Newton's second law in terms of the tangential components of the force and acceleration, we write

$$F_t = m a_t \quad \text{or} \quad F_t = m \frac{dv}{dt},$$

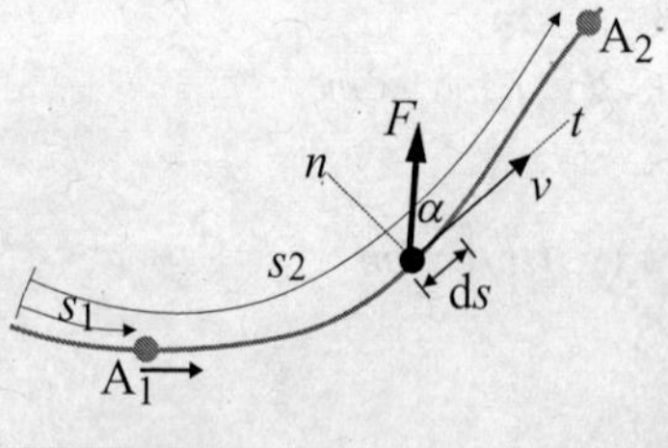

FIGURE 11.10 |

where v is the velocity of the particle along tangent.

$$F_t = m \frac{dv}{dt} = m \frac{dv}{ds} \frac{ds}{dt}$$

or

$$F_t = mv \frac{dv}{ds}$$

(*Note*: Normal force does no work since the particles cannot be displaced in the normal direction.)

or

$$F_t \, ds = mv \, dv$$

Integration from A_1, where $s = s_1$ and $v = v_1$, to A_2, where $s = s_2$ and $v = v_2$, we write

$$\int F_t ds = \int_{v_1}^{v_2} mv \, dv = \frac{1}{2} m v_2^2 - \frac{1}{2} m v_1^2$$

or $$\int F_t ds = \int F \cos \alpha ds = \int_{v_1}^{v_2} mv dv = \frac{1}{2} mv_2^2 - \frac{1}{2} mv_1^2$$

$$U_{1-2} = \frac{1}{2} mv_2^2 - \frac{1}{2} mv_1^2$$

This equation represents the principle of work and energy for the particle. The term on the left is the sum of the work done by all the forces acting on the particle as it moves from A_1 to A_2. The two terms on the right side, which are of the form $T = (1/2)mv^2$, define the particle's final and initial kinetic energy:

$$U_{1-2} = T_2 - T_1$$

The work U_{1-2} of the external forces that act on a particle (or rigid body) as it moves from position 1 to position 2 is equal to the corresponding change in kinetic energy, $T_2 - T_1$, of the body.

CONSERVATIVE AND NON-CONSERVATIVE SYSTEMS

The work that is performed on a system to move it from one configuration to another may depend on the path on which the system moves. This is certainly true for a system in which sliding friction occurs. For example, if we drag a block along a path on the floor, the work we perform on the block depends not only on where we start and stop, but also on the path on the floor. If the work performed on the system depends on the path from one configuration to another, the system is said to be a non-conservative system.

In contrast, a conservative system is one for which the work that is performed on the system in moving it from one configuration to another does not depend on the path. For example, if we lift a body in a gravity field, the work we perform on the body depends only on the change in elevation h of the body, that is, the final elevation minus the initial elevation and not on any path. Hence, gravitation system is conservative.

Conservative Force

A force is said to be conservative if the work done by the force on a system that moves between two configurations is independent of the path the system takes.

Non-conservative Force

A force is said to be non-conservative if the work done by a force on a system depends on the path the system follows.

CONSERVATION OF ENERGY

When a rigid body, or a system of rigid bodies, moves under the action of conservative forces, the sum of the kinetic energy and the potential energy of the system remains constant.

$$T + V = \text{constant} \qquad \text{or} \qquad \Delta T + \Delta V = 0$$

or $$T_1 + V_1 = T_2 + V_2$$

In general, if a body is subjected to both gravitational and elastic forces, the total potential energy is expressed as a potential function V represented as the algebraic sum:

$$V = V_g + V_e$$

V depends on the location of the system with respect to the selected datum.

Consider a conservative system, say, a particle moving from position 1 to position 2. As the particle moves from position 1 to position-2 it undergoes a change in potential energy, $\Delta V = V_2 - V_1$, where V_1 and V_2 are the potential energies in positions 1 and 2, respectively. As the particle is displaced from position 1 to position 2, let U_{1-2} be the work of the forces that act on the particle. The work U_{1-2} does not depend on the path or the speed with which the particle moves on the path. Consequently, $U_{1-2} = V_1 - V_2 = -\Delta V$. Then, the law of kinetic energy ($U_{1-2} = \Delta T$). yields $-\Delta V = \Delta T$, where $\Delta T = T_2 - T_1$ is change in kinetic energy from position 1 to position 2. This relationship may be written as

$$\Delta T + \Delta V = 0$$

$$T + V = \text{constant},$$

where $T + V$ is called the mechanical energy of the particle.

SOLVED PROBLEMS

PROBLEM 11.1

A 10 kg cylinder of diameter 2 m rolls without slipping under the action of a 100 N force. A spring is attached to a cord that is wound around the cylinder [Fig. 11.1(a)]. What is the speed of the cylinder after it has moved 50 cm? The spring is unstretched when the 100 N force is applied.

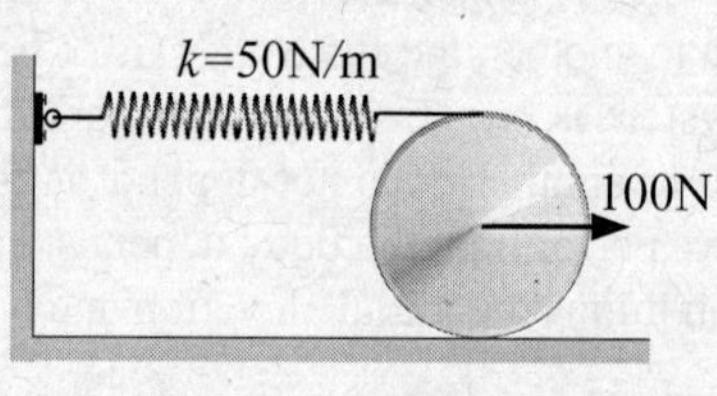

FIGURE 11.1(a) |

Solution. The two positions for the cylinder, when it is located at its initial position 1 and when the cylinder has moved 50 cm to the final position 2, are shown in Figure 11.1(b).

When the cylinder is in position 1:

Since the cylinder is at rest in the initial position 1 so that $v = 0$ and $\omega = 0$. Thus

Kinetic energy $\quad T_1 = 0$

and

Elastic potential energy of the spring $\quad V_1 = 0$

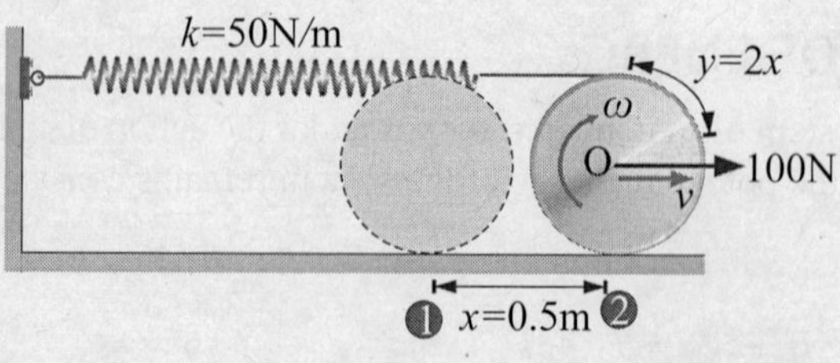

FIGURE 11.1(b) |

When the cylinder is in position 2:

Kinetic energy

$$T_2 = \frac{1}{2}mv^2 + \frac{1}{2}I_o\omega^2,$$

where v is the velocity and ω is the angular velocity of the cylinder,

or $$T_2 = \frac{1}{2}mv^2 + \frac{1}{2}\times\frac{1}{2}mr^2\left(\frac{v}{r}\right)^2 = \frac{3}{4}mv^2$$

Elastic potential energy of spring

$$V_2 = \frac{1}{2}ky^2$$

Since $y = 2x$, $V_2 = 2kx^2$

Work done by the force is $U_{1-2} = Fx$,

$$U_{1-2} = 100 \times \frac{50}{100} = 50\text{J}$$

The work–energy equation gives

$$\{T_1 + V_1\} + U_{1-2} = \{T_2 + V_2\} \Rightarrow 0 + 0 + 50 = \frac{3}{4}mv^2 + 2kx^2$$

$$50 = \frac{3}{4}\times 10v^2 + 2\times 50\times 0.5^2$$

or $$v = 1.82\text{m/s}$$

PROBLEM 11.2

It is desired that a 50 kg weight, when released from rest in the position shown, has no velocity after dropping 7 m to the platform below [Fig. 11.2(a)]. Specify the proper mass m of the counter balancing weight.

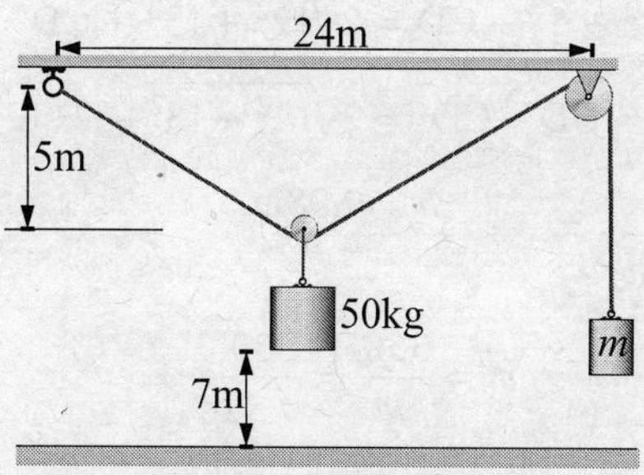

FIGURE 11.2(a) |

Solution. Choosing the datum as shown in Figure 11.2(b). Two diagrams for the system at their initial and final positions are shown.

The system is released from rest so that $(v_P)_1 = 0$. Thus

Kinetic energy $T_1 = 0$.

When the system is in position 2, velocities of P and Q become zero. Thus

Kinetic energy $T_2 = 0$.

The change in potential energy of P is

$$(\Delta V_g)_P = -50(9.81)\times 7$$

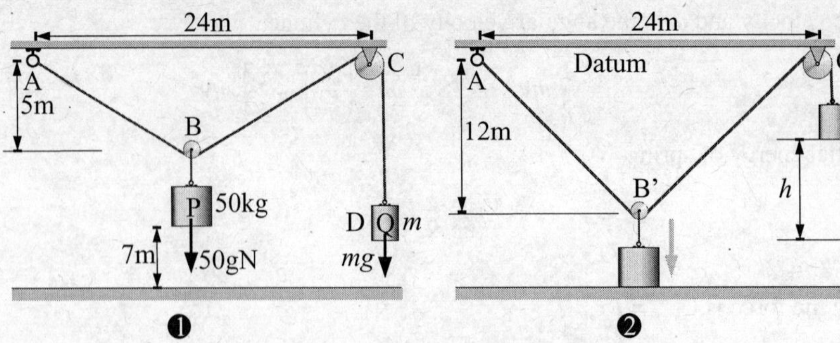

FIGURE 11.2(b) |

The change in potential energy of Q is

$$(\Delta V_g)_Q = mgh$$

Since $T + V$ is a constant for conservative system so that $\Delta T + \Delta V = 0$.

or
$$0 - 50(9.81) \times 7 + mgh = 0$$

or
$$mh = \frac{50(9.81) \times 7}{9.81} = 350 \qquad \text{(i)}$$

From the geometry of the figure length of the rope in position 1 is equal to length of the rope in position 2.

$$2AB + CD = 2AB' + CD'$$

or
$$2\sqrt{12^2 + 5^2} + CD = 2\sqrt{12^2 + 13^2} + CD'$$

or
$$h = CD - CD' = 2\sqrt{12^2 + 13^2} - 2\sqrt{12^2 + 5^2}$$

$$h = 9.38\text{m} \qquad \text{(ii)}$$

From Equations (i) and (ii), we have

$$m = \frac{350}{h} = \frac{350}{9.38}$$

$$m = 37.3\text{kg}$$

PROBLEM 11.3

A heavy uniform rope of length $2l$ hangs over a small smooth pulley, the length $l + a$ being at one side and $l - a$ at the other [Fig. 11.3(a)]. If the end of the shorter portion be held and then let go, show that the rope will slip off the pulley in time

$$t = \sqrt{\frac{l}{g}} \log_e \left\{ \frac{l + \sqrt{l^2 - a^2}}{a} \right\}$$

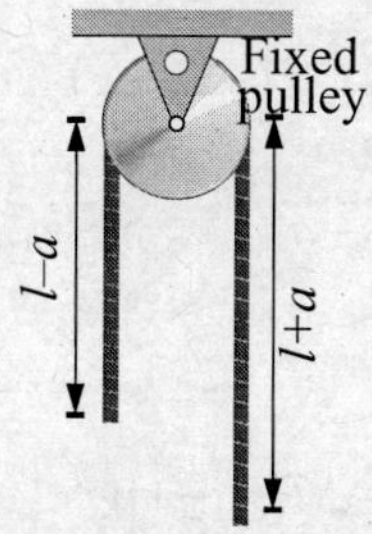

FIGURE 11.3(a) |

Solution. When the lengths of rope at the two sides are $l + a + x$ and $l - a - x$, the depth of the centre of gravity of the rope below the pulley is $\overline{y}_2$ [Fig. 11.3(b)]. where

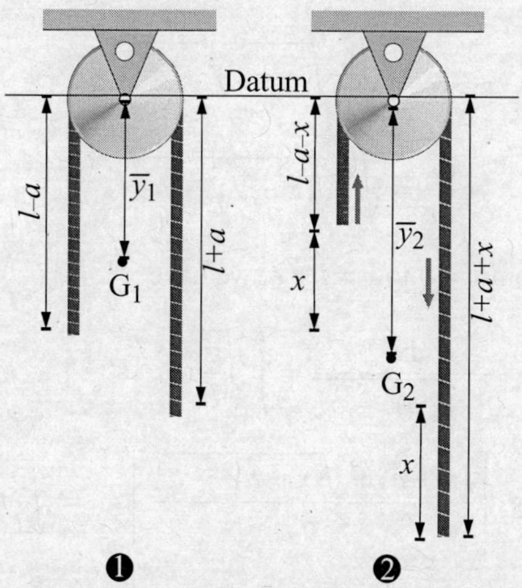

FIGURE 11.3(b) |

$$\overline{y}_2 = \frac{\left(\frac{m}{2l}\right)(l-a-x)(l-a-x)/2 + \left(\frac{m}{2l}\right)(l+a+x)(l+a+x)/2}{\left(\frac{m}{2l}\right)(l-a-x) + \left(\frac{m}{2l}\right)(l+a+x)}$$

or $$\overline{y}_2 = \frac{\left(\frac{m}{2l}\right)(l-a-x)^2/2 + \left(\frac{m}{2l}\right)(l+a+x)^2/2}{m} = \frac{(l-a-x)^2 + (l+a+x)^2}{4l}$$

Initial depth of centre of gravity is

$$\overline{y}_1 = \frac{1}{4l}\left[(l-a)^2 + (l+a)^2\right]$$

Hence change in potential energy is

$$\Delta V = -mg\left(\overline{y}_2 - \overline{y}_1\right)$$

$$= \frac{mg}{4l}\left[(l-a-x)^2 + (l+a+x)^2 - (l-a)^2 - (l+a)^2\right]$$

$$\Delta V = -\frac{mg}{2l}\left[(x+a)^2 - a^2\right]$$

Change of kinetic energy

$$\Delta T = \frac{1}{2}mv^2$$

Applying the principle of conservation of energy

$$\Delta T + \Delta V = 0 \Rightarrow \frac{1}{2}mv^2 - \frac{mg}{2l}\left[(x+a)^2 - a^2\right] = 0$$

or
$$v = \sqrt{\frac{g}{l}}\sqrt{(x+a)^2 - a^2}$$

or
$$\frac{dx}{dt} = \sqrt{\frac{g}{l}}\sqrt{(x+a)^2 - a^2}$$

Integrating above equation from $x = 0$ to $x = l - a$, we have

$$\int_0^{l-a} \frac{dx}{\sqrt{(x+a)^2 - a^2}} = \int_0^{l} \sqrt{\frac{g}{l}}\,dt$$

or
$$\log_e\left[\frac{(x+a) + \sqrt{(x+a)^2 - a^2}}{a}\right]_0^{l-a} = \sqrt{\frac{g}{l}}t,$$

from which

$$t = \sqrt{\frac{l}{g}}\log\frac{l + \sqrt{l^2 - a^2}}{a}$$

PROBLEM 11.4

A chain of length l is released from rest in a smooth tube with $y = 0$ [Fig. 11.4(a)]. Determine the velocity v of the chain in terms of y.

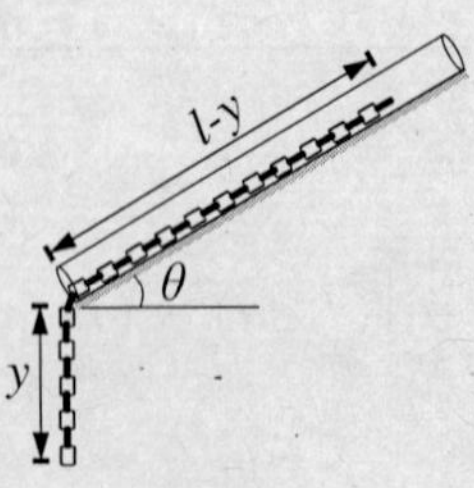

FIGURE 11.4(a) |.

Solution. Let λ be the mass per unit length of chain.

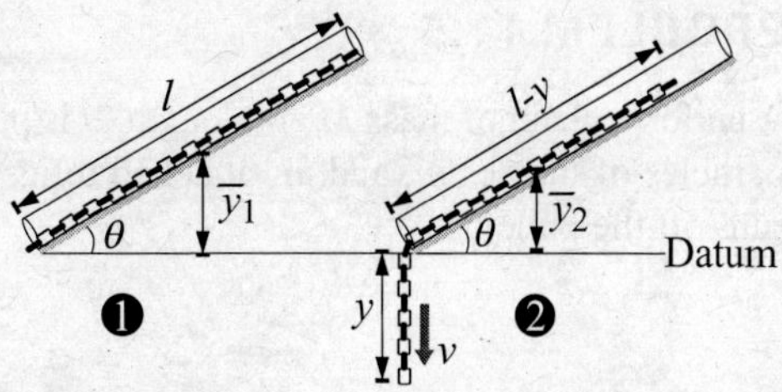

FIGURE 11.4(b) |

Position I

Vertical position of centre of gravity of chain is [(Fig. 11.4(b)]

$$\bar{y}_1 = \frac{l}{2}\sin\theta,$$

therefore, potential energy

$$V_1 = mg\bar{y}_1$$

$$V_1 = \lambda lg\frac{l}{2}\sin\theta$$

Initially, chain is released from rest so that $v = 0$. Thus, kinetic energy $T_1 = 0$.

Position II

Vertical position of centre of gravity of chain

$$\bar{y}_2 = \frac{\lambda(l-y)\dfrac{l-y}{2}\sin\theta - \lambda y \times \dfrac{y}{2}}{\lambda(l-y)+\lambda y}$$

or $$\bar{y}_2 = \frac{(l-y)^2\sin\theta - y^2}{2l}$$

Potential energy

$$V_2 = mg\bar{y}_2$$

$$V_2 = \lambda lg\left[\frac{(l-y)^2\sin\theta - y^2}{2l}\right]$$

Kinetic energy

$$T_2 = \frac{1}{2}mv^2 = \frac{1}{2}\lambda lv^2$$

Applying the principle of conservation of energy between positions 1 and 2, we have

$$T_1 + V_1 = T_2 + V_2$$

or $$\lambda lg\frac{l}{2}\sin\theta + 0 = \lambda lg\left[\frac{(l-y)^2\sin\theta - y^2}{2l}\right] + \frac{1}{2}m\lambda v^2$$

or $$v^2 = gl\sin\theta - \frac{g}{l}\left[(l-y)^2\sin\theta - y^2\right]$$

or $$v^2 = gl\sin\theta - gl\sin\theta + 2gy\sin\theta - \frac{gy^2}{l}\sin\theta + \frac{gy^2}{l}$$

$$v = \sqrt{2gy\left[\sin\theta + \frac{y}{2l}(1-\sin\theta)\right]}$$

PROBLEM 11.5

A uniform chain of mass M and length $2l$ is placed symmetrically over a smooth fixed pulley and has particles of masses m_1 and m_2 attached to its ends [Fig. 11.5(a)]. Find the velocity of chain when it runs off the pulley.

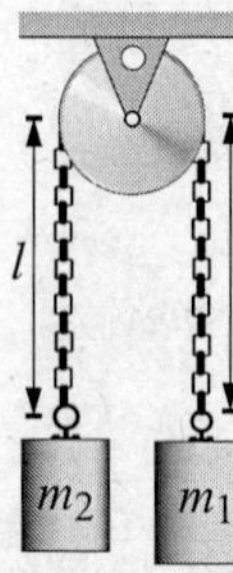

FIGURE 11.5(a) |

Solution.

Position 1

Initially the system is released from rest so that $v = 0$. Thus kinetic energy $T_1 = 0$. Potential energy

$$V_1 = (\Sigma m)\, g\bar{y}_1,$$

where $\bar{y}_1$ is the vertical depth of centre of mass of the system from datum, which is

$$\bar{y}_1 = \frac{(m_1 + m_2)l + \left(\dfrac{Ml}{2}\right)}{m_1 + m_2 + M}$$

$$V_1 = (m_1 + m_2 + M)g\left[\frac{(m_1 + m_2)l + \left(\dfrac{Ml}{2}\right)}{m_1 + m_2 + M}\right]$$

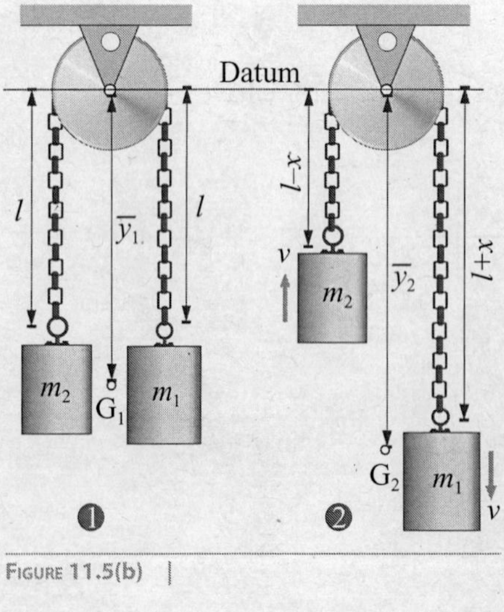

FIGURE 11.5(b) |

Position 2

Kinetic energy

$$T_2 = \frac{1}{2}(m_1 + m_1 + M)\,v^2$$

When the lengths on the two sides are $l + x$ and $l - x$, the depth of the centre of gravity of the chain below the pulleys is $\bar{y}_2$ [Fig. 11.5(b)], where

$$\bar{y}_2 = \frac{m_1(l+x) + \left(\dfrac{M}{2l}\right)(l+x)\times\left[\dfrac{(l+x)}{2}\right] + m_2(l-x) + \left(\dfrac{M}{2l}\right)(l-x)\times\left[\dfrac{(l+x)}{2}\right]}{m_1 + \left(\dfrac{M}{2l}\right)(l+x) + m_2 + \left(\dfrac{M}{2l}\right)(l-x)}$$

or
$$\bar{y}_2 = \frac{(m_1 + m_2)l + (m_1 - m_2)x + \left(\frac{M}{2l}\right)(l^2 + x^2)}{m_1 + m_2 + M}$$

Potential energy

$$V_2 = (\Sigma m)\, g\bar{y}_2$$

$$V_2 = (m_1 + m_1 + M)g\left[\frac{(m_1 + m_2)l + (m_1 - m_2)x + \left(\frac{M}{2l}\right)(l^2 + x^2)}{m_1 + m_2 + M}\right]$$

Applying the principle of conservation of energy,

$$T_1 + V_1 = T_2 + V_2 \Rightarrow -(V_2 - V_1) = (T_2 - T_1)$$

From which

$$(m_1 + m_1 + M)g\left(\frac{(m_1 + m_2)l + (m_1 - m_2)x + \left(\frac{M}{2l}\right)(l^2 + x^2)}{m_1 + m_2 + M} - \frac{(m_1 + m_2)l + \left(\frac{M}{2l}\right)}{m_1 + m_2 + M}\right)$$

$$= \frac{1}{2}(m_1 + m_2 + M)v^2$$

On simplifying, we get

$$v = \sqrt{2g\frac{\left[(m_1 - m_2)x + \left(\frac{Mx^2}{2l}\right)\right]}{m_1 + m_2 + M}}$$

Velocity of the chain when it runs off the pulley, substituting $x = l$,

$$v = \sqrt{\frac{gl\left[M + 2(m_1 - m_2)\right]}{m_1 + m_2 + M}}$$

PROBLEM 11.6

A particle of mass m is at rest A when it is slightly displaced and allowed to slide down the cylindrical surface of radius r [Fig. 11.6(a)]. Neglecting friction, find (a) the velocity of the particle as a function of the angle θ, and (b) the value of θ when the particle leaves the surface.

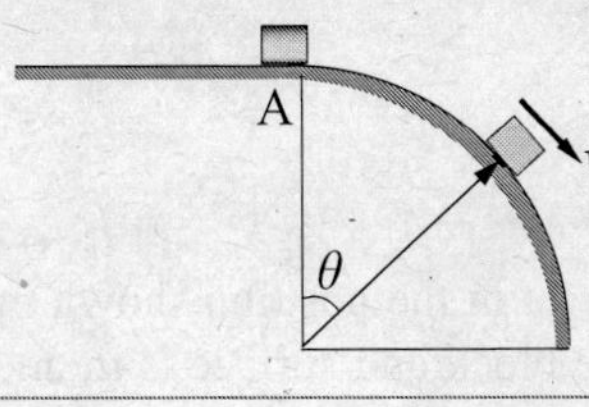

FIGURE 11.6(a) |

Solution. Two positions of the particle are shown in Figure 11.6(b). For convenience, the datum, which is horizontally fixed, passes through point 1.

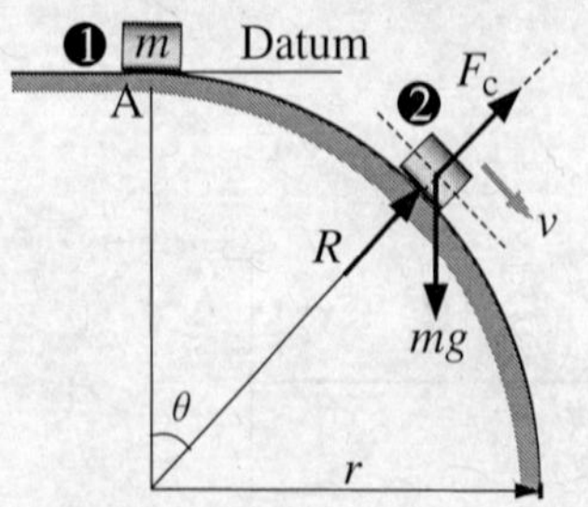

FIGURE 11.6(b)

When the particle is in **position 1**

Potential energy of particle $V_1 = 0$,

and

Kinetic energy of particle $T_1 = 0$.

When the particle is in **position 2**

Potential energy of particle $V_2 = -mg(r - r\cos\theta)$

and

Kinetic energy of particle $T_2 = (1/2)mv^2$,

where v is the velocity of particle at angle θ.

Applying the principle of conservation of energy,

$$T_1 + V_1 = T_2 + V_2$$

or

$$0 + 0 = \frac{1}{2}mv^2 - mg(r - r\cos\theta)$$

or

$$v^2 = 2gr(1 - \cos\theta) \qquad \text{(i)}$$

$$\boxed{v = \sqrt{2gr(1 - \cos\theta)}}$$

(b) Figure 11.6(b) shows the free body diagram of the particle. Resolving the forces along normal to the surface

$$\sum F_n = 0 \Rightarrow R + \frac{mv^2}{r} - mg\cos\theta = 0$$

$$R = mg\cos\theta - \frac{mv^2}{r}$$

Particle leaves the surface if $R = 0$, therefore

$$0 = mg\cos\theta - \frac{mv^2}{r}$$

$$y^2 = gr\cos\theta \qquad \text{(ii)}$$

From Equations (i) and (ii), we have

$$2gr(1 - \cos\theta) = gr\cos\theta$$

or

$$2g - 2g\cos\theta = g\cos\theta$$

or

$$\theta = \cos^{-1}\frac{2}{3}$$

$$\boxed{\theta = 48.18^\circ}$$

PROBLEM 11.7

A block of mass m is released from rest in the position shown in Figure 11.7(a). The coefficient of static and kinetic friction between the block and surface is μ, and the free length of the spring is l_0. Find the maximum displacement of the block.

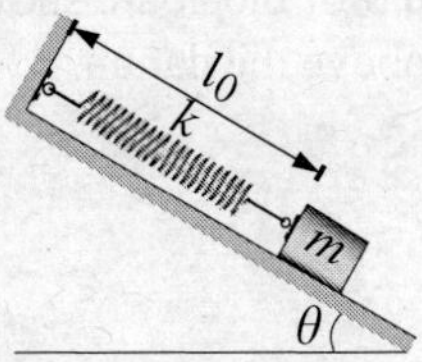

FIGURE 11.7(a)

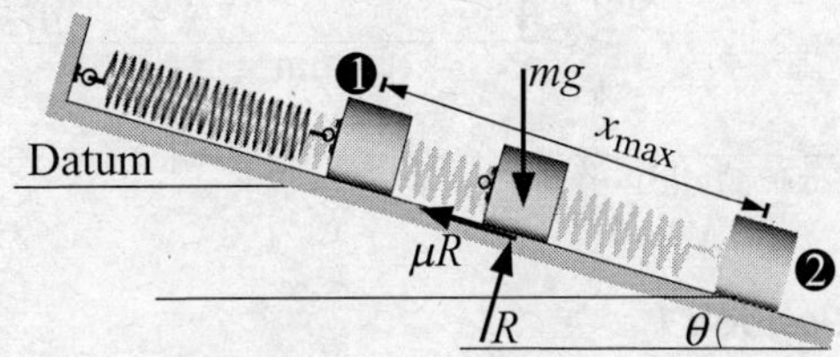

FIGURE 11.7(b)

Solution. We consider the system formed by the spring and block. Let the datum for potential energy be the initial position 1 [Fig. 11.7(b)]. Therefore, the potential energy of the system in position 1 is

$$V_1 = (V_1)_g + (V_1)_s = 0 + 0 = 0$$

and kinetic energy $T_1 = 0$.

When the system is at position 2:

At the condition of maximum deformation x_{max} of the spring, the block is momentarily at rest, which make $T_2 = 0$ and potential energy

$$V_2 = (V_2)_g + (V_2)_s = -mgx_{max}\sin\theta + \frac{1}{2}kx_{max}^2$$

Work against friction force

$$U_{1-2} = -F_f x_{max} = -\mu R x_{max}$$

or

$$U_{1-2} = -\mu mg\cos\theta x_{max}$$

Applying the work–energy equation, we have

$$T_1 + V_1 + U_{1-2} = T_2 + V_2$$

$$0 - \mu mg\cos\theta\, x_{max} = 0 - mgx_{max}\sin\theta + \frac{1}{2}kx_{max}^2$$

or

$$x_{max} = \frac{2mg}{k}(\sin\theta - \mu\cos\theta)$$

PROBLEM 11.8

The coefficient of kinetic friction between the block and the surface is μ [Fig. 11.8(a)]. If the block is released from rest at A, determine (a) the speed of the block at B and (b) the distance x that the block travels on the horizontal surface before coming to rest at C.

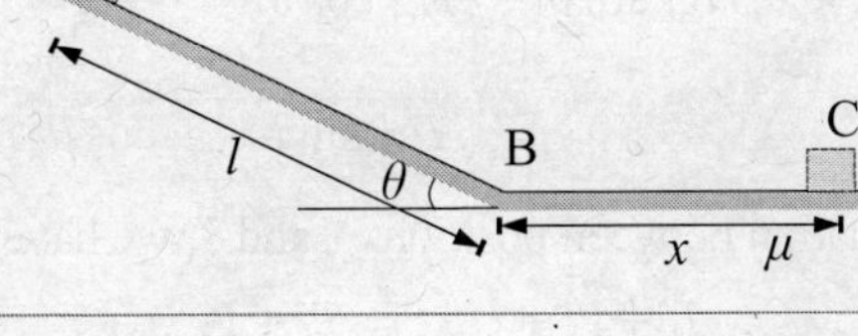

FIGURE 11.8(a)

Solution. Three positions 1, 2 and 3 of block are shown in Figure 11.8(b). When the block is in position 1, the block is located $l \sin\theta$ above the datum.

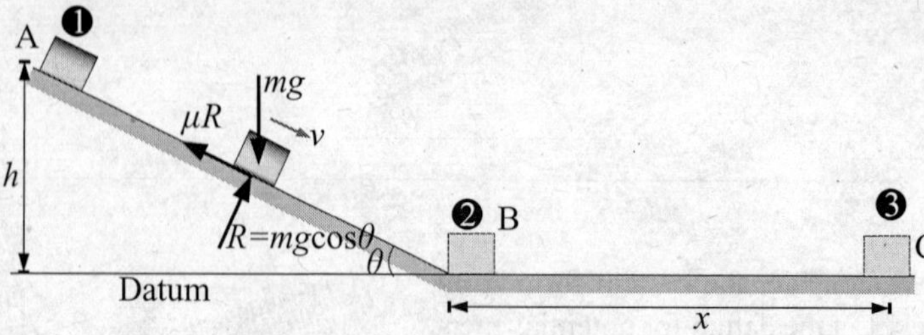

FIGURE 11.8(b) |

Position 1

$$\text{Kinetic energy } T_1 = 0$$

$$\text{Potential energy } V_1 = mgh = mgl \sin\theta$$

Position 2

$$\text{Kinetic energy } T_2 = \frac{1}{2} m v_2^2$$

$$\text{Potential energy } V_2 = 0$$

Position 3

$$\text{Kinetic energy } T_3 = \frac{1}{2} m v_3^2$$

$$\text{Potential energy } V_3 = 0$$

During the motion, friction force does negative work.
Work against friction force between positions 1 and 2

$$U_{1-2} = -\mu mg \cos\theta l$$

Work against friction force between positions 2 and 3

$$U_{2-3} = -\mu mgx$$

(a) Applying work–energy equation between positions 1 and 2, we have

$$T_1 + V_1 + U_{1-2} = T_2 + V_2$$

$$0 + mgl \sin\theta - \mu mg \cos\theta l = 0 + \frac{1}{2} m v_2^2$$

or $$v_2 = \sqrt{2gl(\sin\theta - \mu\cos\theta)}$$

(b) Applying work–energy equation between positions 1 and 3, we have

$$T_1 + V_1 + U_{1-3} = T_3 + V_3$$

$$0 + mgl\sin\theta - \mu mg\cos\theta l - \mu mgx = 0 + \frac{1}{2}mv_3^2$$

$$v_3 = \sqrt{2g(l\sin\theta - \mu l\cos\theta - \mu x)}$$

$$v_3 = 0 = \sqrt{2g(l\sin\theta - \mu l\cos\theta - \mu x)}$$

from which

$$x = \frac{1}{\mu}[\sin\theta - \mu\cos\theta]$$

PROBLEM 11.9

Determine the velocity of slider B [Fig. 11.9(a)] in terms of θ, if the system is released from rest with $\theta = 45°$. The sliders have equal masses. Assume friction is negligible.

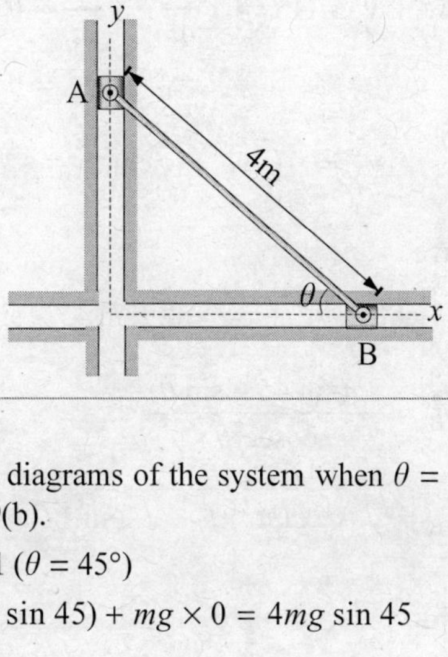

FIGURE 11.9(a)

Solution. Two kinematic diagrams of the system when $\theta = 45°$ (position 1) and an arbitrary angle θ are shown in Figure 11.9(b).

When the system is in **position 1** ($\theta = 45°$)

Potential energy $V_1 = mg(4\sin 45) + mg \times 0 = 4mg\sin 45$

Kinetic energy $T_1 = 0$

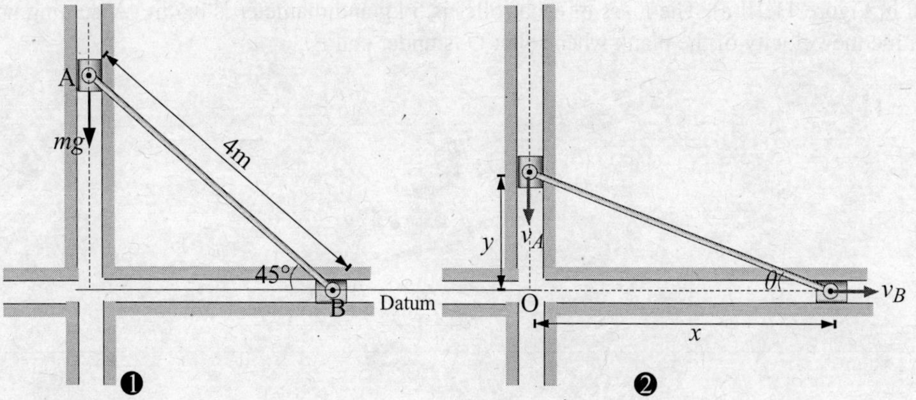

FIGURE 11.9(b)

When the system is in **position 2**

Potential energy $V_1 = mg(4 \sin \theta) + mg \times 0 = 4mg \sin \theta$

Kinetic energy $T_2 = \frac{1}{2} mv_A^2 + \frac{1}{2} mv_B^2$

where v_A and v_B are the velocities of blocks A and B, respectively.

Applying the principle of conservation of energy between positions 1 and 2, we have

$$T_1 + V_1 = T_2 + V_2$$

or
$$4mg \sin 45 = \frac{1}{2} mv_A^2 + \frac{1}{2} mv_B^2 + 4mg \sin \theta$$

or
$$v_A^2 + v_B^2 = 8g(\sin 45 - \sin \theta) \tag{i}$$

From the geometry of the figure

$$x^2 + y^2 = 4^2 \Rightarrow x\frac{dx}{dt} + y\frac{dy}{dt} = 0$$

or
$$xv_B + yv_A = 0$$

or
$$v_A = -\frac{x}{y} v_B = -\cot \theta v_B \tag{ii}$$

From equations (i) and (ii), we have

$$(-\cot \theta v_B)^2 + v_B^2 = 8g(\sin 45 - \sin \theta)$$

or
$$v_B^2 = \frac{8g(\sin 45 - \sin \theta)}{\text{cosec}^2\theta}$$

$$v_B = \sqrt{4\sqrt{2}g(\sin^2 \theta - \sqrt{2}(\sin^3 \theta)}$$

PROBLEM 11.10

Plank AB of mass 10 kg and length 6 m is released from rest with two rollers on an inclined plane as shown in Figure 11.10(a). The mass of each roller is 5 kg and diameter is 0.5 m. Assuming no slip, determine the velocity of the plank when roller Q is under end B.

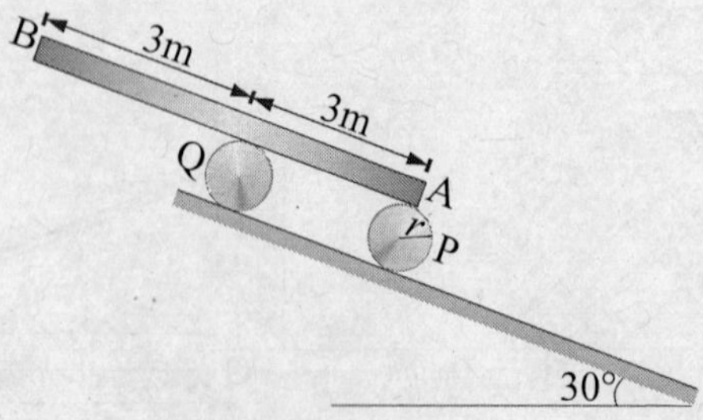

FIGURE 11.10(a) |

Solution. Let the datums for potential energy of rollers and slab be its initial position. Figure 11.10(b) shows the two positions of the system.

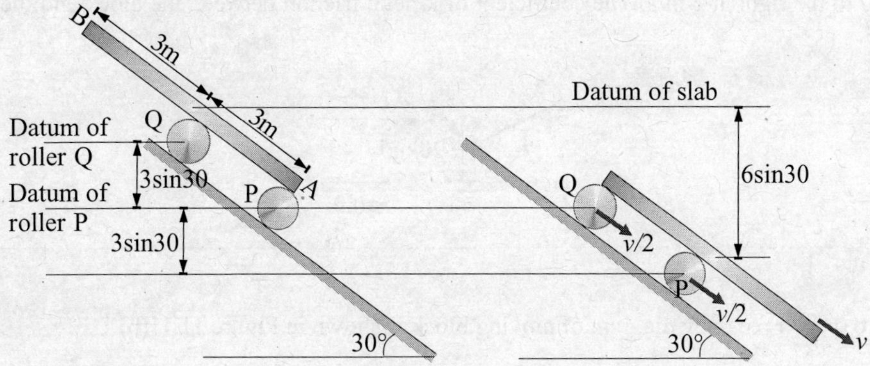

FIGURE 11.10(b) |

Change in potential energy of slab $(\Delta V)_{slab} = -10 \times 9.81 \times 6 \sin 30$

$$(\Delta V)_{slab} = -294.3$$

Change in potential energy of the roller

$$(\Delta V)_{rollers} = -2(5 \times 9.81 \times 3 \sin 30)$$

$$(\Delta V)_{rollers} = -147.15$$

Change in kinetic energy of the slab

$$(\Delta T)_{slab} = \frac{1}{2} \times 10 \times v_{slab}^2 = 5v_{slab}^2$$

Change in kinetic energy of the roller

$$(\Delta T)_{rollers} = 2\left(\frac{1}{2} mv_{roller}^2 + \frac{1}{2} \times I \times \omega_{roller}^2\right) = \frac{3}{2} mv_{roller}^2$$

$$(\Delta T)_{rollers} = \frac{3}{2} \times 5v_{roller}^2 = 7.5v_{roller}^2$$

Since $T + V$ is constant for a conservative system so that $\Delta T + \Delta T = 0$, therefore

$$-294.3 - 147.15 + 5v_{slab}^2 + 7.5v_{roller}^2 = 0$$

Since $v_{slab} = 2v_{roller}$

$$-294.3 - 147.15 + 5v_{slab}^2 + 7.5\left(\frac{v_{slab}}{2}\right)^2 = 0$$

or

$$\frac{27.5}{4} v_{slab}^2 = 441.45$$

from which

$$v_{slab} = 8.01 \text{m/s}$$

PROBLEM 11.11

A 20 kg block is subjected to a force having a constant direction and a magnitude $F = 50x^2$, where x is in meters [Fig. 11.11(a)]. Determine the speed of the block after it slides 3 m. When $x = 0$, the block is moving to the right at 2 m/s. The coefficient of kinetic friction between the block and the surface is $\mu = 0.3$.

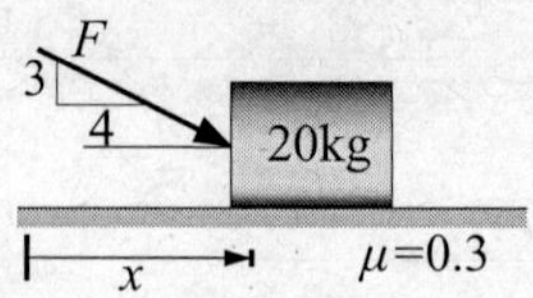

FIGURE 11.11(a) |

Solution. Free body diagram of moving block is shown in Figure 11.11(b).

$$\sum F_y = 0 \Rightarrow R - F\sin\theta - mg = 0$$

or $$R = F\sin\theta + mg = 50x^2 \times \frac{3}{5} + 20 \times 9.81$$

or $$R = 30x^2 + 196.2$$

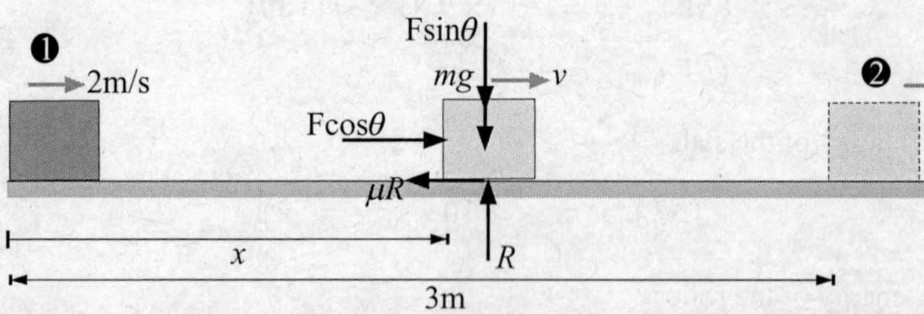

FIGURE 11.11(b) |

Net force along the direction of motion

$$F_H = (F\cos\theta - \mu R) = 50x^2 \times \frac{4}{5} - 0.3(30\,x^2 + 196.2)$$

or $$F_H = 31x^2 - 58.86$$

Applying work−energy equation

Work done by force F_H, U_{1-2} is equal to change kinetic energy ΔT:

$$\int_0^3 F_H dx = \frac{1}{2}mv_2^2 - \frac{1}{2}mv_1^2$$

or $$\int_0^3 (31x^2 - 58.86)dx = \frac{1}{2}mv_2^2 - \frac{1}{2}mv_1^2$$

or $$31 \times \frac{3^3}{3} - 58.86 \times 3 = \frac{1}{2} \times 20v_2^2 - \frac{1}{2} \times 20 \times 2^2$$

$$v_2 = 3.77\text{m/s}$$

PROBLEM 11.12

A 5 kg block is attached to a cable as shown in Figure 11.12(a). If the cable is cut, determine (a) the maximum compression of the spring, (b) the maximum velocity of the block and (c) draw the graph between velocity of the block and x.

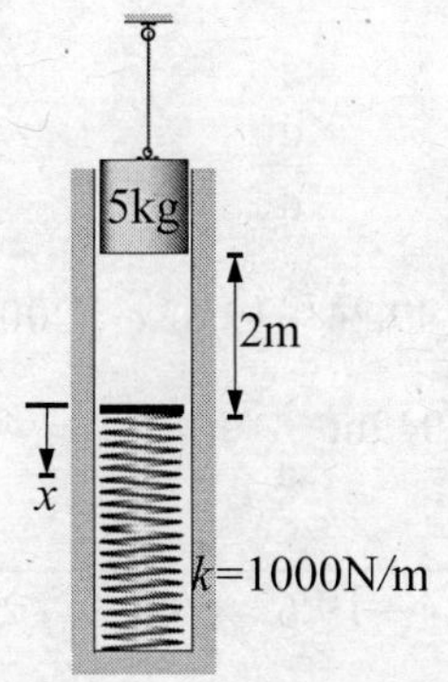

FIGURE 11.12(a)

Solution. Let the datum for potential energy of the block be its initial position and spring is undeformed. Therefore, the initial value of the potential energy of the block and spring is $V_1 = 0$. After the block has moved vertically a distance $h + x$, the spring has compressed by x [Fig. 11.12(b)]. Hence, the potential energy of the block and the spring is

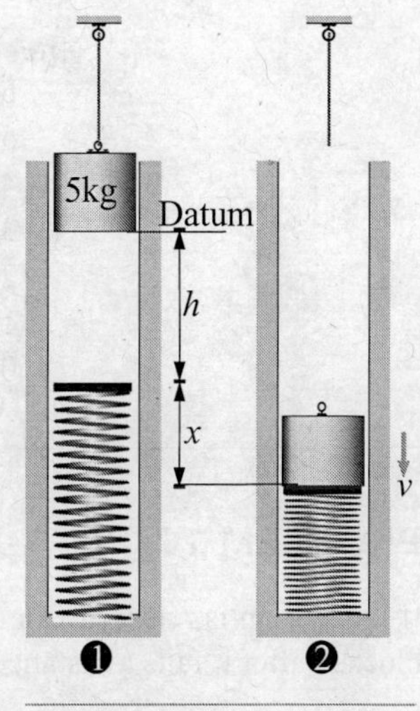

FIGURE 11.12(b)

$$V_2 = \frac{1}{2}kx^2 - mg(h + x)$$

The kinetic energy of the block in position 2 is

$$T_2 = \frac{1}{2}mv^2$$

Initially, $T_1 + V_1 = 0$; since $T + V$ is constant for a conservative system, $T_1 + V_1 = T_2 + V_2$:

$$0 = \frac{1}{2}mv^2 - mg(h + x) + \frac{1}{2}kx^2$$

or

$$\frac{1}{2}mv^2 + \frac{1}{2}kx^2 - mg(h + x) = 0$$

Substituting $m = 5$ kg, $k = 1000$ N/m and $h = 2$ m, we have

$$\frac{1}{2}5v^2 + \frac{1}{2}1000x^2 - 5 \times 9.81(2 + x) = 0$$

or

$$v = \sqrt{39.24 + 19.62x - 200x^2}$$

(a) $v = 0$, at the maximum deformation of the spring

$$200x^2 - 19.62x - 39.24 = 0$$

Solving the quadratic equation, we have

$$x = \frac{19.62 \pm \sqrt{(-19.62)^2 + 4 \times 200 \times 39.24}}{2 \times 200}$$

$$x = 0.49\text{m or } x = 49\text{cm}$$

(b) Condition for maximum velocity

$$\frac{dv}{dx} = 0$$

or
$$\frac{d}{dx}(39.24 + 19.62x - 200x^2) = 0$$

$$x = 0.049\text{m} \quad \text{or} \quad x = 4.9\text{cm}$$

The maximum velocity of the block is

$$v_{max} = \sqrt{39.24 + 19.62 \times 0.049 - 200 \times 0.049^2}$$

$$v_{max} = 6.3\text{m/s}$$

(c) Figure 11.12(c) shows the graph between v and x

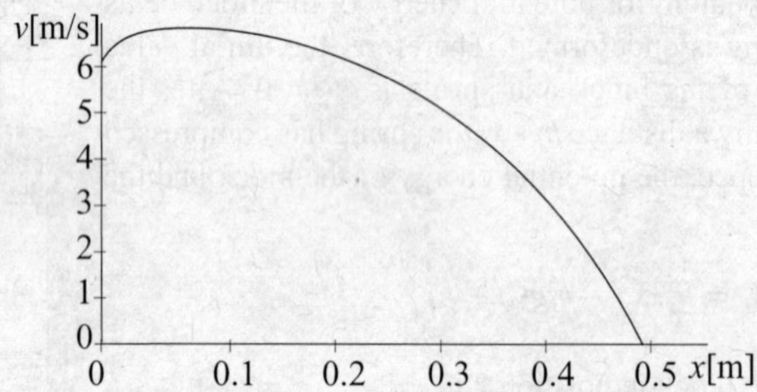

FIGURE 11.12(c) |

PROBLEM 11.13

If the system is released from rest in the configuration shown in Figure 11.13(a) find the velocity of block A after it falls a distance 1.5 m. Neglect the mass of pulley.

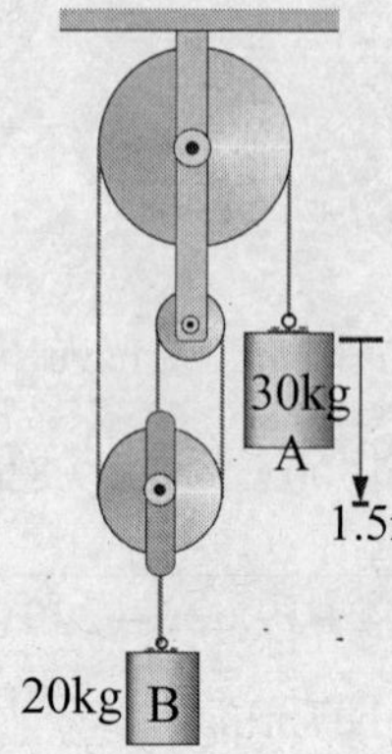

FIGURE 11.13(a) |

Solution. Let the datum for potential energy of blocks A and B be its initial position. The two positions of the system are shown in Figure 11.13(b).

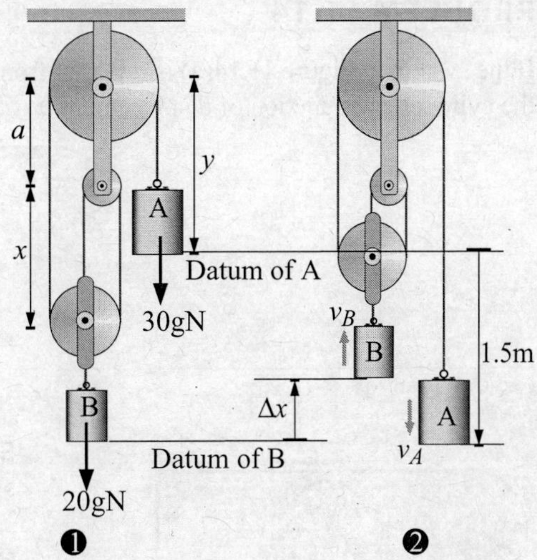

FIGURE 11.13(b)

Change in potential energy of block A

$$(\Delta V)_A = -30 \times 9.81 \times 1.5$$

Change in potential energy of block B

$$(\Delta V)_B = 20 \times 9.81 \times h$$

Therefore,

$$\Delta V = (\Delta V)_A + (\Delta V)_B = -30 \times 9.81 \times 1.5 + 20 \times 9.81 \times h \quad \text{(i)}$$

Change in potential energy of block A

$$(\Delta T)_A = \frac{1}{2} \times 30 v_A^2$$

Change in potential energy of block b

$$(\Delta T)_B = \frac{1}{2} \times 30 v_B^2$$

Therefore,

$$\Delta T = (\Delta T)_A + (\Delta T)_B = \frac{1}{2} \times 30 v_A^2 + \frac{1}{2} \times 30 v_B^2 \quad \text{(ii)}$$

Since $T + V$ is constant for conservative system, $\Delta T + \Delta V = 0$. Thus,

$$\frac{1}{2} \times 30 v_A^2 + \frac{1}{2} \times 30 v_B^2 - 30 \times 9.81 \times 1.5 + 20 \times 9.81 \times h = 0$$

or

$$15 v_A^2 + 15 v_B^2 = 441.45 - 196.2h \quad \text{(iii)}$$

From the geometry of the figure, length of the rope

$$l = y + 3x + a + arc$$

$$y + 3x = \text{constant} \Rightarrow v_A = -3v_B \text{ and}$$

$$\Delta y + 3\Delta x = 0 \Rightarrow h \downarrow = \Delta x = \frac{\Delta y}{3} = \frac{1.5}{3} = 0.5\text{m}\uparrow$$

From Equation (iii), we have

$$15 v_A^2 + 10\left(-\frac{v_A}{3}\right)^2 = 441.45 - 196.2 \times 0.5$$

$$v_A = 4.61 \text{m/s}$$

PROBLEM 11.14

If the system in Figure 11.14(a) is released from rest in the configuration shown, find the velocity v of the cylinder as a function of displacement h.

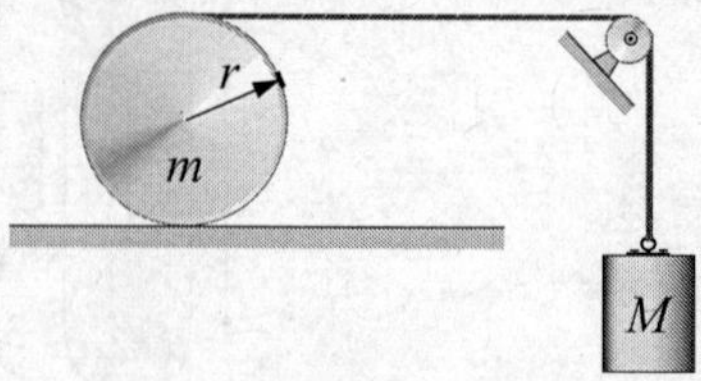

FIGURE 11.14(a)

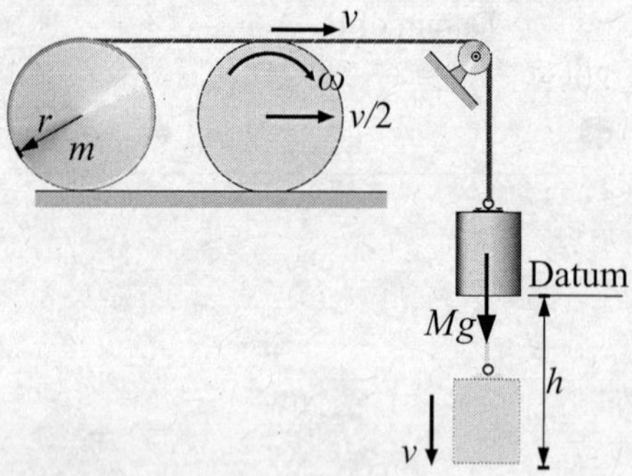

FIGURE 11.14(b)

Solution. Positions 1 and 2 of system are shown in Figure 11.14(b).

Position 1

Potential energy $V_1 = 0$

Kinetic energy $T_1 = 0$

Position 2

Potential energy $V_2 = -Mgh$

Kinetic energy $T_2 = \frac{1}{2}Mv^2 + \frac{1}{2}m\left(\frac{v}{2}\right)^2 + \frac{1}{2}I\omega^2$

or $$T_2 = \frac{1}{2}Mv^2 + \frac{1}{2}m\left(\frac{v}{2}\right)^2 + \frac{1}{2}\left(\frac{1}{2}mr^2\right)\left(\frac{v/2}{r}\right)^2$$

or $$T_2 = \frac{1}{2}Mv^2 + \frac{1}{8}mv^2 + \frac{1}{16}mv^2$$

Applying the principle of conservation of energy,

$$V_1 + T_1 = V_2 + T_2$$

or $$0 = -Mgh + \frac{1}{2}Mv^2 + \frac{1}{8}mv^2 + \frac{1}{16}mv^2$$

From which

$$v = \sqrt{\frac{16Mgh}{8M + 3m}}$$

PROBLEM 11.15

The dynamometer shown in Figure 11.15(a) is used to measure the power output of an engine. When the engine is operating at 300 rpm, the readings of the two spring scales are 60 N and 150 N, respectively. Determine the power being developed by the engine.

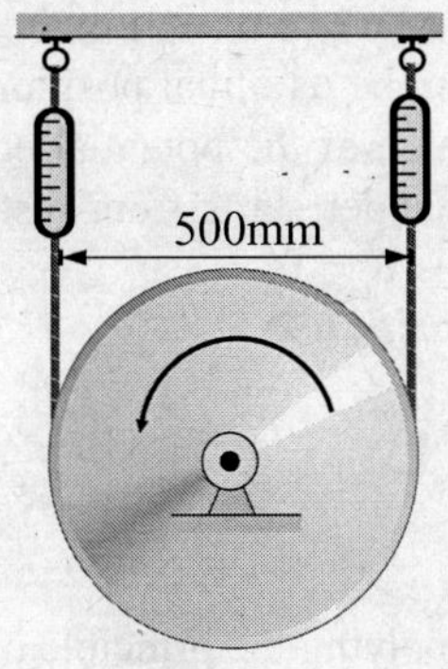

FIGURE 11.15(a) |

Solution. Given: $F_1 = 150$ N, $F_2 = 60$ N, rpm = 300 and $D = 500$ mm.
Angular velocity of brake-drum is

$$\omega = \frac{2\pi N}{60}$$

$$\omega = \frac{2\pi \times 300}{60} = 10\pi \text{rad/s}$$

Peripheral velocity

$$v = \omega r = 10\pi \times \frac{500}{2 \times 1000}$$

$$v = 7.85 \text{m/s}$$

Power developed by the engine

$$P = (\Sigma F)v$$

$$P = (F_1 - F_2)v = (150 - 60) \times 7.85$$

$$P = 706.5 \text{W}$$

PROBLEM 11.16

If the system in Figure 11.16(a) is released from rest in the configuration shown, find the velocity v of the cylinder A after it falls by a distance h. Neglect the friction.

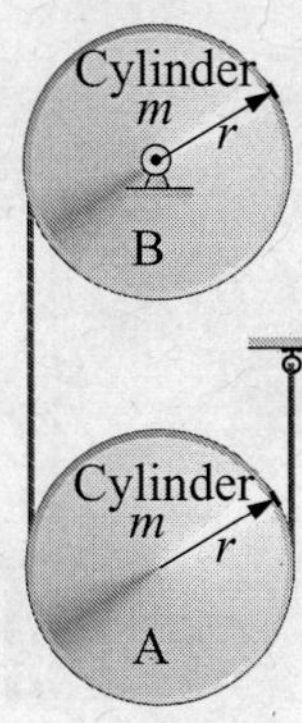

FIGURE 11.16(a) |

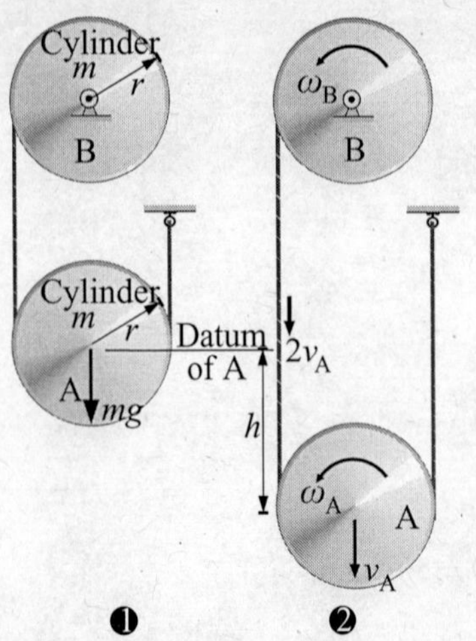

FIGURE 11.16(b) |

Solution. Let the datum for potential energy of the system be its initial position [Fig. 11.16(b)]. Therefore, the initial value of the potential energy of the system $V_1 = 0$. Since the cylinder starts from rest, the initial kinetic energy in position 1 is $T_1 = 0$.

Position 2

$$T_2 = \frac{1}{2}(I_B)_o\omega_B^2 + \left[\frac{1}{2}(I_A)_o\omega_A^2 + \frac{1}{2}mv_A^2\right]$$

$$V_2 = -mgh$$

Applying the principle of conservation of energy,

$$V_1 + T_1 = V_2 + T_2$$

or
$$0 = -mgh + \frac{1}{2}(I_B)_o\omega_B^2 + \left[\frac{1}{2}(I_A)_o\omega_A^2 + \frac{1}{2}mv_A^2\right]$$

or
$$\frac{1}{2}\times\frac{1}{2}mr^2\omega_B^2 + \left[\frac{1}{2}\times\frac{1}{2}mr^2\omega_A^2 + \frac{1}{2}mv_A^2\right] = mgh$$

since
$$\omega_B = 2\omega_A = 2\left(\frac{v_A}{r}\right),$$

$$\frac{1}{2}\times\frac{1}{2}mr^2\left(2\times\frac{v_A}{r}\right)^2 + \left[\frac{1}{2}\times\frac{1}{2}mr^2\left(\frac{v_A}{r}\right)^2 + \frac{1}{2}mv_A^2\right] = mgh$$

or
$$mv_A^2 + \frac{1}{4}mv_A^2 + \frac{1}{2}mv_A^2 = mgh$$

From which

$$v_A = \sqrt{\frac{4gh}{7}}$$

PROBLEM 11.17

If the system in Figure 11.17(a) is released from rest in the configuration shown, find the velocity v of the falling weight A as a function of its displacement h. The mass of the pulley is m and radius of gyration is k.

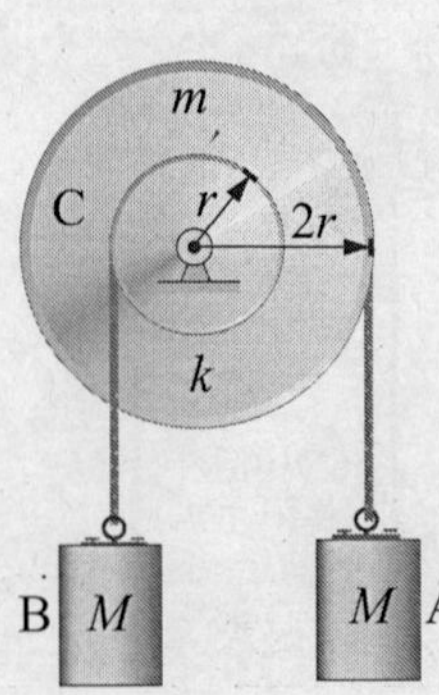

FIGURE 11.17(a) |

Solution. Let the datum for potential energy of the weights A and B be its initial position [Fig. 11.17(b)]. Therefore, the initial value of the potential energy of the weight is

$$V_1 = (V_A)_1 + (V_B)_1 = 0$$

After weight A has moved a distance h down and weight B has moved a distance h_1 up, the potential energy of the system is

$$V_2 = (V_A)_2 + (V_B)_2 = -Mgh + Mgh_1$$

Since the weights A and B start from rest, the initial kinetic energy in position 1 is $T_1 = 0$.

The kinetic energy of system in position 2 is

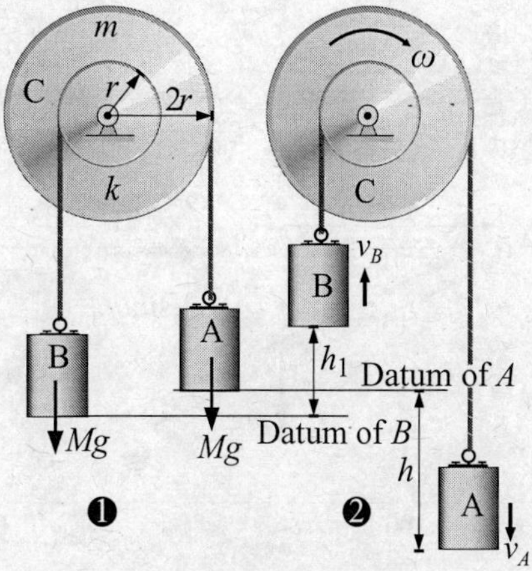

FIGURE 11.17(b)

$$T_2 = \frac{1}{2} Mv_A^2 + \frac{1}{2} Mv_B^2 + \frac{1}{2} \omega^2$$

where v_A is the velocity of weight A, v_B is the velocity of weight B and ω is the angular velocity of the pulley.

Now applying the principle of conservation of energy,

$$T_1 + V_1 = T_2 + V_2$$

or

$$0 = \frac{1}{2} Mv_A^2 + \frac{1}{2} Mv_B^2 + \frac{1}{2} \omega^2 - Mgh + Mgh_1 \qquad \text{(i)}$$

h and h_1 may be related as

$$\theta = \frac{h}{2r} = \frac{h_1}{r} \Rightarrow h = 2h_1$$

v_A and v_B may be related as

$$\omega = \frac{v_A}{2r} = \frac{v_B}{r} \Rightarrow v_A = 2v_B$$

Equation (i) may be written as

$$\frac{1}{2} Mv_A^2 + \frac{1}{2} M\left(\frac{v_A}{2}\right)^2 + \frac{1}{2} mk^2\left(\frac{v_A}{2r}\right)^2 = Mgh - Mg\left(\frac{h}{2}\right)$$

From which

$$v_A = \sqrt{\frac{4Mgh}{5M + m\left(\frac{k}{r}\right)^2}}$$

PROBLEM 11.18

A slender prismatic bar of mass m and length l is released from rest in the horizontal position and allowed to swing about A under the influence of gravity [Fig. 11.18(a)]. Find the angular velocity of the bar as a function of θ.

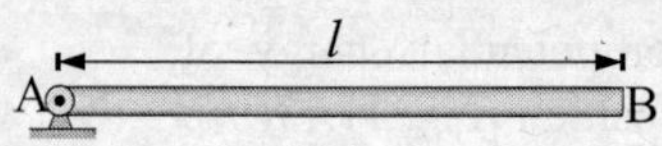

FIGURE 11.18(a) |

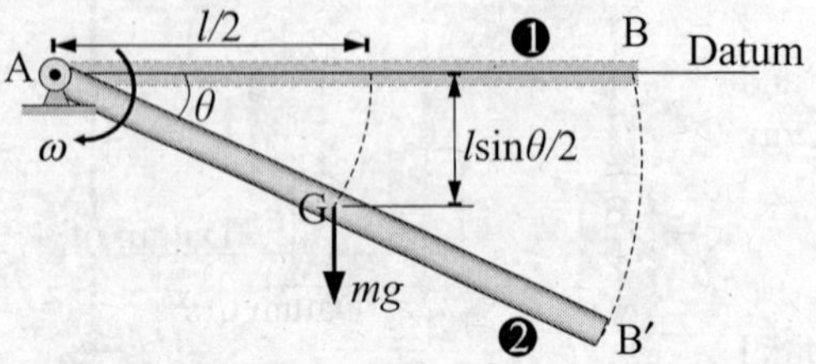

FIGURE 11.18(b) |

Solution. Let the datum for potential energy of the rod be its initial position. Therefore, the initial value of the potential energy of the rod is $V_1 = 0$. [Fig. 11.18(b)]. After the rod has moved by an angle θ, its centre of mass has descended a distance $(l/2)\sin\theta$. Hence, the potential energy of the rod after moving angle θ down is

$$V_2 = -mg\frac{l}{2}\sin\theta$$

Since the rod starts from rest, the initial kinetic energy is $T_1 = 0$.

The kinetic energy of rod at position 2 is

$$T_2 = \frac{1}{2}I_A\omega^2,$$

where I_A is the moment of inertia about A and ω angular velocity of the rod. Initially $T_1 + V_1 = 0$, since $T + V$ is constant for a conservative system, we have

$$T_2 + V_2 = 0 \Rightarrow \frac{1}{2}I_A\omega^2 - mg\frac{l}{2}\sin\theta = 0$$

or
$$\omega = \sqrt{\frac{mgl\sin\theta}{I_A}} = \sqrt{\frac{mgl\sin\theta}{\left(\frac{1}{3}\right)ml^2}}$$

$$\omega = \sqrt{\frac{3g}{l}\sin\theta}$$

PROBLEM 11.19

When a slender 20 kg bar AB is horizontal, it is at rest and the spring is unstretched [Fig. 11.19(a)]. Determine the stiffness k of the spring so that the motion of the bar is momentarily stopped when it has rotated downward 90°.

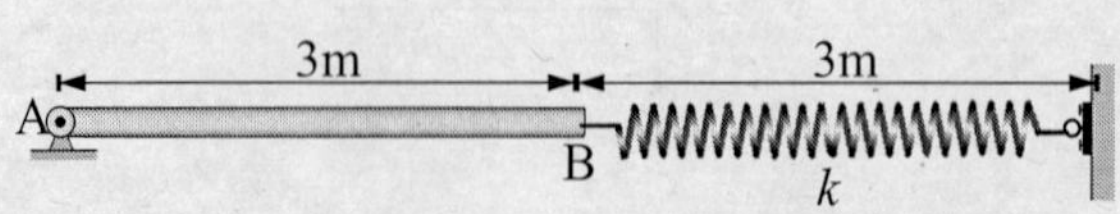

FIGURE 11.19(a) |

Solution. The datum for zero gravitational potential energy is conveniently taken through A as shown in Figure 11.19(b).

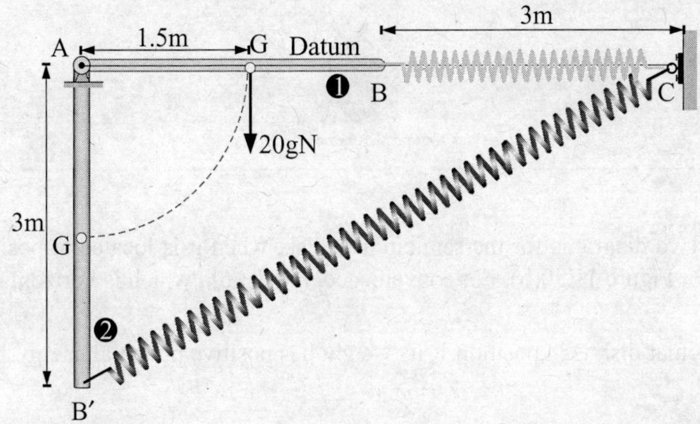

FIGURE 11.19(b) |

Position 1

Since the rod is released from rest, kinetic energy is

$$T_1 = 0$$

Gravitational potential energy $(V_g)_1 = 0$ and

Elastic (spring) potential energy $(V_S)_1 = 0$

Position 2

Given that rod is momentarily stopped so that kinetic energy $T_2 = 0$ and

Gravitational potential energy $(V_g)_2 = -20(9.81) \times 1.5$.

Elastic (spring) potential energy

$$(V_S)_2 = \frac{1}{2}k\delta^2 = \frac{1}{2}k(CB' - CB)^2$$

Applying the principle of conservation of energy

$$T_1 + (V_g)_1 + (V_S)_1 = T_2 + (V_g)_2 + (V_S)_2$$

$$0 + 0 + 0 = 0 - 20(9.81) \times 1.5 + k(CB' - CB)^2$$

or $$\frac{1}{2}k(\sqrt{6^2 + 3^2} - 3)^2 = 20(9.81) \times 1.5$$

or $$k = \frac{2 \times 20(9.81) \times 1.5}{(\sqrt{6^2 + 3^2} - 3)^2}$$

From which

$$k = 42.99\text{N/m}$$

PROBLEM 11.20

A semicircular disk of mass m and radius r is released from rest in the position shown in Figure 11.20(a). Assuming that the semicircular disk rolls without sliding, determine its angular velocity after it has rolled through 90°.

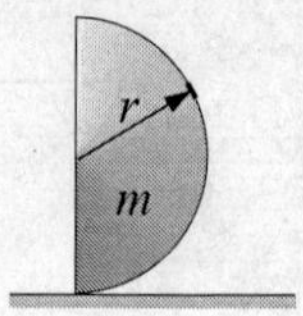

FIGURE 11.20(a) |

Solution. Two diagrams for the semicircular disk, when it is located at position 1 and position 2 are shown in Figure 11.20(b). For convenience, the datum, which is horizontally fixed, passes through point A.

When the semicircular disk is in position 1, its weight has positive potential energy. Thus,

$$V_1 = mgr$$

When the disk is in position 2, the potential energy of disk is

$$V_2 = mg\left(r - \frac{4r}{3\pi}\right)$$

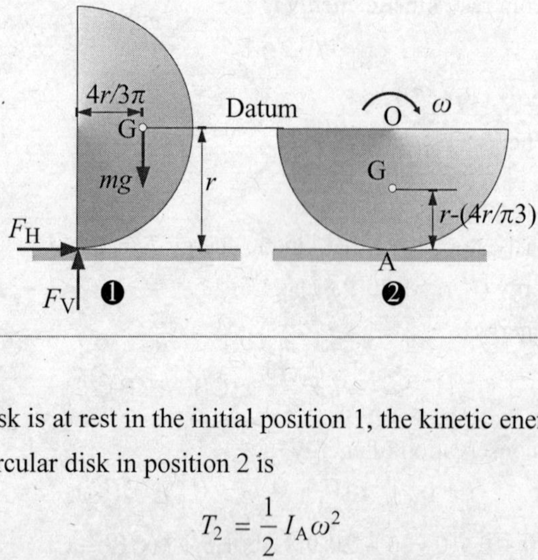

FIGURE 11.20(b) |

Since the semicircular disk is at rest in the initial position 1, the kinetic energy $T_1 = 0$.

Kinetic energy of semicircular disk in position 2 is

$$T_2 = \frac{1}{2} I_A \omega^2$$

Applying the principle of conservation of energy,

$$V_1 + T_1 = V_2 + T_2$$

$$mgr + 0 + mg\left(r - \frac{4r}{3\pi}\right) + \frac{1}{2} I_A \omega^2$$

$$\omega = \sqrt{\frac{2mg\left(\frac{4r}{3\pi}\right)}{I_A}} \qquad \text{(i)}$$

Moment of inertia of semicircular disk about O is

$$I_o = \frac{1}{2} mr^2$$

By applying theorem of parallel axis

$$I_G = I_o - m\text{OG}^2$$

and moment of inertia about A

$$I_A = I_G + m\text{AG}^2$$

$$I_A = I_o - m\text{OG}^2 + m\text{AG}^2$$

or $$I_A = \frac{1}{2}mr^2 - m\left(\frac{4r}{3\pi}\right)^2 + m\left(r - \frac{4r}{3\pi}\right)^2$$

or $$I_A = \frac{(9\pi - 16)}{6\pi}mr^2 \tag{ii}$$

From Equations (i) and (ii), we have

or $$\omega = \sqrt{\frac{2mg\left(\frac{4r}{3\pi}\right)}{I_A}} = \sqrt{\frac{2mg\left(\frac{4r}{3\pi}\right)}{\left[\frac{(9\pi - 16)}{6\pi}\right]mr^2}}$$

or $$\omega = \frac{16g}{(9\pi - 16)r}$$

PROBLEM 11.21

A rope is wrapped around a circular body of mass m, radius r and radius of gyration k and released from rest as shown in Figure 11.21(a). Find the speed of its centre O after it has moved h down from rest.

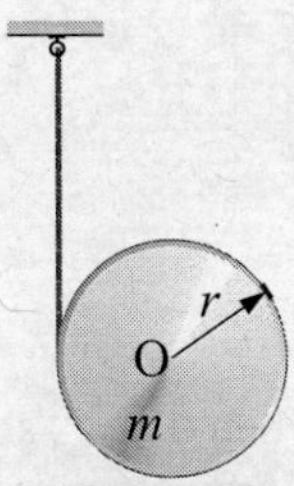

FIGURE 11.21(a) |

Solution. A body is released from rest so that $v = 0$ and $\omega = 0$ so that $T_1 = 0$. When the body is in position 2, the centre of mass G is located below the datum so that gravitational potential energy is negative [Fig. 11.21(b)]. Thus

$$V_2 = -mgh$$

The body has general plane motion, so that its kinetic energy is

$$T_2 = \frac{1}{2}mv^2 + \frac{1}{2}I\omega^2,$$

where v is the linear velocity of O and ω is the angular velocity of the body in position 2.

Applying the principle of conservation of energy ($V_1 + T_1 = V_2 + T_2$),

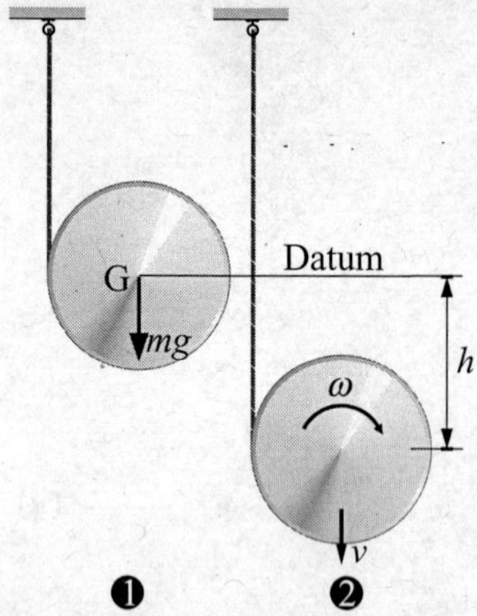

FIGURE 11.21(b)

$$0 + 0 = -mgh + \frac{1}{2}mv^2 + \frac{1}{2}I\omega^2$$

or

$$\frac{1}{2}mv^2 + \frac{1}{2}I\omega^2 = mgh$$

or

$$\frac{1}{2}mv^2 + \frac{1}{2} \times \frac{1}{2}mk^2\left(\frac{v}{r}\right)^2 = mgh$$

$$v = \sqrt{\frac{2gh}{1 + \left(\frac{v}{r}\right)^2}}$$

If a is the acceleration of the body, we have

$$v^2 = u^2 + 2as \Rightarrow a = \frac{v^2}{2h} = \frac{1}{2h} \times \frac{2gh}{1 + \left(\frac{k}{r}\right)^2}$$

$$a = \frac{g}{1 + \left(\frac{k}{r}\right)^2}$$

If t is the time taken to descend h, we have

$$s = ut + \frac{1}{2}at^2 \quad \text{or} \quad t = \sqrt{\frac{2h}{a}} = \sqrt{\frac{\frac{2h}{g}}{1 + \left(\frac{k}{r}\right)^2}}$$

$$t = \sqrt{\frac{2h}{g}\left[1 + \left(\frac{k}{r}\right)^2\right]}$$

PROBLEM 11.22

A bar is released from rest in the position (when $\theta = 0°$) shown in Figure 11.22(a) on a smooth plane. Determine the angular velocity of the bar as a function of θ.

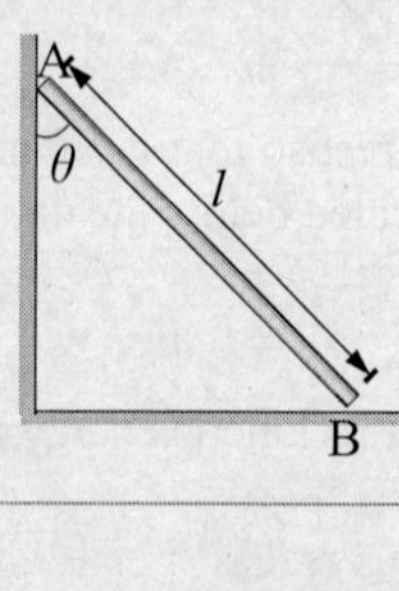

FIGURE 11.22(a)

Solution. Figure 11.22 (b) shows the bar in the release ($\theta = 0$) position 1, and in an arbitrary position 2 defined by an angle θ. Using the horizontal plane as the datum for potential energy, the potential energies of the bar in the two positions are

$$V_1 = mg\frac{l}{2}$$

$$V_2 = mg\frac{l}{2}\cos\theta$$

Initial kinetic energy $T_1 = 0$

and

Kinetic energy in position 2

$$T_2 = \frac{1}{2}I_o\omega^2$$

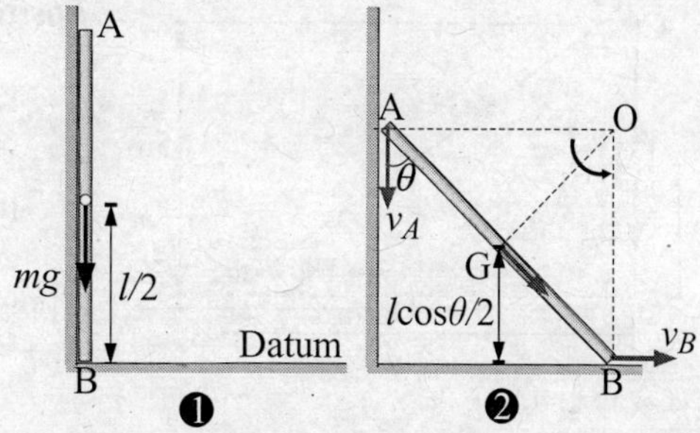

FIGURE 11.22(b) |

O is the point about which the bar rotates (it is also called instantaneous centre).

or
$$T_2 = \frac{1}{2}(I_G + m\text{OG}^2)\omega^2$$

Applying the principle of conservation of energy,

$$T_1 + V_1 = T_2 + V_2$$

or
$$mg\frac{l}{2} + 0 = mg\frac{l}{2}\cos\theta + \frac{1}{2}(I_G + m\text{OG}^2)\omega^2$$

or
$$mg\frac{l}{2}(1-\cos\theta) = \frac{1}{2}\left[\frac{1}{12}ml^2 + m\left(\frac{l}{2}\right)^2\right]\omega^2$$

$$\omega = \sqrt{\frac{3g}{l}(1-\cos\theta)}$$

PROBLEM 11.23

Two blocks A and B are connected by a string that has a length of 14 m and passes over a small pulley C [Fig. 11.23(a)]. If the system is released from rest when $x = 3$ m, determine the velocity of A and B when A hits the floor.

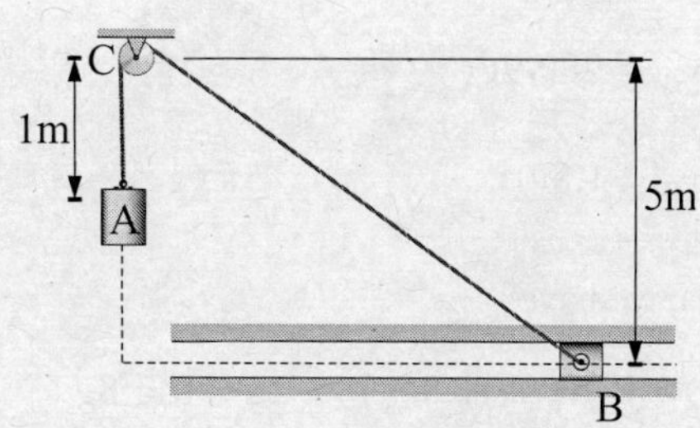

FIGURE 11.23(a) |

Solution.

Position I [Fig. 11.23(a)]

Kinetic energy $T_1 = 0$

and

Potential energy $V_1 = mg \times 4 + 0$

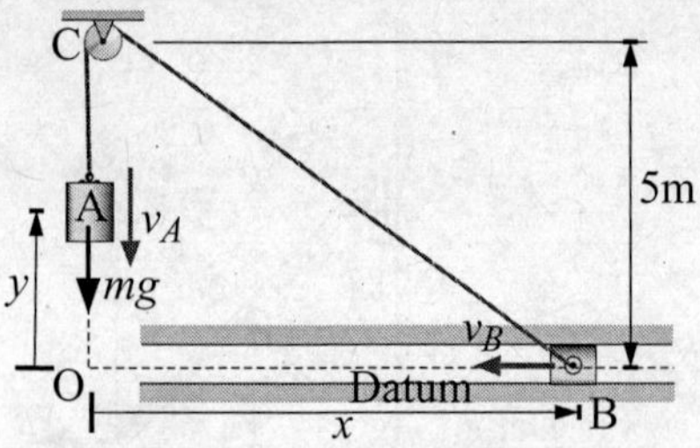

FIGURE 11.23(b)

Position II [Fig. 11.23(b)]

$$\text{Kinetic energy } T_2 = \frac{1}{2}mv_A^2 + \frac{1}{2}mv_B^2$$

$$\text{Potential energy } V_2 = mgy$$

Applying the principle of conservation of energy ($T_1 + V_1 = T_2 + V_2$),

$$mg \times 4 = \frac{1}{2}mv_A^2 + \frac{1}{2}mv_B^2 + mgy$$

or $$v_A^2 + v_B^2 + 2gy = 8g \qquad \text{(i)}$$

From the geometry, length of the rope

$$14 = \sqrt{x^2 + (5-y)^2} + (5-y)$$

or $$9 = \sqrt{x^2 + y^2 - 10y + 25} - y \qquad \text{(ii)}$$

at $$y = 0; 9 = \sqrt{x^2 + 25} \Rightarrow x = 7.48\text{m}$$

Differentiate equation (ii) with respect to t, we have

$$0 = \frac{2xv_B + 2yv_A - 10v_A}{2\sqrt{x^2 + y^2 - 10y + 25}} - v_A = 0$$

When block A hits the ground, substituting $y = 0$ and $x = 7.48$m,

$$0 = \frac{2 \times 7.48v_B - 10v_A}{2\sqrt{(7.48)^2 + 25}} - v_A = 0$$

or $$15.62v_B = 28.99v_A$$

or $$v_B = 1.8v_A \qquad \text{(iii)}$$

From Equations (i), (ii) and (iii), we have

$$v_A^2 + (1.8v_A)^2 + 2g \times 0 = 8g$$

or $$v_A = \sqrt{\frac{8 \times 9.87}{1 + (1.8)^2}}$$

$$v_A = 4.30\text{m/s}$$

From the equation,

$$v_B = 1.8 \times 4.30$$

$$v_B = 7.74 \text{m/s}$$

CHAPTER HIGHLIGHTS

- The work done by a force on a moving body is the product of the force and the distance moved in the direction of the force. $U_{1-2} = \int_{r_1}^{r_2} \mathbf{F} \cdot dr$

  Work of a constant force in Rectilinear Motion $U_{1-2} = F \cos \theta (x_2 - x_1)$

  Work of the Weight $U_{1-2} = -W \Delta y$, Work of a spring force $U_{1-2} = \frac{1}{2} k \left(x_2^2 - x_1^2\right)$

  Work of a Couple $U = \int_{\theta_1}^{\theta_2} M d\theta$
- Power is the time rate at which work is performed. $P = \frac{dU}{dt}$, $P = \mathbf{F} \cdot \mathbf{v}$ and $P = \tau \cdot \omega$
- Efficiency of a machine is the ratio of the output of useful power produced by the machine to the input of power supplied to the machine. $\eta = \frac{\text{Power output}}{\text{Power iutput}}$
- Energy is defined as the capacity to do work. Mechanical energy can be classified into three types of energy (i) kinetic energy, (ii) potential energy and (iii) strain energy.
- Kinetic Energy of Rigid body $T = \frac{1}{2} m v_G{}^2 + \frac{1}{2} I_G \omega^2$
- The work done by a system of forces acting on a body during a displacement is equal to the change in kinetic energy of the body during the same displacement.

$$U_{1-2} = \frac{1}{2} m v_2{}^2 - \frac{1}{2} m v_1{}^2$$

- Conservative force: A force is said to be *conservative* if the work done by the force on a system that moves between two configurations is independent of the path the system takes.
- Nonconservative Force: A force is said to be *nonconservative* if the work done by a force on a system depends on the path the system follows.
- Conservation of Energy: When a rigid body, or a system of rigid bodies, moves under the action of conservative forces, the sum of the kinetic energy and the potential energy of the system remains constant. $T + V =$ constant or $\Delta T + \Delta V = 0$

REVIEW QUESTIONS

Q1. Write the work−energy equation for a curvilinear motion of a particle.

Q2. What do you mean by conservative and non-conservative systems?

Q3. State the law of conservation of energy.

Q4. Derive an expression for kinetic energy of a rigid body.

Q5. Define power and efficiency, and their formula.

Q6. Using the principle of conservation of mechanical energy, write the equation for the angular velocity of a simple pendulum. By differentiating this equation, obtain the equation for the angular acceleration of the pendulum.

$$\left[\begin{array}{l} \text{Ans; } \omega = \sqrt{\dfrac{2g}{l}(\cos\theta - \cos\theta_o)},\ a = \dfrac{2g}{l}\sin\theta, \\ \text{where } \theta_0 \text{ is maximum angle the pendulum arm forms with the viertical} \end{array}\right].$$

Q7. A particle P moves on the path $x = t, y = t^2, z = t^3$, where t is a parameter. The motion is resisted by a force F, defined by its (x, y, z) projection

$$F_x = -3\frac{dx}{dt},\ F_y = -3\frac{dx}{dt},\ F_z = -3\frac{dz}{dt},$$

where length is in meters and force is in Newton. Calculate the work the resting force performs on P during the interval from $t = 1$ to $t = 3$

[Ans; −1417 J].

Q8. A weight W is suspended by a spring with constant k. (i) Derive a formula for the total potential energy V of the system relative to the position of the unstretched spring. (ii) Separate the total potential energy into potential energies of internal and external forces.

Q9. A force pulls a body along a horizontal floor. The force varies with distance x, as shown in Figure 11.Q9. (i) Determine the work done by the force in moving the body from $x = 0$ to $x = 8$ m. (ii) If the body weighs 10 N and the coefficient of sliding friction is 0.20, determine the speed of the body when it has slid 3 m, assuming that it starts from rest. (iii) Determine the speed of the body when it has slid 8 m. (iv) Determine the total distance the body travels before it again comes to rest.

[Ans; (i) work = 27 J, (ii) speed = 5.12 m/s, (iii) speed = 4.65 m/s, (iv) distance = 13.5 *m*].

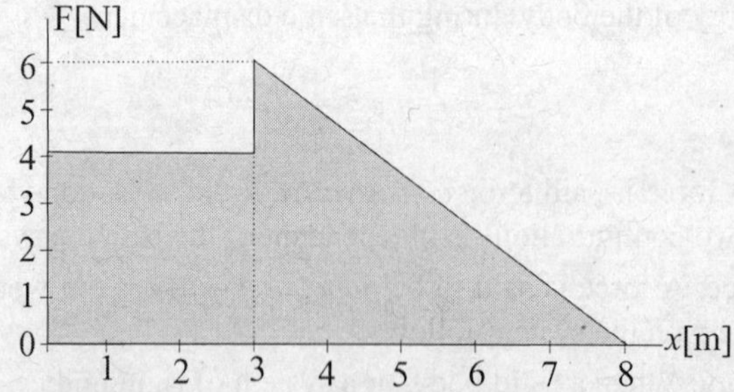

FIGURE 11.Q9 |

Q10. Three crates that rest on a horizontal floor are connected by light cords as shown in Figure 11.Q10. They move to the right under the action of a force $F = 18$ N. The coefficient of kinetic friction μ_k between the crates and the floor is 0.30. (i) Determine the work done on each of the crates as the force pulls the crates 4 m to the right. (ii) Determine the total work performed on the crates during the time. (iii) Determine the change in speed of the crates during the time they slide 4 m. (iv) What net power is being generated by the forces acting on the crates?

[Ans; (i) 0, (ii) 0, (iii) 0, (iv) 0].

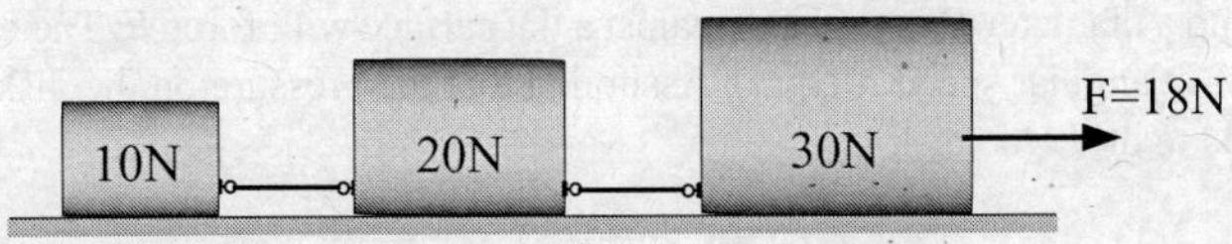

FIGURE 11.Q10 |

Q11. A thin uniform rod is welded to a hoop of radius 150 mm as shown in Figure 11.Q11. The mass of the hoop is half the mass of the rod. In position A, the hoop rolls with an angular velocity of 4 rad/s. What is the angular velocity of the hoop in position B?

[Ans; 11.75 rad/s].

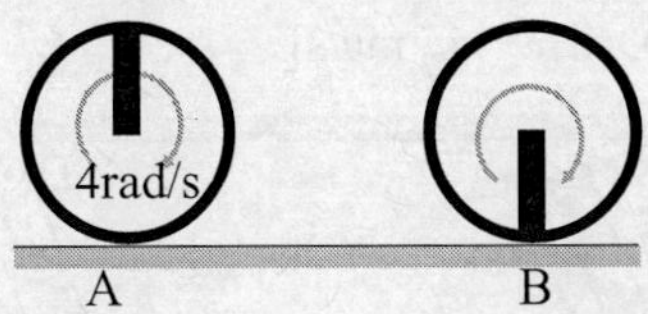

FIGURE 11.Q11 |

Q12. A block whose mass is 6 kg rests on a frictionless horizontal surface as shown in Figure 11.Q12. In this equilibrium position, the block is attached to an unstretched nonlinear spring whose force–extension relation is $F = 400 \sin 10x$ N, where x in metres, and $10x$ is in radians. The block is moved a distance $x = 0.10$ m to the right and released from rest.

(i) Determine the velocity of the block at the instant it passes through its equilibrium position. (ii) Determine the velocity of the block at $x = 0.10$ m.

[Ans; (i) 2.48 m/s, (ii) 0].

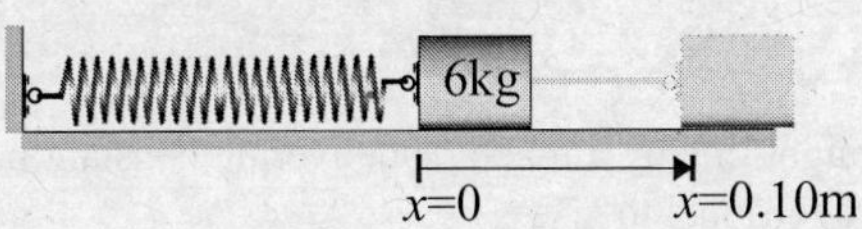

FIGURE 11.Q12 |

Q13. A pulley–cord system shown in Figure 11.Q13. The light cord is inextensible, and pulley A is frictionless and of negligible mass. Mass m_1 is a smooth cylindrical body. Assume that mass m_2 is sufficiently large that when the system is released from rest, m_2 moves downward. Show that when m_2 has moved downward a distance x, the velocities v_1 and v_2 of the masses m_1 and m_2 are

$$v_1 = \sqrt{\frac{(2m_2 - m_1)gx}{4m_2 + m_1}}, \quad v_2 = 2\sqrt{\frac{(2m_2 - m_1)gx}{4m_2 + m_1}}$$

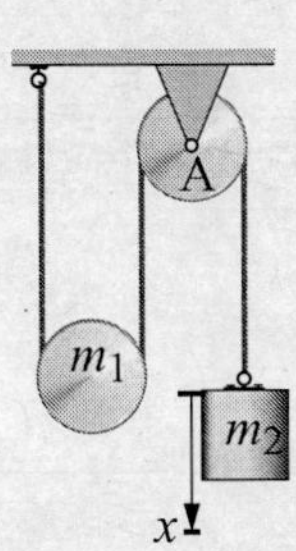

FIGURE 11.Q13 |

Q14. A rotating disk with a diameter D is pressed against a flat surface with force F. The coefficient of friction is μ. The disk rotates at angular speed N (rpm). Assuming that the pressure on the disk is uniform, show that the power required to drive the disk is

$$P = \frac{\pi\mu FDN}{90}$$

Q15. Each arm in the frame ABCD has a weight of 40 N, as shown in Figure 11.Q15. Joints A, B, C and D are frictionless. (i) Take $\theta = 0$ as the datum of potential energy. Determine the potential energy of the system as a function of θ. (*ii*) If the frame is released from rest at $\theta = 90°$, determine the maximum angular speed of the frame.

[Ans; (i) $U = 63.75(1 - \cos\theta)$, (ii) 9.147 rad/s].

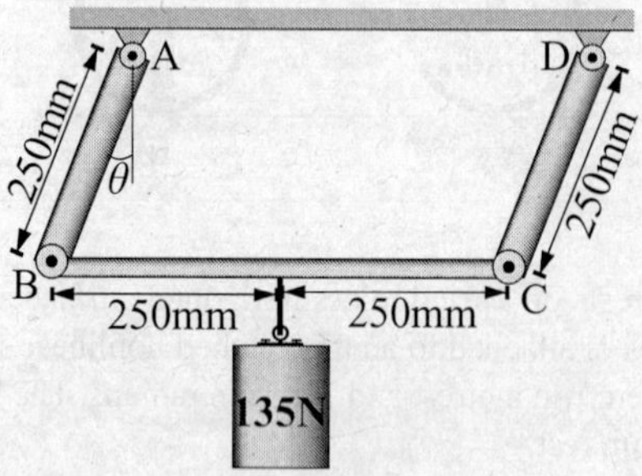

FIGURE 11.Q15 |

MULTIPLE CHOICE QUESTIONS

Q1. A body is moved along a straight line by a machine delivering constant power. The distance by which the body is moved in time t is proportional to

(A) $t^{1/2}$ (B) $t^{3/4}$ (C) $t^{3/2}$ (D) t^2

[Ans: C]

Q2. A man is pulling a bucket of water up to the roof of a building of 6 m height. The total weight of the rope is 20 N and the weight of the bucket with water is 100 N. The work done by the man is

(A) 720 N m (B) 420 N m (C) 660 N m (D) 600 N m

[Ans: C]

Q3. A uniform chain of length L and mass M is lying on a smooth table and one-third of its length is hanging vertically downward over the edge of a table (Fig. 11.MQ3). If g is acceleration due to gravity, the work required to pull the hanging part on to the table is

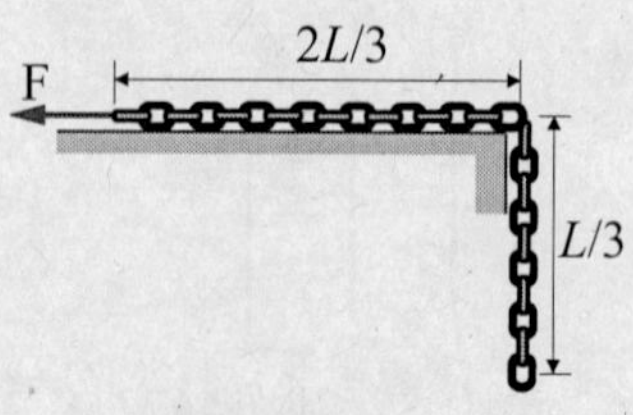

FIGURE 11.MQ3 |

(A) MgL (B) $\frac{1}{3}MgL$ (C) $\frac{1}{9}MgL$ (D) $\frac{1}{18}MgL$

[Ans: D]

Q4. A vehicle is in motion with a velocity v along a straight horizontal road. If the coefficient of friction between the wheels and the road is μ, then the shortest distance in which the vehicle can stop is

(A) $\frac{v}{2\mu g}$ (B) $\frac{v^2}{2\mu g}$ (C) $\sqrt{\frac{v^2}{2\mu g}}$ (D) $\left(\frac{v^2}{2\mu g}\right)^2$

[Ans: B]

Q5. Two objects A and B are thrown upward simultaneously with the same speed. The mass of A is greater than the mass of B. Suppose the air exerts a constant and equal force of resistance on the two bodies.

(A) The two bodies will reach the same height (B) A will go higher than B
(C) B will go higher than A (D) None of these.

[Ans: B]

Q6. A cord is wrapped around a cylinder of radius r and mass m as shown in Figure 11.MQ6. If the cylinder is released from rest, the velocity of the cylinder after it has moved through a distance h will be

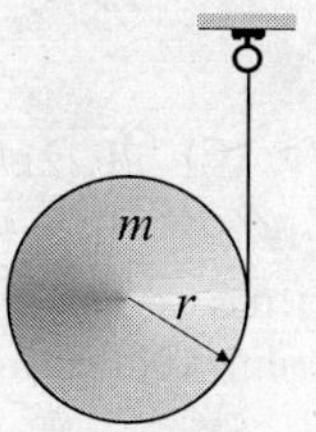

FIGURE 11.MQ6 |

(A) $\sqrt{2gh}$ (B) $\sqrt{gh}$ (C) $\sqrt{\frac{4}{3}gh}$ (D) $\sqrt{\frac{1}{3}gh}$

[Ans: C]

Q7. A particle P is projected from the earth's surface at a latitude 45° with an escape velocity v = 11. 2 km/s (Fig. 11.MQ7). The velocity direction makes an angle θ with the local vertical. The particle will escape the earth's gravitational field,

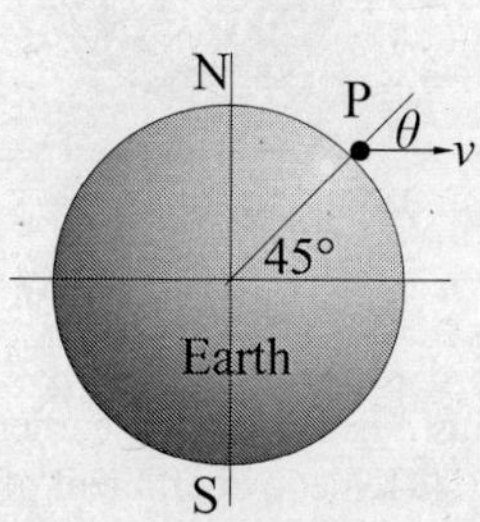

FIGURE 11.MQ7 |

(A) only when $\theta = 0$ (B) only when $\theta = 45°$
(C) only when $\theta = 90°$ (D) irrespective of the value of θ

[Ans: D]

Q8. A truck weighing 150 kN and travelling at 2 m/s impacts with a buffer spring which compresses 1.25 cm per 10 kN (Fig. 11.MQ8). The maximum compression of the spring is

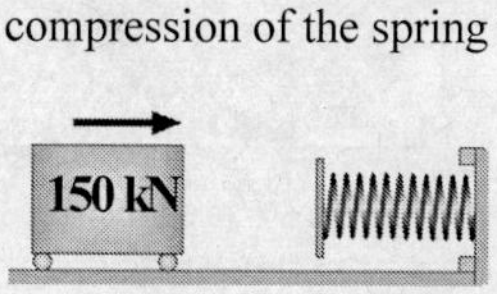

FIGURE 11.MQ8 |

(A) 20.00 cm (B) 22.85 cm (C) 27.66 cm (D) 30.00 cm

[Ans: C]

Q9. The system shown in Figure 11.MQ9 is released from rest. Neglecting the rotational inertia of body A, the velocity of body B as a function of the distance fall h is

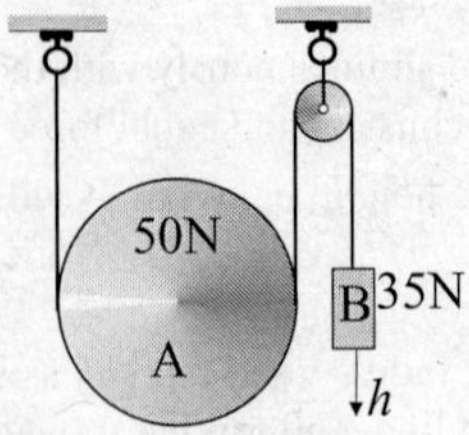

FIGURE 11.MQ9 |

(A) $\sqrt{0.42gh}$ (B) $\sqrt{gh}$ (C) $\sqrt{1.22gh}$ (D) None of these

[Ans: A]

Q10. The dynamometer shown in Figure 11.MQ10 is used to measure the power output of a small turbine. When the turbine is operating at 240 rpm, the readings of the two spring scales are 50 and 100 kN, respectively. The power developed by the turbine is

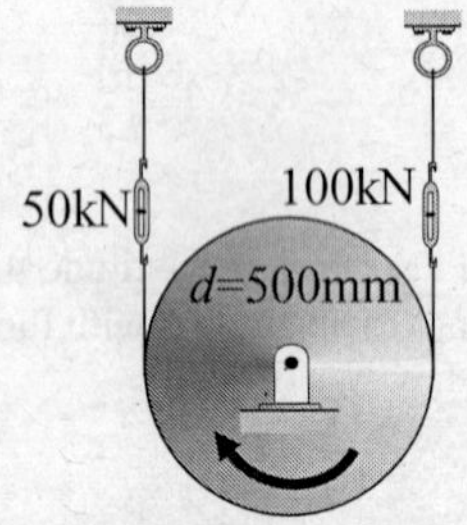

FIGURE 11.MQ10

(A) 204 kW (B) 117 kW (C) 314 kW (D) None of these

[Ans: C]

Q11. A uniform sphere of mass m and radius r is placed in a corner with an initial counter-clockwise angular velocity ω as shown in Figure 11.MQ11. Kinetic coefficient of friction between sphere and both surfaces is μ. The number of rotations before the sphere comes to rest is

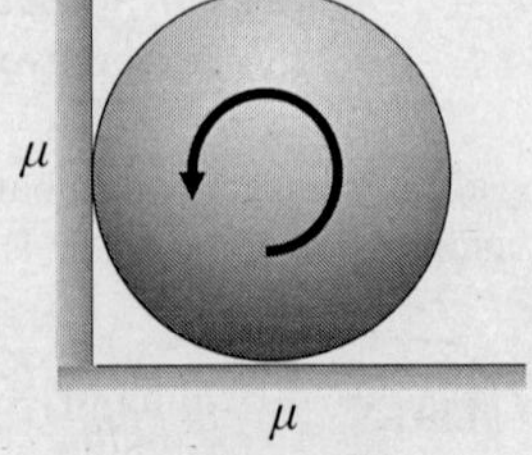

FIGURE 11.MQ11 |

(A) $n = \dfrac{\omega^2 r(1+\mu^2)}{10\mu g\pi(1+\mu)}$ (B) $n = \dfrac{\omega^2 r(1+\mu^2)}{8\mu g\pi(1+\mu)}$

(C) $n = \dfrac{\omega^2 r(1+\mu^2)}{8\mu g\pi}$ (D) None of these

[Ans: A]

Q12. The simple pendulum shown in Figure 11.MQ12 is released from rest at A with a horizontal string that swings downward under the influence of gravity. The tension in the string at an angle θ is

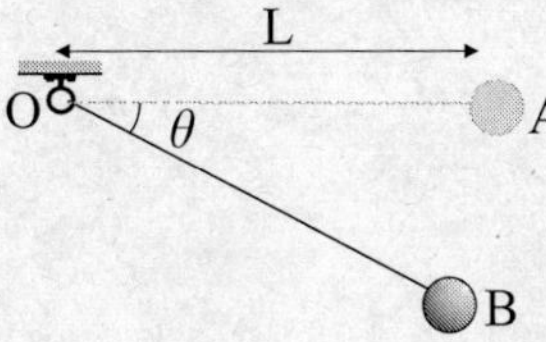

FIGURE 11.MQ12 |

(A) $T = mg\cos\theta$ (B) $T = 2mg\sin\theta$ (C) $T = 3mg\sin\theta$ (D) None of these

[Ans: C]

Q13. An ejector mechanism consists of a helical compression spring having a spring constant of $k = 98.1$ kN/m (Fig. 11.MQ13). It is pre-compressed by 100 mm from its free state. If it is used to eject a mass of 100 kg held on it, the mass will move up through a distance of

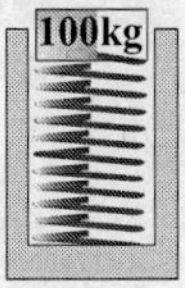

FIGURE 11.MQ13 |

(A) 100 mm (B) 500 mm (C) 981 mm (D) 1000 mm

[Ans: B]

Q14. A block has a mass of 20 kg is attached with two springs and is released from rest when $h = 0$ (Fig. 11.MQ14). The springs are unstreched when $h = 0$. The velocity of the block when it is at $h = 3$ m is

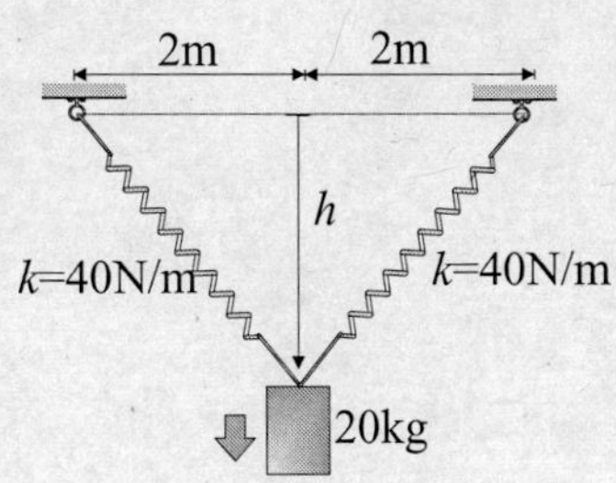

FIGURE 11.MQ14

(A) 2.47 m/s (B) 3.37 m/s (C) 6. 97 m/s (D) None of these

[Ans: C]

Q15. A small block of mass m slides without friction on a cylindrical surface of radius R (Fig. 11.MQ15). If the block is released from rest at $\theta = 0$, then the reaction between block and surface at θ is

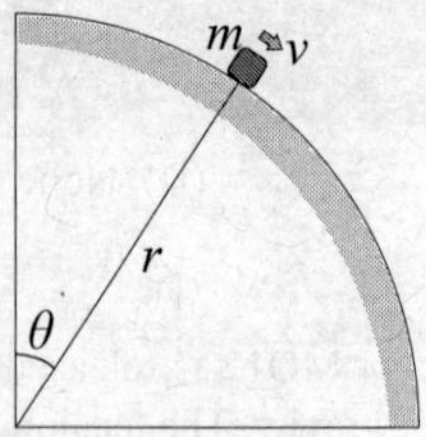

FIGURE 11.MQ15 |

(A) $mg(\cos\theta - 2)$ (B) $mg\cos\theta$ (C) $mg\sin\theta$ (D) $mg(2\cos\theta - 3)$

[Ans: D]

12

Impulse and Momentum

LINEAR MOMENTUM

It is defined as the product of the mass and the velocity of a particle. It is a vector quantity.

$$\mathbf{P} = m\mathbf{v}$$

PRINCIPLE OF MOMENTUM FOR A PARTICLE

Consider a particle of mass m acted upon by a force F. The equation of motion can be written as

$$\mathbf{F} = m\boldsymbol{a} = m\frac{\mathrm{d}\mathbf{v}}{\mathrm{d}t}$$

$$\mathbf{F} = \frac{\mathrm{d}(m\mathbf{v})}{\mathrm{d}t},$$

where m is constant.

Newton called mv the momentum of the particle

$$\mathbf{F} = \frac{\mathrm{d}\mathbf{P}}{\mathrm{d}t}$$

The force that acts on a particle is equal to the time rate of change of the momentum of the particle. This is called the principle of momentum for a particle.

If the resultant force that acts on a particle for a given time interval is zero, momentum is conserved.

$$\mathbf{F} = 0 = \frac{\mathrm{d}\mathbf{P}}{\mathrm{d}t},$$

$$\mathbf{P} = \text{constant}$$

If mass of the system is not constant, we write

$$\mathbf{F} = \frac{\mathrm{d}(m\mathbf{v})}{\mathrm{d}t} = m\frac{\mathrm{d}\mathbf{v}}{\mathrm{d}t} + \mathbf{v}\frac{\mathrm{d}m}{\mathrm{d}t}$$

For a system of particles, the total momentum of a system of particles is equal to the vector sum of the momentum of the individual particles.

$$\sum p = p_1 + p_2 + p_3 + \cdots$$

IMPULSE

It is defined as the product of force and time. Impulse is a vector quantity.

$$\mathbf{I} = \mathbf{F}\Delta t$$

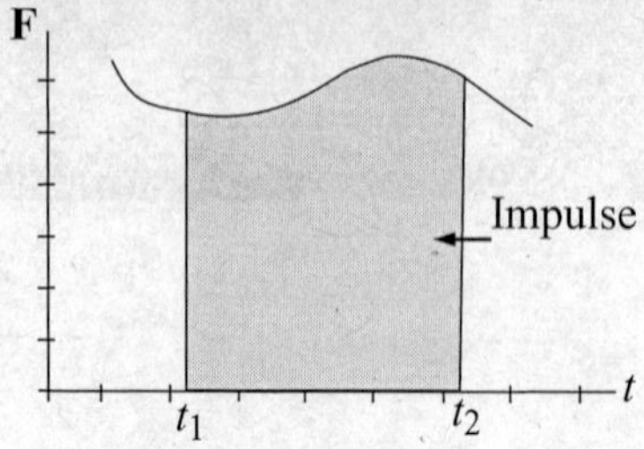

FIGURE 12.1

If force **F** varies with time, then the impulse **I** of a force **F** in a time interval $\Delta t = t_2 - t_1$ is the integral of **F** from time t_1 to t_2 (Fig. 12.1). We write

$$\mathbf{I} = \int_{t_1}^{t_2} \mathbf{F}\,\mathrm{d}t$$

PRINCIPLE OF LINEAR IMPULSE AND MOMENTUM OF A PARTICLE

The impulse of a force that acts on a particle in the time interval $t_2 - t_1$ is equal to the change of momentum of the particle in the same time interval (Fig. 12.2).

Consider a particle of mass m acted upon by a force **F**. Newton's second law may be expressed as

$$\mathbf{F} = \frac{\mathrm{d}(m\mathbf{v})}{\mathrm{d}t}$$

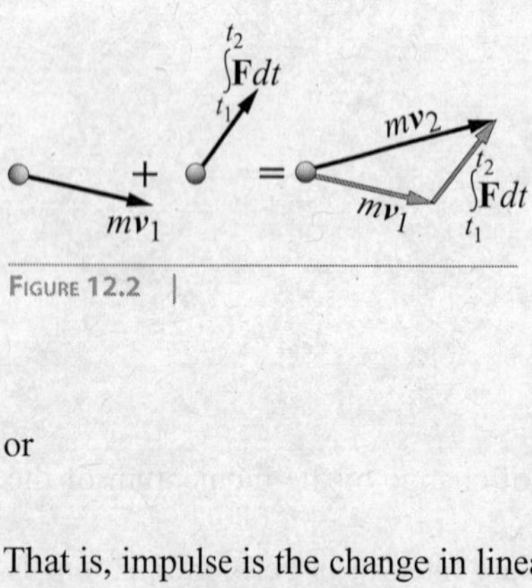

FIGURE 12.2

Multiplying both sides by $\mathrm{d}t$ and integrating from $t = t_1$ where $v = v_1$ to $t = t_2$ where $v = v_2$, we write

$$\int_{t_1}^{t_2} \mathbf{F}\,\mathrm{d}t = m\int_{v_1}^{v_2} \mathrm{d}\mathbf{v}$$

or

$$\int_{t_1}^{t_2} \mathbf{F}\,\mathrm{d}t = m\mathbf{v}_2 - m\mathbf{v}_1$$

That is, impulse is the change in linear momentum.

Transposing the last term

$$m\mathbf{v}_1 + \int_{t_1}^{t_2} \mathbf{F}\,\mathrm{d}t = m\mathbf{v}_2$$

That is the sum of initial momentum and impulse is the final momentum.

If the vectors $\mathbf{v}_1$, $\mathbf{v}_2$ and **F** are resolved into their x, y and z components, we can write the following three scalar equations:

$$m(v_x)_1 + \int_{t_1}^{t_2} \mathbf{F}_x\,\mathrm{d}t = mv(v_x)_2$$

$$m(v_y)_1 + \int_{t_1}^{t_2} \mathbf{F}_y\,\mathrm{d}t = m(v_y)_2$$

$$m(v_z)_1 + \int_{t_1}^{t_2} F_z \mathrm{d}t = m(v_z)_2$$

These equations represent the principle of linear impulse and momentum for a particle in the x, y and z directions, respectively.

CONSERVATION OF LINEAR MOMENTUM

If the resultant force on a particle is zero during an interval of time, linear momentum of the particle remains constant.

The impulse-momentum equation is

$$m\boldsymbol{v}_1 + \int_{t_1}^{t_2} \mathbf{F}\mathrm{d}t = m\boldsymbol{v}_2$$

Resultant force $\mathbf{F} = 0$, we have

$$m\boldsymbol{v}_1 = m\boldsymbol{v}_2$$

PRINCIPLE OF LINEAR IMPULSE AND MOMENTUM FOR A SYSTEM OF PARTICLES

Figure 12.3 shows a system of particles (or rigid body). Consider an ith particle of system of mass m_i. The equation of motion is

$$\mathbf{F}_i = m_i \frac{\mathrm{d}\mathbf{v}_i}{\mathrm{d}t}$$

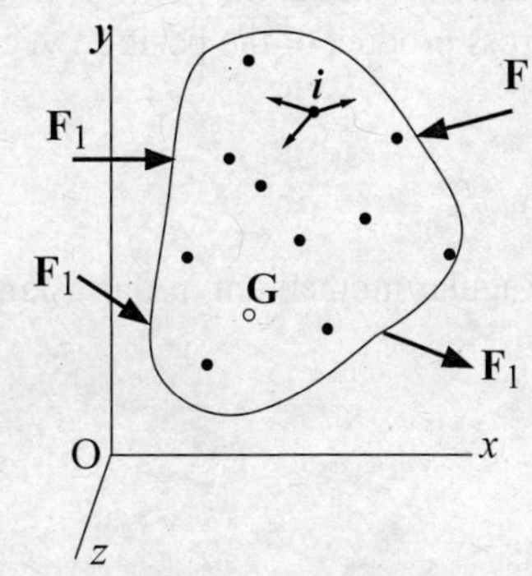

FIGURE 12.3 |

For a whole body we may write

$$\sum \mathbf{F}_i = \sum m_i \frac{\mathrm{d}\mathbf{v}_i}{\mathrm{d}t}$$

ΣF_i represents only the sum of all the external forces acting on the system of particles. (Internal forces between particles do not appear with this summation, since by Newton's third law they occur in equal but opposite collinear pairs and, therefore, cancel out.)

Resultant of external forces

$$\mathbf{F} = \sum \mathbf{F}_i = \sum m_i \frac{\mathrm{d}\mathbf{v}_i}{\mathrm{d}t}$$

Multiplying both sides by $\mathrm{d}t$ and integrating from $t = t_1$, where $v_i = (v_i)_1$ to $t = t_2$ and where $v_i = (v_i)_2$, yields

$$\int_{t_1}^{t_2} \mathbf{F}\mathrm{d}t = \sum m_i[(\mathbf{v}_i)_2 - (\mathbf{v}_i)_1]$$

or

$$\sum m_i(\mathbf{v}_i)_1 + \int_{t_1}^{t_2} \mathbf{F}\mathrm{d}t = \sum m_i(\mathbf{v}_i)_2$$

By the definition of centre of mass

$$mr_G = \sum m_i r_i \Rightarrow m\mathbf{v}_G = m_i \mathbf{v}_i$$

where $m = \Sigma m_i$,

$$\int_{t_1}^{t_2} \mathbf{F}dt = m[(\mathbf{v}_G)_2 - (\mathbf{v}_G)_1]$$

$$m(\mathbf{v}_G)_1 + \int_{t_1}^{t_2} \mathbf{F}dt = m(\mathbf{v}_G)_2$$

The change in linear momentum of a system of particles in any time interval is equal to the linear impulse, over the same time interval, of the resultant of the external forces that act on the body.

If $F = 0$ in any time interval, the above equation reduces to

$$m(\mathbf{v}_G)_1 = m(\mathbf{v}_G)_2 \Rightarrow (\mathbf{v}_G)_1 = (\mathbf{v}_G)_2$$

This equation is referred to as the conservation of linear momentum, which indicates velocity $\mathbf{v}_G$ of centre of mass does not change when no external impulses are applied to the system.

ANGULAR MOMENTUM

The angular momentum of a particle about point O is defined as the moment of the linear momentum vector $m\mathbf{v}$ about the point O. In other words, the angular momentum of a particle about a point is the cross product of the position vector of the particle and the linear momentum of that particle.

$$\mathbf{H}_O = \mathbf{r} \times m\mathbf{v}$$

or

$$\mathbf{H}_O = rm\mathbf{v} \sin\theta$$

Angular momentum in Cartesian coordinates

$$\mathbf{H}_O = \begin{bmatrix} i & j & k \\ x & y & z \\ mv_x & mv_y & mv_z \end{bmatrix}$$

LAW OF ANGULAR MOMENTUM

Consider a particle of mass m. If $\mathbf{F}$ represents the resultant of all forces acting on the particle (Fig. 12.4). The moment of forces about the point O is

$$\mathbf{M}_O = \mathbf{r} \times \mathbf{F}$$

$$\mathbf{M}_O = \mathbf{r} \times m\dot{\mathbf{v}}$$

where, by Newton's second law, $\mathbf{F} = m\dot{\mathbf{v}}$.

Now, differentiate angular momentum equation $\mathbf{H}_O = \mathbf{r} \times m\mathbf{v}$ with respect to time:

$$\mathbf{H}_O = \dot{\mathbf{r}} \times m\mathbf{v} + \mathbf{r} \times m\dot{\mathbf{v}}$$

or

$$\mathbf{H}_O = \mathbf{v} \times m\mathbf{v} + \mathbf{r} \times m\dot{\mathbf{v}}$$

where $\dot{\mathbf{r}} = \mathbf{v}$.

The term $\mathbf{v} \times m\mathbf{v}$ is zero:

$$\mathbf{H}_O = \mathbf{r} \times m\dot{\mathbf{v}}$$

Substituting into the expression for $\mathbf{M}_O$ gives

$$\mathbf{M}_O = \dot{\mathbf{H}}_O \quad \text{or} \quad \mathbf{M}_O = \frac{d\mathbf{H}_O}{dt}$$

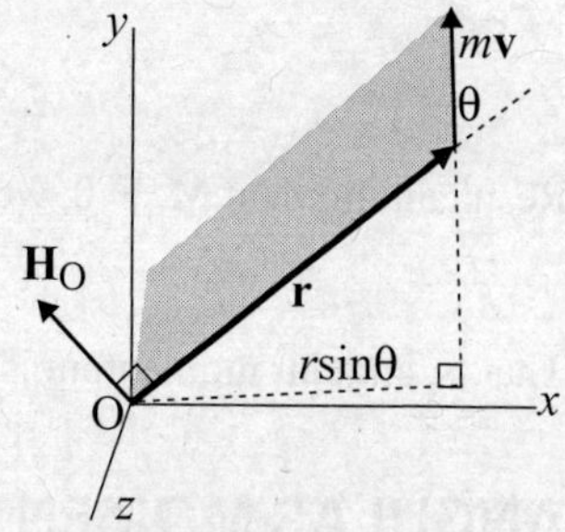

FIGURE 12.4 |

This is the equation of rotational motion. This equation states that the resultant moment about point O of all the forces acting on the particle is equal to the time rate of change of the particle's angular momentum about O.

ANGULAR IMPULSE

The product of moment and time is defined as angular impulse. It is a vector quantity;

$$\text{angular impulse} = \mathbf{M}_O \Delta t$$

If the moment $\mathbf{M}_O$ varies with time then the angular impulse of moment $\mathbf{M}_O$ in a time interval $\Delta t = t_2 - t_1$ is the integral of $\mathbf{M}_O$ from time t_1 to t_2;

$$\text{angular impulse} = \int_{t_1}^{t_2} \mathbf{M}_O \, dt$$

PRINCIPLE OF ANGULAR MOMENTUM AND IMPULSE

The equation of rotational motion is written as

$$\mathbf{M}_O = \frac{d\mathbf{H}_O}{dt}$$

Multiplying by dt and integrating from t_1 to t_2, we have

$$\int_1^2 \mathbf{M}_O \, dt = \int_1^2 d\mathbf{H}_O$$

or

$$\int_1^2 \mathbf{M}_O \, dt = (\mathbf{H}_O)_2 - (\mathbf{H}_O)_1$$

Angular impulse is equal to the change in angular momentum. The change of the angular momentum of a particle in a given time interval is equal to the angular impulse of the sum of the moments of the external forces that act on the particle during the same time interval.

CONSERVATION OF ANGULAR MOMENTUM

If the resultant moment acting on a particle is zero during an interval of time, angular momentum of the particle remains constant, we write angular impulse momentum equation

$$\int_1^2 \mathbf{M}_O dt = (\mathbf{H}_O)_2 - (\mathbf{H}_O)_1$$

Resultant moment $\mathbf{M}_O = 0$, we have

$$(\mathbf{H}_O)_2 = (\mathbf{H}_O)_1$$

That is, angular momentum is a constant.

ANGULAR MOMENTUM OF RIGID BODY

Consider a rigid body of mass m, which is subjected to general plane motion (Fig. 12.5). At the instant shown, the arbitrary point O has a velocity $\mathbf{v}_o$ and the body has an angular velocity ω. Velocity of ith particle of mass dm is

$$\mathbf{v}_i = \mathbf{v}_o + \mathbf{v}_{i/0} = \mathbf{v}_0 + \omega \times \mathbf{r}$$

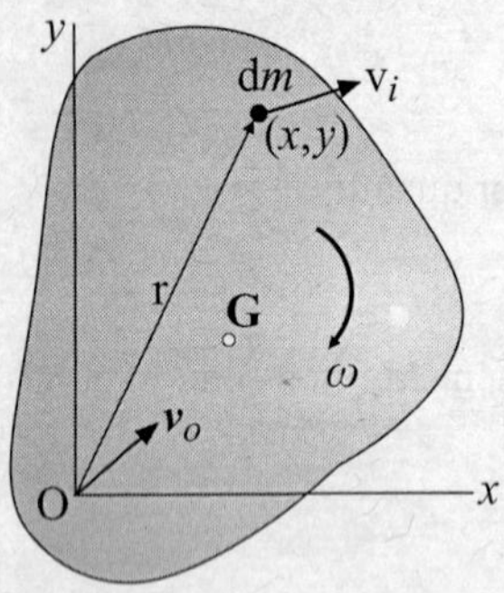

FIGURE 12.5

The angular momentum of particle i about point O is

$$d\mathbf{H}_O = \mathbf{r} \times dm\mathbf{v}_i$$

or

$$d\mathbf{H}_O = \mathbf{r} \times dm(\mathbf{v}_o + \omega \times \mathbf{r})$$

Substituting

$$\mathbf{r} = xi + yj,\ \mathbf{v}_o = (v_o)_x i + (v_o)_y j \text{ and } \omega = \omega k,$$

we have

$$d\mathbf{H}_O = (xi + yj) \times dm[(v_o)_x i + (v_o)_y j + \omega k \times (xi + yj)]$$

On simplifying, we get

$$d\mathbf{H}_O = -dmy(v_o)_x + dmx(v_o)_y + \omega dm(x^2 + y^2)$$

For whole body, integrating the above equation, we have

$$\mathbf{H}_O = -(v_o)_x \int y\, dm + (v_o)_y \int x\, dm + \omega \int r^2 dm$$

Substituting

$$\int y dm = \bar{y}m,\ \int x dm = \bar{x}m, \text{ and } \int r^2\, dm = I_o$$

we have

$$\mathbf{H}_O = -(v_o)_x \bar{y}m + (v_o)_y \bar{x}m + \omega I_o$$

This equation reduces to a simpler form if point O coincides with the mass centre G for the body, in which case $\bar{x} = \bar{y} = 0$, and therefore

$$\mathbf{H}_G = I_G \omega$$

ROTATION ABOUT FIXED AXIS

A body is rotating about a fixed axis passing through O (Fig. 12.6). Hence, the angular momentum about the fixed axis O is

$$\mathbf{H}_O = -0\bar{y}m + 0\bar{x}m + \omega I_o = I_o\omega$$

$$\mathbf{H}_O = I_o\omega$$

or

$$\mathbf{H}_O = (I_G + mr_G^2)\omega = I_G\omega + m\omega r_G \times r_G$$

or

$$\mathbf{H}_O = I_G\omega + mv_G \times r_G$$

$$\mathbf{H}_O = \mathbf{H}_G + (mv_G)r_G$$

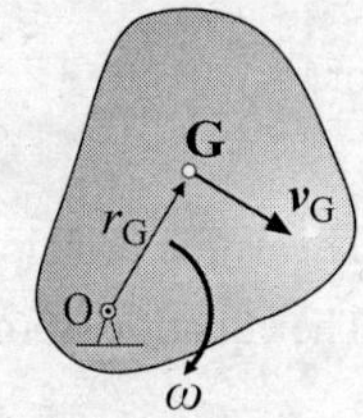

FIGURE 12.6

GENERAL PLANE MOTION

When a rigid body is subjected to general plane motion (Fig. 12.7) then linear momentum of body becomes

$$\boldsymbol{p} = m\mathbf{v}_G$$

Angular momentum about a point A

$$\mathbf{H}_A = \mathbf{H}_G + (mv_G)d$$

$$\mathbf{H}_A = I_G\,\omega + (mv_G)d$$

Angular momentum about centre of mass G is

$$\mathbf{H}_G = I_G\,\omega$$

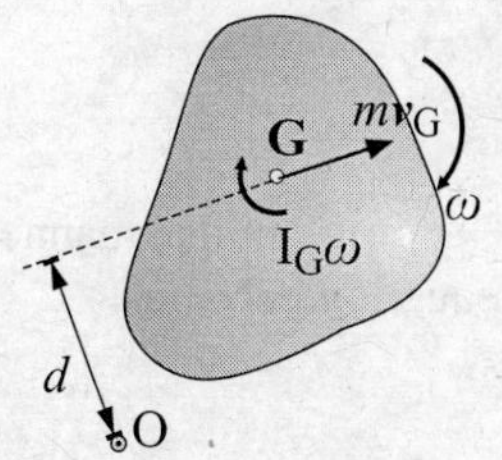

FIGURE 12.7

PRINCIPLE OF IMPULSE AND MOMENTUM FOR A RIGID BODY

Consider a rigid body subjected to external forces and moments as shown in Figure 12.8. The equation of translation for a rigid body can be written as

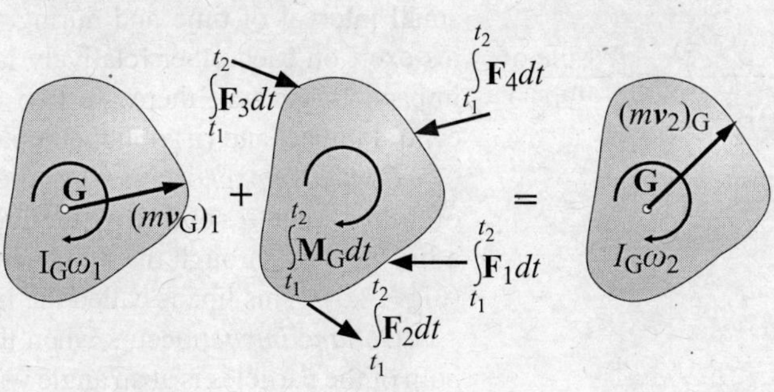

FIGURE 12.8

$$\sum \mathbf{F} = \frac{\mathrm{d}}{\mathrm{d}t}(m\,\mathbf{v}_G)$$

Multiplying both sides by dt and integrating from $t = t_1$ where $v_G = (v_G)_1$ to $t = t_2$ where $v_G = (v_G)_2$, yields

$$\sum \int_{t_1}^{t_2} \mathbf{F}\mathrm{d}t = m \int_{(v_G)_1}^{(v_G)_2} \mathrm{d}\mathbf{v}_G$$

or $$m(\mathbf{v}_G)_1 + \sum \int_{t_1}^{t_2} F\,dt = m(\mathbf{v}_G)_2$$

This is the principle of linear impulse and momentum for rigid body.

The equation of rotation motion for a rigid body can be written as

$$\sum \mathbf{M}_G = \frac{d\mathbf{H}_G}{dt} = \frac{d}{dt}(I_G\omega)$$

Multiplying both sides by dt and integrating from $t = t_1$ where $\omega = \omega_1$ to $t = t_2$ where $\omega = \omega_2$ yields

$$\sum \int_{t_1}^{t_2} \mathbf{M}_G dt = I_G \int_{\omega_1}^{\omega_2} d\omega$$

$$I_G\omega_1 + \sum \int_{t_1}^{t_2} M_G dt = I_G\omega_2$$

In a similar manner, for rotation about a fixed axis passing through the point O, the above equation becomes

$$I_o\omega_1 + \sum \int_{t_1}^{t_2} M_o dt = I_o\omega_2$$

This is the principle of angular impulse and momentum for a rigid body.

IMPACT

A collision between two bodies that occurs in a very small interval of time and during which the two bodies exert on each other relatively large forces is called impact. In general, there are two types of impact: (i) central impact and (ii) oblique impact.

Central impact occurs when the direction of motion of the mass centres of the two colliding particles is along a line passing through the mass centres of the particles (Fig. 12.9). This line is called the line of impact.

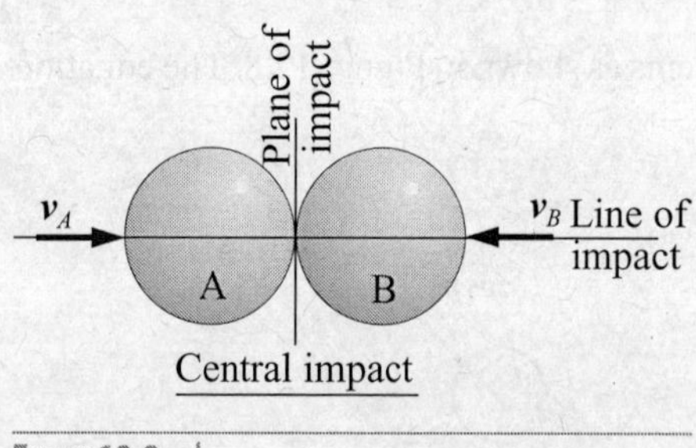

FIGURE 12.9 |

Oblique impact occurs when the motion of one or both of the particles is at an angle with the line of impact as shown in Figure 12.10.

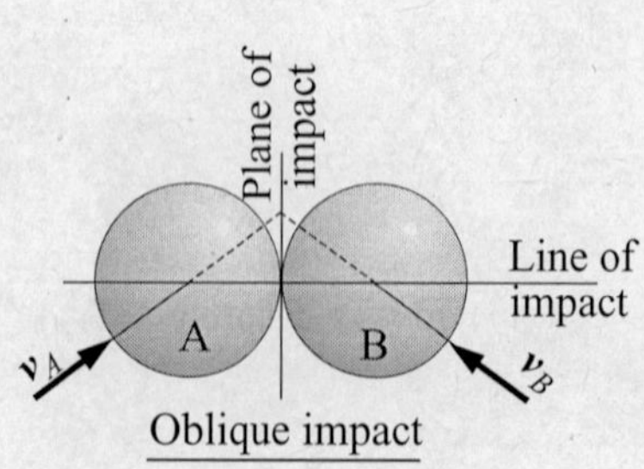

FIGURE 12.10 |

CENTRAL IMPACT

Consider two bodies A and B, of masses m_A and m_B, moving in a same straight line with velocities u_A and u_B (Fig. 12.11). Let $u_A > u_B$, so that the bodies collide(or impact). Under the impact, the bodies will deform and at the end of the period of deformation, they will have the same velocity u. Afterward a period of restitution occurs, in which case the material from which the particles are made will either return to its original shape or remain permanently deformed (Fig. 12.12). The equal but opposite restitution impulse $\int Rdt$ pulses the bodies apart from one another. In reality, impulse is always greater than that of restitution, i.e.,

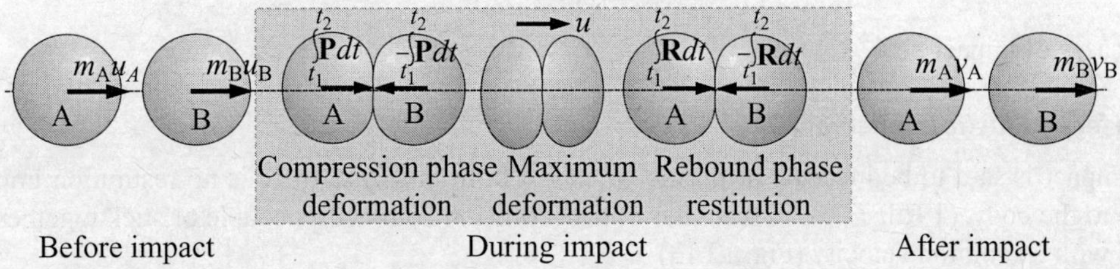

FIGURE 12.11 |

$$\int \mathbf{P}dt > \int \mathbf{R}dt$$

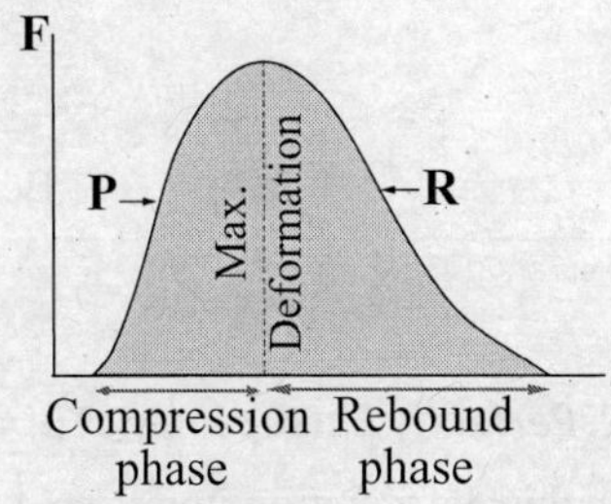

FIGURE 12.12 |

In the system having two bodies A and B, we note that there is no external force. Thus, the total momentum of the two bodies is conserved, and we write

linear momentum of the system before impact is equal to linear momentum of the system after impact:

$$m_A u_A + m_B u_B = m_A v_A + m_B v_B$$

COEFFICIENT OF RESTITUTION

It is the ratio the restitution impulse to the deformation impulse. Mathematically,

$$e = \frac{\int \mathbf{R}dt}{\int \mathbf{P}dt}$$

Applying principle of linear impulse-momentum for body A (Fig. 12.11), before impact to maximum deformation, $m_A u_A - \int \mathbf{P}dt = m_A u$

$$\int \mathbf{P}dt = m_A(u_A - u)$$

Maximum deformation to after impact

$$m_A u - \int \mathbf{R}dt = m_A v_A$$

$$\int \mathbf{R}dt = m_A(u - v_A)$$

Thus,

$$e = \frac{\int \mathbf{R}dt}{\int \mathbf{P}dt} = \frac{m_A(u - v_A)}{m_A(u_A - u)} \quad \text{or} \quad e = \frac{u - v_A}{u_A - u}$$

In a similar manner, we can establish e by considering body B, $e = \dfrac{v_B - u}{u - u_B}$.

If the unknown u is eliminates from above equations, we obtain

$$e = \frac{v_B - v_A}{u_A - u_B}$$

$$v_B - v_A = e(u_A - u_B)$$

Two cases of impact

1. Perfectly Plastic Impact (e = 0)

The impact is said to be inelastic or plastic when $e = 0$. In this case, there is no restitution impulse given to the body ($\int \mathbf{R}dt = 0$), so that after impact collision both bodies couple or stick together and move with a common velocity (Fig. 12.13).

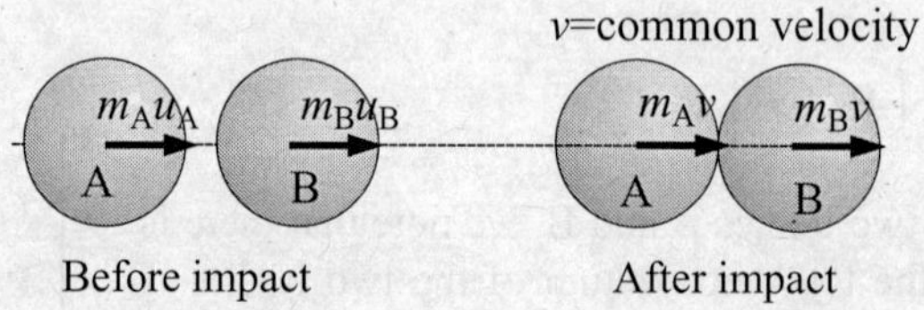

FIGURE 12.13 |

2. Perfectly Elastic Impact (e = 1)

In this deformation impulses $\int \mathbf{P}dt$ is equal and opposite to the restitution $\int \mathbf{R}dt$. Thus,

$$e = 1 = \frac{v_B - v_A}{u_A - u_B}$$

or

$$u_A + v_A = u_B + v_B$$

The momentum equation is

$$m_A u_A + m_B u_B = m_A v_A + m_B v_B$$

or

$$m_A(u_A - v_A) = m_B(v_B - u_B)$$

Multiplying the above equations $m_A(u_A - v_A) = m_B(v_B - u_B)$ and $u_A + v_A = u_B + v_B$, we obtain

$$m_A (u_A^2 - v_A^2) = m_B(v_B^2 - u_B^2)$$

Dividing both sides by 2 and simplifying,

$$\frac{1}{2} m_A u_A^2 + \frac{1}{2} m_B u_B^2 = \frac{1}{2} m_A v_A^2 + \frac{1}{2} m_B v_B^2,$$

which expresses that the kinetic energy of the particle is conserved.

Oblique Impact

Consider two smooth surface bodies A and B of mass m_A and m_B, which are moving in the directions θ_1 and θ_2 as shown in Figure 12.14 with velocities u_A and u_B. Since the velocities v_A and v_B of the bodies after impact are unknown in direction as well as in magnitude, their determination will require the use of four independent equations.

We observe that only impulses exerted on the bodies during the impact are due to internal forces directed along the line of impact, i.e., along x-axis. It follows that

1. Linear momentum of the system is conserved along the line of impact, so that

$$m_A(u_A)_x + m_B(u_B)_x = m_A(v_A)_x + m_B(v_B)_x$$

2. The coefficient of restitution,

$$e = \frac{(v_B)_x - (v_A)_x}{(u_A)_x - (u_B)_x}$$

3. Momentum of particle A is conserved along y-axis, perpendicular to the line of impact, since no impulse acts on particle A in this direction,

$$m_A(u_A)_y = m_A(v_A)_y \Rightarrow (u_A)_y = (v_A)_y$$

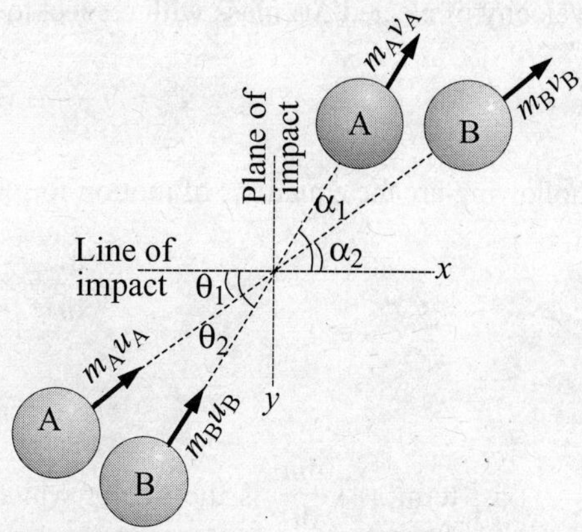

FIGURE 12.14

4. Momentum of particle B is conserved along y-axis, perpendicular to the line of impact, since no impulse acts on particle B in this direction,

$$m_B(u_B)_y = m_B(v_B)_y \Rightarrow (u_B)_y = (v_B)_y$$

SYSTEM OF VARIABLE MASS

Losing Mass

So far, we have discussed motion of a system in which the whole mass of the system remains same and does not undergo any change. There are instances where the mass changes with time and are of very important. Such as rocket in which the gases are continuously exhausted from a jet and thus, while moving mass of rocket continuously decreases.

Consider a system of total mass m moving with a velocity v in a fixed reference frame, under an external force **F**. After a time interval of Δt the system has changed to as shown in Figure 12.15. A mass Δm is ejected with a velocity u (absolute) in the opposite direction to the remaining mass $(m - \Delta m)$. Now the remaining mass $(m - \Delta m)$ travels with a velocity $v + \Delta v$. Applying the impulse-momentum equation

$$mv + \mathbf{F}\Delta t = (m - \Delta m)\,(v + \Delta v) + \Delta\, mu,$$

where F is the resultant of external force.

or

$$\mathbf{F}\Delta t = m\Delta v - \Delta m\,(v - u) - \Delta m\,\Delta v$$

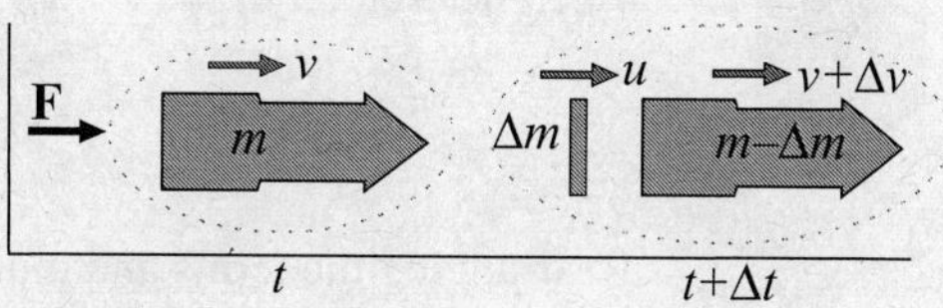

FIGURE 12.15

Neglecting the last term $\Delta m\,\Delta v$ and dividing by Δt, we have

$$\mathbf{F} = m\frac{\Delta v}{\Delta t} - (v - u)\frac{\Delta m}{\Delta t}$$

Taking the limit $\Delta t \to 0$ yields

$$\mathbf{F} = m\frac{dv}{dt} - (v - u)\frac{dm}{dt}$$

Velocity of ejected Δm mass with respect to remaining mass $(m - \Delta m)$ is $v_{rel} = (v - u)$:

$$\mathbf{F} = m\frac{dv}{dt} - v_{rel}\frac{dm}{dt}$$

Following are the equations of motion for variable (loose and gain) mass

$$\mathbf{F} = m\frac{dv}{dt} - v_{rel}\frac{dm}{dt} \quad \text{(losing mass)}$$

$$\mathbf{F} = m\frac{dv}{dt} + v_{rel}\frac{dm}{dt} \quad \text{(gaining mass)}$$

Last term $v_{rel}\frac{dm}{dt}$ is the rate at which the momentum is being transferred between the mass ejected and the remaining mass. Since the rate of change of momentum is defined as force, the term $v_{rel}\frac{dm}{dt}$ can be called the force exerted on the system by the mass that is ejected and is called the force of reaction (thrust) T. Thrust (reaction) is

$$T = v_{rel}\frac{dm}{dt}$$

Rocket

Consider a rocket moving upward in gravitation field (Fig. 12.16). Thus, the external force acting on rocket

$$F = -mg$$

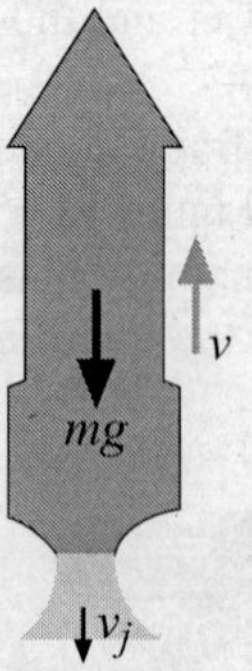

FIGURE 12.16 |

Applying equation of motion for a losing mass system

$$\mathbf{F} = m\frac{dv}{dt} - (v - u)\frac{dm}{dt}$$

or

$$-mg = m\frac{dv}{dt} - (v - u)\frac{dm}{dt},$$

where jet velocity $v_j = (v - u)$

$$-mg = m\frac{dv}{dt} - v_j\frac{dm}{dt}$$

Rearranging the terms and multiplying by dt, we have

$$dv = -g dt + v_j \frac{dm}{m}$$

Let at $t = 0$, mass of rocket $m = m_o$ and velocity $v = v_o$, we have

$$\int_{v_o}^{v} dv = -g\int_{0}^{t} dt + v_j\int_{m_o}^{m}\frac{dm}{m}$$

$$v = v_o - gt + v_j \log_e \frac{m}{m_o}$$

SOLVED PROBLEMS

PROBLEM 12.1

A spring is connected to two blocks and compressed 20 cm as shown in Figure 12.1(a). If the system is released on a smooth horizontal plane, what will be the velocity of each block when the spring is at again its normal length? The length of the uncompressed spring is 60 cm and spring constant is 500 N/m.

FIGURE 12.1(a)

Solution. Since the horizontal impulse of the entire system is zero, the total linear momentum remains zero. Thus,

$$m_A v_A + m_B v_B = 0$$

or

$$2v_A + 3v_B = 0, \tag{i}$$

where v_A and v_B are the velocities of block A and B when the spring is again its normal length [Fig. 12.1(b)].

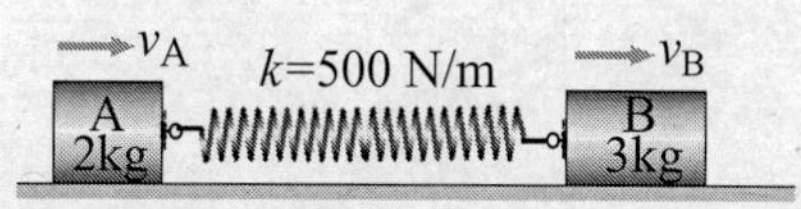

FIGURE 12.1(b)

The work done by the spring in expanding to its original length is

$$U = \int_0^{0.2} 500(0.2 - x)dx = 500\left[(0.2)^2 - \frac{(0.2)^2}{2}\right]$$

or

$$U = 10 \text{ J}$$

Kinetic energy of both blocks is

$$T = \frac{1}{2} \times 2v_A^2 + \frac{1}{2} \times 3v_B^2 = v_A^2 + 1.5v_B^2$$

Applying the principle of conservation of energy ($U = T$),

$$v_A^2 + 1.5v_B^2 = 10 \tag{ii}$$

Solving for v_A and v_B from Equations (i) and (ii), we have

$$v_A = -2.25 \text{ m/s}$$

$$v_B = 1.5 \text{ m/s}$$

PROBLEM 12.2

A 100 g bullet is fired at two blocks resting on a surface where the coefficient of kinetic friction is 0.5. The bullet passes through a 10 kg block and lodges in a 8 kg block. The blocks slide the distances shown in Figure 12.2(a). Calculate the initial velocity u of the bullet.

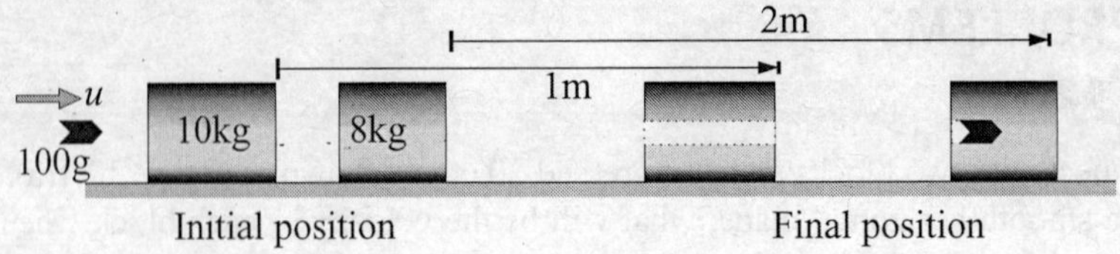

FIGURE 12.2(a) |

Solution. Consider the bullet, 10 kg block and 8 kg block as a system [Fig. 12.2(b)]. Let v_1 and v_2 be the velocities of blocks instant after impact. Applying the conservation of linear momentum,

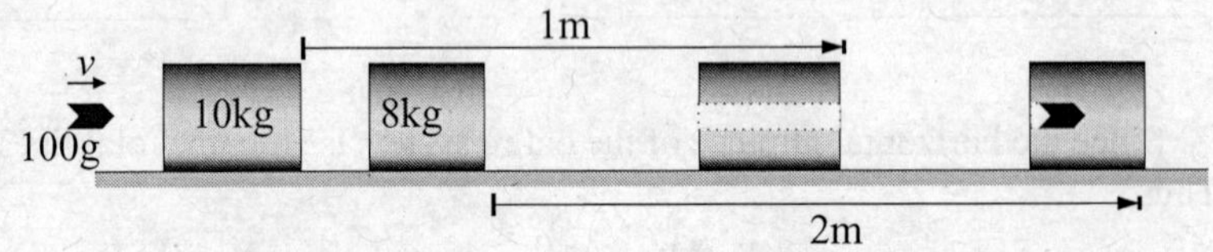

FIGURE 12.2(b) |

$$\frac{100}{1000}u = 10v_1 + \left(8 + \frac{100}{1000}\right)v_2$$

$$u = 100v_1 + 81v_2 \tag{i}$$

Consider the 10 kg block [Fig. 12.2(c)].

Applying the work-energy principle, we have kinetic energy is equal to work done against friction force:

$$\frac{1}{2} \times 10 \times v_1^2 = 0.5 \times 10 \times 9.81 \times 1 \qquad \left(\frac{1}{2}mv^2 = \mu mgd\right)$$

or $$v_1 = \sqrt{\frac{2 \times 0.5 \times 10 \times 9.81 \times 1}{10}} = 3.13$$

or $$v_1 = 3.13 \text{ m/s}$$

Now consider 8 kg block [Fig. 12.2(c)].

Applying work-energy principle, we have

$$\frac{1}{2} \times (8 + 0.1) \times v_2^2 = 0.5 \times (8 + 0.1) \times 9.81 \times 2$$

or $$v_2 = \sqrt{\frac{2 \times 0.5 \times 8.1 \times 9.81 \times 2}{8.1}} = 4.429$$

or $$v_2 = 4.42 \text{ m/s}$$

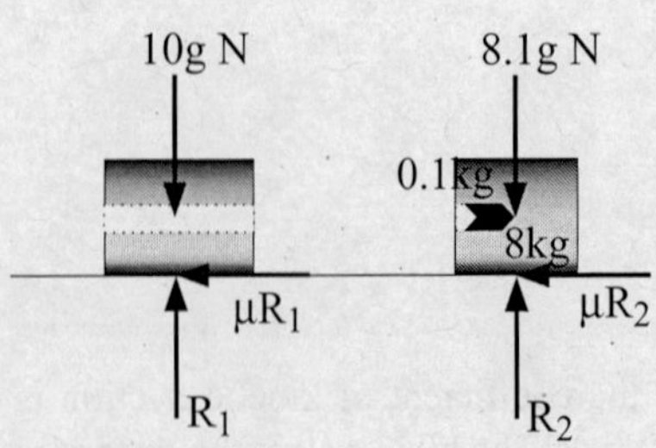

FIGURE 12.2(c) |

Substituting $v_1 = 3.13$ m/s and $v_2 = 4.42$ m/s into the Equation (i), we have

$$u = 100 \times 3.13 + 81 \times 4.42$$

$$u = 671.02 \text{ m/s}$$

PROBLEM 12.3

A ball moving with a velocity of 2 m/s hits against a smooth horizontal plane. Its direction of motion is inclined at an angle of 30° to the plane [Fig. 12.3(a)]. If the coefficient of restitution is 0.8, find the velocity of the ball after hitting the plane.

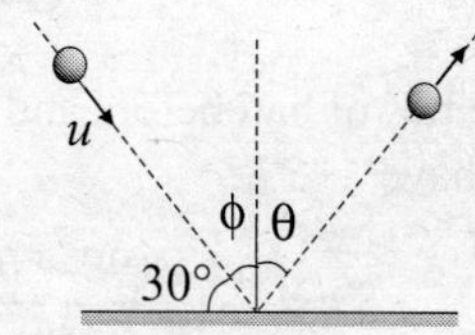

FIGURE 12.3(a)

Solution. Velocity before impact $u = 2$ m/s, and $\phi = 90 - 30 = 60°$.

Let the velocity after the impact be v at the angle θ with the normal to the plane. Since there will be no change in the component to the velocity parallel to the plane (plane of impact), we have

$$v \sin \theta = u \sin \phi$$

or
$$v \sin \theta = 2 \sin 60$$

or
$$v \sin \theta = \sqrt{3} \tag{i}$$

By using the coefficient of restitution equation, the velocity after impact in normal direction is

$$= e \times \text{velocity before impact in normal direction}$$

or
$$v \cos \theta = e \times u \cos \phi$$

or
$$v \cos \theta = 0.8 \times 2 \cos 60$$

or
$$v \cos \theta = 0.8 \tag{ii}$$

Dividing Equation (i) by Equation (ii), we have

$$\tan \theta = \frac{\sqrt{3}}{0.8} = 2.16$$

$$\theta = 65.2°$$

From Equation (i), we have

$$v = \frac{\sqrt{3}}{\sin 65.2} = 1.907$$

$$v = 1.907 \text{ m/s}$$

PROBLEM 12.4

A ball thrown from position A against a smooth vertical circular wall rebounds and hits position C at the other end of the diameter through A [Fig. 12.4(a)]. Find the coefficient of restitution.

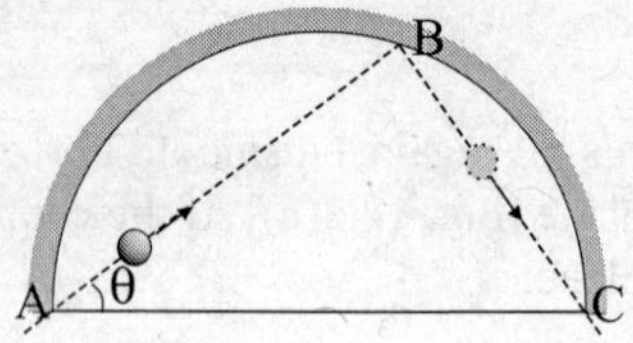

FIGURE 12.4(a) |

Solution. Let u and v be the velocities of ball before and after impact [Fig. 12.4(b)]. Applying equation of coefficient of restitution, we have

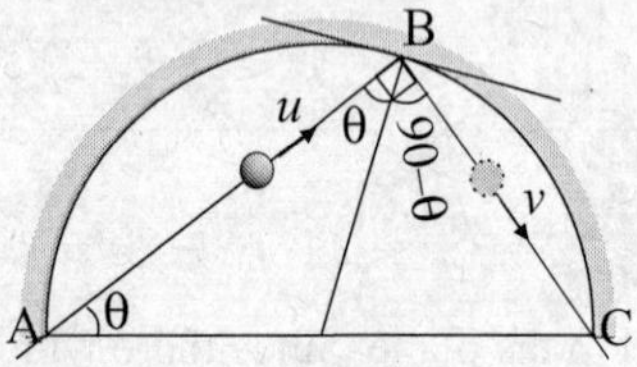

FIGURE 12.4(b) |

$$e = \frac{v\cos(90-\theta)}{u\cos\theta}$$

or $$e = \frac{v}{u}\tan\theta \quad \text{(i)}$$

Applying the equation of impact along plane of impact,

$$u\sin\theta = v\sin(90-\theta)$$

or $$\frac{v}{u} = \tan\theta \quad \text{(ii)}$$

From Equations (i) and (ii), we have

$$e = \tan\theta \times \tan\theta$$

$$e = \tan^2\theta$$

PROBLEM 12.5

A ball dropped from a certain height h on a horizontal plane, bounces up and down and finally comes to rest. What is the total distance covered and total time taken?

Solution. If u be the velocity of the ball then $u = \sqrt{2gh}$. If v be the velocity of the rebound after the first impact, then $v = eu = e\sqrt{2gh}$. The ball goes up with this velocity to a height h_1, so that

$$h_1 = \frac{v^2}{2g} = \frac{e^2 2gh}{2g} = e^2 h$$

and it takes a time

$$t_1 = \frac{v}{g} = \frac{e\sqrt{2gh}}{g} = e\sqrt{\frac{2h}{g}}$$

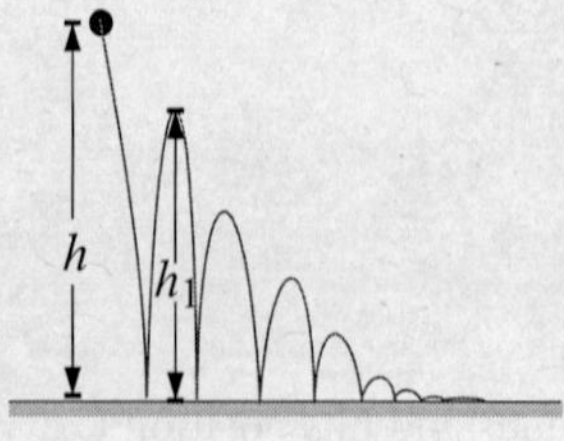

FIGURE 12.5(a) |

The ball again drops and strikes the plane with velocity v and rebounds with a velocity ev [Fig. 12.5(a)]. and goes to a height

$$h_2 = \frac{(ev)^2}{2g} = \frac{e^2}{2g}(e\sqrt{2gh})^2 = e^4 h,$$

after a time $$\frac{ev}{g} = \frac{ee\sqrt{2gh}}{g} = e^2\sqrt{\frac{2h}{g}}$$

and so on. Therefore, the total distance covered

$$S = h + 2h_1 + 2h_2 + 2h_3 + \cdots$$

or
$$S = h + 2e^2h + 2e^4h + 2e^6h + \cdots$$

or
$$S = h(1 + 2e^2 + 2e^4 + 2e^6 + \cdots)$$

or
$$S = h\left(1 + \frac{2e^2}{1-e^2}\right) = h\left(\frac{1+e^2}{1-e^2}\right)$$

$$\boxed{S = h\left(\frac{1+e^2}{1-e^2}\right)}$$

Total time
$$T = t + 2t_1 + 2t_2 + 2t_3 + \cdots$$

or
$$T = \sqrt{\frac{2h}{g}} + 2e\sqrt{\frac{2h}{g}} + 2e^2\sqrt{\frac{2h}{g}} + \cdots$$

or
$$T = \sqrt{\frac{2h}{g}}(1 + 2e + 2e^2 + 2e^3 + \cdots)$$

or
$$T = \sqrt{\frac{2h}{g}}\left(1 + \frac{2e}{1-e}\right)$$

$$\boxed{T = \sqrt{\frac{2h}{g}}\left(\frac{1+e}{1-e}\right)}$$

PROBLEM 12.6

Sand drops continuously from a hopper on to a moving belt as shown in Figure 12.6(a). What force and power are required to keep the belt moving at a constant speed?

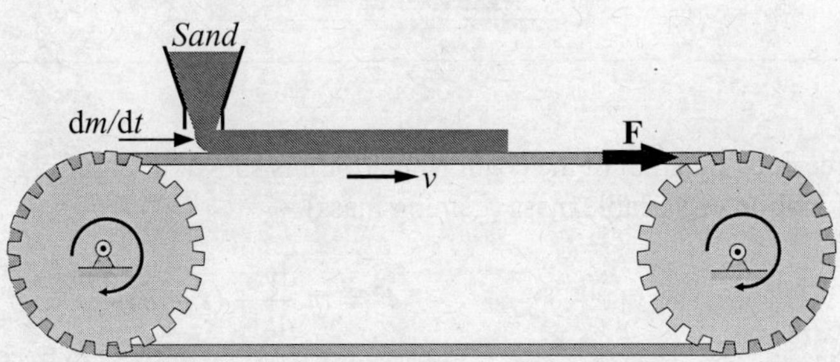

FIGURE 12.6(a) |

Solution. In this problem, we consider the system of varying mass (gaining mass). Applying the equation of motion for varying (increasing) mass

$$\mathbf{F} = m\frac{\mathrm{d}v}{\mathrm{d}t} + (v - u)\frac{\mathrm{d}m}{\mathrm{d}t}$$

Given: velocity of belt is constant $\frac{dv}{dt} = 0$, velocity of sand in the direction of belt $u = 0$.

Thus, force applied on the belt to maintain a constant speed is

$$\mathbf{F} = m \times 0 + (v - 0)\frac{dm}{dt}$$

$$\mathbf{F} = v\frac{dm}{dt}$$

The power requirement to run the belt is

$$\mathbf{P} = \mathbf{F}v = \left(v\frac{dm}{dt}\right)v$$

Power $$\mathbf{P} = v^2\frac{dm}{dt}$$

PROBLEM 12.7

The upper end of an open-link chain is lowered at a constant speed v by the force P [Fig. 12.7(a)]. The length of chain is l and mass per unit length is ρ. Determine the reading R of the platform scale in terms of x.

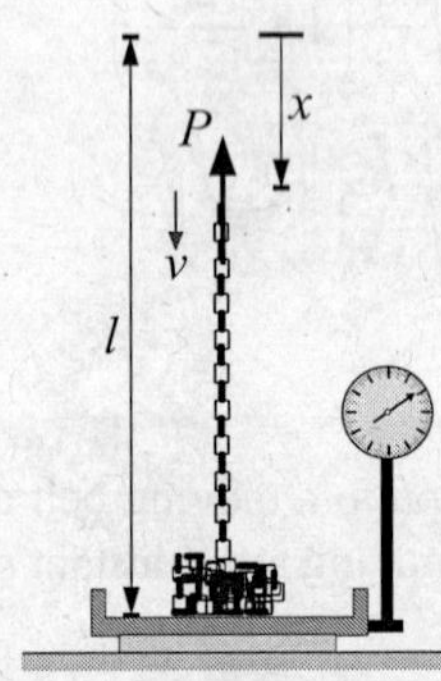

FIGURE 12.7(a) |

Solution. Free body diagram of the chain on platform is shown in Figure 12.7(b). The equation of motion of variable mass (gaining mass)

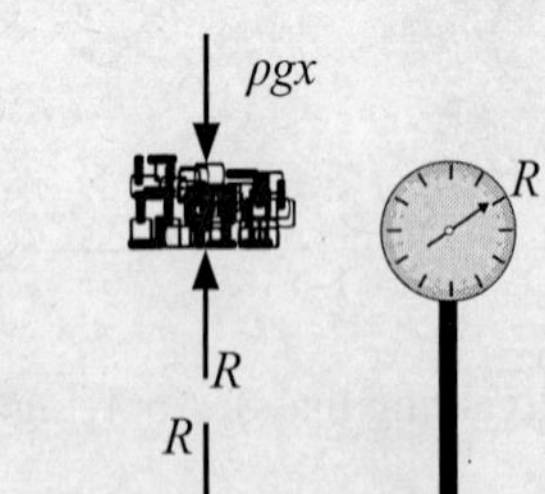

FIGURE 12.7(b) |

$$F = m\frac{dv}{dt} + (v - u)\frac{dm}{dt}$$

The rate of increase of mass is $\frac{dm}{dt} = \rho v$, and $u = 0$. Therefore,

$$R - \rho gx = m \times 0 + (v - 0) \times \rho v$$

or $$R = \rho gx + \rho v^2$$

reading of scale, $R = \rho gx + \rho v^2$

PROBLEM 12.8

The end of an open-link chain is being pulled horizontally along the surface by a constant force P [Fig. 12.8(a)]. If the coefficient of friction between the chain and the surface is μ, and mass per unit length is ρ, determine the acceleration a of the chain in terms of x.

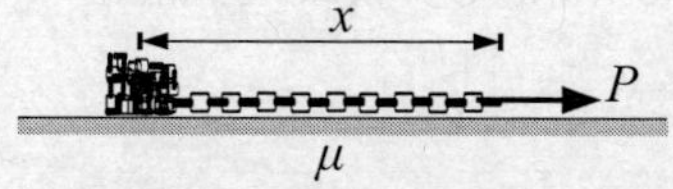

FIGURE 12.8(a)

Solution. The free body diagram of the moving open-link chain is shown in Figure 12.8(b). From the figure, net external force on the moving chain is

$$F = P - \mu\rho g x$$

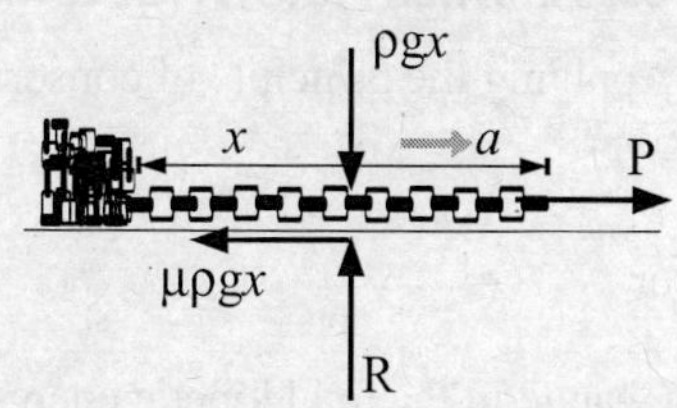

FIGURE 12.8(b)

The rate of increase of mass is $\frac{dm}{dt} = \rho v$, and $u = 0$. Thus, the equation of motion for variable mass (increasing mass) is (note that due to open-link chain there is no reaction between stationary and moving chains)

or
$$F = m\frac{dv}{dt} + (v - u)\frac{dm}{dt}$$

or
$$P - \mu\rho g x = \rho x a + v \times \rho v$$

or
$$P - \mu\rho g x = \rho x a + \rho v^2$$

or
$$a = \frac{P - \mu\rho g x - \rho v^2}{\rho x}$$

Acceleration of moving part

$$a = \frac{\rho}{\rho x} - \mu g - \frac{v^2}{x}$$

PROBLEM 12.9

A shot of mass m penetrates a thickness t of a fixed plate of mass M [Fig. 12.9(a)]. Show that if the plate is free to move, the thickness penetrated is

$$\frac{M}{M + m}t$$

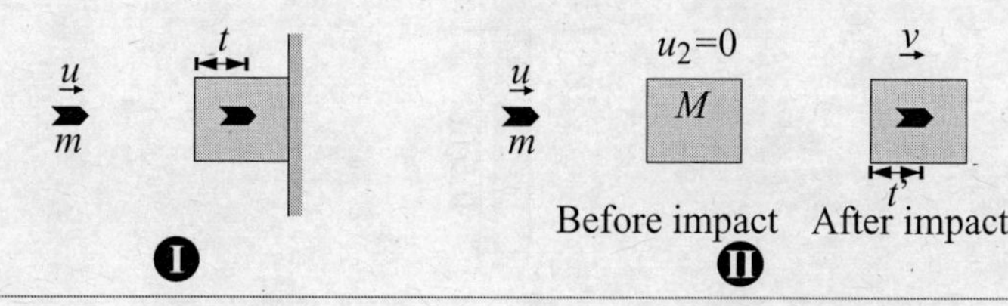

FIGURE 12.9(a)

Solution. Let u be the velocity of the shot, R be the average resistance of the penetration, t be the thickness of penetration in case I, t' be the thickness of penetration in case II, and v be the common velocity of shot and the plate case II.

Case I: When Plate is Fixed

Kinetic energy of the shot is equal to work done in penetration.

$$\frac{1}{2}mu^2 = R \times t$$

$$t = \frac{mu^2}{2R} \tag{i}$$

Case II: When Plate is Free to Move

Applying the principle of conservation of momentum, we have

$$mu = (M + m)v$$

or

$$v = \frac{m}{M+m}u \tag{ii}$$

Change in the total kinetic energy is equal to work done in penetration:

$$\frac{1}{2}mu^2 - \frac{1}{2}(M+m)v^2 = Rt'$$

Substituting the value of v from Equation (ii), we have

$$\frac{mu^2}{2} - \frac{1}{2}(M+m)\left(\frac{mu}{M+m}\right)^2 = Rt'$$

or

$$t' = \frac{mu^2}{2R}\left(\frac{M}{M+m}\right) \tag{iii}$$

From Equations (i) and (iii), we have $t' = t\dfrac{M}{M+m}$

PROBLEM 12.10

A uniform rod of mass 5 kg and length 1 m is suspended from end O as shown in Figure 12.10(a). If a ball of mass 1 kg is thrown at the rod and strikes its centre with a velocity of 9 m/s, determine the angular velocity of the rod just after impact. The coefficient of restitution is $e = 0.6$.

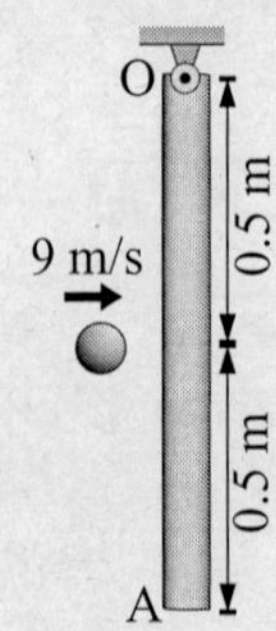

FIGURE 12.10(a) |

Solution. Consider the ball and rod as a system [Fig. 12.10(b)]. The moment of momentum is conserved about O. Therefore,

$$(H_o)_1 = (H_o)_2$$

or $\quad m_B(u_B)_1 \times 0.5 = m_B(v_B)_2 \times 0.5 + m_R(v_G)_2 \times 0.5 + I_G\omega_2$

or $\quad 1 \times 9 \times 0.5 = 1(v_B)_2 \times 0.5 + 5(v_G)_2 \times 0.5 + \frac{1}{12}5 \times 1^2 \times \omega_2$

or $\quad 4.5 = 0.5(v_B)_2 + 2.5(v_G)_2 + 0.4166\omega_2$

since $\quad (v_G)_2 = \omega_2 \times 0.5$

$$4.5 = 0.5(v_B)_2 + 2.5(v_G)_2 + 1.66\omega_2$$

Using the coefficient of restitution equation

$$e = \frac{(v_G)_2 - (v_B)_2}{(u_B)_1 - (u_G)_1} \Rightarrow 0.6 = \frac{(v_G)_2 - (v_B)_2}{9 - 0}$$

or $\quad 0.6 = \frac{0.5\omega_2 - (v_B)_2}{9}$

or $\quad 0.5\omega_2 - (v_B)_2 = 5.4$

Solving Equations (i) and (ii), we have

$$\omega_2 = 3.76\text{rad/s}$$

$$(v_B)_2 = -3.51\text{m/s}$$

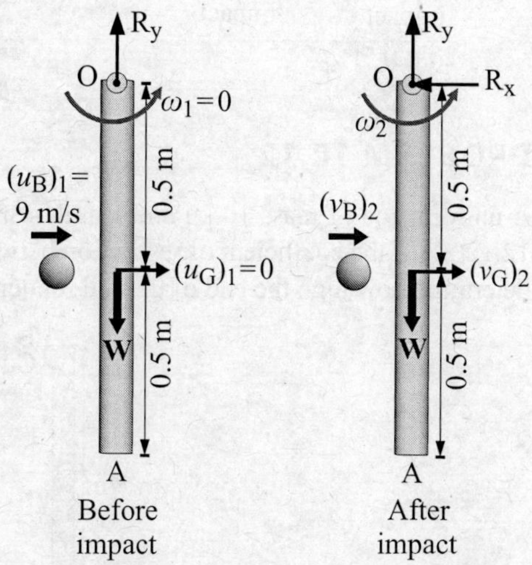

FIGURE 12.10(b) |

PROBLEM 12.11

A 10 g bullet has a horizontal velocity of 1000 m/s as it strikes the 12 kg rod AB, which is suspended from point A and is initially at rest [Fig. 12.11(a)]. Calculate the angular velocity ω which the rod with its embedded bullet has acquired immediately after impact.

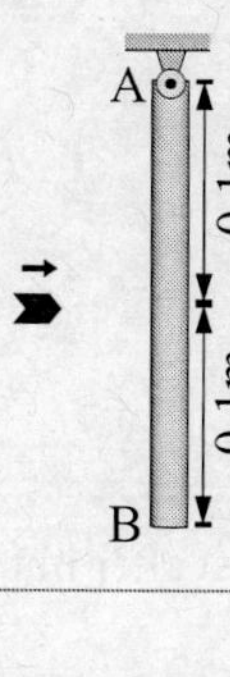

FIGURE 12.11(a) |

Solution. We consider the rod and bullet as a system. The angular momentum is conserved about A, since the impulsive force between rod and bullet is internal [Fig. 12.11(b)]. We have

$$(H_A)_1 = (H_A)_2$$

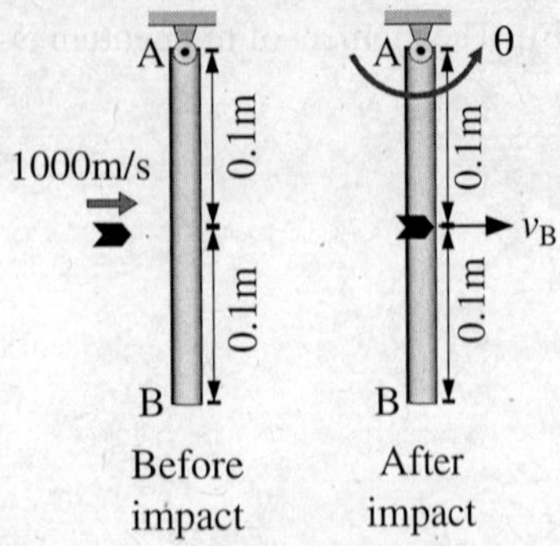

FIGURE 12.11(b) |

or
$$\frac{10}{1000} \times 1000 \times 0.1 = \frac{10}{1000} v_b \times 0.1 + I_A \omega,$$

where ω is the angular velocity instantly after impact.

Since
$$v_B = \omega \times 0.1,$$

or
$$1 = \frac{10}{1000}(0.1)^2 \omega + \frac{1}{3} \times 12 \times (0.2)^2 \omega$$

or
$$1 = 0.16\omega \Rightarrow \omega = 6.25$$

$$\omega = 6.25 \text{ rad/s}$$

PROBLEM 12.12

A uniform rod of mass 10 kg and length 4 m is released in the vertical position as shown in Figure 12.12(a). If the coefficient of restitution between the floor and the cushion at the end of the rod is 0.7, determine how high the end of the rod rebounds after impact with the floor.

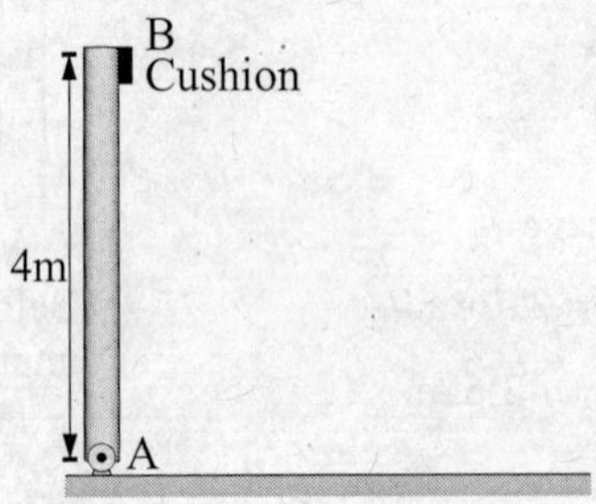

FIGURE 12.12(a) |

Solution. Three diagrams for the rod at different positions are shown in Figure 12.12(b). For the convenience, the datum, which is horizontally fixed, passes through point A.

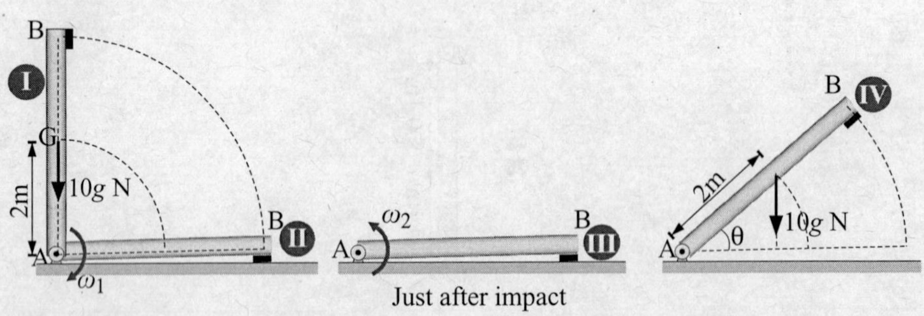

FIGURE 12.12(b) |

When the rod is in position I, the rod's weight has positive potential energy. Thus,

$$V_1 = 10 \times 9.81 \times \frac{4}{2}$$

or
$$V_1 = 196.2\text{J}$$

Kinetic energy $T_1 = 0$.

When the rod is in position II, the potential energy

$$V_2 = 0$$

Kinetic energy

$$T_2 = \frac{1}{2} I_A \omega_1^2 = \frac{1}{2} \times \frac{10 \times 4^2}{3} \times \omega_1^2$$

$$T_2 = 26.66\omega_1^2$$

Applying the principle of conservation of energy,

$$V_1 + T_1 = V_2 + T_2$$

or $$26.66\omega_1^2 = 196.2$$

Angular velocity of the rod instantly after impact is

$$\omega_1 = 2.71 \text{rad/s}$$

Therefore, the velocity of the cushion instantly before impact is

$$(v_B)_1 = \omega_1 \times 4 = 1084 \text{m/s}$$

Let $(v_B)_2$ be the velocity of the cushion instantly after impact. Applying the coefficient of restitution equation,

$$e = 0.7 = \frac{(v_B)_2}{(v_B)_1} \Rightarrow (v_B)_2 = 0.7 \times 10.84$$

$$(v_B)_2 = 7.58 \text{m/s}$$

Therefore, angular velocity of the rod just after impact $\omega_2 \frac{(v_B)_2}{AB} = \frac{7.58}{4}$

$$\omega_2 = 1.89 \text{rad/s}$$

Now applying the principle of conservation of energy between positions III and IV, we have

$$\frac{1}{2} \times I_A \omega_2^2 = mg\Delta h \Rightarrow \frac{1}{2} \times \frac{10 \times 4^2}{3} \times (1.89)^2$$

$$= 10 \times 9.81 \times 2 \sin \theta$$

From which

$$\theta = 29.38^\circ$$

From the geometry of the figure, height of cushion in position IV,

$$h = 4 \sin 29.38$$

$$h = 1.96 \text{ m}$$

PROBLEM 12.13

In Figure 12.13(a), a block of mass m is attached to a cable which is wrapped around a uniform disk of mass M. The block is released from rest and falls through a distance h before the cable becomes taut. Derive an expression for the velocity of the block and the angular velocity of the disk immediately after the impact. Assume that the impact is (a) perfectly plastic and (b) perfectly elastic.

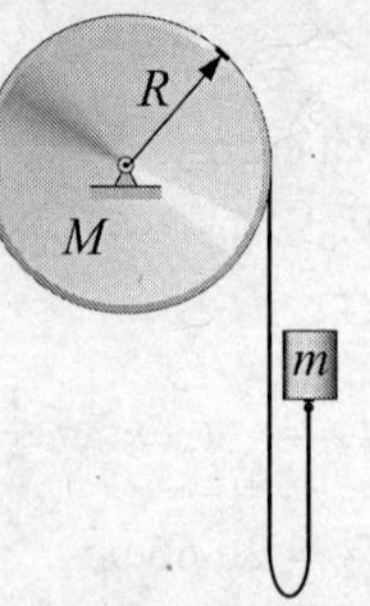

FIGURE 12.13(a) |

Solution. Velocity of block A instantly before the cable becomes tight $(u_A)_1 = \sqrt{2gh}$ (by applying the principle of conservation of energy). The velocity of block A and disk just after the impact is shown in Figure 12.13(b). The angular momentum is conserved about O.

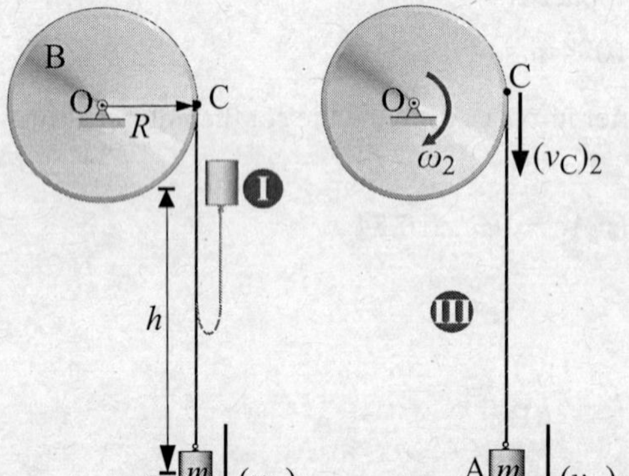

FIGURE 12.13(b)

$$(H_O)_1 = (H_O)_2$$

or $$m(u_A)_1 \times R = I_O\omega_2 + m(v_A)_2 \times R$$

or $$m\sqrt{2gh} \times R = \frac{1}{2}MR^2\omega_2 + m(v_A)_2 \times R$$

or $$m\sqrt{2gh} \times R = \frac{1}{2}MR^2\frac{(v_C)_2}{R} + m(v_A)_2 \times R$$

or $$M(v_C)_2 + 2m(v_A)_2 = 2m\sqrt{2gh} \qquad \text{(i)}$$

Applying the equation of coefficient of restitution

$$e = \frac{(v_C)_2 - (v_A)_2}{(u_A)_1 - (u_C)_1}$$

or $$e = \frac{(v_C)_2 - (v_A)_2}{\sqrt{2gh}}$$

$$(v_C)_2 - (v_A)_2 = e\sqrt{2gh} \qquad \text{(ii)}$$

Solving Equations (i) and (ii), we have

$$(v_C)_2 = \frac{2m(1+e)}{M+2m}\sqrt{2gh}$$

and

$$(v_A)_2 = \frac{2m - Me}{M+2m}\sqrt{2gh}$$

Angular velocity of disk $$\omega_2 = \frac{(v_C)_2}{R} = \frac{2m(1+e)}{M+2m}\frac{\sqrt{2gh}}{R}$$

(a) The impact is perfectly plastic; $e = 0$

$$\omega_2 = \frac{2m}{M + 2m}\frac{\sqrt{2gh}}{R}$$

and

$$(v_A)_2 = \frac{2m}{M + 2m}\sqrt{2gh}$$

(b) The impact is perfectly elastic, $e = 1$

$$\omega_2 = \frac{4m}{M + 2m}\frac{\sqrt{2gh}}{R}$$

and

$$(v_A)_2 = \frac{2m - M}{2m + M}\sqrt{2gh}$$

PROBLEM 12.14

Two homogeneous disks are free to turn in frictionless bearings in the fixed vertical frame shown in Figure 12.14(a). The upper disk A is turning with angular velocity ω clockwise when it is dropped onto the lower disk B. Assuming that the coefficient of friction μ is constant, determine the time until slipping stops. What are the angular velocities at that time?

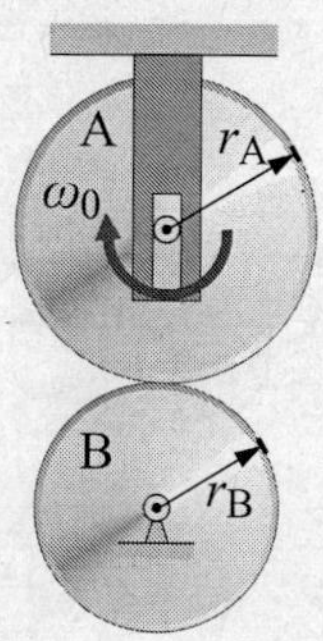

FIGURE 12.14(a) |

Solution. Free body diagram of disks A and B are shown in Figure 12.14(b).

Consider Disk A

Applying the principle of angular momentum and impulse,

$$(I_o)_A\omega_o - \mathbf{F}r_A t = (I_o)_A\omega_A,$$

or

$$(I_o)_A\omega_o - \mu m_A g r_A t = (I_o)_A\omega_A, \qquad \text{(i)}$$

where F is the friction force; $F = \mu m_A g$ during disks slips.

Consider Disk B

Applying the principle of angular momentum and impulse,

$$-\mathbf{F}r_B t = -(I_o)_B\omega_B$$

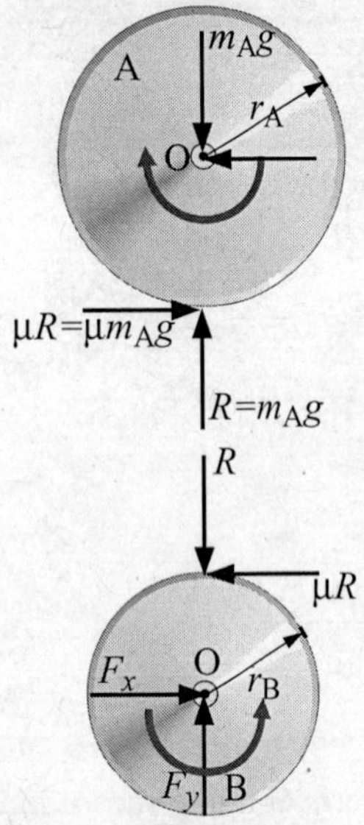

FIGURE 12.14(b) |

or
$$\mu m_A g r_B t = (I_o)_B \omega_B \tag{ii}$$

From Equations (i) and (ii), we have

$$\frac{r_A}{r_B} = \frac{(I_o)_A}{(I_o)_B}\left(\frac{\omega_o - \omega_A}{\omega_B}\right),$$

Substituting

$$(I_o)_A = \frac{1}{2} m_A r_A^2 \quad \text{and} \quad (I_o)_B = \frac{1}{2} m_B r_B^2,$$

we have

$$\frac{m_B r_B}{m_A r_A} = \left(\frac{\omega_o}{\omega_B} - \frac{\omega_A}{\omega_B}\right) \tag{iii}$$

Slipping stops when peripheral velocity of discs A and B becomes equal,

$$v = \omega_A r_A = \omega_B r_B$$

or
$$\frac{\omega_A}{\omega_B} = \frac{r_B}{r_A} \tag{iv}$$

From Equations (iii) and (iv), we have

$$\frac{m_B r_B}{m_A r_A} = \frac{\omega_o}{\omega_B} - \frac{r_B}{r_A}$$

From which

$$\omega_B = \omega_o \left(\frac{m_A}{m_A + m_B}\right)\left(\frac{r_A}{r_B}\right)$$

and

$$\omega_A = \frac{r_B}{r_A} \times \omega_o \left(\frac{m_A}{m_A + m_B}\right)\left(\frac{r_A}{r_B}\right)$$

$$\omega_A = \omega_o \left(\frac{m_A}{m_A + m_B}\right)$$

From Equation (ii),

$$t = \frac{(I_o)_B \omega_B}{\mu m_A g r_B}$$

or

$$t = \frac{\left(\frac{1}{2}\right) m_B r_B^2}{\mu m_A g r_B} \omega_o \left(\frac{m_A}{m_A + m_B}\right)\left(\frac{r_A}{r_B}\right)$$

From which

$$t = \frac{\omega_o r_A}{2\mu g\left[1 + \left(\frac{m_A}{m_B}\right)\right]}$$

PROBLEM 12.15

A uniform rectangular block is sliding on a smooth horizontal surface with a velocity v when it strikes a small step in the surface [Fig. 12.15(a)]. Determine the minimum velocity v that will permit the block to pivot about the edge of the step and just reach the standing position with no velocity.

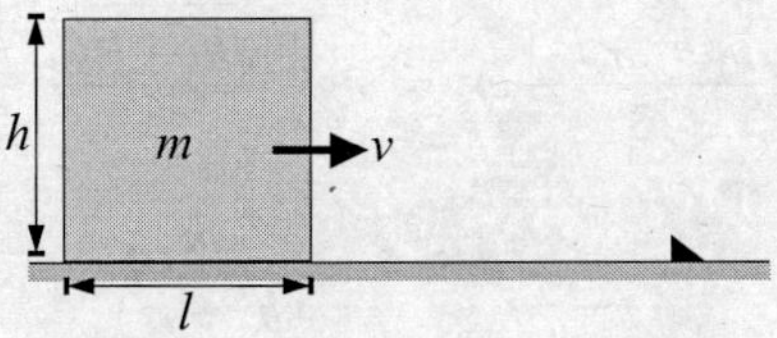

FIGURE 12.15(a) |

Solution. Initial angular momentum of the block about O just before impact

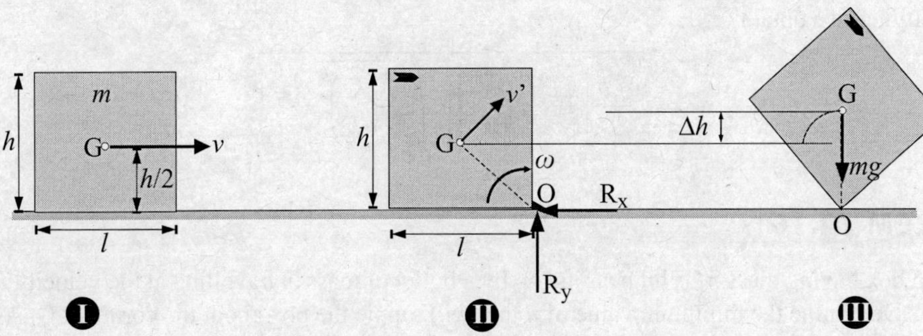

FIGURE 12.15(b) |

$$(H_o)_1 = mv \times \frac{h}{2}$$

The velocity of the centre of mass G immediately after impact is v' and angular velocity is $\omega = v'/OG$.

Angular momentum about O just after impact when the block is starting its rotation about O is $(H_o)_2 = I_o\omega$.

Angular momentum is conserved about O,

$$(H_o)_1 = (H_o)_2 \Rightarrow mv \times \frac{h}{2} = I_o\omega$$

or
$$mv \times \frac{h}{2} = \frac{m(h^2 + l^2)}{3}\omega$$

$$\omega = \frac{3}{2} \times \frac{vh}{h^2 + l^2}$$

This angular velocity will be sufficient to raise the block just in position III if the kinetic energy of rotation equals the increase in potential energy. Thus

$$\frac{1}{2} I_o\omega^2 = mg\Delta h$$

From the geometry of Figure 12.15(b),

Substituting $$\Delta h = \left(\frac{\sqrt{h^2 + l^2}}{2} - \frac{h}{2} \right)$$

$$\frac{1}{2} \times \frac{m(h^2 + l^2)}{3} \omega^2 = mg \left(\frac{\sqrt{h_2 + l^2}}{2} - \frac{h}{2} \right)$$

Substituting $$\omega = \frac{3}{2} \times \frac{vh}{(h^2 + l^2)}, \text{ we have}$$

$$\frac{1}{2} \times \frac{m(h^2 + l^2)}{3} \left[\frac{3}{2} \times \frac{vh}{(h^2 + l^2)} \right]^2 = mg \left(\frac{\sqrt{h^2 + l^2}}{2} - \frac{h}{2} \right)$$

On simplifying, we obtain

$$v = \sqrt{\frac{4gh}{3} \left[1 + \left(\frac{l}{h} \right)^2 \right] \left(\sqrt{1 + \left(\frac{l}{h} \right)^2} - 1 \right)}$$

PROBLEM 12.16

A uniform box having mass M is hit near its top by a bullet of mass m travelling at the velocity u [Fig. 12.16(a)]. Determine the minimum value of u that will topple the box about the corner at O. Assume that the bullet gets embedded and that the box does not bounce out of the corner ($M >> m$)

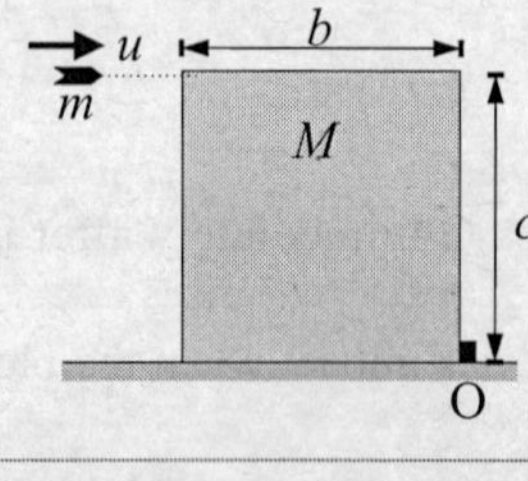

FIGURE 12.16(a) |

Solution. The initial angular momentum of bullet about O just before impact is $H_O = muc$ [Fig. 12.16(b)]. The velocity of the centre of mass G immediately after impact is v', and the angular velocity is $\omega = v'/\text{GO}$. The angular momentum about O instantly after impact when the block is starting its rotation about O is

$$H_o = I_o \omega = \frac{m(b^2 + c^2)}{3} \omega$$

Conservation of angular momentum gives

$$muc = \frac{M(b^2 + c^2)}{3} \omega$$

or $$\omega = \frac{3mc}{M(b^2 + c^2)} u$$

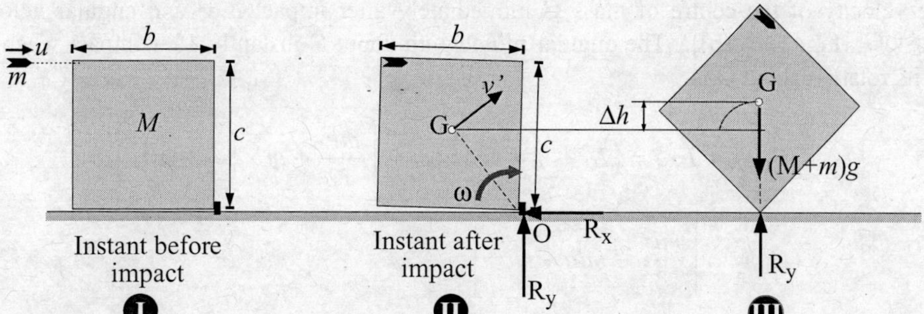

FIGURE 12.16(b) |

This angular velocity will be sufficient to raise the block just in position (III) if the kinetic energy of rotation equals the increase in potential energy

$$\frac{1}{2} I_o \omega^2 = Mg\Delta h$$

Substituting the value of ω, we have

$$\frac{1}{2} \times \frac{M(b^2 + c^2)}{3} \left(\frac{3mc}{M(b^2 + c^2)} u \right)^2 = Mg \left[\sqrt{\left(\frac{b}{2}\right)^2 + \left(\frac{c}{2}\right)^2} - \frac{c}{2} \right].$$

From which

$$u = \frac{M}{m} \sqrt{\frac{gc}{3} \left[1 + \left(\frac{b}{c}\right)^2 \right] \times \left[\sqrt{\left\{ 1 + \left(\frac{b}{c}\right)^2 \right\}} - 1 \right]}$$

PROBLEM 12.17

A uniform rod of mass m and length l is released from a height h in the horizontal position as shown in Figure 12.17(a). If point O of the rod becomes attached to the pivot at C upon impact, determine the angular velocity ω of the bar immediately after impact.

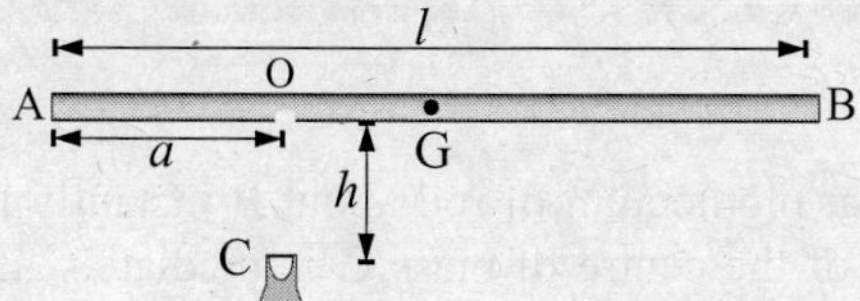

FIGURE 12.17(a) |

Solution. Velocity of the rod instantly before impact $v = \sqrt{2gh}$.

Initial angular momentum of rod about C instant before impact

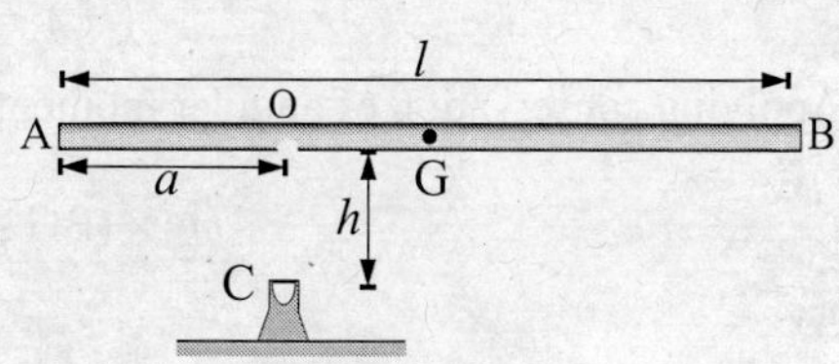

FIGURE 12.17(b) |

$$(H_C)_1 = mv \times OG = m\sqrt{2gh} \times \left(\frac{l}{2} - a \right)$$

The velocity of the centre of mass G immediately after impact is v' and angular velocity is $\omega = v' / \text{OG}$ [Fig. 12.17(b)]. The angular momentum about C instantly after impact when rod is starting its rotation about O is

$$(H_\text{C})_2 = I_\text{o}\omega = \left(I_\text{G} + m \times \text{OG}^2\right)\omega = \left[\frac{ml^2}{12} + m \times \left(\frac{l}{2} - a\right)^2\right]\omega$$

$$(H_\text{C})_2 = \left[\frac{ml^2}{3} - mla + ma^2\right]\omega$$

Applying the conservation of angular momentum about C

$$(H_\text{C})_1 = (H_\text{C})_2$$

or $$m\sqrt{2gh} \times \left(\frac{l}{2} - a\right) = \left[\frac{ml^2}{3} - mla + ma^2\right]\omega$$

$$\omega = \frac{\left[\left(\frac{l}{2}\right) - a\right]\sqrt{2gh}}{\left[\left(\frac{l^2}{3}\right) - la + a^2\right]}$$

PROBLEM 12.18

A uniform rod having mass 10 kg and length 2 m is moving to right on frictionless rollers at A and B with a velocity v when the roller at B strikes the obstruction C without rebounding [Fig. 12.18(a)]. Determine the minimum velocity v for which the rod will reach the vertical position after the impact.

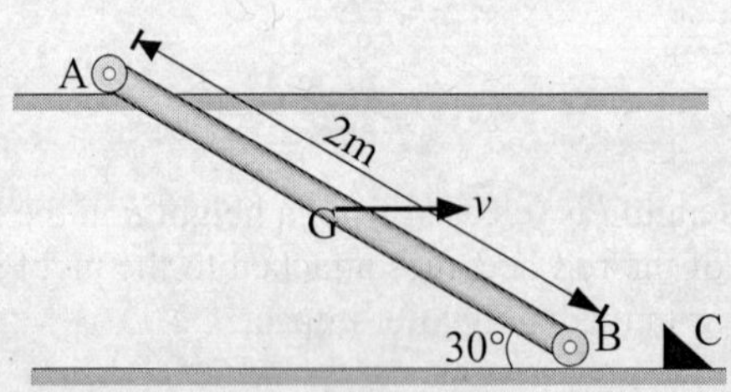

FIGURE 12.18(a) |

Solution. Initial angular momentum of rod about B instantly just before impact is $(H_\text{B})_1 = mv(\text{BG} \sin 30)$. The velocity of the centre of mass G immediately after impact is v' and angular velocity is $\omega = v' / \text{BG}$ [Fig. 12.18(b)]. The angular momentum about B instant after the impact is

$$(H_\text{B})_2 = mv' \times \text{BG} + I_\text{G}\omega = I_\text{B}\omega$$

Applying conservation of angular momentum about B,

$$mv \times (\text{BG} \sin 30) = I_\text{B}\omega$$

or $$10v \times \sin 30 = \frac{1}{3} 10 \times 2^2 \omega$$

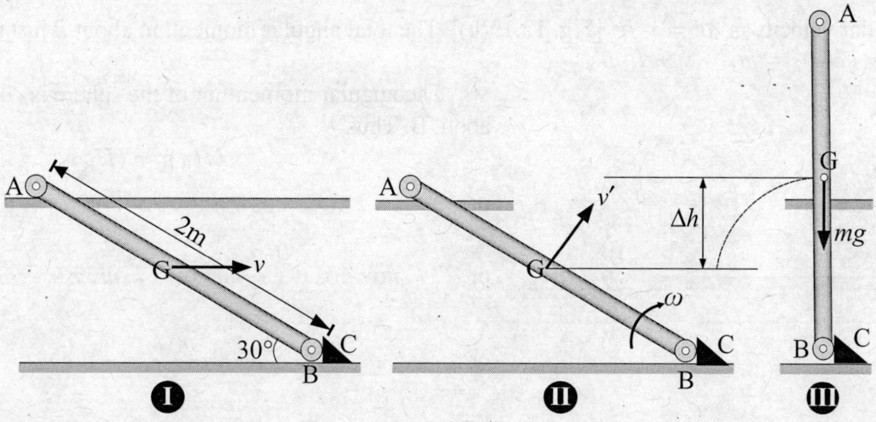

FIGURE 12.18(b)

or
$$5v = 13.3\omega$$

Applying the conservation of energy between position II and position III, we have

$$\frac{1}{2} I_B\omega^2 = mg\Delta h$$

or
$$\frac{1}{2}\times\frac{1}{3}10\times 2^2\omega^2 = 10\times 9.81(1-\sin 30)$$

or
$$\omega = 2.71\text{rad/s}$$

or
$$v = \frac{13.3}{5}\omega$$

or
$$v = \frac{13.3}{5}\times 2.71 = 7.2$$

$$v = 7.2\text{m/s}$$

PROBLEM 12.19

A homogenous sphere which rolls with a velocity v without slipping encounters an abrupt change in direction of its motion as it rolls onto the incline θ [Fig. 12.19(a)]. Determine the new velocity v' of the centre of the sphere as it starts up the incline. Also find the percentage energy loss.

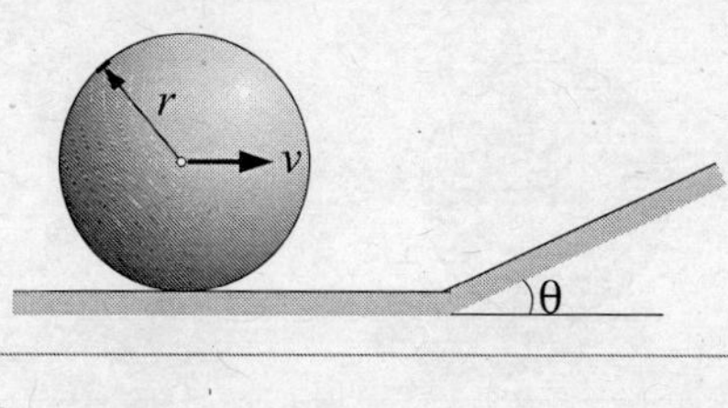

FIGURE 12.19(a)

Solution. The initial total moment of momentum of the sphere about B instantly before impact is $(H_B)_1 = mv\times r\cos\theta + I_G\omega$. The velocity of the centre of mass G immediately after impact is v'

and angular velocity is $\omega' = v'/r$ [Fig. 12.19(b)]. The total angular momentum about B just after the impact is $(H_B)_2 = mv' \times r + I_G\omega'$.

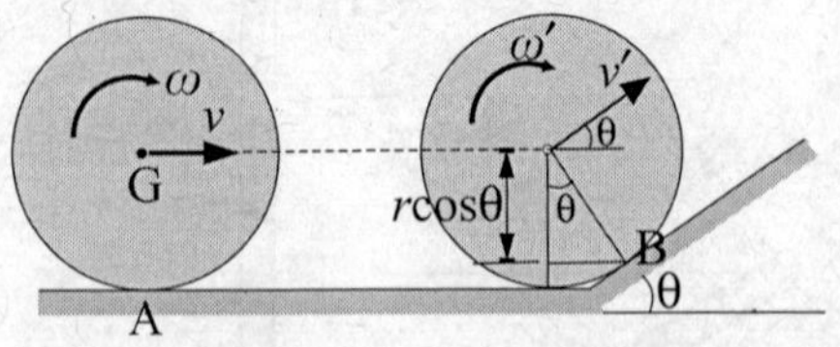

FIGURE 12.19(b) |

The angular momentum of the sphere is conserved about B. Thus

$$(H_B)_1 = (H_B)_2$$

or $$mv \times r\cos\theta + I_G\omega = mv' \times r + I_G\omega'$$

or $$mvr\cos\theta + \frac{2}{5}mr^2 \times \frac{v}{r} = mv'r + \frac{2}{5}mr^2 \times \frac{v'}{r}$$

or $$(2 + 5\cos\theta)\frac{mvr}{5} = \frac{7}{5}mv'r$$

$$v' = \frac{v}{7}(2 + 5\cos\theta)$$

The percentage loss of energy is

$$\frac{\Delta E}{E} = \frac{\left(\frac{1}{2}mv^2 + \frac{1}{2}I_G\omega^2\right) - \left(\frac{1}{2}mv'^2 + \frac{1}{2}I_G\omega'^2\right)}{\frac{1}{2}mv^2 + \frac{1}{2}I_G\omega^2}$$

Substituting

$$I_G = \frac{2}{5}mr^2, \omega' = \frac{v'}{r} \text{ and } v' = \frac{v}{7}(2 + 5\cos\theta),$$

we obtain

$$\frac{\Delta E}{E} = 1 - \left(\frac{2 + 5\cos\theta}{7}\right)^2$$

$$\frac{\Delta E}{E} \times 100 = \frac{45 - 20\cos\theta - 25\cos^2\theta}{49}$$

$$\% \text{ loss of energy} = \frac{45 - 20\cos\theta - 25\cos^2\theta}{49}$$

PROBLEM 12.20

Determine the minimum velocity v which the roller must have to just roll over the obstruction. The radius of gyration of the roller is k, and it is assumed that the roller does not slip [Fig. 12.20(a)].

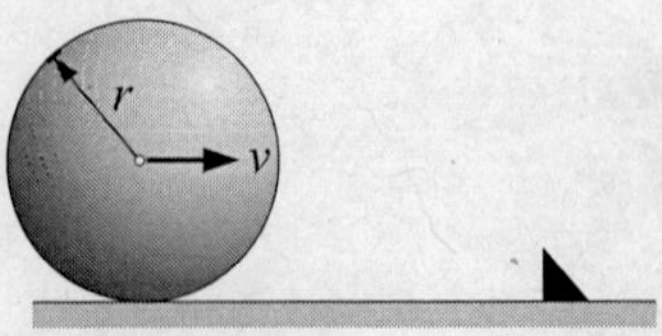

FIGURE 12.20(a) |

Solution. The initial total momentum of the roller about A just before impact is $(H_A)_1 = mv(r - h) + I_G\omega$ The velocity of the centre of mass immediately after impact is v' and the angular velocity is $\omega' = v'/r$ [Fig. 12.20(b)]. The momentum about A just after impact is $(H_A)_2 = mv'r + I_G\omega'$.

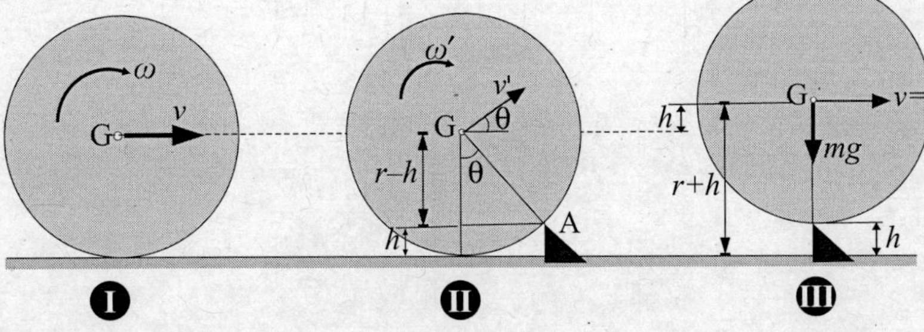

FIGURE 12.20(b) |

Applying the conservation of angular momentum about A,

$$(H_A)_1 = (H_A)_2$$

or
$$mv(r-h) + I_G\omega = mv'r + I_G\omega'$$

or
$$mv(r-h) + mk^2\frac{v}{r} = mv'r + \frac{mk^2v'}{r}$$

or
$$\left[r - h + \frac{k^2}{r}\right]mv = \left(r + \frac{k^2}{r}\right)mv'$$

$$v' = \frac{\left[r - h + \left(\frac{k^2}{r}\right)\right]v}{r + \left(\frac{k^2}{r}\right)} \quad \text{(i)}$$

This velocity will be sufficient to raise the roller position III if the kinetic energy equals the increase in potential energy. Thus,

or
$$\frac{1}{2}mv'^2 + \frac{1}{2}I\omega'^2 = mgh + [mg(r+h) - mgr]$$

or
$$\frac{1}{2}mv'^2 + \frac{1}{2}mk^2\frac{v'^2}{r^2} = mgh$$

or
$$\left(1 + \frac{k^2}{r^2}\right)v'^2 = 2gh$$

$$v' = \sqrt{\frac{2gh}{1 + \left(\frac{k^2}{r^2}\right)}} \quad \text{(ii)}$$

Equating Equations (i) and (ii), we have

$$\left[\frac{r-h+\left(\frac{k^2}{r}\right)}{r+\left(\frac{k^2}{r}\right)}\right]v=\sqrt{\frac{2gh}{1+\left(\frac{k^2}{r^2}\right)}}$$

From which

$$v=\left[\frac{r+\left(\frac{k^2}{r}\right)}{r-h+\left(\frac{k^2}{r}\right)}\right]\sqrt{\frac{2gh}{1+\left(\frac{k^2}{r^2}\right)}}$$

PROBLEM 12.21

A particle of mass m is released on the smooth inside wall of a vertical cylindrical tank at A with a velocity v_o with an angle β with the horizontal tangent. When the particle reaches at B at a distance h below A, determine an expression for the angle θ made by its velocity with the horizontal tangent at B.

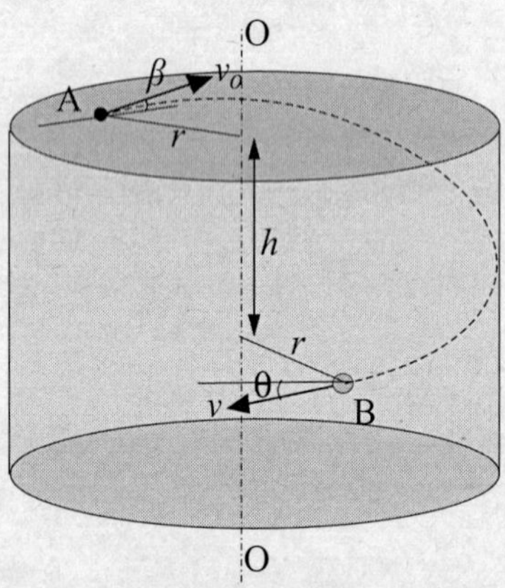

FIGURE 12.21(a) |

Solution. Since direction of reaction between particle and wall passes through OO and weight parallel to OO, angular momentum is conserved about OO [Fig. 12.21(a)]. Thus

$$(H_o)_A=(H_o)_B$$

or

$$mv_o\cos\beta\times r=mv\cos\theta r$$

or

$$v=\frac{v_o\cos\beta}{\cos\theta} \qquad \text{(i)}$$

Also, energy is conserved $(V_1+T_1=V_2+T_2)$,

$$\frac{1}{2}mv_o^2+mgh=\frac{1}{2}mv^2$$

or

$$v_o^2+2gh=v^2$$

or

$$v=\sqrt{v_o^2+2gh} \qquad \text{(ii)}$$

Eliminating v from Equations (i) and (ii), we have

$$\frac{v_o\cos\beta}{\cos\theta}=\sqrt{v_o^2+2gh}$$

$$\cos\theta=\frac{v_o\cos\beta}{\sqrt{v_o^2+2gh}}=\frac{\cos\beta}{\sqrt{1+\left(\frac{2gh}{v_o^2}\right)}}$$

$$\theta=\cos^{-1}\frac{\cos\beta}{\sqrt{1+\left(\frac{2gh}{v_o^2}\right)}}$$

PROBLEM 12.22

A small block having a mass of 0.1 kg is given a horizontal velocity 0.6 m/s tangent to the horizontal rim of smooth hemispherical bowl as shown in Figure 12.22(a). It slides along the smooth surface. When it descends to $h = 0.2$ m, determine its speed and the angle of descent θ, that is, the angle measured from the horizontal to the tangent of the path.

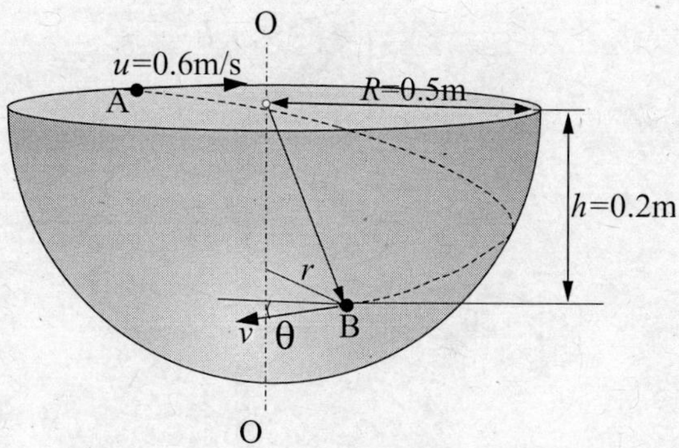

FIGURE 12.22(a)

Solution. Angular momentum of block about OO is conserved:

$$(H_A)_{oo} = (H_B)_{oo}$$

or

$$muR = mv \cos \theta r$$

or

$$uR = vr \cos \theta \tag{i}$$

Applying the principle of conservation of energy between A and B,

$$\frac{1}{2} mu^2 + mgh = \frac{1}{2} mv^2$$

or

$$v = \sqrt{u^2 + 2gh} \tag{ii}$$

From the figure

$$h^2 + r^2 = R^2$$

$$r = \sqrt{R^2 - h^2} \tag{iii}$$

Eliminating v and r from Equations (i), (ii) and (iii), we have

$$\cos \theta = \frac{uR}{\sqrt{u^2 + 2gh}\sqrt{R^2 - h^2}}$$

$$\theta = \cos^{-1} \frac{1}{\sqrt{1 + \left(\frac{2gh}{u^2}\right)}\sqrt{1 - \left(\frac{h^2}{R^2}\right)}}$$

Substituting $u = 0.6$ m/s, $R = 0.5$ m and $h = 0.2$ m,

$$\theta = \cos^{-1} 0.397$$

$$\theta = 66.58°$$

PROBLEM 12.23

Water of density 1000 kg/m^3 issues from two 40 mm diameter holes in the bucket with a velocity 3 m/s in the directions shown in Figure 12.23(a). Determine the force P required to give the bucket an upward acceleration of 0.5 m/s^2 from rest if it contains 40 kg water at that time. The empty bucket has a mass of 0.6 kg.

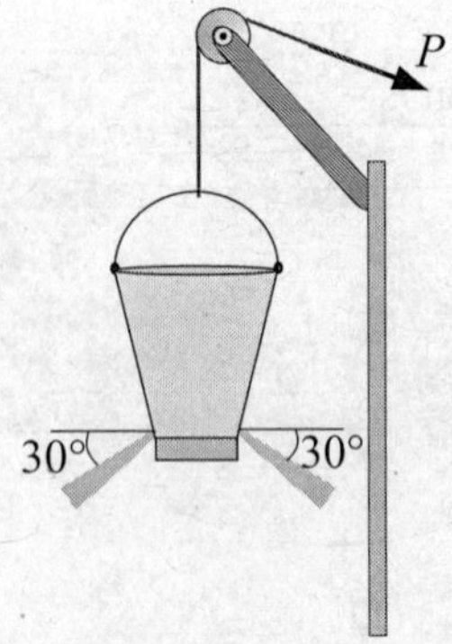

FIGURE 12.23(a) |

Solution. Figure 12.23(b) shows the free body diagram of bucket.

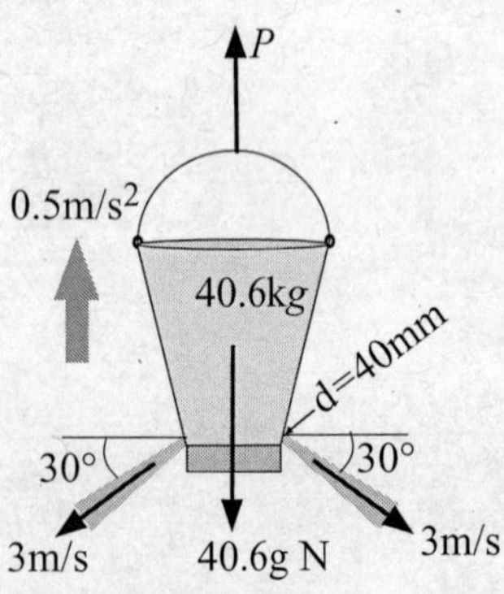

FIGURE 12.23(b) |

Net force in vertical direction

$$F = P - 40.6\text{ g}$$

Velocity of jet in vertical direction

$$(v_j)_y = 3 \sin 30 \text{m/s}$$

Mass flow rate of water from holes

$$\frac{dm}{dt} = \rho av$$

$$= 1000 \times 2 \times \frac{\pi}{4}\left(\frac{40}{1000}\right)^2 \times 3$$

Mass of water with bucket m = 40.6 kg and given acceleration of bucket

$$\frac{dv}{dt} = 0.5 \text{m/s}^2$$

Applying the equation of motion for a bucket (losing mass) in vertical direction,

$$F_y = m\frac{dv}{dt} - v_j\frac{dm}{dt}$$

$$P - 40.6g = 40.6 \times 0.5 - 3 \sin 30 \times 1000 \times 2 \times \frac{\pi}{4}\left(\frac{40}{1000}\right)^2 \times 3$$

or

$$P = 40.6 \times 9.81 + 40.6 \times 0.5 - 3 \times \frac{1}{2} \times 1000 \times 2 \times \frac{\pi}{4}\left(\frac{40}{1000}\right)^2 \times 3$$

$$P = 429.88\text{N}$$

PROBLEM 12.24

A rocket in vertical flight has a mass m_o (including fuel) at lift-off and consumes fuel at the constant rate $\dot{m}$ [Fig. 12.24(a)]. If velocity v_j of the exhaust gases relative to the rocket is constant, derive an expression for the velocity of the rocket as a function of time t.

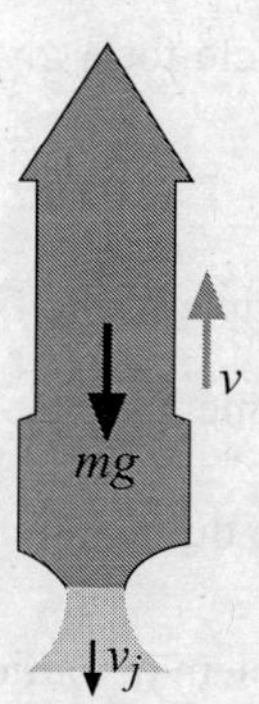

FIGURE 12.24(a) |

Solution. The equation of motion for variable mass (for rocket)

$$F = m\frac{dv}{dt} - v_j\frac{dm}{dt}$$

The force acting on rocket is

$$F = -mg,$$

$$-mg = m\frac{dv}{dt} - v_j\frac{dm}{dt}$$

The mass of the rocket at time t after lift-off is $m = m_0 - \dot{m}t$. Substituting in the above equation yields

$$-(m_0 - \dot{m}t)g = (m_0 - \dot{m}t)\frac{dv}{dt} - v_j\frac{d}{dt}(m_0 - \dot{m}t)$$

or
$$(m_0 - \dot{m}t)\frac{dv}{dt} = v_j\dot{m} - (m_0 - \dot{m}t)g$$

or
$$\frac{dv}{dt} = v_j\frac{\dot{m}}{(m_0 - \dot{m}t)} - g$$

Integration with respect to time gives

$$v = -v_j \log_e(m_0 - \dot{m}t) - gt + c$$

The initial condition $v = 0$ when $t = 0$, we get

$$c = v_j \log_e m_0$$

From which

$$v = v_j \log_e \frac{m_0}{(m_0 - \dot{m}t)} - gt$$

CHAPTER HIGHLIGHTS

- Linear Momentum is the product of the mass and the velocity of the particle. $\boldsymbol{p} = m\boldsymbol{v}$
- The force that acts on a particle is equal to the time rate of change of the momentum of the particle. $\boldsymbol{F} = \dfrac{\mathrm{d}\boldsymbol{p}}{\mathrm{d}t}$
- If the resultant force that acts on a particle for a given time interval is zero, momentum is conserved.
- Impulse is the product of force and time. $\mathbf{I} = \int_{t_1}^{t_2} \boldsymbol{F}\,\mathrm{d}t$
- The impulse of force that acts on a particle in the time interval $t_2 - t_1$ is equal to the change of momentum of the particle in the same time interval. $\int_{t_1}^{t_2} \boldsymbol{F}\,\mathrm{d}t = mv_2 - mv_1$
- If the resultant force on a particle is zero during an interval of time, linear momentum of particle remains constant.
- The change in linear momentum of a system of particles in any time interval is equal to the linear impulse, over the same time interval, of the resultant of the external forces that act on the body. $\int_{t_1}^{t_2} \boldsymbol{F}\,\mathrm{d}t = m(\boldsymbol{v}_G)_2 - m(\boldsymbol{v}_G)_1$
- The angular momentum of a particle about a point is the cross product of the position vector of the particle and the linear momentum of the particle. $\mathbf{H}_O = r \times m\boldsymbol{v}$
- The resultant moment about point O of all the forces acting on the particle is equal to the time rate of change of the particle's angular momentum about O. $\mathbf{M}_O = \dfrac{\mathrm{d}\mathbf{H}_O}{\mathrm{d}t}$
- Angular Impulse is the product of moment and time. Angular impulse $= \int_{t_1}^{t_2} \mathbf{M}_O\mathrm{d}t$
- The change of the angular momentum of a particle in given time interval is equal to the angular impulse of the sum of the moments of the external forces that act on the particle during same time interval. $\int_1^2 \mathbf{M}_O\mathrm{d}t = (\mathbf{H}_O)_2 - (\mathbf{H}_O)_1$
- If the resultant moment acting on a particle is zero during an interval of time, angular momentum of particle remains constant.
- Angular Momentum of Rigid Body $\mathbf{H}_G = \mathrm{I}_G\,\omega$.
- Principle of angular impulse and momentum for rigid body is $I_o\omega_1 + \sum \int_{t_1}^{t_2} \mathbf{M}_o\mathrm{d}t = I_o\omega_2$.
- Impact is a collision between two bodies which occurs in a very small interval of time, and during which the two bodies exert on each other relatively large forces. There are two type of impact (*i*) central impact (*ii*) oblique impact.
- Coefficient of Restitution is the ratio of restitution impulse to deformation impulse. $e = \dfrac{\int \mathbf{R}\mathrm{d}t}{\int \mathbf{P}\mathrm{d}t}$ where $e = 0$ for Perfectly Plastic Impact and $e = 1$ for Perfectly Elastic Impact
- Equation of motion for variable mass system is

$$\mathbf{F} = m\frac{\mathrm{d}v}{\mathrm{d}t} \pm (v - u)\frac{\mathrm{d}m}{\mathrm{d}t} \qquad \text{(–ve for lossing mass, +ve for gaining mass)}$$

Thrust (Reaction) $\mathrm{T} = v_{rel}\dfrac{\mathrm{d}m}{\mathrm{d}t}$

REVIEW QUESTIONS

Q1. State the theorem of principle of momentum. From it, obtain the law of conservation of momentum.

Q2. Define impulse. Derive the equation that relates impulse to the change of momentum of a system.

Q3. State the theorem of moment of momentum. From it, obtain the law of conservation of moment of momentum.

Q4. Define the coefficient of restitution. Two bodies are having direct impact. Find an expression for the coefficient of restitution in terms of initial and final velocities of the two bodies.

Q5. What is impact? What are the kinds of impact?

Q6. Obtain an expression for the velocity attained by a rocket starting from rest, at a given time t.

Q7. The position vector of a particle of mass 5 kg is given by $\mathbf{r} = (2t^2 i + 3j)m$ where t is in seconds. Determine the angular momentum and the torque acting on the particle about the origin.

[Ans; $60tk$ kg m^2/s, $-60k$ N m].

Q8. The force F_x acting on a 2 kg particle varies in time as shown in Figure 12.Q8. Find (i) the impulse of the force, (ii) the final velocity of the particle if it is initially at rest and (iii) the final velocity of the particle if it is initially moving along the x-axis with a velocity of −2 m/s.

[Ans; (i) 12.0 kg m/s, (ii) 6.0 m/s, (iii) 4.0 m/s].

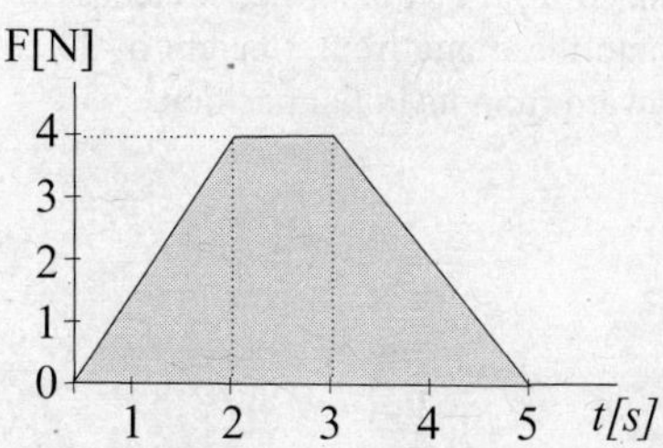

FIGURE 12.Q8

Q9. A 3 kg steel ball strikes a massive wall with a speed of 10 m/s at an angle of 60° with the surface. It bounces off with the same speed and angle (Fig. 12.Q9). If the ball is in contact with the wall for 0.20 s, what is the average force exerted on the ball by the wall?

[Ans; 260 N].

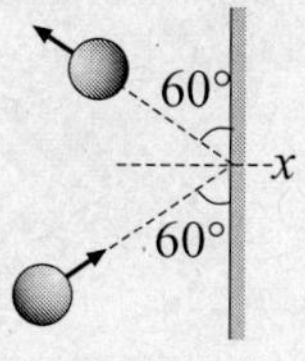

FIGURE 12.Q9

Q10. A 5 g bullet is fired horizontally into a 1.20 kg wooden block resting on a horizontal surface. The coefficient of kinetic friction between block and surface is 0.20. The bullet remains embedded in the block, which is observed to slide 0.230 m along the surface before stopping. What was the initial speed of the bullet?

[Ans; 229 m/s].

Q11. A mass m is attached to a cord passing through a small hole in a frictionless, horizontal surface as shown in Figure 12.Q11. The mass is initially orbiting in a circle of radius r_o with velocity v_o. The cord is then slowly pulled from below, decreasing the radius of the circle to r. (i) What is the velocity of the mass when the radius is r? (ii) Find the tension in the cord as a function of r. (iii) How much work is done in moving m from r_o to r?

$$\left[\text{Ans; (i) } v = \frac{v_o r_o}{r}, \text{ (ii) } T = \frac{m v_o^2 r_o^2}{r^3}, \text{ (iii) } U = \frac{1}{2} m v_o^2 \left(\frac{r_o^2}{r^2} - 1\right)\right].$$

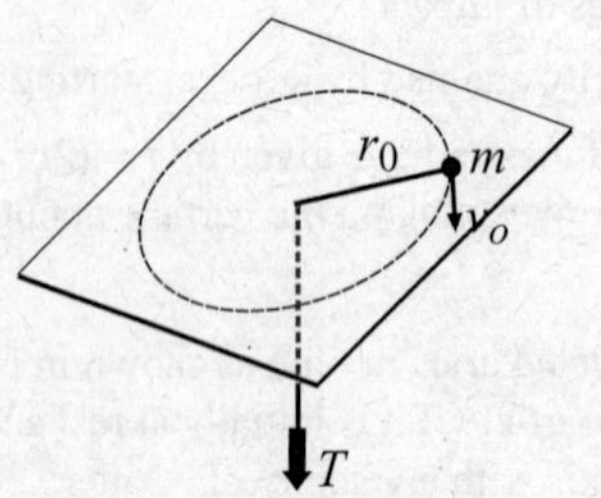

FIGURE 12.Q11 |

Q12. A 0.150 kg frame, when suspended from a coil spring, stretches the spring by 0.050 m. A 0.20 kg lump of putty is dropped from rest onto the frame from a height of 30 cm (Fig. 12.Q12). Find the maximum distance the frame moves downward from its initial position.

[Ans; 23.2 cm].

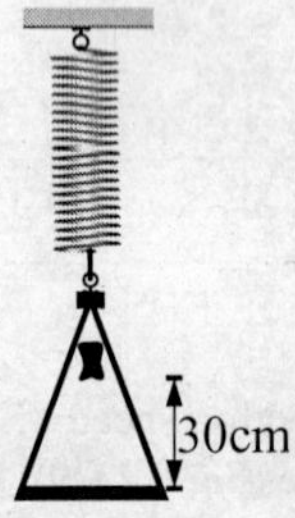

FIGURE 12.Q12 |

Q13. Consider a frictionless track ABC as shown in Figure 12.Q13. A block of mass $m_1 = 5$ kg is released from A. It makes a head-on elastic collision with a block of mass $m_2 = 10$ kg at B, initially at rest. Calculate the maximum height to which m_1 will rise after the collision.

[Ans; 0.556 m].

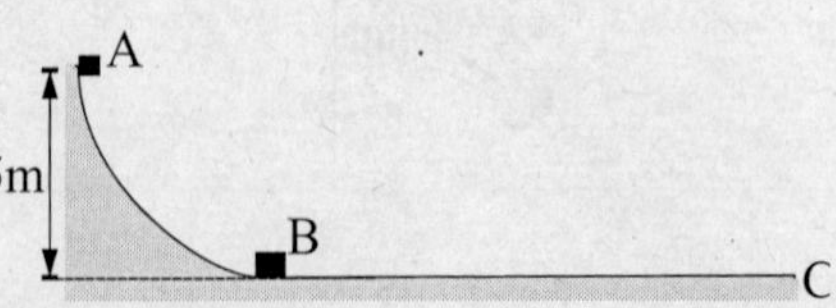

FIGURE 12.Q13 |

Q14. A rocket engine consumes fuel 80 kg/s. If the exhaust velocity is 2.5×10^3 m/s, calculate the thrust on the rocket.

[Ans; 200 kN].

Q15. Fuel aboard a rocket has a density of 1.4×10^3 kg/m^3 and is ejected with a speed of 3000 m/s. If the engine is to provide a thrust of 2500 kN, what volume of fuel must be burned per second.

[Ans; 0.595 m^3/s].

Q16. A 5 g bullet moving with an initial speed of 400 m/s is fired into and passes through a 1 kg block, as in Figure 12.Q16. The block, initially at rest on a frictionless, horizontal surface, is connected to a spring of force constant 900 N/m. If the block moves a distance of 5 cm to the right after impact, find (i) the speed at which the bullet emerges from the block and (ii) the energy lost in collision.

[Ans; (i) 100 m/s, (ii) 374 J].

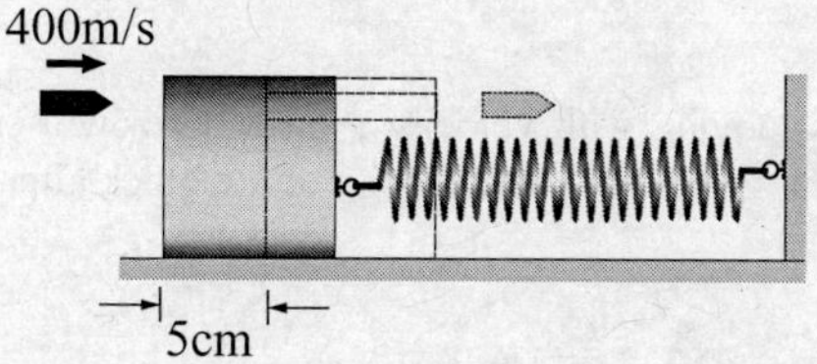

FIGURE 12.Q16 |

Q17. A chain of length L and total mass M is released from rest with its lower end just touching the top of a table as shown in Figure 12.Q17. Find the force of the table on the chain after the chain has fallen through a distance x, as shown in the figure. (Assume each link comes to rest the instant it reaches the table.)

$$\left[\text{Ans; } \frac{3Mgx}{L}\right].$$

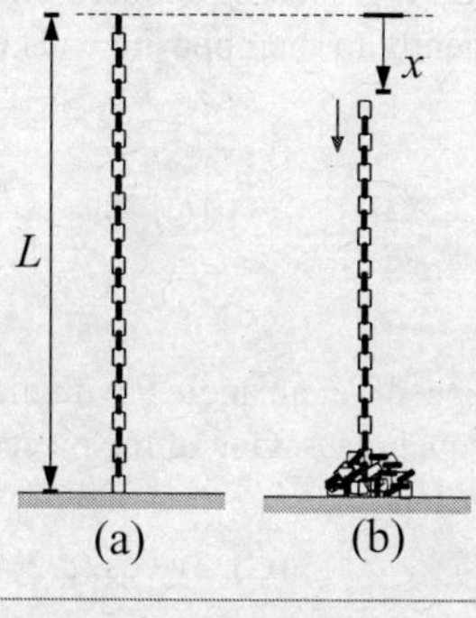

FIGURE 12.Q17 |

MULTIPLE CHOICE QUESTIONS

Q1. A ball, of mass m, strikes the ground at an angle α as shown in Figure 12.MQ1 and is deflected at angle β. The coefficient of restitution will be

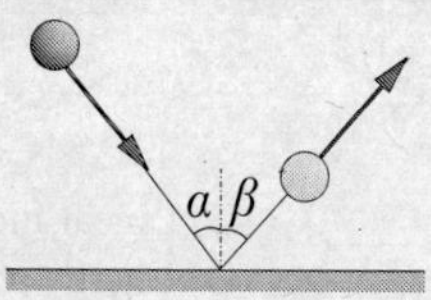

FIGURE 12.MQ1 |

(A) $\dfrac{\tan \alpha}{\tan \beta}$ (B) $\dfrac{\cos \alpha}{\cos \beta}$ (C) $\dfrac{\sin \alpha}{\sin \beta}$ (D) $\dfrac{\tan \beta}{\tan \alpha}$

[Ans: A]

Q2. A ball falls from a height of 1 m, hits the ground and rebounds with half its velocity just before impact. Then after rising, it falls and hits the ground and again rebounds with its velocity just before impact, and so on. The total distance travelled by the ball till it comes to rest on the ground is

(A) 2 m (B) $\frac{5}{3}$ m (C) $\frac{5}{2}$ m (D) $\frac{5}{4}$ m

[Ans: B]

Q3. A cubical block of side a is moving with velocity v on a horizontal smooth plane as shown in Figure 12.MQ3. It hits a ridge at point O. The angular velocity of the block after it hits O is

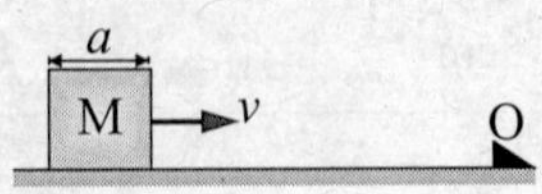

FIGURE 12.MQ3 |

(A) $\dfrac{3v}{4a}$ (B) $\dfrac{3v}{2a}$ (C) $\dfrac{2v}{3a}$ (D) zero

[Ans: A]

Q4. A thin circular ring of mass M and radius R is rotating about its axis with a constant angular velocity ω. Two objects each of mass m are attached gently to the opposite ends of a diameter of the ring. The wheel now rotates with an angular velocity

(A) $\omega \dfrac{M}{(m+M)}$ (B) $\omega \dfrac{(M-2m)}{(M+2m)}$ (C) $\omega \dfrac{M}{(M+2m)}$ (D) $\omega \dfrac{(M+2m)}{M}$

[Ans: C]

Q5. A shell is fired from the canon with a speed v at an angle θ with the horizontal direction. At the highest point in its path, it explodes into pieces of equal mass. One of the pieces retraces its path to the canon. The speed of other piece immediately after explosion is

(A) $3v \cos \theta$ (B) $2v \cos \theta$ (C) $3v \cos (\theta/2)$ (D) $(3/2)^{0.5} v \cos \theta$

[Ans: A]

Q6. A wheel of centroidal radius of gyration k is rolling on a horizontal surface with a constant velocity (Fig. 12.MQ6). It comes across an obstruction of height h, and, because of its rolling speed, just overcomes the obstruction. To determine v, one should use the principle(s) of conservation of

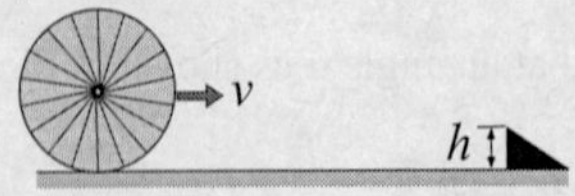

FIGURE 12.MQ6 |

(A) Energy (B) Linear momentum
(C) Energy and linear momentum (D) Energy and angular momentum

[Ans: D]

Q7. A bullet of mass 0.1 kg is fired with a velocity of u from a gun of mass 10 kg. The ratio of kinetic energy of bullet and gun is

(A) 1 (B) 10 (C) 100 (D) 1000

[Ans: C]

Q8. A bullet of mass m travels at a very high velocity v (as shown in Fig. 12.MQ8) and gets embedded inside a block of mass M initially at rest on a rough horizontal floor. The block with the bullet is seen to move a distance S along the floor. Assuming μ to be the coefficient of kinetic friction between the block and the floor and g the acceleration due to gravity, what is the velocity v of the bullet?

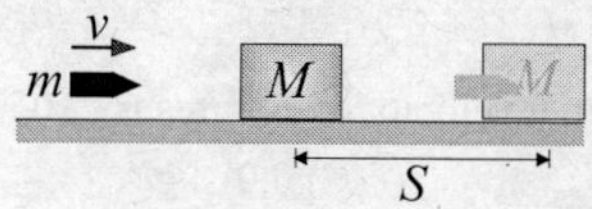

FIGURE 12.MQ8 |

(A) $\frac{(M+m)}{M}\sqrt{2\mu gS}$ m/s (B) $(M-m)\sqrt{2\mu gS}$ m/s

(C) $\mu(M+m)\sqrt{2gS}$ m/s (D) $M\sqrt{2\mu gS}$ m/s

[Ans: A]

Q9. When any two elastic bodies collide and move in the same direction, the velocity of separation is

(A) independent of their initial velocities (B) the difference of their initial velocities
(C) the sum of their initial velocities (D) the average of their initial velocities

[Ans: B]

Q10. A ball moving with a momentum of 15 kg m/s strikes against a wall at an angle of 45° and is reflected at the same angle. The impulse is

(A) 0 (B) 15.1 kg m/s (C) 21.2 kg m/s (D) 30 kg m/s

[Ans: C]

Q11. A bullet is fired from a gun. The force on the bullet is given by $F = 600 - 2 \times 10^5 t$, where F is in Newton and t in seconds. The force on the bullet becomes zero as soon as it leans the barrel. The average impulse imparted to the bullet is

(A) 0.9 N s (B) 1.8 N s (C) 4.9 N s (D) 9 N s

[Ans: A]

Q12. A shot of mass 0.2 kg fired at the bob of a pendulum of mass 1 kg hits it with a velocity of 20 m/s and remains stuck on it after impact (Fig. 12.MQ12). After the impact, the bob will swing from its initial position to an approximate height h of

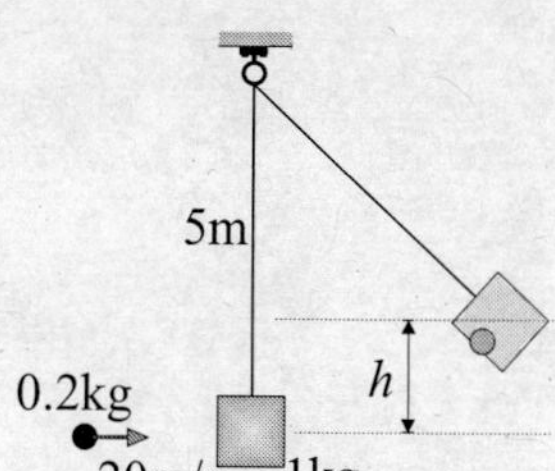

FIGURE 12.MQ12 |

(A) 10 m (B) 4.08 m (C) 3.39 m (D) 0.56 m

[Ans: D]

Q13. An elastic ball hits a horizontal floor vertically with a speed of u and is allowed to strike the floor twice again after which it rebounds from the floor with a velocity $(27/64)u$. Neglecting the resistance of air, the coefficient of elasticity between the ball and the floor is

(A) $\frac{4}{5}$ (B) $\frac{3}{4}$ (C) $\frac{2}{3}$ (D) $\frac{9}{16}$

[Ans: B]

Q14. A 500 kg rocket is set for firing vertically. The exhaust speed 800 m/s. To give an initial upward acceleration of 20 m/s^2, the amount of gas ejected per second to supply the needed thrust will be

(A) 127.5 kg/s (B) 137 kg/s (C) 155.5 kg/s (D) 186.3 kg/s

[Ans: D]

Q15. If the force on a rocket moving with a velocity of 300 m/s is 210 N, then the rate of combustion of fuel is

(A) 0.7 kg/s (B) 1.4 kg/s (C) 0.07 kg/s (D) 10.7 kg/s

[Ans: A]

13 Vibrations

INTRODUCTION

A *vibration* is the motion of a body or system of connected bodies that oscillates about a position of equilibrium. In general, there are two types of vibration, *free* and *forced*. When the motion is maintained by the restoring forces only, the vibration is said to be a *free vibration*. When a periodic force is applied to the system, the resulting motions are described as a *forced vibration*. Free and forced vibrations may be either damped or undamped. When the effects of internal or external friction may be neglected, the vibration is said to be *undamped free vibration*. However, all vibrations are actually damped to some degree. The amplitude of the damped free vibration slowly decreases until, after a certain time, the motion comes to a stop.

DEGREES OF FREEDOM

The number of independent coordinates required to describe a vibratory system is known as its degree of freedom. A simple spring-mass system and a simple pendulum oscillating in one plane [part (a) of Fig. 13.1] are the examples of systems with a single degree of freedom. A two-mass two-spring system constrained to move in one direction and a double pendulum belong to systems with two degrees of freedom [part (b) of Fig. 13.1]. A system which has continuously distributed mass, such as a string stretched between two supports [part (c) of Fig. 13.1], has infinite degrees of freedom, as such the system is equivalent to an infinite number of masses concentrated at different points.

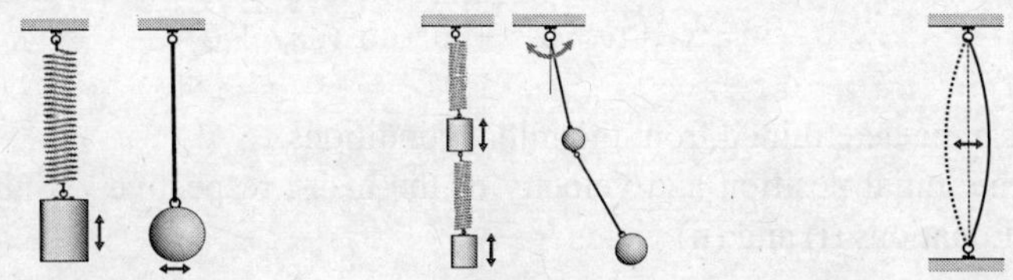

FIGURE 13.1

FREE VIBRATIONS OF PARTICLES

Figure 13.1 shows a mass m that is suspended from an ideal spring of stiffness k. If we consider only the vertical movement, there is only one degree of freedom, represented by the position coordinate x, measured downward from the equilibrium position of the mass. When the mass is at the equilibrium position, the elongation of the spring, called the static extension, is

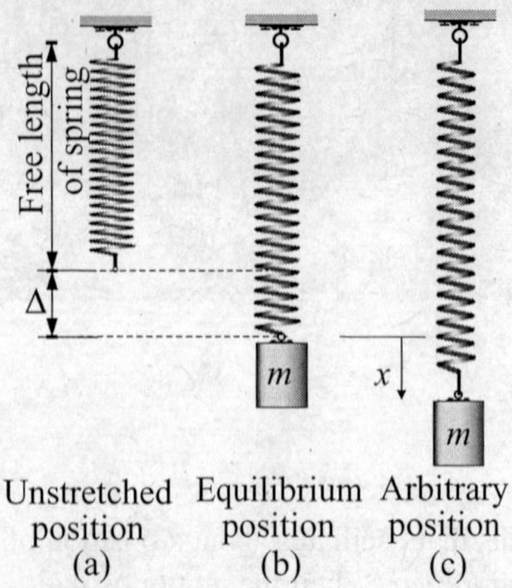

FIGURE 13.2 |

$$\Delta = \frac{mg}{k}$$

Figure 13.2 shows the free body diagram of the mass is at arbitrary, non-equilibrium position. We note that the forces acting on the particle are its weight and the force exerted by the spring which, in this position, has a magnitude $F_{spring} = k(\Delta + T)$.

The equation of motion of the mass is

$$\Sigma Fx = ma_x \Rightarrow mg - k(\Delta + x) = m\ddot{x}$$

Because $mg = k\Delta$, this equation reduces to

$$m\ddot{x} + kx = 0,$$

which can be written as

$$\ddot{x} + \omega_n{}^2 x = 0, \qquad \text{(i)}$$

where

$$\omega_n = \sqrt{\frac{k}{m}}$$

The motion defined by Equation (i) is called simple harmonic motion. It is characterized by the fact that the acceleration is proportional to the displacement and is of opposite direction. The solution of the second-order linear differential equation in Equation (i) is

$$x = A\cos\omega_n t + B\sin\omega_n t$$

Or, equivalently

$$x = \sqrt{A^2 + B^2}\,\sin(\omega_n t + \phi) \qquad \text{where } \phi = \tan^{-1}\left(\frac{A}{B}\right)$$

Velocity of mass

$$v = \dot{x} = \omega_n\sqrt{A^2 + B^2}\,\cos(\omega_n t + \phi) \qquad \text{(ii)}$$

and constant A and B can be determined from the initial conditions.

Let x_0 and v_0 be the initial position and velocity of the mass, respectively. Substituting $x = x_0$, $\dot{x} = v_0$ and $t = 0$ into Equations (i) and (ii) yields

$$A = x_0, \qquad B = \frac{v_0}{\omega_n}, \qquad \phi = \tan^{-1}\left(\frac{x_0\omega_n}{v_0}\right)$$

Therefore, the motion is described by

$$x = x_0\cos\omega_n t + \frac{v_0}{\omega_n}\sin\omega_n t$$

or

$$x = \sqrt{x_0{}^2 + \left(\frac{v_0}{\omega_n}\right)^2}\,\sin\left[\omega_n t + \tan^{-1}\left(\frac{x_0\omega_n}{v_0}\right)\right]$$

GRAPHICAL REPRESENTATION OF MOTION

The motion may be represented graphically, as in Figure 13.3, where x is seen to be the projection onto a vertical axis of the rotating vector of length $\sqrt{x_0{}^2 + \left(\frac{v_0}{\omega_n}\right)^2}$. The vector rotates at a constant angular velocity ω_n, which is called the natural circular frequency and has the units rad/s. As shown in the figure, τ denotes the period of the motion, that is, the time taken by one complete cycle of the motion. Therefore, $\omega_n T = 2\pi$, which gives

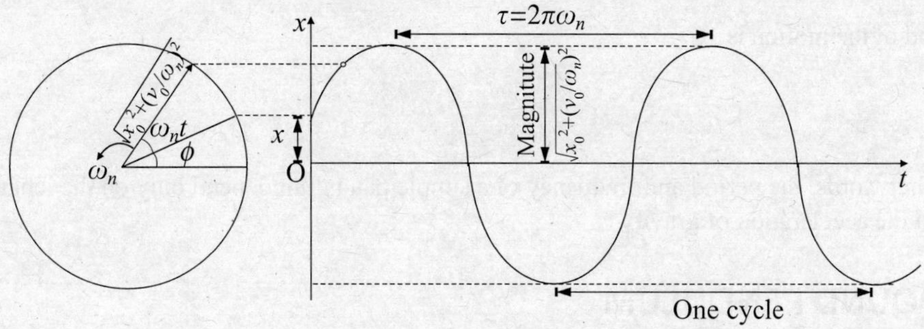

FIGURE 13.3 |

$$\text{period} = \tau = \frac{2\pi}{\omega_n}$$

The frequency of the motion is the number of cycles completed per unit time

$$\text{frequency} = f = \frac{1}{\tau} = \frac{\omega_n}{2\pi}$$

SIMPLE PENDULUM

The simple pendulum consists of a point mass m suspended by a light string of length l, where the upper end of the string is fixed as shown in Figure 13.4. The motion occurs in a vertical plane and is driven by the force of gravity. We shall show that the motion is that of a simple harmonic oscillator, provided the angle θ that the pendulum makes with the vertical is small.

The forces acting on the mass are the tension T and the weight mg. The tangential component of the weight, $mg \sin \theta$, always acts toward O, opposite the displacement. Therefore, the tangential force is a restoring force, and we can write the equation of motion in the tangential direction

$$F_t = -mg \sin \theta = m \frac{d^2 s}{dt^2},$$

where s is the displacement measured along the arc and the minus sign indicates that F_t acts toward the equilibrium position. Since $s = l\theta$, this equation reduces to

$$ml \frac{d^2 \theta}{dt^2} = -mg \sin \theta$$

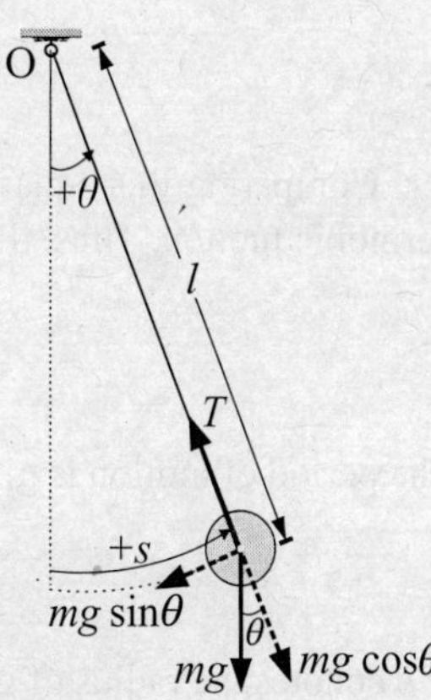

FIGURE 13.4 |

If we assume that θ is small, we can use the approximation $\sin\theta \approx \theta$, where θ is measured in radians. Therefore, the equation of motion becomes

$$\frac{d^2\theta}{dt^2} + \frac{g}{l}\theta = 0$$

Since this equation is in the standard form, the circular frequency is

$$\omega_n = \sqrt{\frac{g}{l}}$$

The period of the motion is

$$\tau = \frac{2\pi}{\omega_n} = 2\pi\sqrt{\frac{g}{l}}$$

In other words, the period and frequency of a simple pendulum depend only on the length of the string and the acceleration of gravity.

COMPOUND PENDULUM

A compound pendulum consists of any rigid body suspended from a fixed axis that does not pass through the body's centre of gravity. The body will oscillate when displaced from its equilibrium position. Consider a rigid body pivoted at a point O that is at a distance l from the centre of gravity, as shown in Figure 13.5. The restoring torque about O is provided by the force of gravity (mg), and its magnitude is ($mg \sin\theta$) l. Therefore, we can write the equation of motion

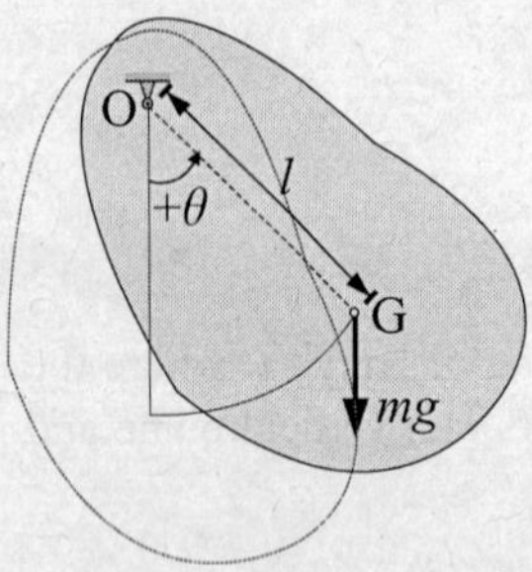

FIGURE 13.5 |

$$-(mg\sin\theta)l = I_o\frac{d^2\theta}{dt^2},$$

where $[M = I\ddot{\theta}]$ and I_o is the moment of inertia of body about O.

The negative sign shows that the restoring torque is clockwise when the displacement is counter-clockwise, and vice versa.

If we again assume that θ is small, then the approximation $\sin\theta = \theta$ is valid and the equation of motion reduces to

$$\frac{d^2\theta}{dt^2} + \frac{mgl}{I_o}\theta = 0$$

Comparing this equation with equation $\ddot{x} + \omega_n^2 x = 0$, which is the "standard form" for simple harmonic motion. Thus, the circular frequency is given by

$$\omega_n = \sqrt{\frac{mgl}{I_o}}$$

The period of motion is given by

$$\tau = \frac{2\pi}{\omega_n} = 2\pi\sqrt{\frac{I_o}{mgl}}$$

If k denotes the radius of gyration of the body about G, we have

$$I_o = I_G + ml^2 \quad \text{or} \quad I_o = mk^2 + ml^2$$

Then the above equation becomes

$$\tau = 2\pi\sqrt{\frac{l + \frac{k^2}{l}}{g}}$$

Minimum value of period

$$\tau_{\min} = 2\pi\sqrt{\frac{2k}{g}} \quad \text{at} \quad l = k$$

TORSIONAL PENDULUM

Figure 13.6 shows a rigid body suspended by a wire attached at the top to a fixed support. When the body is twisted through some small angle θ, the twisted wire exerts a restoring torque on the body proportional to the angular displacement. That is

$$M = -k_\theta \theta,$$

where k_θ is called the torsional constant of the wire.

torsion equation, $$\frac{T}{J} = \frac{G\theta}{l} \Rightarrow k_\theta = \frac{T}{\theta} = \frac{GJ}{l}$$

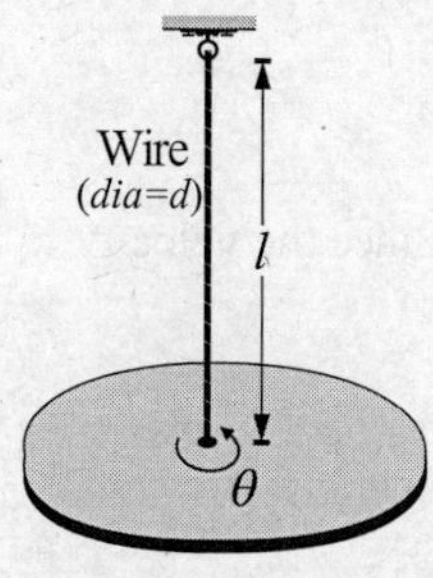

FIGURE 13.6

Applying Newton's second law for rotational motion gives

$$-k_\theta \theta = I\frac{d^2\theta}{dt^2}$$

or $$\frac{d^2\theta}{dt^2} + \frac{k_\theta}{I}\theta = 0$$

Since this equation is in the "standard form", the circular frequency is $\omega_n = \sqrt{\frac{k_\theta}{I}}$ and the period

$$\tau = 2\pi\sqrt{\frac{I}{k_\theta}}$$

This system is called a torsional pendulum.

ENERGY METHOD

In a conservative system, the total energy is constant, and the differential equation of motion can also be established by the principle of conservation of energy. For the undamped free vibration system, the energy is partly potential and partly kinetic. The kinetic energy (KE) is stored in the mass by virtue of its velocity, whereas the potential energy (PE) is stored in the form of strain energy in elastic deformation or work done in a force field such as gravity. The total energy being constant, its rate of change is zero:

$$\text{KE} + \text{PE} = \text{constant},$$

$$\frac{d}{dt}(\text{KE} + \text{PE}) = 0$$

To illustrate this alternative approach, consider the block and spring shown in Figure 13.7 when the block is displaced by an arbitrary amount x from the equilibrium position, the kinetic energy is

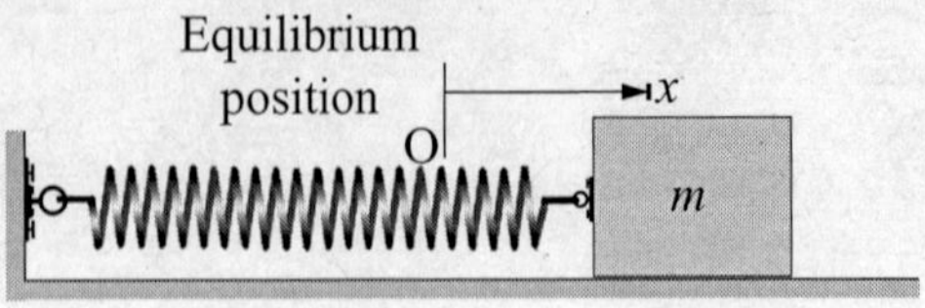

FIGURE 13.7 |

$$KE = \frac{1}{2} m\dot{x}^2$$

and the potential energy is

$$PE = \frac{1}{2} kx^2$$

By the conservation of energy equation

$$\frac{d}{dt}(KE + PE) = 0$$

$$\frac{d}{dt}\left(\frac{1}{2} m\dot{x}^2 + \frac{1}{2} kx^2\right) = 0$$

or

$$m\dot{x}\ddot{x} + k\dot{x}x = 0$$

or

$$\left(\ddot{x} + \frac{k}{m} x\right)\dot{x} = 0$$

Since the velocity $\dot{x}$ is not always zero in a vibrating system,

$$\ddot{x} + \frac{k}{m} x = 0$$

or

$$\ddot{x} + \omega_n^2 x = 0,$$

which is the same as Equation (i).

If the energy equation is written for a system of connected bodies, the natural frequency, or the equation of motion can also be determined by time differentiation.

SOLVED PROBLEMS

PROBLEM 13.1

Determine the period τ for small oscillations of an assembly composed of two light bars and two particles, each of mass m, as shown in Fig. 13.1(a).

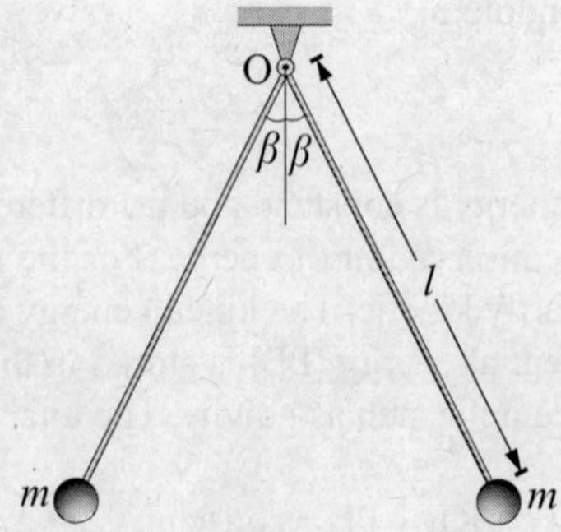

FIGURE 13.1(a) |

Solution. The moment of inertia of the system about O is

$$I_o = ml^2 + ml^2 = 2ml^2$$

Let the system be displaced through a small angle θ, as shown in Figure 13.1(b).

Then restoring moment about O is

$-2mg \sin\theta \times OG = -2mg\,\theta \times l \cos\beta$,

since θ is small.

Therefore, the equation of motion is

$$I_o\ddot{\theta} = -2mg\theta \times l \cos\beta$$

or

$$2ml^2\ddot{\theta} + 2mgl\cos\beta\theta = 0,$$

which reduces to

$$\ddot{\theta} + \frac{g\cos\beta}{l}\theta = 0$$

Hence,

$$\omega_n = \sqrt{\frac{g\cos\beta}{l}}$$

Therefore,

$$\tau = \frac{2\pi}{\omega_n} = 2\pi\sqrt{\frac{l}{g\cos\beta}}$$

$$\tau = 2\pi\sqrt{\frac{l}{g\cos\beta}}$$

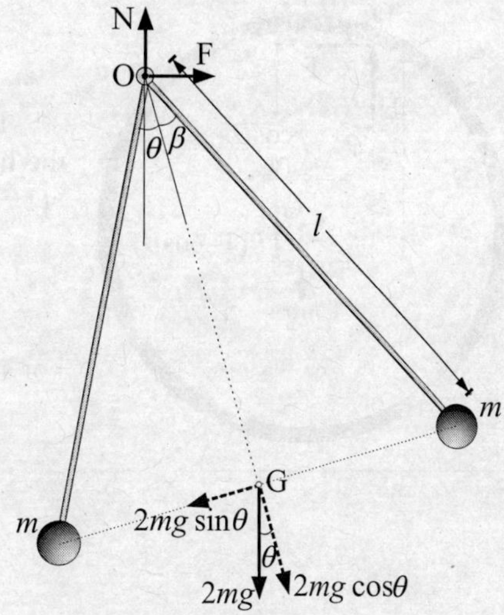

FIGURE 13.1(b) |

PROBLEM 13.2

A thin hoop of mass m and radius r is supported by a peg at O, as shown in Figure 13.2(a). Determine the period of oscillation for small amplitudes of swing.

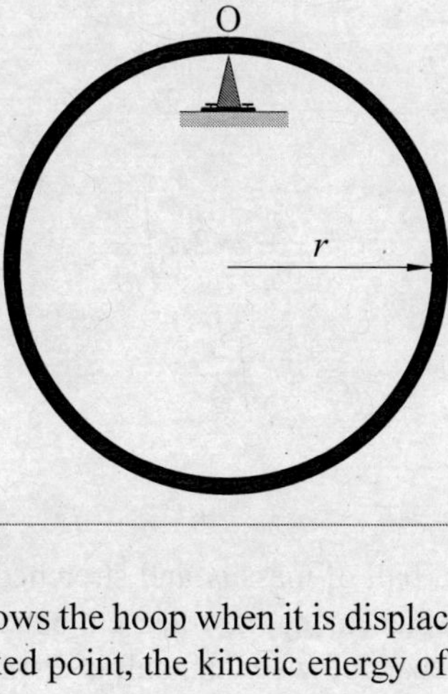

FIGURE 13.2(a) |

Solution. Figure 13.2(b) shows the hoop when it is displaced by a small angle θ from the equilibrium position. Because O is a fixed point, the kinetic energy of the hoop may be written as

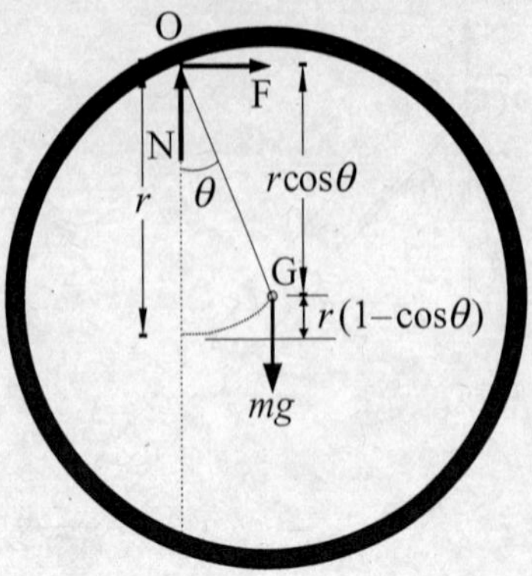

FIGURE 13.2(b)

$$KE = \frac{1}{2} I_0\omega^2 = \frac{1}{2}(mr^2 + mr^2)\dot{\theta}^2 = mr^2\dot{\theta}^2$$

If a horizontal datum is placed through centre of gravity of the hoop when $\theta = 0$, then the centre of gravity moves upward $r(1-\cos\theta)$ in the displaced position. The potential energy is

$$PE = mgr(1 - \cos\theta) = mgr \times 2\sin^2\frac{\theta}{2}$$

For small angle θ,

$$\sin\frac{\theta}{2} \approx \frac{\theta}{2}$$

$$PE = \frac{1}{2} mgr\theta^2$$

The total energy in the system is

$$KE + PE = mr^2\dot{\theta}^2 + \frac{1}{2} mgr\theta^2$$

The condition for conservation of energy

$$\frac{d}{dt}(KE + PE) = \frac{d}{dt}\left(mr^2\dot{\theta}^2 + \frac{1}{2} mgr\theta^2\right) = 0$$

$$mr^2 2\dot{\theta}\ddot{\theta} + mgr\theta\dot{\theta} = 0$$

$$mr\dot{\theta}(2r\ddot{\theta} + g\theta) = 0$$

Since $\dot{\theta}$ is not always equal to zero, from the terms in parentheses

$$2r\ddot{\theta} + g\theta = 0 \Rightarrow \ddot{\theta} + \frac{g}{2r}\theta = 0$$

Hence, the natural frequency

$$\omega_n = \sqrt{\frac{g}{2r}}$$

So that

$$\tau = \frac{2\pi}{\omega_n} = 2\pi\sqrt{\frac{2r}{g}}$$

$$\tau = 2\pi\sqrt{\frac{2r}{g}}$$

PROBLEM 13.3

Two elastic cords are connected to a ball of mass m and stretched to initial tension T [Fig. 13.3(a)]. If the ball is given a small lateral displacement and released, determine the frequency of the ensuing vibration (assume that T remains unchanged when the ball is displaced).

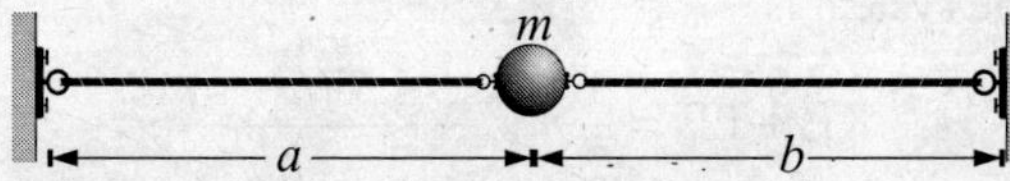

FIGURE 13.3(a) |

Solution. The free body diagram of the ball when it is displaced by an arbitrary distance y is shown in Figure 13.3(b). Because the change in the tension can be neglected if y is small enough, the forces in the strings are equal T.

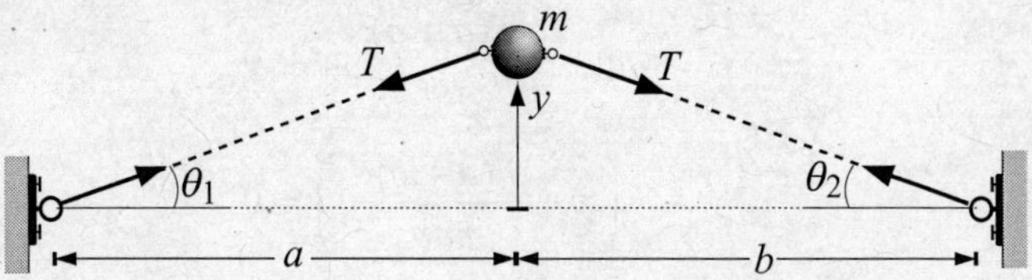

FIGURE 13.3(b) |

The equation of motion in y is

$$\sum F_y = m\ddot{y} \Rightarrow -(T\sin\theta_1 + T\sin\theta_2) = m\ddot{y}$$

or

$$m\ddot{y} + (T\sin\theta_1 + T\sin\theta_2) = 0$$

Using the small angle approximation, $\sin\theta_1 \approx \dfrac{y}{a}$, $\sin\theta_2 \approx \dfrac{y}{b}$, the equation of motion for the ball may be written as

$$m\ddot{y} + T\left(\frac{y}{a} + \frac{y}{b}\right) = 0$$

or

$$\ddot{y} + \frac{T(a+b)}{mab}\,y = 0$$

Hence,

$$\omega_n = \sqrt{\frac{T(a+b)}{mab}}$$

Energy Method

The kinetic energy of the ball is

$$\text{KE} = \frac{1}{2}m\dot{y}^2$$

The potential energy of elastic cords is

$$\text{PE} = \frac{1}{2}(T\sin\theta_1 + T\sin\theta_2)y$$

For small displacement y, $\sin\theta_1 = \dfrac{y}{a}$ and $\sin\theta_2 = \dfrac{y}{b}$,

$$\text{PE} = \frac{1}{2}\times\frac{T(a+b)}{ab}\,y^2$$

Therefore, total energy of the system is

$$\text{KE} + \text{PE} = \frac{1}{2} m\dot{y}^2 + \frac{1}{2}\frac{T(a+b)}{ab} y^2$$

Total energy remains constant, $\frac{\text{d}}{\text{d}t}(\text{KE} + \text{PE}) = 0,$

$$\frac{\text{d}}{\text{d}t}\left(\frac{1}{2} m\dot{y}^2 + \frac{1}{2}\frac{T(a+b)}{ab} y^2\right) = 0$$

or

$$m\dot{y}\ddot{y} + \frac{T(a+b)}{ab} y\dot{y} = 0$$

or

$$\dot{y}\left(\ddot{y} + \frac{T(a+b)}{mab} y\right) = 0$$

Since $\dot{y}$ is not always zero,

$$\ddot{y} + \frac{T(a+b)}{mab} y = 0$$

Hence,

$$\omega_n = \sqrt{\frac{T(a+b)}{mab}}$$

PROBLEM 13.4

A simply supported beam with a concentrated load m acting on the mid-span is shown in Figure 13.4(a). If the mass of the beam is negligible compared to the mass acting, find the natural frequency of the system.

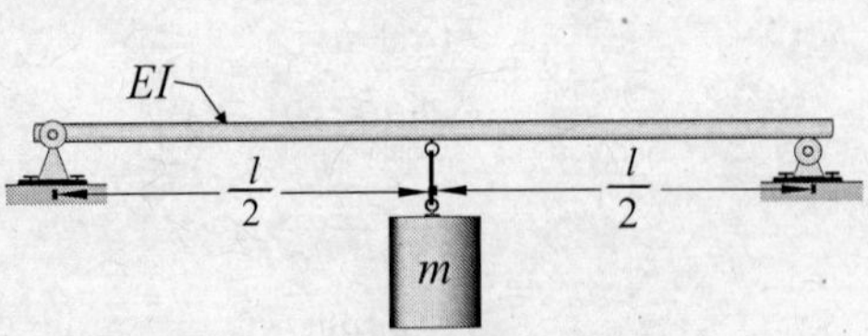

FIGURE 13.4(a) |

Solution. From the strength of material, the deflection at centre of a simply supported beam due to a concentrated load P at the centre of the beam [Fig. 13.4(b)] is given by

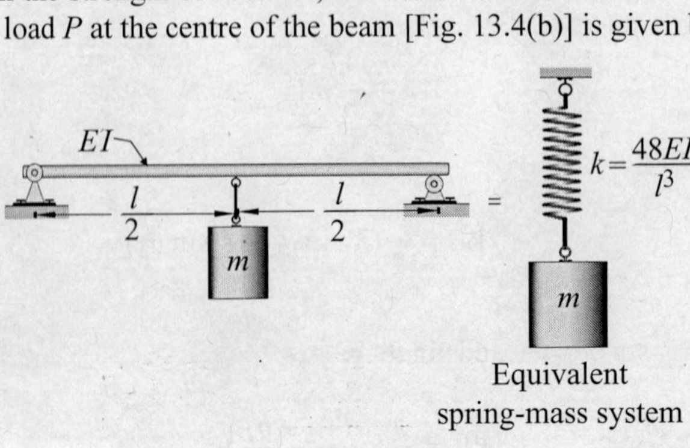

FIGURE 13.4(b) |

$$\delta = \frac{Pl^3}{48EI}$$

Thus, the stiffness of the beam is

$$k = \frac{P}{\delta} = \frac{48EI}{l^3}$$

Hence, the natural frequency of the system is

$$\omega_n = \sqrt{\frac{k}{m}} = \sqrt{\frac{48EI}{ml^3}}$$

$$\omega_n = \sqrt{\frac{48EI}{ml^3}}$$

PROBLEM 13.5

A square plate of mass m and side a is suspended at its corner by the pin O, as shown in Figure 13.5(a). Determine the natural period of oscillation if it is displaced by a small amount and released.

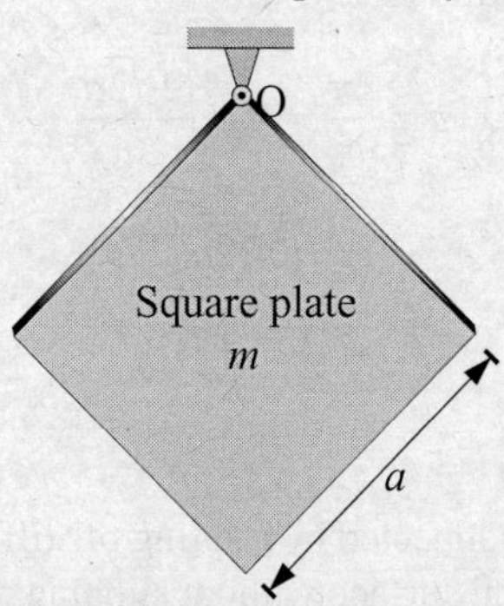

FIGURE 13.5(a)

Solution. The free body diagram of the plate when it is displaced by small angle θ from vertical is shown in Figure 13.5(b).

Summing moments about the point O

$$\sum M_o = I_o\ddot{\theta} \Rightarrow -mg \sin\theta \times \text{OG} = I_o\ddot{\theta},$$

which, on rearranging the terms, becomes

$$I_o\ddot{\theta} + mg \sin\theta \times \text{OG} = 0,$$

where

$$\text{OG} = \frac{a}{\sqrt{2}}$$

Using the small angle approximation $\sin\theta \approx \theta$, the equation of motion may be written as

$$I_o\ddot{\theta} + mg\frac{a}{\sqrt{2}} \times \theta = 0$$

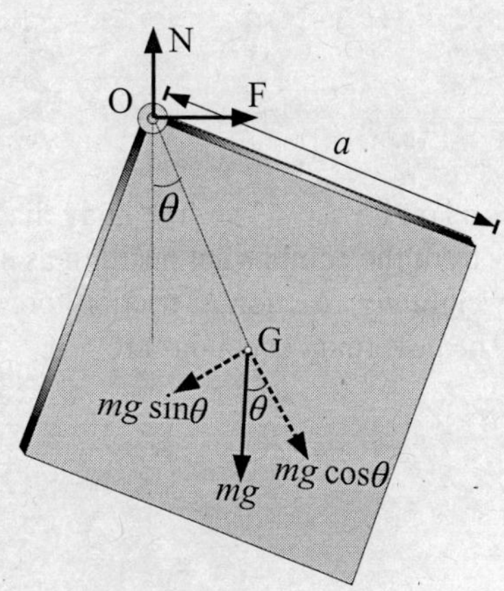

FIGURE 13.5(b)

or
$$\ddot{\theta} + \frac{mg \times a}{I_o\sqrt{2}}\theta = 0$$

Hence, the natural frequency is

$$\omega_n = \sqrt{\frac{mga}{I_o\sqrt{2}}}$$

The moment of inertia of the plate about O is

$$I_o = \frac{1}{6}ma^2 + m \times \left(\frac{a}{\sqrt{2}}\right)^2 = \frac{2}{3}ma^2$$

$$\omega_n = \sqrt{\frac{mga}{\frac{2}{3}ma^2\sqrt{2}}} = \sqrt{\frac{3g}{2\sqrt{2}a}}$$

Therefore, the natural period of vibration is

$$\tau = \frac{2\pi}{\omega_n} = 2\pi\sqrt{\frac{2\sqrt{2}a}{3g}}$$

$$\tau = 2\pi\sqrt{\frac{2\sqrt{2}a}{3g}}$$

PROBLEM 13.6

A cylinder of mass m and radius r is connected by a spring of stiffness k as shown in Figure 13.6(a). If it is free to roll on the rough horizontal surface without slipping, find its natural frequency.

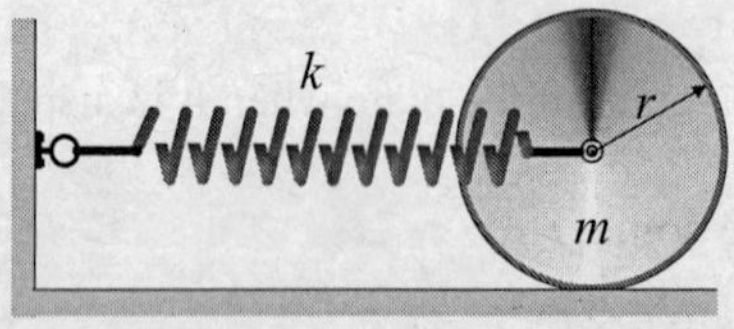

FIGURE 13.6(a)

Solution. The free body diagram of the cylinder is constructed for an arbitrary displacement x from the equilibrium position as shown in Figure 13.6(b). The forces acting on the cylinder are its weight mg, reaction N, friction force F, and spring force kx ($=kr\theta$).
The governing equations are

$$\Sigma F_x = ma_x \Rightarrow -kx + F = m\ddot{x}$$

$$\Sigma F_y = 0 \Rightarrow mg = N$$

$$\Sigma M_o = I_o\ddot{\theta} \Rightarrow -Fr = I_o\ddot{\theta}$$

Eliminating F from the above equations, we have

$$-kx + \frac{-I_o\ddot{\theta}}{r} = m\ddot{x}$$

or $$-kr\theta - \frac{I_o\ddot{\theta}}{r} = mr\ddot{\theta} \quad (x = r\theta),$$

which, on rearranging the terms, becomes

$$\ddot{\theta} + \frac{kr^2}{(mr^2 + I_o)}\theta = 0$$

The natural frequency is determined from differential equation as

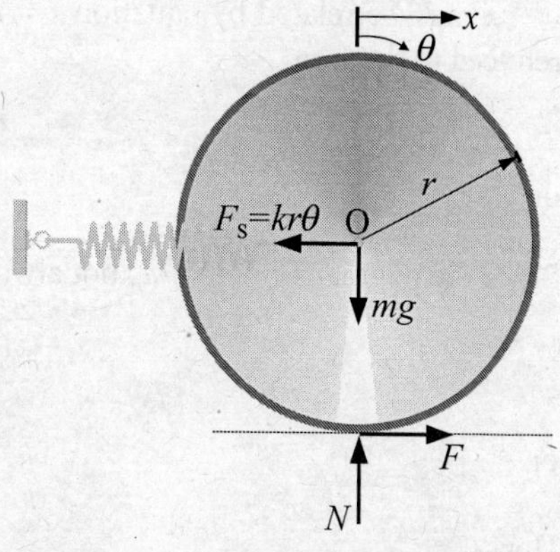

FIGURE 13.6(b) |

$$\omega_n = \sqrt{\frac{kr^2}{(mr^2 + I_o)}}$$

$$= \sqrt{\frac{kr^2}{mr^2 + \left(\frac{1}{2}\right)mr^2}} = \sqrt{\frac{2k}{3m}} \qquad \left(I_o = \frac{1}{2}mr^2\right)$$

$$\omega_n = \sqrt{\frac{2k}{3m}}$$

Energy Method

Figure 13.6(b) shows the cylinder when it is displaced by a small angle θ from the equilibrium. The kinetic energy of the cylinder may be written as

$$\text{KE} = \frac{1}{2}m\dot{x}^2 + \frac{1}{2}I_o\dot{\theta}^2$$

The potential energy is

$$\text{PE} = \frac{1}{2}kx^2$$

Total energy of the system is

$$\text{KE} + \text{PE} = \frac{1}{2}m\dot{x}^2 + \frac{1}{2}I_o\dot{\theta}^2 + \frac{1}{2}kx^2$$

Total energy of the system remains constant,

$$\frac{\text{d}}{\text{d}t}(\text{KE} + \text{PE}) = 0,$$

$$\frac{\text{d}}{\text{d}t}\left(\frac{1}{2}m\dot{x}^2 + \frac{1}{2}I_o\dot{\theta}^2 + \frac{1}{2}kx^2\right) = 0$$

$$m\dot{x}\ddot{x} + I_o\dot{\theta}\ddot{\theta} + kx\dot{x} = 0$$

x may be related by equation $x = r\theta$, so that $\dot{x} = r\dot{\theta}$ and $\ddot{x} = r\ddot{\theta}$. Hence the above equation is reduced to the form

$$mr^2\dot{\theta}\ddot{\theta} + I_o\dot{\theta}\ddot{\theta} + kr^2\theta\dot{\theta} = 0$$

or

$$\dot{\theta}[(mr^2 + I_o)\ddot{\theta} + kr^2\theta] = 0$$

Since the angular velocity $\dot{\theta}$ is not always zero,

$$(mr^2 + I_o)\ddot{\theta} + kr^2\theta = 0$$

or

$$\ddot{\theta} + \frac{kr^2}{mr^2 + I_o}\theta = 0$$

Hence,

$$\omega_n = \sqrt{\frac{kr^2}{mr^2 + I_o}} = \sqrt{\frac{kr^2}{mr^2 + \frac{1}{2}mr^2}} \qquad \left[I_o = \frac{1}{2}mr^2\right]$$

$$\omega_n = \sqrt{\frac{2k}{3m}}$$

PROBLEM 13.7

A uniform cylinder of length l and mass m having a cross-section area A is suspended with its vertical length from a fixed point by a mass-less spring, such that it is half submerged in a liquid of density ρ. at equilibrium position [Fig. 13.7(a)]. When the cylinder is given a small downward push and released, it starts oscillating vertically with small amplitude. Determine the natural frequency of the cylinder.

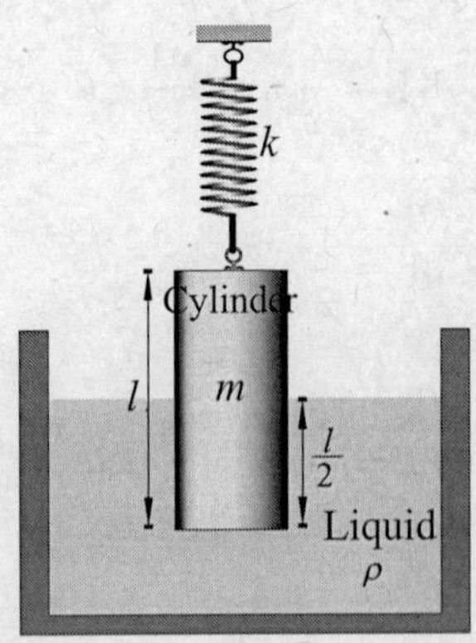

FIGURE 13.7(a) |

Solution. The free body diagram of the cylinder is constructed for an arbitrary displacement x from the equilibrium position as shown in Figure 13.7(b). The forces acting on the cylinder are its weight mg, spring force $k(x_{st} + x)$ and buoyant force F_b. The differential equation of the motion in the x-direction is

$$\sum F_x = m\ddot{x} \Rightarrow mg - k(x_{st} + x) - F_b = m\ddot{x}$$

$$mg - k(x_{st} + x) - \rho gA\left(\frac{l}{2} + x\right) = m\ddot{x},$$

which, on rearranging the terms, becomes

$$m\ddot{x} + \left(kx_{st} + \rho gA\frac{l}{2} - mg\right) + kx + \rho gAx = 0$$

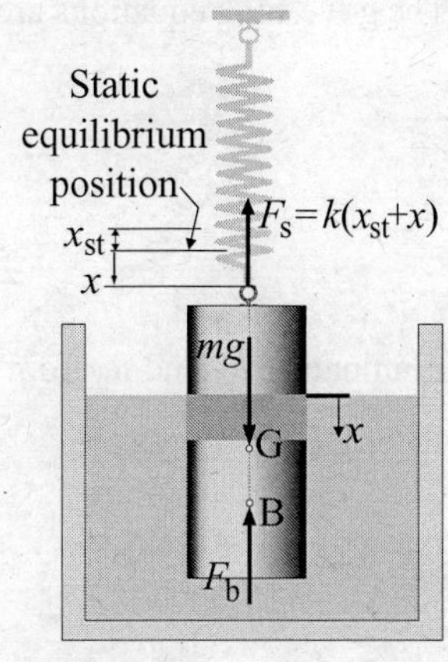

FIGURE 13.7(b) |

When the cylinder is at static equilibrium, upward force F_b and spring force kx_{st} is balanced by the downward force, mg, i.e., $kx_{st} + F_{b_{st}} = mg \Rightarrow kx_{st} + \rho gA\frac{l}{2} = mg$, the above equation becomes

$$m\ddot{x} + (k + \rho gA)x = 0$$

or

$$\ddot{x} + \frac{(k + \rho gA)}{m}x = 0$$

From which the natural frequency is determined as

$$\omega_n = \sqrt{\frac{(k + \rho gA)}{m}}$$

PROBLEM 13.8

A uniform metallic plate of mass m is supported on two identical cylindrical rollers rotating in opposite directions at the same angular speed as shown in Figure 13.8(a). When the centre of gravity of the plate is initially displaced, the block will be set into simple harmonic motion. Find the natural frequency of the plate. The coefficient of friction between plate and rollers is. μ

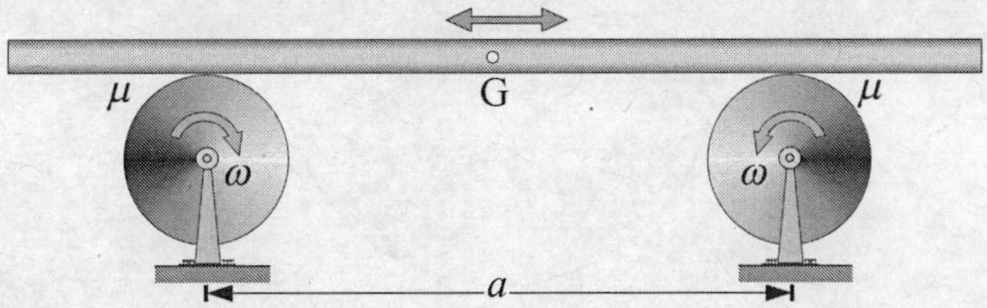

FIGURE 13.8(a) |

Solution. The free body diagram of the plate is constructed for an arbitrary displacement x from the central position as shown in Figure 13.8(b).

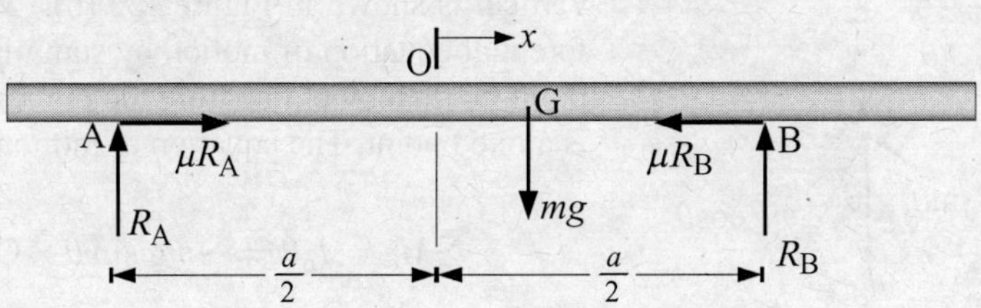

FIGURE 13.8(b) |

The governing equations are

$$\sum F_x = m\ddot{x} \Rightarrow \mu R_A - \mu R_B = m\ddot{x}$$

$$\Sigma F_y = 0 \Rightarrow R_A + R_B = mg$$

$$\Sigma M_G = 0 \Rightarrow R_B\left(\frac{a}{2} - x\right) - R_A\left(\frac{a}{2} + x\right) = 0$$

Eliminating R_A and R_B from the first equation, we have

$$\mu\left[\frac{mg\left(\frac{a}{2} - x\right)}{a} - \frac{mg\left(\frac{a}{2} + x\right)}{a}\right] = m\ddot{x},$$

which reduces to

$$m\ddot{x} + \frac{2\mu mg}{a}x = 0$$

or

$$\ddot{x} + \frac{2\mu g}{a}x = 0$$

From which the natural frequency is determined as

$$\omega_n = \sqrt{\frac{2\mu g}{a}}$$

PROBLEM 13.9

A semicircular disc of mass m and radius r pivots freely about a horizontal axis through O as shown in Figure 13.9(a). Determine the natural frequency ω_n of small oscillations about the equilibrium position.

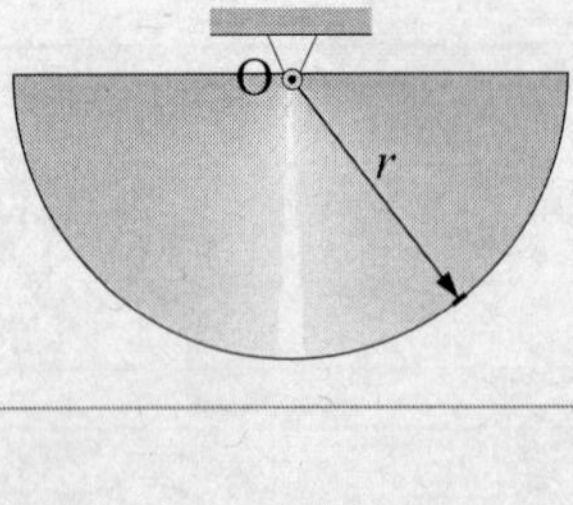

FIGURE 13.9(a) |

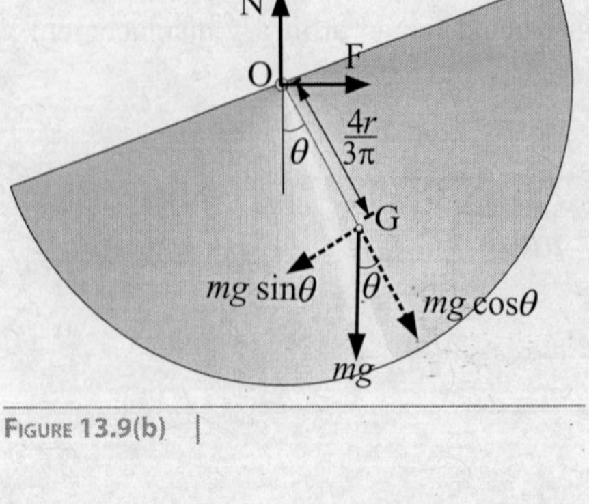

FIGURE 13.9(b) |

Solution. The free body diagram of the semicircular disc when it is displaced by a small angle θ from the vertical is shown in Figure 13.9(b). We derive the differential equation of motion by summing moment about point O, thereby eliminating the pin reaction, because O is a fixed point. The equation of motion is

$$\Sigma M_o = I_o\ddot{\theta} \Rightarrow -mg\sin\theta \times \text{OG} = I_o\ddot{\theta}$$

which, on rearranging the terms, becomes

$$I_o\ddot{\theta} + mg\sin\theta \times \frac{4r}{3\pi} = 0 \qquad \left[\text{OG} = \frac{4r}{3\pi}\right]$$

and assuming small θ, leads to

$$\frac{1}{2}mr^2\,\ddot{\theta} + mg \times \frac{4r}{3\pi} \times \theta = 0$$

or

$$\ddot{\theta} + \frac{8g}{3\pi r}\theta = 0$$

The natural frequency is determined from differential equation as

$$\omega_n = \sqrt{\frac{8g}{3\pi r}}$$

PROBLEM 13.10

Two cities on the surface of the earth are joined by a straight, smooth underground tunnel of length l. A body is released into the tunnel from one city [Fig. 13.10(a)]. How much time will it take to reach the other city? Given $g = 9.81$ m/s^2.

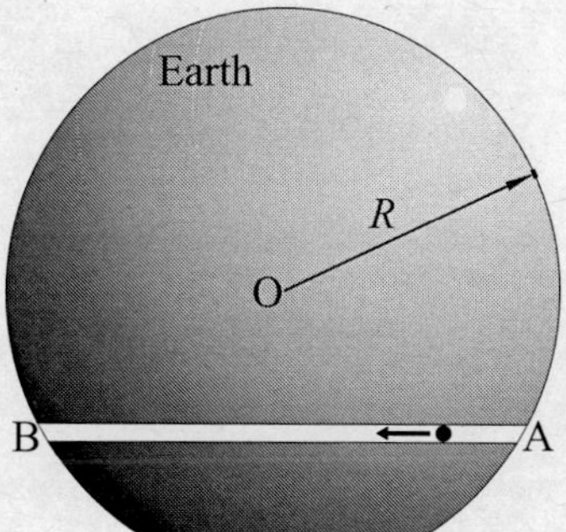

FIGURE 13.10(a) |

Solution. Let A and B be the two cities on the surface joined by a straight smooth underground tunnel AB.

Let the body of mass m, released from a city, be at a point P inside the tunnel at a given instant such that it is at a distance x from O′ and OP = r [Fig. 13.10(b)]. The differential equation of motion along x-direction is

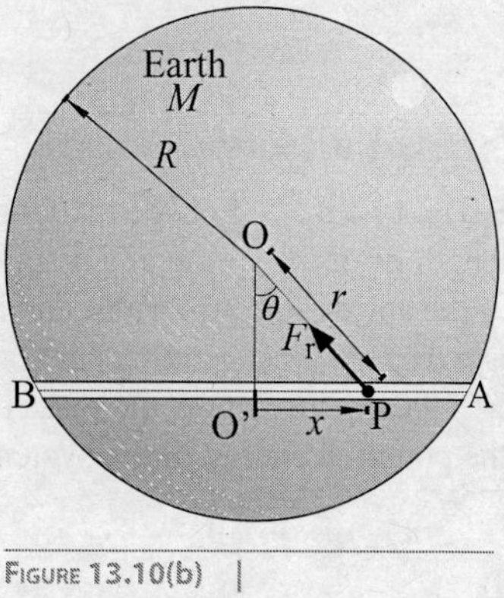

FIGURE 13.10(b) |

$$m\ddot{x} = -F_x$$

or

$$m\ddot{x} = -F_r \sin\theta$$

or

$$m\ddot{x} = -\left(\frac{GM'm}{r^2}\right)\sin\theta$$

or $$m\ddot{x} = -\left(\frac{GMmr^3}{R^3 \times r^2}\right) \times \frac{x}{r} \qquad \left(\text{where } M' = \frac{M}{R^3} r^3\right)$$

or $$\ddot{x} + \frac{GM}{R^3} x = 0$$

or $$\ddot{x} + \frac{g}{R} x = 0 \qquad \left(\text{where } g = \frac{GM}{R^2}\right)$$

Hence,

$$\omega_n = \sqrt{\frac{g}{R}}$$

So that

$$\tau = \frac{2\pi}{\omega_n} = 2\pi\sqrt{\frac{R}{g}} = 84.5\text{min}$$

Time taken for reaching from A to B is

$$T = \frac{\tau}{2} = \pi\sqrt{\frac{R}{g}} = 42.2\text{min}$$

$$T = \pi\sqrt{\frac{R}{g}} = 42.2\text{min}$$

PROBLEM 13.11

Determine the effect of the mass of the spring on the natural frequency of the spring-mass system shown in Figure 13.11(a).

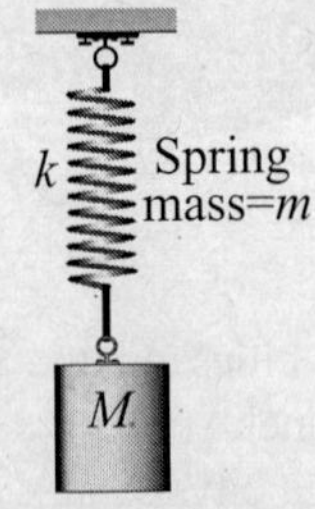

FIGURE 13.11(a) |

Solution. Let l be the length of the spring under equilibrium condition. Consider an element of length ds of the spring at a distance s from the support [Fig. 13.11(b)]. If ρ is the mass per unit length of the spring in equilibrium condition, then the mass of the spring $m = \rho l$ and the mass of the element is ρ ds.

At any instant, let the mass be displaced from the equilibrium position through a distance x. Then the potential energy of the system is

$$\text{PE} = \frac{1}{2} kx^2$$

The kinetic energy of the system at this instant consists of kinetic energy of mass M plus the kinetic energy of the spring is

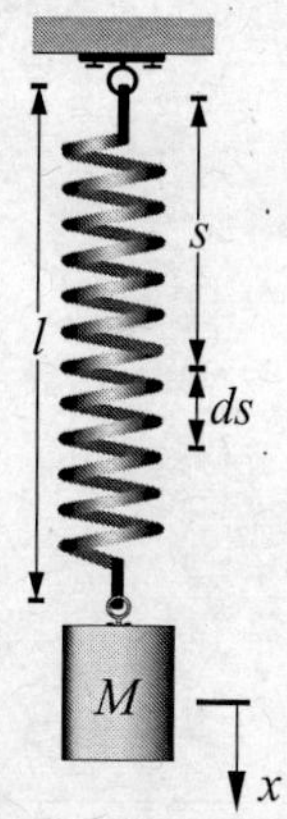

FIGURE 13.11(b)

$$KE = KE_{mass} + KE_{spring}$$

$$= \frac{1}{2} M\dot{x}^2 + \int_0^l \frac{1}{2}(\rho ds)\dot{s}^2$$

$$= \frac{1}{2} M\dot{x}^2 + \int_0^l \frac{1}{2}(\rho ds)\left(\frac{s}{l}\dot{x}\right)^2$$

$$= \frac{1}{2} M\dot{x}^2 + \frac{1}{2} \times \frac{1}{3}\rho l\dot{x}^2$$

$$= \frac{1}{2}\left(M + \frac{m}{3}\right)\dot{x}^2 \qquad (m = \text{mass of spring} = \rho l)$$

Total energy for the system is, therefore

$$PE + KE = \frac{1}{2}kx^2 + \frac{1}{2}\left(M + \frac{m}{3}\right)\dot{x}^2 = \text{constant}$$

The condition for conservation of energy is

$$\frac{d}{dt}(KE + PE) = 0,$$

$$\frac{d}{dt}\left\{\frac{1}{2}kx^2 + \frac{1}{2}\left(M + \frac{m}{3}\right)\dot{x}^2\right\} = 0$$

or

$$kx\dot{x} + \left(M + \frac{m}{3}\right)\dot{x}\ddot{x} = 0$$

or

$$\dot{x}\left[kx + \left(M + \frac{m}{3}\right)\ddot{x}\right] = 0$$

Since $\dot{x}$ is not always equal to zero,

$$\left(M + \frac{m}{3}\right)\ddot{x} + kx = 0$$

Hence, the natural frequency

$$\omega_n = \sqrt{\frac{k}{\left(M + \dfrac{m}{3}\right)}}$$

PROBLEM 13.12

A uniform bar of length l may oscillate about a hinge O located at a distance c from the mass centre G of the bar [Fig. 13.12(a)]. (a) Determine the distance c for which the period of oscillation is minimum. Draw the graph between τ vs. c.

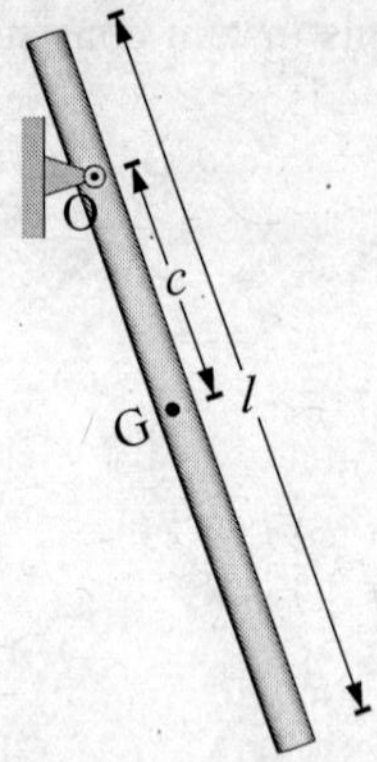

FIGURE 13.12(a) |

Solution. The moment of inertia of the bar about O,

$$I_o = \frac{ml^2}{12} + mc^2$$

or $$I_o = m\left(\frac{l^2}{12} + c^2\right)$$

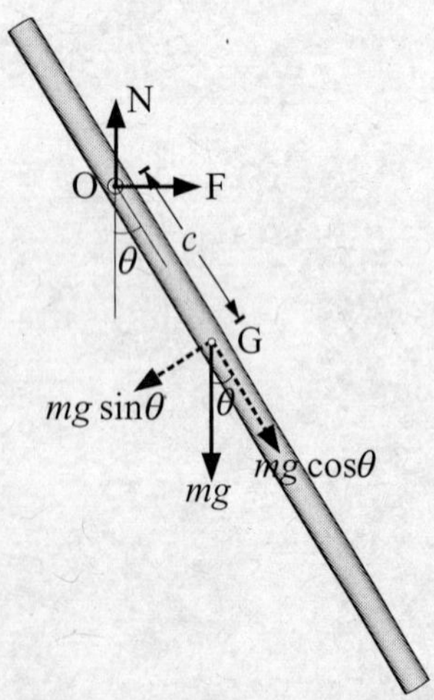

FIGURE 13.12(b) |

Let the bar be displaced through a small angle θ from the vertical equilibrium position as shown in Figure 13.12(b).

Then restoring moment about O

$$M_o = -mg\sin\theta \times \text{OG}$$

or $$M_o = -mg\theta \times \text{OG}, \qquad \text{since } \theta \text{ is small}$$

Therefore, the equation of motion for bar is

$$I_o\ddot{\theta} = -mg\theta \times \text{OG}$$

or $$m\left(\frac{l^2}{12} + c^2\right)\ddot{\theta} + mg\theta \times c = 0$$

or $$\ddot{\theta} + \frac{gc}{\left(\frac{l^2}{12} + c^2\right)}\theta = 0$$

So that

$$\omega_n = \sqrt{\frac{gc}{\left(\frac{l^2}{12} + c^2\right)}}$$

Thus

$$\tau = \frac{2\pi}{\omega_n} = 2\pi\sqrt{\frac{\left(\frac{l^2}{12c} + c\right)}{g}}$$

The condition for maximum/minimum period of oscillation,

$$\frac{d(\tau)}{dc} = 0,$$

$$\frac{d}{dc}\left(2\pi\sqrt{\frac{\left(\frac{l^2}{12c} + c\right)}{g}}\right) = 0 \Rightarrow -\frac{l^2}{12c^2} + 1 = 0$$

$$\Rightarrow c = \pm\frac{l}{\sqrt{12}}$$

$$\tau_{\min} = \tau_{c=\pm l/\sqrt{12}} = 2\pi\sqrt{\frac{l}{\sqrt{3}g}}$$

The graph between τ vs. c is plotted in Figure 13.12(c).

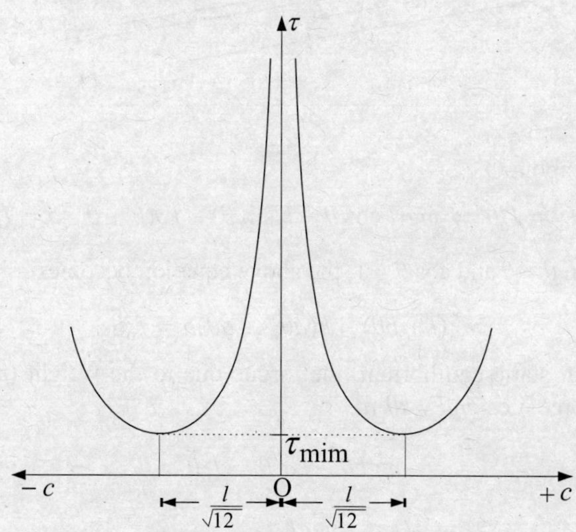

Graph between τ and c

FIGURE 13.12(c) |

PROBLEM 13.13

The small mass m is attached to the end of the L-shaped rod of negligible mass. The spring tension is adjusted so that the system is in equilibrium in the position shown in Figure 13.13(a). Determine the expression for the natural frequency of the system.

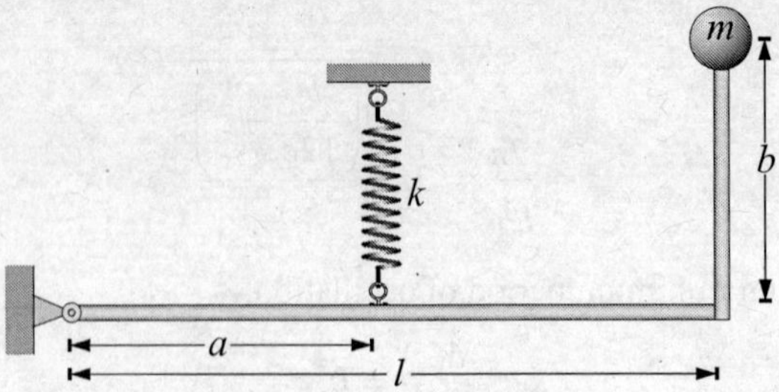

FIGURE 13.13(a) |

Solution. The free body diagram of the rod is constructed for an arbitrary angular displacement θ from the equilibrium position as shown in Figure 13.13(b).

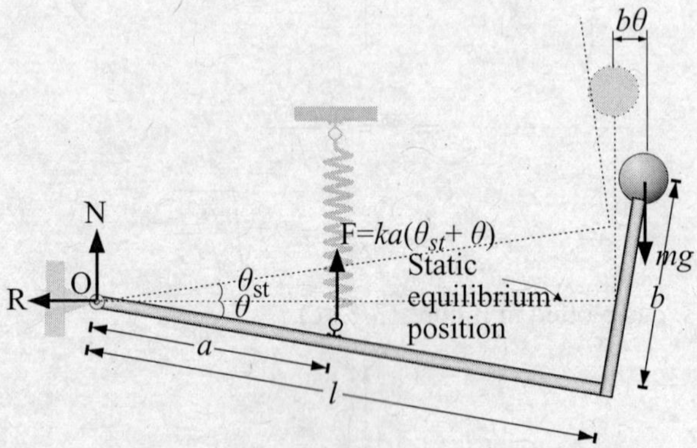

FIGURE 13.13(b) |

Summing moments about O

$$\Sigma M_o = I_o\ddot{\theta} \Rightarrow mg(l\cos\theta + b\sin\theta) - ka(\theta + \theta_{st})a = I_o\ddot{\theta}$$

For the small angle θ, $\sin\theta \approx \theta$ and $\cos\theta = 1$, the above equation becomes

$$mg(l + b\theta) - k(a\theta_{st} + a\theta)a = I_o\ddot{\theta}$$

When the mass is at static equilibrium, the torque due to the weight (mg) is balanced by the moment due to spring force, i.e., $mgl = k\theta_{st}a^2$,

$$mgb\theta - ka^2\theta = I_o\ddot{\theta}$$

or

$$\ddot{\theta} + \frac{(ka^2 - mgb)}{I_o}\theta = 0$$

Hence, the natural frequency of vibration

$$\omega_n = \sqrt{\frac{ka^2 - mgb}{I_o}}$$

$$\boxed{\omega_n = \sqrt{\frac{ka^2 - mgb}{I_o}}}$$

System does not oscillate if

$$ka^2 - mgb \le 0 \Rightarrow b \ge \frac{ka^2}{mg}$$

PROBLEM 13.14

A circular plate of mass m is supported by three symmetrically placed cords of length l, as shown in Figure 13.14(a). If the plate is given a slight rotation about a vertical axis through its centre and released, determine the natural period of oscillation.

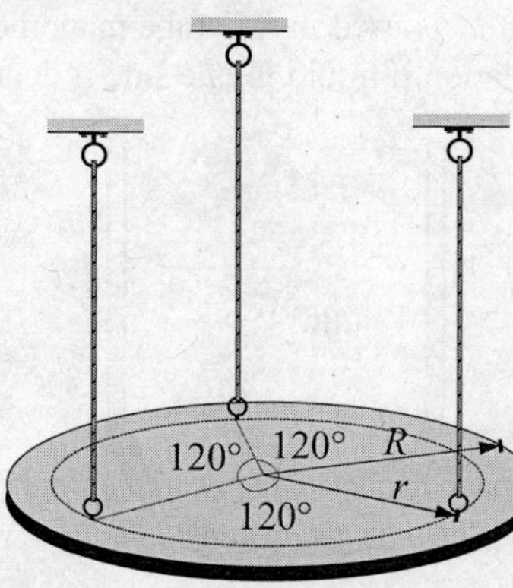

FIGURE 13.14(a) |

Solution. Figure 13.14(b) shows the free body diagram of the plate when its is displaced by a small angle θ about centre line. The tension in each cord is $T = \frac{1}{3}mg$. The horizontal component of the tension is $T \sin\phi \left(= \frac{1}{3} mg \sin\phi\right)$.

The moment equation about the vertical centre line

$$\Sigma M_o = I_o\ddot{\theta} \Rightarrow -3T \sin\phi \times r = I_o\ddot{\theta}$$

$$I_o\ddot{\theta} + 3T \sin\phi \times r = 0$$

For the small angle ϕ, $\sin\phi \approx \phi$ and $r\theta = l\phi$.

$$I_o\ddot{\theta} + 3 \times \frac{mg}{3} \times \frac{r\theta}{l} \times r = 0$$

or $$\frac{1}{2}mR^2\ddot{\theta} + \frac{mgr^2}{l}\theta = 0 \qquad \left(I_o = \frac{1}{2}mR^2\right)$$

or $$\ddot{\theta} + \frac{2gr^2}{lR^2}\theta = 0$$

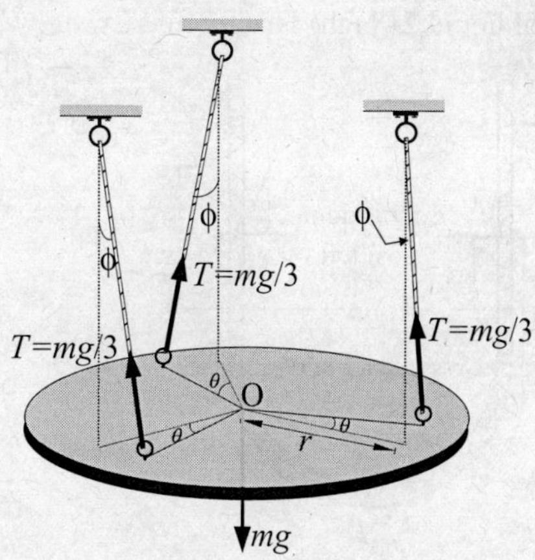

FIGURE 13.14(b) |

Hence,

$$\omega_n = \sqrt{\frac{2gr^2}{lR^2}}$$

So that

$$\tau = \frac{2\pi}{\omega_n} = 2\pi\sqrt{\frac{IR^2}{2gr^2}} = 2\pi\frac{R}{r}\sqrt{\frac{l}{2g}}$$

$$\boxed{\tau = 2\pi\frac{R}{r}\sqrt{\frac{l}{2g}}}$$

PROBLEM 13.15

A liquid of density ρ and total length l is used in a U-tube manometer as shown in Figure 13.15(a). What will be the frequency of oscillation if liquid in one side is depressed somewhat and released.

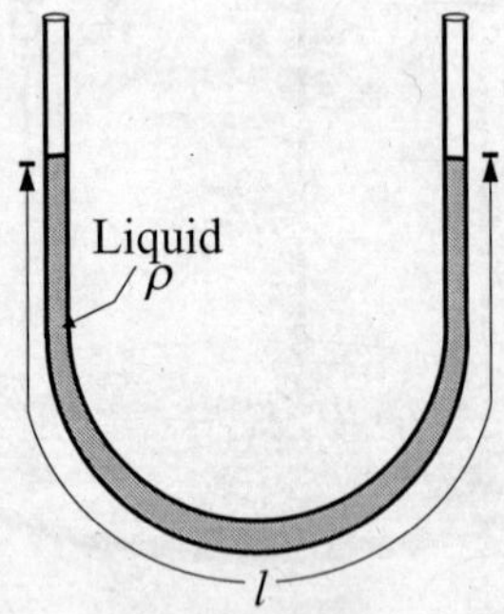

FIGURE 13.15(a) |

Solution. Figure 13.15(b) shows when water is displaced by a distance x below the equilibrium position in the left column and, of course, a distance x above the equilibrium position in the right column. The unbalanced force tending to restore equilibrium is the gravitational force and a column of liquid $2x$ high. This force is $2x\rho ag$, where a is the cross-section of the liquid.

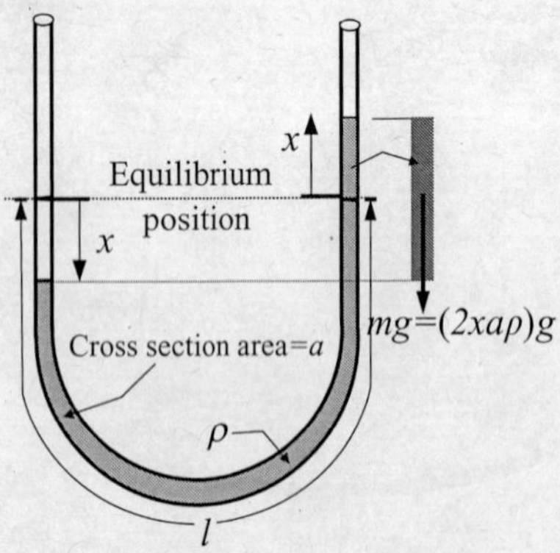

FIGURE 13.15(b) |

The governing equation is

$$m\ddot{x} = -(2x\rho a)g,$$

Where $m = \rho al$ = total mass of liquid.

$$\rho al\ddot{x} + (2\rho ag)x = 0$$

or

$$\ddot{x} + \frac{2g}{l}x = 0$$

Hence, the natural frequency is

$$\boxed{\omega_n = \sqrt{\frac{2g}{l}}}$$

Energy Method

The total energy of the liquid consists of kinetic energy and potential energy. Kinetic energy of the system is

$$\text{KE} = \frac{1}{2} m\dot{x}^2 = \frac{1}{2}(\rho al)\dot{x}^2$$

Potential energy of system is

$$\text{PE} = \int_0^x F\,\text{d}x = \int_0^x (2xa\rho)g\,\text{d}x = ax^2\rho g$$

Total energy for the system is, therefore,

$$\text{KE} + \text{PE} = \frac{1}{2}(al\rho)\dot{x}^2 + ax^2\rho g$$

The condition for conservation of energy

$$\frac{\text{d}}{\text{d}t}(\text{KE} + \text{PE}) = 0$$

$$\frac{\text{d}}{\text{d}t}\left[\frac{1}{2}(al\rho)\dot{x}^2 + a\rho x^2 g\right] = 0$$

or

$$(al\rho)\dot{x}\ddot{x} + 2a\rho g x\dot{x} = 0$$

or

$$\rho a\dot{x}[l\ddot{x} + 2gx] = 0$$

Since $\dot{x}$ is not always equal to zero,

$$l\ddot{x} + 2gx = 0 \Rightarrow \ddot{x} + \frac{2g}{l}x = 0$$

Hence,

$$\omega_n = \sqrt{\frac{2g}{l}}$$

PROBLEM 13.16

Find the natural frequency of vibration of the half solid cylinder shown in Figure 13.16(a), when slightly displaced from the equilibrium position and released.

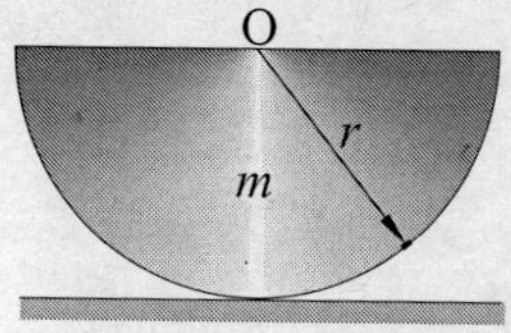

FIGURE 13.16(a) |

Solution. The free body diagram of the half cylinder when it is displaced by a small angle θ from the vertical is shown in Figure 13.16(b).

The moment equation about P is

$$\sum M_P = I_P\ddot{\theta} \Rightarrow -mg \times \text{OG} \sin\theta = I_P\ddot{\theta},$$

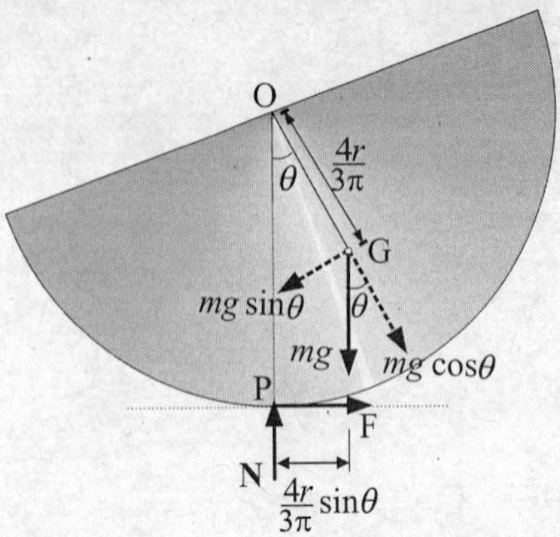

FIGURE 13.16(b) |

which, on rearranging the terms, becomes

$$I_P\ddot{\theta} + mg \times \frac{4r}{3\pi}\sin\theta = 0,$$

and assuming θ is small leads to

$$I_P\ddot{\theta} + mg \times \frac{4r}{3\pi} \times \theta = 0$$

or
$$\ddot{\theta} + \frac{4mgr}{3\pi I_P}\theta = 0$$

The natural frequency is determined from differential equation as

$$\omega_n = \sqrt{\frac{4mgr}{3\pi I_P}}$$

Mass moment of inertia about P is I_p

$$I_P = I_G + m(GP)^2 = [I_o - m(OG)^2] + m(GP)^2, \qquad [I_o = I_G + m(OG)^2],$$

$$= \frac{1}{2}mr^2 - m\left(\frac{4r}{3\pi}\right)^2 + m\left(r - \frac{4r}{3\pi}\right)^2, \qquad \left(I_o = \frac{1}{2}mr^2, OG = \frac{4r}{3\pi}, GP \approx r - \frac{4r}{3\pi}\right),$$

$$= mr^2\left(\frac{3}{2} - \frac{8}{3\pi}\right)$$

$$I_P = mr^2\left(\frac{3}{2} - \frac{8}{3\pi}\right)$$

Hence, the natural frequency

$$\omega_n = \sqrt{\frac{4g}{3\pi r\left(\frac{3}{2} - \frac{8}{3\pi}\right)}}$$

$$\omega_n = \sqrt{\frac{8g}{r(9\pi - 16)}}$$

Energy Method

Figure 13.16(c) shows the half cylinder when it is displaced by a small angle θ from the equilibrium position. The kinetic energy of the half cylinder may be written as

$$KE = \frac{1}{2}I_P\dot{\theta}^2$$

If a horizontal datum is placed through centre of gravity of the hoop when $\theta = 0$, then the centre of gravity moves upward OG$(1 - \cos\theta)$ in the displaced position. The potential energy is

$$PE = mg \times OG \times (1 - \cos\theta)$$

$$= mg \times \frac{4r}{3\pi} \times (1 - \cos\theta)$$

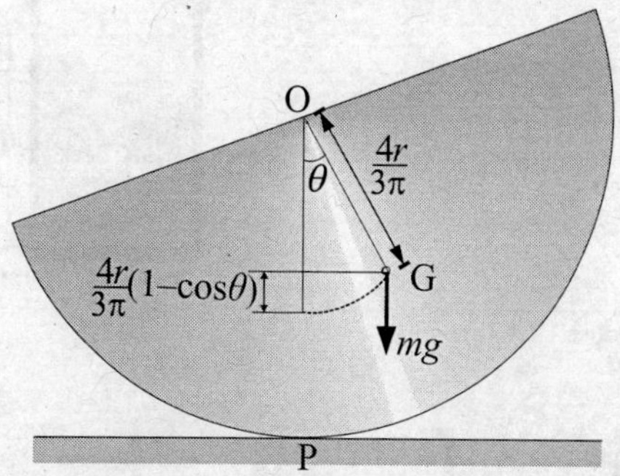

FIGURE 13.16(c) |

The total energy of the system is, therefore

$$KE + PE = \frac{1}{2} I_P \dot{\theta}^2 + mg \times \frac{4r}{3\pi}(1 - \cos\theta)$$

Total energy of the system remain constant:

$$\frac{d}{dt}(KE + PE) = 0$$

$$\frac{d}{dt}\left[\frac{1}{2} I_P \dot{\theta}^2 + mg\frac{4r}{3\pi}(1 - \cos\theta)\right] = 0$$

or

$$I_P \dot{\theta}\ddot{\theta} + mg\frac{4r}{3\pi}\sin\theta\dot{\theta} = 0$$

For small angle θ, $\sin\theta \approx \theta$.

$$\dot{\theta}\left(I_P \ddot{\theta} + mg\frac{4r}{3\pi}\theta\right) = 0$$

or

$$\ddot{\theta} + \frac{4mgr}{3\pi I_P}\theta = 0$$

Hence,

$$\omega_n = \sqrt{\frac{4mgr}{3\pi I_P}}$$

or

$$\omega_n = \sqrt{\frac{4g}{3\pi r\left(\frac{3}{2} - \frac{8}{3\pi}\right)}}$$

$$= \sqrt{\frac{8g}{r(9\pi - 16)}} \qquad \left[I_P = mr^2\left(\frac{3}{2} - \frac{8}{3\pi}\right)\right]$$

$$\omega_n = \sqrt{\frac{8g}{r(9\pi - 16)}}$$

PROBLEM 13.17

A pipe of cross-sectional area a_p and length l connects two tanks of cross-sectional area a_1 and a_2, as shown in Figure 13.17(a). Find the natural period of oscillation of the liquid between the two tanks.

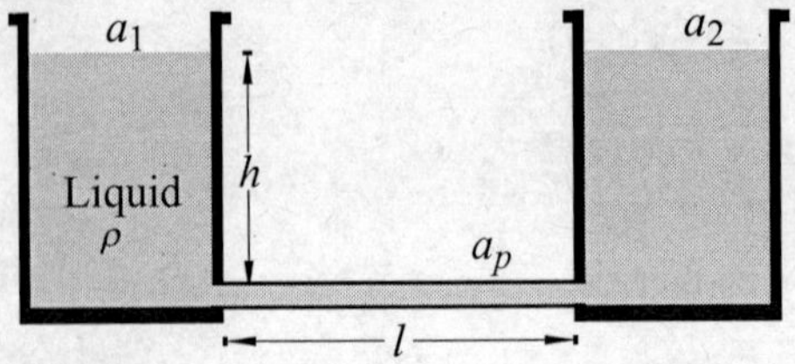

FIGURE 13.17(a) |

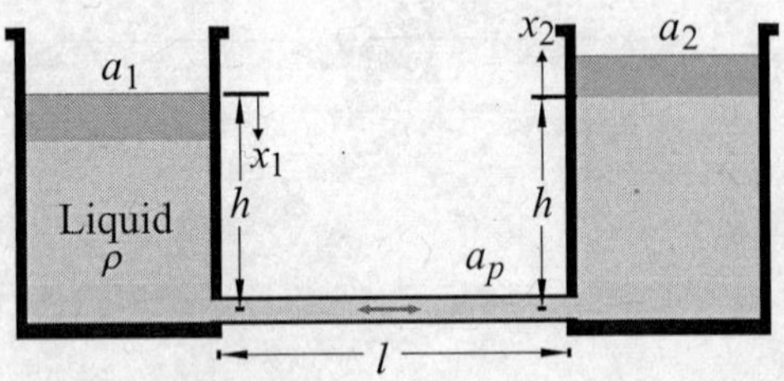

FIGURE 13.17(b) |

Solution. Figure 13.17(b) shows arbitrary vertical displacement x_1 below from the equilibrium position in the left hand tank and, of course, a displacement $x_2 = \left(\dfrac{a_1}{a_2}\right) x_1$ above the equilibrium position in the right hand tank, and the displacement results in a velocity $\dot{x}_p$ of the fluid in the pipe.

The kinetic energy of the fluid at this instant is

$$\text{KE} = \frac{1}{2} m_1 \dot{x}_1^{\,2} + \frac{1}{2} m_2 \dot{x}_2^{\,2} + \frac{1}{2} m_p \dot{x}_p^{\,2}$$

But

$$\dot{m} = \rho a_1 \dot{x}_1 = \rho a_2 \dot{x}_2 = \rho a_p \dot{x}_p \Rightarrow \dot{x}_2 = \frac{a_1}{a_2}\dot{x}_1,\ \dot{x}_p = \frac{a_1}{a_p}\dot{x}_1,$$

and the above equation becomes

$$\text{KE} = \frac{1}{2}\left[m_1 + m_2 \left(\frac{a_1}{a_2}\right)^2 + m_p \left(\frac{a_1}{a_p}\right)^2 \right] \dot{x}_1^{\,2}$$

$$\text{KE} = \frac{1}{2}\left[\rho a_1 h + \rho a_2 h \left(\frac{a_1}{a_2}\right)^2 + \rho a_p l \left(\frac{a_1}{a_p}\right)^2 \right] \dot{x}_1^{\,2}$$

The potential energy of the fluid

$$\text{PE} = \int_0^{x_2} F\,\text{d}x_2 = mg(\bar{y}_2 - \bar{y}_1),$$

where m is the total mass of the fluid, $\bar{y}_2$ is the vertical distance of centre of gravity from datum at the displaced position and $\bar{y}_1$ is the vertical distance of centre of gravity from datum at the equilibrium position.

$$\text{PE} = \int_0^{x_2} F\,\text{d}x_2 = \int_0^{x_2} \rho a_2 (x_1 + x_2) g\,\text{d}x_2$$

$$= \int_0^{x_2} \rho a_2 \left(x_1 + \frac{a_1}{a_2} x_1 \right) g \times \frac{a_1}{a_2}\,\text{d}x_1 \qquad \left[x_2 = \frac{a_1}{a_2} x_1 \right]$$

$$\text{PE} = \frac{\rho a_1 g}{2}\left(1+\frac{a_1}{a_2}\right)x_1^2$$

Total energy remains constant;

$$\frac{\mathrm{d}}{\mathrm{d}t}(\text{KE} + \text{PE}) = 0$$

$$\frac{\mathrm{d}}{\mathrm{d}t}\left[\frac{1}{2}\left\{m_1 + m_2\left(\frac{a_1}{a_2}\right)^2 + m_\mathrm{p}\left(\frac{a_1}{a_\mathrm{p}}\right)^2\right\}\dot{x}_1^2 + \frac{1}{2}\rho a_1 g\left(1+\frac{a_1}{a_2}\right)x_1^2\right] = 0$$

or
$$\frac{1}{2}\left\{m_1 + m_2\left(\frac{a_1}{a_2}\right)^2 + m_\mathrm{p}\left(\frac{a_1}{a_\mathrm{p}}\right)^2\right\}2\dot{x}_1\ddot{x}_1 + \frac{1}{2}\rho a_1 g\left(1+\frac{a_1}{a_2}\right)2x_1\dot{x}_1 = 0$$

or
$$\dot{x}_1\left[\left\{m_1 + m_2\left(\frac{a_1}{a_2}\right)^2 + m_\mathrm{p}\left(\frac{a_1}{a_\mathrm{p}}\right)^2\right\}\ddot{x}_1 + \rho a_1 g\left(1+\frac{a_1}{a_2}\right)x_1\right] = 0$$

Since $\dot{x}$ is not always zero

$$\left\{m_1 + m_2\left(\frac{a_1}{a_2}\right)^2 + m_\mathrm{p}\left(\frac{a_1}{a_\mathrm{p}}\right)^2\right\}\ddot{x}_1 + \rho a_1 g\left(1+\frac{a_1}{a_2}\right)x_1 = 0$$

or
$$\ddot{x}_1 + \frac{\rho a_1 g\left(1+\dfrac{a_1}{a_2}\right)}{\rho a_1 h + \rho a_2 h\left(\dfrac{a_1}{a_2}\right)^2 + \rho a_\mathrm{p} l\left(\dfrac{a_1}{a_\mathrm{p}}\right)^2}x_1 = 0;$$

where $m_1 = \rho a_1 h,\ m_2 = \rho a_2 h,\ m_\mathrm{p} = \rho a_\mathrm{p} l$

Hence, the natural frequency

$$\omega_\mathrm{n} = \sqrt{\frac{a_1 g\left(1+\dfrac{a_1}{a_2}\right)}{a_1 h + a_2 h\left(\dfrac{a_1}{a_2}\right)^2 + a_\mathrm{p} l\left(\dfrac{a_1}{a_\mathrm{p}}\right)^2}}$$

PROBLEM 13.18

A circular body of mass m, and radius r and radius of gyration k rolls without slipping on a cylindrical surface of radius R as shown in Figure 13.18(a). Determine its natural frequency for small oscillation about the lowest point.

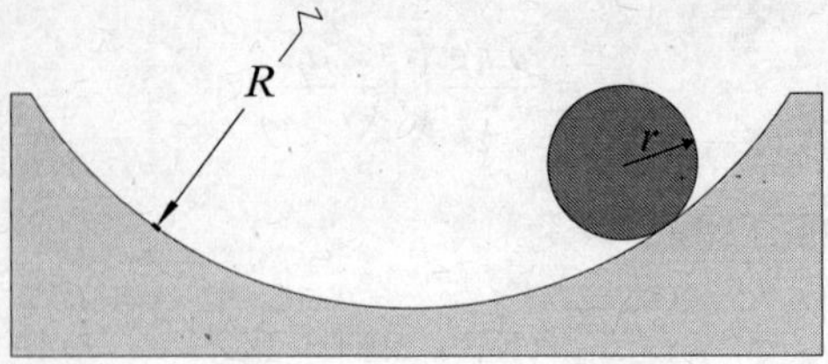

FIGURE 13.18(a) |

Solution. The free body diagram of the body is constructed for an arbitrary displacement θ from the equilibrium position, as shown in Figure 13.18(b). The forces acting of the body are its weight mg, reaction N and friction force F.

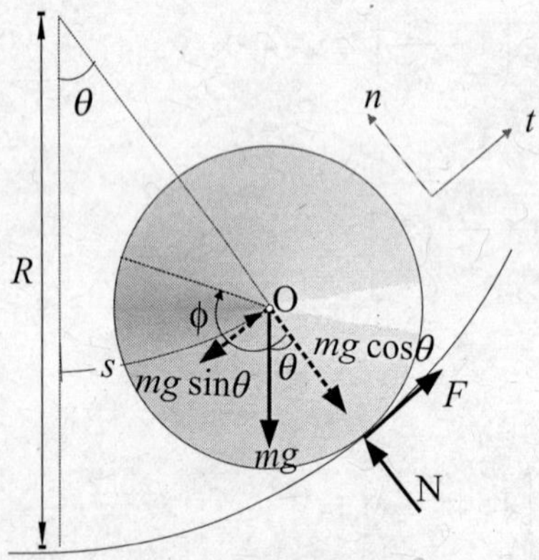

FIGURE 13.18(b) |

The governing equations are

$$\sum F_{\mathrm{T}} = m\ddot{s} \Rightarrow -mg\sin\theta + F = m\ddot{s}$$

$$\sum F_{\mathrm{N}} = 0 \Rightarrow mg\cos\theta = N$$

$$\sum M_{\mathrm{o}} = I_{\mathrm{o}}\ddot{\phi} \Rightarrow -Fr = I_{\mathrm{o}}\ddot{\phi}$$

$$(R-r)\theta = r\phi \Rightarrow \ddot{\phi} = \frac{(R-r)}{r}\ddot{\theta}$$

Eliminating F from first equation

$$-mg\sin\theta - \frac{I_{\mathrm{o}}\ddot{\phi}}{r} = m\ddot{s}$$

and assuming small θ, leads to

$$-mg\theta - \frac{I_{\mathrm{o}}\ddot{\phi}}{r} = m\ddot{s}$$

or

$$-mg\theta - \frac{I_{\mathrm{o}}}{r}\frac{(R-r)}{r}\ddot{\theta} = m(R-r)\ddot{\theta} \qquad [\ddot{s} = (R-r)\ddot{\theta}],$$

which, on rearranging the terms, becomes

$$(R-r)\left(m + \frac{I_{\mathrm{o}}}{r^2}\right)\ddot{\theta} + mg\theta = 0$$

or

$$\ddot{\theta} + \frac{mg}{(R-r)\left(m + \frac{I_{\mathrm{o}}}{r^2}\right)}\theta = 0$$

The natural frequency is determined from differential equation as

$$\omega_{\mathrm{n}} = \sqrt{\frac{mg}{(R-r)\left(m + \frac{I_{\mathrm{o}}}{r^2}\right)}}$$

$$\omega_n = \sqrt{\frac{g}{(R-r)\left(1+\frac{k^2}{r^2}\right)}}$$

$$(\omega_n)_{\text{cylinder}} = (\omega_n)_{\text{disc}} = \sqrt{\frac{2g}{3(R-r)}},$$

$$(\omega_n)_{\text{sphere}} = \sqrt{\frac{5g}{7(R-r)}},$$

$$(\omega_n)_{\text{ring}} = \sqrt{\frac{g}{2(R-r)}}$$

Energy Method

The total energy of the body consists of kinetic energy (rotational and translation) and potential energy.

Translation $$\text{KE} = \frac{1}{2}m[(R-r)\dot{\theta}]^2$$

Rotation $$\text{KE} = \frac{1}{2}I_o\dot{\phi}^2 \qquad \left[\dot{\phi} = \frac{(R-r)}{r}\dot{\theta}\right]$$

Thus, total kinetic energy for the body is

$$\text{KE} = \frac{1}{2}m[(R-r)\dot{\theta}]^2 + \frac{1}{2}I_o\left[\left(\frac{R}{r}-1\right)\dot{\theta}\right]^2$$

Potential energy is

$$\text{PE} = mg(R-r)(1-\cos\theta)$$

Total energy of the system is [Fig. 13.18(c)], therefore,

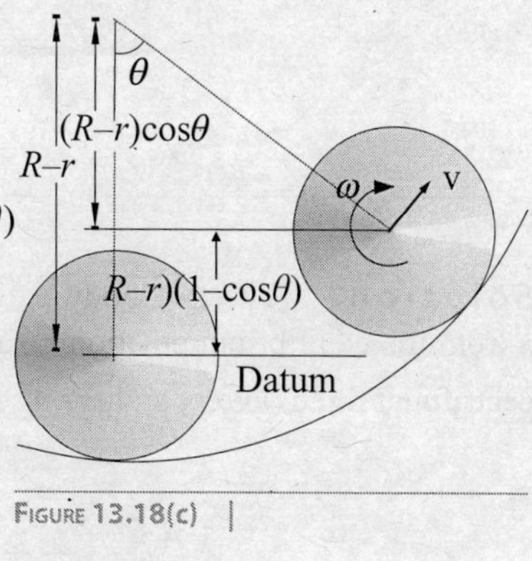

FIGURE 13.18(c) |

KE + PE

$$= \frac{1}{2}m[(R-r)\dot{\theta}]^2 + \frac{1}{2}I_o\left[\left(\frac{R}{r}-1\right)\dot{\theta}\right]^2 + mg(R-r)(1-\cos\theta)$$

The condition for conservation of energy is

$$\frac{\mathrm{d}}{\mathrm{d}t}(\text{KE}+\text{PE}) = 0$$

That is,

$$m(R-r)^2\dot{\theta}\ddot{\theta} + I_o\left(\frac{R}{r}-1\right)^2\dot{\theta}\ddot{\theta} + mg(R-r)\sin\theta\dot{\theta} = 0$$

For a small angle θ of oscillation, $\sin\theta \approx \theta$, the equation of motion becomes

or $$(R-r)^2\dot{\theta}\left[\left(m+\frac{I_o}{r^2}\right)\ddot{\theta} + \frac{mg}{R-r}\theta\right] = 0$$

Since $\dot{\theta}$ is not always equal to zero,

$$\left[\left(m+\frac{I_o}{r^2}\right)\ddot{\theta}+\frac{mg}{R-r}\theta\right]=0$$

or

$$\ddot{\theta}+\frac{mg}{(R-r)\left(m+\dfrac{I_o}{r^2}\right)}\theta=0$$

Hence,

$$\omega_n=\sqrt{\frac{mg}{(R-r)\left(m+\dfrac{I_o}{r^2}\right)}}\qquad [I_o=mk^2]$$

$$\omega_n=\sqrt{\frac{g}{(R-r)\left(1+\dfrac{k^2}{r^2}\right)}}$$

PROBLEM 13.19

A sleeve of mass m is fixed between two identical springs of stiffness k each, as shown in Figure 13.19(a). The sleeve can slide without friction over a horizontal bar AB. The arrangement rotates with a constant angular velocity ω about a vertical axis passing through the middle of the bar. Find the period of small oscillations of the sleeve. At what values of ω will there be no oscillations of the sleeve?

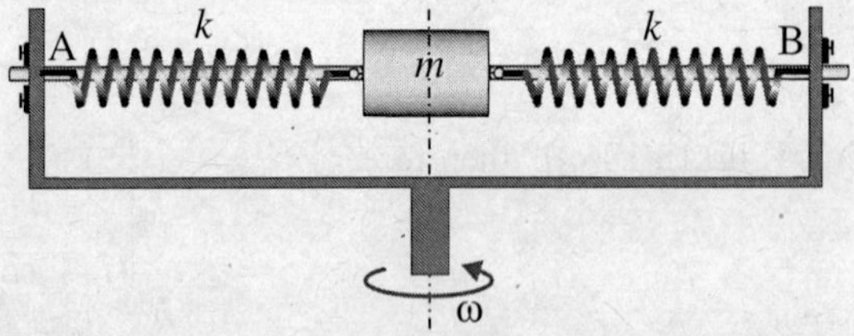

FIGURE 13.19(a) |

Solution. The free body diagram of the sleeve is constructed for an arbitrary displacement x from the equilibrium position O, as shown in Figure 13.19(b). The forces acting on the sleeve are centrifugal force ($m\omega_2 x$) and spring forces $\left(\dfrac{k}{2}x\right)$.

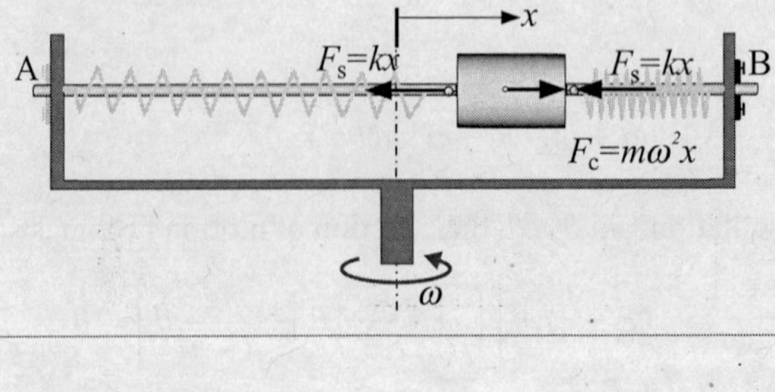

FIGURE 13.19(b) |

The differential equation of motion in the x-direction is

$$\sum Fx = m\ddot{x} \Rightarrow -kx + m\omega^2 x - kx = m\ddot{x}$$

or $$m\ddot{x} + 2kx - m\omega^2 x = 0$$

or $$\ddot{x} + \left(\frac{2k}{m} - \omega^2\right)x = 0$$

Hence,

$$\omega_n = \sqrt{\frac{2k}{m} - \omega^2}$$

So that

$$\tau = \frac{2\pi}{\omega_n} = \frac{2\pi}{\sqrt{\frac{2k}{m} - \omega^2}}$$

Sleeve will not oscillate if

$$\frac{2k}{m} - \omega^2 \le 0 \Rightarrow \omega \ge \sqrt{\frac{2k}{m}}$$

$$\omega \ge \sqrt{\frac{2k}{m}}$$

PROBLEM 13.20

A weight of mass m is suspended by means of a spring k_2 from the end of a rigid mass-less beam as shown in Figure 13.20(a). It is also supported in a horizontal position by a spring k_1 attached to the frame. Determine the natural frequency ω_n of the system.

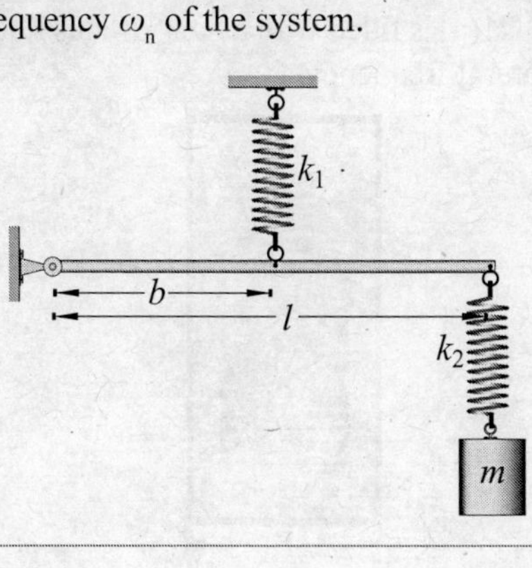

FIGURE 13.20(a) |

Solution. Figure 13.20(b) shows a method for calculating the equivalent spring stiffness. First, a vertical force F is applied at B, and the vertical movement δ is computed, the equivalent spring stiffness $k_{eq} = F/\delta$. The total displacement δ is equal to the stretch δ_C of spring k_2 plus the deflection δ_B of end B due to stretch δ_A of spring k_1.

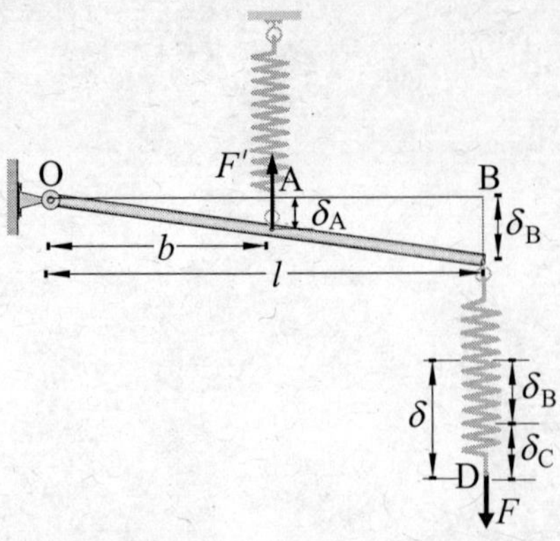

FIGURE 13.20(b) |

Hence,

$$\delta = \delta_B + \delta_C$$

$$= \frac{\delta_A}{b} \times l + \delta_C = \frac{F'}{k_1} \times \frac{l}{b} + \frac{F}{k_2}$$

$$\delta = \frac{F}{k_1} \times \frac{l}{b} \times \frac{l}{b} + \frac{F}{k_2} \qquad \left[\text{where } F' = F\frac{l}{b}\right]$$

$$\delta = F\left(\frac{l^2}{b^2} \times \frac{1}{k_1} + \frac{1}{k_2}\right)$$

or

$$k_{eq} = \frac{F}{\delta} = \frac{1}{\left(\frac{l^2}{b^2} \times \frac{1}{k_1} + \frac{1}{k_2}\right)}$$

Hence, the natural frequency

$$\omega_n = \sqrt{\frac{k_{eq}}{m}} = \sqrt{\frac{1}{m} \times \frac{1}{\left(\frac{l^2}{b^2} \times \frac{1}{k_1} + \frac{1}{k_2}\right)}}$$

$$\omega_n = \sqrt{\frac{k_1 k_2}{m\left[k_1 + \left(\frac{l}{b}\right)^2 k_2\right]}}$$

PROBLEM 13.21

A vibrometer shown in Figure 13.21(a) is filled with a non-viscous liquid. If m is the mass of the liquid displaced by the body, find its natural frequency.

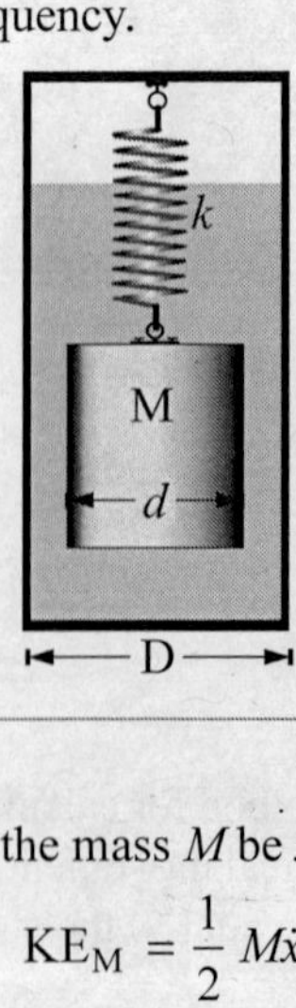

FIGURE 13.21(a) |

Solution. Let the displacement of the mass M be x [Fig. 13.21(b)]. The kinetic energy of M is

$$KE_M = \frac{1}{2} M\dot{x}^2$$

Let the velocity gained by liquid be $\dot{y}$. From the continuity equation, we have rate of liquid going up is equal to rate of liquid displaced:

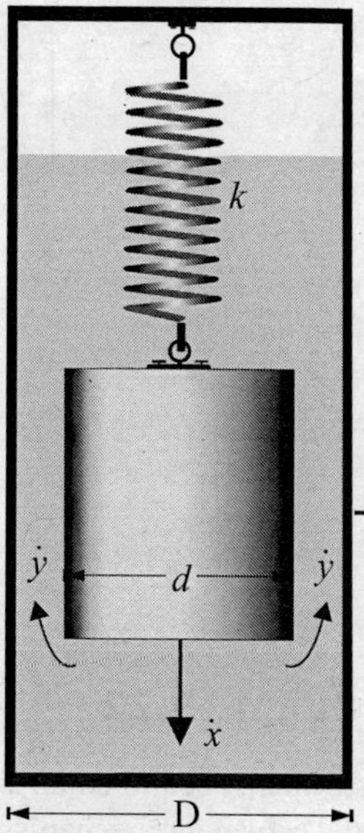

FIGURE 13.21(b) |

$$\rho\left(\frac{\pi}{4}D^2 - \frac{\pi}{4}d^2\right)\dot{y} = \rho\left(\frac{\pi}{4}d^2\right)\dot{x},$$

where ρ is the density of liquid.

From which,

$$\dot{y} = \frac{d^2}{D^2 - d^2}\,\dot{x}$$

The kinetic energy of the liquid is

$$\text{KE}_{\text{liquid}} = \frac{1}{2}m\dot{y}^2$$

Substituting $\dot{y}$, the kinetic energy of the liquid

$$\text{KE}_{\text{liquid}} = \frac{1}{2}m\left(\frac{d^2}{D^2 - d^2}\right)^2\dot{x}^2$$

The total kinetic energy,

$$\text{KE} = \text{KE}_{\text{M}} + \text{KE}_{\text{liquid}} = \frac{1}{2}M\dot{x}^2 + \frac{1}{2}m\left(\frac{d^2}{D^2 - d^2}\right)^2\dot{x}^2$$

The potential energy,

$$\text{PE} = \frac{1}{2}kx^2$$

Therefore, the total energy

$$\text{KE} + \text{PE} = \frac{1}{2}M\dot{x}^2 + \frac{1}{2}m\left(\frac{d^2}{D^2 - d^2}\right)^2\dot{x}^2 + \frac{1}{2}kx^2$$

The condition for conservation of energy,

$$\frac{\text{d}}{\text{d}t}(\text{KE} + \text{PE}) = 0$$

$$\frac{\text{d}}{\text{d}t}\left[\frac{1}{2}M\dot{x}^2 + \frac{1}{2}m\left(\frac{d^2}{D^2 - d^2}\right)^2\dot{x}^2 + \frac{1}{2}kx^2\right] = 0$$

or $$M\dot{x}\ddot{x} + m\left(\frac{d^2}{D^2 - d^2}\right)^2\dot{x}\ddot{x} + kx\dot{x} = 0$$

or $$\dot{x}\left[\left\{M + m\left(\frac{d^2}{D^2 - d^2}\right)^2\right\}\ddot{x} + kx\right] = 0 \Rightarrow \dot{x} \neq 0$$

Since

$$\left[M + m\left(\frac{d^2}{D^2 - d^2}\right)^2\right]\ddot{x} + kx = 0$$

or

$$\ddot{x} + \frac{k}{M + m\left(\dfrac{d^2}{D^2 - d^2}\right)^2}x = 0$$

Hence,

$$\omega_n = \sqrt{\frac{k}{M + m\left(\dfrac{d^2}{D^2 - d^2}\right)^2}}$$

PROBLEM 13.22

An instrument used to count vertical oscillation of a transmission line consists of a seismic pendulum and an escapement mechanism as shown in Figure 13.22(a). Find the natural frequency of vibration of the instrument. Neglect the mass of rod.

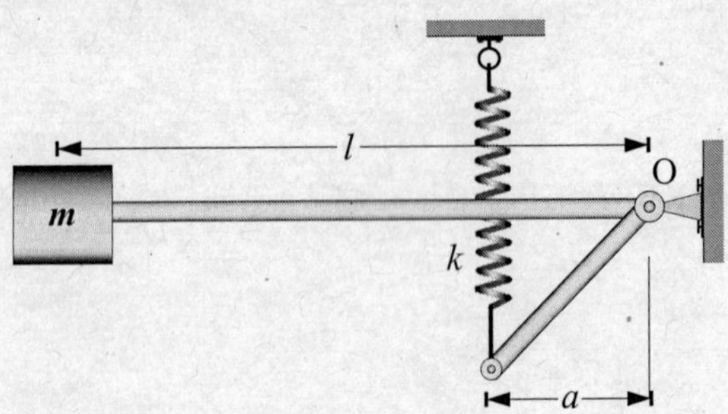

FIGURE 13.22(a) |

Solution. The free body diagram is constructed for an arbitrary displacement θ from the equilibrium position, as shown in Figure 13.22(b).

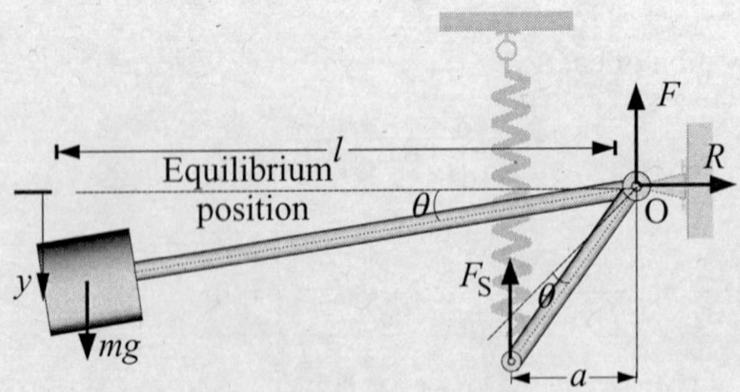

FIGURE 13.22(b) |

Summing moments about O,

$$\sum M_o = I_o\ddot{\theta} \Rightarrow mg \times l - F_s \times a = I_o\ddot{\theta}$$

or

$$mg \times l - k(a\theta + a\theta_{st}) \times a = I_o\ddot{\theta}$$

When the mass is at static equilibrium, the torque due to the weight (mg) is balanced by the moment due to spring force (F_s), i.e.,

$$mgl = k\theta_{st}a^2$$

$$-ka^2\theta = ml^2\ddot{\theta}$$

or

$$\ddot{\theta} + \frac{ka^2}{ml^2}\theta = 0$$

Hence, the natural frequency

$$\omega_n = \sqrt{\frac{ka^2}{ml^2}}$$

$$\omega_n = \frac{a}{l}\sqrt{\frac{k}{m}}$$

PROBLEM 13.23

A cylinder of mass m and cross-sectional area A floats in a liquid of density ρ as shown in Figure 13.23(a). What will be the frequency of oscillation if the cylinder is depressed somewhat and released? Neglect the damping effect of the liquid.

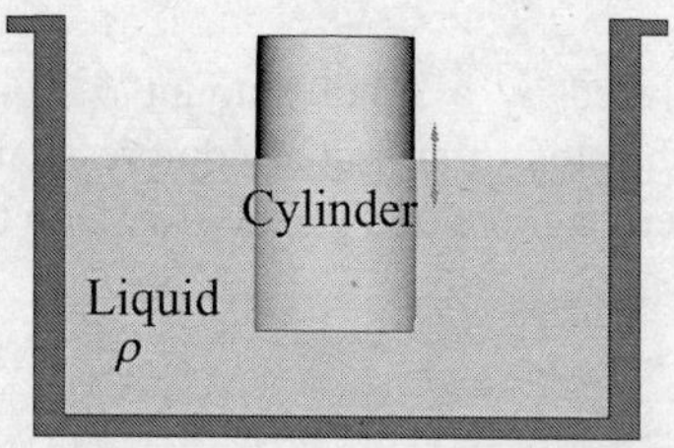

FIGURE 13.23(a)

Solution. The free body diagram of the cylinder when it is displaced by an arbitrary distance x below its equilibrium position is shown in Figure 13.23(b). It is acted upon by a buoyant force equal in magnitude to the gravitational force on the displaced liquid. Unbalanced force, which is up for a downward displacement, is equated to the product of the mass and its acceleration. Then the equation of motion is

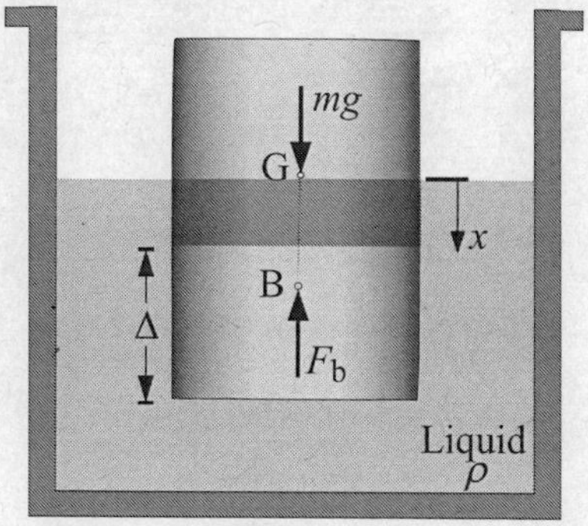

FIGURE 13.23(b)

$$\Sigma F_y = m\ddot{x} \Rightarrow -(F_b - mg) = m\ddot{x},$$

where F_b is the buoyant force and is equal to $\rho A(\Delta + x)g$:

$$mg - \rho A(\Delta + x)g = m\ddot{x}$$

When the cylinder is at static equilibrium, the weight of the cylinder is balanced by the buoyant force, i.e., $mg = \rho A\Delta g$, the equation becomes

$$-\rho Axg = m\ddot{x}$$

or

$$\ddot{x} + \frac{\rho Ag}{m}x = 0$$

Hence, the natural frequency of the system is

$$\omega_n = \sqrt{\frac{\rho A g}{m}}$$

PROBLEM 13.24

The bent rod shown in Figure 13.24(a) has a negligible mass and supports a collar of mass m at its end. Determine the natural period of vibration of the system.

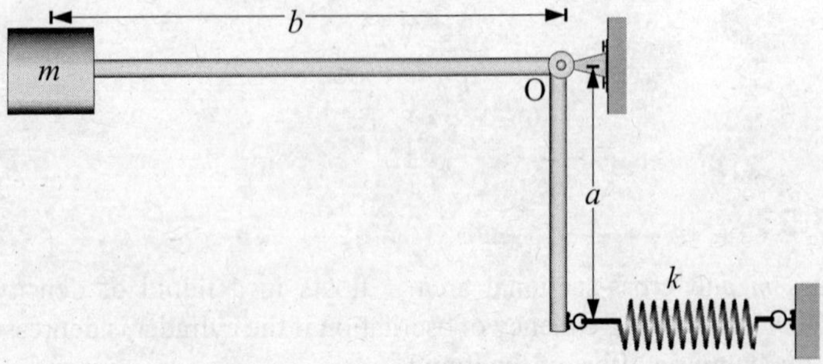

FIGURE 13.24(a) |

Solution. The rod is displaced by a small amount θ from the equilibrium position [Fig. 13.24(b)]. Since the spring is subjected to an initial compression of x_{st} for equilibrium, when the displacement $x > x_{st}$, the spring exerts a force of $F_s = kx - kx_{st}$ on the rod.

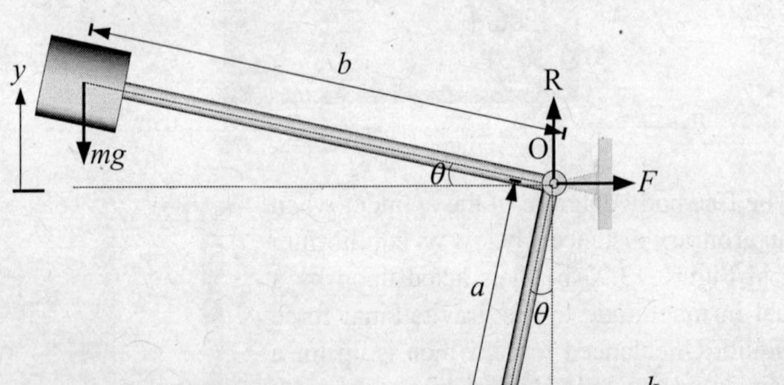

FIGURE 13.24(b) |

Summing moment about point O to eliminate the unknown pin reaction, the moment equation is

$$\Sigma M_o = I_o\ddot{\theta} \Rightarrow -(F_s \times a + mg \times b) = I_o\ddot{\theta}$$

or

$$-[(kx - kx_{st}) \times a + mg \times b] = I_o\ddot{\theta}$$

or

$$mb^2\ddot{\theta} + kxa + (mgb - kx_{st}a) = 0 \qquad (I_o = mb^2)$$

Under the static equilibrium, $mgb = kx_{st}a$, the above equation becomes

$$mb^2\ddot{\theta} + kxa = 0$$

or

$$\ddot{\theta} + \frac{ka^2}{mb^2}\theta = 0 \qquad (x = a\theta)$$

Hence, the natural frequency

$$\omega_n = \sqrt{\frac{ka^2}{mb^2}}$$

Therefore, the natural period of vibration is

$$\tau = \frac{2\pi}{\omega_n} = 2\pi\sqrt{\frac{mb^2}{ka^2}}$$

$$\tau = 2\pi\frac{b}{a}\sqrt{\frac{m}{k}}$$

PROBLEM 13.25

A block of mass m is suspended from a cord that passes over a disc of mass M and radius r as shown in Figure 13.25(a). The spring has a stiffness k. Determine the natural period of vibration for the system.

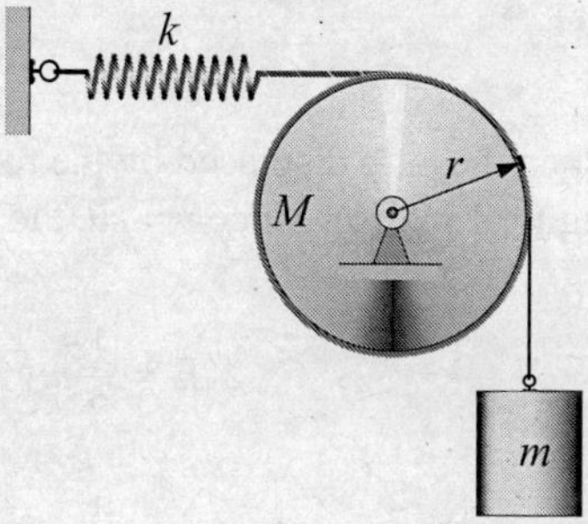

FIGURE 13.25(a) |

Solution. The system consists of the disc, which undergoes a rotation defined by an angle θ, and the block which translates by an amount y [Fig. 13.25(b)]. $I_o\ddot{\theta}$ acts in the direction of positive θ, and consequently ma_y acts downward in the direction of positive y.

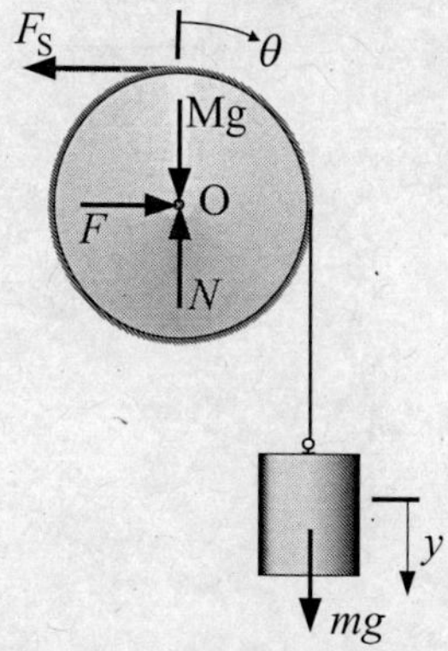

FIGURE 13.25(b) |

Summing moments about O to eliminate the reactions F and N,

$$\sum M_o = I_o\ddot{\theta} \Rightarrow mg \times r - F_s \times r = I_o\ddot{\theta} + m\ddot{y} \times r$$

or

$$(mg - F_s)r = \frac{1}{2}Mr^2\ddot{\theta} + m\ddot{y}r$$

A small angular displacement θ of the disc causes the block to lower by an amount $y = r\theta$; hence, $\ddot{y} = r\ddot{\theta}$. When $\theta = 0$, the spring force required for equilibrium of the disc is mg, acting to the left. For position θ, the spring force is $F_s = kr\theta + mg$. Substituting these into the above equation, we have

$$(mg - kr\theta - mg) \times r = \frac{1}{2} Mr^2\ddot{\theta} + mr\ddot{\theta}r,$$

which, on rearranging the terms, becomes

or
$$\left(\frac{1}{2} M + m\right) r^2\ddot{\theta} + kr^2\theta = 0$$

or
$$\ddot{\theta} + \frac{2k}{M + 2m}\theta = 0$$

Hence, the natural frequency of vibration

$$\omega_n = \sqrt{\frac{2k}{M + 2m}}$$

Therefore, the natural period of vibration is

$$\tau = \frac{2\pi}{\omega_n} = 2\pi\sqrt{\frac{M + 2m}{2k}}$$

$$\tau = 2\pi\sqrt{\frac{M + 2m}{2k}}$$

Energy Method

A diagram of the block and disc, when they are displaced by the respective amounts y and θ from the equilibrium position is shown in Figure 13.25(c). Since $y = r\theta$, the kinetic energy of the system is

$$\begin{aligned} \text{KE} &= \frac{1}{2} m\dot{y}^2 + \frac{1}{2} I_o\omega^2 \\ &= \frac{1}{2} m(r\dot{\theta})^2 + \frac{1}{2} \times \frac{1}{2} Mr^2\dot{\theta}^2 \\ &= \frac{1}{2}\left(m + \frac{M}{2}\right) r^2\dot{\theta}^2 \end{aligned}$$

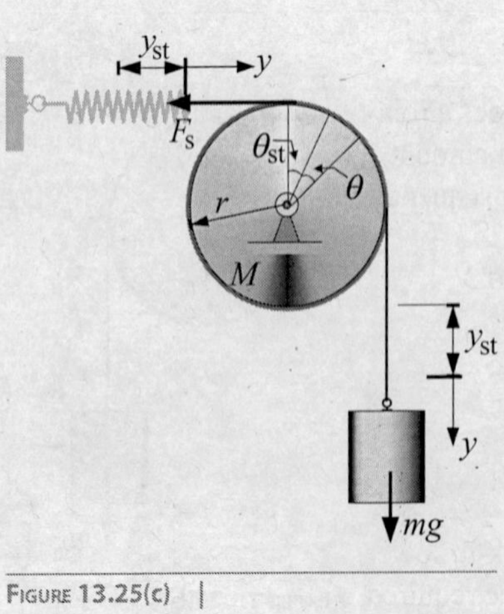

FIGURE 13.25(c) |

Establishing the datum at the equilibrium position of the block and realizing that the spring stretches y_{st} for equilibrium, we can write the potential energy as

$$\text{PE} = \frac{1}{2} k(y_{st} + y)^2 - mgy$$

$$\text{PE} = \frac{1}{2} k(r\theta_{st} + r\theta)^2 - mgr\theta$$

Total energy of the system is

$$\text{KE} + \text{PE} = \frac{1}{2} \times \left(m + \frac{M}{2} \right) r^2 \dot{\theta}^2 + \frac{1}{2} kr^2 (\theta_{\text{st}} + \theta)^2 - mgr\theta$$

Because energy is conserved,

$$\text{KE} + \text{PE} = \text{constant} \quad \text{or} \quad \frac{\text{d}}{\text{d}t}(\text{KE} + \text{PE}) = 0$$

or

$$\frac{\text{d}}{\text{d}t} \left[\frac{1}{2} \left(m + \frac{M}{2} \right) r^2 \dot{\theta}^2 + \frac{1}{2} kr^2 (\theta_{\text{st}} + \theta)^2 - mgr\theta \right] = 0$$

or

$$\left[\left(m + \frac{M}{2} \right) r^2 \dot{\theta} \ddot{\theta} + kr^2 (\theta_{\text{st}} + \theta) \dot{\theta} - mgr\dot{\theta} \right] = 0$$

or

$$r\dot{\theta} \left[\left(m + \frac{M}{2} \right) r\ddot{\theta} + kr\theta_{\text{st}} + kr\theta - mg \right] = 0$$

Under static equilibrium $kr\theta_{\text{st}} = mg$, the above equation becomes

$$r\dot{\theta} \left[\left(m + \frac{M}{2} \right) r\ddot{\theta} + kr\theta \right] = 0$$

Since $\dot{\theta}$ is not always equal to zero in vibrating system, $r\dot{\theta} \neq 0$,

$$\left[\left(m + \frac{M}{2} \right) r\ddot{\theta} + kr\theta \right] = 0$$

or

$$\ddot{\theta} + \frac{2k}{(2m + M)} \theta = 0$$

Hence, the natural frequency of the system is

$$\omega_{\text{n}} = \sqrt{\frac{2k}{(2m + M)}}$$

Therefore, the natural period of vibration is

$$\tau = \frac{2\pi}{\omega_{\text{n}}} = 2\pi \sqrt{\frac{2m + M}{2k}}$$

$$\boxed{\tau = 2\pi \sqrt{\frac{2m + M}{2k}}}$$

PROBLEM 13.26

Two springs of stiffness k_1 and k_2 support a block of mass m as shown in Figure 13.26(a). Determine the natural frequency of the system. Neglect the mass of the rod.

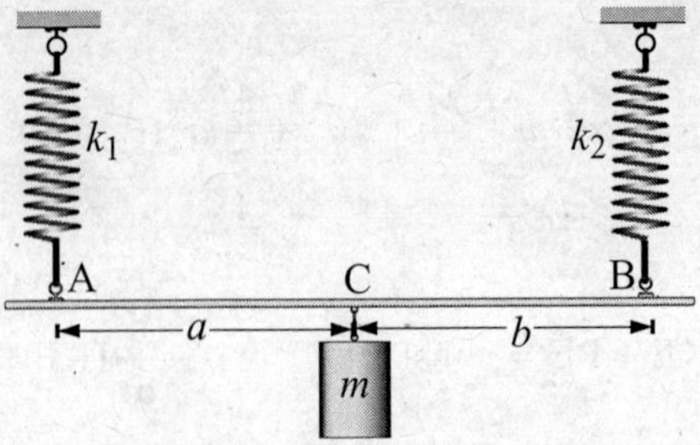

FIGURE 13.26(a) |

Solution. This problem may be solved by substitution k_{eq} into the equation

$$\omega_n = \sqrt{\frac{k_{eq}}{m}}$$

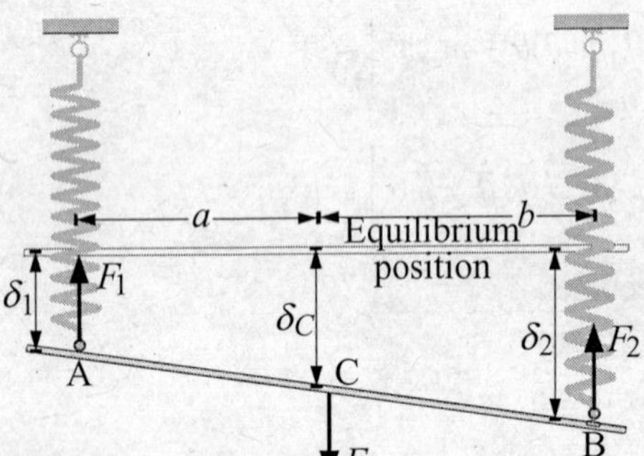

FIGURE 13.26(b) |

Figure 13.26(b) shows a method for calculating the equivalent spring stiffness. First, the vertical force **F** is applied at C, and vertical displacement δ_c is computed. A force F applied at C results in forces $Fb/(a+b)$ and $Fa/(a+b)$ at points A and B, respectively.

The displacements of point A and B, therefore are

$$\delta_1 = \frac{F_1}{k_1} = \frac{Fb}{(a+b)k_1}$$

$$\delta_2 = \frac{F_2}{k_2} = \frac{Fa}{(a+b)k_2}$$

Resultant displacement δ_c of point C is

$$\delta_c = \delta_1 + \frac{\delta_2 - \delta_1}{a+b}a = \frac{\delta_2 a + \delta_1 b}{a+b}$$

$$\delta_c = \frac{F}{(a+b)^2}\left[\frac{b^2}{k_1} + \frac{a^2}{k_2}\right]$$

Equivalent spring stiffness

$$k_{eq} = \frac{F}{\delta_c} = \frac{(a+b)^2}{\left(\dfrac{b^2}{k_1} + \dfrac{a^2}{k_2}\right)}$$

Hence, the natural frequency

$$\omega_n = \sqrt{\frac{K_{eq}}{m}} = \sqrt{\frac{(a+b)^2}{m\left(\dfrac{b^2}{k_1} + \dfrac{a^2}{k_2}\right)}}$$

$$\omega_n = \sqrt{\frac{(a+b)^2 k_1 k_2}{m(b^2 k_2 + a^2 k_1)}}$$

PROBLEM 13.27

Two masses m_1 and m_2 are connected by a spring of stiffness k and placed on a smooth horizontal surface, as shown in Figure 13.27(a). The masses are displaced slightly in opposite directions and released. Determine the frequency of oscillations.

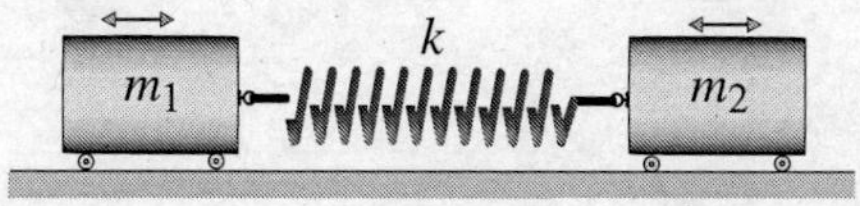

FIGURE 13.27(a) |

Solution. Let l be the normal length of the spring and let at any given instant the coordinate of the masses be x_1 and x_2. Then the extension of spring

$$x = (x_2 - x_1) - l$$

The force F exerted by the spring on the two bodies are obviously equal in magnitude but opposite in sign, as indicated in Figure 13.27(b), with the magnitude of each being kx.

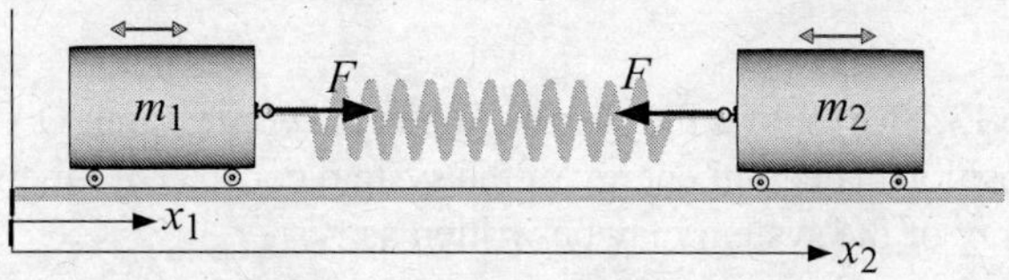

FIGURE 13.27(b) |

The equation of motions for the masses m_1 and m_2 are, respectively,

$$\sum F = m_1\ddot{x}_1 \Rightarrow F = m_1\ddot{x}_1$$

or

$$kx = m_1\ddot{x} \tag{i}$$

$$\sum F = m_2\ddot{x}_2 \Rightarrow F = m_2\ddot{x}_2$$

or

$$-kx = m_2\ddot{x}_2 \tag{ii}$$

Multiplying Equation (i) by m_2 and Equation (ii) by m_1, and then subtracting Equation (i) by Equation (ii),we have

$$m_1m_2\ddot{x}_2 - m_1m_2\ddot{x}_1 = -km_1x - km_2x$$

or

$$m_1m_2(\ddot{x}_2 - \ddot{x}_1) = -k(m_1 + m_2)x$$

or

$$(\ddot{x}_2 - \ddot{x}_1) + \frac{k(m_1 + m_2)}{m_1m_2}x = 0$$

or

$$\ddot{x} + \frac{k(m_1 + m_2)}{m_1m_2}x = 0 \qquad [x = (x_2 - x_1) - l \Rightarrow \ddot{x} = (\ddot{x}_2 - \ddot{x}_1)]$$

Hence,

$$\omega_n = \sqrt{\frac{k(m_1 + m_2)}{m_1m_2}}$$

$$\omega_n = \sqrt{\frac{k}{m_{eq}}}$$

where equivalent mass $m_{eq} = \dfrac{m_1m_2}{m_1 + m_2}$.

PROBLEM 13.28

The link system consists of two similar slender bars and a solid cylinder shown in Figure 13.28(a). Find the natural frequency of vibration of the system.

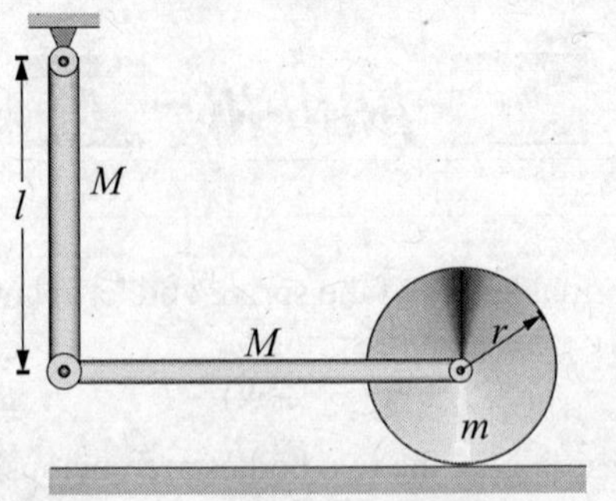

FIGURE 13.28(a) |

Solution. Figure 13.28(b) shows the system. When the vertical link OA is displaced by a small angle θ from vertical position. The total energy of the system consists of kinetic energy and potential energy. The kinetic energy of the system may be written as

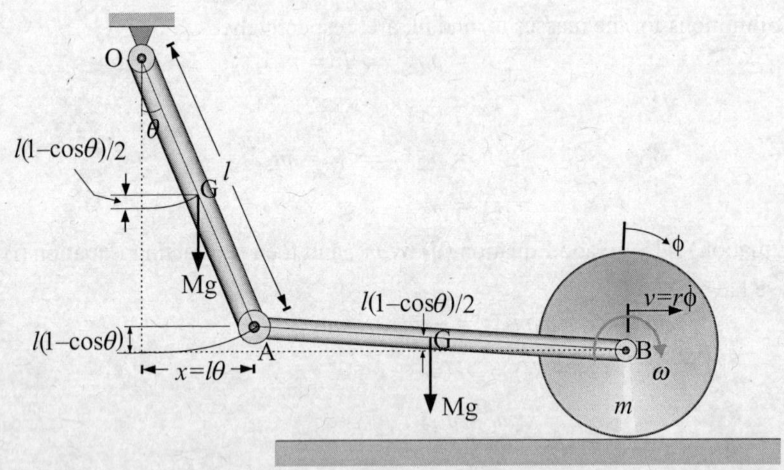

FIGURE 13.28(b) |

$$KE = KE_{OA} + KE_{AB} + (KE_{rot} + KE_{trans})_{cyl}$$

$$= \frac{1}{2}(I_o)_{OA}\dot{\theta}^2 + \frac{1}{2}M\dot{x}^2 + \left(\frac{1}{2}I_B\dot{\phi}^2 + \frac{1}{2}m\dot{x}^2\right)$$

We have

$$x = l\theta = r\phi \Rightarrow l\dot{\theta} = r\dot{\phi}$$

$$\text{KE} = \frac{1}{2}\frac{ml^2}{3}\dot{\theta}^2 + \frac{1}{2}M(l\dot{\theta})^2 + \frac{1}{2}\times\frac{1}{2}mr^2\left(\frac{l\dot{\theta}}{r}\right)^2 + \frac{1}{2}m(l\dot{\theta})^2$$

$$= \frac{1}{6}Ml^2\dot{\theta}^2 + \frac{1}{2}Ml^2\dot{\theta}^2 + \frac{1}{4}Ml^2\dot{\theta}^2 + \frac{1}{2}ml^2\dot{\theta}^2$$

$$\text{KE} = \left(\frac{2}{3}M + \frac{3}{4}m\right)l^2\dot{\theta}^2 = \frac{(8M + 9m)}{12}l^2\dot{\theta}^2$$

When computing the gravitational potential energy, we note from the figure that the centre of gravity of link OA has moved upward through a distance $\frac{l}{2}(1 - \cos\theta)$ and centre of gravity of the link AB has moved upward through a distance $l(1 - \cos\theta)/2$. The potential energy of system is, therefore,

$$\text{PE} = Mg\frac{l}{2}(1 - \cos\theta) + Mg\frac{(1 - \cos\theta)}{2}$$

$$\text{PE} = Mgl(1 - \cos\theta)$$

Total energy of the system is, therefore,

$$\text{KE} + \text{PE} = \frac{(8M + 9m)}{12}l^2\dot{\theta}^2 + Mgl(1 - \cos\theta)$$

The condition for conservation of energy is

$$\frac{\mathrm{d}}{\mathrm{d}t}(\text{KE} + \text{PE}) = 0$$

$$\frac{\mathrm{d}}{\mathrm{d}t}\left[\frac{(8M + 9m)}{12}l^2\dot{\theta}^2 + Mgl(1 - \cos\theta)\right] = 0$$

$$\frac{(8M + 9m)}{6}l^2\dot{\theta}\ddot{\theta} + Mgl\sin\theta\dot{\theta} = 0$$

For oscillation of small amplitude, we may replace $\sin\theta \approx \theta$:

$$\frac{(8M + 9m)}{6}l^2\dot{\theta}\ddot{\theta} + Mgl\theta\dot{\theta} = 0$$

$$l\dot{\theta}\left[\frac{(8M + 9m)}{6}l\ddot{\theta} + Mg\theta\right] = 0$$

Since the $\dot{\theta}$ is not always equal to zero in vibrating system

$$\frac{(8M + 9m)}{6}l\ddot{\theta} + Mg\theta = 0$$

or

$$\ddot{\theta} + \frac{6Mg}{(8M + 9m)l}\theta = 0$$

Hence,

$$\omega_n = \sqrt{\frac{6Mg}{(8M + 9m)l}}$$

PROBLEM 13.29

A slender uniform rod is balanced on a rough cylindrical surface as shown in Figure 13.29(a). If the rod is tipped slightly and released, determine the frequency of the resulting oscillation.

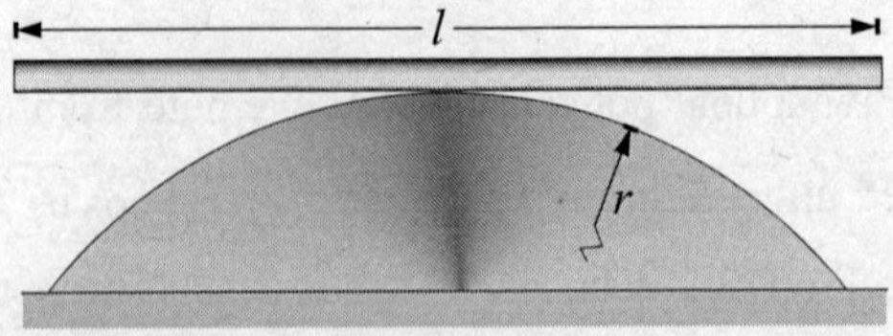

FIGURE 13.29(a)

Solution. Let the rod be displaced through a small angle θ from the equilibrium position, as shown in Figure 13.29(b).

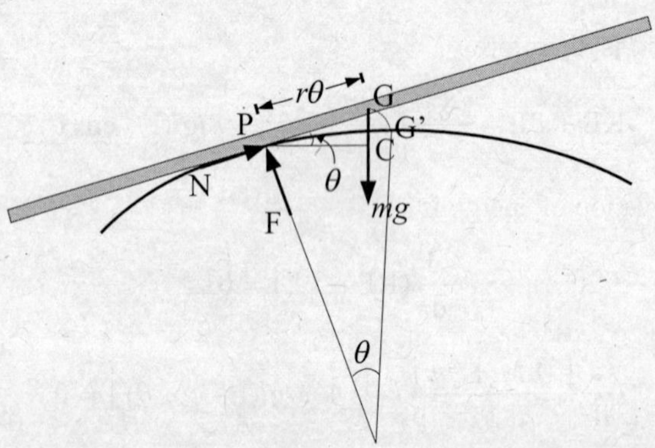

FIGURE 13.29(b)

Then restoring moment about contact point P

$$M_p = -mg\text{PC}$$

$$= -mg\text{PG}\cos\theta \qquad (\text{arcPG}' = r\theta = \text{PG})$$

$$= -mgr\theta \qquad (\text{since } \theta \text{ is small, } \cos\theta = 1)$$

Therefore, the equation of motion for rod is

$$I_p\ddot{\theta} = -mgr\theta$$

or

$$\frac{ml^2}{12}\ddot{\theta} + mgr\theta = 0 \qquad \left(I_p = \frac{ml}{12}\right),$$

which reduces to

$$\ddot{\theta} + \frac{12gr}{l^2}\theta = 0$$

Hence,

$$\omega_n = \sqrt{\frac{12gr}{l^2}}$$

So that

$$\tau = \frac{2\pi}{\omega_n} = 2\pi\sqrt{\frac{l^2}{12gr}}$$

$$\boxed{\tau = 2\pi\sqrt{\frac{l^2}{12gr}}}$$

PROBLEM 13.30

Two uniform discs of mass m each are attached to a rod of mass M, and a spring of constant k is attached to the rod as shown in Figure 13.30(a). Knowing that the discs roll without sliding, determine the frequency of vibration of the system.

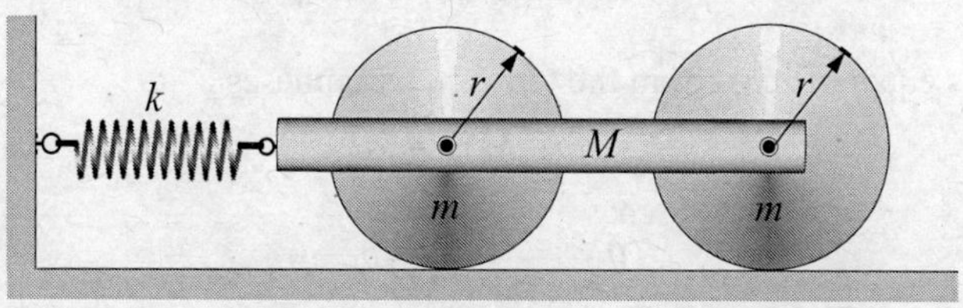

FIGURE 13.30(a) |

Solution. A diagram of the rod and discs when they are displaced by x from the equilibrium position is shown in Figure 13.30(b).

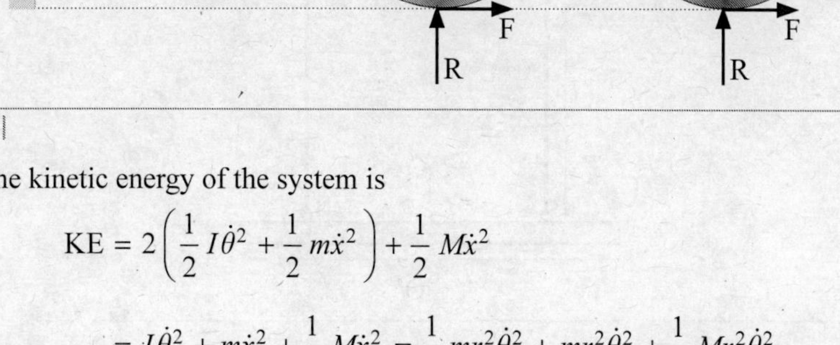

FIGURE 13.30(b) |

Since $x = r\theta$, the kinetic energy of the system is

$$\text{KE} = 2\left(\frac{1}{2}I\dot{\theta}^2 + \frac{1}{2}m\dot{x}^2\right) + \frac{1}{2}M\dot{x}^2$$

$$= I\dot{\theta}^2 + m\dot{x}^2 + \frac{1}{2}M\dot{x}^2 = \frac{1}{2}mr^2\dot{\theta}^2 + mr^2\dot{\theta}^2 + \frac{1}{2}Mr^2\dot{\theta}^2$$

$$= \frac{1}{2}(3m + M)r^2\dot{\theta}^2$$

The potential energy of the system is

$$\text{PE} = \frac{1}{2}kx^2$$
$$= \frac{1}{2}kr^2\theta^2 \qquad (x = r\theta)$$

The total energy for the system is, therefore,

$$\text{KE} + \text{PE} = \frac{1}{2}(3m + M)r^2\dot{\theta}^2 + \frac{1}{2}kr^2\theta^2$$

The condition for conservation of energy is $\frac{\text{d}}{\text{d}t}(\text{KE} + \text{PE}) = 0.$

$$\frac{\text{d}}{\text{d}t}\left(\frac{1}{2}(3m + M)r^2\dot{\theta}^2 + \frac{1}{2}kr^2\theta^2\right) = 0$$

or

$$(3m + M)r^2\dot{\theta}\ddot{\theta} + kr^2\theta\dot{\theta} = 0$$

or

$$r^2\dot{\theta}[(3m + M)\ddot{\theta} + k\theta] = 0$$

Since $\dot{\theta}$ is not always equal to zero, from the terms in parentheses

$$(3m + M)\ddot{\theta} + k\theta = 0$$

or

$$\ddot{\theta} + \frac{k}{(3m + M)}\theta = 0$$

Hence,

$$\omega_n = \sqrt{\frac{k}{3m + M}}$$

PROBLEM 13.31

A uniform bar of length L and mass m is suspended symmetrically by two strings, as shown in Figure 13.31(a). Determine the expression for the natural frequency of the rod, oscillating about the vertical axis O–O′.

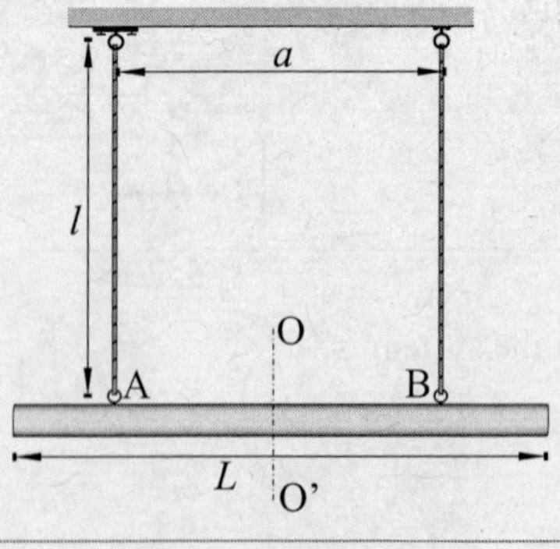

FIGURE 13.31(a) |

Solution. Figure 13.31(b) shows the free body diagram of the bar when it is displaced by a small angle θ about the centre line O–O′ and the tension in each string $T = mg/2$. The horizontal component of the tension is

$$T \sin \phi = \frac{1}{2} mg \sin \phi$$

The moment equation about the vertical centre line is

$$\sum M_o = I_o \ddot{\theta} \Rightarrow -2T \sin \phi \times \text{OA} = I_o \ddot{\theta}$$

or $$I_o \ddot{\theta} + 2T \times \frac{a}{2} \times \sin \phi = 0$$

For a small angle ϕ, $\sin \phi \approx \phi$ and

$$l\phi = \frac{a}{2} \theta \Rightarrow \phi = \frac{a}{2l} \theta$$

The equation becomes

$$I_o \ddot{\theta} + Ta\phi = 0$$

or $$\frac{mL^2}{12} \ddot{\theta} + \frac{mg}{2} \times a \times \frac{a}{2l} \theta = 0, \qquad \left[I_o = \frac{mL^2}{12} \right]$$

or $$\ddot{\theta} + \frac{3ga^2}{l \times L^2} \theta = 0$$

Hence,

$$\omega_n = \sqrt{\frac{3ga^2}{l \times L^2}} = \frac{a}{L} \sqrt{\frac{3g}{l}}$$

$$\omega_n = \frac{a}{L} \sqrt{\frac{3g}{l}}$$

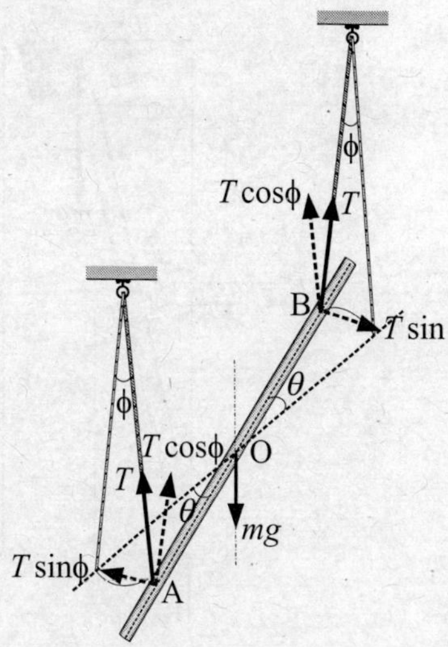

FIGURE 13.31(b)

PROBLEM 13.32

Find the natural frequency of vibration of the system shown in Figure 13.32(a) for small amplitudes. If k, l and m are fixed, determine the value of a for which the system will not vibrate.

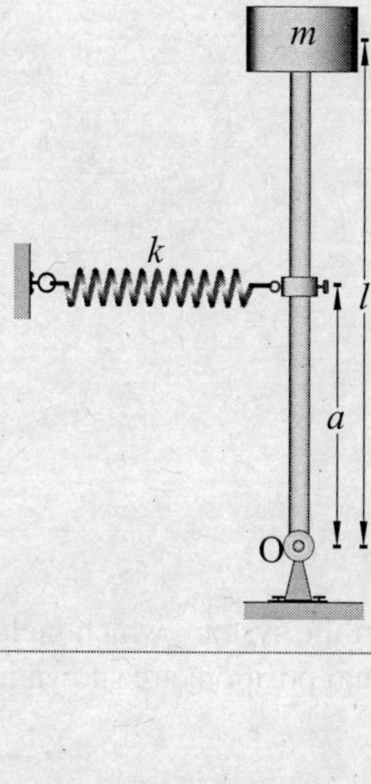

FIGURE 13.32(a)

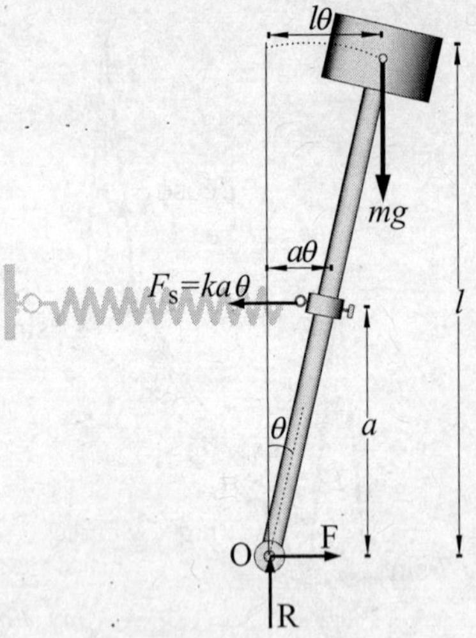

FIGURE 13.32(b) |

Solution. Figure 13.32(b) shows the free body diagram of the system when it is displaced by a small angle θ from the vertical equilibrium position.

Applying the moment equation about O,

$$\Sigma M_o = I_o \ddot{\theta}$$

$$-F_s \times a + mg \times l\theta = I_o \ddot{\theta}$$

Spring force $F_s = ka\theta$; the equation reduces to

$$I_o \ddot{\theta} + (ka^2 - mgl)\theta = 0$$

or $$\ddot{\theta} + \frac{(ka^2 - mgl)}{I_o}\theta = 0$$

Hence,

$$\omega_n = \sqrt{\frac{ka^2 - mgl}{I_o}}$$

System will not vibrate if $(ka^2 - mgl) \leq 0$

or $$a^2 \leq \frac{mgl}{k} \Rightarrow a \leq \sqrt{\frac{mgl}{k}}$$

$$a \leq \sqrt{\frac{mgl}{k}}$$

PROBLEM 13.33

Figure 13.33(a) shows a mass m hanging from a cord attached to a homogeneous circular disc of mass M and radius R. To restrain the disc from rotating, a spring of stiffness k is attached at a radius r from the disc's centre. Determine the natural frequency of oscillation of the system when the mass is displaced from the equilibrium position.

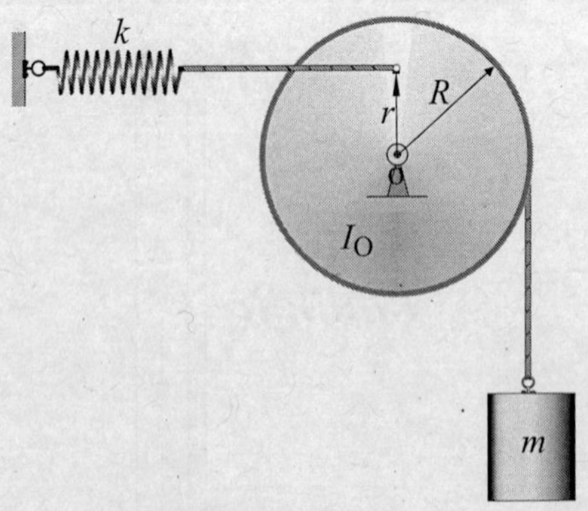

FIGURE 13.33(a)

Solution. Free body diagrams of the system, which includes the block and the disc at an arbitrary displacement x from the equilibrium position, are shown in Figure 13.33(b).

The equations of motion for the mass m and the disc are, respectively,

$$\sum F = m\ddot{x} \Rightarrow mg - T = m\ddot{x}$$

$$\sum M_o = I_o\ddot{\theta} \Rightarrow TR - kr^2(\theta + \theta_{st}) = I_o\ddot{\theta}$$

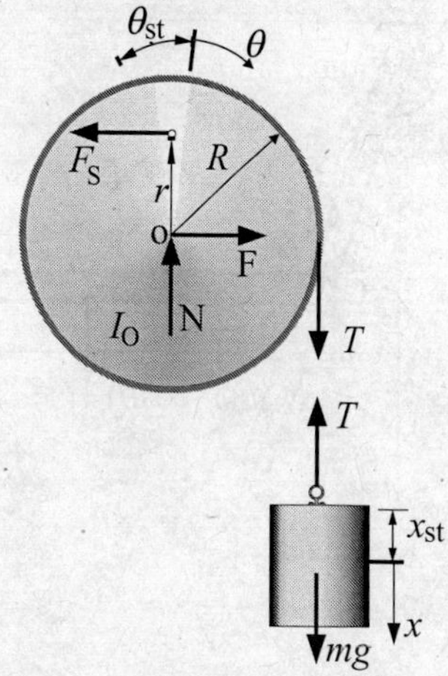

FIGURE 13.33(b) |

Eliminating T from the above equations, we have

$$R(mg - m\ddot{x}) - kr^2(\theta + \theta_{st}) = I_o\ddot{\theta}$$

Replacing $\ddot{x}$ by $R\ddot{\theta}$, the equation becomes

$$mgR - mR^2\ddot{\theta} - kr^2\theta - kr^2\theta_{st} = I_o\ddot{\theta}$$

Rearrange the terms,

$$(mR^2 + I_o)\ddot{\theta} + kr^2\theta + kr^2\theta_{st} - mgR = 0$$

When the mass is at static equilibrium, the torque due to the weight of the mass is balanced by the moment due to the restoring force in the spring, i.e.,

$$mgR = T_o r = kr^2\theta_{st} \text{ (where } T_o = kr^2\theta_{st})$$

$$(mR^2 + I_o)\ddot{\theta} + kr^2\theta = 0$$

or

$$\ddot{\theta} + \frac{kr^2}{mR^2 + I_o}\theta = 0$$

Hence,

$$\omega_n = \sqrt{\frac{kr^2}{I_o + mR^2}}$$

PROBLEM 13.34

Three identical rods are connected as shown in Figure 13.34(a). Determine the frequency of small oscillation of the system.

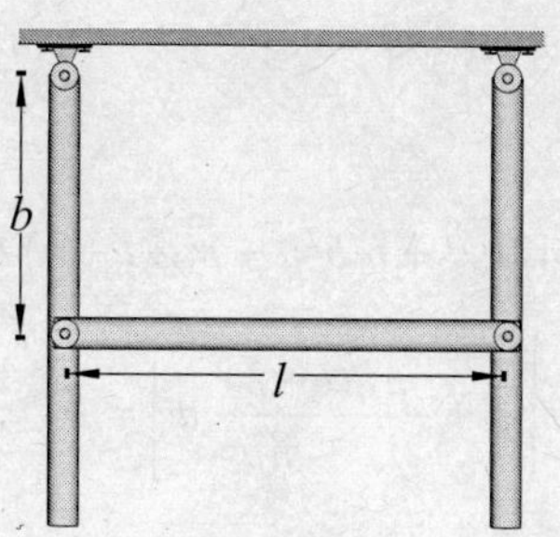

FIGURE 13.34(a) |

Solution. Figure 13.34(b) shows the frame when it is displaced by a small angle θ from the equilibrium position. The kinetic energy of the frame may be written as

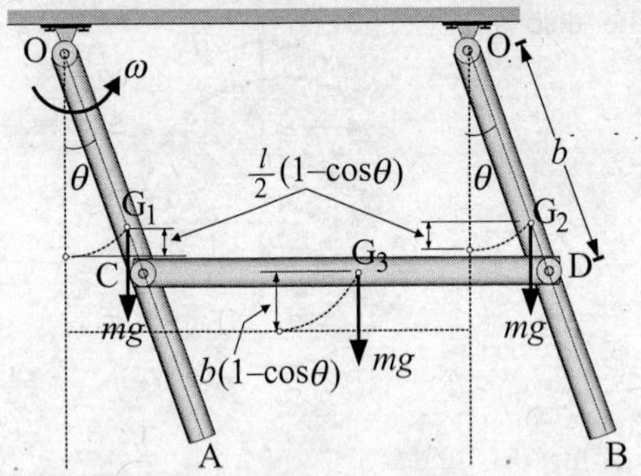

FIGURE 13.34(b) |

$$KE = KE_{link\ OA} + KE_{link\ OB} + KE_{link\ CD}$$

$$= 2 \times KE_{link\ OA} + KE_{link\ CD}$$

$$KE = 2\left(\frac{1}{2} I_o \dot{\theta}^2\right) + \frac{1}{2} mv^2$$

$$= I_o \dot{\theta}^2 + \frac{1}{2} m(b\dot{\theta})^2$$

If a horizontal datum is displaced through the centre of gravity of the links when $\theta = 0$, then the centre of gravity of vertical links moves upward $(l/2)(1 - \cos\theta)$ and centre of gravity of horizontal links moves upward $b(1 - \cos\theta)$.

The potential energy of system is

$$PE = 2\left[mg\frac{l}{2}(1 - \cos\theta)\right] + mgb(1 - \cos\theta)$$

$$= mg(l + b)(1 - \cos\theta)$$

The total energy of the system is

$$KE + PE = I_o\dot{\theta}^2 + \frac{1}{2} m(b\dot{\theta})^2 + mg(l + b)(1 - \cos\theta)$$

The condition for conservation of energy,

$$\frac{d}{dt}[KE + PE] = 0$$

$$\frac{d}{dt}\left[I_o\dot{\theta}^2 + \frac{1}{2} mb^2\dot{\theta}^2 + mg(l + b)(1 - \cos\theta)\right] = 0$$

or

$$2I_o\dot{\theta}\ddot{\theta} + mb^2\dot{\theta}\ddot{\theta} + mg(l + b)\sin\theta\dot{\theta} = 0$$

For a small angle θ, $\sin\theta \approx \theta$

$$\dot{\theta}[2I_o\ddot{\theta} + mb^2\ddot{\theta} + mg(l + b)\theta] = 0$$

or

$$\dot{\theta}\left[\ddot{\theta} + \frac{mg(l + b)}{2I_o + mb^2}\theta\right] = 0$$

Since $\dot{\theta}$ is not always equal to zero in a vibrating system

$$\ddot{\theta} + \frac{mg(l + b)}{2I_o + mb^2}\theta = 0$$

or $$\ddot{\theta} + \frac{3g(l+b)}{2l^2 + 3b^2}\theta = 0 \qquad \left(I_o = \frac{1}{3}ml^2\right)$$

Hence,

$$\omega_n = \sqrt{\frac{3g(l+b)}{2l^2 + 3b^2}}$$

PROBLEM 13.35

Two uniform rods AB and CD, each of mass m and length l, are attached to two gears as shown in Figure 13.35(a). Neglecting the mass of the gears, determine the period of small oscillations of the system.

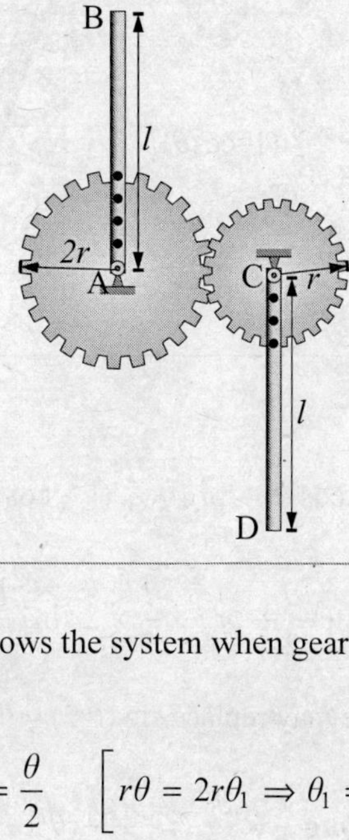

FIGURE 13.35(a) |

Solution. Figure 13.35(b) shows the system when gear Q has rotated through a small angle θ from the equilibrium position.

Since

$$\theta_1 = \frac{\theta}{2} \qquad \left[r\theta = 2r\theta_1 \Rightarrow \theta_1 = \frac{\theta}{2}\right]$$

the kinetic energy of the system is

$$\text{KE} = \text{KE}_{AB} + \text{KE}_{CD} = \frac{1}{2}(I_{AB})_A\dot{\theta}_1^2 + \frac{1}{2}(I_{CD})_C\dot{\theta}^2$$

$$= \frac{1}{2} \times \frac{1}{3}ml^2\left(\frac{\dot{\theta}}{2}\right)^2 + \frac{1}{2} \times \frac{1}{3}ml^2\dot{\theta}^2 = \frac{5}{24}ml^2\dot{\theta}^2$$

$$\text{KE} = \frac{5}{24}ml^2\dot{\theta}^2.$$

When computing the gravitational potential energy, we note from Figure 13.35(b) that the centre of gravity of the lower rod CD has moved upward through a distance $\frac{l}{2}(1 - \cos\theta)$, and centre the of gravity of the upper rod AB has moved downward through the distance $\frac{l}{2}(1 - \cos\theta_1)$. The potential energy is

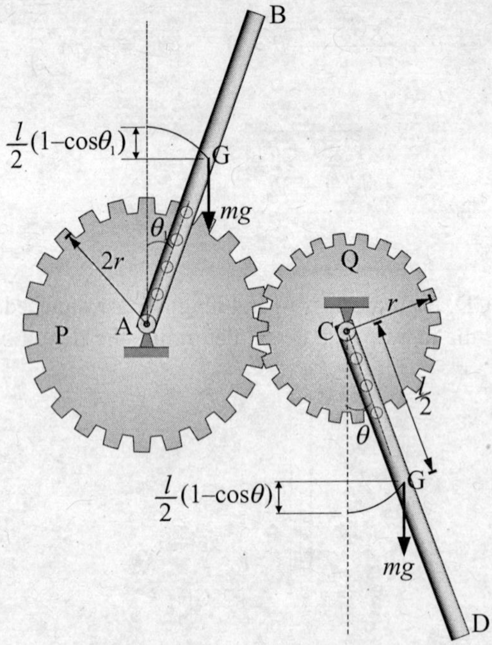

FIGURE 13.35(b) |

$$\mathrm{PE} = +mg \times \frac{l}{2}(1 - \cos\theta) - mg \times \frac{l}{2}(1 - \cos\theta_1)$$

$$\mathrm{PE} = +\ mg \times \frac{l}{2} 2\sin^2\frac{\theta}{2} - mg \times \frac{l}{2} \times 2\sin^2\frac{\theta_1}{2}$$

For oscillations of small amplitude, we may replace $\sin(\theta/2) \approx \theta/2$ and $\sin(\theta_1/2) \approx \theta_1/2$,

$$\mathrm{PE} = mg\frac{l}{2} \times \frac{\theta^2}{2} - mg\frac{l}{2} \times \frac{\theta_1^2}{2} = \frac{mgl}{4}(\theta^2 - \theta_1^2)$$

$$= \frac{mgl}{4}\left(\theta^2 - \frac{\theta^2}{4}\right) = \frac{3}{16} mgl\theta^2 \qquad \left(r\theta = 2r\theta_1 \Rightarrow \theta_1 = \frac{\theta}{2}\right)$$

Total energy of the system is, therefore,

$$\mathrm{KE} + \mathrm{PE} = \frac{5}{24} ml^2\dot{\theta}^2 + \frac{3}{16} mgl\theta^2$$

The condition for conservation of energy, $\frac{\mathrm{d}}{\mathrm{d}t}(\mathrm{KE} + \mathrm{PE}) = 0$

$$\frac{\mathrm{d}}{\mathrm{d}t}\left(\frac{5}{24} ml^2\dot{\theta}^2 + \frac{3}{16} mgl\theta^2\right) = 0$$

or
$$\frac{5}{12}ml^2\ddot{\theta}\dot{\theta}+\frac{3}{8}mgl\theta\dot{\theta}=0$$

or
$$ml\dot{\theta}\left(\frac{5}{12}l\ddot{\theta}+\frac{3}{8}g\theta\right)=0$$

Since the $\dot{\theta}$ is not always equal to zero, from the terms in parenthesis, we have

$$\frac{5}{12}l\ddot{\theta}+\frac{3}{8}g\theta=0$$

or
$$\ddot{\theta}+\frac{9g}{10l}\theta=0$$

Hence,
$$\omega_{\text{n}}=\sqrt{\frac{9g}{10l}}$$

So that
$$\tau=\frac{2\pi}{\omega_{\text{n}}}=\frac{2\pi}{\sqrt{\frac{9g}{10l}}}$$

$$\tau=2\pi\sqrt{\frac{10l}{9g}}$$

PROBLEM 13.36

Find the natural frequency of oscillation of spring-mass system as shown in Figure 13.36(a).

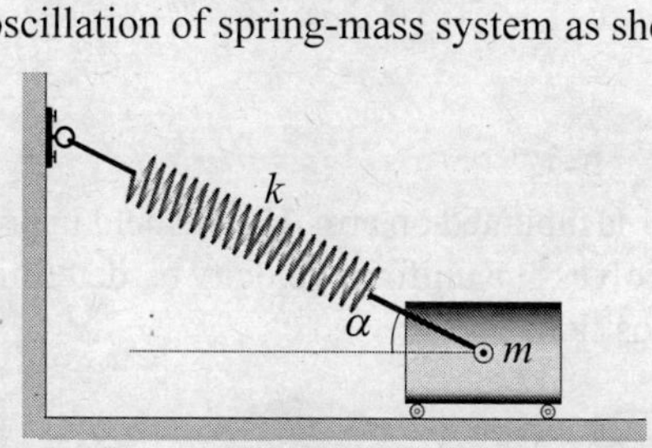

FIGURE 13.36(a) |

Solution. The free body diagram of the mass is constructed for an arbitrary displacement x from equilibrium position, as shown in Figure 13.36(b). The spring is deformed by an amount PQ $= x\cos a$ along its axis. Spring force along the spring axis is

$$F_{\text{s}}=k(x\cos a)$$

The component of F_{s} along the direction of motion of mass is

$$F_x=F_{\text{s}}\cos a$$

or
$$F_x=kx\cos^2 a$$

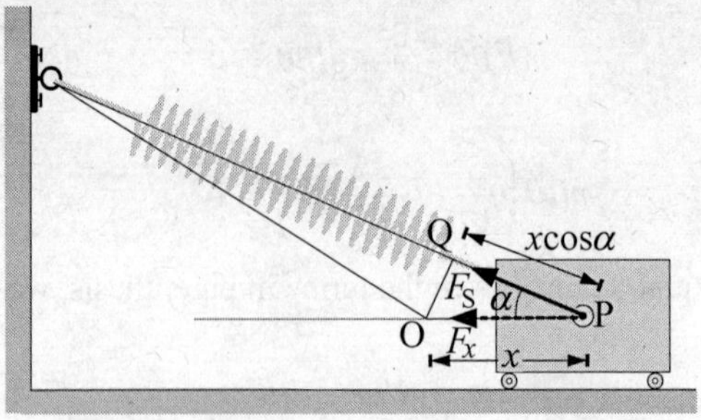

FIGURE 13.36(b) |

The equation of motion of the mass m is

$$m\ddot{x} = -F_x$$

or

$$m\ddot{x} = -kx\cos^2 a$$

or

$$\ddot{x} + \frac{k\cos^2 a}{m}x = 0$$

From the above equation, it may be noted that the equivalent stiffness k_{eq} of a spring that makes an angle α with the axis of motion is given as

$$k_{eq} = k\cos^2 a$$

Hence, the natural frequency is

$$\omega_n = \sqrt{\frac{k\cos^2 a}{m}}$$

PROBLEM 13.37

A cylinder of mass m and radius r is mounted on arm AB and held in position by two springs as shown in Figure 13.37(a). If the arm revolves at a uniform velocity ω, determine the natural frequency of the cylinder about the equilibrium position.

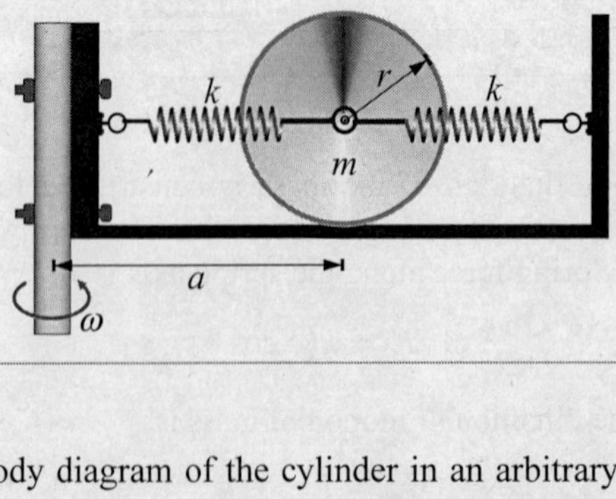

FIGURE 13.37(a) |

Solution. The free body diagram of the cylinder in an arbitrary position is shown in Figure 13.37(b).

The forces acting on the cylinder are its weight mg, centrifugal force $m\omega^2(a + x)$, reaction N, friction force F, and two spring forces F_s.

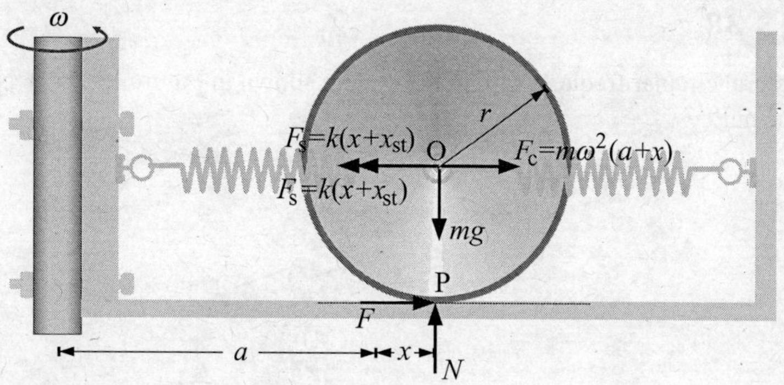

FIGURE 13.37(b) |

The governing equations are

$$\sum F_x = ma_x \Rightarrow -2k(x + x_{st}) + m\omega^2(a + x) + F = m\ddot{x}$$

$$\sum F_y = 0 \Rightarrow mg = N$$

$$\sum M_o = I_o\ddot{\theta} \Rightarrow -Fr = I_o\ddot{\theta}$$

Eliminating F from the above equations, we have

$$-2k(x + x_{st}) + m\omega^2(a + x) + \frac{-I_o\ddot{\theta}}{r} = m\ddot{x}$$

When the cylinder is at static equilibrium, the centrifugal force ($m\omega^2 a$) is balanced by the spring forces ($2kx_{st}$), i.e., $m\omega^2 a = 2kx_{st}$:

$$-2kx + m\omega^2 x - \frac{I_o\ddot{\theta}}{r} = m\ddot{x}$$

x may be related to θ by the equation $x = r\theta$, so that $\ddot{x} = r\ddot{\theta}$:

$$-2kr\theta + m\omega^2 r\theta - \frac{I_o\ddot{\theta}}{r} = mr\ddot{\theta},$$

which, on rearranging the terms, becomes

$$\ddot{\theta} + \frac{(2k - m\omega^2)r^2}{(mr^2 + I_o)}\theta = 0$$

The natural frequency is determined from differential equation as

$$\omega_n = \sqrt{\frac{(2k - m\omega^2)r^2}{(mr^2 + I_o)}} \qquad \left[I_o = \frac{1}{2}mr^2\right]$$

$$\omega = \sqrt{\frac{2(2k - m\omega^2)}{3m}}$$

PROBLEM 13.38

Determine the natural circular frequency ω_n of the system shown in Figure 13.38(a). Neglect the mass and friction of the pulleys.

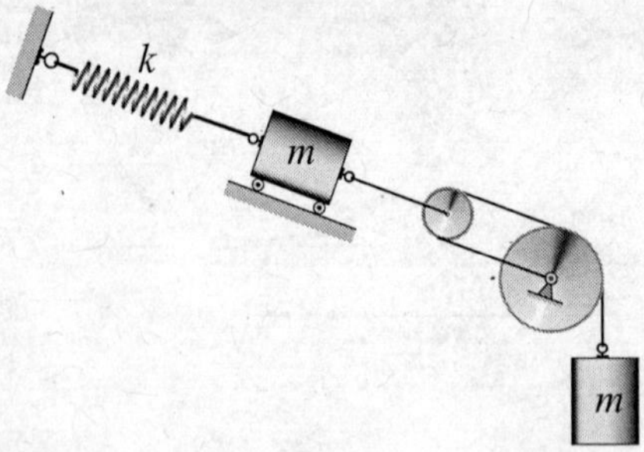

FIGURE 13.38(a) |

Solution. Let T be the tension in the cord connected to mass A, and x_1 and x be the displacements of masses A and B, respectively, from the static equilibrium, as shown in Figure 13.38(b).

Then the equation of motion for mass A is

$$mg - T = m\ddot{x}_1$$

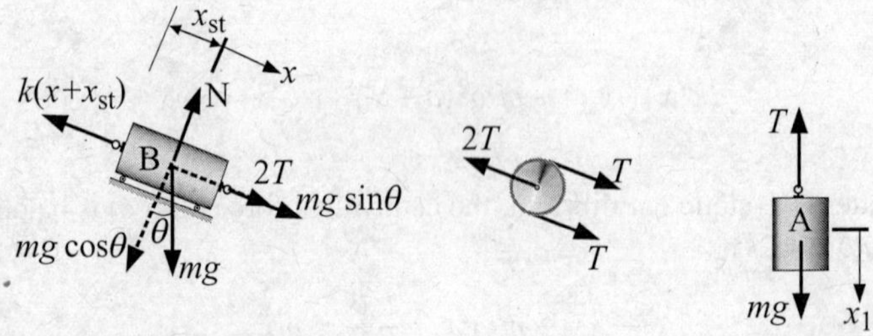

FIGURE 13.38(b) |

x_1 may be related by x by the equation $x_1 = 2x$, so that $\ddot{x}_1 = 2\ddot{x}$:

$$mg - T = m(2\ddot{x}) \tag{i}$$

$$\left(x_1 = 2x \Rightarrow \ddot{x}_1 = 2\ddot{x}\right)$$

and the equation of motion for mass B is

$$mg\sin\theta + 2T - k(x + x_{st}) = 5m\ddot{x} \tag{ii}$$

Eliminating T from Equations (i) and (ii), we have

$$2mg + mg\sin\theta - k(x + x_{st}) = 5m\ddot{x} \tag{iii}$$

Under the static equilibrium

$$T_{st} = mg \tag{iv}$$

and

$$mg\sin\theta + 2T_{st} = kx_{st} \tag{v}$$

Eliminating T_{st} from Equations (iv) and (v), we have

$$mg\sin\theta + 2mg = kx_{st}$$

Then Equation (iii) reduces to

$$-kx = 5m\ddot{x}$$

or

$$\ddot{x} + \frac{k}{5m}x = 0$$

Hence,

$$\ddot{x} + \frac{k}{5m}x = 0$$

$$\omega_n = \sqrt{\frac{k}{5m}}$$

PROBLEM 13.39

A block of mass m is suspended from a cord that passes over a disc as shown in Figure 13.39(a). If the spring has a stiffness k, determine the natural period of vibration for the system.

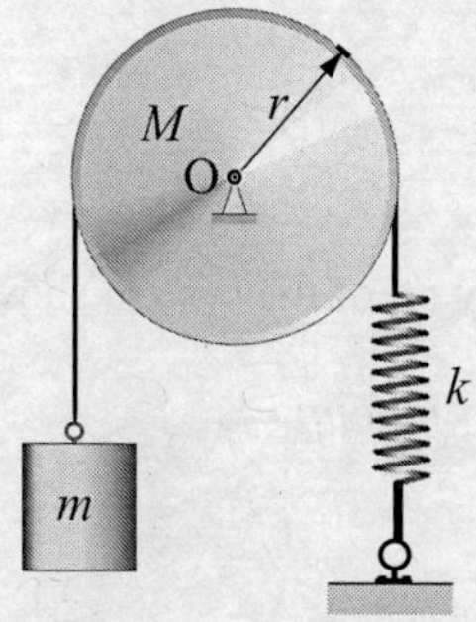

FIGURE 13.39(a) |

Solution. Let T be the tension in the rope, and x and θ be the displacements of the mass and pulley, respectively, from the static equilibrium position, as shown in Figure 13.39(b).

Then the equation of motion of the mass is

$$mg - T = m\ddot{x}$$

and the equation of motion of the pulley is

$$Tr - k(x + x_{st})r = I_o\ddot{\theta}$$

Eliminating T from the above equations

$$mg - k(x + x_{st}) = \frac{I_o}{r}\ddot{\theta} + m\ddot{x}$$

or

$$mg - k(x + x_{st}) = \frac{I_o}{r^2}\ddot{x} + m\ddot{x} \qquad \left(\ddot{\theta} = \frac{\ddot{x}}{r}\right)$$

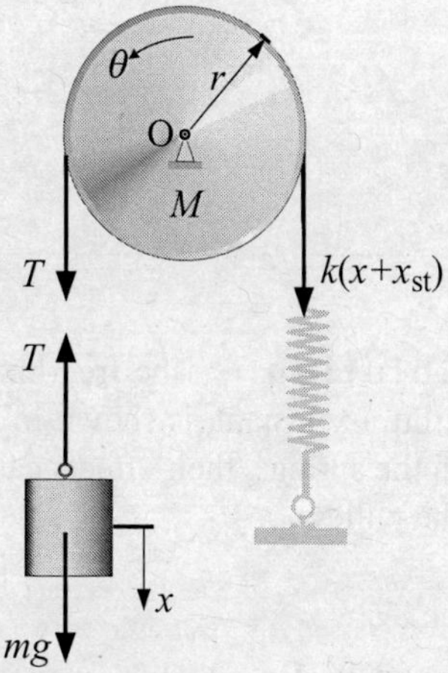

FIGURE 13.39(b) |

Under the static equilibrium, $mg = kx_{st}$, the equation reduces to

$$\left(\frac{I_o}{r^2} + m\right)\ddot{x} = -kx$$

or

$$\ddot{x} + \frac{k}{\left(\frac{I_o}{r^2}\right) + m}x = 0$$

So that

$$\omega_n = \sqrt{\frac{k}{\left(\frac{I_o}{r^2}\right) + m}} = \sqrt{\frac{2k}{(M + 2m)}} \qquad \left(I_o = \frac{1}{2}Mr^2\right)$$

Thus,

$$\tau = \frac{2\pi}{\omega_n} = 2\pi\sqrt{\frac{M + 2m}{2k}}$$

$$\boxed{\tau = 2\pi\sqrt{\frac{M + 2m}{2k}}}$$

PROBLEM 13.40

Determine the natural circular frequency ω_n of the system shown in Figure 13.40(a).

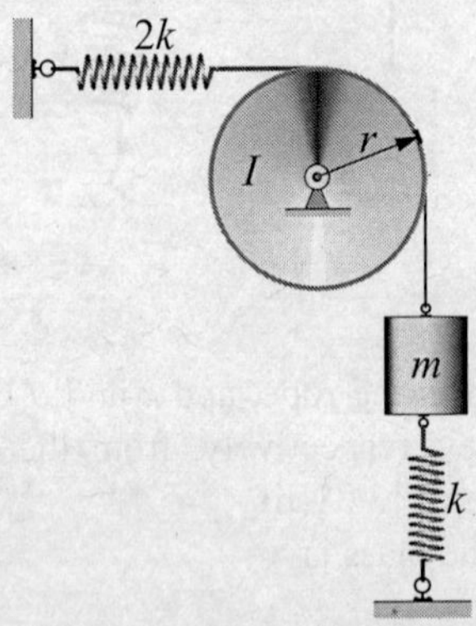

FIGURE 13.40(a) |

Solution. The free body diagram of the system, which includes a block and a pulley at an arbitrary instant, is shown in Figure 13.40(b). It is noted that since gravity leads to static deflection in the springs, their effects cancel in the differential equation. Summing moments about the centre of the pulley

$$(\Sigma M_o)_{ext} = (\Sigma M_o)_{eff},$$

which leads to

$$(-kx)r - (2kx)r = (m\ddot{x})r + I_o\ddot{\theta}$$

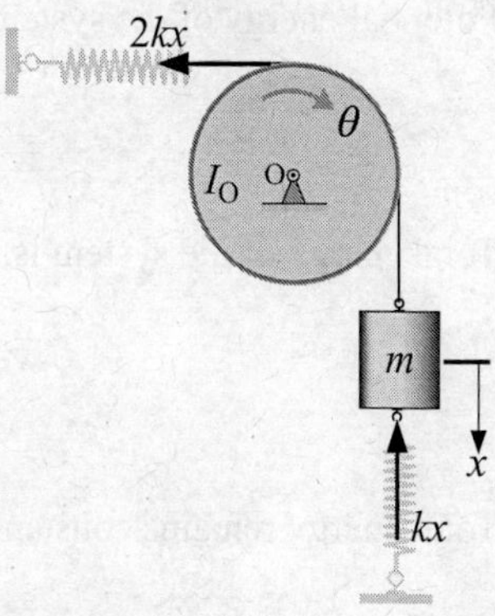

FIGURE 13.40(b) |

or $$\left(mr\ddot{x} + I_o\frac{\ddot{x}}{r}\right) + 3kxr = 0 \qquad \left(x = r\theta \Rightarrow \ddot{\theta} = \frac{\ddot{x}}{r}\right)$$

or $$\left(m + \frac{I_o}{r^2}\right)\ddot{x} + 3kx = 0$$

or $$\ddot{x} + \frac{3k}{m + \left(\frac{I_o}{r^2}\right)}x = 0$$

The natural frequency is determined from the differential equation as

$$\omega_n = \sqrt{\frac{3k}{m + \left(\frac{I_o}{r^2}\right)}}$$

PROBLEM 13.41

Figure 13.41(a) shows a homogeneous disc of mass M and radius r, supporting a load of mass m by an inextensible cord, and the disc is supported by a spring of stiffness k. Find the frequency of oscillations of the system.

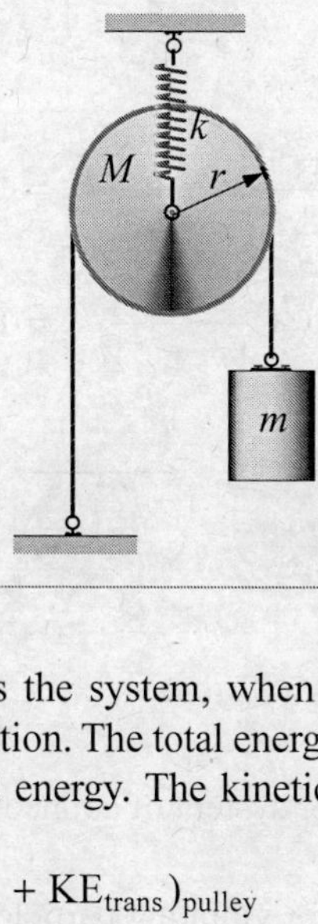

FIGURE 13.41(a) |

Solution. Figure 13.41(b) shows the system, when the mass m is displaced by x from the equilibrium position. The total energy of the system consists of kinetic energy and potential energy. The kinetic energy of the system may be written as

$$KE = KE_m + (KE_{rot} + KE_{trans})_{pulley}$$

$$= \frac{1}{2}m\dot{x}^2 + \left(\frac{1}{2}I_o\dot{\theta}^2 + \frac{1}{2}M\dot{y}^2\right)$$

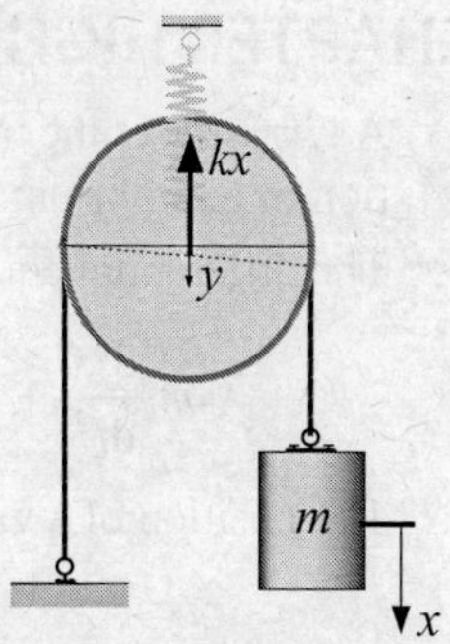

FIGURE 13.41(b) |

For any small displacement $\theta, y = r\theta$ and $y = x/2$ so that $\dot{y} = \dot{x}/2, \dot{\theta} = \dot{x}/r$;

$$KE = \frac{1}{2}m\dot{x}^2 + \frac{1}{2}\left(\frac{Mr^2}{2}\right)\left(\frac{\dot{y}}{r}\right)^2 + \frac{1}{2}M\left(\frac{\dot{x}}{2}\right)^2$$

$$= \frac{(8m + 3M)}{16}\dot{x}^2$$

Potential energy of the system

$$\text{PE} = \frac{1}{2}ky^2 = \frac{1}{2}k\left(\frac{x}{2}\right)^2 = \frac{1}{8}kx^2$$

Total energy of the system is, therefore,

$$\text{KE} + \text{PE} = \frac{8m + 3M}{16}\dot{x}^2 + \frac{1}{8}kx^2$$

Total energy remains constant; $\frac{\mathrm{d}}{\mathrm{d}t}(\text{KE} + \text{PE}) = 0$

$$\frac{\mathrm{d}}{\mathrm{d}t}\left[\frac{(8m + 3M)}{16}\dot{x}^2 + \frac{1}{8}kx^2\right] = 0$$

or

$$\frac{(8m + 3M)}{16}2\dot{x}\ddot{x} + \frac{1}{8}k2x\dot{x} = 0$$

or

$$\dot{x}\left[\frac{(8m + 3M)}{8}\ddot{x} + \frac{1}{4}kx\right] = 0$$

Since $\dot{x}$ is not always zero,

$$\frac{(8m + 3M)}{8}\ddot{x} + \frac{1}{4}kx = 0$$

or

$$\ddot{x} + \frac{2k}{8m + 3M}x = 0$$

Hence,

$$\omega_n = \sqrt{\frac{2k}{8m + 3M}}$$

CHAPTER HIGHLIGHTS

- A *vibration* is the motion of a body or system of connected bodies which oscillates about a position of equilibrium.
- The differential equation of motion for vibrating particle is

$$m\frac{\mathrm{d}^2x}{\mathrm{d}t^2} + kx = 0 \quad \text{or} \quad \frac{\mathrm{d}^2x}{\mathrm{d}t^2} + \omega_n^2x = 0 \quad \text{Where } \omega_n = \sqrt{\frac{k}{m}}$$

- The position of a vibrating particle varies periodically in time according to the relation

$$x = \sqrt{x_0^2 + \left(\frac{v_0}{\omega_n}\right)^2}\sin\left[\omega_n t + \tan^{-1}\left(\frac{x_0\omega_n}{v_0}\right)\right]$$

where x_0 and v_0 are the initial position and velocity of the mass, respectively.

- The differential equation of motion for compound pendulum $\dfrac{d^2\theta}{dt^2} + \dfrac{mgl}{I_0}\theta = 0$,

  where $\omega_n = \dfrac{2\pi}{\tau} = \sqrt{\dfrac{mgl}{I_o}} \Rightarrow \tau = 2\pi\sqrt{\dfrac{1+\dfrac{k^2}{l}}{g}}$ and $\tau_{min} = 2\pi\sqrt{\dfrac{2k}{g}}$ at $l = k$.

- The differential equation of motion for torsional pendulum is

  $\dfrac{d^2\theta}{dt^2} + \dfrac{k_\theta}{I}\theta = 0$, where k_θ torsional constant of the wire.

- Total energy is conserved in undamped free vibration system. KE + PE = *constant* or $\dfrac{d}{dt}(KE + PE) = 0$ which can be reduced to standard form $\dfrac{d^2x}{dt^2} + \omega_n^2 x = 0$.

REVIEW QUESTIONS

Q1. What is vibration?

Q2. Explain the degrees of freedom with suitable examples.

Q3. Obtain an expression for the motion of a particle in simple harmonic motion.

Q4. Give the differential equation of motion for a compound pendulum when it performs an oscillatory motion about the fixed axis through a point of support.

Q5. Give the expressions for the equivalent length of a compound pendulum.

Q6. When will we get the maximum frequency of oscillation of a compound pendulum?

Q7. Give an expression for frequency of torsional pendulum.

Q8. Obtain an expression for the motion of a particle by energy method.

Q9. Determine the natural frequency of a spring-mass system, taking into account the mass of the spring.

Q10. Determine the period τ for small oscillations of an assembly composed of a light circular frame and two particles, each of mass m as shown in Figure 13.Q10.

$\left[\text{Ans; } \tau = 2\pi\sqrt{\dfrac{2r}{g}}\right]$.

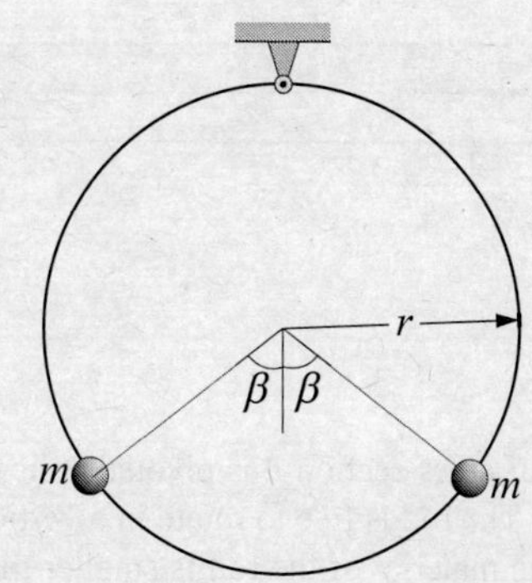

FIGURE 13.Q10 |

Q11. Determine the natural frequency of the mass m on the end of a cantilever beam of negligible mass shown in Figure 13.Q11.

$$\left[\text{Ans; } \omega_n = \sqrt{3\frac{EI}{ml^3}}\right].$$

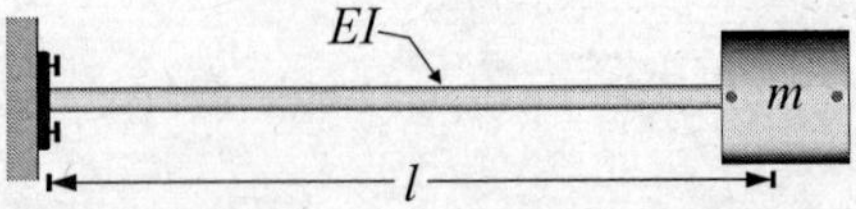

FIGURE 13.Q11 |

Q12. A thin rectangular plate is bent into a semicircular cylinder shell as shown in Figure 13.Q12. Determine its period of oscillation if it is allowed to rock on a horizontal surface.

$$\left[\text{Ans; } \tau = 2\pi\sqrt{\frac{r(\pi-2)}{g}}\right].$$

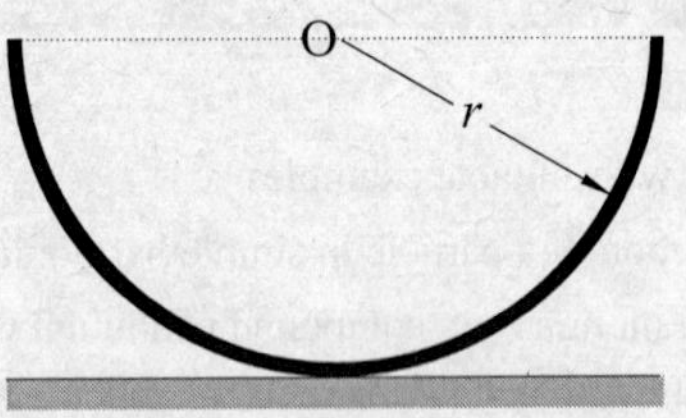

FIGURE 13.Q12 |

Q13. A uniform rod of mass m is supported by pin O. If the rod is given a small displacement and released, it starts oscillating. Determine the natural period of vibration. The springs are unstretched when the rod is in the position shown in Figure 13.Q13.

$$\left[\text{Ans; } \tau = \pi\sqrt{\frac{2m}{3k}}\right].$$

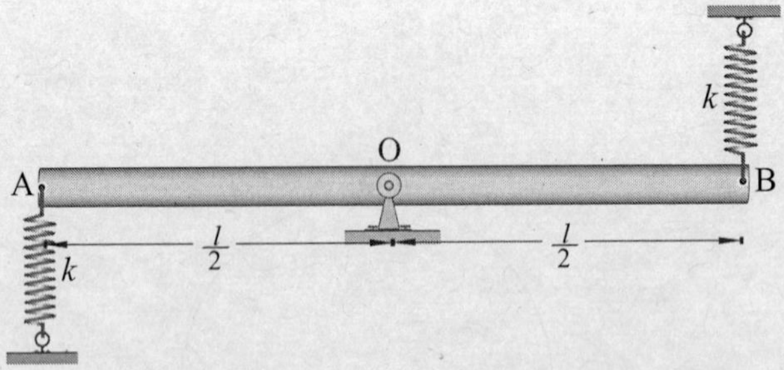

FIGURE 13.Q13 |

Q14. A thin rod of length l and area of cross-section A is pivoted at its lowest point O inside a non-viscous liquid as shown in Figure 13.Q14. The rod is free to rotate in a vertical plane about a horizontal axis passing through O. The density ρ_1 of the material of the rod is smaller than the density ρ_2 of the liquid. The rod is displaced by a small angle θ from its equilibrium position and then released, and it starts oscillating. Determine the natural frequency.

$$\left[\text{Ans; } \omega_n = \sqrt{\frac{3(\rho_2 - \rho_1)g}{2\rho_1 l}}\right].$$

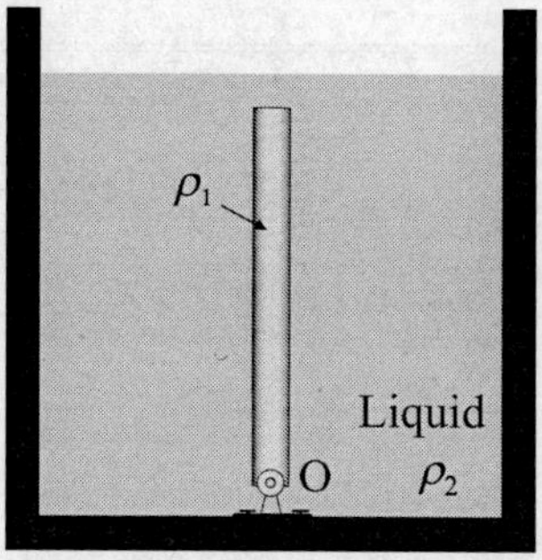

FIGURE 13.Q14 |

Q15. A small sphere of mass m is attached to one end of a weightless stiff rod that is rigidly connected to the centre of a homogeneous cylinder of radius R as shown in Figure 13.Q15. If the cylinder rolls without slipping, determine the natural frequency of oscillation of the system.

$$\left[\text{Ans; } \omega_n = \sqrt{\frac{2mgL}{3MR^2 + 2m(L - R^2)}}\right].$$

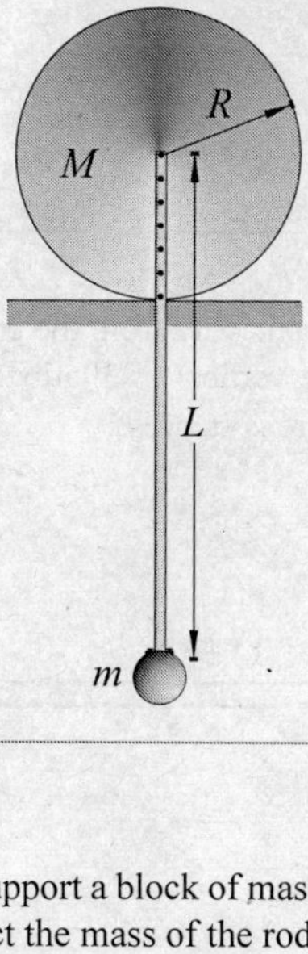

FIGURE 13.Q15 |

Q16. Three springs of stiffness k_1, k_2 and k_3 support a block of mass m as shown in Figure 13.Q16. Determine the natural frequency of the system. Neglect the mass of the rod.

$$\left[\text{Ans; } \omega_n = \sqrt{\frac{4k_1k_2k_3}{m(4k_1k_2 + k_1k_3 + k_2k_3)}}\right].$$

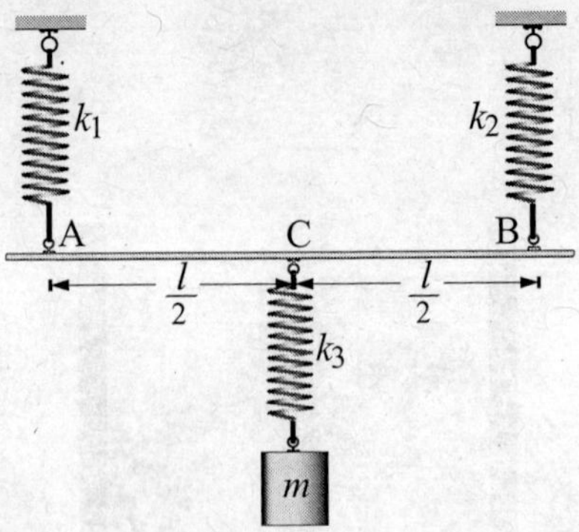

FIGURE 13.Q16 |

Q17. A thin disc of radius r may oscillate about O located as shown at a distance l from the centre as shown in Figure 13.Q17. (a) Determine the period of small oscillations if $l = r$. (b) Determine the second value of l for which the period of oscillation is the same as that found in part (a). (c) Draw the graph between τv/sl.

$$\left[\text{Ans; (a) } \tau = 2\pi\sqrt{\frac{(r^2/2l) + l}{g}}, \text{(b) } l = \frac{r}{2}\right].$$

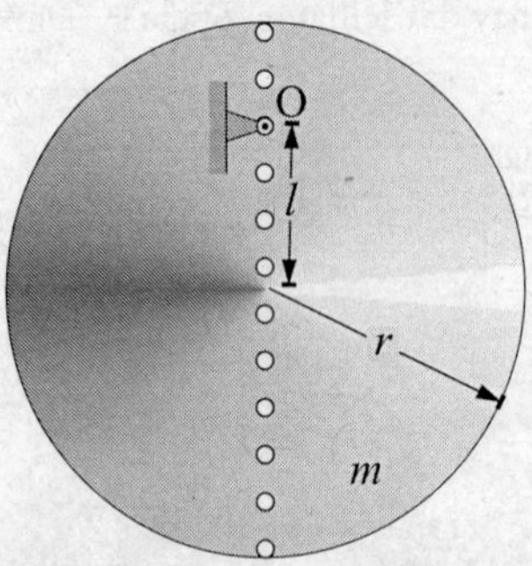

FIGURE 13.Q17 |

Q18. A load of mass m is suspended by an elastic wire from the end of a cantilever as shown in Figure 13.Q18. Determine the natural frequency for the vertical oscillations of the system. Assume that the modulus of elasticity for the cantilever beam and wire is same, E.

$$\left[\text{Ans; } \omega_n = \sqrt{\frac{3EIAg}{W(AL^3 + 3Il)}}\right].$$

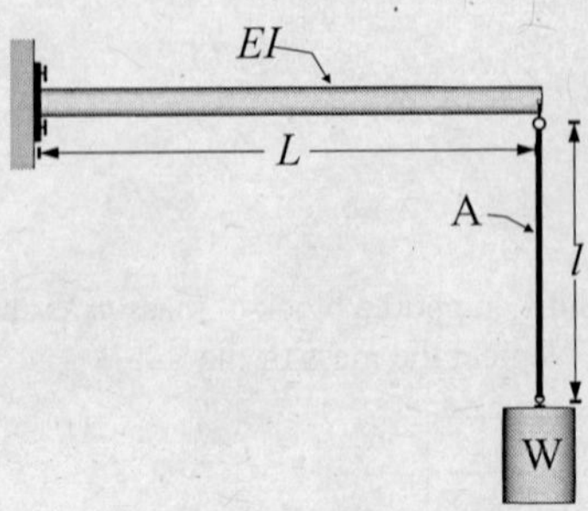

FIGURE 13.Q18 |

Q19. Imagine a hollow pipe going all the way through the earth from the north pole to the south pole along its rotation axis as shown in Figure 13.Q19. Assuming the earth to be a homogeneous ball and neglecting the air drag, (a) find the equation of motion of a body falling down into the hollow pipe and (b) how long does it take the body to reach the other end of the shaft.

$$\left[\text{Ans; } T = \pi\sqrt{\frac{3}{4\pi G\rho}}\right].$$

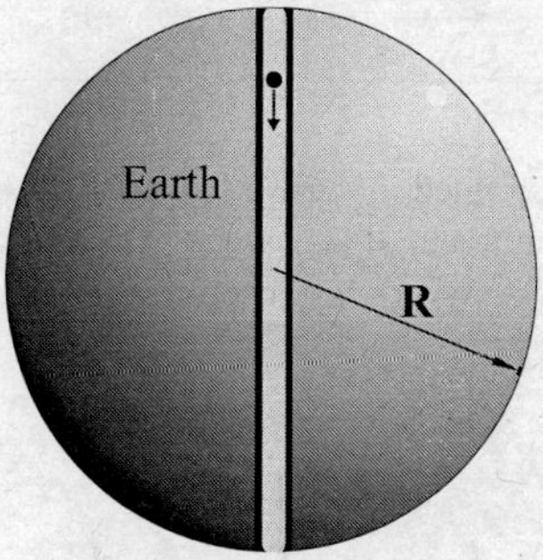

FIGURE 13.Q19 |

Q20. Two uniform rods, each of mass m and length l, are welded together to form the T-shaped assembly shown in Figure 13.Q20. Determine the natural frequency of oscillation.

$$\left[\text{Ans; } \omega_n = \sqrt{\frac{18g}{17l}}\right].$$

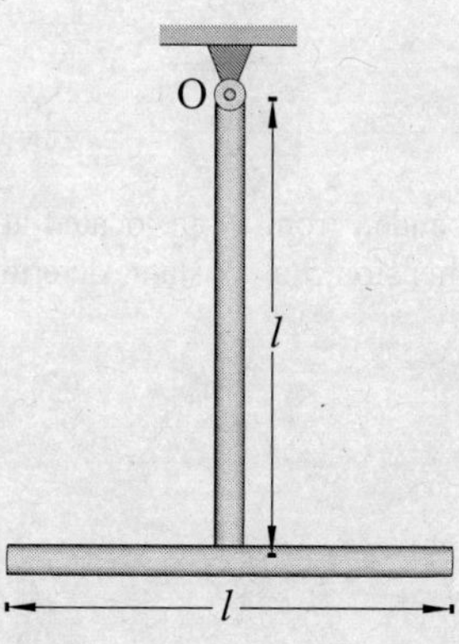

FIGURE 13.Q20 |

Q21. Three uniform thin rods, each of mass m and length l, are welded together to form an equilateral triangle assembly, as shown in Figure 13.Q21. Determine the frequency of small oscillations about O.

$$\left[\text{Ans; } \omega_n = \sqrt{\frac{2g}{\sqrt{3}l}}\right].$$

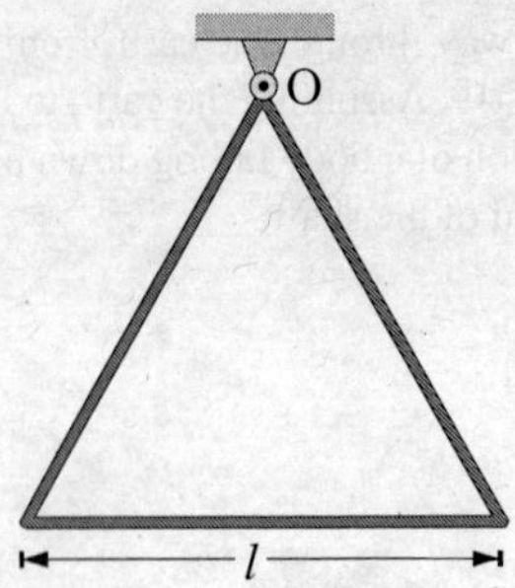

FIGURE 13.Q21 |

Q22. A rod of mass m and length l is restrained in the vertical position by two identical springs having a spring constant k. Find a vertical load P in terms of m, l and k, which causes the rod to have a natural frequency of oscillation about O approaching zero for small oscillations (Fig. 13.Q22).

$$\left[\text{Ans; } P = 2kl - \frac{mg}{2}\right].$$

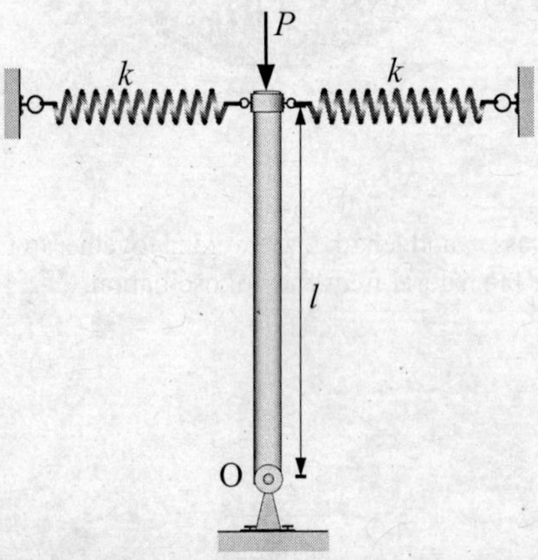

FIGURE 13.Q22 |

Q23. A uniform rectangular plate is suspended from a pin located at the mid-point of one edge as shown in Figure 13.Q23. Considering the dimension b a constant, determine the ratio c/b for which the period of oscillation of the plate is minimum.

$$\left[\text{Ans; } \tau_{\min} = 2\pi\sqrt{\frac{2b}{3g}}\right].$$

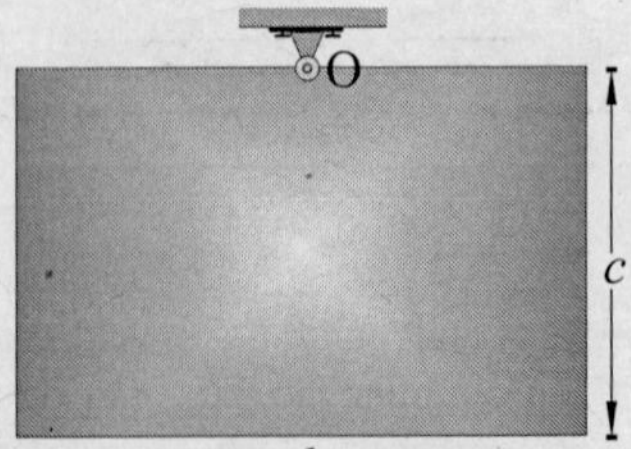

FIGURE 13.Q23 |

Q24. A homogeneous wire of length $2l$ is bent as shown in Figure 13.Q24 and allowed to oscillate about a frictionless pin at O. Determine the period of small oscillations.

$$\left[\text{Ans; } \tau = 2\pi\sqrt{\frac{2l}{3g\cos\beta}}\right].$$

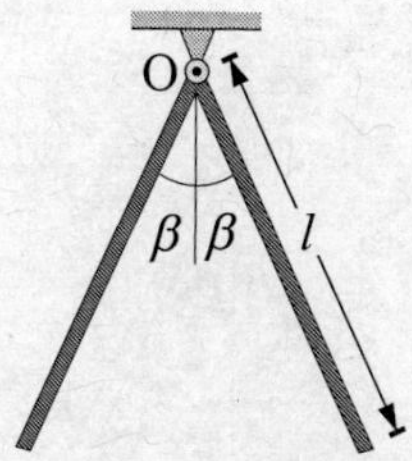

FIGURE 13.Q24 |

Q25. A rectangular block of mass m is resting on top of a semi-cylindrical surface as shown in Figure 13.Q25. If the block is slightly tipped at one end, find its frequency of oscillation.

$$\left[\text{Ans; } \omega_n = \sqrt{\frac{12g\left(r - \frac{h}{2}\right)}{4h^2 + l^2}}\right].$$

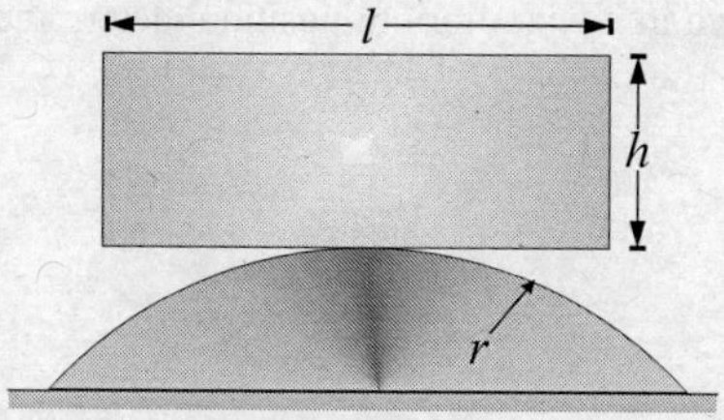

FIGURE 13.Q25

Q26. A bar AB of length l rolls on wheels of negligible weight on a circular path of radius r as shown in Figure 13.Q26. Determine the frequency of small oscillations of the system. Assume that the bar moves in the vertical plane and is displaced slightly from its equilibrium.

$$\left[\text{Ans; } \omega_n = \sqrt{\frac{g\sqrt{r^2 - \frac{l^2}{4}}}{r^2 - \frac{l^2}{6}}}\right].$$

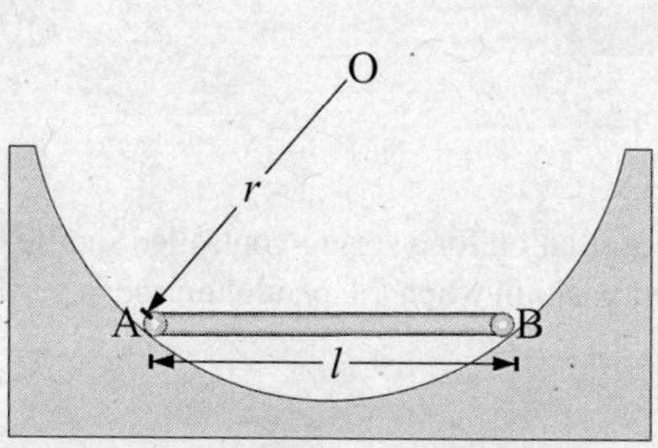

FIGURE 13.Q26 |

Q27. Determine the natural frequency of the uniform circular sector in terms of r, β and g. Assume oscillations of small amplitude about the pin at O (Fig. 13.Q27).

$$\left[\text{Ans; } \omega_n = \sqrt{\frac{4g \sin \beta}{3r\beta}}\right].$$

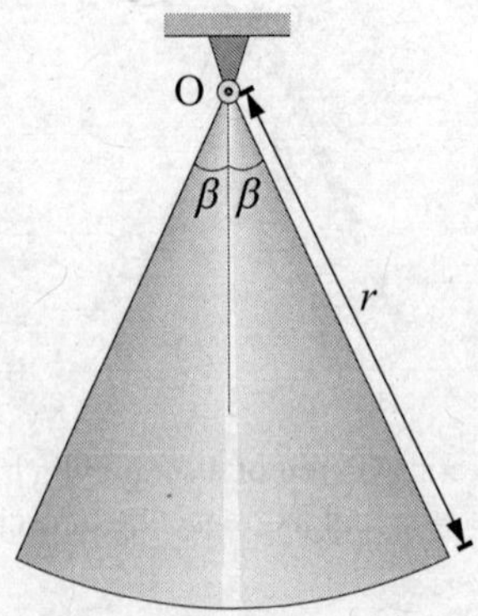

FIGURE 13.Q27 |

Q28. A uniform slender rod of length $5R$ and mass m is secured to the uniform disc of radius R and mass M (Fig. 13.Q28). If the system is shown in its equilibrium position, determine the natural frequency ω_n for small oscillations about the pivot O.

$$\left[\text{Ans; } \omega_n = \sqrt{\frac{18k}{M + 10m}}\right].$$

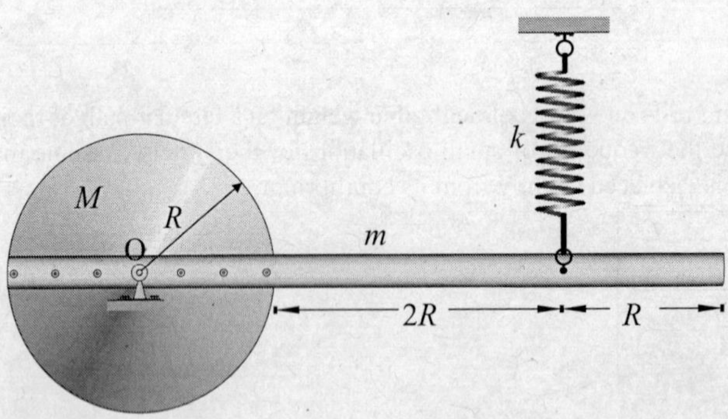

FIGURE 13.Q28 |

Q29. Find the natural frequency of oscillation for a spring-controlled simple pendulum shown in Figure 13.Q29. The spring is in its unstretched position when the pendulum rod is vertical. Neglect the mass of rod.

$$\left[\text{Ans; } \omega_n = \sqrt{\frac{g}{l} + \frac{k}{m}\left(\frac{a}{l}\right)^2}\right].$$

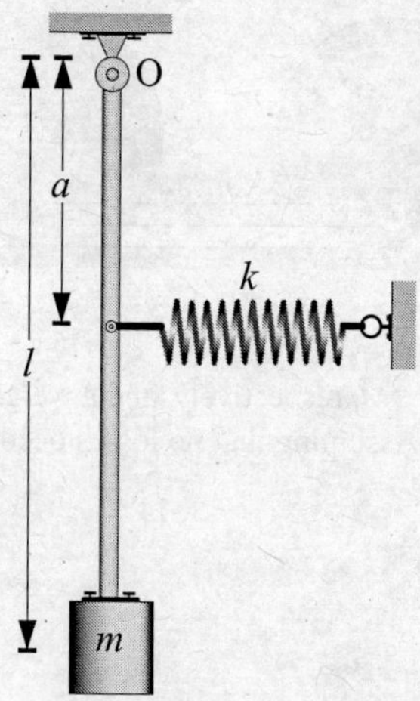

FIGURE 13.Q29 |

Q30. A weight of mass m is fixed to the end of a T-rod assembly. If both springs are unstretched when the assembly is in the position shown in Figure 13.Q30, determine the period of vibration for the system when it is displaced slightly and released. Neglect the size of the block and the mass of the rods.

$$\left[\text{Ans; } \tau = \sqrt{\frac{1}{\left(\frac{g}{l} + \frac{k}{2m}\right)}}\right].$$

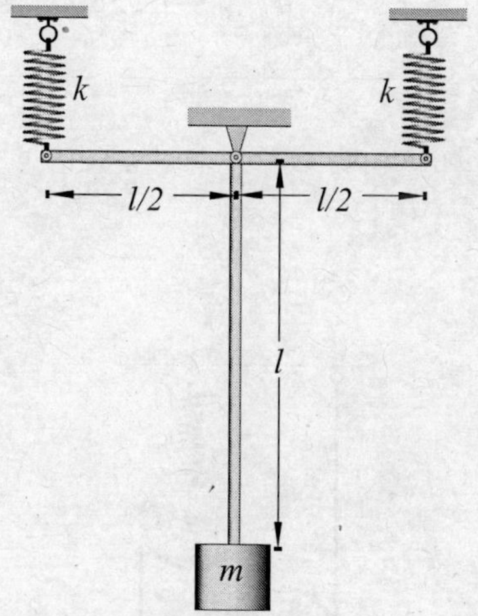

FIGURE 13.Q30 |

Q31. Derive the expression for the natural frequency ωn of the system composed of two homogeneous circular cylinders, each of mass M and the connecting link AB of mass m (Fig. 13.Q31).

$$\left[\text{Ans; } \omega_n = \sqrt{\frac{mgr}{3MR^2 + m(R - r)^2}}\right].$$

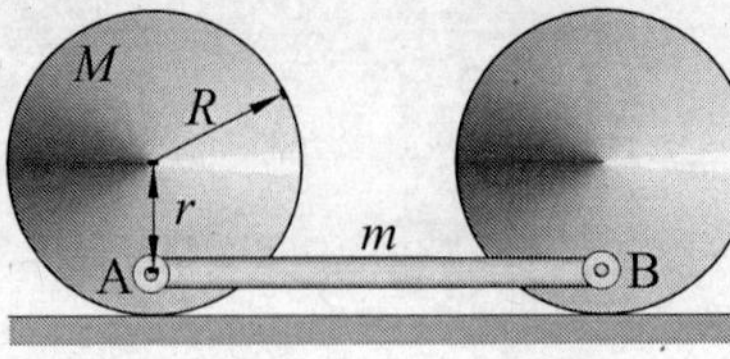

FIGURE 13.Q31 |

Q32. Gears A and B are of masses M and $3M$, respectively, and a weight C of mass m is attached to the teeth of gear B as shown in Figure 13.Q32. Assuming that no losses occur between gears, determine the period of small oscillations of the system.

$$\left[\text{Ans; } \tau = 2\pi\sqrt{\frac{(4M + 2m)r}{mg}}\right].$$

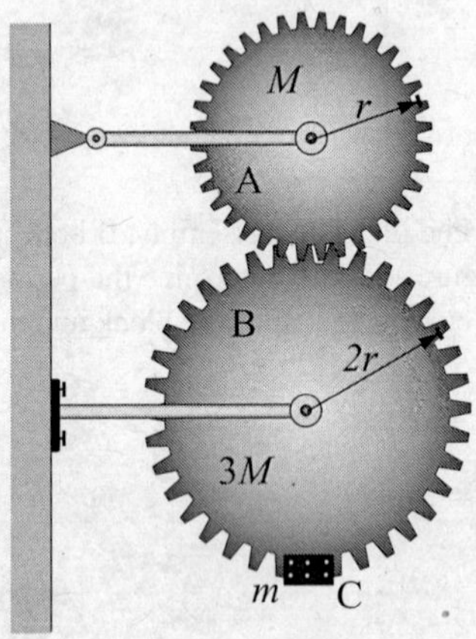

FIGURE 13.Q32 |

Q33. Find the natural frequency of vibration of a system of pulleys shown in Figure 13.Q33. Neglect the weight of pulleys.

$$\left[\text{Ans; } \omega_n = \sqrt{\frac{k_a k_b}{4m(k_a + k_b)}}\right].$$

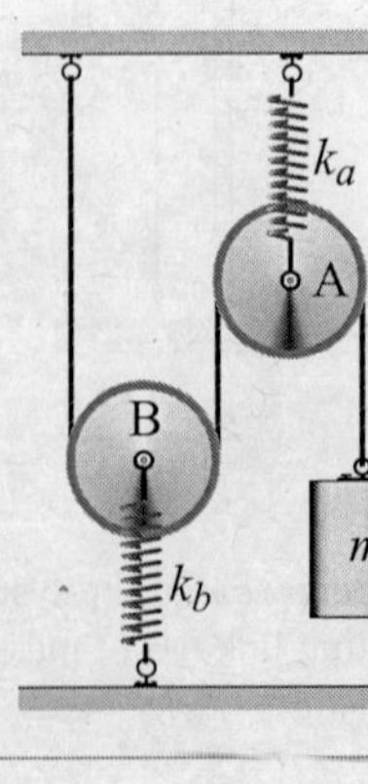

FIGURE 13.Q33 |

Q34. Neglecting mass of links, find the natural frequency of the mass m shown in Figure 13.Q34.

$$\left[\text{Ans; } \omega_n = \left(\frac{a}{a+b}\right)^2 \sqrt{\frac{k}{m}}\right].$$

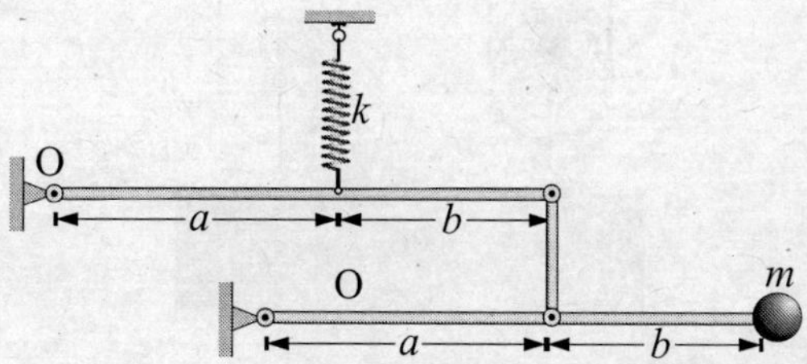

FIGURE 13.Q34 |

Q35. A homogeneous solid cylinder of mass m is linked by a spring constant k and is resting on an inclined plane as shown in Figure 13.Q35. If it rolls without slipping, write its equation of vibration.

$$\left[\text{Ans; } \ddot{x} + \left(\frac{2k}{3m}\right)x = 0\right].$$

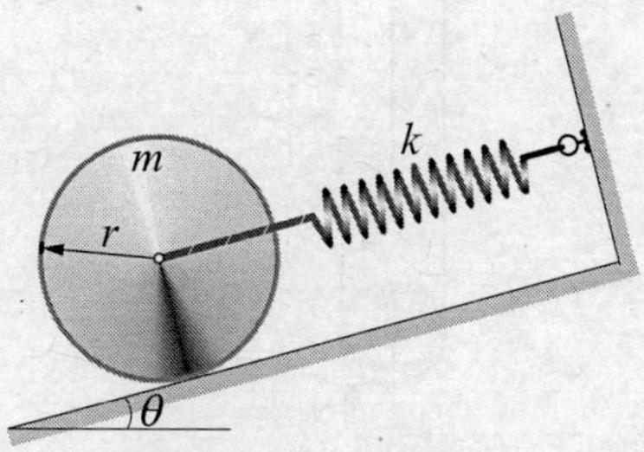

FIGURE 13.Q35 |

Q36. A sphere of mass m is attached to a rod of negligible mass as shown in Figure 13.Q36. Determine the natural frequency of vibration of the sphere. Neglect the size of the sphere.

$$\left[\text{Ans; } \omega_n = \sqrt{\frac{k}{4m}}\right].$$

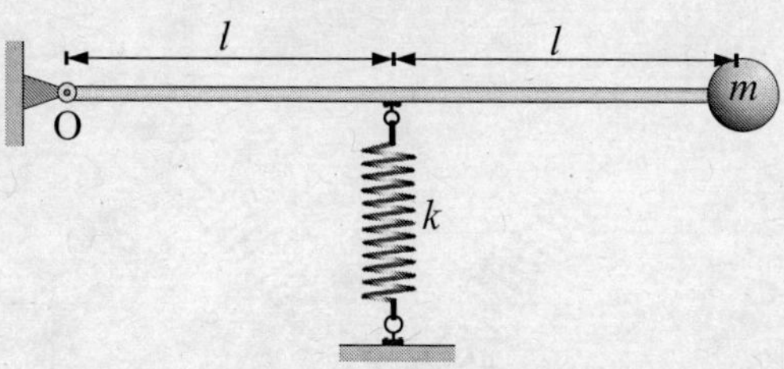

FIGURE 13.Q36 |

Q37. Determine the period of vertical oscillation of system shown in Figure 13.Q37. Each spring has stiffness k and the mass of pulleys may be neglected.

$$\left[\text{Ans; } \omega_n = \sqrt{\frac{8k}{m}}\right].$$

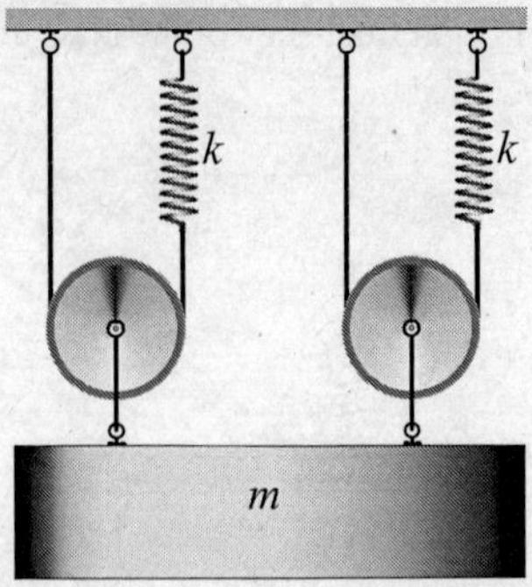

FIGURE 13.Q37 |

Q38. Neglecting the mass of pulleys, find the natural frequency of the system shown in Figure 13.Q38.

$$\left[\text{Ans; } \omega_n = \sqrt{\frac{64k}{m}}\right].$$

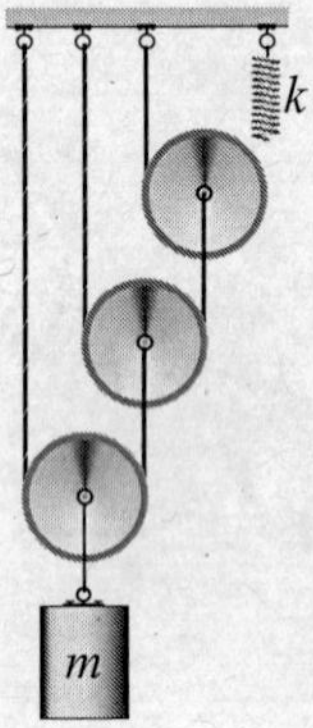

FIGURE 13.Q38 |

Q39. A bar of mass m is suspended from two springs such that when it is in equilibrium, the springs make an angle of 45° with the horizontal as shown in Figure 13.Q39. Determine the natural period of vibration, if the bar is pulled down a short distance and released. Each spring has a stiffness k.

$$\left[\text{Ans; } \omega_n = \sqrt{\frac{k}{m}}\right].$$

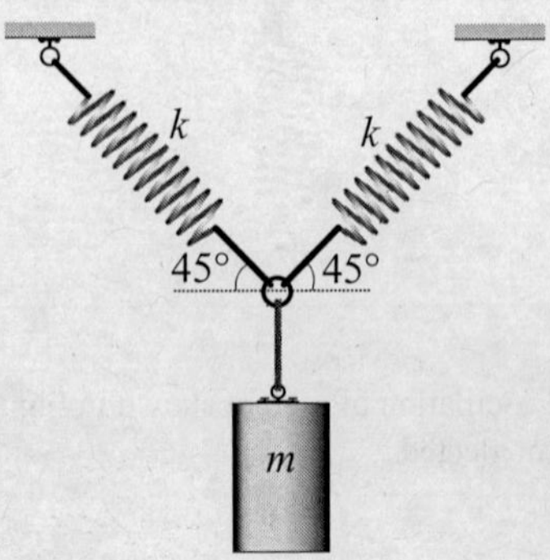

FIGURE 13.Q39

Q40. Determine the natural frequency of the system shown in Figure 13.Q40. Neglect the mass of the links and lever.

$$\left[\text{Ans; } \omega_n = \sqrt{\frac{k_1 k_2}{m[k_1 + k_2(a/l)^2]}}\right].$$

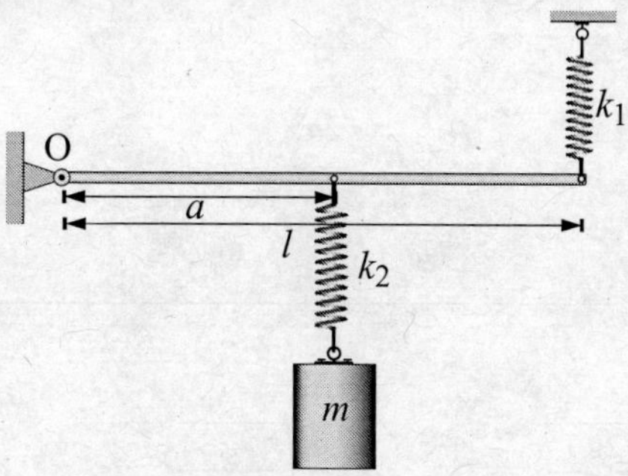

FIGURE 13.Q40 |

Q41. Determine the natural frequency of the system shown in Figure 13.Q41. Neglect the mass of the links and lever.

$$\left[\text{Ans; } \omega_n = \sqrt{\frac{k_1 a^2 + k_2 b^2}{m_1 a^2 + m_2 b^2}}\right].$$

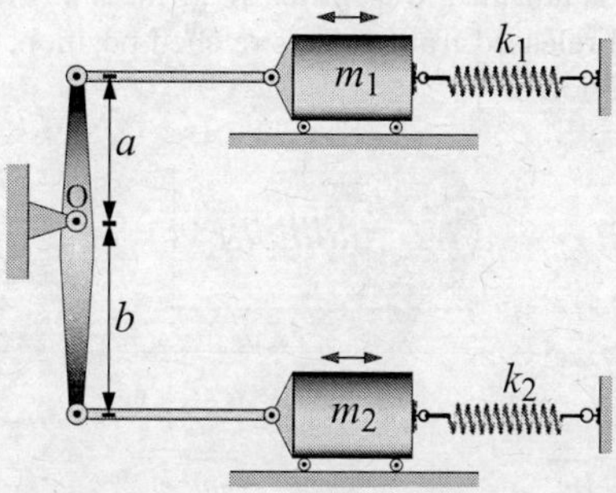

FIGURE 13.Q41 |

MULTIPLE CHOICE QUESTIONS

Q1. A uniform stiff rod of length 300 mm having a weight of 300 N is pivoted at one end and connected to a spring at the other end (Fig. 13.MQ1). For keeping the rod vertical in a stable position, the minimum value of spring constant k needed is

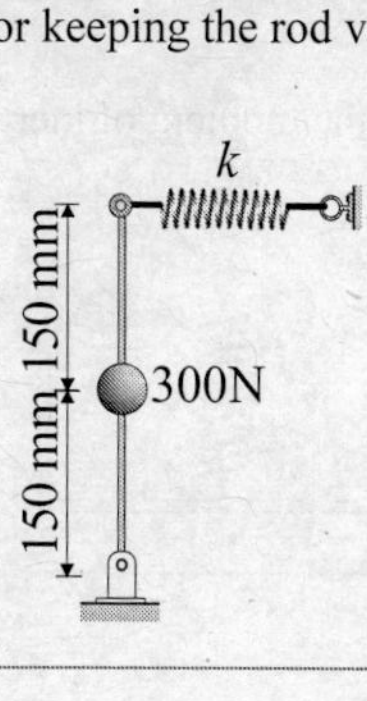

FIGURE 13.MQ1 |

(A) 300 N/m (B) 400 N/m (C) 500 N/m (D) 1000 N/m

[Ans: C]

Q2. The assembly shown in Figure 13.MQ2 is composed of two mass-less rods of length L with two particles, each of mass m. The natural frequency ω_n of this assembly for small oscillation is

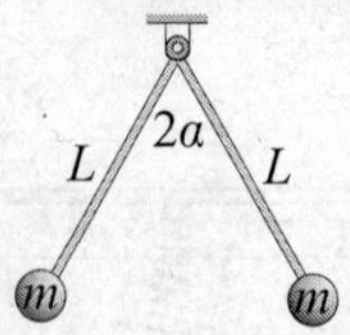

FIGURE 13.MQ2 |

(A) $\sqrt{\frac{g}{L}}$ (B) $\sqrt{\frac{2g}{L\cos a}}$ (C) $\sqrt{\frac{g}{L\cos a}}$ (D) $\sqrt{\frac{g\cos a}{L}}$

[Ans: D]

Q3. A particle oscillates on the x-axis according to the equation $x = A + B\sin\omega t$. The amplitude of motion is

(A) A (B) B (C) $A + B$ (D) $\sqrt{A^2 + B^2}$

[Ans: B]

Q4. A disc of mass m and radius r is attached to a spring of stiffness k (Fig. 13.MQ4). During its motion the disc rolls on the ground. When released from some stretched position, the centre of the disc will execute harmonic motion with a time period of

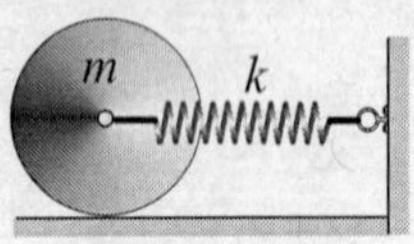

FIGURE 13.MQ4 |

(A) $2\pi\sqrt{\frac{m}{2k}}$ (B) $2\pi\sqrt{\frac{m}{k}}$ (C) $2\pi\sqrt{\frac{3m}{2k}}$ (D) $2\pi\sqrt{\frac{2m}{k}}$

[Ans: C]

Q5. A particle moves on the x-axis according to the equation $x = A\sin_2\omega t$. The motion is simple harmonic

(A) with amplitude A (B) with amplitude $2A$

(C) with time period $\frac{2\pi}{\omega}$ (D) with time period $\frac{\pi}{\omega}$

[Ans: D]

Q6. For the system shown in Figure 13.MQ6, the moment of inertia of the weight W and the bar about the pivot point is I_0. The system will vibrate when

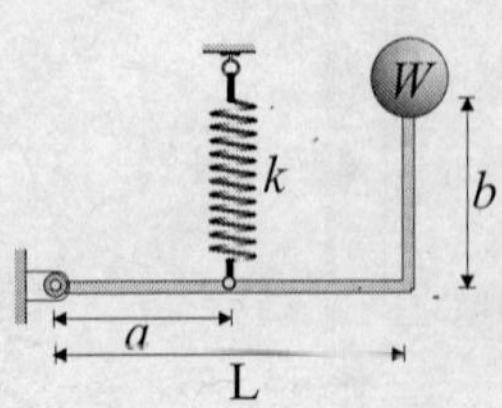

FIGURE 13.MQ6 |

(A) $b < \frac{ka^2}{W}$ (B) $b = \frac{ka^2}{W}$ (C) $b > \frac{ka^2}{W}$ (D) $b = 0$

[Ans: A]

Q7. A uniform rod of length l and mass m is suspended from one of its ends in vertical plane. The natural frequency of this rod for small amplitude

(A) $\frac{1}{2\pi}\sqrt{\frac{g}{l}}$ (B) $\frac{1}{2\pi}\sqrt{\frac{g}{3l}}$ (C) $\frac{1}{2\pi}\sqrt{\frac{2g}{3l}}$ (D) $\frac{1}{2\pi}\sqrt{\frac{3g}{2l}}$

[Ans: D]

Q8. A particle of unit mass oscillates in a straight path under a force of attraction proportional to its distance from a fixed point O. If its velocity at distances a and b from O be v_1 and v_2, respectively, then the period of oscillation is

(A) $2\pi\sqrt{\frac{b^2 - a^2}{v_1^2 - v_2^2}}$ (B) $2\pi\sqrt{\frac{v_1^2 - v_2^2}{b^2 - a^2}}$ (C) $2\pi\sqrt{\frac{b^2 + a^2}{v_1^2 - v_2^2}}$ (D) $2\pi\sqrt{\frac{b^2}{v_2^2} + \frac{a^2}{v_1^2}}$

[Ans: A]

Q9. A system is shown in Figure 13.MQ9. The bar AB is assumed to be rigid and weightless. The natural frequency (f_n) of vibration of the system is given by

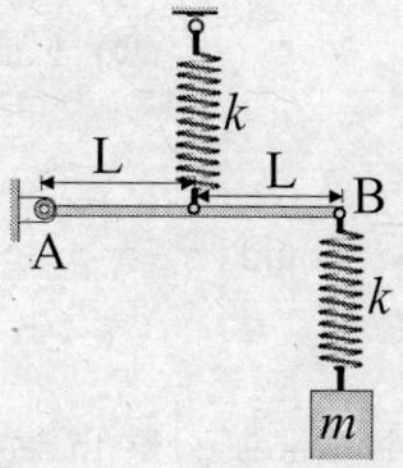

FIGURE 13.MQ9 |

(A) $f_n = \frac{1}{2\pi}\sqrt{\frac{k}{5m}}$ (B) $f_n = \frac{1}{2\pi}\sqrt{\frac{k}{2m}}$ (C) $f_n = \frac{1}{2\pi}\sqrt{\frac{k}{m}}$ (D) $f_n = \frac{1}{2\pi}\sqrt{\frac{2}{km}}$

[Ans: A]

Q10. A solid cylinder of height h and uniform material relative density β floats in water with its axis vertical [Fig. 13.MQ10]. Disregarding friction, the period of oscillation of the cylinder, when it is vertically depressed in water and released, will be

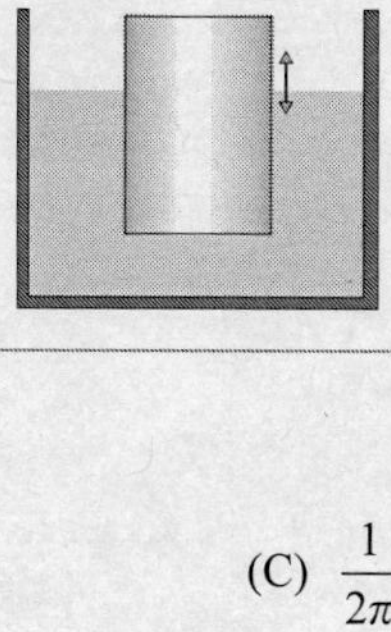

FIGURE 13.MQ10 |

(A) $\frac{1}{2\pi}\sqrt{\frac{\beta g}{h}}$ (B) $2\pi\sqrt{\frac{\beta g}{h}}$ (C) $\frac{1}{2\pi}\sqrt{\frac{g}{h\beta}}$ (D) $\frac{1}{2\pi}\sqrt{\frac{h}{\beta g}}$

[Ans: B]

Q11. A glass tube bent into a circle of radius R is rigidly fixed in the vertical plane, with the top ends open to the atmosphere. A liquid of density ρ is filled in the tube so that the liquid column forms a semicircle. If the liquid is disturbed slightly, the time period of oscillations will be

(A) $\sqrt{\dfrac{4\pi^3 R}{g}}$ (B) $\sqrt{\dfrac{3\pi^3 R}{g}}$ (C) $\sqrt{\dfrac{\pi R}{g}}$ (D) $\sqrt{\dfrac{2\pi^3 R}{g}}$

[Ans: D]

Q12. The displacement x of a particle at a time t is given by $x = 5 \sin 2t$, where x is in metres and t is in seconds. A simple pendulum has the same period as the particle when the length of the pendulum is

(A) 10.0 m (B) 5.0 m (C) 2.5 m (D) 2.0 m

[Ans: B]

Q13. If the differential equation of a particle executing simple harmonic motion about a point is $\dfrac{d^2x}{dt^2} + \mu x = 0$, where μ is some constant of proportionality, then consider the following statements:

1. Frequency of oscillation is $\sqrt{\mu}/2\pi$.
2. Maximum velocity of the particle is $\sqrt{\mu}a$, where a is the distance of the mean position from the point from where it starts moving.
3. If the point from which the particle starts moving is altered, there will be change in the time period of oscillation.

Of these statements

(A) 1, 2 and 3 are correct (B) 1 and 2 are correct

(C) 2 and 3 are correct (D) 1 and 3 are correct

[Ans: B]

Q14. Distance x covered by a particle in time t is given by $x = 2 \cos a^2 t - \sin (2a - 1)t$, where $(a - 1)$. If motion is required to be simple harmonic, then a should be

(A) $1 \pm \sqrt{3}$ (B) $-1 \pm \sqrt{2}$ (C) $1 \pm \sqrt{5}$ (D) $-1 \pm \sqrt{3}$

[Ans: B]

Q15. A particle is oscillating in a straight line is expressed by the equation $x = a \cos nt + b \sin nt$. The time period of particle is

(A) $\dfrac{2\pi}{n}$ (B) a (C) b (D) $\sqrt{a^2 + b^2}$

[Ans: A]

Index